# Lecture Notes in Computer Science 3059

Commenced Publication in 1973
Founding and Former Series Editors:
Gerhard Goos, Juris Hartmanis, and Jan van Leeuwen

**Springer**
*Berlin*
*Heidelberg*
*New York*
*Hong Kong*
*London*
*Milan*
*Paris*
*Tokyo*

Celso C. Ribeiro   Simone L. Martins (Eds.)

# Experimental and Efficient Algorithms

Third International Workshop, WEA 2004
Angra dos Reis, Brazil, May 25-28, 2004
Proceedings

Springer

Volume Editors

Celso C. Ribeiro
Simone L. Martins
Universidade Federal Fluminense
Department of Computer Science
Niterói, RJ 24210-240, Brazil
E-mail: {celso,simone}@ic.uff.br

Library of Congress Control Number: 2004105538

CR Subject Classification (1998): F.2.1-2, E.1, G.1-2, I.3.5, I.2.8

ISSN 0302-9743
ISBN 3-540-22067-4 Springer-Verlag Berlin Heidelberg New York

Springer-Verlag is a part of Springer Science+Business Media

springeronline.com

© Springer-Verlag Berlin Heidelberg 2004
Printed in Germany

Typesetting: Camera-ready by author, data conversion by PTP-Berlin, Protago-TeX-Production GmbH
Printed on acid-free paper     SPIN: 11008408     06/3142     5 4 3 2 1 0

# Preface

The Third International Workshop on Experimental and Efficient Algorithms (WEA 2004) was held in Angra dos Reis (Brazil), May 25–28, 2004.

The WEA workshops are sponsored by the European Association for Theoretical Computer Science (EATCS). They are intended to provide an international forum for researchers in the areas of design, analysis, and experimental evaluation of algorithms. The two preceding workshops in this series were held in Riga (Latvia, 2001) and Ascona (Switzerland, 2003).

This proceedings volume comprises 40 contributed papers selected by the Program Committee along with the extended abstracts of the invited lectures presented by Richard Karp (University of California at Berkeley, USA), Giuseppe Italiano (University of Rome "Tor Vergata", Italy), and Christos Kaklamanis (University of Patras, Greece).

As the organizer and chair of this wokshop, I would like to thank all the authors who generously supported this project by submitting their papers for publication in this volume. I am also grateful to the invited lecturers, who kindly accepted our invitation.

For their dedication and collaboration in the refereeing procedure, I would like also to express my gratitude to the members of the Program Committee: E. Amaldi (Italy), J. Blazewicz (Poland), V.-D. Cung (France), U. Derigs (Germany), J. Diaz (Spain), M. Gendreau (Canada), A. Goldberg (USA), P. Hansen (Canada), T. Ibaraki (Japan), K. Jansen (Germany), S. Martello (Italy), C.C. McGeoch (USA), L.S. Ochi (Brazil), M.G.C. Resende (USA), J. Rolim (Switzerland), S. Skiena (USA), M. Sniedovich (Australia), C.C. Souza (Brazil), P. Spirakis (Greece), D. Trystram (France), and S. Voss (Germany). I am also grateful to the anonymous referees who assisted the Program Committee in the selection of the papers to be included in this publication.

The idea of organizing WEA 2004 in Brazil grew out of a few meetings with José Rolim (University of Geneva, Switzerland). His encouragement and close collaboration at different stages of this project were fundamental for the success of the workshop. The support of EATCS and Alfred Hofmann (Springer-Verlag) were also appreciated.

I am thankful to the Department of Computer Science of *Universidade Federal Fluminense* (Niterói, Brazil) for fostering the environment in which this workshop was organized. I am particularly indebted to Simone Martins for her invaluable support and collaboration in the editorial work involved in the preparation of the final camera-ready copy of this volume.

Angra dos Reis (Brazil), May 2004                    Celso C. Ribeiro (Chair)

# Table of Contents

# A Hybrid Bin-Packing Heuristic to Multiprocessor Scheduling

Adriana C.F. Alvim[1] and Celso C. Ribeiro[2]

[1] Universidade Federal do Rio de Janeiro, COPPE, Programa de Engenharia de
Produção, PO Box 68507, Rio de Janeiro 21945-970, Brazil.
`alvim@inf.puc-rio.br`
[2] Universidade Federal Fluminense, Department of Computer Science, Rua Passo da
Pátria 156, Niterói, RJ 24210-240, Brazil.
`celso@inf.puc-rio.br`

**Abstract.** The multiprocessor scheduling problem consists in scheduling a set of tasks with known processing times into a set of identical processors so as to minimize their makespan, i.e., the maximum processing time over all processors. We propose a new heuristic for solving the multiprocessor scheduling problem, based on a hybrid heuristic to the bin packing problem. Computational results illustrating the effectiveness of this approach are reported and compared with those obtained by other heuristics.

## 1 Introduction

Let $T = \{T_1, \ldots, T_n\}$ be a set of $n$ tasks with processing times $t_i$, $i = 1, \ldots, n$, to be processed by a set $P = \{P_1, \ldots, P_m\}$ of $m \geq 2$ identical processors. We assume the processing times are nonnegative integers satisfying $t_1 \geq t_2 \geq \ldots \geq t_n$. Each processor can handle at most one task at any given time and preemption is not possible. We denote by $A_j$ the set formed by the indices of the tasks assigned to processor $P_j$ and by $t(P_j) = \sum_{i \in A_j} t_i$ its overall processing time, $j = 1, \ldots, n$. A solution is represented by the lists of tasks assigned to each processor. The makespan of a solution $S = (A_1, \ldots, A_m)$ is given by $C_{\max}(S) = \max_{j=1,\ldots,m} t(P_j)$.

The multiprocessor scheduling problem $P\|C_{\max}$ consists in finding an optimal assignment of tasks to processors, so as to minimize their makespan, see e.g. [5,16,20]. $P\|C_{\max}$ is NP-hard [4,14]. We denote the optimal makespan by $C^*_{\max}$. Minimizing the schedule length is important since it leads to the maximization of the processor utilization factor [3].

There is a duality relation [5,17,18,24] between $P\|C_{\max}$ and the bin packing problem (BP), which consists in finding the minimum number of bins of a given capacity $C$ which are necessary to accommodate $n$ items with weights $t_1, \ldots, t_n$.

The worst case performance ratio $r(H)$ of a given heuristic H for $P\|C_{\max}$ is defined as the maximum value of the ratio $H(I)/C^*_{\max}(I)$ over all instances $I$, where $C^*_{\max}(I)$ is the optimal makespan of instance $I$ and $H(I)$ is the makespan of the solution computed by heuristic H. The longest processing time (LPT)

C.C. Ribeiro and S.L. Martins (Eds.): WEA 2004, LNCS 3059, pp. 1–13, 2004.

heuristic of Graham [15] finds an approximate solution to $P\|C_{\max}$ in time $O(n \log n + n \log m)$, with $r(\text{LPT}) = 4/3 - 1/3m$. The MULTIFIT heuristic proposed by Coffman et al. [6] explores the duality relation with the bin packing problem, searching by binary search the minimum processing time (or bin capacity) such that the solution obtained by the FFD heuristic [19,21] to pack the $n$ tasks (or items) makes use of at most $m$ processors (or bins). It can be shown that if MULTIFIT is applied $k$ times, then it runs in time $O(n \log n + kn \log m)$ and $r(\text{MULTIFIT}) = 1.22 + 2^{-k}$. Friesen [13] subsequently improved this ratio to $1.20 + 2^{-k}$. Yue [25] further improved it to $13/11$, which is tight. Finn and Horowitz [9] proposed the 0/1-INTERCHANGE heuristic running in time $O(n \log m)$, with worst case performance ratio equal to 2. Variants and extensions of the above heuristics can be found in the literature.

The duality relation between the bin packing problem and $P\|C_{\max}$ was also used by Alvim and Ribeiro [1] to derive a hybrid improvement heuristic to the former. This heuristic is explored in this paper in the context of $P\|C_{\max}$. Lower and upper bounds used by the heuristic are described in Section 2. The main steps of the hybrid improvement heuristic to multiprocessor scheduling are presented in Section 3. Numerical results illustrating the effectiveness of the proposed algorithm are reported in Section 4. Concluding remarks are made in the last section.

## 2    Lower and Upper Bounds

The lower bound $L_1 = \max\left\{\left\lceil \frac{1}{m} \sum_{i=1}^{n} p_i \right\rceil, \max_{i=1,\ldots,n}\{p_i\}\right\}$ proposed by McNaughton [22] establishes that the optimal makespan cannot be smaller than the maximum between the average processing time over all processors and the longest duration over all tasks. This bound can be further improved to $L_2 = \max\left\{L_1, p_m + p_{m+1}\right\}$.

Dell'Amico and Martello [7] proposed the lower bound $L_3$. They also showed that the combination of lower and upper bounds to the makespan makes it possible to derive lower and upper bounds to the number of tasks assigned to each processor, leading to a new lower bound $L_\vartheta$. Bounds $L_2, L_3$, and $L_\vartheta$ are used in the heuristic described in the next section.

We used three construction procedures for building feasible solutions and computing upper bounds to $P\|C_{\max}$:

- Construction heuristic H1: Originally proposed in [1,2], it is similar to the Multi-Subset heuristic in [7]. It considers one processor at a time. The longest yet unassigned task is assigned to the current processor. Next, assign to this same processor a subset of the yet unassigned tasks such that the sum of their processing times is as close as possible to a given limit to the makespan. The polynomial-time approximation scheme MTSS(3) of Martello and Toth [21] is used in this step. The remaining unassigned tasks are considered one by one in non-increasing order of their processing times. Each of them is assigned to the processor with the smallest load.

- Construction heuristic H2: Hochbaum and Shmoys [17] proposed a new approach to constructing approximation algorithms, called dual approximation algorithms. The goal is to find superoptimal, but infeasible, solutions. They showed that finding an $\epsilon$-approximation to $P\|C_{\max}$ is equivalent to finding an $\epsilon$-dual-approximation to BP. For the latter, an $\epsilon$-dual-approximation algorithm constructs a solution in which the number of bins is at most the optimal number of bins and each bin is filled with at most $1+\epsilon$ (bin capacity $C = 1$ and item weights $t_i$ scaled by $t_i/C$). In particular, they proposed a scheme for $\epsilon = 1/5$. Given a lower bound $L$ and an upper bound $U$ to $C^*_{\max}$, we obtain by binary search the smallest value $C$ such that $L \leq C \leq U$ and the 1/5-dual-approximation solution to BP uses no more than $m$ bins. This approach characterizes a $1/5 + 2^{-k}$-approximation algorithm to $P\|C_{\max}$, where $k$ is the number of iterations of the search. At the end of the search, the value of $C$ gives the $L_{HS}$ lower bound to the makespan.
- Construction heuristic H3: This is the longest processing time heuristic LPT [15]. Tasks are ordered in non-increasing order of their processing times. Each task is assigned to the processor with the smallest total processing time.

França et al. [10] proposed algorithm 3-PHASE based on the idea of balancing the load between pair of processors. Hübscher and Glover [18] also explored the relation between $P\|C_{\max}$ and BP, proposing a tabu search algorithm using 2-exchanges and influential diversification. Dell'Amico and Martello [7] developed a branch-and-bound algorithm to exactly solve $P\|C_{\max}$. They also obtained new lower bounds. Scholl and Voss [24] considered two versions of the simple assembly line balancing problem. If the precedence constraints are not taken into account, these two versions correspond to BP and $P\|C_{\max}$. Fatemi-Ghomi and Jolai-Ghazvini [8] proposed a local search algorithm using a neighborhood defined by exchanges of pairs of tasks in different processors. Frangioni et al. [12] proposed new local search algorithms for the minimum makespan processor scheduling problem, which perform multiple exchanges of jobs among machines. The latter are modelled as special disjoint cycles and paths in a suitably defined improvement graph. Several algorithms for searching the neighborhood are suggested and computational experiments are performed for the case of identical processors.

## 3   Hybrid Improvement Heuristic to $P\|C_{m\,ax}$

The hybrid improvement heuristic to $P\|C_{\max}$ is described in this section. The core of this procedure is formed by the construction, redistribution, and improvement phases, as illustrated by the pseudo-code of procedure C+R+I in Figure 1. It depends on two parameters: the target makespan Target and the maximum number of iterations MaxIterations performed by the tabu search procedure used in the improvement phase. The loop in lines 1–8 makes three attempts to build a feasible solution $S$ to the bin packing problem defined by the processing times $t_i, i = 1, \ldots, n$, with bin capacity Target, using exactly $m$ bins. Each of the heuristics H1, H2, and H3 is used at each attempt in line 2. If $S$ is feasible to the associated bin packing problem, then it is returned in line 3. Otherwise, load

```
procedure C+R+I(Target, MaxIterations);
1   for k = 1, 2, 3 do
2       Build a solution S = {A₁, ..., Aₘ} to P‖Cₘ ₐₓ using heuristic Hk;
3       if Cₘ ₐₓ(S) ≤ Target then return S;
4       S ← Redistribution(S);
5       if Cₘ ₐₓ(S) ≤ Target then return S;
6       S ← TabuSearch(S, MaxIterations);
7       if Cₘ ₐₓ(S) ≤ Target then return S;
8   end
9   return S;
end C+R+I
```

**Fig. 1.** Pseudo-code of the core `C+R+I` procedure.

redistribution is performed in line 4 to improve processor usability and the modified solution $S$ is returned in line 5 if it is feasible to the bin packing problem. Finally, a tabu search procedure is applied in line 6 as an attempt to knock down the makespan of the current solution and the modified solution $S$ is returned in line 7 if it is feasible to the bin packing problem. Detailed descriptions of the redistribution and improvement phases are reported in [1].

The pseudo-code of the complete hybrid improvement heuristic HI_PCmax to $P\|C_{\max}$ is given in Figure 2. An initial solution $S$ is built in line 1 using heuristic H3. The lower bound $L_2$ is computed in line 2. If the current lower and upper bounds coincide, then solution $S$ is returned in line 3. The lower bound $L_3$ is computed in line 4 and the current lower bound is updated. If the current lower and upper bounds coincide, then solution $S$ is returned in line 5. The lower bound $L_\vartheta$ is computed in line 6 and the current lower bound is updated. If the current lower and upper bounds coincide, then solution $S$ is returned in line 7. A new solution $S'$ is built in line 8 using heuristic H2. The currently best solution and the current upper bound are updated in line 9, while the current lower bound is updated in line 10. If the current lower and upper bounds coincide, then the currently best solution $S$ is returned in line 11. A new solution $S'$ is built in line 12 using heuristic H1. The currently best solution and the current upper bound are updated in line 13. If the current lower and upper bounds coincide, then the currently best solution $S$ is returned in line 14. At this point, $UB$ is the upper bound associated with the currently best known solution $S$ to $P\|C_{\max}$ and $LB$ is an unattained makespan. The core procedure `C+R+I` makes an attempt to build a solution with makespan equal to the current lower bound in line 15. The currently best solution and the current upper bound are updated in line 16. If the current lower and upper bounds coincide, then the currently best solution $S$ is returned in line 17. The loop in lines 18–23 implements a binary search strategy seeking for progressively better solutions. The target makespan $C_{\max} = \lfloor (LB + UB)/2 \rfloor$ is set in line 19. Let $S'$ be the solution obtained by the core procedure `C+R+I` applied in line 20 using $C_{\max}$ as the target makespan. If its makespan is at least as good as the target makespan

$C_{\max}$, then the current upper bound $UB$ and the currently best solution $S$ are updated in line 21. Otherwise, the unattained makespan $LB$ is updated in line 22, since the core procedure C+R+I was not able to find a feasible solution with the target makespan. The best solution found $S$ is returned in line 24.

The core procedure C+R+I is applied at two different points: once in line 15 using the lower bound $LB$ as the target makespan and in line 20 at each iteration of the binary search strategy using $C_{\max}$ as the target makespan. This implementation follows the same EBS (binary search with prespecified entry point) scheme suggested in [24]. Computational experiments have shown that it is able to find better solutions in smaller computation times than other variants which do not explore the binary search strategy or do not make a preliminary attempt to build a solution using $LB$ as the target makespan.

## 4   Computational Experiments

All computational experiments were performed on a 2.4 GHz AMD XP machine with 512 MB of RAM memory.

---

**procedure HI_PCmax(MaxIterations);**
1   Compute a solution $S$ using heuristic H3 and set $UB \leftarrow C_{\max}(S)$;
2   Compute the lower bound $L_2$ and set $LB \leftarrow L_2$;
3   **if** $LB = UB$ **then return** $S$;
4   Compute $L_3$ using binary search in the interval $[LB, UB]$ and set $LB \leftarrow L_3$;
5   **if** $LB = UB$ **then return** $S$;
6   Compute $L_\vartheta$ using $LB$ and $UB$ and update $LB \leftarrow \max\{LB, L_\vartheta\}$;
7   **if** $LB = UB$ **then return** $S$;
8   Compute a solution $S'$ and the lower bound $L_{HS}$ using heuristic H2;
9   **if** $C_{\max}(S') < UB$ **then set** $UB \leftarrow C_{\max}(S')$ and $S \leftarrow S'$;
10  Update $LB \leftarrow \max\{LB, L_{HS}\}$;
11  **if** $LB = UB$ **then return** $S$
12  Compute a solution $S'$ using heuristic H1;
13  **if** $C_{\max}(S') < UB$ **then set** $UB \leftarrow C_{\max}(S')$ and $S \leftarrow S'$;
14  **if** $LB = UB$ **then return** $S$;
15  $S' \leftarrow$ C+R+I($LB$, MaxIterations);
16  **if** $C_{\max}(S') < UB$ **then set** $UB \leftarrow C_{\max}(S')$ and $S \leftarrow S'$;
17  **if** $LB = UB$ **then return** $S$;
18  **while** $LB < UB - 1$ **do**
19      $C_{\max} \leftarrow \lfloor (LB + UB)/2 \rfloor$;
20      $S' \leftarrow$ C+R+I($C_{\max}$, MaxIterations);
21      **if** $C_{\max}(S') \leq C_{\max}$ **then set** $UB \leftarrow C_{\max}(S')$ and $S \leftarrow S'$;
22      **else** $LB \leftarrow C_{\max}$;
23  **end**
24  **return** $S$;
**end HI_PCmax**

---

**Fig. 2.** Pseudo-code of the hybrid improvement procedure to $P\|C_{\max}$.

Algorithms HI_PCmax and LPT were coded in C and compiled with version 2.95.2 of the gcc compiler with the optimization flag -O3. The maximum number of iterations during the tabu search improvement phase is set as MaxIterations = 1000. We compare the new heuristic HI_PCmax with the 3-PHASE heuristic of França et al. [10], the branch-and-bound algorithm B&B of Dell'Amico and Martello [7], and the ME multi-exchange algorithms of Frangioni et al. [12]. The code of algorithm B&B [7] was provided by Dell'Amico and Martello.

## 4.1   Test Problems

We considered two families of test problems: uniform and non-uniform. In these families, the number of processors $m$ takes values in $\{5, 10, 25\}$ and the number of tasks $n$ takes values in $\{50, 100, 500, 1000\}$ (the combination $m = 5$ with $n = 10$ is also tested). The processing times $t_i, i = 1, \ldots, n$ are randomly generated in the intervals $[1, 100]$, $[1, 1000]$, and $[1, 10000]$. Ten instances are generated for each of the 39 classes defined by each combination of $m$, $n$ and processing time interval.

The two families differ by the distribution of the processing times. The instances in the first family were generated by França et al. [10] with processing times uniformly distributed in each interval. The generator developed by Frangioni et al. [11] for the second family was obtained from [23]. For a given interval $[a, b]$ of processing times, with $a = 1$ and $b \in \{100, 1000, 10000\}$, their generator selects 98% of the processing times from a uniform distribution in the interval $[0.9(b - a), b]$ and the remaining 2% in the interval $[a, 0.2(b - a)]$.

## 4.2   Search Strategy

In the first computational experiment, we compare three different search methods that can be used with HI_PCmax.

In all three methods the core procedure C+R+I is used to check whether there exists a solution with a certain target makespan in the interval $[LB, UB]$. In the lower bound method (LBM), we start the search using $LB$ as the target makespan, which is progressively increased by one. In the binary search method (BSM), the interval $[LB, UB]$ is progressively bisected by setting $\lfloor (LB + UB)/2 \rfloor$ as the target makespan. The binary search with prespecified entry point method (EBS) is that used in the pseudo-code in Figure 2. It follows the same general strategy of BSM, but starts using $LB$ as the first target makespan.

Table 1 presents the results observed with each search method. For each family of test problems, we report the following statistics for the 130 instances with the same interval for the processing times: the total computation time in seconds, the maximum and the average number of executions of the core procedure C+R+I. These results show that EBS is consistently better than the other methods: the same solutions are always found by the three methods, with significantly smaller total times and fewer executions of C+R+I in the case of EBS.

**Table 1.** Search strategies: LBM, BSM, and EBS.

| | | Search method | | | | | | | | |
|---|---|---|---|---|---|---|---|---|---|---|
| | | LBM | | | BSM | | | EBS | | |
| Instances | $t_i$ | time (s) | max | avg | time (s) | max | avg | time (s) | max | avg |
| uniform | $[1, 100]$ | 0.17 | 1 | 1.00 | 0.17 | 1 | 1.00 | 0.14 | 1 | 1.00 |
| | $[1, 1000]$ | 3.02 | 14 | 2.88 | 2.16 | 6 | 3.38 | 1.98 | 6 | 2.38 |
| | $[1, 10000]$ | 33.42 | 190 | 1.71 | 11.77 | 10 | 6.50 | 11.67 | 10 | 6.23 |
| non-uniform | $[1, 100]$ | 16.07 | 8 | 1.73 | 21.56 | 5 | 3.56 | 14.02 | 4 | 1.47 |
| | $[1, 1000]$ | 47.08 | 26 | 1.77 | 54.14 | 9 | 6.06 | 21.94 | 6 | 1.15 |
| | $[1, 10000]$ | 499.77 | 254 | 9.18 | 491.21 | 12 | 8.98 | 83.11 | 10 | 2.02 |

### 4.3  Phases

In this experiment, we investigate the effect of the preprocessing, construction, redistribution, and improvement phases. Four variants of the hybrid improvement procedure HI_PCmax are created:

– Variant P: only lines 1–4 corresponding to the preprocessing phase of the pseudo-code in Figure 2 are executed.
– Variant P+C: the core procedure C+R+I is implemented without the redistribution and improvement phases.
– Variant P+C+R: the core procedure C+R+I is implemented without the improvement phase.
– Variant P+C+R+I: the core procedure C+R+I is fully implemented with all phases, corresponding to the complete HI_PCmax procedure itself.

Table 2 shows the results obtained with each variant. The differences between corresponding columns associated with consecutive variants give a picture of the effect of each additional phase. For each family of test poblems and for each interval of processing times, we report the number of optimal solutions found and the total computation time in seconds over all 130 instances.

**Table 2.** Phases: preprocessing, construction, redistribution, and improvement.

| | | Variants | | | | | | | |
|---|---|---|---|---|---|---|---|---|---|
| | | P | | C | | C+R | | C+R+I | |
| Instances | $p_i$ | opt. | time(s) | opt. | time (s) | opt. | time (s) | opt. | time (s) |
| uniform | $[1, 100]$ | 130 | 0.06 | 130 | 0.06 | 130 | 0.07 | 130 | 0.14 |
| | $[1, 1000]$ | 122 | 1.37 | 123 | 1.38 | 125 | 1.59 | 126 | 1.98 |
| | $[1, 10000]$ | 101 | 2.50 | 103 | 2.77 | 110 | 4.70 | 110 | 11.67 |
| | | 353 | | 356 | | 365 | | 366 | |
| non-uniform | $[1, 100]$ | 71 | 0.48 | 71 | 1.17 | 85 | 20.51 | 120 | 14.02 |
| | $[1, 1000]$ | 65 | 0.62 | 65 | 1.90 | 70 | 40.76 | 128 | 21.94 |
| | $[1, 10000]$ | 68 | 1.08 | 68 | 4.44 | 70 | 87.10 | 121 | 83.11 |
| | | 204 | | 204 | | 225 | | 369 | |

These results show that the uniform instances are relatively easy and the construction, redistribution, and improvement phases do not bring significative benefits in terms of solution quality or computation times. We notice that 90.5% of the 390 uniform instances are already solved to optimality after the preprocessing phase. The three additional phases allow solving only 13 other instances to optimality, at the cost of multiplying the total computation time by a factor of almost five. This picture is quite different for the non-uniform instances. In this case, only 204 out of the 390 test problems (52.3%) are solved to optimality after the preprocessing phase. The complete procedure with all phases made it possible to solve 165 additional instances to optimality.

In consequence of the order in which the three heuristics are applied in the preprocessing phase (lines 1, 8, and 12 of the pseudo-code in Figure 2), of the 557 optimal solutions found after this phase, 171 were obtained with the LPT heuristic H3, one with the $1/5 + 2^{-k}$-approximation heuristic H2, and 385 with the construction heuristic H1 proposed in Section 2. However, we note that if the lower bound $\max(L_2, L_3, L_\vartheta, L_{HS})$ is used, then heuristic H1 alone is capable of finding 556 out of the 557 optimal solutions obtained during the preprocessing phase. This shows that H1 is indeed a very useful fast heuristic to $P\|C_{\max}$.

## 4.4   Comparison with Other Approaches

In this final set of experiments, we compare the hybrid improvement heuristic HI_PCmax with the list scheduling algorithm LPT [15], the 3-PHASE heuristic of França et al. [10], and the branch-and-bound algorithm B&B of Dell'Amico and Martello [7] with the number of backtracks set at 4000 as suggested by the authors, as well as with the best solution found by the multi-exchange (ME) algorithms.

**Table 3.** Comparative results, uniform instances, $t_i \in [1, 100]$.

| | | LPT | | B&B | | | HI_PCmax | | | 3-PHASE |
|---|---|---|---|---|---|---|---|---|---|---|
| $m$ | $n$ | error | opt | error | opt | time (s) | error | opt | time (s) | error |
| 5 | 10 | 3.54e-03 | 9 | 0 | 10 | - | 0 | 10 | - | 0.018 |
| 5 | 50 | 4.58e-03 | 1 | 0 | 10 | - | 0 | 10 | - | 0.000 |
| 5 | 100 | 8.81e-04 | 4 | 0 | 10 | - | 0 | 10 | - | 0.000 |
| 5 | 500 | 0 | 10 | 0 | 10 | - | 0 | 10 | - | 0.000 |
| 5 | 1000 | 0 | 10 | 0 | 10 | - | 0 | 10 | - | 0.000 |
| 10 | 50 | 1.56e-02 | 0 | 0 | 10 | - | 0 | 10 | - | 0.002 |
| 10 | 100 | 3.64e-03 | 1 | 0 | 10 | - | 0 | 10 | - | 0.002 |
| 10 | 500 | 1.20e-04 | 7 | 0 | 10 | - | 0 | 10 | - | 0.000 |
| 10 | 1000 | 0 | 10 | 0 | 10 | - | 0 | 10 | - | 0.000 |
| 25 | 50 | 8.61e-03 | 6 | 0 | 10 | - | 0 | 10 | 0.01 | 0.011 |
| 25 | 100 | 2.37e-02 | 0 | 0 | 10 | - | 0 | 10 | - | 0.003 |
| 25 | 500 | 9.04e-04 | 4 | 0 | 10 | - | 0 | 10 | - | 0.000 |
| 25 | 1000 | 0 | 10 | 0 | 10 | - | 0 | 10 | - | 0.000 |

'-' is used to indicate negligible computation times.

**Table 4.** Comparative results, uniform instances, $t_i \in [1, 1000]$.

| | | LPT | | B&B | | | HI_PCmax | | | 3-PHASE |
|---|---|---|---|---|---|---|---|---|---|---|
| $m$ | $n$ | error | opt | error | opt | time (s) | error | opt | time (s) | error |
| 5 | 10 | 0 | 10 | 0 | 10 | - | 0 | 10 | - | 0.010 |
| 5 | 50 | 3.33e-03 | 0 | 0 | 10 | - | 0 | 10 | - | 0.001 |
| 5 | 100 | 1.02e-03 | 0 | 0 | 10 | - | 0 | 10 | - | 0.000 |
| 5 | 500 | 4.63e-05 | 1 | 0 | 10 | - | 0 | 10 | 0.02 | 0.000 |
| 5 | 1000 | 7.97e-06 | 5 | 0 | 10 | - | 0 | 10 | 0.05 | 0.000 |
| 10 | 50 | 1.61e-02 | 0 | 8.17e-05 | 8 | 0.01 | 7.74e-05 | 8 | 0.02 | 0.002 |
| 10 | 100 | 4.04e-03 | 0 | 0 | 10 | - | 0 | 10 | - | 0.000 |
| 10 | 500 | 2.21e-04 | 0 | 0 | 10 | - | 0 | 10 | 0.01 | 0.000 |
| 10 | 1000 | 4.42e-05 | 3 | 0 | 10 | - | 0 | 10 | 0.04 | 0.000 |
| 25 | 50 | 1.06e-02 | 4 | 0 | 10 | - | 0 | 10 | 0.02 | 0.011 |
| 25 | 100 | 3.15e-02 | 0 | 2.07e-04 | 8 | 0.11 | 9.93e-05 | 8 | 0.04 | 0.003 |
| 25 | 500 | 1.41e-03 | 0 | 0 | 10 | - | 0 | 10 | - | 0.000 |
| 25 | 1000 | 2.75e-04 | 0 | 0 | 10 | - | 0 | 10 | 0.01 | 0.000 |

'-' is used to indicate negligible computation times.

**Table 5.** Comparative results, uniform instances, $t_i \in [1, 10000]$.

| | | LPT | | B&B | | | HI_PCmax | | | 3-PHASE |
|---|---|---|---|---|---|---|---|---|---|---|
| $m$ | $n$ | error | opt | error | opt | time (s) | error | opt | time (s) | error |
| 5 | 10 | 6.48e-03 | 9 | 0 | 10 | - | 0 | 10 | 0.03 | 0.013 |
| 5 | 50 | 5.92e-03 | 0 | 9.26e-06 | 7 | 0.01 | 0 | 10 | 0.03 | 0.000 |
| 5 | 100 | 1.41e-03 | 0 | 0 | 10 | - | 0 | 10 | - | 0.000 |
| 5 | 500 | 4.87e-05 | 0 | 0 | 10 | - | 0 | 10 | 0.01 | 0.000 |
| 5 | 1000 | 1.02e-05 | 0 | 0 | 10 | - | 0 | 10 | 0.14 | 0.000 |
| 10 | 50 | 2.45e-02 | 0 | 1.14e-03 | 0 | 0.18 | 1.03e-04 | 0 | 0.25 | 0.004 |
| 10 | 100 | 4.82e-03 | 0 | 0 | 10 | - | 0 | 10 | - | 0.000 |
| 10 | 500 | 2.34e-04 | 0 | 0 | 10 | - | 0 | 10 | 0.01 | 0.000 |
| 10 | 1000 | 6.31e-05 | 0 | 0 | 10 | - | 0 | 10 | 0.04 | 0.000 |
| 25 | 50 | 7.25e-03 | 6 | 0 | 10 | - | 0 | 10 | 0.03 | 0.004 |
| 25 | 100 | 2.76e-02 | 0 | 4.49e-03 | 0 | 0.74 | 3.47e-04 | 0 | 0.59 | 0.001 |
| 25 | 500 | 1.00e-03 | 0 | 0 | 10 | - | 0 | 10 | 0.01 | $a$ |
| 25 | 1000 | 3.12e-04 | 0 | 0 | 10 | - | 0 | 10 | 0.02 | 0.000 |

'-' is used to indicate negligible computation times.
[a] not reported in [10].

Tables 3 to 5 report the results obtained by heuristics LPT, B&B, HI_PCmax, and 3-PHASE for the uniform instances. Tables 6 to 8 give the results obtained by heuristics LPT, B&B, HI_PCmax, and ME for the non-uniform instances. The following statistics over all ten test instances with the same combination of $m$ and $n$ are reported: (a) average relative errors with respect to the best lower bound for algorithms LPT, B&B, and HI_PCmax; (b) average relative errors reported in [10] for algorithm 3-PHASE; (c) average relative errors reported in [12] for the best among the solutions obtained by ME algorithms 1-SPT, 1-BPT, and K-SPT; (d) number of optimal solutions found by LPT, B&B, and HI_PCmax; (e) average computation times observed for LPT, B&B, and HI_PCmax on a 2.4 GHz AMD XP machine; and (f) average computation times reported in [12]

**Table 6.** Comparative results, non-uniform instances, $t_i \in [1, 100]$.

| | | LPT | | B&B | | | HI_PCmax | | | ME | |
|---|---|---|---|---|---|---|---|---|---|---|---|
| $m$ | $n$ | error | opt | error | opt | time (s) | error | opt | time (s) | error | time (s) |
| 5 | 10 | 0 | 10 | 0 | 10 | - | 0 | 10 | - | 0 | - |
| 5 | 50 | 9.37e-03 | 0 | 7.24e-03 | 0 | 0.11 | 0 | 10 | 0.03 | 8.58e-03 | 0.01 |
| 5 | 100 | 1.67e-02 | 0 | 0 | 10 | - | 0 | 10 | 0.01 | 5.31e-05 | 0.02 |
| 5 | 500 | 5.86e-04 | 0 | 0 | 10 | 0.01 | 0 | 10 | 0.01 | 0 | 1.01 |
| 5 | 1000 | 1.44e-04 | 0 | 0 | 10 | - | 0 | 10 | 0.02 | 0 | 9.71 |
| 10 | 50 | 1.13e-02 | 0 | 8.34e-03 | 2 | 0.23 | 7.28e-03 | 4 | 0.30 | 1.57e-02 | - |
| 10 | 100 | 9.02e-03 | 0 | 7.96e-03 | 0 | 0.21 | 2.13e-04 | 8 | 0.22 | 5.09e-03 | 0.04 |
| 10 | 500 | 1.00e-02 | 0 | 0 | 10 | 0.02 | 0 | 10 | - | 2.13e-05 | 3.26 |
| 10 | 1000 | 3.83e-04 | 1 | 0 | 10 | 0.01 | 0 | 10 | - | 0 | 17.14 |
| 25 | 50 | 0 | 10 | 0 | 10 | - | 0 | 10 | - | 0 | 0.01 |
| 25 | 100 | 4.77e-03 | 0 | 2.65e-03 | 3 | 1.12 | 1.34e-03 | 8 | 0.55 | 9.85e-03 | 0.04 |
| 25 | 500 | 9.79e-03 | 0 | 1.60e-04 | 8 | 1.63 | 0 | 10 | 0.08 | 2.12e-04 | 3.29 |
| 25 | 1000 | 9.30e-03 | 0 | 1.17e-03 | 4 | 9.56 | 0 | 10 | 0.18 | 7.97e-05 | 36.59 |

'-' is used to indicate negligible computation times.

**Table 7.** Comparative results, non-uniform instances, $t_i \in [1, 1000]$.

| | | LPT | | B&B | | | HI_PCmax | | | ME | |
|---|---|---|---|---|---|---|---|---|---|---|---|
| $m$ | $n$ | error | opt | error | opt | time (s) | error | opt | time (s) | error | time (s) |
| 5 | 10 | 0 | 10 | 0 | 10 | - | 0 | 10 | - | 0 | - |
| 5 | 50 | 8.82e-03 | 0 | 7.94e-03 | 0 | 0.20 | 0 | 10 | 0.03 | 8.92e-03 | 0.01 |
| 5 | 100 | 1.70e-02 | 0 | 1.55e-04 | 9 | 0.02 | 0 | 10 | 0.02 | 7.43e-05 | 0.06 |
| 5 | 500 | 6.35e-04 | 0 | 0 | 10 | - | 0 | 10 | - | 0 | 1.39 |
| 5 | 1000 | 1.78e-04 | 0 | 0 | 10 | - | 0 | 10 | 0.02 | 0 | 14.80 |
| 10 | 50 | 4.08e-03 | 0 | 1.57e-03 | 0 | 0.42 | 0 | 10 | - | 1.46e-02 | 0.01 |
| 10 | 100 | 8.66e-03 | 0 | 8.18e-03 | 0 | 0.39 | 0 | 10 | 0.35 | 4.64e-03 | 0.07 |
| 10 | 500 | 1.01e-02 | 0 | 0 | 10 | 0.01 | 0 | 10 | 0.07 | 0 | 5.61 |
| 10 | 1000 | 4.12e-04 | 0 | 0 | 10 | 0.11 | 0 | 10 | 0.06 | 0 | 14.61 |
| 25 | 50 | 0 | 10 | 0 | 10 | - | 0 | 10 | - | 0 | 0.01 |
| 25 | 100 | 4.67e-03 | 0 | 3.80e-03 | 0 | 1.06 | 1.34e-03 | 8 | 1.08 | 7.93e-03 | 0.08 |
| 25 | 500 | 1.00e-02 | 0 | 1.39e-03 | 1 | 5.53 | 0 | 10 | 0.13 | 4.25e-05 | 15.39 |
| 25 | 1000 | 9.38e-03 | 0 | 7.97e-06 | 9 | 0.73 | 0 | 10 | 0.43 | 7.97e-06 | 138.21 |

'-' is used to indicate negligible computation times.

for the best ME algorithm on a 400 MHz Pentium II with 256 Mbytes of RAM memory.

Most uniform instances are easy and can be solved in negligible computation times. Tables 3 to 5 show that HI_PCmax found better solutions than LPT, B&B, and 3-PHASE for all classes of test problems. Only four instances in Table 4 and 20 instances in Table 5 were not solved to optimality by HI_PCmax. B&B outperformed LPT and 3-PHASE, but found slightly fewer optimal solutions and consequently slightly larger relative errors than HI_PCmax. We also notice that the uniform test instances get harder when the range of the processing times increase.

The non-uniform test instances are clearly more difficult that the uniform. Once again, HI_PCmax outperformed the other algorithms considered in Tables 6 to 8 in terms of solution quality and computation times. This conclusion

**Table 8.** Comparative results, non-uniform instances, $t_i \in [1, 10000]$.

| | | LPT | | B&B | | | HI_PCmax | | | ME | |
|---|---|---|---|---|---|---|---|---|---|---|---|
| $m$ | $n$ | error | opt | error | opt | time (s) | error | opt | time (s) | error | time (s) |
| 5 | 10 | 0 | 10 | 0 | 10 | - | 0 | 10 | - | 0 | - |
| 5 | 50 | 8.79e-03 | 0 | 8.11e-03 | 0 | 0.28 | 0 | 10 | 0.02 | 8.95e-03 | 0.01 |
| 5 | 100 | 1.70e-02 | 0 | 1.00e-04 | 8 | 0.05 | 0 | 10 | 0.01 | 5.78e-05 | 0.09 |
| 5 | 500 | 6.42e-04 | 0 | 0 | 10 | - | 0 | 10 | - | 0 | 1.97 |
| 5 | 1000 | 1.78e-04 | 0 | 0 | 10 | - | 0 | 10 | 0.03 | 0 | 13.88 |
| 10 | 50 | 4.12e-03 | 0 | 2.02e-03 | 0 | 0.54 | 4.22e-06 | 8 | 0.35 | 1.46e-02 | 0.01 |
| 10 | 100 | 8.61e-03 | 0 | 8.28e-03 | 0 | 0.52 | 0 | 10 | 0.24 | 4.63e-03 | 0.15 |
| 10 | 500 | 1.02e-02 | 0 | 0 | 10 | 0.02 | 0 | 10 | 0.01 | 1.06e+00 | 7.99 |
| 10 | 1000 | 4.10e-04 | 0 | 0 | 10 | 0.01 | 0 | 10 | 0.02 | 0 | 15.57 |
| 25 | 50 | 0 | 10 | 0 | 10 | - | 0 | 10 | - | 0 | 0.01 |
| 25 | 100 | 4.73e-03 | 0 | 4.16e-03 | 0 | 1.34 | 1.35e-03 | 3 | 4.29 | 7.76e-03 | 0.14 |
| 25 | 500 | 1.01e-02 | 0 | 7.23e-04 | 1 | 12.70 | 0 | 10 | 3.13 | 1.91e-05 | 20.37 |
| 25 | 1000 | 9.40e-03 | 0 | 1.86e-06 | 8 | 5.55 | 0 | 10 | 0.22 | 5.31e-06 | 195.88 |

'-' is used to indicate negligible computation times.

is particularly true if one compares the results observed for the largest test instances with $m = 25$ and $n \geq 250$.

Table 9 summarizes the main results obtained by algorithms HI_PCmax and B&B on the same computational environment. For each group of test problems and for each algorithm, it indicates the number of optimal solutions found over the 130 instances, the average and maximum absolute errors, the average and maximum relative errors, and the average and maximum computation times. The superiority of HI_PCmax is clear for the non-uniform instances. It not only found better solutions, but also in smaller computation times.

**Table 9.** Comparative results: HI_PCmax vs. B&B.

| | | HI_PCmax | | | | | |
|---|---|---|---|---|---|---|---|
| | | opt | absolute error | | relative error | | time (s) | |
| Instances | $t_i \in$ | | avg | max | avg | max | avg | max |
| uniform | $[1, 100]$ | 130 | 0.00 | 0 | 0 | 0 | 0 | 0.09 |
| | $[1, 1000]$ | 126 | 0.03 | 1 | 1.36e-05 | 5.15e-04 | 0.02 | 0.19 |
| | $[1, 10000]$ | 110 | 0.75 | 12 | 3.46e-05 | 5.83e-04 | 0.09 | 0.77 |
| non-uniform | $[1, 100]$ | 120 | 0.32 | 7 | 6.79e-04 | 1.50e-02 | 0.11 | 3.76 |
| | $[1, 1000]$ | 128 | 0.38 | 25 | 1.03e-04 | 6.71e-03 | 0.17 | 9.49 |
| | $[1, 10000]$ | 121 | 3.90 | 253 | 1.04e-04 | 6.75e-03 | 0.64 | 19.62 |

| | | B&B | | | | | |
|---|---|---|---|---|---|---|---|
| | | opt | absolute error | | relative error | | time (s) | |
| Instances | $t_i \in$ | | avg | max | avg | max | avg | max |
| uniform | $[1, 100]$ | 130 | 0.00 | 0 | 0 | 0 | 0 | 0.01 |
| | $[1, 1000]$ | 126 | 0.05 | 3 | 2.22e-05 | 1.55e-03 | 0.01 | 0.75 |
| | $[1, 10000]$ | 107 | 9.31 | 173 | 4.33e-04 | 8.30e-03 | 0.07 | 1.12 |
| non-uniform | $[1, 100]$ | 87 | 1.84 | 20 | 2.12e-03 | 1.50e-02 | 0.99 | 35.41 |
| | $[1, 1000]$ | 79 | 15.59 | 152 | 1.77e-03 | 9.42e-03 | 0.65 | 9.99 |
| | $[1, 10000]$ | 77 | 150.01 | 880 | 1.80e-03 | 9.73e-03 | 1.62 | 23.60 |

# 5    Concluding Remarks

We proposed a new strategy for solving the multiprocessor scheduling problem, based on the application of a hybrid improvement heuristic to the bin packing problem. We also presented a new, quick construction heuristic, combining the LPT rule with the solution of subset sum problems.

The construction heuristic revealed itself as a very effective approximate algorithm and found optimal solutions for a large number of test problems. The improvement heuristic outperformed the other approximate algorithms in the literature, in terms of solution quality and computation times. The computational results are particularly good in the case of non-uniform test instances.

**Acknowledgments:** The authors are grateful to M. Dell'Amico for having kindly provided the code of B&B algorithm used in the computational experiments. We are also thankful to P. França for making available the instances of the uniform family.

# References

1. A.C.F. Alvim, C.C. Ribeiro, F. Glover, and D.J. Aloise, "A hybrid improvement heuristic for the one-dimensional bin packing problem", *Journal of Heuristics*, 2004, to appear.

2. A.C.F. Alvim, *Uma heurística híbrida de melhoria para o problema de bin packing e sua aplicação ao problema de escalonamento de tarefas*, Doctorate thesis, Catholic University of Rio de Janeiro, Department of Computer Science, Rio de Janeiro, 2003.

3. J. Błażewicz, "Selected topics in scheduling theory", in *Surveys in Combinatorial Optimization* (G. Laporte, S. Martello, M. Minoux, and C.C. Ribeiro, eds.), pages 1–60, North-Holland, 1987.

4. J.L. Bruno, E.G. Coffman Jr., and R. Sethi, "Scheduling independent tasks to reduce mean finishing time", *Communications of the ACM* 17 (1974), 382–387.

5. T. Cheng and C. Sin, "A state-of-the-art review of parallel-machine scheduling research", *European Journal of Operational Research* 47 (1990), 271–292.

6. E.G. Coffman Jr., M.R. Garey, and D.S. Johnson, "An application of bin-packing to multiprocessor scheduling", *SIAM Journal on Computing* 7 (1978), 1–17.

7. M. Dell'Amico and S. Martello, "Optimal scheduling of tasks on identical parallel processors", *ORSA Journal on Computing* 7 (1995), 191–200.

8. S.M. Fatemi-Ghomi and F. Jolai-Ghazvini, "A pairwise interchange algorithm for parallel machine scheduling", *Production Planning and Control* 9 (1998), 685–689.

9. G. Finn and E. Horowitz, "A linear time approximation algorithm for multiprocessor scheduling", *BIT* 19 (1979), 312–320.

10. P.M. França, M. Gendreau, G. Laporte, and F.M. Müller, "A composite heuristic for the identical parallel machine scheduling problem with minimum makespan objective", *Computers Ops. Research* 21 (1994), 205–210.

11. A. Frangioni, M. G. Scutellà, and E. Necciari, "Multi-exchange algorithms for the minimum makespan machine scheduling problem", Report TR-99-22, Dipartimento di Informatica, Università di Pisa, Pisa, 1999.

12. A. Frangioni, E. Necciari, and M. G. Scutellà, "A multi-exchange neighborhood for minimum makespan machine scheduling problems", *Journal of Combinatorial Optimization*, to appear.
13. D.K. Friesen, "Tighter bounds for the MULTIFIT processor scheduling algorithm", *SIAM Journal on Computing* 13 (1984), 170–181.
14. M.R. Garey and D.S. Johnson, *Computers and Intractability: A Guide to the Theory of NP-Completeness*, W.H. Freeman and Company, 1979.
15. R.L. Graham, "Bounds on multiprocessing timing anomalies", *SIAM Journal of Applied Mathematics* 17 (1969),416–429.
16. R.L. Graham, E.L. Lawler, J.K. Lenstra, and A.H.G. Rinnooy Kan, "Optimization and approximation in deterministic sequencing and scheduling: A survey", *Annals of Discrete Mathematics* 5 (1979), 287–326.
17. D.S. Hochbaum and D. B. Shmoys, "Using dual approximation algorithms for scheduling problems: Theoretical and practical results", *Journal of the ACM* 34 (1987), 144–162.
18. R. Hübscher and F. Glover, "Applying tabu search with influential diversification to multiprocessor scheduling", *Computers and Operations Research* 21 (1994), 877–884.
19. D.S. Johnson, A. Demers, J.D. Ullman, M.R. Garey, and R.L. Graham, "Worst case performance bounds for simple one-dimensional packing algorithms", *SIAM Journal on Computing* 3 (1974), 299–325.
20. E.L. Lawler, J. K. Lenstra, A.H.G. Rinnooy Kan, and D.B. Shmoys, "Sequencing and scheduling: Algorithms and complexity", in *Logistics of Production and Inventory: Handbooks in Operations Research and Management Science* (S.C. Graves, P.H. Zipkin, and A.H.G. Rinnooy Kan, eds.), 445–522, North-Holland, 1993.
21. S. Martello and P. Toth, *Knapsack Problems: Algorithms and Computer Implementations*, Wiley, 1990.
22. McNaughton, "Scheduling with deadlines and loss functions", *Management Science* 6 (1959), 1–12.
23. E. Necciari,"Istances of machine scheduling problems". Online document available at http://www.di.unipi.it/di/groups/optimize/Data/MS.html, last visited on November 21, 2001.
24. A. Scholl and S. Voss, "Simple assembly line balancing - Heuristic approaches",*Journal of Heuristics* 2 (1996), 217–244.
25. M. Yue, "On the exact upper bound for the MULTIFIT processor scheduling algorithm", *Annals of Operations Research* 24 (1990), 233–259.

# Efficient Edge-Swapping Heuristics for Finding Minimum Fundamental Cycle Bases

Edoardo Amaldi, Leo Liberti, Nelson Maculan\*, and Francesco Maffioli

DEI, Politecnico di Milano, Piazza L. da Vinci 32, 20133 Milano, Italy.
{amaldi,liberti,maculan,maffioli}@elet.polimi.it

**Abstract.** The problem of finding a fundamental cycle basis with minimum total cost in a graph is NP-hard. Since fundamental cycle bases correspond to spanning trees, we propose new heuristics (local search and metaheuristics) in which edge swaps are iteratively applied to a current spanning tree. Structural properties that make the heuristics efficient are established. We also present a mixed integer programming formulation of the problem whose linear relaxation yields tighter lower bounds than known formulations. Computational results obtained with our algorithms are compared with those from existing constructive heuristics on several types of graphs.

## 1 Introduction

Let $G = (V, E)$ be a simple, undirected graph with $n$ nodes and $m$ edges, weighted by a non-negative cost function $w : E \to \mathbb{R}^+$. A *cycle* is a subset $C$ of $E$ such that every node of $V$ is incident with an even number of edges in $C$. Since an elementary cycle is a connected cycle such that at most two edges are incident to any node, cycles can be viewed as the (possibly empty) union of edge-disjoint elementary cycles. If cycles are considered as edge-incidence binary vectors in $\{0,1\}^{|E|}$, it is well-known that the cycles of a graph form a vector space over $GF(2)$. A set of cycles is a *cycle basis* if it is a basis in this cycle vector space associated to $G$. The cost of a cycle is the sum of the costs of all edges contained in the cycle. The cost of a set of cycles is the sum of the costs of all cycles in the set. Given any spanning tree of $G$ characterized by an edge set $T \subseteq E$, the edges in $T$ are called *branches* of the tree, and those in $E \backslash T$ (the co-tree) are called the *chords* of $G$ with respect to $T$. Any chord uniquely identifies a cycle consisting of the chord itself and the unique path in $T$ connecting the two nodes incident on the chord. These $m - n + 1$ cycles are called *fundamental cycles* and they form a *Fundamental Cycle Basis* (FCB) of $G$ with respect to $T$. It turns out [1] that a cycle basis is fundamental if and only if each cycle in the basis contains at least one edge which is not contained in any other cycle in the basis. In this paper we consider the problem of finding Minimum Fundamental Cycle Bases (MIN FCB) in graphs, that is FCBs with minimum

---

\* On academic leave from COPPE, Universidade Federal do Rio de Janeiro, Brazil, e-mail: maculan@cos.ufrj.br.

C.C. Ribeiro and S.L. Martins (Eds.): WEA 2004, LNCS 3059, pp. 14–29, 2004.
© Springer-Verlag Berlin Heidelberg 2004

total cost. Since the cycle space of a graph is the direct sum of the cycle spaces of its biconnected components, we assume that $G$ is biconnected, i.e., $G$ contains at least two edge-disjoint paths between any pair of nodes.

Cycle bases have been used in the field of electrical networks since the time of Kirchoff [2]. Fundamental cycle bases can be uniquely identified by their corresponding spanning trees, and can therefore be represented in a highly compact manner. Besides the above-mentioned characterization, Sysło established several structural results concerning FCBs [3,1,4]. For example, two spanning trees whose symmetric difference is a collection of 2-paths (paths where each node, excluding the endpoints, has degree 2) give rise to the same FCB [1]. Although the problem of finding a minimum cycle basis can be solved in polynomial time (see [5] and the recent improvement [6]), requiring fundamentality makes the problem NP-hard [7]. In fact, it does not admit a polynomial-time approximation scheme (PTAS) unless P=NP; that is, under the same assumption there exists no polynomial-time algorithm that guarantees a solution within a factor of $1 + \varepsilon$ for every instance and for any $\varepsilon > 0$ [8]. In the same work, a $4 + \varepsilon$ approximation algorithm is presented for complete graphs, and a $2^{O(\sqrt{\log n \log \log n})}$ approximation algorithm for arbitrary graphs.

Interest in minimum FCBs arises in a variety of application fields, such as electrical circuit testing [9], periodic timetable planning [10] and generating minimal perfect hash functions [11].

The paper is organized as follows. In Section 2 we describe a local search algorithm in which the spanning tree associated to the current FCB is iteratively modified by performing edge swaps, and we establish structural results that make its implementation efficient. In Section 3 the same type of edge swaps is adopted within two metaheuristic schemes, namely a variable neighbourhood search and a tabu search. To provide lower bounds on the cost of optimal solutions, a new mixed integer programming (MIP) formulation of the problem is presented in Section 4. Computational results are reported and discussed in Section 5.

## 2   Edge-Swapping Local Search

In our local search algorithm for the MIN FCB problem, we start from the spanning tree associated to an initial FCB. At each iteration we swap a branch of the current spanning tree with one of its chords until the cost cannot be further decreased, i.e., a local minimum is found.

### 2.1   Initial Feasible Solutions

Initial solutions are obtained by applying a very fast "tree-growing" procedure [12], where a spanning tree and its corresponding FCB are grown by adding nodes to the tree according to predefined criteria. The adaptation of Paton's procedure to the MIN FCB problem proceeds as follows. The node set of the initial tree $V_T$ only contains a root node $v_0$, and the set $X$ of nodes to be examined is initialized at $V$. At each step a node $u \in X \cap V_T$ (not yet examined) is selected according to a predefined ordering. For all nodes $z$ adjacent to $u$, if $z \notin V_T$, the edge $\{z, u\}$ is included in $T$ (the edge is selected), the node $z$ is added to $V_T$ and the node $u$ is removed from $X$. Nodes to be examined are selected according to non-increasing degree and, to break ties, to increasing edge star costs. The resulting order tends to maximize the chances of finding very short fundamental cycles early in the process. The performance of this tree-growing procedure is comparable to other existing tree-growing techniques [7,11].

## 2.2   Edge Swap

Using edge swaps to search the solution space of the MIN FCB problem is a good strategy, since all spanning trees of a graph can be obtained from any initial spanning tree by the repeated application of edge swaps [13]. Consider any given spanning tree $T$ of $G$. For each branch $e$ of $T$, the removal of $e$ from $T$ induces the partition of the node set $V$ into two subsets $S_T^e$ and $\bar{S}_T^e$. Denote by $\delta_T^e$ the *fundamental cut* of $G$ induced by the branch $e$ of $T$, i.e., $\delta_T^e = \delta(S_T^e) = \{\{u, v\} \in E \mid u \in S_T^e, v \in \bar{S}_T^e\}$. For any chord $f \in \delta_e^T$, let $\pi = (e, f)$ be the edge swap which consists in removing the branch $e$ from $T$ while adding $f$ to $T$. Denote by $\pi T$ the resulting spanning tree.

---

Let $T$ be the initial spanning tree constructed as in Section 2.1;
**loop**
    $\Delta_{\mathrm{opt}} := 0$;
    initialize $\pi_{\mathrm{opt}}$ to the identity;
    **for all** $e \in T$
        **for all** $f \in \delta_T^e$ with $f \neq e$
            $\pi := (e, f)$;
            **if** $\Delta_\pi \geq \Delta_{\mathrm{opt}}$ **then**
                $\pi_{\mathrm{opt}} = \pi$;
                $\Delta_{\mathrm{opt}} = \Delta_\pi$;
            **end if**
        **end for**
    **end for**
    **if** $\pi_{\mathrm{opt}}$ is not the identity **then**
        $T := \pi_{\mathrm{opt}} T$;
    **end if**
**until** $\pi_{\mathrm{opt}}$ is the identity

---

**Fig. 1.** Local search algorithm for the MIN FCB problem.

For any spanning tree $T$, let $\mathcal{C}(T)$ be the set of cycles in the FCB associated to $T$, and let $w(\mathcal{C}(T))$ denote the FCB total cost (the function $w$ is extended to sets of edges in the obvious way: $w(F) = \sum_{f \in F} w(f)$ for any $F \subseteq E$). We are interested in finding edge swaps $\pi = (e, f)$, where $e$ is a branch of the current tree $T$ and $f$ is a chord in the fundamental cut $\delta_T^e$ induced by $e$, such that $w(\mathcal{C}(\pi T)) < w(\mathcal{C}(T))$. For each branch $e$ and chord $f$, the cost difference $\Delta_\pi = w(\mathcal{C}(T)) - w(\mathcal{C}(\pi T))$ must be computed. Let $\Delta_{\text{opt}}$ be the largest such $\Delta_\pi$, and $\pi_{\text{opt}}$ be the corresponding edge swap. If $\Delta_{\text{opt}} \leq 0$ we let $\pi_{\text{opt}}$ be the identity permutation.

The local search algorithm based on this edge-swapping operation is summarized in Fig. 1.

## 2.3   Efficient Implementation

Given the high worst-case computational complexity of each iteration of the basic local search procedure (see Section 2.4), an efficient implementation is of foremost importance. Since applying an edge swap to a spanning tree may change the fundamental cycles and cut structure considerably, efficient procedures are needed to determine the cuts $\delta_{\pi T}^e$ for all $e \in \pi T$, and to compute $\Delta_\pi$ from the data at the previous iteration, namely from $T$, $\pi$ and the cuts $\delta_T^e$, for $e \in T$.

**Edge swap effect on the cuts.** In this subsection we prove that any edge swap $\pi = (e, f)$ applied to a spanning tree $T$, where $e \in T$ and $f \in \delta_T^e$, changes a cut $\delta_T^h$ if and only if $f$ is also in $\delta_T^h$. Furthermore, $\pi(\delta_T^h) = \delta_{\pi T}^h$ is the symmetric difference $\delta_T^h \triangle \delta_T^e$. This makes it easy to maintain data structures relative to the cuts that can be updated efficiently when $\pi$ is applied to $T$.

For each pair of nodes $u, v \in V$ let $\langle u, v \rangle$ be the unique path in $T$ from $u$ to $v$. Let $e = \{u_e, v_e\} \in T$ be an edge of the spanning tree and $c = \{u_c, v_c\} \notin T$ be a chord, where the respective endpoints $u_e, v_e, u_c, v_c$ are nodes in $V$. Let $p_1(e, c) = \langle u_e, u_c \rangle$, $p_2(e, c) = \langle u_e, v_c \rangle$, $p_3(e, c) = \langle v_e, u_c \rangle$, $p_4(e, c) = \langle v_e, v_c \rangle$ and $P_T(e, c) = \{p_i(e, c) \subseteq T \mid i = 1, \dots, 4\}$. Note that exactly two paths in $P_T(e, c)$ do not contain $e$. Let $\bar{P}_T(e, c)$ denote the subset of $P_T(e, c)$ composed of those two paths not containing $e$. Let $P_T^*(e, c)$ be whichever of the sets $\{p_1(e, c), p_4(e, c)\}$, $\{p_2(e, c), p_3(e, c)\}$ has shortest total path length in $T$ (see Fig. 2). In the sequel, with a slight abuse of notation, we shall sometimes say that an edge belongs to a set of nodes, meaning that its endpoints belong to that set of nodes. For a path $p$ and a node set $N \subseteq V(G)$ we say $p \subseteq N$ if the edges of $p$ are in the edge set $E(G_N)$ (i.e., the edges of the subgraph of $G$ induced by $N$). Furthermore, we shall say that a path connects two edges $e, f$ if it connects an endpoint of $e$ to an endpoint of $f$.

**Lemma 2.1.** *For any branch $e \in T$ and chord $c \in E \backslash T$, we have $c \in \delta_T^e$ if and only if $\bar{P}_T(e, c) = P_T^*(e, c)$.*

*Proof.* First assume that $c \in \delta_T^e$. Denoting by $u_c, v_c$ the endpoints of $c$ and by $u_e, v_e$ those of $e$, we can assume w.l.o.g. that $u_c, v_c, u_e, v_e$ are labeled so that $u_c, u_e \in S_T^e$ and $v_c, v_e \in \bar{S}_T^e$. Since there is a unique shortest path $q$ in $T$ connecting $u_c$ to $v_c$ with $u_c \in S_T^e$, $v_c \in \bar{S}_T^e$, then $e \in p$. Thus, there are unique shortest sub-paths $q_1, q_2$ of $q$ such that $q_1 = \langle u_e, u_c \rangle$, $q_2 = \langle v_e, v_c \rangle$ and $q_1 \subseteq$

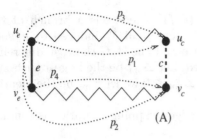 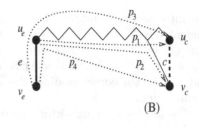

**Fig. 2.** (A) If $c$ is in the fundamental cut induced by $e$, $\bar{P}_T(e,c) = P_T^*(e,c) = \{p_1, p_4\}$. Otherwise, up to symmetries, we have the situation depicted in (B) where $\bar{P}_T(e,c) \neq P_T^*(e,c)$.

$S_T^e$, $q_2 \subseteq \bar{S}_T^e$. Hence $P_T^*(e,c) = \{q_1, q_2\} = \bar{P}_T(e,c)$. Conversely, let $P_T^*(e,c) = \{q_1, q_2\}$, and assume that $e \notin q_1, q_2$. Since either $q_1 \subseteq S_T^e$ and $q_2 \subseteq \bar{S}_T^e$ or vice versa, the endpoints of $c$ are separated by the cut $\delta_T^e$, i.e., $c \in \delta_T^e$. □

Let $T$ be a spanning tree for $G = (V, E)$ and $\pi = (e, f)$ an edge swap with $e \in T, f \in \delta_T^e$ and $f \neq e$. First we note that the cut in $G$ induced by $e$ of $T$ is the same as the cut induced by $f$ of $\pi T$.

**Proposition 2.2.** $\pi(\delta_T^e) = \delta_{\pi T}^f$.

*Proof.* Since $f \in \delta_T^e$, swapping $e$ with $f$ does not modify the partitions that induce the cuts, i.e., $S_T^e = S_{\pi T}^f$. □

Second, we show that the cuts that do not contain $f$ are not affected by $\pi$.

**Proposition 2.3.** *For each $h \in T$ such that $h \neq e$, and $f \notin \delta_T^h$, we have $\pi(\delta_T^h) = \delta_T^h$.*

*Proof.* Let $g \in \delta_T^h$. By Lemma 2.1, the shortest paths $p_1^T, p_2^T$ from the endpoints of $h$ to the endpoints of $g$ do not contain $h$. We shall consider three possibilities. (1) In the case where $e$ and $f$ do not belong either to $p_1^T$ or $p_2^T$ we obtain trivially that $\bar{P}_{\pi T}(e,c) = \bar{P}_T(e,c) = P_T^*(e,c) = P_{\pi T}^*(e,c)$ and hence the result. (2) Assume now that $e \in p_1^T$, and that both $e, f$ are in $S_T^h$. The permutation $\pi$ changes $p_1^T$ so that $f \in p_1^{\pi T}$, whilst $p_2^{\pi T} = p_2^T$. Now $p_1^{\pi T}$ is shortest because it is the unique path in $\pi T$ connecting the endpoints of $p_1^T$, and since $h \notin p_1^{\pi T}, p_2^{\pi T}$ because $\pi$ does not affect $h$, we obtain $\bar{P}_{\pi T}(e,c) = P_{\pi T}^*(e,c)$. (3) Suppose that $e \in p_1^T$, $e \in S_T^h$ and $f \in \bar{S}_T^h$. Since $f \in \delta_T^e$, by Lemma 2.1 there are shortest paths $q_1^T, q_2^T$ connecting the endpoints of $e$ and $f$ such that $q_1^T \subseteq S_T^e$, $q_2^T \subseteq \bar{S}_T^e$. Since $e \in S_T^h$, $f \in \bar{S}_T^h$ and $T$ is tree, there is an $i$ in $\{1, 2\}$ such that $h \in q_i^T$ (say, w.l.o.g. that $i = 1$); let $q_1^T = r_1^T \cup \{h\} \cup r_2^T$, where $r_1^T \subseteq S_T^h$ connects $h$ and $e$, and $r_2^T \subseteq \bar{S}_T^h$ connects $h$ and $f$. Let $q^T = r_1^T \cup \{e\} \cup q_2^T$, then $q^T$ is the unique path in $S_T^h$ connecting $h$ and $f$. Since $r_2^T$ connects $h$ and $f$ in $\bar{S}_T^h$, we must conclude that $f \in \delta_T^h$, which is a contradiction. □

Third, we prove that any cut containing $f$ is mapped by the edge swap $\pi = (e, f)$ to its symmetric difference with the cut induced by $e$ of $T$.

**Theorem 2.4.** *For each $h \in T$ such that $h \neq e$ and $f \in \delta_T^h$, we have $\pi(\delta_T^h) = \delta_T^h \triangle \delta_T^e$.*

Due to its length, the proof is given in the Appendix.

**Edge swap effect on the cycles.** In order to compute $\Delta_\pi$ efficiently, we have to determine how an edge swap $\pi = (e, f)$ affects the FCB corresponding to the tree $T$. For each chord $h$ of $G$ with respect to $T$, let $\gamma_T^h$ be the unique fundamental cycle induced by $h$.

**Fact** *If $h \notin \delta_T^e$, then $\gamma_T^h$ is unchanged by $\pi$.*

The next result characterizes the way $\pi$ acts on the cycles that are changed by the edge swap $\pi$.

**Theorem 2.5.** *If $h \in \delta_T^e$, then $\gamma_{\pi T}^h = \pi(\gamma_T^h) = \gamma_T^h \triangle \gamma_T^f$, where $\gamma_T^f$ is the fundamental cycle in $T$ corresponding to the chord $f$.*

*Proof.* We need the two following claims.

*Claim 1.* For all $h \in \delta_T^e$ such that $h \neq e$, $\gamma_T^h \cap \delta_T^e = \{e, h\}$.
*Proof.* Since $\gamma_T^h$ is the simple cycle consisting of $h$ and the unique path in $T$ connecting the endpoints of $h$ through $e$, the only edges both in the cycle and in the cut of $e$ are $e$ and $h$.

*Claim 2.* For all pairs of chords $g, h \in \delta_T^e$ such that $g \neq h$ there exists a unique simple cycle $\gamma \subseteq G$ such that $g \in \gamma$, $h \in \gamma$, and $\gamma \backslash \{g, h\} \subseteq T$.
*Proof.* Let $g = (g_1, g_2)$, $h = (h_1, h_2)$ and assume w.l.o.g. $g_1, h_1, g_2, h_2$ are labeled so that $g_1, h_1 \in S_T^e$ and $g_2, h_2 \in \bar{S}_T^e$. Since there exist unique paths $p \subseteq T$ connecting $g_1, h_1$ and $q \subseteq T$ connecting $g_2, h_2$, the edge subset $\gamma = \{g, h\} \cup p \cup q$ is a cycle with the required properties. Assume now that there is another cycle $\gamma'$ with the required properties. Then $\gamma'$ defines paths $p', q'$ connecting respectively $g_1, h_1$ and $g_2, h_2$ in $T$. Since $T$ is a spanning tree, $p = p'$ and $q = q'$; thus $\gamma' = \gamma$.

Consider the cycle $\gamma = \gamma_T^h \triangle \gamma_T^f$. By definition, $e \in \gamma_T^h$, $e \in \gamma_T^f$, $h \in \gamma_T^h$, $f \in \gamma_T^f$. Since $h, f \in \delta_T^e$, by Claim 1 $h \notin \gamma_T^f$ and $f \notin \gamma_T^h$. Thus $h \in \gamma$, $f \in \gamma$, $e \notin \gamma$. Consider now $\pi(\gamma_T^h)$. Since $e \in \gamma_T^h$ and $\pi = (e, f)$, $f \in \pi(\gamma_T^h)$. Furthermore, since $\pi$ fixes $h$, $h \in \pi(\gamma_T^h)$. Hence, by Claim 2, we have that $\pi(\gamma_T^h) = \gamma = \gamma_T^h \triangle \gamma_T^f$.  $\square$

## 2.4  Computational Complexity

We first evaluate the complexity of applying an edge swap to a given spanning tree and of computing the fundamental cut and cycle structures in a basic implementation. Computing the cost of a FCB given the associated spanning tree $T$ is $O(mn)$, since there are $m - n + 1$ chords of $G$ relative to $T$ and each one of the corresponding fundamental cycles contains at most $n$ edges. To select the best edge swap available at any given iteration, one has to evaluate the FCB cost for all the swaps involving one of the $n - 1$ branches $e \in T$ and one of the (at most $m - n + 1$) chords $f \in \delta_T^e$. Since computing a fundamental cut requires $O(m)$, the total complexity for a single edge swap is $O(m^3 n^2)$.

In the efficient implementation described in Section 2.3, fundamental cuts and cycles are computed by using symmetric differences of edge sets, which require linear time in the size of the sets. Since there are $m$ fundamental cycles of size at most $n$, and $n$ fundamental cuts of size at most $m$, updating the fundamental cut and cycle structures after the application of an edge swap $(e, f)$ requires $O(mn)$. Doing this for each branch of the tree and for each chord in the fundamental cut induced by the branch, leads to an $O(m^2 n^2)$ total complexity.

It is worth pointing out that computational experiments show larger speed-ups in the average running times (with respect to the basic implementation) than those suggested by the worst-case analysis.

## 2.5   Edge Sampling

The efficient implementation of the local search algorithm described in Fig. 1 is still computationally intensive, since at each iteration all pairs of tree branches $e$ and chords $f \in \delta_T^e$ must be considered to select the best available edge swap. Ideally, we would like to test the edge swap only for a small subset of pairs $e, f$ while minimizing the chances of missing pairs which yield large cost decreases.

**Fig. 3.** All edge weights are equal to 1 and the numbers indicated on the chords correspond to the costs of the corresponding fundamental cycles. The cut on the left has a difference between the cheapest and the most expensive cycles of $10 - 4 = 6$; after the edge swap the difference amounts to $6 - 4 = 2$.

A good strategy is to focus on branches inducing fundamental cuts whose edges define fundamental cycles with "unbalanced" costs, i.e., with a large difference between the cheapest and the most expensive of those fundamental cycles. See Fig. 3 for a simple example. This is formalized in terms of an order $<_b$ on the tree branches. For branches $e_1, e_2 \in T$, we have $e_1 <_b e_2$ if the difference between the maximum and minimum fundamental cycle costs deriving from edges in $\delta_T^{e_1}$ is smaller than that deriving from edges in $\delta_T^{e_2}$. Computational experience suggests that branches that appear to be larger according to the above order tend to be involved in edge swaps leading to largest decreases in the FCB cost.

This strategy can be easily adapted to sampling by ordering the branches of the current spanning tree as above and by testing the candidate edge swaps only for the first $\sigma$ fraction of the branches, where $0 < \sigma \leq 1$ is an arbitrary sampling constant.

# 3  Metaheuristics

To go beyond the scope of local search and try to escape from local minima, we have implemented and tested two well-known metaheuristics: variable neighbourhood search (VNS) [14] and tabu search (TS) [15].

## 3.1  Variable Neighbourhood Search

In VNS one attempts to escape from a local minimum $x'$ by choosing another random starting point in increasingly larger neighbourhoods of $x'$. If the cost of the local minimum $x''$ obtained by applying the local search from $x'$ is smaller than the cost of $x'$, then $x''$ becomes the new best local minimum and the neighbourhood size is reset to its minimal value. This procedure is repeated until a given termination condition is met.

For the MIN FCB problem, given a locally optimal spanning tree $T'$ (obtained by applying the local search of Fig. 1), we consider a neighbourhood of size $p$ consisting of all those spanning trees $T$ that can be reached from $T'$ by applying $p$ consecutive edge swaps. A random solution in a neighbourhood of size $p$ is then obtained by generating a sequence of $p$ random edge swaps and applying it to $T'$.

## 3.2  Tabu Search

Our implementation of tabu search includes diversification steps à la VNS (vTS). In order to escape from local minima, an edge swap that worsens the FCB cost is applied to the current solution and inserted in a tabu list. If all possible edge swaps are tabu or a pre-determined number of successive non-improving moves is exceeded, $t$ random edge swaps are applied to the current spanning tree. The number $t$ increases until a pre-determined limit is reached, and is then re-set to 1. The procedure runs until a given termination condition is met.

Other TS variants were tested. In particular, we implemented a "pure" TS (pTS) with no diversification, and a fine-grained TS (fTS) where, instead of forbidding moves (edge swaps), feasible solutions are forbidden by exploiting the fact that spanning trees can be stored in a very compact form. We also implemented a TS variant with the above-mentioned diversification steps where pTS tabu moves and fTS tabu solutions are alternatively considered. Although the results are comparable on most test instances, vTS performs best on average. Computational experiments indicate that diversification is more important than intensification when searching the MIN FCB solution space with our type of edge swaps.

## 4  Lower Bounds

A standard way to derive a lower bound on the cost of the optimal solutions of a combinatorial optimization problem (and thus to estimate heuristics performance) is to solve a linear relaxation of a (mixed) integer programming formulation. Three different integer programming formulations were discussed in [16].

We now describe an improved formulation that uses non-simultaneous flows on arcs to ensure that the cycle basis is fundamental. Consider a biconnected graph $G = (V, E)$ with a non-negative cost $w_{ij}$ assigned to each edge $\{i, j\} \in E$. For each node $v \in V$, $\delta(v)$ denotes the node star of $v$, i.e., the set of all edges incident to $v$. Let $G_0 = (V, A)$ be the directed graph associated with $G$, namely $A = \{(i, j), (j, i) | \{i, j\} \in E\}$. We use two sets of decision variables. For each edge $\{k, l\} \in E$, the variable $x_{ij}^{kl} \geq 0$ represents the flow through arc $(i, j) \in A$ from $k$ to $l$. Moreover, for each edge $\{i, j\} \in E$, the variable $z_{ij}$ is equal to 1 if edge $\{i, j\}$ is in the spanning tree of $G$, and equal to 0 otherwise. For each pair of arcs $(i, j) \in A$ and $(j, i) \in A$, we define $w_{ji} = w_{ij}$.

The following MIP formulation of the MIN FCB problem provides much tighter bounds than those considered in [16]:

$$\min \sum_{\{k,l\}\in E} \sum_{(i,j)\in A} w_{ij} x_{ij}^{kl} + \sum_{\{i,j\}\in E} (1 - 2z_{ij})w_{ij} \tag{1}$$

$$\sum_{j\in\delta(k)} (x_{kj}^{kl} - x_{jk}^{kl}) = 1 \qquad \forall \{k,l\} \in E \tag{2}$$

$$\sum_{j\in\delta(i)} (x_{ij}^{kl} - x_{ji}^{kl}) = 0 \qquad \forall \{k,l\} \in E,\ \forall i \in V \backslash \{k,l\} \tag{3}$$

$$x_{ij}^{kl} \leq z_{ij} \qquad \forall \{k,l\} \in E,\ \forall \{i,j\} \in E \tag{4}$$

$$x_{ji}^{kl} \leq z_{ij} \qquad \forall \{k,l\} \in E,\ \forall \{i,j\} \in E \tag{5}$$

$$\sum_{\{i,j\}\in E} z_{ij} = n - 1 \tag{6}$$

$$x_{ij}^{kl} \geq 0 \qquad \forall \{k,l\} \in E,\ \forall (i,j) \in A$$

$$z_{ij} \in \{0,1\} \qquad \forall \{i,j\} \in E.$$

For each edge $\{k, l\} \in E$, a path $p$ from $k$ to $l$ is represented by a unit of flow through each arc $(i, j)$ in $p$. In other words, a unit of flow exists node $k$ and enters node $l$ after going through all other (possible) nodes in $p$. For each edge $\{k, l\} \in E$, the flow balance constraints (2) and (3) account for a directed path connecting nodes $k$ and $l$. Note that the flow balance constraint for node $l$ is implied by constraints (2) and (3). Since constraints (4) and (5) require that $z_{ij} = 1$ for every edge $\{i, j\}$ contained in some path (namely with a strictly positive flow), the $z$ variables define a connected subgraph of $G$. Finally, constraint (6) ensures that the connected subgraph defined by the $z$ variables is a spanning tree. The objective function (1) adds the cost of the path associated to every edge $\{k, l\} \in E$ and the cost of all tree chords, and subtracts from it the cost of the tree branches (which are counted when considering the path for every edge $\{k, l\}$).

Besides the quality of the linear relaxation bounds, the main shortcoming of this formulation is that it contains a large number of variables and constraints and hence its solution becomes cumbersome for the largest instances.

## 5  Some Computational Results

Our edge-swapping local search algorithm and metaheuristics have been implemented in C++ and tested on three types of unweighted and weighted graphs. CPU times refer to a Pentium 4 2.66 GHz processor with 1 GB RAM running Linux.

### 5.1  Unweighted Mesh Graphs

One of the most challenging testbeds for the MIN FCB problem is given by the square $n \times n$ mesh graphs with unit costs on the edges. This is due to the large number of symmetries in these graphs, which bring about many different spanning trees with identical associated FCB costs. Uniform cost square mesh graphs have $n^2$ nodes and $2n(n-1)$ edges. Table 1 reports the FCB costs and corresponding CPU times of the solutions found with: the local search algorithm (LS) of Fig. 1, the variant with edge sampling (Section 2.5), the NT-heuristic cited in [11], the VNS and tabu search versions described in Section 3. For LS with edge sampling, computational experiments indicate that a sampling constant of 0.1 leads to a good trade-off between solution quality and CPU time for this type of graphs. The lower bounds in the last column correspond to the cost of a non-fundamental minimal cycle basis, that is to four times the number of cycles in a basis: $4(m - n + 1)$. For this particular type of graphs, the linear relaxation of the MIP formulation provides exactly the same lower bounds.

### 5.2  Random Euclidean Graphs

To asses the performance of our edge-swapping heuristics on weighted graphs, we have generated simple random biconnected graphs. The nodes are positioned uniformly at random on a $20 \times 20$ square centered at the origin. Between each pair of nodes an edge is generated with probability $p$, with $0 < p < 1$. The cost of an edge is equal to the Euclidean distance between its adjacent nodes. For each $n$ in $\{10, 20, 30, 40, 50\}$ and $p$ in $\{0.2, 0.4, 0.6, 0.8\}$, we have generated a random graph of size $n$ with edge probability $p$.

Table 2 reports the results obtained with the edge-swapping heuristics (pure local search and metaheuristics) on these random graphs. The first two columns indicate the performance of LS in terms of FCB cost and CPU time. The next two columns correspond to the lower bounds obtained by partially solving the MIP formulation of Section 4. The third and fourth two-column groups indicate the performances of VNS and TS. There was enough available data to ascribe

**Table 1.** Computational results (FCB cost and CPU times (h:mm:ss)) for $n \times n$ mesh graphs having unit edge costs, obtained with different heuristics. The VNS and TS metaheuristics were run for 10 minutes (after finding the first local optimum). Values marked with * denote an improved value with respect to LS. Lower bounds on the optimal value are reported in the last column.

| | LS | | LS with edge sampling | | NT [11] | | VNS | TS | Bound |
|---|---|---|---|---|---|---|---|---|---|
| $n$ | Cost | Time | Cost | Time | Cost | Time | Cost | Cost | Cost |
| 5 | 72 | 0:00:00 | 74 | 0:00:00 | 78 | 0:00:00 | 72 | 72 | 64 |
| 10 | 474 | 0:00:00 | 524 | 0:00:00 | 518 | 0:00:00 | 466* | 466* | 324 |
| 15 | 1318 | 0:00:00 | 1430 | 0:00:00 | 1588 | 0:00:00 | 1280* | 1276* | 784 |
| 20 | 2608 | 0:00:03 | 3186 | 0:00:00 | 3636 | 0:00:00 | 2572* | 2590* | 1444 |
| 25 | 4592 | 0:00:16 | 5152 | 0:00:02 | 6452 | 0:00:00 | 4464* | 4430* | 2304 |
| 30 | 6956 | 0:00:47 | 8488 | 0:00:03 | 11638 | 0:00:00 | 6900* | 6882* | 3364 |
| 35 | 10012 | 0:02:19 | 11662 | 0:00:08 | 16776 | 0:00:00 | 9982* | 9964* | 4624 |
| 40 | 13548 | 0:06:34 | 15924 | 0:00:26 | 28100 | 0:00:01 | 13524* | 13534* | 6084 |
| 45 | 18100 | 0:14:22 | 22602 | 0:01:00 | 35744 | 0:00:01 | 18100 | 18100 | 7744 |
| 50 | 23026 | 0:31:04 | 33274 | 0:01:10 | 48254 | 0:00:03 | 23026 | 23552 | 9604 |

some statistical significance to the average percentage gap between heuristic and lower bounding values (8.19%), and its reassuringly low standard deviation (± 5.15%). The maximum frequency value is also a rather low value (6%). It is worth pointing out that the lower bounds obtained by solving the linear relaxation of the formulation presented in Section 4 are generally much tighter than those derived from the formulations considered in [16].

## 5.3   Weighted Graphs from Periodic Timetabling

An interesting application of MIN FCB arises in periodic timetabling for transportation systems. To design the timetables of the Berlin underground, Liebchen and Möhring [10] consider the mathematical programming model based on the Periodic Event Scheduling Problem (PESP) [17] and the associated graph $G$ in which nodes correspond to events. Since the number of integer variables in the model can be minimized by identifying an FCB of $G$ and the number of discrete values that each integer variable can take is proportional to the total FCB cost, good models for the PESP problem can be obtained by looking for minimum fundamental cycle bases of the corresponding graph $G$.

Due to the way the edge costs are determined, the MIN FCB instances arising from this application have a high degree of symmetry. Such instances are difficult because, at any given heuristic iteration, a very large number of edge swaps may lead to FCBs with the same cost. Notice that this is generally not the case for weighted graphs with uniformly distributed edge costs. The results reported in Table 3 for instance timtab2, which is available from MIPLIB (http://miplib.zib.de) and contains 88 nodes and 316 edges, are promising. According to practical modeling requirements, certain edges are mandatory and must belong to the spanning tree associated to the MIN FCB solution. The

**Table 2.** Computational results (FCB costs, CPU times (mm:ss), lower bounds) for Euclidean random graphs. Lower bound values marked with $^\dagger$ denote an optimal solution (MIP solved to optimality). FCB costs are marked with $^\dagger$ when the metaheuristic improved on the value found by LS. Missing values are due to excessive CPU timings.

| $p = 0.2$ | | | | | | | | |
|---|---|---|---|---|---|---|---|---|
| $n$ | LS | CPU time | Bound | CPU time | VNS | CPU time | TS | CPU time |
| 10 | $216.698^\dagger$ | 0 | $216.698^\dagger$ | 0 | $216.698^\dagger$ | 0 | $216.698^\dagger$ | 0 |
| 20 | $1052.38^\dagger$ | 0 | $1052.38^\dagger$ | 0:56 | $1052.38^\dagger$ | 0 | $1052.38^\dagger$ | 0 |
| 30 | 3315.89 | 0 | 2750.92 | 0:28 | $3111.71^*$ | 0:14 | 3315.89 | 0 |
| 40 | 4634.04 | 0 | 4065.187 | 16:58 | $4504.84^*$ | 0:22 | $4633.45^*$ | 0 |
| 50 | 7007.34 | 0:01 | 6448.711 | 2:38:51 | $6991.53^*$ | 1:11 | 7007.34 | 0:02 |

| $p = 0.4$ | | | | | | | | |
|---|---|---|---|---|---|---|---|---|
| $n$ | LS | CPU time | Bound | CPU time | VNS | CPU time | TS | CPU time |
| 10 | 472.599 | 0 | $459.305^\dagger$ | 0:02 | $459.305^\dagger$ | 0 | 472.599 | 0 |
| 20 | 2021.82 | 0 | 1894.747 | 0:08 | $2021.37^*$ | 0:04 | $2021.37^*$ | 0 |
| 30 | 4467.13 | 0 | 4265.6 | 22:56 | $4455.2^*$ | 0:29 | $4455.2^*$ | 0:01 |
| 40 | 7685.97 | 0:01 | - | - | $7648^*$ | 1:46 | $7684.53^*$ | 0:02 |
| 50 | 11096.8 | 0:05 | - | - | $11022.8^*$ | 9:32 | $11073.4^*$ | 0:12 |

| $p = 0.6$ | | | | | | | | |
|---|---|---|---|---|---|---|---|---|
| $n$ | LS | CPU time | Bound | CPU time | VNS | CPU time | TS | CPU time |
| 10 | 581.525 | 0 | $547.406^\dagger$ | 0:08 | $547.406^\dagger$ | 0 | $547.406^\dagger$ | 0 |
| 20 | 2776.22 | 0 | 2627.558 | 0:59 | $2756.6^*$ | 0:08 | $2756.6^*$ | 0 |
| 30 | 7031.2 | 0 | 6445.83 | 39:32 | $6979.15^*$ | 1:13 | 7031.2 | 0:03 |
| 40 | 11686.0 | 0:02 | - | - | $11513^*$ | 6:40 | $11683.4^*$ | 0:04 |
| 50 | 19387.3 | 0:10 | - | - | $19174.1^*$ | 7:06 | $19174.1^*$ | 1:06 |

| $p = 0.8$ | | | | | | | | |
|---|---|---|---|---|---|---|---|---|
| $n$ | LS | CPU time | Bound | CPU time | VNS | CPU time | TS | CPU time |
| 10 | 992.866 | 0 | $775.838^\dagger$ | 0:26 | $775.838^\dagger$ | 0 | $775.838^\dagger$ | 0 |
| 20 | 3478.11 | 0 | 3164.9 | 2:31 | $3383.45^*$ | 0:13 | $3383.45^*$ | 0:02 |
| 30 | 8971.78 | 0:01 | 7823.848 | 1:43:05 | $8384.32^*$ | 2:42 | $8930.17^*$ | 0:02 |
| 40 | 14946.4 | 0:07 | - | - | $14870.7^*$ | 5:30 | $14902.2^*$ | 0:16 |
| 50 | 25349.9 | 0:12 | - | - | $25061.2^*$ | 31:55 | $25245.5^*$ | 0:53 |

above-mentioned instance contains 80 mandatory edges out of 87 tree branches, and most of the these 80 fixed edges have very high costs. As shown in Table 3 (instance `liebchen-fixed`), this additional condition obviously leads to FCBs with substantially larger costs.

## 6   Concluding Remarks

We described and investigated new heuristics, based on edge swaps, for tackling the MIN FCB problem. Compared to existing tree-growing procedures, our local search algorithm and simple implementation of the VNS and Tabu search metaheuristics look very promising, even though computationally more intensive. We established structural results that allow an efficient implementation of

**Table 3.** Computational results for Liebchen's instance. Missing values are due to a missing implementation of the corresponding algorithm which deals with mandatory spanning tree edges. Values marked with * correspond to an improvement with respect to the LS solution.

| Instance | Local search FCB cost | Time | NT [11] FCB cost | VNS FCB cost | Time | TS FCB cost | Time | Lower bound FCB cost |
|---|---|---|---|---|---|---|---|---|
| liebchen | 40520 | 0.7s | 50265 | 39801* | 30s | 39841* | 30s | 31220.534 |
| liebchen-fixed | 46072 | 0.13s | - | - | - | 46002* | 30s | 39907.96 |

the proposed edge swaps. We also presented a new MIP formulation whose linear relaxation provides tighter lower bounds than known formulations on several classes of graphs.

# References

1. Sysło, M.: On some problems related to fundamental cycle sets of a graph. In Chartrand, R., ed.: Theory of Applications of Graphs, New York, Wiley (1981) 577–588
2. Kirchhoff, G.: Über die auflösung der gleichungen, auf welche man bei der untersuchungen der linearen verteilung galvanisher ströme geführt wird. Poggendorf Annalen Physik **72** (1847) 497–508
3. Sysło, M.: On cycle bases of a graph. Networks **9** (1979) 123–132
4. Sysło, M.: On the fundamental cycle set graph. IEEE Transactions on Circuits and Systems **29** (1982) 136–138
5. Horton, J.: A polynomial-time algorithm to find the shortest cycle basis of a graph. SIAM Journal of Computing **16** (1987) 358–366
6. Amaldi, E., Rizzi, R.: Personal communication. (2003)
7. Deo, N., Prabhu, G., Krishnamoorthy, M.: Algorithms for generating fundamental cycles in a graph. ACM Transactions on Mathematical Software **8** (1982) 26–42
8. Galbiati, G., Amaldi, E.: On the approximability of the minimum fundamental cycle basis problem. In Jansen, K., Solis-Oba, R., eds.: Approximation and Online Algorithms: First international workshop WAOA 2003, Lecture Notes in Computer Science. Volume 1909., Springer-Verlag (2004) 151–164
9. Brambilla, A., Premoli, A.: Rigorous event-driven (RED) analysis of large-scale nonlinear RC circuits. IEEE Transactions on Circuits and Systems–I: Fundamental Theory and Applications **48** (2001) 938–946
10. Liebchen, C., Möhring, R.H.: A case study in periodic timetabling. In Wagner, D., ed.: Electronic Notes in Theoretical Computer Science. Volume 66., Elsevier (2002)
11. Deo, N., Kumar, N., Parsons, J.: Minimum-length fundamental-cycle set problem: New heuristics and an empirical investigation. Congressus Numerantium **107** (1995) 141–154
12. Paton, K.: An algorithm for finding a fundamental set of cycles of a graph. Communications of the ACM **12** (1969) 514–518
13. Shioura, A., Tamura, A., Uno, T.: An optimal algorithm for scanning all spanning trees of undirected graphs. SIAM Journal of Computing **26** (1997) 678–692

14. Hansen, P., Mladenović, N.: Variable neighbourhood search. In Glover, F., Kochen-berger, G., eds.: Handbook of Metaheuristics, Dordrecht, Kluwer (2003)
15. Hertz, A., Taillard, E., de Werra, D.: Tabu search. In Aarts, E., Lenstra, J., eds.: Local Search in Combinatorial Optimization, Chichester, Wiley (1997) 121–136
16. Liberti, L., Amaldi, E., Maculan, N., Maffioli, F.: Mathematical models and a constructive heuristic for finding minimum fundamental cycle bases. Submitted to Yugoslav Journal of Operations Research (2003)
17. Serafini, P., Ukovich, W.: A mathematical model for periodic scheduling problems. SIAM Journal of Discrete Mathematics **2** (1989) 550–581

## Appendix: Proof of Theorem 2.4

To establish that, for each $h \in T$ such that $h \neq e$ and $f \in \delta_T^h$, we have $\pi(\delta_T^h) = \delta_T^h \triangle \delta_T^e$, we proceed in four steps.

We prove that: (1) $g \in \delta_T^h \cap \delta_T^e \Rightarrow g \notin \pi(\delta_T^h)$, (2) $g \notin \delta_T^h \cup \delta_T^e \Rightarrow g \notin \pi(\delta_T^h)$, (3) $g \in \delta_T^h \setminus \delta^e \Rightarrow g \in \pi(\delta_T^h)$, and (4) $g \in \delta^e \setminus \delta_T^h \Rightarrow g \in \pi(\delta_T^h)$.

When there is no ambiguity, $\delta_T^e$ is written $\delta^e$.

**Claim 1:** $g \in \delta^h \cap \delta^e \Rightarrow g \notin \pi(\delta^h)$.
*Proof.* Since $g \in \delta^h$ there are shortest paths $p_1^T, p_2^T$ connecting $g, h$ and not

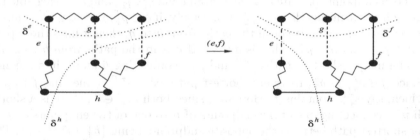

**Fig. 4.** Claim 1: If $g \in \delta^h \cap \delta^e$ then $g \notin \pi(\delta^h)$.

containing $h$, such that $p_1^T \subseteq S_T^h$ and $p_2^T \subseteq \bar{S}_T^h$. Since $g \in \delta^e$, either $p_1^T$ or $p_2^T$ must contain $e$, but not both. Assume w.l.o.g. that $e \in p_1$, $e \notin p_2$ (i.e., $e \in S_T^h$). Thus $\pi$ sends $p_1^T$ to a path $p_1^{\pi T}$ containing $f$, whereas $p_2^{\pi T} = p_2^T$. Thus $P_{\pi T}^*(h, g) = \{p_1^{\pi T}, p_2^{\pi T}\}$. Since $f \in \delta_T^h$, there exist shortest paths $q_1^T \subseteq S_T^h$ and $q_2^T \subseteq \bar{S}_T^h$ connecting $f, h$ and not containing $h$. Because $q_2^T \subseteq \bar{S}_T^h$, $e \notin q_2^T$. In $\pi T$, $q_2^T$ can be extended to a path $q^{\pi T} = q_2^T \cup \{f\}$. By Proposition 2.2 $g \in \delta_{\pi T}^f$. Thus in $\pi T$ there exist paths $r_1^{\pi T}$ in $S_{\pi T}^f = S_T^e$ and $r_2^{\pi T}$ in $\bar{S}_{\pi T}^f = \bar{S}_T^e$ connecting $f, g$ and not containing $f$. Notice that the path $q^{\pi T} \cup r_1^{\pi T}$ connects the endpoint of $h$ in $\bar{S}_T^h$ and $g$, and $p_2^{\pi T}$ connects the same endpoint of $h$ with the opposite endpoint of $g$. Thus $p_1^{\pi T} = \{h\} \cup q^{\pi T} \cup r_1^{\pi T}$, which means that $h \in p_1^{\pi T}$, i.e., $P_{\pi T}^*(h, g) \neq \bar{P}_{\pi T}(h, g)$. By Lemma 2.1, the claim is proved.

**Claim 2**: $g \notin \delta^h \cup \delta^e \Rightarrow g \notin \pi(\delta^h)$.

*Proof.* By hypothesis, $g \in S_T^e \cap S_T^h$ or $g \in \bar{S}_T^e \cap \bar{S}_T^h$. Assume the former w.l.o.g. Let

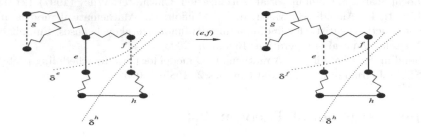

**Fig. 5.** Claim 2: If $g \notin \delta^h \cup \delta^e$ then $g \notin \pi(\delta^h)$.

$p_1^T, p_2^T$ be unique shortest paths from $h$ to $g$, and assume $h \in p_1^T$. Thus $p_2^T \subseteq S_T^h$. Since $g \notin \delta^e$, there are unique shortest paths $q_1^T, q_2^T$ from $e$ to $g$ such that $e$ is in one of them, assume $e \in q_1^T$. Since both $h, e \in T$, there is a shortest path $r^T$ between one of the endpoints of $h$ and one of the endpoints of $e$, while the opposite endpoints are linked by the path $\{h\} \cup r^T \cup \{e\}$. Suppose $e \in p_1^T$. Then since both endpoints of $g$ are reachable from $e$ via $q_1^T, q_2^T$, and $e$ is reachable from $h$ through $r^T$, it means that $e \in p_2^T$. Conversely, if $e \notin p_1^T$, then $e \notin p_2^T$. Thus, we consider two cases. If $e$ is not in the paths from $h$ to $g$, then $\pi$ fixes those paths, i.e., $h \in p_1^{\pi T}$ and $h \notin p_2^{\pi T}$, that is $g \notin \delta^h$. If $e$ is in the paths from $h$ to $g$, then both the unique shortest paths $p_1^{\pi T}$ and $p_2^{\pi T}$ connecting $h$ and $g$ in $\pi T$ contain $f$. Since $g \notin \delta_{\pi T}^f = \delta_T^e$, there are shortest paths $s_1^{\pi T}, s_2^{\pi T}$ connecting $f$ to $g$ one of which, say $s_1^{\pi T}$, contains $f$. Moreover, since both $h, f \in T$, there is a shortest path $u^{\pi T}$ connecting one of the endpoints of $h$ to one of the endpoints of $f$, the other shortest path between the opposite endpoints being $\{h\} \cup u^{\pi T} \cup \{f\}$. Thus, either $p_1^{\pi T} = u^{\pi T} \cup s_1^{\pi T}$ and $p_2^{\pi T} = \{h\} \cup u^{\pi T} \cup \{f\} \cup s_2^{\pi T}$, or $p_1^{\pi T} = u^{\pi T} \cup \{f\} \cup s_2^{\pi T}$ and $p_2^{\pi T} = \{h\} \cup u^{\pi T} \cup s_1^{\pi T}$. Either way, one of the paths contains $h$. By Lemma 2.1, the claim is proved.

**Claim 3**: $g \in \delta^h \setminus \delta^e \Rightarrow g \in \pi(\delta^h)$.

*Proof.* Since $g \in \delta^h$, there are shortest paths $p_1^T \subseteq S_T^h, p_2^T \subseteq \bar{S}_T^h$ connecting $h$ and $g$, none of which contains $h$. Assume w.l.o.g. $e \in S_T^h$. Suppose $e \in p_1^T$, say $p_1^T = q^T \cup \{e\} \cup r^T$. Consider $s^T = q^T \cup \{h\} \cup p_2^T$ and $r^T$. These are a pair of shortest paths connecting $e$ and $g$ such that $e$ does not belong to either; i.e., $g \in \delta_T^e$, which contradicts the hypothesis. Thus $e \notin p_1^T$, i.e., $\pi$ fixes paths $\pi_1^T, \pi_2^T$; thus $\bar{P}_{\pi T}(h, g) = \bar{P}_T(h, g) = P_T^*(h, g) = P_{\pi T}^*(h, g)$, which proves the claim.

**Claim 4**: $g \in \delta^e \setminus \delta^h \Rightarrow g \in \pi(\delta^h)$.

*Proof.* First consider the case where $e, g \in S_T^h$. Since $g \notin \delta_T^h$, the shortest paths $p_1^T, p_2^T$ connecting $h, g$ are such that one of them contains $h$, say $h \in p_1^T$, whilst $p_2^T \subseteq S_T^h$. Since $g \in \delta^e$, there are shortest paths $q_1^T, q_2^T$ entirely in $S_T^h$, connecting $e, g$ such that neither contains $e$. Since both $e, h \in T$ there is a shortest path

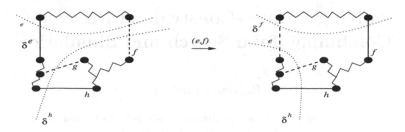

**Fig. 6.** Claim 3: If $g \in \delta^h$ and $g \notin \delta^e$, then $g \in \pi(\delta^h)$.

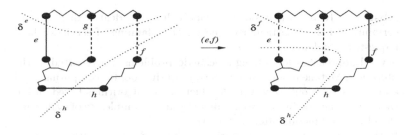

**Fig. 7.** Claim 4: If $g \notin \delta^h$ and $g \in \delta^e$, then $g \in \pi(\delta^h)$.

$r^T \subseteq S_T^h$ connecting an endpoint of $h$ to an endpoint of $e$, the opposite endpoints being joined by $\{h\} \cup r^T \cup \{e\}$. Thus w.l.o.g. $p_1^T = \{h\} \cup r^T \cup \{e\} \cup q_1^T$. Since the path $r^T \cup q_2^T$ does not include $e$ and connects $h, g$, $e \notin p_2^T$. Thus $p_2^{\pi T} = p_2^T$ and $h \notin p_2^{\pi T}$. On the other hand $\pi$ sends $p_1^T$ to a unique shortest path $p_1^{\pi T}$ connecting $h, g$ that includes $f$. Since $f \in \delta_T^h$, there are shortest paths $s_1^T \subseteq S_T^h, s_2^T \subseteq \bar{S}_T^h$ that do not include $h$. Since $p_2^T \subseteq S_T^h$, $s_1^T$ may only touch the same endpoint of $h$ as $p_2^T$. Thus the endpoint of $h$ touched by $p_1^T$ also originates $s_2^T$. Since $s_2^T \subseteq \bar{S}_T^h$, $h \notin s_2^T$. Since $p_1^{\pi T}$ joins $h, g$, contains $f$ and is shortest, $p_1^{\pi T} = s_1^T \cup \{f\} \cup u$, where $u^{\pi T}$ is a shortest path from $f$ to $g$ (which exists because by Proposition 2.2 $g \in \delta_{\pi T}^f$), which shows that $h \notin p_1^{\pi T}$. Thus by Lemma 2.1, $g \in \pi(\delta^h)$. The second possible case is that $e \in S_T^h, g \in \bar{S}_T^h$. Since $g \in \delta_T^e$ there are shortest paths $p_1^T, p_2^T$ connecting $e, g$ such that neither includes $e$. Assume w.l.o.g. $h \in S_T^e$. Since $e, g$ are partitioned by $\delta_T^h$, exactly one of $p_1^T, p_2^T$ includes $h$ (say $h \in p_1^T$, which implies $p_1^T \subseteq S_T^h$). Let $q_1^T$ be the sub-path of $p_1^T$ joining $h$ and $g$ and not including $h$, and let $r^T$ be the sub-path of $p_1^T$ joining $h$ and $e$ and not including $h$. Let $q_2^T = r^T \cup \{e\} \cup p_2^T$. We have that $q_2$ is a shortest path joining $h, g$ not including $h$. Thus $\bar{P}_T(h, g) = \{q_1^T, q_2^T\} = P_T^*(h, g)$, and by Lemma 2.1 $g \in \delta_T^h$, which is a contradiction. $\qquad \square$

# Solving Chance-Constrained Programs Combining Tabu Search and Simulation

Roberto Aringhieri

DTI, University of Milan, via Bramante 65, I-26013 Crema, Italy.
roberto.aringhieri@unimi.it

**Abstract.** Real world problems usually have to deal with some uncertainties. This is particularly true for the planning of services whose requests are unknown *a priori*.

Several approaches for solving stochastic problems are reported in the literature. Metaheuristics seem to be a powerful tool for computing good and robust solutions. However, the efficiency of algorithms based on Local Search, such as Tabu Search, suffers from the complexity of evaluating the objective function after each move.

In this paper, we propose alternative methods of dealing with uncertainties which are suitable to be implemented within a Tabu Search framework.

## 1 Introduction

Consider the following *deterministic linear program*:

$$\textbf{LP} : \min \quad \sum_j c_j x_j$$

$$\text{s.t.} \quad \sum_j a_{ij} x_j \le b_i, \qquad i = 1, \dots, m \tag{1a}$$

$$x_j \ge 0, \qquad j = 1, \dots, n. \tag{1b}$$

The *cost* coefficients $c_j$, the *technological* coefficients $a_{ij}$, and the right-hand side values $b_i$ are the problem parameters. In practical applications, any or all of these parameters may not be precisely defined. When some of these parameters are modelled as random variables, a stochastic problem arises.

When $c_j$, $a_{ij}$ or $b_i$ are random variables having a known joint probability distribution, $z$ is also a random variable. When only the coefficients $c_j$ are random, the problem can be formulated as the minimization of the *expected value* of $z$. Otherwise, a stochastic programming approach must be used. Two main variants of stochastic programming (see e.g. [4]) are the *stochastic programming with recourse* and the *chance-constrained programming*.

Charnes and Cooper [5] proposed to replace constraints (1a) with a number of probabilistic constraints. Denoting with $\mathbb{P}(.)$, the probability of an event, we

C.C. Ribeiro and S.L. Martins (Eds.): WEA 2004, LNCS 3059, pp. 30–41, 2004.

consider the probability that constraint $i$ is satisfied, i.e. :

$$\mathbb{P}\left(\sum_j a_{ij}x_j \leq b_i\right).$$

Let $\alpha_i$ be the maximum allowable probability that constraint $i$ is violated, a **chance-constrained programming** formulation of **LP**, say **CCP**, can be obtained by replacing (1a) with the following *chance constraints*:

$$\mathbb{P}\left(\sum_j a_{ij}x_j \leq b_i\right) \geq 1 - \alpha_i, \quad i = 1, \ldots, m. \tag{2}$$

When all $c_j$'s are known, this formulation minimizes $z$ while forbidding the constraints to exceed certain threshold values $\alpha_i$.

Moving constraints (2) to the objective function via Lagrangean multipliers, we obtain the following stochastic program:

$$\textbf{SPR} : \min \quad \sum_j c_j x_j + \sum_i \lambda_i \, \mathbb{P}\left(\sum_j a_{ij}x_j > b_i\right) \tag{3}$$

$$\text{s.t.} \quad x_j \geq 0, \qquad\qquad\qquad j = 1, \ldots, n.$$

Using the Lagrangean multipliers, **SPR** directly considers the *cost of recourse*, i.e. the cost of bringing a violated constraint to feasibility.

When the deterministic mathematical model involves also binary and/or integer variables, as in many real applications, the complexity of the associated stochastic program increases. To this purpose, several solution approaches are reported in literature such as the Integer $L$-Shaped Method [12], heuristics (see e.g. those for Stochastic Vehicle Routing Problem [7,8]), methods based on *a priori optimization* [3] or *sample-average approximation* [13].

In this paper we propose a new algorithmic approach which combines Tabu Search and simulation for Chance-Constrained Programming. Glover and Kelly have described the benefits of applying Simulation to the solution of $\mathcal{NP}$-hard problems [9]. Tabu Search [10] is a well-known metaheuristic algorithm which has proved effective in a great number of applications.

The paper is organized as follows. The motivations and the basic idea of combining Tabu Search and Simulation are presented and discussed in section 2. In section 3 we introduce two optimization problems which are used to evaluate the efficiency of the proposed algorithms. Section 4 describes the two problems. Section 5 reports about the planning and the results of the computational experiments. Finally, ongoing work is discussed in section 6.

## 2   Motivations and Basic Ideas

In the following, we refer to the general model **IP** derived from **LP** setting the $x_j$'s to be integer.

Tabu Search (TS) explores the solution space by moving at each iteration to the best neighbor of the current solution, even if the objective function is not improved. In order to avoid cycling, each move is stored in a list of *tabu* moves for a number of iterations: a tabu move is avoided until it is deleted from the list. A basic scheme of TS algorithm, say BTS, is depicted in Algorithm 1.

---

**Algorithm 1 BTS**

$k := 1$;
$x :=$ InitialSol();
**while** (**not** STOP) **do**
  $N(x) :=$ NeighborhoodOf($x$);
  $N^{(k)}(x) := N(x) \setminus T^{(k)}(x) \cup A^{(k)}(x)$;
  $x' :=$ BestOf($N^{(k)}(x)$);
  $x := x'$;
  $k := k + 1$;
**end while**

---

Note that: $T^{(k)}(x)$ is the set of *tabu solutions* generated by $x$ using tabu moves at iteration $k$; $A^{(k)}(x)$ is the set of tabu solution which are evaluated since they respect some *aspiration criteria* (e.g. their objective function value improves that of current best solution). The algorithm usually stops after a given number of iterations or after a number of not improving iterations.

From a computational point of view, the computation of $N(x)$ and its evaluation (the choice of the best move) are the most time consuming components. For instance, a linear running time to evaluate the objective function of a single move is usually considered acceptable for a deterministic problem.

Unfortunately, this is not always true for stochastic programs. If we consider both **SPR** and **CCP** models, we observe that a move evaluation requires to compute a quite complex probability function. For instance, in [8], the authors proposed a **SPR** formulation for Vehicle Routing Problem with Stochastic Demands and Customers: the proposed TS algorithm, TABUSTOCH, requires at least $O(n^3)$ to evaluate a single move, where $n$ is the number of demand locations. More generally, the evaluation of a new move involves probability and, at least, two stages of computation [4].

Our main concern is to reduce the computational complexity required for neighborhood exploration by introducing simulation methods within TS framework for solving a **CCP** programs.

The idea is based on a different way of dealing with random parameters: instead of computing directly the probability function, which is computationally expensive, we use simulation to evaluate random parameters. Then, we use these simulated random parameters, within the TS framework, in order to avoid moves which lead to unfeasible solutions, i.e. moves which make unfeasible the chance-constraints (2).

For the sake of simplicity, we assume that only the $b_i$'s are random. Clearly, the following remarks can be extended straightforwardly to the other problem parameters. In order to simulate random parameters, we introduce the following notation:

$$x^{(k)} = \text{the variable } x \text{ at } k\text{-th iteration,}$$
$$\widetilde{b}_i^{(t)} = \text{the } t\text{-th simulated value of } b_i \quad (t = 1, \dots, T),$$
$$\delta_i^{(k,t)} = \widetilde{b}_i^{(t)} - \sum_j a_{ij} x_j^{(k)} \quad (t = 1, \dots, T).$$

Let $S_i^{(k)}$ be given as follows:

$$S_i^{(k)} = \sum_{t=1}^{T} S_i^{(k,t)}, \text{ where } S_i^{(k,t)} = \begin{cases} 1 & \text{if } \delta_i^{(k,t)} > 0 \\ 0 & \text{otherwise} \end{cases}. \tag{4}$$

The value of $S_i^{(k)}$ counts the number of *successes* for the $i$-th constraint (i.e. the constraint is satisfied) The value $\bar{S}_i^{(k)} = \dfrac{S_i^{(k)}}{T}$ estimates the probability of constraint $i$ to be satisfied at iteration $k$

Taking into account **CCP** models, we are interested in computing solutions such that the chance-constraints (2) are satisfied for a given probability.

In this case, we can introduce a concept similar to that of a tabu move. The idea is to avoid all the moves leading to solutions which make unfeasible the respective chance-constraint. More formally, a move is *probably tabu* at iteration $k$ if

$$\bar{S}_i^{(k)} < 1 - \alpha_i, \quad i = 1, \dots, m. \tag{5}$$

Let $P^{(k)}(x)$ be the set of probably tabu solutions generated by $x$ at iteration $k$. Then, the corresponding TS algorithm, say SIMTS-CCP, can be obtained from Algorithm 1 by modifying the computation of $N^{(k)}(x)$ as

$$\bar{N}^{(k)}(x) := N(x) \setminus T^{(k)}(x) \setminus P^{(k)}(x) \cup A^{(k)}(x). \tag{6}$$

The SIMTS-CCP procedure is sketched in Algorithm 2.

Finally, TS offers to the researchers a great flexibility. For instance, a common practice is to use a *simple* neighborhood structure and a penalized objective function to take into account the unfeasibility when some constraints are violated (see e.g. [6]). A general form of penalized function can be the following:

$$z + \sum_i \beta_i p_i(x) \tag{7}$$

where $\beta_i > 0$ is usually a self-adjusting penalty coefficient and $p_i(x) \geq 0$ is a measure of how much the $i$-th constraint is unfeasible.

---

**Algorithm 2** SIMTS-CCP

$k := 1$;
$x := $ InitialSol();
**while** (**not** STOP) **do**
   $N(x) := $ NeighborhoodOf$(x)$;
   $\bar{N}^{(k)}(x) := N(x) \setminus T^{(k)}(x) \setminus P^{(k)}(x) \cup A^{(k)}(x)$;
   $x' := $ BestOf$(N^{(k)}(x))$;
   $x := x'$;
   $k := k + 1$;
**end while**

---

In the same way, we can adapt the function in (7) to take into account the unfeasibility of chance-constraints. For instance, the $p_i(x)$ function for the $i$-th constraint (2) can have the following general form:

$$p_i(x) = 1 - \alpha_i - \mathbb{P}\left(\sum_j a_{ij}x_j \leq b_i\right), \quad i = 1, \ldots, m. \tag{8}$$

## 3   Test Problems

In order to test the SIMTS-CCP algorithm, we will consider two $\mathcal{NP}$-hard optimization problems arising in the design of telecommunication networks based on SONET technology. For this class of problems, extensive computational experiences have been made and efficient tabu search algorithms are available [1].

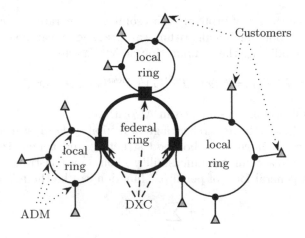

**Fig. 1.** A SONET network with DXC

The SONET network is a collection of rings connecting a set of customer sites. Each customer needs to transmit, receive and relay a given traffic with a subset of the other customers. Add-Drop Multiplexers (ADM) and Digital Cross Connectors (DXC) are the technologies allowing the connection between customers and rings. Since they are very expensive, the main concern is to reduce the number of DXCs used.

Two main topologies for the design of a SONET network are available. The first topology consists in the assignment of each customer to exactly one ring by using one ADM and allowing connection between different rings through a unique *federal ring* composed by one DXC for each connected ring. The objective of this problem, say SRAP and depicted in Figure 1, is to minimize the number of DXCs.

Under some distance requirements, a second topology is possible: the use of a federal ring is avoided assigning each traffic between two different customers to only one ring. In this case, each customer can belong to different rings. The objective of this problem, say IDP and depicted in Figure 2, is to minimize the number of ADMs. For further details see e.g. [2,1,11].

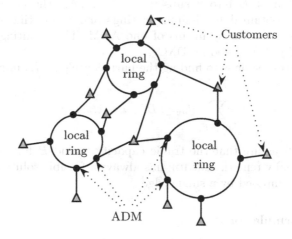

**Fig. 2.** A SONET network without DXC

In order to formulate these problems, we consider the undirected graph $G = (V, E)$: the node set $V$ ($|V| = n$) contains one node for each customer; the edge set $E$ has an edge $[u, v]$ for each pair of customers $u, v$ such that the amount of traffic $d_{uv}$ between $u$ and $v$ is greater than 0 and $d_{uv} = d_{vu}, \forall u, v \in V, u \neq v$. Given a subset of edges $E_i \subset E$, let $V(E_i) \subseteq V$ be the subset of nodes induced by $E_i$, i.e. $V(E_i) = \{u, v \in V : [u, v] \in E_i\}$.

## SRAP Formulation

Given a partition of $V$ into $r$ subsets $V_1, V_2, \ldots V_r$, the corresponding SRAP network is obtained by defining $r$ local rings, connecting each customer of subset $V_i$ to the $i$-th local ring, and one federal ring, connecting the $r$ local rings by using $r$ DXCs. The resulting network uses $n$ ADMs and $r$ DXCs.

Solving SRAP corresponds to finding the partition $V_1, \ldots V_r$ minimizing $r$, and such that

$$\sum_{\substack{u \in V_i}} \sum_{\substack{v \in V \\ v \neq u}} d_{uv} \leq B, \quad i = 1, \ldots, r \tag{9a}$$

$$\sum_{i=1}^{r-1} \sum_{j=i+1}^{r} \sum_{u \in V_i} \sum_{v \in V_j} d_{uv} \leq B \tag{9b}$$

Constraints (9a) and (9b) impose, respectively, that the capacity bound $B$ is satisfied for each local ring and for the federal ring.

## IDP Formulation

Given a partition of $E$ into $r$ subsets $E_1, E_2, \ldots E_r$, the corresponding IDP network can be obtained by defining $r$ rings and connecting each customer of $V(E_i)$ to the $i$-th ring by means of one ADM. The resulting network uses $\varphi = \sum_{i=1}^{r} |V(E_i)|$ ADMs and no DXC.

Solving IDP corresponds to finding the partition $E_1, \ldots E_r$ minimizing $\varphi$ and such that

$$\sum_{[u,v] \in E_i} d_{uv} \leq B, \quad i = 1, \ldots, r \tag{10}$$

Constraints (10) assure that the traffic capacity bound $B$ for each ring is not exceeded. we finally remark that IDP has always a feasible solution, e.g. the one with $|E|$ rings composed by a single edge.

## Stochastic Formulations

The stochastic version of SRAP and IDP considers the demand $d_{uv}$ as random parameters. The corresponding chance-constrained programs are obtained by replacing constraints (9a) and (9b) with

$$\mathbb{P} \left( \sum_{\substack{u \in V_i}} \sum_{\substack{v \in V \\ v \neq u}} d_{uv} \leq B \right) \geq 1 - \alpha_i, \quad i = 1, \ldots, r \tag{11a}$$

$$\mathbb{P} \left( \sum_{i=1}^{r-1} \sum_{j=i+1}^{r} \sum_{u \in V_i} \sum_{v \in V_j} d_{uv} \leq B \right) \geq 1 - \alpha_0 \tag{11b}$$

for SRAP, and constraints (10) with

$$\mathbb{P}\left(\sum_{[u,v]\in E_i} d_{uv} \leq B\right) \geq 1 - \alpha_i, \quad i = 1, \ldots, r \tag{12}$$

for IDP.

## 4   The Algorithms for SRAP and IDP

In [1] the authors proposed a short-term memory TS guided by a variable objective function: the main idea of the variable objective function $z_v$ is to lead the search within the solution space from unfeasible solutions to feasible ones, as in Table 1.

**Table 1.** Variable objective function $z_v$

|  | SRAP | | | IDP | |
|---|---|---|---|---|---|
|  | feas. sol | not feas. sol |  | feas. sol | not feas. sol |
| feas. sol | $k\,B + BN$ | $(k+1)\widetilde{BN}$ | feas. sol | $\varphi\,B + M$ | $2\varphi\,\widetilde{M}$ |
| not feas. sol | $k\,B$ | $n\,\widetilde{BN}$ | not feas. sol | $\varphi\,B$ | $2\varphi\,\widetilde{M}$ |
| where | $BN$ is the maximum value between maximum ring and federal ring capacities | | | | |
|  | $\widetilde{BN}$ is the value of $BN$ associated to an unfeasible solution | | | | |
|  | $M$ is the maximum ring capacity value | | | | |
|  | $\widetilde{M}$ is the value of $M$ associated to an unfeasible solution | | | | |

A diversification strategy is implemented by varying multiple neighborhoods during the search. More specifically, DMN uses mainly a neighborhood based on moving one customer or demand at a time in such a way that the receiving ring does not exceed the bound $B$. Otherwise, if $B$ is exceeded, we consider also the option of switching two customers or demands belonging to two different rings. After $\Delta$ consecutive non improving iterations, a second neighborhood is used for few moves. During this phase, DMN empties a ring by moving its elements (customers or demands respectively for SRAP and IDP) to the other rings disregarding the capacity constraint while locally minimizing the objective function.

To turn the computation of each move efficient, some data structures, representing the traffic within and outside a ring, are maintained along the computation.

The whole algorithm is called Diversification by Multiple Neighborhoods, say DMN. For a more detailed description of the whole algorithm, refer to the description given in [1].

## The simts-ccp Algorithms

In order to devise the SIMTS-CCP algorithms for our problems, we need to implement the evaluation of chance-constraints through the generation of random parameters $d_{uv}$.

As described in section 2, we need $T$ observations of $S_i^{(k,t)}$ to evaluate the value of $\bar{S}_i^{(k)}$ defined in (4). Moreover, each $S_i^{(k,t)}$ needs the generation of all $d_{uv}$ traffic demands values according to their probability distribution function.

---

**Algorithm 3** Computation of $\bar{S}_i^{(k)}$

---

  **for all** $t = 1, \dots, T$ **do**
    generate random $d_{uv}, \forall u, v \in V, u \neq v$;
    **for** $i = 1, \dots, r$ **do** compute $\delta_i^{(k,t)}$;
  **end for**
  **for** $i = 1, \dots, r$ **do** compute $\bar{S}_i^{(k)}$;

---

Algorithm 3 describes, with more details, the evaluation of $\bar{S}_i^{(k)}$. Since the number of traffic demands $d_{uv}$ is $O(n^2)$, the complexity of the computation of $\bar{S}_i^{(k)}$ values is $O(T \times n^2)$. Note that the complexity of $\delta_i^{(k,t)}$ can be reduced employing the current traffic values available in the traffic data structures.

Here we propose two SIMTS-CCP algorithms derived from DMN. The basic idea is to add the computation of $\bar{S}_i^{(k)}$ to the original framework of DMN at the end of neighborhood exploration: starting from solution $x$, the algorithms generate each possible move $x'$ as in DMN using the mean value of $d_{uv}$; then, the stochastic feasibility (respecting to the chance-constraints) is tested through the computation of $\bar{S}_i^{(k)}$. The algorithms differ in how the test is used to reject, or not, solution $x'$.

The first one, called DMN-STOCH-1, is the simplest one: each move not belonging to $\bar{N}^{(k)}(x)$, defined in (6), is avoided. In other words, DMN-STOCH-1 allows only moves which satisfy the chance-constraints. Note that the objective function remains the one in Table 1.

**Table 2.** Penalized variable objective function $\bar{z}_v$

| | SRAP | | | IDP | |
| --- | --- | --- | --- | --- | --- |
| | feas. sol | not feas. sol | | feas. sol | not feas. sol |
| feas. sol | $z_v + Bs(x)$ | $z_v + \widetilde{BN}s(x)$ | feas. sol | $z_v + Bs(x)$ | $z_v + \widetilde{M}s(x)$ |
| not feas. sol | $z_v + Bs(x)$ | $z_v + \widetilde{BN}s(x)$ | not feas. sol | $z_v + Bs(x)$ | $z_v + \widetilde{M}s(x)$ |

On the contrary, the second one, say DMN-STOCH-2, allows moves in $P^{(k)}(x)$ but penalizes them using an objective function which also measures how unfeasible the chance-constraints are. Referring to the general form reported in (7) and (8), our penalized objective function $\bar{z}_v$ is depicted in Table 2 where $s(x) = \sum_i p_i(x)$ and

$$p_i(x) = \begin{cases} 100(1 - \alpha_i - \bar{S}_i^{(k)}) & \text{if } \bar{S}_i^{(k)} < 1 - \alpha_i \\ 0 & \text{otherwise} \end{cases}.$$

# 5 Preliminary Computational Results

In this section, we report the planning of computational experiments and the preliminary results.

In our computational experiments, we used the well-known Marsaglia-Zaman generator [14,15], say RANMAR. This algorithm is the best known random number generator available since it passes *all* of the tests for random number generators. The algorithm is a combination of a Fibonacci sequence (with lags of 97 and 33, and operation *"subtraction plus one, modulo one"*) and an *"arithmetic sequence"* (using subtraction), which gives a period of $2^{144}$. It is completely portable and gives an identical sequence on all machines having at least 24 bit mantissas in the floating point representation.

**Table 3.** DMN: best computational results.

| | $z$ | # opt | gaps | | | | avg. time in ms | |
|---|---|---|---|---|---|---|---|---|
| | | | # 1 | # 2 | # 3 | # >3 | optimal | all |
| SRAP | $z_v$ | 118 | 0 | 0 | 0 | 0 | 60.4 | 60.4 |
| IDP | $z_v$ | 129 | 23 | 2 | 0 | 0 | 808.9 | 846.4 |

## Results of the Deterministic Version

Considering the set of 160 benchmark instances generated by Goldschmidt *et al.* [11], DMN solves to optimality all the 118 instances, which are known to be feasible for SRAP, with an average computing time of 60 mseconds. On the same benchmark but considering IDP, for which the optimal solution value is known only for 154 instances, DMN solves 129 instances to optimality, 23 instances with a gap of 1 and the remaining two instances with a gap of 2. The average computing time is 850 mseconds. The overall results are reported in Table 3.

## Results of the Stochastic Version

We have tested our algorithms on the same benchmark varying the parameters in the following ranges: $T \in \{50, 75, 100\}$, $\alpha_i = \alpha \in \{0.3, 0.2, 0.1\}$ and maintaining unaltered those giving the best result for the deterministic version (see [1]).

**Table 4.** DMN-STOCH-1: best and worst computational results.

| type of | results | # opt | gaps | | | | avg. time in ms | |
|---|---|---|---|---|---|---|---|---|
| | | | # 1 | # 2 | # 3 | # >3 | optimal | all |
| SRAP | best | 103 | 3 | 3 | 4 | 5 | 341.4 | 397.1 |
| SRAP | worst | 99 | 3 | 8 | 1 | 7 | 412.3 | 441.3 |
| IDP | best | 105 | 27 | 8 | 4 | 10 | 3127.3 | 3301.5 |
| IDP | worst | 101 | 25 | 10 | 3 | 15 | 3108.2 | 3309.9 |

The comparisons are made with the optimal value of the deterministic version, that is the values obtained using the mean value of traffic demands. Our tests try to investigate how far the solution computed by the SIMTS-CCP is from the one computed by its deterministic version.

**Table 5.** DMN-STOCH-2: best and worst computational results.

| type of | results | # opt | gaps | | | | avg. time in ms | |
|---|---|---|---|---|---|---|---|---|
| | | | # 1 | # 2 | # 3 | # >3 | optimal | all |
| SRAP | best | 106 | 4 | 5 | 2 | 1 | 394.1 | 432.7 |
| SRAP | worst | 104 | 5 | 4 | 2 | 3 | 443.5 | 457.9 |
| IDP | best | 118 | 22 | 6 | 5 | 3 | 3212.3 | 3287.2 |
| IDP | worst | 108 | 25 | 8 | 4 | 9 | 3459.1 | 3501.4 |

The results, reported in Table 4 and 5, show the impact of $\bar{S}_i^{(k)}$ computation: although the increase in the average computation time is quite remarkable with respect the deterministic version, we observe that the quality of solutions computed by both algorithms is acceptable.

## 6   Conclusions and Further Work

The paper addresses the problem of solving chance-constrained optimization problems combining Tabu Search and Simulation. After a brief introduction to stochastic programming, the class of SIMTS-CPP algorithms is proposed. The reported computational results show that the solutions computed have a quality comparable to those computed by the deterministic version. Also the increase in the average running time is acceptable.

Further work will be mainly concerned with two topics. The first one is the extension of computational experiments regarding SIMTS-CCP. The second one concerns the study of a similar algorithm for **SPR** problems.

**Acknowledgments.** The author wishes to thank Paola Valpreda and Roberto Cordone for their help during the proofreading process.

# References

1. R. Aringhieri and M. Dell'Amico. Comparing Intensification and Diversification Strategies for the SONET Network Design Problem. Technical report, DISMI, 2003. submitted to Journal of Heuristics.
2. R. Aringhieri, M. Dell'Amico, and L. Grasselli. Solution of the SONET Ring Assignment Problem with capacity constraints. Technical Report 12, DISMI, 2001. To appear in "Adaptive Memory and Evolution: Tabu Search and Scatter Search", Eds. C. Rego and B. Alidaee.
3. D.J. Bertsimas, P. Jaillet, and A. R. Odoni. A Priori Optimization. *Operations Research*, 38(6):1019–1033, 1990.
4. J.R. Birge and F. Louveaux. *Introduction to Stochastic Programming*. Springer, 1997.
5. A. Charnes and W. Cooper. Chance-constrained programming. *Management Science*, 6:73–79, 1959.
6. J-F. Cordeau, M. Gendreau, G. Laporte, J-Y Potvin, and F. Semet. A guide to vehicle routing heuristics. *Journal of the Operational Research Society*, 53:512–522, 2002.
7. M. Dror and P. Trudeau. Stochastic vehicle routing with modified savings algorithm. *European Journal of Operational Research*, 23:228–235, 1986.
8. M. Gendreau, G. Laporte, and R. Seguin. A Tabu Search heuristic for the Vehicle Routing Problem with Stochastic Demands and Customers. *Operations Research*, 44(3):469–447, 1996.
9. F. Glover and J. Kelly. New Advanced combining Optimization and Simulation. In *Airo 1999 Proceedings*.
10. F. Glover and M. Laguna. *Tabu Search*. Kluwer Academic Publishers, Boston, 1997.
11. O. Goldschmidt, A. Laugier, and E. V. Olinick. SONET/SDH Ring Assignment with Capacity Constraints. *Discrete Applied Mathematics*, (129):99–128, 2003.
12. G. Laporte and F. V. Louveaux. The Integer *L*-Shaped Method for Stochastic Integer Problems with Complete Recourse. *Operations Research Letters*, 13:133–142, 1993.
13. J. Linderoth, A. Shapiro, and S. Wright. The Empirical Behavior of Sampling Methods for Stochastic Programming. Technical report, Computer Science Department, University of Wisconsin-Madison, 2001.
14. G. Marsaglia and A. Zaman. Toward a Universal Random Number Generator. Technical Report FSU-SCRI-87-50, Florida State University, 1987.
15. G. Marsaglia and A. Zaman. A New Class of Random Number Generators. *Annals of Applied Probability*, 3(3):462–480, 1991.

# An Algorithm to Identify Clusters of Solutions in Multimodal Optimisation

Pedro J. Ballester and Jonathan N. Carter

Imperial College London, Department of Earth Science and Engineering, RSM
Building, Exhibition Road, London SW7 2AZ, UK.
{p.ballester,j.n.carter}@imperial.ac.uk

**Abstract.** Clustering can be used to identify groups of similar solutions
in Multimodal Optimisation. However, a poor clustering quality reduces
the benefit of this application. The vast majority of clustering methods
in literature operate by resorting to a priori assumptions about the data,
such as the number of cluster or cluster radius. Clusters are forced to
conform to these assumptions, which may not be valid for the considered
population. The latter can have a huge negative impact on the clustering quality. In this paper, we apply a clustering method that does not
require a priori knowledge. We demonstrate the effectiveness and efficiency of the method on real and synthetic data sets emulating solutions
in Multimodal Optimisation problems.

## 1 Introduction

Many real-world optimisation problems, particularly in engineering design, have
a number of key features in common: the parameters are real numbers; there are
many of these parameters; and they interact in highly non-linear ways, which
leads to many local optima in the objective function. These optima represent
solutions of distinct quality to the presented problem. In Multimodal Optimisation, one is interested in finding the global optimum, but also alternative good
local optima (ie. diverse high quality solutions). There are two main reasons to
seek for more than one optimum. First, real-world functions do not come without errors, which distort the fitness landscape. Therefore, global optima may not
correspond to the true best solution. This uncertainty is usually addressed by
considering multiple good optima. Also, the best solution represented by a global
optimum may be impossible to implement from the engineering point of view. In
this case, an alternative good solution could be considered for implementation.

Once a suitable search method is available, an ensemble of diverse, high
quality solutions are obtained. Within this ensemble, there are usually several
groups of solutions, each group representing a different optimum. In other words,
the ensemble of solutions is distributed into clusters. Clustering can be defined
as the partition of a data set (ensemble of solutions) into groups named clusters
(part of the ensemble associated with an optimum). Data points (individuals)
within each cluster are similar (or close) to each other while being dissimilar to
the remaining points in the set.

C.C. Ribeiro and S.L. Martins (Eds.): WEA 2004, LNCS 3059, pp. 42–56, 2004.
© Springer-Verlag Berlin Heidelberg 2004

The identification of these clusters of solutions is useful for several reasons. First, at the implementation stage, we may want to consider distinct solutions instead of implementing similar solutions. This is specially convenient when the implementation costs time and money. On the other hand, the uncertainty associated with one solution can be estimated by studying all the similar solutions (ie. those in the same cluster). Also, one may want to optimise the obtained solutions further. This is done by using a faster (but less good at searching) optimiser on the region defined by all solutions. With the boundaries provided by the clustering algorithm, you could do the same but for each cluster (ie. for each different high performance region). Lastly, an understanding of the solutions is needed. In real world engineering design, for instance, it is common to have many variables and a number of objectives. Packham and Parmee [8] claim that in this context it is extremely difficult to understand the possible interactions between variables and between variables and objectives. These authors pointed out the convenience of finding the location and distribution of different High Performance regions (ie. regions containing high quality solutions).

There are many clustering algorithms proposed in the literature (a good review has been written by Haldiki et al. [7]). The vast majority of clustering algorithms operate by using some sort of a priori assumptions about the data, such as the cluster densities, sizes or number of clusters. In these cases, clusters are forced to conform to these assumptions, which may not be valid for the data set. This can have a huge negative impact on the clustering quality. In addition, there is a shortage in the literature of effective clustering methods for high dimensional data. This problem is known as the Curse of Dimensionality [6] in Clustering. This issue is discussed extensively in [1].

In this work, we apply a new algorithm [2] for clustering data sets obtained by the application of a search and optimisation method. The method is known as CHIDID (Clustering HIgh DImensional Data) and it is aimed at full dimensional clustering (ie. all components have a clustering tendency). To use this algorithm it is not necessary to provide a priori knowledge related to the expected clustering behaviour (eg. cluster radius or number of clusters). The only input needed is the data set to be analysed. The algorithm does contain two tuning parameters. However, extensive experiments, not reported in this paper, lead us to believe that the performance of the algorithm is largely independent of the values of these two parameters, which we believe are suitable for most data sets and are used for all our work. CHIDID scales linearly with the number of dimensions and clusters. The output is the assigment of data points to clusters (the number of clusters is automatically found).

The rest of this paper is organised as follows. We begin by describing the clustering method in Sect. 2. Section 3 explains how the test data sets are generated. The analysis of the results is discussed in Sect. 4. Lastly, Sect. 5 presents the conclusions.

## 2   Clustering Method

We describe our clustering method in three stages. We start by introducing the notation used. Next, the clustering criterion is presented as a procedure used to find the cluster in which a given data point is included. Finally, the clustering algorithm is introduced. The operation of this algorithm is based on the iterative application of the clustering criterion until all clusters and outliers are found.

### 2.1   Notation

We regard the population as a collection of N distinct vectors, or data points in a M-dimensional space, over which the clustering task is performed. We represent this data set as

$$\Omega = \{x^i\}_{i=1}^N, \text{ with } x^i \in \mathrm{R}^M \; \forall i \in I = \{1, \ldots, N\} \; . \tag{1}$$

Clustering is the process by which a partition is performed on the data set. Points belonging to the same cluster are similar to each other, but dissimilar to those belonging to other clusters. This process results in $C$ pairwise disjoint clusters, whose union is the input data set. That is

$$\Omega = \bigcup_{k=1}^C \Omega_k, \text{ with } \Omega_k = \{x^i\}_{i \in I_k} \; . \tag{2}$$

where $\Omega_k$ is the $k^{th}$ cluster, $I_k$ contains the indices for the data points included in the $k^{th}$ cluster and $\Omega_k \bigcap \Omega_l = \emptyset \; \forall k \neq l$ with $k, l \in K = \{1, \ldots, C\}$.

An outlier is defined as a data point which is not similar to any other point in the data set. Hence, an outlier can be regarded as a cluster formed by a single point.

Proximity (also known as similarity) is the measure used to quantify the degree of similarity between data points. A low value of the proximity between two points means these points are similar. Conversely, a high value of their proximity implies that the points are dissimilar. In this work, the Manhattan distance is adopted as the proximity measure. Thus the proximity of $x^i$ with respect to $x^l$ is calculated as

$$p_{il} = \sum_{j=1}^M |x_j^l - x_j^i| \; . \tag{3}$$

In practice, the value range of a given component can be very large. This would dominate the contribution of the components with much smaller value ranges. In this case, we would recommend scaling all components to the same range so that every component contributes equally to the proximity measure. However, in this paper the scaling will not be necessary since all components will be in the same interval.

Based on the proximity measure, a criterion is needed to determine if two points should be considered members of the same cluster. The quality of the clustering directly depends on the choice of criterion. In this work, clusters are regarded as regions, in the full dimensional space, which are densely populated with data points, and which are surrounded by regions of lower density. The clustering criterion must serve to identify these dense regions in the data space.

## 2.2   Clustering Criterion

As we previously described, points belonging to the same cluster have a low proximity between them, and high proximity when compared with points belonging to other clusters. The proximity $\{p_{il}\}_{l=1}^{N}$ from an arbitrary point $x^i$ to the rest of the data set (ie. $x^i$ with $l \neq i$) should split into two groups, one group of similar low values and the other of significantly higher values. This first group corresponds to the cluster to which $x^i$ belongs to. The goal of the clustering criterion is to determine the cluster cutoff in an unsupervised manner. The criterion is also expected to identify outliers as points which are not similar to any other point in the data set.

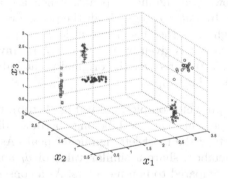

**Fig. 1.** Example data set with N= 150, M= 3 and C= 5

In order to illustrate the operation of the clustering criterion, consider the data set shown in Fig. 1. It contains 150 points unevenly distributed in 5 clusters in a 3-dimensional space. Let us take a point from one of the clusters (say the $k^{th}$) in the example, which will be referred to as the cluster representative and denoted by $x^{i_k}$. We apply the following procedure to determine which are its cluster members:

1. Calculate the sequence $\{p_{i_k l}\}_{l=1}^{N}$, using (3), and sort it in increasing order to get $\{d_l\}_{l=1}^{N}$. The sorting operation is represented as a correspondence between sets of point indices, that is, $S : I \rightarrow L$. The plot of $d_l$ can be found

at the bottom part of Fig. 2. Note that $d_1 = 0$ corresponds to $p_{i_k i_k}$, $d_2$ is the nearest neighbour to $x^{i_k}$ and so on.

2. Define the relative linear density up to the $l^{th}$ data point as $\theta_l = \frac{l}{N} \frac{\beta + d_N}{\beta + d_l}$ and compute $\{\theta_l\}_{l=1}^N$. This expression is equivalent to the cumulative number of points divided by an estimation of the cluster size relative to the same ratio applied over all N points. $\beta$ is a parameter that will be discussed later (we recommend that $\beta = \frac{2}{3}d_2$).

3. Define the relative linear density increment for the $l^{th}$ point as $\triangle \theta_l = |\theta_l - \theta_{l-1}|$ and calculate the sequence $\{\triangle \theta_l\}_{l=2}^N$. The plot of $\triangle \theta_l$ is presented at the top part of Fig. 2. If $\triangle \theta_l$ is high then it will be unlikely that the $l^{th}$ point forms part of the cluster.

4. Calculate $l_1 = \{l \in \{2, \ldots, N\} \mid \triangle \theta_l = max(\{\triangle \theta_{l'}\}_{l'=2}^N)\}$ (ie. the position of the highest value of $\triangle \theta_l$ in Fig. 2). Define $(l_1 - 1)$ as the provisional cluster cutoff, and it means that only the $l_1 - 2$ closer points to the cluster representative would be admitted as members of the cluster.

5. Define the significant peaks in $\{\triangle \theta_l\}_{l=2}^N$ as those above the mean by $\alpha$ times the standard deviation, ie. $\{l \in \{2, \ldots, N\} \mid \triangle \theta_l > \overline{\triangle \theta_l} + \alpha \ \sigma_{\triangle \theta_l}\}$. $\alpha$ is a parameter that will be discussed later (we recommend $\alpha = 5$).

6. Identify the significant peak with the lowest value of $l$, that is, the most left significant peak and take it as the definitive cluster cutoff $l_c$. In Fig. 2 (top), the example shows two significant peaks, which are above the horizontal dotted line given by $\overline{\triangle \theta_l} + \alpha \ \sigma_{\triangle \theta_l}$. If all the peaks are below $\overline{\triangle \theta_l} + \alpha \ \sigma_{\triangle \theta_l}$, then take the highest as the cluster cutoff, ie. $l_c = l_1 - 1$.

7. Finally, the $k^{th}$ cluster is given by $L_k = \{1, ..., l_c\}_k$. Invert the sorting operation $(I_k = S^{-1}(L_k))$ to recover the original point indices which define the $k^{th}$ cluster as $\Omega_k = \{x^i\}_{i \in I_k}$.

Under the $\triangle \theta_l$ representation pictured in Fig. 2 (top), the natural cluster to which the cluster representative $l = 1$ belongs becomes distinguishable as the data points to the left of the most left significant peak. As has been previously discussed, cluster members share a similar value of $d_l$ among themselves and are dissimilar when compared to non-members. As a consequence, the sequence $\{\triangle \theta_l\}_{l=2}^{l_c}$ contain low values, whereas $\triangle \theta_{l_c+1}$ is higher in comparison. Nevertheless, $\triangle \theta_{l_c+1}$ is not necessarily the highest in the sequence $\{\triangle \theta_l\}_{l=2}^N$ and therefore Step 6 is required to localise $\triangle \theta_{l_c+1}$ as the most inner peak. A decrease in the value of $\alpha$ will result in clusters with a lower proximity between them, in other words, more restrictive clusters. In the light of these considerations, $\alpha$ can be regarded as a threshold for the clustering resolution. Despite the fact that $\alpha$ plays a relevant role in data sets containing few points and dimensions, as the number of points and dimensions increase the plot of $\triangle \theta_l$ tend to show a single sharp peak corresponding to the cluster cutoff. It suffices to take any high value of $\alpha$.

Another issue is the role of $\beta$ in the definition of $\theta_l$ at Step 2. $d_l$ is a better estimation of the cluster diameter than $\beta + d_l$. However, the latter is preferred since with the former neither $\theta_1$ nor $\triangle \theta_2$ would be defined due to the zero value of $d_1$. These quantities lack meaning in the sense that a density cannot be defined

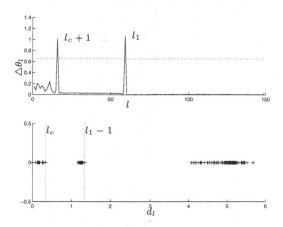

**Fig. 2.** Characteristic $\{\triangle\theta_l\}_{l=2}^N$ (top) and $\{d_l\}_{l=1}^N$ (bottom, represented by '+' signs) plots for a given point in the example data set

for only one point. Without $\triangle\theta_2$, it would not be possible to find out whether the cluster representative is an outlier or not. A suitable value of $\beta$ will vary depending on the cluster and hence it is preferable to link it to an estimation of the proximity between points in the cluster. We choose to fix $\beta = \gamma d_2$ with $0 < \gamma < 1$ (a high $\gamma$ implies a higher tendency to include outliers in clusters). In appendix A, $\gamma$ is shown to be a threshold for outlier detection. Note that an additional advantage of posing $\beta$ in this form is that it becomes negligible with respect to high values of $d_l$ and hence ensures the density effect of $\theta_l$. We set $\gamma = \frac{2}{3}$ throughout this work.

Finally, the detection of outliers in very high dimensional spaces is a difficult task that requires a preliminary test in order to ensure the effectiveness of the outlier detection for every value of $\gamma$. If this additional test, placed before Step 2, concludes that the cluster representative is an outlier, that is, $l_c = 1$ then Steps 2 to 6 are skipped. This test consists in checking whether the proximity from the cluster representative to the nearest neighbour $d_2$ is the highest of all consecutive differences in $d_l$, $\{\triangle d_l\}_{l=2}^N$, calculated as $\triangle d_l = |d_l - d_{l-1}|$. Note that this constitutes a sufficient condition for the cluster representative to be an outlier since it implies that the considered point is not similar to its nearest neighbour. This issue is explained further in [1].

## 2.3 Clustering Algorithm

In the previous section we have presented a criterion to determine the cluster to which a given data point belongs. We describe now an algorithm based on the designed criterion to carry out the clustering of the whole data set, whose flow diagram is presented in Fig. 3.

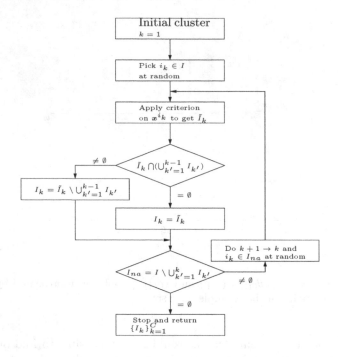

**Fig. 3.** Flow diagram of CHIDID

1. Randomly choose a point, $x^{i_k}$, from the data set, that has not been allocated to a cluster.
2. Apply the clustering criterion *over the whole data set* to determine the cluster associated to $x^{i_k}$.
3. In case that the proposed cluster, $\tilde{I}_k$, contains any points already included in any of the $k-1$ previous clusters, delete them from the proposed cluster. The points form the definitive $k^{th}$ cluster given by the index vector $I_k$.
4. Form the non-allocated index vector $I_{na} = I \setminus (\bigcup_{k'=1}^{k} I_{k'})$. If $I_{na} \neq \emptyset$, pass to next cluster by randomly choosing a point $i_{k+1 \rightarrow k}$ from $I_{na}$ and go back to Step 2. Otherwise stop the algorithm and return $\{I_k\}_{k=1}^{C}$, which constitutes the solution described in (2).

Let us go back to the example in Fig. 1 to illustrate the algorithm. Step 1 picks at random a point to initiate the clustering process. Apply Step 2 to find the associated cluster to the selected point. Step 3 deletes no points. In Step 4, $I_{na} \neq \emptyset$, since $I_{na}$ contains points corresponding to the four clusters yet to be discovered. Therefore another point is selected at random among those in $I_{na}$. We repeat Steps 1, 2 and 3, but this time Step 3 checks whether the actual cluster intersects with the previous one or not, after which the second cluster is formed. The latter procedure is repeated until all points have been allocated to their respective clusters.

As clusters can be arbitrarily close, it is possible to have points from different clusters within a proximity $d_{l_c}$ from the cluster representative. In such cases, the algorithm is likely to merge all these natural clusters together in a single output cluster. The aim of Step 3 is to ensure that formed clusters are pairwise disjoint. This is achieved by removing from the current cluster those points already allocated in previously found clusters. However, it must be highlighted that a merge is a very unlikely event. Firstly, a merge can only happen if the representative is located in the cluster region closest to the neighbouring cluster. Even in low dimensions, few points have this relative location within a cluster. Therefore, the likelihood of selecting one of these problematic representatives is also low. Secondly, if neighbouring clusters present a significant difference in the value of just one component then a merge cannot occur. Thus, while it is strictly possible, the likelihood of having a merge quickly diminishes as the dimensionality increases.

CHIDID is a method aiming at full dimensional clustering (ie. all components exhibit clustering tendency). It is expected to be able to handle data containing a few components without clustering tendency. However, the contribution of these irrelevant components to the proximity measure must be much smaller than that of the components with clustering tendency. In this work, we restrict the performance tests to data sets without irrelevant components. This issue will be discussed further when generating the test data sets in Sect. 3.2.

Since $\alpha$ and $\gamma$ are just thresholds, the only input to the algorithm is the data set containing the population. No a priori knowledge is required. The output is the assignment of data points (solutions) to clusters (optima or high performance regions), whose number is automatically found. The performance of the algorithm does not rely on finding suitable values for the parameters. A wide range of threshold values will provide the same clustering results if clusters and outliers are clearly identifiable. The algorithm naturally identifies outliers as byproduct of its execution. Within this approach, an outlier is found whenever the additional outlier test is positive or the criterion determines that the cluster representative is actually the only member of the cluster.

CHIDID carries an inexpensive computational burden. It is expected to run quickly, even with large high dimensional data sets. There are several reasons for this behaviour. First, the algorithm only visits populated regions. This constitutes a significant advantage in high dimensional spaces. Second, it scales linearly with $M$, the dimensionality of the space. Finally, it only calculates $C$ times $N$ distances between points, where $C$ is the number of found clusters and outliers.

# 3 Generation of Test Data Sets

## 3.1 Synthetic Data Sets

These data sets emulate the output of an ideal search and optimisation method on a multimodal optimisation problem. The assumed structure is most of data points distributed into clusters, with a small subset points as outliers.

The generation of these synthetic data sets is beneficial for two reasons. First, emulated outputs can be designed with very diverse characteristics. This is important because, in practice, there is not a unique form for these outputs. Second, it allows us to test the clustering algorithm with data sets of higher dimensionality than those achievable by current search methods.

It is reasonable to assume that each cluster of solutions is confined within the basin of attraction of a different function minimum (we restrict to minimisation without loss of generality). The Rastrigin function is suitable to generate such synthetic outputs because the location and extent of its minima is known. It is defined as

$$f(\boldsymbol{x}) = MA + \sum_{j=1}^{M}(x_j^2 - A\cos(2\pi x_j)) \ . \tag{4}$$

This function is highly multimodal. It has a local minimum at any point with every component $x_j$ taking an integer value $k_j$ and a single global minimum, at $\boldsymbol{x} = \boldsymbol{0}$. The function has a parabolic term and sinusoidal term. The parameter $A$ controls the relative importance between them (we set $A = 10$).

We generate the first synthetic data set (SD1) as follows. It contains $N_c = 100$ data points distributed into clusters, and $N_o = 10$ outliers with each component selected at random within $[-2.5, 2.5]$. The total number of points is $N = N_c + N_o$ and the number of variables is $M = 20$. Points are distributed among five clusters in the following way: $N_1 = 0.4N_c$, $N_2 = N_3 = 0.2N_c$ and $N_4 = N_5 = 0.1N_c$. Each cluster is associated with a minimum, whose components are picked at random among $c_j^k \in \{-2, -1, 0, 1, 2\}$. The basin of attraction of the $c^k$ minimum is defined by the interval $[c_j^k - 0.25, c_j^k + 0.25]$. The size of the interval is given by half the distance between consecutive function minima. Point components in each cluster are generated at random within the basin of attraction of the corresponding minimum.

In addition, we want to consider data sets with very diverse characteristics to check the performance robustness of the clustering method. We start by describing an automatic way to construct data sets. This is based on the previous data set, but with two variants. First, each point component is now generated within $[c_j^k - \triangle x, c_j^k + \triangle x]$, where $\triangle x$ is a random number between $(0, 1/2)$. Second, the number of points of each cluster $N_k$ is now determined as follows. We first determine $r_k$, which controls the proportion of points in cluster $k$, $r_k$ is drawn from a uniform distribution in the range $(1, 5)$. Next, we determine the number of points $N_k$ in cluster $k$ using the formula $N_k = N_c r_k / \sum_{k'=1}^{C} r_{k'}$, where $C$ is the number of generated clusters. Note that each of these data sets contain clusters of different sizes, cardinalities, densities and relative distances between them.

Finally, we define eight groups of data sets from all possible combinations of $N_c = 100, 1000$; $M = 10, 100$ and $C = 3, 7$. For all of them we add $N_o = 10$ randomly generated outliers within $[-2.5, 2.5]$. For each group $(N_c, M, C)$, we create ten realisations of the data set. We will refer to these data sets groups as SD2 to SD9. All these data sets are summarised in Table 1.

**Table 1.** Test data sets. $N_c$ is number of points in clusters, $M$ is the dimensionality and $C$ is the number of clusters. In the real data sets, RD1 and RD2, $C$ is unknown.

| Set | SD1 | SD2 | SD3 | SD4 | SD5 | SD6 | SD7 | SD8 | SD9 | RD1 | RD2 |
|-----|-----|-----|-----|-----|-----|-----|-----|-----|-----|-----|-----|
| $N_c$ | 100 | 100 | 100 | 100 | 100 | 1000 | 1000 | 1000 | 1000 | 177 | 203 |
| $M$ | 20 | 10 | 100 | 10 | 100 | 10 | 100 | 10 | 100 | 5 | 7 |
| $C$ | 5 | 3 | 3 | 7 | 7 | 3 | 7 | 3 | 7 | ? | ? |

## 3.2   Real Data Sets

We generate two real data sets (RD1 and RD2) by applying a search method on a analytical function with four global minimum. This function is used in [5] and defined as

$$f(\boldsymbol{x}) = A \left[ 1 - \sum_{k=1}^{C_G} \exp\left(\frac{x_1 - a^k}{\sigma_k^2}\right) \right] + B \frac{d(\boldsymbol{x}, \boldsymbol{b}(x_1))}{\sqrt{M}} . \tag{5}$$

with

$$\{\sigma_k\}_{k=1}^{C_G} = 0.5 \qquad a^k = \begin{cases} 2 & \text{if } k = 1 \\ 5 & \text{if } k = 2 \\ 7 & \text{if } k = 3 \\ 9 & \text{if } k = 4 \end{cases} \qquad b(x_1) = \begin{cases} 2 & \text{if } 0 < x_1 \leq 4 \\ 5 & \text{if } 4 < x_1 \leq 6 \\ 7 & \text{if } 6 < x_1 \leq 8.5 \\ 9 & \text{if } 8.5 < x_1 \leq 10 \end{cases}$$

where $d(\cdot, \cdot)$ is the euclidean distance, $C_G = 4$ (number of global minima), $M$ is the dimensionality, $A = 5$ and $B = 1$. As a search method, we use a Real-parameter Genetic Algorithm (GA) [3] [4] that has been modified [5] to be effective at finding multiple good minima. In brief the details are: a steady-state population of size N, parents are selected randomly (without reference to their fitness), crossover is performed using a self-adaptive parent-centric operator, and culling is carried out using a form of probabilistic tournament replacement involving NREP individuals.

The output of this GA is formed with all individuals that enter the population during the run. Thus, many of them are not optimised. From the clustering point of view, these points contain many irrelevant components which harm the clustering. Inevitably, one cannot find clusters where there are not clusters to find. Therefore, a preprocessing is needed to include most of the points with clustering tendency. The preprocessing consists in defining a threshold value for the objective function and include in the data set all points in the outputs below that threshold. The chosen threshold is $f(\boldsymbol{x}) < 1$.

The first data set, RD1, is obtained by applying the GA (N= 100 and NREP= 30, using 40,000 function evaluations) on the 5-dimensional instance of the function. RD2 comes from the application of GA (N= 150 and NREP= 40, using 200,000 function evaluations) on the 7-dimensional instance of the function.

## 4   Analysis of the Results

In the experiments performed in this study, we set the resolution of the clustering as $\alpha = 5$ and the threshold for outlier detection as $\gamma = \frac{2}{3}$. We required only one algorithm run for each data set to provide the presented results. These experiments were performed in an Intel Pentium IV 1.5GHz, with memory of 256MB.

Firstly, the clustering was performed on all the synthetic data sets described in the previous section (SD1 to SD9). The correct clustering was achieved for all of them, meaning that the correct assignment of individuals to clusters was found. Outliers were also identified correctly. No merge, as defined in Sect. 2.3, occurred. Also, due to their linkage to cluster properties, a single choice of parameters was valid for all experiments. The high variety of test data sets demonstrates the robustness of the method. The highest CPU time used was 0.8 seconds (with SD9).

Figure 4 provides a visualisation of the obtained clustering quality with SD1. The five clusters in the data set are correctly revealed by CHIDID, which has correctly assigned data points to clusters as well as discard the outliers.

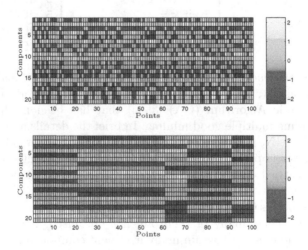

**Fig. 4.** Results for SD1. The upper plot shows the input data set without outliers, with colors corresponding to the component value. The bottom plot shows the five clusters found with CHIDID (outliers were correctly identified and are not included)

Likewise, Fig. 5 shows equally good performance on SD5, a data set with the same number of points as SD1 but a much higher dimensionality (M= 100 instead of M= 20).

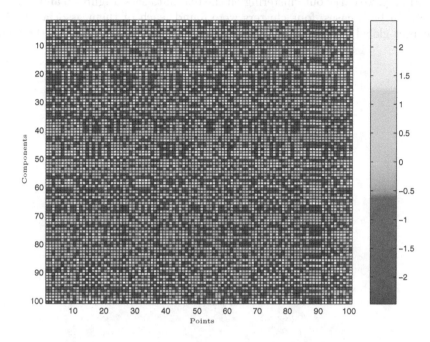

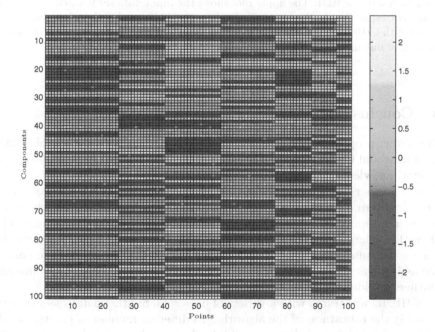

**Fig. 5.** Results for SD5. The upper plot shows the input data set without outliers, with colors corresponding to the component value. The bottom plot shows the seven clusters found with CHIDID (outliers were correctly identified and are not included)

Finally, we carry out clustering on the real data sets. Figures 6 and 7, respectively. In both cases, four clusters were found, each of them associated with a different global optimum. The quality of the clustering is clear from the figures.

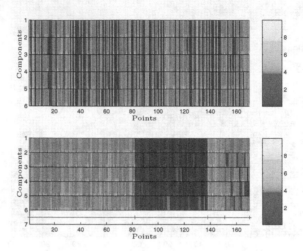

**Fig. 6.** Results for RD1. The upper plot shows the input data set without outliers, with colors corresponding to the component value. The bottom plot shows the four clusters found with CHIDID (outliers were correctly discriminated and are not included). The horizontal line underneath the plot marks the separation between clusters

# 5   Conclusions

This paper proposes a new algorithm (CHIDID) to identify clusters of solutions in Multimodal Optimisation. To use this algorithm it is not necessary to provide a priori knowledge related to the expected clustering behaviour (eg. cluster radius or number of clusters). The only input needed is the data set to be analysed. The algorithm does contain two tuning parameters. However, extensive experiments, not reported in this paper, lead us to believe that the performance of the algorithm is largely independent of the values of these two parameters, which we believe are suitable for most data sets and are used in this work. The output is the assignment of data points to clusters, whose number is found automatically. Outliers are identified as a byproduct of the algorithm execution.

CHIDID was tested with a variety of data sets. Synthetic data sets were used to study the robustness of the algorithm to different number of points, variables and clusters. Also, real data sets, obtained by applying a GA on an analytical function, were considered. CHIDID has been shown to be efficient and effective at clustering all these data sets.

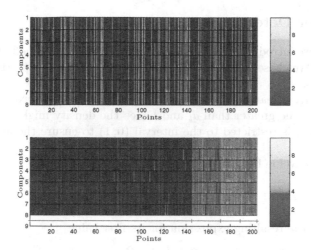

**Fig. 7.** Results for RD2. The upper plot shows the input data set without outliers, with colors corresponding to the component value. The bottom plot shows the four clusters found with CHIDID (outliers were correctly discriminated and are not included). The horizontal line underneath the plot marks the separation between clusters

# References

1. P. J. Ballester. *The use of Genetic Algorithms to Improve Reservoir Characterisation*. PhD thesis, Department of Earth Science and Engineering, Imperial College London, 2004.
2. P. J. Ballester and J. N. Carter. Method for managing a database, 2003. UK Patent Application filed on 5th August 2003.
3. P. J. Ballester and J. N. Carter. Real-parameter genetic algorithms for finding multiple optimal solutions in multi-modal optimization. In *Genetic and Evolutionary Computation Conference, Lecture Notes in Computer Science 2723*, July 2003.
4. P. J. Ballester and J. N. Carter. An effective real-parameter genetic algorithms for multimodal optimization. In Ian C. Parmee, editor, *Proceedings of the Adaptive Computing in Design and Manufacture VI*, April 2004. In Press.
5. P. J. Ballester and J. N. Carter. Tackling an inverse problem from the petroleum industry with a genetic algorithm for sampling. In *the 2004 Genetic and Evolutionary Computation COnference (GECCO-2004), Seatle, Washington, U.S.A.*, June 2004. In Press.
6. R. Bellman. *Adaptive control processes : a guided tour*. Princeton University Press, 1961.
7. M. Halkidi, Y. Batistakis, and M. Vazirgiannis. Clustering algorithms and validity measures. In *Proceedings Thirteenth International Conference on Scientific and Statistical Database Management*, pages 3–22. IEEE Computer Soc, Los Alamitos, 2001.
8. I. S. J. Packham and I. C. Parmee. Data analysis and visualisation of cluster-oriented genetic algorithm output. In *Proceedings of the 2000 IEEE International Conference on Information Visualization*, pages 173–178. IEEE Service Center, 2000.

## A     Outlier Detection

In Sect. 2.2, we introduced the parameter $\gamma$ as a threshold aimed at discriminating outliers. We chose to fix $\beta = \gamma d_2$ with $0 < \gamma < 1$. In this section, we justify these settings.

The relative linear density was defined as $\theta_l = \frac{l}{N}\frac{\beta + d_N}{\beta + d_l}$. Note that in case $\gamma > 1$, $\beta$ might be greater than $d_l$ and hence the density might be notably distorted. Thus, $\gamma$ is restricted to the interval $(0,1)$ to ensure the accuracy of the density measure.

Next, we develop the relative linear density increment as

$$\triangle\theta_l \equiv |\theta_l - \theta_{l-1}| = \frac{\beta + d_N}{N}\left|\frac{l}{\beta + d_l} - \frac{l-1}{\beta + d_{l-1}}\right| . \tag{6}$$

and evaluate the resulting expression for its first value

$$\triangle\theta_2 = \frac{\beta + d_N}{N}\left|\frac{2}{\beta + d_2} - \frac{1}{\beta}\right| . \tag{7}$$

From the latter equation, if $\beta \to 0$ (ie. $\gamma \to 0$) then $\triangle\theta_2 \to +\infty$ and thus $\triangle\theta_2$ will be the highest of all $\{\triangle\theta_l\}_{l=2}^N$. This would make the representative to appear always as an outlier. By contrast, if $\beta \to d_2$ (ie. $\gamma \to 1$) then $\triangle\theta_2 \to 0$ and thus $\triangle\theta_2$ will be the lowest of all $\{\triangle\theta_l\}_{l=2}^N$. This situation corresponds to the representative being never an outlier.

We next substitute $\beta = \gamma d_2$ in (7) to give

$$\triangle\theta_2 = \frac{\gamma d_2 + d_N}{N d_2}f(\gamma), \text{ with } f(\gamma) \equiv \frac{1 - \gamma}{(1 + \gamma)\gamma} . \tag{8}$$

Since $0 < \gamma < 1$, it is clear that $f(\gamma)$ controls the tendency of regarding a representative as an outlier. Figure 8 presents the plot of $f(\gamma)$ against $\gamma$. The figure shows that a higher value of $\gamma$ implies a higher tendency to include outliers in clusters.

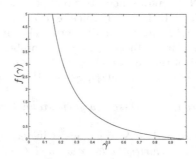

**Fig. 8.** Plot of $f(\gamma)$ against $\gamma$

# On an Experimental Algorithm for Revenue Management for Cargo Airlines

Paul Bartodziej and Ulrich Derigs

Department of Information Systems and Operations Research, University of Cologne,
Pohligstr. 1, 50696 Cologne, Germany.
{bartodziej,derigs}@winfors.uni-koeln.de

**Abstract.** In this paper we present a set of algorithms which represent different implementations of strategies for revenue management for air cargo airlines. The problem is to decide on the acceptance of a booking request given a set of fixed accepted requests and expected further demand. The strategies are based on a planning model for air cargo routing which determines the maximal contribution to profit for the capacities of a given flight schedule. This planning model associates an optimal set of so-called itineraries/paths to the set of possible origin-destination pairs according to given yield values. In mathematical terms the model is the path flow formulation of a special multi-commodity flow problem. This model is solved by intelligently applying column generation.

## 1 Introduction

Revenue Management (RM) deals with the problem of effectively using perishable resources or products in businesses or markets with high fixed cost and low margins which are price-segmentable like airline, hotel, car rental, broadcasting etc. (see Cross [1997] and Mc Gill et. al [1999]. Revenue management has to be supported by forecasting systems and optimization systems. In this paper we focus solely on the optimization aspect.

RM has its origin and has found broad application in the airline business and here especially in passenger flight revenue management. Only recently the concepts which have been developed for the passenger sector have been adequately modified and transferred to the cargo sector. Here an immediate transfer is not feasible since the two businesses although at first sight offering a similar kind of transportation service face differences in product offering and in production which have consequences for capacity planning and revenue management. While the demand in the passenger sector is one-dimensional (seat) and kind of smooth, the demand in the air cargo sector is multi-dimensional (weight, volume, classification) and lumpy. On the other side, capacity in the airline business is fixed while in the cargo business we face stochastic capacity when using idle belly capacities of passenger flights for instance.

Yet, the most subtle difference is as follows. In the passenger business the service which customers are booking is a concrete itinerary i.e. a sequence of

C.C. Ribeiro and S.L. Martins (Eds.): WEA 2004, LNCS 3059, pp. 57–71, 2004.
© Springer-Verlag Berlin Heidelberg 2004

flight connections, so-called legs leading the passenger from an origin to a destination. Thus a passenger booking a flight from Cologne to Rio de Janeiro via Frankfurt on a given date has to be given seat availability on two specific flights: (CGN,FRA) and (FRA,GIG). In the cargo business the customers book a transportation capacity from an origin to a destination, so-called O&D's at a certain service level i.e. within a certain time-window and the airline has some degree of freedom to assign booked requests to concrete flights later. For the example above the cargo client would book the connection (CGN,GIG) and he is not especially interested in the actual path or routing for his package of goods. This difference leads to different planning models.

Antes et. al. [1998] have developed a model and introduced a decision support system for evaluating alternative schedules by optimally assigning demand to flights. The model leads to a special kind of multi-commodity flow problem.

In this paper we show and analyze how this model can be adapted to aid in solving the operational problems faced in revenue management. The paper is organized as follows. In section 2 we introduce the problem and the model for transportation planning in the air cargo business and in section 3 we show how this model can be solved by column generation. In section 4 we describe how this model can be extended to the problem of revenue management from a conceptual point of view. In section 5 we propose several alternative strategies for implementing the model into the revenue management process and in section 6 we present first computational results for realistic problem scenarios.

## 2    The Basic Planning Model for Air Cargo Routing

A cargo airline offers the conceptually simple service to transport a certain amount of goods from an origin to a destination within a certain time interval at a given price. For this purpose the airline keeps a fleet of aircrafts and/or leases capacities from other companies, especially ground capacities for transporting goods to, from, and between airports. The tactical planning problem of such airlines is to design a (weekly) flight schedule which allows the most profitable service of the unknown market demand for the next period. Then, on the operational level the requests have to be assigned to concrete flights based on the given flight schedule.

From the verbal description of an air-cargo service given in the introduction we can identify three basic entity types in our world of discourse: AIRPORT, PRODUCT the class of all conceptually different "product types",and, TIME being a tuple of three components: day of the week, hour and minute.

The flight schedule can be described as a set of so-called legs. Here a *leg* is a direct flight between two airports offered at the same time every week during the planning period. We model the object class LEG as a 4-ary (recursive) relationship-type between AIRPORT playing the role of an origin (from) and the destination (to), respectively, and TIME playing the role of a departure and arrival time, respectively.

Every flight schedule defines a so-called *time-space network* $G = (V, E)$ with $V$ the set of nodes representing the airports at a certain point of time. $E$ is composed from two subsets, a set of *flight arcs* which represent the legs and connect the associated airports at departure and arrival time, respectively, and a set of *ground arcs* which connect two nodes representing the same airport at two consecutive points in time. Associated with every leg/arc $e \in E$ is the *weight capacity* $u_e$ measured in kg, the *volume capacity* $v_e$ measured in cubic meter and the operating cost $c_e$ measured in in \$.

Market-demand can be described as a set of estimated or booked service requests. Such a request has several attributes with the origin and destination airport being the dominant characteristic. Therefore, these request objects are commonly referred to as "O&D's" or "O&D-pairs". Conceptually the object class OD can be modelled as a complex 5-ary (recursive) relationship-type between AIRPORT playing the role of an origin and a destination, respectively, TIME playing the role of an availability and due time, respectively, and PRODUCT.

Then within the concept of network flow, the different O&D's define different commodities which have to be routed through this network (see Ahuja et. al. [1993]). For the following models we abstract in our notation and use the symbol $k$ for representing an O&D-commodity and $K$ is the set of all commodities/O&D's. Associated with a commodity/O&D $k \in K$ is a specific demand $b^k$ measured in kg, a value $d^k$ giving the volume per kg and a freight rate or yield value $y^k$ measured in \$ per unit of commodity $k$.

The so-called *path-flow model* of the multi-commodity flow problem, i.e. the problem to construct optimal assignments of O&D's/commodities to legs/arcs is based on the obvious fact that any unit transported from an origin to a destination has taken a sequence of legs (possibly only one leg) a so-called path or *itinerary* connecting the origin node with the destination node in the network. A path or itinerary for an O&D/commodity k is a sequence $p = (l_1, \ldots, l_{r(p)})$ of legs $l_i \in LEG, r(p) \geq 1$ with the following properties

$$origin(l_1) = origin(k)$$
$$destination(l_i) = origin(l_{i+1}) \qquad i = 1, \ldots, r(p) - 1$$
$$destination(r(p)) = destination(k)$$

A path $p$ is called *k-feasible* if additional requirements are fulfilled which vary with the problem definition. Here we consider several types of constraints which concern due dates, transfer times and product compatibility. Note that feasibility of paths is checked outside the decision model. For every O&D/commodity $k$ we denote by $P^k$ the set of $k$-feasible itineraries. A path may be feasible for many different O&D's. In our model we have to distinguish these roles and consider multiple copies of the same path/the same legs assigned to different commodities/O&D's. The relation between arcs and itineraries is represented in a binary indicator $\delta : L \times S \to \{0, 1\}$

$$\delta(l, p) := \begin{cases} 1 \text{ if leg l is contained in path p} \\ 0 \text{ } else \end{cases}$$

Given an itinerary $p \in P^k$ we can easily calculate $c^k(p) := \sum_{e \in E} c_e^k \delta_e(p) = \sum_{e \in p} c_e^k$ the operating cost as well as $y^k(p) := y^k - \sum_{e \in p} c_e$ the yield per kg of commodity $k$ which is transported over $p$. This calculation as well as the construction of the set $P^k$ is done outside our model using a so-called "connection builder" and then fed as input data into the model. For the model we introduce for every $p \in P^k$ a decision-variable $f(p)$ giving the amount (in kg) transported via $p$.

Now the planning problem is to select the optimal combination of paths giving maximal contribution to profit which leads to a standard linear (multi-commodity flow) program.

**Planning model (P)**

$$\max \sum_{k \in K} \sum_{p \in P^k} y^k(p) f(p) \tag{1}$$

$$\text{s.t.} \sum_{k \in K} \sum_{p \in P^k} \delta_e(p) f(p) \leq u_e \quad \forall e \in E \tag{2}$$

$$\sum_{k \in K} \sum_{p \in P^k} \delta_e(p) d^k f(p) \leq v_e \quad \forall e \in E \tag{3}$$

$$\sum_{p \in P^k} f(p) \leq b^k \quad \forall k \in K \tag{4}$$

$$f(p) \geq 0 \quad \forall k \in K \quad \forall p \in P^k \tag{5}$$

The advantage of the itinerary-based model over leg-based flow models is the possibility to consider rather general and complicated constraints for feasibility of transportation in the path-construction phase via the connection builder, i.e. keeping this knowledge away from the optimization model, thereby reducing the complexity of the optimization phase respectively, allowing the same standard (LP-)solution procedure for a wider range of different planning situations.

Moreover, this approach allows for scaling, i.e. it is not necessary to construct all possible paths beforehand. Working with a "promising subset" of profitable paths only, reduces the size of the problem instance but may lead to solutions which although not optimal in general, are highly acceptable in quality. Finally, an approach called *column generation* allows to generate feasible paths on the run during optimization and thus keeps problem size manageable throughout the optimization process.

## 3   Solving the Planning Model via Column Generation

Column generation goes back to Dantzig and Wolfe [1960] as an approach for solving large linear programs with decomposable structures. It has become the leading optimization technique for solving huge constrained routing and scheduling problems (see Desrochers et al. [1995]). Due to degeneracy problems column

generation often shows unsatisfactory convergence. Recently, so-called stabiliza-
tion approaches have been introduced and studied which accelerate convergence
(see du Merle et. al. [1999]. We did not adapt these advanced techniques since
our objective is not to solve single static instances of a hard optimization problem to near-optimality in reasonable time, but to analyze whether and how the
concepts which have shown to be appropriate for solving air-cargo network design and analysis problems on a tactical level can be applied on the operative in
a dynamic and uncertain environment. The suitability and use of such a common
modelling concept is necessary with respect to consistency of network capacity
planing and revenue control.

In the following we will only outline the column generation approach for solving (P). Let us denote this problem by MP, which stands for master problem.
Now, instead of generating all feasible itineraries $p \in P^k$, $k \in K$ and thus initializing MP only a promising subset of itineraries $R^k$, $k \in K$ is constructed in the
initialization phase. Such a subset can be obtained for instance by constructing and introducing for every O&D the m feasible itineraries with maximum
contribution to profit, where m is a relatively small number.

Since in the model every path or itinerary constitutes a decision variable and
a column in the LP we have generated with this initialisation for every O&D
a subset of columns associated with a subset of the set of all decision variables
defining a subproblem the so-called restricted master problem (RMP). RMP is
again of the (P)– type and thus can be solved by any standard LP-technique,
the simplex method for instance. Solving RMP will generate a feasible freight
flow, since every feasible solution to a restricted problem is also feasible for the
original master problem.

Now the question arises whether the optimal solution for RMP is optimal for
MP, too. Solving MP, by the simplex method for instance, we obtain for every
arc/leg $e$ a shadow price $w_e$ for the weight constraint and $z_e$ for the volume
constraint. The demand constraint associates with every commodity $k$ a shadow
price $\sigma^k$. Based on these prices the so-called "reduced cost coefficient" for an
itinerary $p \in P^k$ is defined as

$$r(p) = y^k - \sum_{e \in p} (c_e + w_e + d^k z_e) - \sigma^k \qquad (6)$$

LP-theory states that in the maximization case a feasible (basic) solution for an
LP is optimal if and only if every non-basic variable has non-positive reduced
cost. Thus, if in an optimal solution for RMP an itinerary $p \in P^k$ has positive
reduced cost $r(p)$ then transporting (one unit of) commodity k on p leads to a
solution for MP which is more profitable than the current optimal solution for
RMP.

Thus in the second phase we calculate the reduced cost values for (all) $p \in
P^k \backslash R^k$ and we check whether exists $p \in P^k$ for a commodity $k$ with $y^k - \sigma^k >
\sum_{e \in p} (c_e + w_e + d^k z_e)$, a so-called promising path. This phase is called $outpricing$.

A common and efficient approach for the outpricing phase is by solving a
shortest path problem for each $k \in K$ i.e. to determine the shortest path from

the origin of $k$ to the destination of $k$ with respect to the modified arc-costs $(c_e + w_e + d^k z_e)$. If the shortest path length exceeds $y^k - \sigma^k$ then no promising path exists. If this holds for all $k \in K$, this indicates that the optimal solution to RMP is optimal for MP. Otherwise we define additional variables for all or some itineraries for which $r(p) > 0$ holds and generate the associated columns to be introduced into RMP, which then has to be (re-) solved again.

The column generation concept which we have described above leaves a great variety of strategies on how to implement the different phases. The basic philosophy of column generation is to keep the restricted master problem RMP rather small to speed up the LP-solution. Yet, keeping RMP small requires to test more columns during the out pricing phase. Thus there is a trade-off which has to be evaluated by testing several options.

Accordingly, two different algorithms T1 and T2 were implemented to examine the trade-off between calls to outpricing and calls LP-(re-)optimition. In T1, RMP is re-solved and the dual prices are updated every time a new path is found while in T2 RMP is re-solved and the dual prices are updated only after the path search has been executed for all commodities.

Another important aspect of applying the path-flow formulation and column generation concept to the problem of generating optimal freight flows is the fact that all constraints on od-feasibility can be checked in the outpricing phase and thus need not be represented in the LP-formulation and tested during LP-solution. This does not only reduce the complexity of the LP, but enables the consideration of arbitrary constraints on the feasibility of itineraries and allows the application of one model to a variety of problem types characterised by different sets of constraints. Note that when the feasibility of itineraries is restricted by constraints, so-called constrained-shortest path problems CSPP have to be solved (see Nickel [2000]).

## 4 Operational Models for Air Cargo Booking and Revenue Management

The planning model can be used to evaluate alternative schedules, to identify capacity bottlenecks, to support static pricing etc. In an operational environment given a flight schedule and pricing information booking requests will occur over time and decisions have to be made concerning the acception or rejection of requests due to limited resources and/or (expected) opportunity costs and appropriate capacity reservations have to be made. Yet, there is no need for the airline to actually book a certain request to a specific itinerary i.e. set of legs. The airline has to ensure only that at any time there exists a feasible routing for all booked requests. This aspect is represented in the following booking model where we distinguish between a set $B$ of booked O&D- requests with $\beta^a$ the demand of a booked request $a \in B$, and a forecasted demand $b^k$ for a set $K$ of O&D's. W.l.o.g. we can assume that $K$ is the set of all O&D's. Then the model determines the maximal contribution to profit subject to the set of accepted/booked requests and the additional expected demand.

**Booking model**

$$\max \sum_{k \in K \cup B} \sum_{p \in P^k} y^k(p) f(p) \tag{7}$$

$$\text{s.t.} \sum_{k \in K} \sum_{p \in P^k} \delta_e(p) f(p) \leq u_e \quad \forall e \in E \tag{8}$$

$$\sum_{k \in K} \sum_{p \in P^k} \delta_e(p) d^k f(p) \leq v_e \quad \forall e \in E \tag{9}$$

$$\sum_{p \in P^k} f(p) \leq b^k \quad \forall k \in K \tag{10}$$

$$\sum_{p \in P^a} f(p) = \beta^a \quad \forall a \in B \tag{11}$$

$$f(p) \geq 0 \quad \forall k \in K \cup B \quad \forall p \in P^k \tag{12}$$

The next model represents the situation in revenue management where the decision on the acceptance of a single booking request $r$ with demand $\beta^r$ for a specific O&D subject to a set $B$ of already booked requests and estimated further demand for the commodities $K$ has to be made.

**Request model**

$$\max \sum_{k \in K \cup B \cup \{r\}} \sum_{p \in P^k} y^k(p) f(p) \tag{13}$$

$$\text{s.t.} \sum_{k \in K} \sum_{p \in P^k} \delta_e(p) f(p) \leq u_e \quad \forall e \in E \tag{14}$$

$$\sum_{k \in K} \sum_{p \in P^k} \delta_e(p) d^k f(p) \leq v_e \quad \forall e \in E \tag{15}$$

$$\sum_{p \in P^k} f(p) \leq b^k \quad \forall k \in K \tag{16}$$

$$\sum_{p \in P^a} f(p) = \beta^a \quad \forall a \in B \tag{17}$$

$$\sum_{p \in P^r} f(p) \leq \beta^r \tag{18}$$

$$f(p) \geq 0 \quad \forall k \in K \cup B \cup \{r\} \quad \forall p \in P^k \tag{19}$$

Note that for both models we assume an (expected) yield $y^k$ associated with every commodity $k \in K \cup B$. This yield has been used to calculate the yield of the paths $p \in P^k$ and thus is contained in the model only implicitly.

The booking model can be used to evaluate the acceptability of a request at a pre-specified yield as well as "dynamic pricing", i.e. the determination of the minimum acceptable yield. Conceptually, the first question can be answered by comparing the value of the optimal solution of the booking model without the actual request and the value of the optimal solution for the request model. We will present several more efficient algorithmic ideas based on computing promising paths in the sequel. Dynamic Pricing can be supported by modifying the procedure for solving the booking model. Here paths with increasing opportunity costs are constructed sequentially until a set of paths with sufficient capacity for fulfilling the additional demand is constructed. Then the opportunity costs of these path give the minimum acceptable yield, a concept which is often called "bid-price". We do not address aspects of dynamic pricing in this paper.

## 5    Strategies for Applying the Request Model in Revenue Management

When incorporating this modelling approach into a (real) revenue management process the specification of expected demand is essential. For this purpose the optimization module has to be integrated with a forecasting module where calibrated forecasting methods update the demand for O&D's. Thus we assume for the following that such a forecasting module triggers the update of $b^k$ and $y^k$ for $k \in K$ in the booking model which then is resolved by the optimization module. This computation is done outside the revenue management process which we describe in the following. Thus for our algorithms we assume that we always work on the optimal solution for a given booking model.

Now, there is one problem with applying and simulating a procedure for handling a series of booking requests. In a realistic environment the forecasted demand would be updated over time through the forecasting module, thus changing the booking model, its optimal freight flow and its optimal dual prices. Yet, there will always exist a feasible solution with respect to the booked requests. Associated with a request $r$ is a commodity $k$ for which a certain forecast $b^k$ is contained in the booking model, and thus $\beta^k$ is "contained" in the forecast $b^k$. Thus, after processing request $r$ the expected demand for commodity $k$ has to be adjusted especially if the demand of requests is kind of lumpy. In our experiments we take this into account by reducing the demand $b^k$ by $\beta^k$ for the next requests.

In the following we first discuss the basic applications of the request model with one additional request only and we assume that this request has to be evaluated in concurrency with booked requests and expected further demand. Such a request is characterized by the following data ( r , $\beta^r$, $y^r$). The basic algorithmic idea for (on-line/real-time) processing is as follows: Start from an optimal primal and dual solution of the booking model without the actual request $r$, called the master model MP, and sequentially determine feasible paths of maximal reduced cost until the demand of the request is fulfilled.

---

**Algorithm S1**

---

let $f$ be the optimal primal solution of the MP

while $\beta^r - \sum\limits_{p \in P^r} f(p) > 0$ // i.e. demand of request r is not fulfilled

do

    determine $p$ the r-feasible path with maximal reduced cost

    if the reduced cost is $\leq 0$ then: reject request r and restore the old MP,
        STOP.

    else

        introduce $p$ into the master problem MP

        (re-)solve MP and read new solution $f$

    end if

end do

accept the request and update the booking model by introducing the request
into B: let B $:=$ B $\cup$ {r} and introduce an equality constraint for r

---

This procedure does not always yield the correct answer. A request r which improves the expected contribution to profit may be rejected since S1 tries to ship the complete demand $\beta^r$ over itineraries with positive reduced costs only. Yet, using some itineraries with negative reduced costs could lead to an improvement with respect to total contribution.

In modification S2 we do not compute the path(s) for serving request $r$ "from scratch" based on the primal and dual solution of MP. Here we make use of the fact, that we may serve already forecasted demand for the same O&D $k$ in the optimal solution of MP and we use the associated itineraries for serving request $r$ . Only in the case that these path capacities are insufficient we would compute additional shortest paths as in S1. Yet, there is one problem that has to be handled appropriately. The itineraries in the optimal MP-solution which are associated with a forecasted demand for a commodity $k$ have been selected based on an expected revenue $y^k$. Thus, at this point of the decision process we have to compare $y^k$ with $y^r$ the actual yield for request r.

In the algorithms formulated so far we perform each outpricing and path determination for a single request on the basis of the "true" dual values and opportunity costs. Thus the master problem has to be resolved after each "augmentation" to update these cost values. In time-critical environments like on-line booking such an exact optimization may be much too costly, and strategies have to be applied which reduce the number of dual price determination, eventually at the cost of not getting maximal contribution to profit. Thus, in algorithm S3 we "freeze" the dual prizes as long as we can find profitable paths and we set the dual value/shadow price of arcs/legs which have become saturated to Big M, a sufficiently large value, which prevents constructing itineraries which use these arcs in the next steps.

Note that applying algorithm S3 we will not always obtain the solution with maximal contribution to profit.

## Algorithm S2

let $PF := \sum\limits_{p \in P^r} f(p)$ and let $RF$ be the additional capacity of paths
$p \in P^r$ with $f(p) > 0$ and let $RZ$ be the additional capacity of paths
$p \in P^r$ in MP with $f(p) = 0$
if $PF + RF + RZ < \beta^r$ // not enough capacity available
then
    apply algorithm S1
else
    if $y^k \leq y^r$ and $PF + RF \geq \beta^r$ //capacity is available and r is profitable
    then
        assign $Q := \{p \in P^k$ with $f(p) > 0\}$ to $r$ modifying $y^k(p)$ using $y^r$
        introduce $r$ into $B$ and introduce an equality constraint for $r$
        accept $r$ and resolve the booking model
    else //capacity available but $r$ may be unprofitable
        reduce $b^k$ by $\beta^r$ and solve the modified booking model,
        let z be the optimal value
        assign $Q := \{p \in P^k$ with $f(p) > 0\}$ to $r$ modifying $y^k(p)$ using $y^r$
        introduce $r$ into $B$ and introduce an equality constraint for $r$
        solve the modified booking model, let z' be the optimal value
        if $z' > z$ accept request $r$
        else reject request $r$ and restore the old MP
    end if
end if

## Algorithm S3

set the remaining demand $rd := \beta^r$
while $rd > 0$ and exists r-feasible path $p$ with positive reduced cost
do **path search for request r**
end while
if $rd > 0$ then // resolve and try again
    solve the booking model with inequality constraint for $\beta^r$
    while $rd > 0$ and exists r-feasible path $p$ with positive reduced cost
    do **path search for request r**
    end while
end if
if $rd = 0$
then
    accept request $r$
    introduce $r$ into $B$ and introduce an equality constraint for $r$
else
    reject the request and restore the old MP
end if

This algorithm makes use of the following subroutine

---

**Algorithm "path search for request r"**

compute $r$-feasible path $p$ of maximal reduced cost
let $q$ be the capacity of $p$
reduce capacity of arcs in $p$ by $\min\{rd, q\}$
reduce $rd$ by $\min\{rd, q\}$
set dual variables of arcs with zero capacity to Big M
introduce $p$ into MP

---

A proper strategy to reduce the need for real-time computing is to accept bookings which are uncritical with respect to capacity and apparently profitable without evaluation and optimization and to update the set of booked requests and construct a feasible freight flow using the booking model in a batch-processing kind of mode. In our fourth strategy (Algorithm S4) we give up the requirement to decide on the acceptance of requests sequentially one by one and we accumulate requests to *blocks* and then decide on which requests are accepted in one single optimization run per block.

---

**Algorithm S4**

---

Phase 1 (collection)

for all requests $r = r(k, \beta^r, y^r)$ in the block do:
    introduce into MP an in-equality constraint for $r$ with right hand side $\beta^r$
end for

Phase 2 (decision)

repeat
    solve MP and update primal solution $f$ and dual variables
    for all requests $r = r(k, \beta^r, y^r)$ in the block do:
        if exists r-feasible path $p$ with positive reduced cost then
            compute r-feasible path $p$ of maximal reduced cost
            introduce $p$ into MP
        end if
    end for
until no path was found in last repeat-loop run
for all requests $r = r(k, \beta^r, y^r)$ in the block do:
    if $\beta^r - \sum\limits_{p \in P^r} f(p) = 0$ // demand of request r is fulfilled
    then
        accept request $r$
        introduce $r$ into $B$ and introduce an equality constraint for $r$
    else
        reject request $r$
        remove the in-quality constraint and all paths found for $r$ from MP
    end if
end for

---

# 6   Computational Results

The algorithms which we have described in section 4 have been implemented in
Microsoft Visual C++ and have been applied to several real-world problems on a
PC with Pentium III Processor with 600 MHz and 256 MB RAM under Windows
98. The LP's were solved using the ILOG-CPLEX 6.5 solver. For solving the
constrained shortest path problems we have used a proprietary code (see Nickel
[2000]).

Our test problems were generated from 3 real world planning problems of a
cargo airline representing specific markets (see: Zils [2000]):

- Problem P10 with 10 airports, 624 legs and 1338 O&D's
- Problem P64 with 64 airports, 1592 legs and 3459 O&Ds
- Problem P79 with 79 airports. 1223 legs and 1170 O&D's'

In our experiment we have used the demand of the planning situation and we
have generated a sequence of requests over time. First we have split the demand
randomly into lumpy requests each having a demand between 20% and 100% of
the total demand. Then we applied a random perturbation in the range of -10%
to +10% to the demand of every request. With this procedure, we obtained a
number of requests which was about twice the number of O&D's (2533, 6521,
and 2206, respectively).

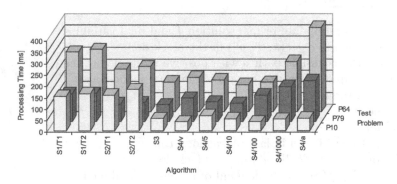

**Fig. 1.** Processing time of the different algorithms

For all runs we used the solution of the planning problem obtained by Algorithm T1 as starting situation. For algorithm S1 and S2 we also performed
test runs based on planning solutions obtained by T2. For algorithm S4 we have
performed tests with different blocksizes of 5, 10, 100 and 1000 requests. Moreover we have generated a test run with a variable blocksize. Here, starting with
a blocksize of 1% of the number of O&D's the size was doubled after the processing of 10 blocks until after 40 blocks the blocksize was held constant. This

experiment should represent the situation that closer to the deadline the number of requests per time unit increases while the decision is made within constant time intervals and thus the number of request per block is increasing. Finally we performed a test (indicated by "a") where all requests were put into one single block.

Figure 1 gives the result on the average processing time per call to the algorithms in ms. For all problem instances algorithm S3 and algorithm S4 with small blocksize have the smallest processing time which indicates that the means to reduce the effort materialize. Comparing algorithms S1 and S2 which focus more on profitability, we can see that using the paths of the MP-solution pays off.

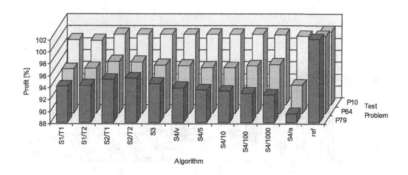

**Fig. 2.** Contribution to profit of different solutions

In Figure 2 we compare the contribution to profit which is obtained, i.e. we state the optimal values as fractions of the planning solution. Note that by our modification of the demand we could have generated instances where the value of the solution in the dynamic revenue management scenario could outperform the solution for the planning problem. Therefore we have taken the value obtained when applying algorithm S4/a before rejecting those requests which cannot be served completely as reference value. This is indicated by "ref". The results show that the algorithms differ with respect to quality significantly. Algorithm S2 which is comparable to S1 with respect to computation time is outperforming S1 and thus preferable. Algorithm S4 even when using small blocksizes is inferior to S3. Thus the assumption that one should postpone decisions is critical. Also, comparing algorithm S3 with algorithms S1 and S2 we see that the quality of S3 is only slightly inferior while on the other hand running time is significantly smaller.

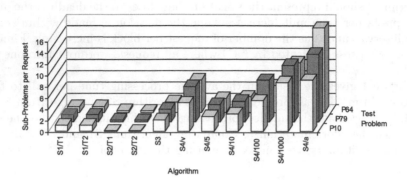

**Fig. 3.** Number of subproblems to be solved per request

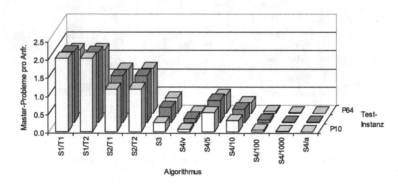

**Fig. 4.** Number of master problems to be solved

To analyse the computational behaviour of the different algorithms we have counted the number of sub-problems i.e. shortest path problems which have to be solved. Figure 3 gives the average number per request. For each algorithm, the numbers do not differ very much for the three test problems. It is significant that algorithm S2 has to compute the smallest number of paths. Figure 4 displays results on the number of master problems which have to be solved, Here we see that algorithms S3 and S4 need to solve significantly less LP's which accounts for the smaller processing time of S3 and S4. Further analysis shows that algorithm S3 (and S4) are more efficient than S2 only in those cases where the master-LP's are large enough, i.e. the number of O&D's is large enough.

After all, algorithm S2 or algorithm S3 seem to be the choice among the algorithms which we have proposed and tested. Here we can say, that with respect to running time the relation between the effort to solve the master LP's and to solve the path problems is crucial. The advantage of S2 with respect to quality becomes more significant if capacities become more scarce. There is no general best algorithm. Especially the running time for solving the LP's becomes unacceptable high for problem instances of larger size. Thus in these realistic and time-critical environments additional means to reduce the size of the master problems should be applied. Here one could reduce the number of paths in the search by limiting the allowable number of legs per path. Another option would be the decomposition of the network. Then the first step in the revenue management decision would be to assign every request to a suitable sub-network.

# References

1. Ahuja R.K.; Magnanti T.L. and J.B. Orlin, (1993): *Network Flows: Theory, Algorithms, and Applications*, Prentice Hall, Englewood Cliffs.
2. Antes J., Campen L., Derigs U., Titze C. and G.-D. Wolle, (1998): *SYNOPSE: a model-based decision support system for the evaluation of flight schedules for cargo airlines*, Decision Support Systems 22, 307-323.
3. Cross R.G., (1997): *Revenue Management: Hard Core Tactics for Market Domination*, Broadway Books, New York.
4. Dantzig G.B. and P. Wolfe, (1960): *Decomposition principles for linear programs*, Operations Research 8, 101-111.
5. Desrosiers J.; Dumas Y.; Solomon M.M. and F. Soumis, (1995):*Time constrained routing and scheduling* in: M.O. Ball et al. /(eds) Network Routing, Handbook in Operations Research and Management Science Vol. 8, North Holland, 35-139.
6. du Merle O.; Villeneuve D.; Desrosiers J. and P. Hansen (1999): *Stabilized column generation*, Discrete Mathematics 194, 229-237.
7. Mc Gill J.I., and G.J. van Ryzin (1999): *Revenue Management: Research Overview and Prospects*, Transportation Science 33, 233-256.
8. Nickel, N.-H.(2000):*Algorithmen zum Constrained Shortest Path Problem*, Technical Report, WINFORS, University of Cologne, Germany.
9. Zils, M. (1999): *AirCargo Scheduling Problem Benchmark Instanzen*, Working Paper, WINFORS, University of Cologne, Germany.

# Cooperation between Branch and Bound and Evolutionary Approaches to Solve a Bi-objective Flow Shop Problem

Matthieu Basseur, Julien Lemesre, Clarisse Dhaenens, and El-Ghazali Talbi

Laboratoire d'Informatique Fondamentale de Lille (LIFL), UMR CNRS 8022,
University of Lille, 59655 Villeneuve d'Asq Cedex, France.
{basseur,lemesre,dhaenens,talbi}@lifl.fr

**Abstract.** Over the years, many techniques have been established to solve NP-Hard Optimization Problems and in particular multiobjective problems. Each of them are efficient on several types of problems or instances. We can distinguish exact methods dedicated to solve small instances, from heuristics - and particularly metaheuristics - that approximate best solutions on large instances. In this article, we firstly present an efficient exact method, called the two-phases method. We apply it to a biobjective Flow Shop Problem to find the optimal set of solutions. Exact methods are limited by the size of the instances, so we propose an original cooperation between this exact method and a Genetic Algorithm to obtain good results on large instances. Results obtained are promising and show that cooperation between antagonist optimization methods could be very efficient.

## 1 Introduction

A large part of real-world optimization problems are of multiobjective nature. In trying to solve Multiobjective Optimization Problems (MOPs), many methods scalarize the objective vector into a single objective. Since several years, interest concerning MOPs area with Pareto approaches always grows. Many of these studies use Evolutionary Algorithms to solve MOPs [4,5,23] and only few approaches propose exact methods such as a classical branch and bound with Pareto approach, an $\epsilon$-constraint method and the two-phases method.

In this paper, we propose to combine the two types of approaches: a metaheuristic and an exact method. Therefore, we firstly present a two-phases method developed to exactly solve a BiObjective Flow Shop Problem (BOFSP) [11]. In order to optimize instances which are too large to be solved exactly, we propose and present cooperation methods between Genetic Algorithms (GAs) and the two-phases method.

In section II, we define MOPs and we present a BOFSP. In section III, we present the two-phases method applied to the BOFSP, and computational results. In section IV, we present cooperation schemes between GA and the two-phases method. Section V presents results on non-solved instances. In the last

C.C. Ribeiro and S.L. Martins (Eds.): WEA 2004, LNCS 3059, pp. 72–86, 2004.
© Springer-Verlag Berlin Heidelberg 2004

section, we discuss on effectiveness of this approach and perspectives of this work.

## 2    A Bi-objective Flow Shop Problem (BOFSP)

### 2.1    Multiobjective Optimization Problems (MOPs)

Although single-objective optimization may have a unique optimal solution, MOPs present a set of optimal solutions which are proposed to a decision maker. So, before presenting BOFSP, we have to describe and define MOPs in a general case. We assume that a solution $x$ to such a problem can be described by a decision vector $(x_1, x_2, ..., x_n)$ in the decision space $X$. A cost function $f : X \rightarrow Y$ evaluates the quality of each solution by assigning it an objective vector $(y_1, y_2, ..., y_p)$ in the objective space $Y$ (see Fig. 1). So, multiobjective optimization consists in finding the solutions in the decision space optimizing (minimizing or maximizing) $p$ objectives.

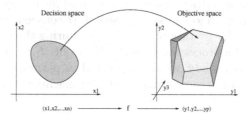

**Fig. 1.** Example of MOP

For the following definitions, we consider the minimization of $p$ objectives. In the case of a single objective optimization, comparison between two solutions $x$ and $x'$ is immediate. For multiobjective optimization, comparing two solutions $x$ and $x'$ is more complex. Here, there exists only a partial order relation, known as the Pareto dominance concept:

**Definition 1.** *A solution $x$ dominates a solution $x'$ if and only if:*

$$\begin{cases} \forall k \in [1..p], f_k(x) \leq f_k(x') \\ \exists k \in [1..p]/f_k(x) < f_k(x') \end{cases}$$

In MOPs, we are looking for Pareto Optimal solutions:

**Definition 2.** *A solution is Pareto optimal if it is not dominated by any other solution of the feasible set.*

The set of optimal solutions in the decision space $X$ is denoted as the Pareto set, and its image in the objective space is the Pareto front. Here we are interested in a apriori approach where we want to find every Pareto solutions.

## 2.2   Flow Shop Problem (FSP)

The FSP is one of the numerous scheduling problems. Flow-shop problem has been widely studied in the literature. Proposed methods for its resolution vary between exact methods, as the branch & bound algorithm [16], specific heuristics [15] and meta-heuristics [14]. However, the majority of works on flow-shop problem studies the problem in its single criterion form and aims mainly to minimize makespan, which is the total completion time. Several bi-objective approaches exist in the literature. Sayin et al. proposed a branch and bound strategy to solve the two-machine flow-shop scheduling problem, minimizing the makespan and the sum of completion times [16]. Sivrikaya-Serifoglu et al. proposed a comparison of branch & bound approaches for minimizing the makespan and a weighted combination of the average flowtime, applied to the two-machine flow-shop problem [17]. Rajendran proposed a specific heuristic to minimize the makespan and the total flowtime [15]. Nagar et al. proposed a survey of the existing multicriteria approaches of scheduling problems [14].

FSP can be presented as a set of $N$ jobs $J_1, J_2, \ldots, J_N$ to be scheduled on $M$ machines. Machines are critical resources: one machine cannot be assigned to two jobs simultaneously. Each job $J_i$ is composed of $M$ consecutive tasks $t_{i1}, \ldots, t_{iM}$, where $t_{ij}$ represents the $j^{th}$ task of the job $J_i$ requiring the machine $m_j$. To each task $t_{ij}$ is associated a processing time $p_{ij}$. Each job $J_i$ must be achieved before its due date $d_i$. In our study, we are interested in permutation FSP where jobs must be scheduled in the same order on all the machines (Fig. 2).

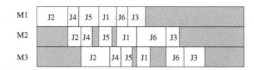

**Fig. 2.** Example of permutation Flow-Shop

In this work, we minimize two objectives: $C_{max}$, the makespan (Total completion time), and $T$, the total tardiness. Each task $t_{ij}$ being scheduled at the time $s_{ij}$, the two objectives can be computed as follows:

$$C_{max} = Max\{s_{iM} + p_{iM} | i \in [1 \ldots N]\}$$

$$T = \sum_{i=1}^{N} [max(0, s_{iM} + p_{iM} - d_i)]$$

In the Graham et. al. notation, this problem is denoted [7]: $F/\text{perm}, d_i/(C_{max}, T)$. $C_{Max}$ minimization has been proven to be NP-hard for more than two machines, in [12]. The total tardiness objective $T$ has been studied only a few times for M machines [9], but total tardiness minimization for one machine has been proven to be NP-hard [6]. The

evaluation of the performances of our algorithm has been realized on some
Taillard benchmarks for the FSP [18], extended to the bi-objective case [20]
(bi-objective benchmarks and results obtained are available on the web at
http://www.lifl.fr/~basseur).

## 3   An Exact Approach to Solve BOFSP: The Two-Phases Method (TPM)

On the Pareto front two types of solutions may be distinguished : the supported
solutions, that may be found thanks to linear combinations of criteria, and non
supported solutions [21]. As supported solutions are the vertices of the convex
hull, they are nearer to the ideal optimal solution, and we can ask why it is
important to look for non supported solutions. Figure 3, shows the importance
of non-supported solutions. It represents the Pareto front for one instance of the
bicriteria permutation flowshop with 20 jobs and 5 machines. This figure shows
that for this example, only two Pareto solutions are supported (the extremes)
and to get a good compromise between the two criteria, it is necessary to choose
one of the non-supported solutions.

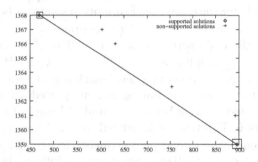

**Fig. 3.** Importance of non supported solutions (Pb: ta_20_5_02)

A lot of heuristic methods exist to solve multicriteria (and bicriteria) problems. In this section we are interested in developing an exact method able to
enumerate all the Pareto solutions for a bicriteria flowshop problem.

A method, called the Two-Phases Method, has been proposed by Ulungu
and Teghem to initially solve a bicriteria assignment problem [21]. This method
is in fact a very general scheme that could be applied to other problems at
certain conditions. It has not yet been very often used for scheduling applications
where the most famous exact method for bicriteria scheduling problems is the
$\epsilon$-constraint approach, proposed by Haimes et al. [8]. This section presents the
application of the scheme of the two-phases method to the bicriteria flow shop
problem under study.

## 3.1   The Two-Phases Method

Here we present the general scheme of the method. It proceeds in two phases. The
first phase finds all the supported solutions and the second all the non-supported
ones.

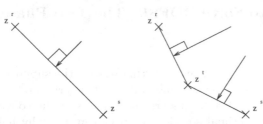

**Fig. 4.** Search direction.     **Fig. 5.** New searches.

**Fig. 6.** Non supported solutions.

- The first phase consists in finding supported solutions with aggregations of
  the two objectives $C_1$ and $C_2$ in the form $\lambda_1 C_1 + \lambda_2 C_2$. It starts to find
  the two extreme efficient solutions that are two supported solutions. Then it
  looks recursively for the existence of supported solutions between two already
  found supported solutions $z^r$ and $z^s$ (we suppose $z_1^{(r)} < z_1^{(s)}$ and $z_2^{(r)} > z_2^{(s)}$)
  according to a direction perpendicular to the line $(z^r\ z^s)$ (see figure 4), while
  defining $\lambda_1$ and $\lambda_2$ as follows: $\lambda_1 = z_2^{(r)} - z_2^{(s)}$, $\lambda_2 = z_1^{(s)} - z_1^{(r)}$. Each new
  supported solution generates two new searches (see figure 5).
- The second phase consists in finding non-supported solutions. Graphically,
  any non-supported solution between $z^r$ and $z^s$ belongs to the triangle represented in figure 6. This triangle is defined by $z^r$, $z^s$ and Y, which is the point
  $[z_1^{(s)}, z_2^{(r)}]$ . Hence, the second phase consists in exploiting all the triangles,
  underlying each pair of adjacent supported solutions, in order to find the
  non-supported solutions.

## 3.2   Applying the Two-Phases Method to a Bicriteria Flow Shop
Problem

The interesting point of the two-phases method, is that it solves exactly a bicriteria problem without studying the whole search space. Hence we want to apply
it to solve BOFSP for which the complete search space is too large to enable a
complete enumeration. But this method is only a general scheme and applying it
to a given problem requires a monocriterion exact method to solve aggregations.

As this scheduling problem (even in its monocriterion form) is NP-Hard, we
decided to develop a branch-and-bound method. A large part of the success of a
branch-and-bound is based on the quality of its lower bounds. As the makespan
minimization has been widely studied, we have adapted an existing bound for

this criterion whereas for the total tardiness we propose a new bound. Details about these bounds may be found in [11].

The search strategy used is "a depth first search" where at each step, the node with the best bound is chosen. Moreover, a large part of the tardiness $(T)$ value is generated by the last scheduled jobs. So the construction of solutions places jobs either at the beginning or at the end of the schedule, in order to have a precise estimation of the final $T$ value fastly.

## 3.3   Improvements of the Two-Phases Method

The two-phases method can be applied to any bicriteria problem. Applying it to scheduling problems allows improvements:

- Search of the extremes: The calculation of the extremes may be very long for scheduling problems as there exists a lot of solutions with the same value for one criterion. Hence, we propose to find extremes in a lexicographic order. A criterion is first optimized and then the second, without degrading the first one.
- Search intervals: The objective of the first phase is to find all the supported solutions in order to reduce the search space of the second phase. But when supported solutions are very near to each other, it is not interesting to look for all of them, as it will be very time consuming. Moreover, in the second phase, the search is, in fact, not reduced to the triangle shown on figure 6 but to the whole rectangle $(z_2^{(r)}, Y, z_1^{(s)}, 0)$. Hence, during the second phase, it is possible to find supported solutions that still exists. Then to avoid uninteresting branch-and-bounds we propose to execute a first phase only between solutions far from each other (a minimal distance is used).

## 3.4   Results

Table 1 presents results obtained with the two-phases method on the studied problems. The first column describes the instances of the problem: ta_number of jobs_number of machines_number of the instance. Then the three following columns indicate computational time with three different versions : the original two-phases method, the method with improvements proposed, and its parallel version[1]. It shows that both, improvements and parallelization allow to solve problems faster. Sequential runs have been executed on a 1.00Ghz machine. The parallel version has been executed on eight 1.1Ghz machines.

---

[1] The parallel version is described in [11]

**Table 1.** Results of two-phases method.

| Instances | Original method | Time With improvements | With parallelization |
|---|---|---|---|
| ta_20_5_01 | 30" | 17" | no need |
| ta_20_5_02 | 15' | 14' | no need |
| ta_20_10_01 | one week | 2 days | 1 day |
| ta_20_10_02 | one week | 2 days | 1 day |
| ta_20_20_01 | Unsolved | Unsolved | few weeks |

# 4 Using the Two-Phases Approach to Approximate the Optimal Pareto Front

## 4.1 Motivations

Exact methods are always limited by the size of the problem. Moreover, when the optimal Pareto front is not reached, these methods do not give good solutions. So, for these problems, heuristics are usually proposed. In this section, we propose to use the adaptation of the TPM to improve Pareto fronts obtained with a heuristic. Firstly, we briefly present the hybrid GA which will cooperate with TPM. Then we propose several cooperation mechanisms between TPM and the hybrid GA.

## 4.2 An Adaptive Genetic/Memetic Algorithm (AGMA)

In order to optimize solutions of FSP, AGMA algorithm has been proposed in [3]. AGMA is firstly a genetic algorithm (GA) which proposes an adaptive selection between mutation operators. Crossover, selection and diversification operators are described in [2]. Moreover, AGMA proposes an original hybrid approach: the search alternates adaptively between a Genetic Algorithm and a Memetic Algorithm (MA). The hybridization works as follows: Let $P_{PO*}$ be the value of the modification rate done on the Pareto front $PO*$ computed on the last generations of the GA. If this value goes below a threshold $\alpha$, the MA is launched on the current GA population. When the MA is over, the Pareto front is updated, and the GA is re-run with the previous population (Algorithm 1).

Computational results presented in [3] show that we have a good approximation of the Pareto front. In order to improve these results, we propose some cooperative schemes between AGMA and TPM.

## 4.3 Cooperation between AGMA and TPM

Recently, interest for cooperation methods grows. A large part of them are hybrid methods, in which a first heuristic gives solution(s) to a second one which upgrades its (their) quality [19]. But different Optimization Methods (OMs) can

---

**Algorithm 1** AGMA algorithm

---

Create an initial population
**while** run time not reached **do**
   Make a GA generation with adaptive mutation
   Update $PO*$ an $P_{PO*}$
   **if** $P < \alpha$ **then**
      /* Make a generation of MA on the population */
      Apply crossover on randomly selected solutions of $PO$ to create a set of new solutions.
      Compute the non-dominated set $PO'$ on these solutions
      **while** New solutions found **do**
         Create the neighborhood $N$ of each solution of $PO'$
         Let PO' be the non-dominated set of $N \bigcup PO'$
      **end while**
      Update $PO*$ an $P_{PO*}$
   **end if**
   Update selection probability of each mutation operator
**end while**

---

cooperate in several ways as shown in figure 7. This cooperation can be sequential ($a$), often called hybridization. The search can also alternate between two OMs ($b$). The alternativity may be decided thanks to thresholds ($c$). Finally a cooperation can be established, with one method integrated in a mechanism of the second one with or without threshold ($d$).

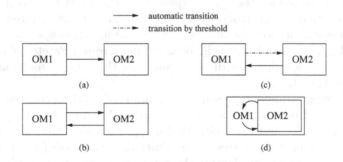

**Fig. 7.** Examples of cooperation scheme

Here we present three cooperation methods that combine the two-phases method (TPM) and the Adaptive Genetic/Memetic Algorithm (AGMA) presented before. The first one is an exact method which uses the Pareto set obtained with AGMA to speed up TPM. But the computational time of TPM still grows exponentially with the size of the instances. So, for the next approaches running on larger problems, we add constraints to TPM, to guide the algorithm despite of the loss of the guaranty to obtain the optimal Pareto set.

These three methods use the cooperation scheme $(a)$. But we can apply these methods with the other cooperation schemes, which are more evolved.

**Approach 1 - An improved exact approach: Using AGMA solutions as initial values:** In this approach, we run the whole two-phases method. For every branch-and-bounds of the TPM, we consider the best solutions given by the meta-heuristic as initial values. Therefore we can cut a lot of nodes of the branch-and-bound and find all optimal solutions with this method.

The time required to solve a given problem is of course smaller if the distance between the initial front (given by the meta-heuristic) and the optimal front is small. If the distance between them is null, the TPM will be used to prove that solutions produced by AGMA are optimal.

Even if this approach reduces the time needed to find the exact Pareto front, it does not allow to increase a lot the size of the problems solved.

**Approach 2 - Using TPM as a Very Large Neighborhood Search (VLNS):** Neighborhood search algorithms (also called local search algorithms) are a wide class of improvement heuristics where at each iteration an improving solution is found by exploring the "neighborhood" of the current solution. Ahuja et. al remark that a critical issue in the design of a neighborhood search is the choice of the neighborhood structure [1]. In a general case, larger is the neighborhood, more efficient is the neighborhood search. So, VLNS algorithms consist in exploring exponential neighborhood in a practical time to get better results. In [1], several exponential neighborhoods techniques are exposed. Here, we propose to use TPM as a VLNS algorithm.

The idea is to reduce the space explored by the TPM by cutting branches when the solution in construction is too far from the initial Pareto solution. An efficient neighborhood operator for FSP is the insertion operator [3]. So, we allow TPM to explore only the neighborhood of an initial Pareto solution which consists of solutions that are distant from less than $\delta_{max}$ insertion operator applications from it:

The following example represents an example of solution construction using VLNS approach from the initial solution $abcdefghij$. In this example two sets of jobs are used: The first one (initialized to $\{\}$) represents the current partially constructed solution and the second one (initialized to $\{abcdefghij\}$) represents jobs that have to be placed. During the solution construction, $\delta$ value (initially set to 0) is incremented for each breaking order with the initial solution. If $\delta = \delta_{max}$, then no more breaking order is allowed, so in this case, only one schedule is explored:
Example (constraint: $\delta_{max} = 2$):
- Initialization: $\{\},\{abcdefghij\}$ (represents: {jobs scheduled},{jobs to be placed})
- We firstly place the two first jobs: $\{ab\}, \{cdefghij\}$. $\delta = 0$.
- Then we place job $g$, so we apply insertion operator on the remaining jobs: $\{abg\}, \{cdefhij\}$. For the moment, the distance $\delta$ between the initial solution and the current solution is 1 (one breaking order).

- Then we place jobs $c$ and $d$: $\{abgcd\}, \{efhij\}$. $\delta = 1$.
- Then we place job $h$: $\{abgcdh\}, \{efij\}$. $\delta = 2$.
- Here, $\delta = \delta_{max}$, so the last jobs have to be scheduled without breaking order. So, the single solution explored is the schedule $\{abgcdhefij\}, \{\}$, with $\delta = 2$. Others possible schedules are too far from the initial solution.

The size of the insertion neighborhood is in $\Theta(N^2)$, so the size of the space explored by TPM may be approximated (for $d_{max} << N$) by $\Theta(N^{2d_{max}})$. Hence, we have to limit $d_{max}$ value, especially on instances with many jobs.

**Approach 3 - A local optimization with TPM:** This third cooperation limits the size of the explored space, while reducing it to a partition of Pareto solutions proposed by AGMA. So TPM is applied on regions of Pareto solutions.

The main goal of this approach is to limit the size of the trees obtained by the TPM, in order to apply this approach to large instances. In this section, we will present a non-exact two-phases method in which we only explore a region of the decisional space. So, we select partitions of each solution obtained by AGMA algorithm. Then we explore all the solutions obtained with modifications of these partitions using the two-phases method. After having explored a partition for all the Pareto set, we extract the new Pareto set from the obtained solutions.

This Simple Partitionning Post Optimization Branch & Bound (SPPOBB) works as follows:

The two-phases method explores the tree by placing jobs either at the beginning or at the end of the schedule. So, if the partition is defined from job $X_i$ to job $X_j$, it places, jobs $0..X_i - 1$ at the beginning of the schedule and jobs $X_j + 1..N$ at the end. Then it explores the remaining solutions of the tree by using the two-phases method technique.

Figure 8 shows an example of hybridization by the two-phases method - it can be applied for other branch and bound methods. In this figure, we consider an initial solution $a, b, \ldots, i, j$, which is on the Pareto front obtained by AGMA algorithm. In this example, $N = 10$, the partition size is 4, and is applied from job number 4 to 7, i.e jobs $d, e, f, g$ in the schedule. The first phase consists in placing the first three jobs at the beginning of the schedule. Then, it places the last three jobs at the end of the schedule (a job $j$ placed in queue is symbolized by $-j$). Then, we apply the two-phases method on the remaining jobs. After cutting several nodes, 5 complete schedules have been explored:

- a,b,c,-j,-i,-h,d,-e,f,g which corresponds to the schedule abcdfgehij
- a,b,c,-j,-i,-h,d,-g,e,f which corresponds to the schedule abcdefghij (the initial solution)
- a,b,c,-j,-i,-h,d,-g,f,e which corresponds to the schedule abcdfeghij
- a,b,c,-j,-i,-h,g,-d,-e,f which corresponds to the schedule abcgfedhij
- a,b,c,-j,-i,-h,g,-d,-f,e which corresponds to the schedule abcgefdhij

**Parameters:**

Different parameters have to be set to have an efficient search without having a too large time expense.

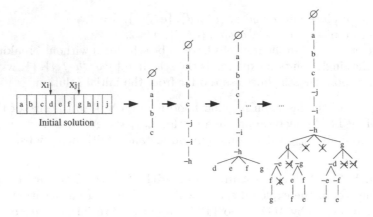

**Fig. 8.** Example: one partition exploration

- Size of the partitions: The cardinality of the Pareto set obtained with AGMA algorithm varies between several tens and two hundred solutions. In order to obtain solutions rapidly, we limit the size of partitions to 15 jobs for the 10-machines instances, and 12 jobs for the 20-machines instances. So each two-phases execution can be solved in several seconds or minutes.
- Number of partitions for each solution: Enough partitions of the complete schedule have to be considered to treat each job at least once by TPM approach. Moreover, it is interesting to superpose consecutive partitions to authorize several moves of a same job during optimization. Then, a job which is early scheduled could be translated at the end of the schedule by successive moves. On the other side, the more partitions we have, the more processing time is needed. So we take 8 partitions for the 50-jobs instances, 16 partitions for the 100-jobs instances and 32 partitions for the 200-jobs instances.

## 5   Results

We test the first approach to prove optimality of Pareto fronts on small instances. This approach reduces the time needed by the TPM to exactly solve these instances. Then we test the last two approaches. Results are comparable on 50 machines instances, but the computational time of the VLNS approach is exponential, so we present here only the results obtained with SPPOBB. However, the other approaches give some perspectives about cooperation mechanisms.

In this part, we firstly present performance indicators to evaluate effectiveness of this approach. Then we apply these indicators to compare the fronts obtained before and after cooperation with SPPOBB.

### 5.1   Quality Assessment of Pareto Set Approximation

Solutions' quality can be assessed in different ways. We can observe graphically progress realized as in figures 9 and 10. Here, we use the contribution metric

[13] to evaluate the proportion of Pareto solutions given by each front, and the S metric, as suggested in [10], to evaluate the dominated area.

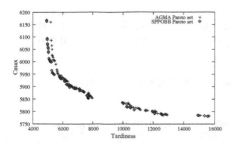

**Fig. 9.** SPPOBB results: instance with 100 jobs and 10 machines.

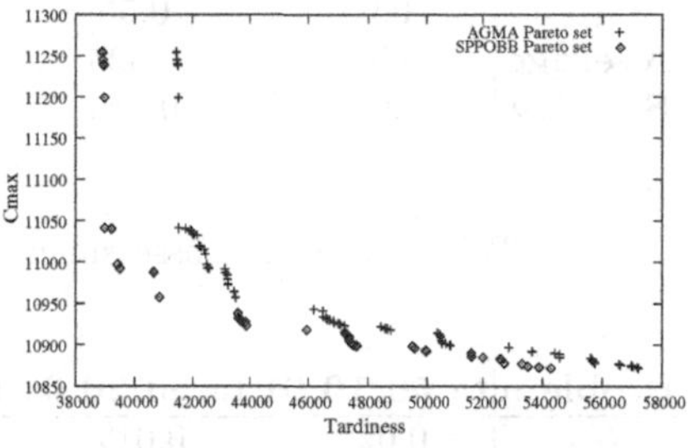

**Fig. 10.** SPPOBB results: instance with 200 jobs and 10 machines.

**Contribution metric:** The contribution of a set of solutions $PO_1$ relatively to a set of solutions $PO_2$ is the ratio of non-dominated solutions produced by $PO_1$ in $PO^*$, where $PO^*$ is the set of Pareto solutions of $PO_1 \cup PO_2$.

- Let $PO$ be the set of solutions in $PO_1 \cap PO_2$.
- Let $W_1$ (resp. $W_2$) be the set of solutions in $PO_1$ (resp. $PO_2$) that dominate some solutions of $PO_2$ (resp. $PO_1$).
- Let $L_1$ (resp. $L_2$) be the set of solutions in $PO_1$ (resp. $PO_2$) that are dominated by some solutions of $PO_2$ (resp. $PO_1$).
- Let $N_1$ (resp. $N_2$) be the other solutions of $PO_1$ (resp. $PO_2$): $N_i = PO_i \setminus (PO \cup W_i \cup L_i)$.

$$Cont(PO_1/PO_2) = \frac{\frac{\|PO\|}{2} + \|W_1\| + \|N_1\|}{\|PO^*\|}$$

**$S$ metric:** A definition of the S metric is given in [22]. Let $PO$ be a non-dominated set of solutions. S metric calculates the hyper-volume of the multi-dimensional region enclosed by $PO$ and a reference point $Z_{ref}$.

Let $PO_1$ and $PO_2$ be two sets of solutions. To evaluate quality of $PO_1$ against $PO_2$, we compute the ratio $(S(PO_1) - S(PO_2))/S(PO_2)$. For the evaluation, the reference point is the one with the worst value on each objective among all the Pareto solutions found over the runs.

### 5.2   Computational Results

We use S and Contribution metrics to compute improvements realized on fronts. Tests were realized for 10 runs per instance, on a 1.6Ghz machine. Tables 2 and 3 show the results obtained for these metrics.

**Table 2.** Quality assessment (contribution metric): C(SPPOBB/AGMA)

| Problem | ta_50_10_01 | ta_50_20_01 | ta_100_10_01 | ta_100_20_01 | ta_200_10_01 |
|---------|-------------|-------------|--------------|--------------|--------------|
| $C_{Min}$ | 0.54 | 0.51 | 0.96 | 0.73 | 1.00 |
| $C_{Max}$ | 0.63 | 0.55 | 1.00 | 0.96 | 1.00 |
| **Average** | 0.594 | 0.525 | 0.986 | 0.876 | 1.000 |
| **Std dev** | 0.026 | 0.015 | 0.015 | 0.062 | 0.000 |

**Table 3.** Quality assessment (S metric): S(SPPOBB)/S(AGMA)

| Problem | ta_50_10_01 | ta_50_20_01 | ta_100_10_01 | ta_100_20_01 | ta_200_10_01 |
|---------|-------------|-------------|--------------|--------------|--------------|
| $S_{Min}$ | 0.02% | 0.01% | 0.75% | 0.28% | 8.35% |
| $S_{Max}$ | 0.46% | 0.27% | 2.10% | 1.92% | 15.57% |
| **Average** | 0.185% | 0.093% | 1.199% | 0.970% | 13.094% |
| **Std dev** | 0.122% | 0.095% | 0.387% | 0.412% | 1.974% |

Table 2 shows that improvements realized on 50*10 and 50*20 instances were small in a general case. In fact we have an average improvement of 18.8 per cent of the initial Pareto set for the 50*10 instance, and 4.8 per cent for the 50*20 instance. For the other problems, a large part of the new Pareto set dominates the initial set of Pareto solutions. Table 3 shows a good progression of the Pareto front for large problems, especially for the 200 jobs* 10 machines instance.

Table 4 shows that the time required to realize the set of two phases is almost regular despite of the branch & bound approach.

**Table 4.** Run time

| Problem | ta_50_10_01 | ta_50_20_01 | ta_100_10_01 | ta_100_20_01 | ta_200_10_01 |
|---------|-------------|-------------|--------------|--------------|--------------|
| $T_{Min}$ | 3h16' | 5h33' | 28h11' | 37h26' | 63h24' |
| $T_{Max}$ | 5h26' | 6h19' | 52h44' | 65h02' | 122h45' |
| **Average** | 4h19' | 5h49' | 35h54' | 50h25' | 90h53' |
| **Std dev** | 42' | 25' | 8h05' | 7h58' | 17h16' |

## 6   Conclusion and Perspectives

In this paper, we have first presented an exact approach and a metaheuristic approach to solve MOPs. These approaches have been applied on a BOFSP. Then we have proposed original approaches to upgrade metaheuristic results by using an exact method i.e. the two-phases method. These approaches were tested, and their effectivenesses were shown by improvements realized on Pareto fronts obtained with AGMA algorithm. These results show the interest of this type of methods, which can be improved by adding other mechanisms to explore a large region of the search space without exploring a great part of the solutions. In the future, cooperation could be made in a hybrid way to combine the partitionning and the VLNS approaches. Another way for cooperation between evolutionary and exact approaches, without considering partitions of optimal solution, is to extract information from these solutions to reduce sufficiently the size of the search space.

# References

1. Ravindra K. Ahuja, Ã-zlem Ergun, James B. Orlin, and Abraham P. Punnen. A survey of very large-scale neighborhood search techniques. *Discrete Appl. Math.*, 123(1-3):75–102, 2002.
2. M. Basseur, F. Seynhaeve, and E-G. Talbi. Design of multi-objective evolutionary algorithms: Application to the flow-shop scheduling problem. In *Congress on Evolutionary Computation CEC'02*, pages 1151–1156, Honolulu, USA, 2002.
3. M. Basseur, F. Seynhaeve, and E-G. Talbi. Adaptive mechanisms for multi-objective evolutionary algorithms. In *Congress on Engineering in System Application CESA'03*, Lille, France, 2003.
4. C. A. Coello Coello, D. A. Van Veldhuizen, and G. B. Lamont. *Evolutionary algorithms for solving Multi-Objective Problems*. Kluwer, New York, 2002.
5. K. Deb. *Multi-objective optimization using evolutionary algorithms*. Wiley, Chichester, UK, 2001.
6. J. Du and J. Y.-T. Leung. Minimizing total tardiness on one machine is np-hard. *Mathematics of operations research*, 15:483–495, 1990.
7. R. L. Graham, E. L. Lawler, J. K. Lenstra, and A. H. G. Rinnooy Kan. Optimization and approximation in deterministic sequencing and scheduling: a survey. In *Annals of Discrete Mathematics*, volume 5, pages 287–326. 1979.
8. Y. Haimes, L. Ladson, and D. Wismer. On a bicriterion formulation of the problems of integrated system identification and system optimization. *IEEE Transaction on system, Man and Cybernetics*, pages 269–297, 1971.
9. Y-D. Kim. Minimizing total tardiness in permutation flowshops. *European Journal of Operational Research*, 33:541–551, 1995.
10. J. D. Knowles and D. W. Corne. On metrics for comparing non-dominated sets. In IEEE Service Center, editor, *Congress on Evolutionary Computation (CEC'2002)*, volume 1, pages 711–716, Piscataway, New Jersey, May 2002.
11. J. Lemesre. Algorithme parallèle exact pour l'optimisation multi-objectif: application à l'ordonnancement. Master's thesis, University of Lille, 2003.
12. J. K. Lenstra, A. H. G. Rinnooy Kan, and P. Brucker. Complexity of machine scheduling problems. *Annals of Discrete Mathematics*, 1:343–362, 1977.
13. H. Meunier, E. G. Talbi, and P. Reininger. A multiobjective genetic algorithm for radio network optimisation. In *CEC*, volume 1, pages 317–324, Piscataway, New Jersey, jul 2000. IEEE Service Center.
14. A. Nagar, J. Haddock, and S. Heragu. Multiple and bicriteria scheduling: A litterature survey. *European journal of operational research*, (81):88–104, 1995.
15. C. Rajendran. Heuristics for scheduling in flowshop with multiple objectives. *European journal of operational research*, (82):540–555, 1995.

16. S. Sayin and S. Karabati. A bicriteria approach to the two-machine flow shop scheduling problem. *European journal of operational research*, (113):435–449, 1999.
17. F. Sivrikaya and G. Ulusoy. A bicriteria two-machine permutation flowshop problem. *European journal of operational research*, (107):414–430, 1998.
18. E. Taillard. Benchmarks for basic scheduling problems. *European Journal of Operations Research*, 64:278–285, 1993.
19. E-G. Talbi. A taxonomy of hybrid metaheuristics. *Journal of Heuristics*, 8:541–564, 2002.
20. E. G. Talbi, M. Rahoual, M. H. Mabed, and C. Dhaenens. A hybrid evolutionary approach for multicriteria optimization problems : Application to the flow shop. In E. Zitzler et al., editors, *Evolutionary Multi-Criterion Optimization*, volume 1993 of *LNCS*, pages 416–428. Springer-Verlag, 2001.
21. M. Visée, J. Teghem, M. Pirlot, and E.L. Ulungu. The two phases method and branch and bound procedures to solve the bi-objective knapsack problem. *Journal of Global Optimization*, Vol. 12:p. 139–155, 1998.
22. E. Zitzler. Evolutionary algorithms for multiobjective optimization: Methods and applications. Master's thesis, Swiss federal Institute of technology (ETH), Zurich, Switzerland, november 1999.
23. E. Zitzler, K. Deb, L. Thiele, C. A. Coello Coello, and D. Corne, editors. *Proceedings of the First International Conference on Evolutionary Multi-Criterion Optimization (EMO 2001)*, volume 1993 of *Lecture Notes in Computer Science*, Berlin, 2001. Springer-Verlag.

# Simple Max-Cut for Split-Indifference Graphs and Graphs with Few $P_4$'s

Hans L. Bodlaender[1], Celina M. H. de Figueiredo[2], Marisa Gutierrez[3], Ton Kloks[4], and Rolf Niedermeier[5]

[1] Institute of Information and Computing Sciences, Utrecht University, Padualaan 14, 3584 CH Utrecht, The Netherlands.
hansb@cs.uu.nl
[2] Instituto de Matemática and COPPE, Universidade Federal do Rio de Janeiro, Caixa Postal 68530, 21945-970 Rio de Janeiro, Brazil.
celina@cos.ufrj.br
[3] Departamento de Matemática, Universidad Nacional de La Plata, C. C. 172, (1900) La Plata, Argentina.
marisa@mate.unlp.edu.ar
[4] klokskloks@zonnet.nl
[5] Wilhelm-Schickard-Institut für Informatik, Universität Tübingen, Sand 13, D-72076 Tübingen, Germany.
niedermr@informatik.uni-tuebingen.de

**Abstract.** The SIMPLE MAX-CUT problem is as follows: given a graph, find a partition of its vertex set into two disjoint sets, such that the number of edges having one endpoint in each set is as large as possible. A split graph is a graph whose vertex set admits a partition into a stable set and a clique. The SIMPLE MAX-CUT decision problem is known to be NP-complete for split graphs. An indifference graph is the intersection graph of a set of unit intervals of the real line. We show that the SIMPLE MAX-CUT problem can be solved in linear time for a graph that is both split and indifference. Moreover, we also show that for each constant $q$, the SIMPLE MAX-CUT problem can be solved in polynomial time for $(q, q - 4)$-graphs. These are graphs for which no set of at most $q$ vertices induces more than $q - 4$ distinct $P_4$'s.

## 1 Introduction

The MAXIMUM CUT problem (or the MAXIMUM BIPARTITE SUBGRAPH problem) asks for a bipartition of the graph (with edge weights) with a total weight as large as possible. In this paper we consider only the simple case, i.e., all edges in the graph have weight one. Then the objective of this SIMPLE MAX-CUT problem is to delete a minimum number of edges such that the resulting graph is bipartite. Making a graph bipartite with few edge deletions has many applications [26]. A very recent one is found in the emerging field of SNP (single nucleotide polymorphism) analysis in computational molecular biology, e.g., see [11,27]. Aiming for efficient algorithms, we only consider the unweighted case since the classes of graphs we consider in this paper contain all complete graphs and the (weighted)

C.C. Ribeiro and S.L. Martins (Eds.): WEA 2004, LNCS 3059, pp. 87–99, 2004.

MAXIMUM CUT problem is NP-complete for every class of graphs containing all complete graphs [21,26].

As SIMPLE MAX-CUT is NP-complete in general, there are basically two lines of research to cope with its computational hardness. First, one may study polynomial-time approximation algorithms (it is known to be approximable within 1.1383, see [13]) or try to develop exact (exponential-time) algorithms (see [15] for an algorithm running in time $2^{m/3} \cdot m^{O(1)}$, where $m$ is the number of edges in the graph). Approximation and exact algorithms both have their drawbacks, i.e., non-optimality of the gained solution or poor running time even for relatively small problem instance sizes. Hence, the second line of research—as pursued in this paper—is to determine and analyze special graph structures that make it possible to solve the problem efficiently *and* optimally. This leads to the study of special graph classes. (Have a look at the classics [14,9] for general information on numerous graph classes.) For example, it was shown that the SIMPLE MAX-CUT problem remains NP-complete for cobipartite graphs, split graphs, and graphs with chromatic number three [6]. On the positive side, the problem can be efficiently solved for cographs [6], linegraphs [1], planar graphs [24,16], and for graphs with bounded treewidth [29].

In this paper we consider two classes of graphs, both of which possess nice decomposition properties which we make use of in the algorithms for SIMPLE MAX-CUT to be described. Also, both graph classes we study are related to split graphs. An indifference graph is the intersection graph of a set of unit intervals of the real line. (See [23] for more information on intersection graphs and their applications in biology and other fields.) A split graph is a graph whose vertex set admits a partition into a stable set and a clique. Ortiz, Maculan, and Szwarcfiter [25] characterized graphs that are both split and indifference in terms of their maximal cliques, and used this characterization to edge-colour those graphs in polynomial time. First, we show that this characterization also leads to a linear-time solution for the SIMPLE MAX-CUT problem for graphs that are both split and indifference.

Second, we study the class of $(q, q - 4)$-graphs (also known as graphs with few $P_4$'s [4] and introduced in [2]). These are graphs for which no set of at most $q$ vertices induces more than $q - 4$ distinct $P_4$'s. (A $P_4$ is a path with four vertices.) In this terminology, the cographs are exactly the $(4, 0)$-graphs. The class of $(5, 1)$-graphs are called $P_4$-sparse graphs. Jamison and Olariu [20] showed that $(q, q - 4)$-graphs allow a nice decomposition tree similar to cographs [20]. This decomposition can be used to find fast solutions for several in general NP-complete problems (see, e.g., [3,22]). Also using this decomposition, we show that the SIMPLE MAX-CUT problem can be solved in polynomial time for $(q, q - 4)$-graphs for every constant $q$.

## 2   Preliminaries

In this paper, $G$ denotes a simple, undirected, finite, connected graph, and $V(G)$ and $E(G)$ are respectively the vertex and edge sets of $G$. The vertex-set size is denoted by $|V(G)| = N$, and $K_N$ denotes the complete graph on $N$ vertices. A *stable set* (or *independent set*) is a set of vertices pairwise non-adjacent in $G$. A *clique* is a set of vertices pairwise adjacent in $G$. A *maximal clique* of $G$ is a clique not properly contained in any other clique. A *subgraph* of $G$ is a graph $H$ with $V(H) \subseteq V(G)$ and $E(H) \subseteq E(G)$. For $X \subseteq V(G)$, we denote by $G[X]$ the *subgraph induced by* $X$, that is, $V(G[X]) = X$ and $E(G[X])$ consists of those edges of $E(G)$ having both ends in $X$.

Given nonempty subsets $X$ and $Y$ of $V(G)$, the symbol $(X, Y)$ denotes the subset $\{xy \in E(G) : x \in X, y \in Y\}$ of $E(G)$. A *cut* $\mathcal{K}$ of a graph $G$ is the set of edges $(S, V(G) \setminus S)$, defined by a subset $S \subseteq V(G)$. We often write $\overline{S}$ instead of $V(G) \setminus S$. We also write $\delta(S)$ for the set of edges with exactly one endpoint in $S$ (and the other endpoint in $V(G) \setminus S$). By $|\mathcal{K}|$ we denote the number of edges in the cut $\mathcal{K}$ and $\ell(\mathcal{K})$ is the number of edges in $E(G) \setminus \mathcal{K}$, i.e., the number of edges that are *lost* by the cut $\mathcal{K}$. A *max-cut* $\mathcal{K}$ is a cut such that $|\mathcal{K}|$ is as large as possible. The (simple) max-cut problem considers the computation of two complementary parameters of a graph $G$: $mc(G) = \max\{|\mathcal{K}| : \mathcal{K}$ is a cut of $G\} = \max_{S \subset V} |\delta(S)|$, the maximum number of edges in a cut of $G$; and $\ell(G) = |E(G)| - mc(G)$, the minimum number of edges lost by a cut of $G$ (making the remaining graph bipartite). Instead of calculating $mc(G)$ directly it is sometimes more convenient to calculate first, for $i = 1, \ldots, n$, the values $mc(G, i) = \max_{S \subset V, |S| = i} |\delta(S)|$.

In the sequel, the following observations will be helpful.

**Remark 1.** *For $K_N$, the complete graph on $N$ vertices, we have:*

- *If $(S, \overline{S})$ is a max-cut of $K_N$, then $|S| = \lfloor \frac{N}{2} \rfloor$;*
- $mc(K_N) = \lfloor \frac{N}{2} \rfloor \cdot \lceil \frac{N}{2} \rceil$.

We say that a max-cut in a complete graph is a *balanced* cut.

**Remark 2.** *Let $H$ be a subgraph of a graph $G$ and let $\mathcal{K}$ be a cut of $G$. If $\ell(\mathcal{K}) = \ell(H)$, then $\mathcal{K}$ is a max-cut of $G$.*

*Proof.* Since $H$ is a subgraph of $G$, any cut $\mathcal{N}$ of $G$ satisfies $\ell(\mathcal{N}) \geq \ell(H) = \ell(\mathcal{K})$. Hence $\mathcal{K}$ is a cut of minimum loss in $G$, in other words, $\mathcal{K}$ is a max-cut of $G$. ∎

**Remark 3.** *Let $|V(G)| = N$ and let $S$ be a subset of $V(G)$ satisfying:*

- $|S| = \lfloor \frac{N}{2} \rfloor$;
- *every vertex of $S$ is adjacent to every vertex of $\overline{S}$.*

*Then $(S, \overline{S})$ is a max-cut of $G$.*

*Proof.* Clearly the cut $(S, \overline{S})$ has $\lfloor \frac{N}{2} \rfloor \cdot \lceil \frac{N}{2} \rceil$ edges, the maximum possible size of a cut in $G$. ∎

The *union* of two graphs $G_1$ and $G_2$, denoted by $G_1 \cup G_2$, is the graph such that $V(G_1 \cup G_2) = V(G_1) \cup V(G_2)$ and $E(G_1 \cup G_2) = E(G_1) \cup E(G_2)$. By way of contrast, $G_1 \setminus G_2$ denotes the subgraph of $G_1$ induced by $V(G_1) \setminus V(G_2)$. The (disjoint) *sum* of two graphs $G_1$ and $G_2$ makes every vertex of $G_1$ adjacent to every vertex of $G_2$.

# 3   Linear-Time Solution for Split-Indifference Graphs

## Some Preliminaries

An *interval graph* is the intersection graph of a set of intervals of the real line (cf. [9,23] for general expositions). In case of unit intervals the graph is called *unit interval, proper interval,* or *indifference graph*. We shall adopt the latter name, to be consistent with the terminology of indifference *orders*, defined next. (For a recent proof that the class of unit interval graphs coincides with that of the proper interval graphs, see [8].) Indifference graphs can be characterized as those interval graphs without an induced claw, (i.e., a $K_{1,3}$). Indifference graphs can also be characterized by a linear order: their vertices can be linearly ordered so that the vertices contained in the same maximal cliques are consecutive [28]. We call such an order an *indifference order*.

A *split graph* is a graph whose vertex set can be partitioned into a stable set and a clique. A *split-indifference graph* is a graph that is both split and indifference. We shall use the following characterization of split-indifference graphs in terms of their maximal cliques due to [25].

**Theorem 1.** *Let $G$ be a connected graph. Then $G$ is a split-indifference graph if and only if*

- $G = K_N$, *or*
- $G = K_m \cup K_n$, *where* $n \geq m > 1$, *and* $K_m \setminus K_n = K_1$, *or*
- $G = K_m \cup K_n \cup K_l$, *where* $n \geq m > 1$, $n \geq l > 1$, *and* $K_m \setminus K_n = K_1$, $K_l \setminus K_n = K_1$. *Moreover,* $V(K_m) \cap V(K_l) = \emptyset$ *or* $V(K_m) \cup V(K_l) = V(G)$.

This characterization was applied to obtain a polynomial-time algorithm to edge colour split-indifference graphs [25]. In the sequel, we show how to apply this characterization to obtain a linear-time algorithm to solve the max-cut problem for split-indifference graphs.

## The Balanced Cut Is not Always Maximal

A natural approach [7] for solving max-cut for indifference graphs is the following. Let $v_1, v_2, \ldots, v_t$ be an indifference order for $G$ and define $\mathcal{K} = (S, \overline{S})$ as follows: Place in $S$ all vertices with odd labels and place in $\overline{S}$ the remaining vertices (i.e., those with even labels). By definition of $\mathcal{K}$ and by Remark 1, $\mathcal{K} \cap E(\mathcal{M})$ is a max-cut of $\mathcal{M}$, for every graph $\mathcal{M}$ induced by a maximal clique of $G$. This natural approach defines a cut that is locally balanced, i.e., it gives a cut that is a max-cut with respect to each maximal clique. The following example shows that $\mathcal{K}$ is not necessarily a max-cut of $G$. Consider the indifference graph $G$ with five (ordered) vertices $v_1, v_2, v_3, v_4, v_5$, where $\{v_1, v_2, v_3, v_4\}$ induce a $K_4$, and $\{v_3, v_4, v_5\}$ induce a $K_3$. Note that the cut $(\{v_1, v_3, v_5\}, \{v_2, v_4\})$ has 5 edges, whereas the cut $(\{v_1, v_2, v_5\}, \{v_3, v_4\})$ has 6 edges. Therefore, this approach works only when the indifference graph $G$ has only one maximal clique, i.e., when $G$ is a complete graph which covers the first point in Theorem 1.

Let $G = K_n \cup K_m$, where $|V(K_n) \cap V(K_m)| = i$. Call $K_i$ the graph induced by the vertices of the *intersection*. We say that a cut $\mathcal{K}$ of $G$ is *compatible* if:

**a)** $\mathcal{K} \cap E(K_n)$ is a max-cut of $K_n$ and $\mathcal{K} \cap E(K_m)$ is a max-cut of $K_m$;

**b)** Among all cuts $\mathcal{K}$ of $G$ satisfying condition a), $|\mathcal{K} \cap E(K_i)|$ is minimal.

Clearly, the cut proposed by the natural approach satisfies condition a) but not necessarily condition b) of the definition of compatible cut. Clearly, for the example above the compatible cut gives the maximum cut. However, our subsequent study of the max-cut problem for graphs with two maximal cliques shows that it is not always possible to define a max-cut which is a compatible cut for the graph. We actually show that there are graphs for which the max-cut is not balanced with respect to any maximal clique of the graph.

In the sequel, we show how to use this approach—considering cuts $\mathcal{K}$ such that locally $\mathcal{K} \cap E(\mathcal{M})$ is a max-cut of $\mathcal{M}$, for every graph $\mathcal{M}$ induced by a maximal clique—to find first a max-cut in a graph with two maximal cliques (which covers the second point in Theorem 1) and then to find a max-cut in a split-indifference graph (by dealing with the third point in Theorem 1).

## Graphs with Two Maximal Cliques

In this section we consider general graphs with precisely two maximal cliques. Note that a graph with precisely two maximal cliques is necessarily an indifference graph but not necessarily a split graph.

**Lemma 1.** *Let $G = K_n \cup K_m$ with $n \geq m > i \geq 1$, where $|V(K_n) \cap V(K_m)| = i$. Call $K_i$ the graph induced by the vertices of the intersection. Let $(S, \overline{S})$ be a cut of $G$. Let $x = |S \cap V(K_i)|$. Suppose $x \leq \lfloor \frac{i}{2} \rfloor$. Then, the maximum value of a cut $(S, \overline{S})$ having $x$ vertices in $S \cap V(K_i)$ is obtained by placing the vertices outside the intersection $K_i$ as follows:*

- *Place in $S$ the largest possible number that is less than or equal to $\lceil \frac{n}{2} \rceil - x$ of vertices of $K_n \setminus K_i$;*
- *Place in $S$ the largest possible number that is less than or equal to $\lceil \frac{m}{2} \rceil - x$ of vertices of $K_m \setminus K_i$.*

*Proof.* Let $\mathcal{N} = (S, \overline{S})$ be a cut of $G$. Since $G$ contains two maximal cliques, i.e., $G = K_n \cup K_m$, with $|V(K_n) \cap V(K_m)| = i$, we may count the number of edges in the cut $\mathcal{N}$ as follows:

$$|\mathcal{N}| = |\mathcal{N} \cap E(K_n)| + |\mathcal{N} \cap E(K_m)| - |\mathcal{N} \cap E(K_i)|.$$

Now because $x = |S \cap V(K_i)|$, we have $|\mathcal{N} \cap E(K_i)| = x(i - x)$. Hence, by placing the vertices outside the intersection $K_i$ as described, we get a cut as close as possible to the balanced cut with respect to both $K_n$ and $K_m$. ∎

By using the notation of Lemma 1, let $M(x)$ be the number of edges of a maximum cut of $G$ having $x$ vertices of $K_i$ in $S$. By Lemma 1, $M(x)$ is well defined as a function of $x$ in the interval $[0, \lfloor \frac{i}{2} \rfloor]$. We consider three cases according to the relation between $i$ and $\lceil \frac{m}{2} \rceil$, and $i$ and $\lceil \frac{n}{2} \rceil$. In each case, our goal is to find the values of $x$ in the interval $[0, \lfloor \frac{i}{2} \rfloor]$ which maximize $M(x)$.

**Case 1:** $i \leq \lceil \frac{m}{2} \rceil \leq \lceil \frac{n}{2} \rceil$ In this case, $x \leq \lceil \frac{m}{2} \rceil$ and $i-x \leq \lceil \frac{m}{2} \rceil$. Hence, vertices outside the intersection can be placed accordingly to get balanced partitions for both $K_n$ and $K_m$. Then $M(x)$ is equal to $M_1(x)$, which is defined as follows: $M_1(x) = \lceil \frac{n}{2} \rceil \lfloor \frac{n}{2} \rfloor + \lceil \frac{m}{2} \rceil \lfloor \frac{m}{2} \rfloor - x(i - x)$. We want to maximize $M_1(x)$ over the interval $[0, \lfloor \frac{i}{2} \rfloor]$. In this case, we have just one maximum, which occurs at $x = 0$.

**Case 2:** $\lceil \frac{m}{2} \rceil < i \leq \lceil \frac{n}{2} \rceil$ In this case, we still have $x \leq \lceil \frac{m}{2} \rceil$, but not necessarily $i - x \leq \lceil \frac{m}{2} \rceil$. If $i - x \leq \lceil \frac{m}{2} \rceil$, then the function $M(x)$ is equal to $M_1(x)$ above. Otherwise, $i - x > \lceil \frac{m}{2} \rceil$, and it is not possible to get a balanced partition with respect to $K_m$: By Lemma 1, the maximum cut in this case is obtained by placing all vertices of $K_m \setminus K_i$ in $S$. Therefore, the function $M(x)$ is

$$M(x) = \begin{cases} M_2(x) = \lceil \frac{n}{2} \rceil \lfloor \frac{n}{2} \rfloor + (m - i)(i - x) & \text{for } 0 \leq x < i - \lceil \frac{m}{2} \rceil \\ M_1(x) = \lceil \frac{n}{2} \rceil \lfloor \frac{n}{2} \rfloor + \lceil \frac{m}{2} \rceil \lfloor \frac{m}{2} \rfloor - x(i - x) & \text{for } i - \lceil \frac{m}{2} \rceil \leq x \leq \lfloor \frac{i}{2} \rfloor \end{cases}$$

It is easy to see that $M(x)$ is a function that is continuous and decreasing with maximum at $x = 0$.

**Case 3:** $\lceil \frac{m}{2} \rceil \leq \lceil \frac{n}{2} \rceil < i$ In this case, we distinguish three intervals for $i - x$ to be in:

If $i - x \leq \lceil \frac{m}{2} \rceil$, then vertices outside the intersection can be placed accordingly to get balanced partitions for both $K_n$ and $K_m$, and $M(x) = M_1(x)$.

If $\lceil \frac{m}{2} \rceil < i - x \leq \lceil \frac{n}{2} \rceil$, then only $K_n$ gets a balanced partition and $M(x) = M_2(x)$.

Finally, if $i - x > \lceil \frac{n}{2} \rceil$, then a maximum cut is obtained by placing all vertices outside the intersection in $S$ and we get a new function $M_3(x)$.

Therefore, a complete description of the function $M(x)$ is

$$M(x) = \begin{cases} M_3(x) = (i - x)(n + m - 2i + x) & \text{for } 0 \leq x < i - \lceil \frac{n}{2} \rceil \\ M_2(x) = \lceil \frac{n}{2} \rceil \lfloor \frac{n}{2} \rfloor + (m - i)(i - x) & \text{for } i - \lceil \frac{n}{2} \rceil \leq x < i - \lceil \frac{m}{2} \rceil \\ M_1(x) = \lceil \frac{n}{2} \rceil \lfloor \frac{n}{2} \rfloor + \lceil \frac{m}{2} \rceil \lfloor \frac{m}{2} \rfloor - x(i - x) & \text{for } i - \lceil \frac{m}{2} \rceil \leq x \leq \lfloor \frac{i}{2} \rfloor \end{cases}$$

Observe that this function is also continuous but not always decreasing. The function $M_3(x)$ is a parabola with apex at $x = \frac{2i - N}{2}$, where $N = m + n - i$ is the total number of vertices of $G$. For this reason, we distinguish two cases, according to the relation between $i$ and $N$, as follows: $M(x)$ has maximum at $x = 0$ when $i \leq \lfloor \frac{N}{2} \rfloor$, and $M(x)$ has maximum at $x = \frac{2i - N}{2}$ when $i > \lfloor \frac{N}{2} \rfloor$. Since $x$ takes values on the interval $[0, \lfloor \frac{i}{2} \rfloor]$, we have two possible values for $x$ in this case: the maximum cut has either $i - \lfloor \frac{N}{2} \rfloor$ or $i - \lceil \frac{N}{2} \rceil$ vertices of $K_i$ in $S$.

In summary, we have shown:

**Theorem 2.** *Let $G = K_n \cup K_m$ with $n \geq m > i \geq 1$, where $|V(K_n) \cap V(K_m)| = i$. Call $K_i$ the graph induced by the vertices of the intersection. Let $v_1, v_2, \ldots, v_N$ be an indifference order of $G$ such that vertices $v_1, v_2, \ldots, v_n$ induce a $K_n$, vertices $v_{n-i}, v_{n-i+1}, \ldots, v_n$ induce a $K_i$ containing the vertices of the intersection, and $v_{n-i}, v_{n-i+1}, \ldots, v_N$ induce a $K_m$. A maximum cut of $G$ is obtained as follows:*

- If $i \leq \lceil \frac{m}{2} \rceil \leq \lceil \frac{n}{2} \rceil$, then the compatible cut $(S, \overline{S})$ that places in $S$ the first $\lceil \frac{n}{2} \rceil$ vertices, and the last $\lceil \frac{m}{2} \rceil$ vertices, contains zero edges of $K_i$, and is a maximum cut of $G$.

- If $\lceil \frac{m}{2} \rceil < i \leq \lceil \frac{n}{2} \rceil$, then the cut $(S, \overline{S})$ that places in $S$ the first $\lceil \frac{n}{2} \rceil$ vertices, and the last $m - i$ vertices, contains zero edges of $K_i$, is not a compatible cut, and is a maximum cut of $G$.

- If $\lceil \frac{m}{2} \rceil \leq \lceil \frac{n}{2} \rceil < i$, then we distinguish two cases. If $i \leq \lfloor \frac{N}{2} \rfloor$, then the cut $(S, \overline{S})$ that places in $S$ the first $n - i$ vertices, and the last $m - i$ vertices, contains zero edges of $K_i$, is not a compatible cut, and is a maximum cut of $G$. If $i > \lfloor \frac{N}{2} \rfloor$, then the cut $(S, \overline{S})$ that places in $\overline{S}$ $\lceil \frac{N}{2} \rceil$ vertices if the intersection is not a compatible cut, and is a maximum cut of $G$.

## Split-Indifference Graphs with Three Maximal Cliques

In this section we consider split-indifference graphs with precisely three maximal cliques. By Theorem 1, any such graph $G = K_m \cup K_n \cup K_l$, with $n \geq m$, $n \geq l$, satisfies $K_m \setminus K_n = \{1\}$, $K_l \setminus K_n = \{t\}$, i.e., the vertex set $V(G) = V(K_n) \cup \{1, t\}$. In other words, we have $|V(G)| = N = n + 2$. In addition, there exists an indifference order for $G$ having vertex 1 first, vertex $t$ last, and the remaining vertices between 1 and $t$.

To obtain a solution for the max-cut problem for a split-indifference graph with precisely three maximal cliques, we shall consider three cases.

**Case 1: vertex 1 is adjacent to at most $\lfloor \frac{n}{2} \rfloor$ vertices or vertex $t$ is adjacent to at most $\lfloor \frac{n}{2} \rfloor$ vertices** In the preceding subsection we studied the case of two maximal cliques. In particular, we got the easy case that if a graph $H = K_n \cup K_m$, with $n \geq m$ and such that $K_m \setminus K_n = \{1\}$, then there exists a max-cut of $H$ that places on the same side the $\lceil \frac{n}{2} \rceil$ vertices that are closer to vertex 1 with respect to the indifference order of $H$, and places vertex 1 and the remaining $\lfloor \frac{n}{2} \rfloor$ vertices on the opposite side.

Now suppose vertex $t$ is adjacent to at most $\lfloor \frac{n}{2} \rfloor$ vertices. Let $(S, V(H) \setminus S)$ be a max-cut of $H$ that places all neighbours of $t$ on the same side $S$. By Remark 2, $(S, V(H) \setminus S \cup \{t\})$ is a max-cut of the entire graph $G$, because $(S, V(H) \setminus S \cup \{t\})$ looses the same number of edges as the cut $(S, V(H) \setminus S)$.

**Case 2: both vertices 1 and $t$ are adjacent to at least $\lceil \frac{n}{2} \rceil$ vertices but there are not $\lfloor \frac{N}{2} \rfloor$ vertices adjacent to both 1 and $t$** Note that every vertex of $K_n$ is adjacent to 1 or to $t$. Let $S$ contain vertex 1 and a set of $\lceil \frac{n}{2} \rceil$ neighbours of $t$ that includes all nonneighbours of 1. The only "missing" edge in the cut $(S, \overline{S})$ is the edge $1t$, an edge not present in $G$. Since there are not $\lfloor \frac{N}{2} \rfloor$ vertices adjacent to both 1 and $t$, it is not possible to define a cut for $G$ larger than $(S, \overline{S})$ by placing vertices 1 and $t$ on the same side.

**Case 3: there exist $\lfloor \frac{N}{2} \rfloor$ vertices adjacent to both 1 and $t$** Let $S$ be a set of $\lfloor \frac{N}{2} \rfloor$ vertices adjacent to both 1 and $t$. Remark 3 justifies $(S, \overline{S})$ to be a max-cut of $G$.

**Theorem 3.** *Let $G$ be a split-indifference graph with three maximal cliques $K_m$, $K_n$, and $K_l$, with $n \geq m$, $n \geq l$, and satisfying $K_m \setminus K_n = \{1\}$, $K_l \setminus K_n = \{t\}$. Let $v_1, v_2, \ldots, v_N$ be an indifference order of $G$ having vertex 1 first, vertex $t$ last. A maximum cut of $G$ is obtained as follows:*

- *If vertex $t$ is adjacent to at most $\lfloor \frac{n}{2} \rfloor$ vertices, then the cut $(S, \overline{S})$ that places in $S$ vertex 1 and the $\lfloor \frac{n}{2} \rfloor$ vertices that are closer to $t$ with respect to the indifference order is a maximum cut of $G$. An analogous result follows if vertex 1 is adjacent to at most $\lfloor \frac{n}{2} \rfloor$ vertices.*
- *If both vertices 1 and $t$ are adjacent to at least $\lceil \frac{n}{2} \rceil$ vertices but there are not $\lfloor \frac{N}{2} \rfloor$ vertices adjacent to both 1 and $t$, then the cut $(S, \overline{S})$ that places in $S$ vertex 1 and the $\lceil \frac{n}{2} \rceil$ vertices that are closer to $t$ with respect to the indifference order is a maximum cut of $G$.*
- *If there exist $\lfloor \frac{N}{2} \rfloor$ vertices adjacent to both 1 and $t$, then the cut $(S, \overline{S})$ that places in $S$ a set of $\lfloor \frac{N}{2} \rfloor$ vertices adjacent to both 1 and $t$ is a maximum cut of $G$.*

Altogether, we obtain the following main result.

**Corollary 1.** SIMPLE MAX-CUT *can be solved in linear time for split-indifference graphs.*

*Proof.* The result directly follows from combining Theorem 1 with Remark 1, Theorem 2, and Theorem 3. ∎

## 4   Polynomial-Time Solution for $(q, q-4)$-Graphs

**Some preliminaries.** A graph is a $(q, t)$-*graph* if no set of at most $q$ vertices induces more than $t$ distinct $P_4$'s. The class of cographs are exactly the $(4, 0)$-graphs, i.e., cographs are graphs without induced $P_4$. The class of so-called $P_4$-sparse graphs coincides with the $(5, 1)$-graphs. The class of $P_4$-sparse graphs was extensively studied in [17,18,19,12].

It was shown in [3] that many problems can be solved efficiently for $(q, q-4)$-graphs for each constant $q$. These results make use of a decomposition theorem which we state below. In this section we show that this decomposition can also be used to solve the SIMPLE MAX-CUT problem. In order to state the decomposition theorem for $(q, q-4)$-graphs we need some preliminaries.

Recall that a *split graph* is a graph of which the vertex set can be split into two sets $K$ and $I$ such that $K$ induces a clique and $I$ induces an independent set in $G$. A *spider* is a split graph consisting of a clique and an independent set of equal size (at least two) such that each vertex of the independent set has precisely one neighbor in the clique and each vertex of the clique has precisely one neighbor in the independent set, or it is the complement of such a graph. We call a spider *thin* if every vertex of the independent set has precisely one neighbor in the clique. A spider is *thick* if every vertex of the independent set is non-adjacent to precisely one vertex of the clique. The smallest spider is a path with four vertices (i.e., a $P_4$) and this spider is at the same time both thick and thin.

The SIMPLE MAX-CUT problem is easy to solve for spiders:

**Remark 4.** *Let $G$ be a thin spider with $2n$ vertices where $n \geq 3$. Then $mc(G) = \lfloor \frac{n^2}{4} \rfloor + n$. If $G$ is a thick spider then $mc(G) = n(n-1)$.*

A graph $G$ is *p-connected* if for every partition into two non-empty sets there is a crossing $P_4$, that is a $P_4$ with vertices in both sets of the partition. The *p-connected components* of a graph are the maximal induced subgraphs which are p-connected. A p-connected graph is *separable* if there is a partition $(V_1, V_2)$ such that every crossing $P_4$ has its midpoints in $V_1$ and its endpoints in $V_2$.

Recall that a *module* is a non-trivial (i.e., not $\emptyset$ or $V$) set of vertices which have equivalent neighborhoods outside the set. The *characteristic* of a graph is obtained by shrinking the non-trivial modules to single vertices. It can be shown (see [2,20]) that a p-connected graph is separable if and only if its characteristic is a split graph.

Our main algorithmic tool is the following structural theorem due to [20].

**Theorem 4.** *For an arbitrary graph $G$ exactly one of the following holds:*

- *$G$ or $\overline{G}$ is disconnected.*
- *There is a unique proper separable p-connected component $H$ of $G$ with separation $(V_1, V_2)$ such that every vertex outside $H$ is adjacent to all vertices of $V_1$ and to none of $V_2$.*
- *$G$ is p-connected.*

Furthermore, the following characterization of p-connectedness for $(q, q-4)$-graphs was obtained in [2] (also see [4]).

**Theorem 5.** *Let $G = (V, E)$ be a $(q, q-4)$-graph which is p-connected. Then either $|V| < q$ or $G$ is a spider.*

Theorem 4 and Theorem 5 lead to a binary *decomposition tree* for $(q, q-4)$-graphs (also see [3] for more details). This decomposition tree can be found in linear time [5]. The leaves of this tree correspond with spiders or graphs with less than $q$ vertices (this reflects the last point of Theorem 4 and Theorem 5). The internal nodes of this tree have one of three possible labels. If the label of an internal node is 0 or 1, then the graph corresponding with this node is the disjoint union or the sum of the graphs corresponding with the children of the node (this reflects the first point of Theorem 4). If the label of the node is 2 (this reflects the second point of Theorem 4), one of the graphs, w.l.o.g. $G_1$, has a separation $(V_1^1, V_1^2)$ and it is either a spider or a graph with less than $q$

vertices of which the characteristic is a split graph (Theorems 4 and 5), and $G_2$ is arbitrary. If $G_1$ is a spider, all vertices of $G_2$ are made adjacent exactly to all vertices of the clique (induced by $V_1^1$) of $G_1$. If $G_1$ is a graph of which the characteristic is a split graph, all vertices of $G_2$ are made adjacent exactly to all vertices (i.e., $V_1^1$) of every clique module of $G_1$.

In the following subsections we briefly describe the method to compute the simple max-cut for graphs with few $P_4$'s. The main idea of the algorithm is that we compute for each node of the decomposition tree all relevant values of $mc(G', i)$, $G'$ being the graph corresponding with this node. The table of values for such a node is computed, given the tables of the children of the node. In the subsequent paragraphs, we discuss the methods to do this, for each of the types of nodes in the decomposition tree. Once we have the table of the root node, i.e., all values $mc(G, i)$, we are done.

**Cographs.** We review the algorithm for the SIMPLE MAX-CUT problem for cographs (i.e., $(4, 0)$-graphs) which was published in [6]. A cograph which is not a single vertex is either the sum or the union of two (smaller) cographs. In other words: cographs have a decomposition tree with all internal nodes labelled 0 or 1.

**Lemma 2.** Let $G = (V, E)$ be the union of $G_1 = (V_1, E_1)$ and $G_2 = (V_2, E_2)$. Then

$$mc(G, i) = \max\{mc(G_1, j) + mc(G_2, i - j) \ : \ 0 \le j \le i \wedge$$
$$|V_1| \ge j \wedge |V_2| \ge i - j\}$$

Let $G = (V, E)$ be the sum of $G_1 = (V_1, E_1)$ and $G_2 = (V_2, E_2)$. Then

$$mc(G, i) = \max\{mc(G_1, j) + mc(G_2, i - j) + j(|V_2| - (i - j)) +$$
$$(|V_1| - j)(i - j) \ : \ 0 \le j \le i \wedge |V_1| \ge j \wedge |V_2| \ge i - j\}$$

**Corollary 2.** There exists an $O(N^2)$ time algorithm to compute the simple max-cut of a cograph.

**$P_4$-sparse graphs.** The decomposition tree (as defined above) for graphs that are $P_4$-sparse has nodes with label 0, 1, or 2 [17]. Note that in the case of label 2 we can assume here that the graph $G_1$ is a spider (see the discussion after Theorems 4 and 5, and [18]). In the lemma below, we assume $G$ is obtained from $G_1$ and $G_2$ as described above by the operation of a 2-labeled node. Let $K$ be the clique and $S$ be the independent set of $G_1$. Let $n_i$ denote the number of vertices of $G_i$. Note that every vertex of $G_2$ is adjacent to every vertex of $K$.

**Lemma 3.** Let $G$, $G_1$, $G_2$, $S$, and $K$ be as above. Let $G_1$ be a thick spider. Then

$$mc(G, i) = \max\{mc(G_2, j) + j(|K| - j') + j'(n_2 - j) +$$
$$(i - j - j')(|K| - 1) \ : \ 0 \le j, j' \le i\}$$

*Let $G_1$ be a thin spider. Then*

$$mc(G, i) = \max\{mc(G_2, j) + j(|K| - j') + j'(n_2 - j) +$$
$$(i - j - j') \; : \; 0 \leq j, j' \leq i\}$$

For $P_4$-sparse graphs (i.e., $(5, 1)$-graphs), Lemmas 2 and 3 are sufficient to compute all the values $mc(G', i)$ for all graphs associated with nodes in the decomposition tree. Thus, we obtain:

**Corollary 3.** *There exists an $O(N^3)$ time algorithm to compute the simple max-cut for a $P_4$-sparse graph.*

**$|V_1| < q$ and the characteristic of $G_1$ is a split graph.** If we have a decomposition tree of $(q, q - 4)$-graphs, then there is one remaining case: $G$ is obtained from $G_1$ and $G_2$ by the operation corresponding to a 2-labeled node, and $G_1$ has less than $q$ vertices. In this case the vertex set of $G_2$ acts as a module, i.e., every vertex of $G_2$ has exactly the same set of neighbors in $G_1$. Let $K$ be the set of vertices of $G_1$ which are adjacent to all vertices of $G_2$.

Let $mc(G_1, j, j')$ be the maximum cut in $G_1$ with exactly $j$ vertices in $K$ and $j'$ vertices in $V_1 - K$. Since $G_1$ is constant size the numbers $mc(G_1, j, j')$ can easily be computed in constant time.

**Lemma 4.** *Let $G$, $G_1$, $G_2$, $K$ be as above. Suppose that $|V_1| < q$ and the characteristic of $G_1$ is a split graph. Then*

$$mc(G, i) = \max\{mc(G_2, j) + mc(G_1, j', i - j - j') +$$
$$j(|K| - j') + j'(n_2 - j) + (i - j - j') \; : \; 0 \leq j, j' \leq i\}$$

Now, with Lemma 4, and Lemmas 2 and 3, we obtain:

**Theorem 6.** *There exists an $O(N^4)$ time algorithm for the* SIMPLE MAX-CUT *problem on $(q, q - 4)$-graphs for each constant $q$.*

## 5  Concluding Remarks

This paper considers two classes of graphs: indifference graphs and $(q, q - 4)$-graphs. Both classes possess nice decomposition properties which we make use of in the described algorithms for SIMPLE MAX-CUT. Also, both graph classes we study are related to split graphs, a class of graphs for which SIMPLE MAX-CUT is known to be hard.

A linear-time algorithm for the recognition of indifference graphs was presented by de Figueiredo et al. [10]. The algorithm partitions in linear time the vertex set of an indifference graph into sets of *twin* vertices, i.e., vertices of the graph that belong to the same set of maximal cliques.

Given a graph $G$ with a bounded number of maximal cliques, the partition of $G$ into sets of twins contains a bounded number $k$ of sets. Hence, we can compute $mc(G)$ in polynomial time, by maximizing a function on $k$ variables $x$, that assume integer values in a limited region of the space, i.e., on a finite domain. This simple argument establishes the polynomial upper bound $O(N^k)$ for the max-cut problem for a class of graphs with a bounded number of maximal cliques.

One goal of this paper was to establish a linear time upper bound for the computation of $mc(G)$ for a split-indifference graph $G$, by computing the value of $mc(G)$ in constant time, given that we can in linear time determine which case of the computation we are in. We leave it as an open problem to extend the proposed solution to the whole class of indifference graphs.

Another goal reached by this paper was to extend to the whole class of $(q, q-4)$-graphs the known solution of SIMPLE MAX-CUT for cographs. We leave it as an open problem to find a more efficient polynomial-time algorithm for the computation of $mc(G)$ for a $(q, q-4)$-graph $G$.

**Acknowledgments:** We are grateful to J. Spinrad for pointing out that the max-cut problem is solvable in polynomial time for any class of graphs with a bounded number of maximal cliques. Parts of this research were done while Celina M. H. de Figueiredo visited Universidad Nacional de La Plata supported by FAPERJ, and Marisa Gutierrez visited Universidade Federal do Rio de Janeiro supported by FOMEC and FAPERJ. Rolf Niedermeier is supported by the DFG, Emmy Noether research group "PIAF" (fixed-parameter algorithms), NI 369/4.

# References

1. C. Arbib. A polynomial characterization of some graph partitioning problem. *Inform. Process. Lett.*, 26:223–230, 1987/1988.
2. L. Babel. On the $P_4$-structure of graphs, *Habilitationsschrift*, Zentrum für Mathematik, Technische Universität München, 1997.
3. L. Babel, T. Kloks, J. Kratochvíl, D. Kratsch, H. Müller, and S. Olariu. Efficient algorithms for graphs with few $P_4$'s. Combinatorics (Prague, 1998). *Discrete Math.*, 235:29–51, 2001.
4. L. Babel and S. Olariu. On the structure of graphs with few $P_4$s. *Discrete Appl. Math.*, 84:1–13, 1998.
5. S. Baumann. A linear algorithm for the homogeneous decomposition of graphs, Report No. M-9615, Zentrum für Mathematik, Technische Universität München, 1996.
6. H. L. Bodlaender and K. Jansen. On the complexity of the maximum cut problem. In *STACS 94*, Lecture Notes in Computer Science 775, 769–780, Springer, Berlin, 1994. Also in *Nordic J. Comput.*, 7(1):14–31, 2000.
7. H. L. Bodlaender, T. Kloks, and R. Niedermeier. Simple max-cut for unit interval graphs and graphs with few $P_4$'s. In *Extended abstracts of the 6th Twente Workshop on Graphs and Combinatorial Optimization* 12–19, 1999. Also in *Electronic Notes in Discrete Mathematics* **3**, 1999.
8. K. P. Bogart and D. B. West. A short proof that 'proper = unit', *Discrete Math.*, 201:21–23, 1999.

9. A. Brandstädt, V. B. Le, and J. P. Spinrad, *Graph Classes: a Survey*. SIAM Monographs on Discrete Mathematics and Applications. Society for Industrial and Applied Mathematics (SIAM), Philadelphia, PA, 1999.
10. C. M. H. de Figueiredo, J. Meidanis, and C. P. de Mello. A linear-time algorithm for proper interval graph recognition. *Inform. Process. Lett.*, 56:179–184, 1995.
11. E. Eskin, E. Halperin, and R. M. Karp. Large scale reconstruction of haplotypes from genotype data. In *RECOMB 2003*, pp. 104–113, ACM Press, 2003.
12. V. Giakoumakis, F. Roussel, and H. Thuillier. On $P_4$-tidy graphs. *Discrete Mathematics and Theoretical Computer Science*, 1:17–41, 1997.
13. M. X. Goemans and D. P. Williamson. Improved approximation algorithms for maximum cut and satisfiability problems using semidefinite programming. *J. ACM*, 42:1115–1145, 1995.
14. M. C. Golumbic, *Algorithmic Graph Theory and Perfect Graphs*, Academic Press, New York, 1980.
15. J. Gramm, E. A. Hirsch, R. Niedermeier, and P. Rossmanith. Worst-case upper bounds for MAX-2-SAT with application to MAX-CUT. *Discrete Appl. Math.*, 130(2):139–155, 2003.
16. F. O. Hadlock. Finding a maximum cut of a planar graph in polynomial time. *SIAM J. Comput.*, 4:221–225, 1975.
17. B. Jamison and S. Olariu. A tree representation for $P_4$-sparse graphs. *Discrete Appl. Math.*, 35:115–129, 1992.
18. B. Jamison and S. Olariu. Recognizing $P_4$-sparse graphs in linear time. *SIAM J. Comput.*, 21:381–406, 1992.
19. B. Jamison and S. Olariu. Linear time optimization algorithms for $P_4$-sparse graphs. *Discrete Appl. Math.*, 61:155–175, 1995.
20. B. Jamison and S. Olariu. p-components and the homogeneous decomposition of graphs. *SIAM J. Discrete Math.*, 8:448–463, 1995.
21. R. M. Karp. Reducibility among combinatorial problems. *Complexity of computation* (R. E. Miller and J. W. Thather eds.), pp. 85–103, 1972.
22. T. Kloks and R. B. Tan. Bandwidth and topological bandwidth of graphs with few $P_4$'s. In 1st Japanese-Hungarian Symposium for Discrete Mathematics and its Applications (Kyoto, 1999). *Discrete Appl. Math.*, 115(1–3):117–133, 2001.
23. T. A. McKee and F. R. Morris. *Topics in Intersection Graph Theory*. SIAM Monographs on Discrete Mathematics and Applications. Society for Industrial and Applied Mathematics (SIAM), Philadelphia, PA, 1999
24. G. I. Orlova and Y. G. Dorfman, Finding the maximal cut in a graph, *Engrg. Cybernetics*, 10:502–504, 1972.
25. C. Ortiz Z., N. Maculan, and J. L. Szwarcfiter. Characterizing and edge-colouring split-indifference graphs. *Discrete Appl. Math.*, 82(1–3):209–217, 1998.
26. S. Poljak and Z. Tuza. Maximum cuts and large bipartite subgraphs. *DIMACS Series in Discrete Mathematics and Theoretical Computer Science* (W. Cook, L. Lovász, and P. Seymour eds.), 20:181–244, Amer. Math. Soc., Providence, RI, 1995.
27. R. Rizzi, V. Bafna, S. Istrail, and G. Lancia. Practical algorithms and fixed-parameter tractability for the single individual SNP haplotypying problem. In *WABI 2002*, Lecture Notes in Computer Science 2452, pp. 29–43, Springer, Berlin, 2002.
28. F. S. Roberts. On the compatibility between a graph and a simple order. *J. Combinatorial Theory Ser. B*, 11:28–38, 1971.
29. T. V. Wimer. *Linear algorithms on k-terminal graphs*, PhD Thesis, Department of Computer Science, Clemson University, South Carolina, 1987.

# A Randomized Heuristic for Scene Recognition by Graph Matching

Maria C. Boeres[1], Celso C. Ribeiro[2], and Isabelle Bloch[3]

[1] Universidade Federal do Espírito Santo, Department of Computer Science, R. Fernando Ferrari, Campus de Goiabeiras, Vitória, ES 29060-970, Brazil.
boeres@inf.ufes.br
[2] Universidade Federal Fluminense, Department of Computer Science, Rua Passo da Pátria 156, Niterói, RJ 24210-240, Brazil.
celso@inf.puc-rio.br
[3] Ecole Nationale Supérieure des Télécommunications, CNRS URA 820, 46 rue Barrault, 75634 Paris Cedex 13, France.
Isabelle.Bloch@enst.fr

**Abstract.** We propose a new strategy for solving the non-bijective graph matching problem in model-based pattern recognition. The search for the best correspondence between a model and an over-segmented image is formulated as a combinatorial optimization problem, defined by the relational attributed graphs representing the model and the image where recognition has to be performed, together with the node and edge similarities between them. A randomized construction algorithm is proposed to build feasible solutions to the problem. Two neighborhood structures and a local search procedure for solution improvement are also proposed. Computational results are presented and discussed, illustrating the effectiveness of the combined approach involving randomized construction and local search.

## 1 Introduction

The recognition and the understanding of complex scenes require not only a detailed description of the objects involved, but also of the spatial relationships between them. Indeed, the diversity of the forms of the same object in different instantiations of a scene, and also the similarities of different objects in the same scene, make relationships between objects of prime importance in order to disambiguate the recognition of objects with similar appearance. Graph based representations are often used for scene representation in image processing [6,9, 11,20,21]. Vertices of the graphs usually represent the objects in the scenes, while their edges represent the relationships between the objects. Relevant information for the recognition is extracted from the scene and represented by relational attributed graphs. In model-based recognition, both the model and the scene are represented by graphs.

The assumption of a bijection between the elements in two instantiations of the same scene is too strong for many problems. Usually, the model has a

C.C. Ribeiro and S.L. Martins (Eds.): WEA 2004, LNCS 3059, pp. 100–113, 2004.
© Springer-Verlag Berlin Heidelberg 2004

schematic aspect. Moreover, the construction of the image graph often relies on segmentation techniques that may fail in accurately segmenting the image into meaningful entities. Therefore, no isomorphism can be expected between both graphs and, in consequence, scene recognition may be better expressed as an non-bijective graph matching problem.

Our motivation comes from an application in medical imaging, in which the goal consists in recognizing brain structures from 3D magnetic resonance images, previously processed by a segmentation method. The model consists of an anatomical atlas. A graph is built from the atlas, in which each node represents exactly one anatomical structure of interest. Edges of this graph represent spatial relationships between the anatomical structures. Inaccuracies constitute one of the main characteristics of the problem. Objects in the image are segmented and all difficulties with object segmentation will be reflected in the representation, such as over-segmentation, unexpected objects found in the scene (pathologies for instance), expected objects not found and deformations of objects [13]. Also, the attributes computed for the image and the model may be imprecise. To illustrate these difficulties, Figure 1 presents slices of three different volumes: (a) a normal brain, (b) a pathological brain with a tumor, and (c) the representation of a brain atlas where each grey level corresponds to a unique connected structure. Middle dark structures (lateral ventricles) are much bigger in (b) than in (a). The white hyper-signal structure (tumor) does not appear in the atlas (c) nor in the normal brain (a). Similar problems occur in other applications, such as aerial or satellite image interpretation using a map, face recognition, and character recognition.

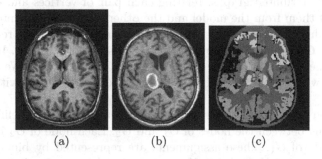

(a)                    (b)                    (c)

**Fig. 1.** Examples of magnetic resonance images: (a) axial slice of a normal brain, (b) axial slice of a pathological brain with a tumor, and (c) axial slice of a brain atlas.

This paper focuses on algorithms for the non-bijective graph matching problem [1,7,10,13,15,17,19], which is defined by the relational attributed graphs representing the model and the over-segmented image, together with the node and edge similarities between their nodes and edges. Section 2 describes our formulation of the search for the best correspondence between the two graphs as a non-bijective graph matching problem. We discuss the nature of the objective function and of the constraints of the graph matching problem. A randomized construction algorithm is proposed in Section 3 to build feasible solutions. Besides the quality of the solutions found, this algorithm may also be used as a robust generator of initial solutions for a GRASP metaheuristic [16] or for population methods such as the genetic algorithm described in [14]. A local search algorithm is proposed in Section 4 to improve the solutions obtained by the construction algorithm. Numerical results obtained with the randomized construction and the local search algorithms are presented and discussed in Section 5. They illustrate the robustness of the construction algorithm and the improvements attained by the local search algorithm in terms of solution quality and object identification. Concluding remarks are presented in the last section.

## 2    Non-bijective Graph Matching

Attributed graphs are widely used in pattern recognition problems. The definition of the attributes and the computation of their values are specific to each application and problem instance. Fuzzy attributed graphs are used for recognition under imprecisions [2,3,4,5,12,13,14,15]. The construction of a fuzzy attributed graph depends on the imperfections of the scene or of the reference model, and on the attributes of the object relations. The common point is that there is always a single vertex for each region of each image. Differences may occur due to the strategy applied for the creation of the edge set, as a result of the chosen attributes or of defects in scene segmentation. Once the graph is built, the next step consists in computing the attributes of vertices and edges. Finally, vertex-similarity and edge-similarity matrices are computed from the values of the attributed graphs, relating each pair of vertices and each pair of edges, one of them from the model and the other from the segmented image.

Two graphs are used to represent the problem: $G_1 = (N_1, E_1)$ represents the model, while $G_2 = (N_2, E_2)$ represents the over-segmented image. In each case, $N_i$ denotes the vertex set and $E_i$ denotes the edge set, $i \in \{1,2\}$. We assume that $|N_1| \leq |N_2|$, which is the case when the image is over-segmented with respect to the model.

A solution to the non-bijective graph matching problem is defined by a set of associations between the nodes of $G_1$ and $G_2$. Each node of $G_2$ is associated with one node of $G_1$. These assignments are represented by binary variables: $x_{ij} = 1$ if nodes $i \in N_1$ and $j \in N_2$ are associated, $x_{ij} = 0$ otherwise. The set $A(i) = \{j \in N_2 \,|\, x_{ij} = 1\}$ denotes the subset of vertices of $N_2$ associated with vertex $i \in N_1$. To ensure that the structure of $G_1$ appears within $G_2$, we favor solutions where a correspondence between edges also implies a correspondence between their extremities (edge association condition). Thus, edge associations are derived from vertex associations, according to the following rule: edge $(a, b) \in E_1$ is associated with all edges $(a', b') \in E_2$ such that (i) $a' \in N_2$ is associated with $a \in N_1$ and $b' \in N_2$ is associated with $b \in N_1$ or (ii) $a' \in N_2$ is associated with $b \in N_1$ and $b' \in N_2$ is associated with $a \in N_1$.

A good matching is a solution in which the associations correspond to high similarity values. Similarity matrices are constructed from similarity values calculated from graph attributes. The choice of these attributes depends on the images. Let $S^v$ (resp. $S^e$) denote an $|N_1| \times |N_2|$ (resp. $|E_1| \times |E_2|$) vertex-similarity (resp. edge-similarity) matrix, where the elements $s^v(i,j)$ (resp. $s^e((i,i'),(j,j')))$ $\in [0,1]$ represent the similarity between vertices (resp. edges) $i \in N_1$ and $j \in N_2$ (resp. $(i,i') \in E_1$ and $(j,j') \in E_2$). The value of any solution is expressed by an objective function, defined for each solution $x$ as

$$f(x) = \frac{\alpha}{|N_1| \cdot |N_2|} f^v(x) + \frac{(1-\alpha)}{|E_1| \cdot |E_2|} f^e(x),$$

with

$$f^v(x) = \sum_{i \in N_1} \sum_{j \in N_2} (1 - |x_{ij} - s^v(i,j)|)$$

and

$$f^e(x) = \sum_{(i,i') \in E_1} \sum_{(j,j') \in E_2} (1 - |\max\{x_{ij}x_{i'j'}, x_{ij'}x_{i'j}\} - s^e((i,i'),(j,j'))|),$$

where $\alpha$ is a parameter used to weight each term of $f$. This function consists of two terms which represent the vertex and edge contributions to the measure of the solution quality associated with each correspondence. Vertex and edge associations with high similarity values are privileged, while those with low similarity values are penalized. The first term represents the average vertex contribution to the correspondence. The second term represents the average edge contribution to the correspondence and acts to enforce the edge association condition. For instance, if $s^e((i,i'),(j,j'))$ is high and there are associations between the extremities of edges $(i,i')$ and $(j,j')$, then $\max\{x_{ij}x_{i'j'}, x_{ij'}x_{i'j}\} = 1$ and the edge contribution is high. On the contrary, if the extremities of edges $(i,i')$ and $(j,j')$ are not associated, then $\max\{x_{ij}x_{i'j'}, x_{ij'}x_{i'j}\} = 0$ and the edge contribution is null. This function behaves appropriately when the image features are well described by the graph attributes.

The search is restricted only to solutions in which each vertex of $N_2$ has to be associated with exactly one vertex of $N_1$. The rationale for this condition is that image segmentation is performed by an appropriate algorithm which preserves the boundaries and, in consequence, avoids situations in which one region of the segmented image is located in the frontier of two adjacent regions of the model: Constraint (1): For every $j \in N_2$, there exists exactly one node $i \in N_1$ such that $x_{ij} = 1$, i.e. $|A^{-1}(j)| = 1$.

The quality of the input data (vertex and edge similarity matrices) is primordial for the identification of the best correspondence. However, as this quality is not always easy to be achieved in real applications, we emphasize some aspects that can be used as additional information to improve the search. Vertices of $G_2$ associated with the same vertex of $G_1$ should be connected among themselves in real situations, since an over-segmentation method can split an object in several smaller pieces, but it does not change the piece positions. Regions of the segmented image corresponding to the same region of the model should necessarily be connected. A good strategy to avoid this type of solution is to restrain the search to correspondences for which each set $A(i)$ of vertices induces a connected subgraph in $G_2$, for every model vertex $i \in N_1$ (connectivity constraint):

Constraint (2): For every $i \in N_1$, the subgraph induced in $G_2(N_2, E_2)$ by $A(i)$ is connected.

Pairs of vertices with null similarity cannot be associated. Such associations are discarded by the constraint below, which strengthens the penalization of associations between vertices with low similarity values induced by the objective function:

Constraint (3): For every $i \in N_1$ and $j \in N_2$, if $s^v(i,j) = 0$, then $x_{ij} = 0$.

Finally, to ensure that all objects of the model appear in the image graph, one additional constraint is imposed:

Constraint (4): For every $i \in N_1$, there exists at least one node $j \in N_2$ such that $(i, j) \in E'$ (i.e., $|A(i)| \geq 1$).

## 3   Randomized Construction Algorithm

The construction algorithm proposed in this section is based on progressively associating a node of $N_1$ with each node of $N_2$, until a feasible solution $x$ is built. The objective function $f(x)$ does not have to be evaluated from scratch at each iteration. Its value is initialized once for all and progressively updated after each new association between a pair of vertices from $N_1$ and $N_2$. Since

$$f^v(x) = \sum_{i \in N_1} \sum_{j \in N_2} (1 - |x_{ij} - s^v(i,j)|) =$$

$$= \sum_{(i,j) \in N_1 \times N_2} (1 - s^v(i,j)) + \sum_{(i,j):x_{ij}=1} (2s^v(i,j) - 1),$$

then $f(x') = f(x) + 2s^v(i,j) - 1$ for any two solutions $x$ and $x'$ that differ only by one additional association between vertices $i \in N_1$ and $j \in N_2$. Similar considerations are used in the evaluation of the term $f^e(x)$, which is increased by $2s^e((a, a'), (b, b')) - 1$ whenever a new pair of edges $(a, a') \in E_1$ and $(b, b') \in E_2$ are associated.

The pseudo-code of the RandomizedConstruction randomized algorithm is given in Figure 2. The algorithm takes as parameters the initial seed, the maximum number $MaxTrials$ of attempts to build a feasible solution before stopping, and the maximum number $MaxSolutions$ of solutions built. We denote by $\Gamma_G(j)$ the nodes adjacent to vertex $j$ in a graph $G$. The number of attempts, the number of solutions built, and the indicator that a feasible solution has been found are initialized in line 1. The optimal value is initialized in line 2. The loop in lines 3–35 performs at most $MaxTrials$ attempts to build at most $MaxSolutions$ solutions. Lines 4–7 prepare and initialize the data for each attempt. The solution $x$, the set $A(i)$ of nodes associated with each node $i \in N_1$, and the node $A^{-1}(j)$ associated with each node $j \in N_2$ are initialized in line 4. The terms $f^v$ and $f^e$ are initialized respectively in lines 5 and 6. A temporary copy $V_2$ of the node set

```
procedure RandomizedConstruction(seed,MaxTrials,MaxSolutions)
1.    trials ← 1, solutions ← 1, and feasible ← .FALSE.;
2.    f* ← −∞;
3.    while trials ≤ MaxTrials and solutions ≤ MaxSolutions do
4.        x ← 0, A(i) ← ∅ ∀i ∈ N₁, and A⁻¹(j) ← ∅ ∀j ∈ N₂;
5.        fᵛ ← ∑₍ᵢ,ⱼ₎∈N₁×N₂(1 − sᵛ(i,j));
6.        fᵉ ← ∑₍₍ᵢ,ᵢ'₎,₍ⱼ,ⱼ'₎₎∈E₁×E₂(1 − sᵉ((i,i'),(j,j')));
7.        V₂ ← N₂;
8.        while V₂ ≠ ∅ do
9.            Randomly select a node j from V₂ and update V₂ ← V₂ − {j};
10.           V₁ ← N₁;
11.           while V₁ ≠ ∅ and A⁻¹(j) = ∅ do
12.               Randomly select a node i from V₁ and update V₁ ← V₁ − {i};
13.               if sᵛ(i,j) > 0 and
                      the graph induced in G₂ by A(i) ∪ {j} is connected
14.               then do
15.                   xᵢⱼ ← 1;
16.                   A(i) ← A(i) ∪ {j} and A⁻¹(j) ← {i};
17.                   fᵛ ← fᵛ + 2sᵥ(i,j) − 1;
18.                   forall i' ∈ Γ_G₁(i) do
19.                       forall j' ∈ Γ_G₂(j) do
20.                           if xᵢ'ⱼ' = 1
21.                           then fᵉ ← fᵉ + 2sᵛ((i,i'),(j,j')) − 1;
22.                       end_forall;
23.                   end_forall;
24.               end_if;
25.           end_while;
26.       end_while;
27.       if A(i) ≠ ∅ ∀i ∈ N₁ and A⁻¹(j) ≠ ∅ ∀j ∈ N₂
28.       then do
29.           feasible ← .TRUE.;
30.           solutions ← solutions + 1;
31.           Compute f ← α/(|N₁| · |N₂|) · fᵛ + (1 − α)/(|E₁| · |E₂|) · fᵉ;
32.           if f > f* then update f* ← f and x* ← x;
33.       end_if;
34.       trials ← trials + 1;
35.   end_while;
36.   return x*, f*;
end RandomizedConstruction.
```

$$\textbf{Fig. 2.} \text{ Pseudo-code of the randomized construction algorithm.}$$

$N_2$ is created in line 7. The loop in lines 8–26 performs one attempt to create a feasible solution and stops after the associations to each node in $V_2$ have been investigated. A node $j \in V_2$ is randomly selected and eliminated from $V_2$ in line 9. A temporary copy $V_1$ of the node set $N_1$ is created in line 10. The loop in lines 11–25 searches for a node in $N_1$ to be associated with node $j \in V_2$. It stops after all possible associations to nodes in $N_1$ have been investigated without success or if one possible association was found. A node $i \in V_1$ is randomly selected and eliminated from $V_1$ in line 12. The algorithm checks in line 13 if node $i$ can be associated with node $j$, i.e., if their similarity is not null and if the graph induced in $G_2$ by $A(i) \cup \{j\}$ is connected. If this is the case, the current solution and its objective function value are updated in lines 14–24. The current solution is updated in lines 15–16. The term $f^v$ corresponding to the node similarities is updated in line 17. The term $f^e$ corresponding to the edge similarities is updated in lines 18–23. The algorithm checks in line 27 if the solution $x$ built in lines 8–26 is feasible, i.e., if there is at least one node of $N_2$ associated with every node of $N_1$ and if there is exactly one node of $N_1$ associated with every node of $N_2$. If this is the case, the indicator that a feasible solution was found is reset in line 29 and the number of feasible solutions built is updated in line 30. The value of the objective function for the new solution is computed in line 31. If the new solution is better than the incumbent, then the latter is updated in line 32. The number of attempts to build a feasible solution is updated in line 34 and a new iteration resumes, until the maximum number of attempts is reached. The best solution found $x^*$ and its objective function value $f^*$ are returned in line 36. In case no feasible solution was found, the returned value is $f^* = -\infty$. The complexity of each attempt to build a feasible solution is $O(|N_1|^2 \cdot |N_2|^2)$.

## 4   Local Search

The solutions generated by a randomized construction algorithm are not necessarily optimal, even with respect to simple neighborhoods. Hence, it is almost always beneficial to apply a local search to attempt to improve each constructed solution. A local search algorithm works in an iterative fashion by successively replacing the current solution by a better solution in the neighborhood of the current solution. It terminates when a local optimum is found, i.e., when no better solution can be found in the neighborhood of the current solution.

We define the neighborhood $N^a(x)$ associated with any solution $x$ as formed by all feasible solutions that can be obtained by the modification of $A^{-1}(j)$ for some $j \in N_2$. For each vertex $j \in N_2$, the candidate set $C(j)$ is formed by all vertices in $N_1$ that can replace the node currently associated with $N_2$, i.e. $C(j) = \{k \in N_1 \mid x'$ is a feasible solution, where $x'_{i\ell} = 1$ if $i = k$ and $\ell = j, x'_{i\ell} = 0$ if $i = A^{-1}(j)$ and $\ell = j, x'_{i\ell} = x_{i\ell}$ otherwise$\}$. The number of solutions within this neighborhood is bounded by $|N_1| \cdot |N_2|$.

The pseudo-code of the local search algorithm LS using a first-improving strategy based on the neighborhood structure $N^a$ defined above is given in Figure 3. The algorithm takes as inputs the solution $x^*$ built by the randomized construction algorithm and its objective function value $f^*$. Initializations are performed in lines 1-2. The loop in lines 3-32 performs the local search and stops at the first local optimum of the objective function with respect to the neighborhood defined by the sets $C(j)$. The control variable is initialized at each local search iteration in line 4. The loop in lines 5-31 considers each node $j$ of graph $G_2$. The replacement of the node $i = A^{-1}(j)$ currently associated with $j$

```
procedure LS(x*, f*)
1.   improvement ← .TRUE.;
2.   Build sets C(j), ∀j ∈ N₂;
3.   while improvement do
4.       improvement ← .FALSE.;
5.       forall j ∈ N₂ while .NOT.improvement do
6.           i ← A⁻¹(j);
7.           forall k ∈ C(j) while .NOT.improvement do
8.               Δᵛ ← 2 · sᵛ(k, j) − 2 · sᵛ(i, j);
9.               Δᵉ ← 0;
10.              forall j' ∈ Γ_{G₂}(j) do
11.                  forall i' ∈ Γ_{G₁}(i) do
12.                      if i' = A⁻¹(j')
13.                      then Δᵉ ← Δᵉ + 1 − 2 · sᵉ((i, i'), (j, j')));
14.                  end_forall;
15.                  forall k' ∈ Γ_{G₁}(k) do
16.                      if k' = A⁻¹(j')
17.                      then Δᵉ ← Δᵉ − 1 + 2 · sᵉ((k, k'), (j, j')));
18.                  end_forall;
19.              end_forall;
20.              Δ ← α/(|N₁| · |N₂|) · Δᵛ + (1 − α)/(|E₁| · |E₂|) · Δᵉ;
21.              if Δ > 0
22.              then do
23.                  improvement ← .TRUE.;
24.                  x*_{kj} ← 1, x*_{ij} ← 0;
25.                  A(i) ← A(i) − {j};
26.                  A(k) ← A(k) ∪ {j};
27.                  f* ← f* + Δ;
28.                  Update sets C(j), ∀j ∈ N₂;
29.              end_if;
30.          end_forall;
31.      end_forall;
32.  end_while;
33.  return x*, f*;
end LS.
```

**Fig. 3.** Pseudo-code of the basic local search algorithm using neighborhood $N^a$.

(line 6) by each node belonging to the candidate set $C(j)$ is investigated in the loop in lines 7-30. The increase in the value of the objective function due to the node similarity contributions is computed in line 8, while that due to the edge similarity contributions is computed in lines 9-19. If the total increase in the objective function value computed in line 20 is strictly positive (line 21), then the current solution and the control variables are updated in lines 22-28. The procedure returns in line 33 the local optimum found and the corresponding solution value. Each local search iteration within neighborhood $N^a$ has complexity $O(|N_1| \cdot |N_2|^2 + |N_2| \cdot |E_2|)$.

We notice that if $A(i) = \{j\}$ for some $i \in N_1$ and $j \in N_2$ (i.e. $|A(i)| = 1$) then $|C(j)| = \emptyset$, because in this case vertex $i$ would not be associated with any other vertex. It can also be the case that a node $j \in A(i)$ cannot be associated with any other node because $A(i) \setminus \{j\}$ induces a non-connected graph in $G_2$. In consequence, in these situation the vertex associated with node $j$ cannot be changed by local search within the neighborhood $N^a$, even if there are other feasible associations. As an attempt to avoid this situation, we define a second neighborhood structure $N^b(x)$ associated with any feasible solution $x$. This is a swap neighborhood, in which the associations of two vertices $j, j' \in N_2$ are exchanged. A solution $x' \in N^b(x)$ if there are two vertices $i', i'' \in N_1$ and two vertices $j', j'' \in N_2$ such that $x_{i'j'} = 1$, $x_{i''j''} = 1$, $x'_{i'j''} = 1$, and $x'_{i''j'} = 1$, with all other associations in solutions $x$ and $x'$ being the same.

Local search within the swap neighborhood $N^b$ has a higher time complexity $O(|N_2|^2 \cdot |E_2|)$ than within neighborhood $N^a$. Also, $|N^b(x)| >> |N^a(x)|$ for any feasible solution $x$. Accordingly, we propose an extended local search procedure LS+ which makes use of both neighborhoods. Whenever the basic local search procedure LS identifies a local optimum $x^*$ with respect to neighborhood $N^a$, the extended procedure starts a local search from the current solution $x^*$ within neighborhood $N^b$. If this solution is also optimal with respect to neighborhood $N^b$, then the extended procedure stops; otherwise algorithm LS resumes from any improving neighbor of $x^*$ within $N^b$.

## 5    Computational Results

The algorithms described in the previous sections were implemented in C and compiled with version 2.96 of the gcc compiler. We used an implementation in C of the random number generator described in [18]. All computational experiments were performed on a 450 MHz Pentium II computer with 128 Mbytes of RAM memory, running under version 7.1 of the Red Hat Linux operating system.

Unlike other problems in the literature, there are no benchmark instances available for the problem studied in this paper. We describe below a subset of seven test instances used in the evaluation of the model and the algorithms proposed in Sections 3 and 4.

Instances GM-5, GM-8, and GM-9 were randomly generated [1], with node and edge similarity values in the interval [0,1]. Instance GM-8 was also used in the computational experiments reported in [1]. Instances GM-5 and GM-8 have isolated nodes: two in the image graph $G_2$ of GM-5 and three in the model graph $G_1$ of GM-8. Instances GM-5a and GM-8a are derived from them, by the introduction of additional edges to connect the isolated nodes.

Instances GM-6 and GM-7 were provided by Perchant and Bengoetxea [12, 14] and built from real images. Instance GM-6 was built from magnetic resonance images of a human brain, as depicted in Figure 4. Instance GM-7 was created for the computational experiments reported in [14] from the 2D images given in Figure 5. The image (a) was over-segmented in 28 regions (c) and compared with a model with only 14 well defined regions (b). The model graph $G_1$ has 14 vertices and 27 edges, while the over-segmented image graph $G_2$ has 28 vertices and 63 edges. Grey levels were used in the computation of node similarities, while distances and adjacencies were used for the computation of edge similarities.

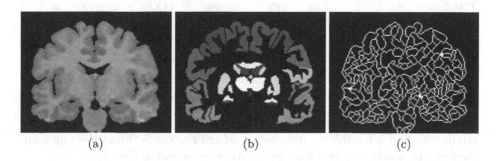

Fig. 4. Instance GM-6: (a) original image, (b) model, and (c) over-segmented image.

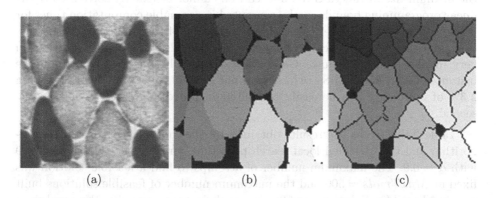

Fig. 5. Cut of a muscle (instance GM-7): (a) original 2D image, (b) model, and (c) over-segmented image.

We summarize the characteristics of instances GM-5 to GM-9 in Table 1. For each instance, we first give the number of nodes and edges of the model and image graphs. We also give the optimal value $f^*$ obtained by the exact integer programming formulation proposed by Duarte [8] using the mixed integer programming solver CPLEX 9.0 and the associated computation time in seconds on a 2.0 GHz Pentium IV computer (whenever available), considering exclusively the vertex contribution to the objective function. In the last two columns, we give the value $f^{LS+}$ of the solution obtained by the randomized construction algorithm followed by the application of the extended local search procedure LS+ and the total computation time in seconds, with the maximum number of

**Table 1.** Characteristics and exact results for instances GM-5 to GM-9.

| Instance | $|N_1|$ | $|E_1|$ | $|N_2|$ | $|E_2|$ | $f^*$ | time (s) | $f^{LS+}$ | time (s) |
|---|---|---|---|---|---|---|---|---|
| GM-5 | 10 | 15 | 30 | 39 | 0.5676 | 7113.34 | 0.5534 | 0.01 |
| GM-5a | 10 | 15 | 30 | 41 | 0.5690 | 2559.45 | 0.5460 | 0.02 |
| GM-6 | 12 | 42 | 95 | 1434 | 0.4294 | 23668.17 | 0.4286 | 2.68 |
| GM-7 | 14 | 27 | 28 | 63 | 0.6999 | 113.84 | 0.6949 | $< 10^{-3}$ |
| GM-8 | 30 | 39 | 100 | 297 | (a) 0.5331 | (a) 4.27 | 0.5209 | 1.02 |
| GM-8a | 30 | 42 | 100 | 297 | (a) 0.5331 | (a) 4.12 | 0.5209 | 1.02 |
| GM-9 | 50 | 88 | 250 | 1681 | – | – | 0.5204 | 42.26 |

(a) linear programming relaxation
–: not available

attemps to find a feasible solution set at $MaxTrials = 500$ and the maximum number of feasible solutions built set at $MaxSolutions = 100$.

The results in Table 1 illustrate the effectiveness of the heuristics proposed in this work. The non-bijective graph matching problem can be exactly solved only for small problems by a state-of-the-art solver such as CPLEX 9.0. Even the medium size instances GM-8 and GM-8a cannot be exactly solved. Only the linear programming bounds can be computed in resonable computation times for both of them. On the other hand, the combination of the randomized construction algorithm with the local search procedure provides high quality solutions in very small computation times. Good approximate solutions for the medium size instances GM-8 and GM-8a (which were not exactly solved by CPLEX) within 2.3% of optimality can be easily computed in processing times as low as one second.

Table 2 illustrates the results obtained by the randomized construction algorithm and the extended local search procedure for instances GM-5 to GM-9 with $\alpha = 0.9$. The maximum number of attempts to find a feasible solution was fixed at $MaxTrials = 500$ and the maximum number of feasible solutions built was fixed at $MaxSolutions = 100$. For each instance, we give the number of attempts necessary to find the first feasible solution, the value $f^{(1)}$ of the first feasible solution found, the number of attempts necessary to find the best among the first 100 feasible solutions built, the value $f^{(100)}$ of the best feasible solution found, and the average computation time per attempt in seconds. The last three columns report statistics for the extended local search algorithm: the number of local search iterations until local optimality, the value $f^{LS+}$ of the best solution found, and the average computation time per iteration in seconds.

The computation time taken by each attempt of the randomized construction algorithm to build a feasible solution is very small, even for the largest instances. The algorithm is very fast and finds the first feasible solution in only a few attempts, except in the cases of the difficult instances with isolated nodes. However, even in the case of the hard instance GM-5, the algorithm managed to find a feasible solution after 297 attempts. For the other instances, the construction algorithm found a feasible solution in very few iterations. Even better solutions can be obtained if additional attempts are performed.

The local search algorithm improved the solutions built by the construction algorithm for all test instances. The average improvement with respect to the value of the solution obtained by the construction algorithm was approximately 3%.

**Table 2.** Results obtained by the randomized construction algorithm and the extended local search procedure with $MaxTrials = 500$ and $MaxSolutions = 100$.

| Instance | first | $f^{(1)}$ | best | $f^{(100)}$ | time (s) | iteration | $f^{LS+}$ | time (s) |
|----------|-------|-----------|------|-------------|----------|-----------|-----------|----------|
| GM-5  | 297 | 0.5168 | 297 | 0.5168 | $< 10^{-3}$ | 19  | 0.5474 | 0.002 |
| GM-5a | 9   | 0.4981 | 417 | 0.5243 | $< 10^{-3}$ | 13  | 0.5434 | 0.002 |
| GM-6  | 1   | 0.4122 | 40  | 0.4168 | 0.001       | 320 | 0.4248 | 0.020 |
| GM-7  | 5   | 0.6182 | 34  | 0.6282 | $< 10^{-3}$ | 12  | 0.6319 | 0.001 |
| GM-8  | 26  | 0.4978 | 292 | 0.5022 | 0.002       | 118 | 0.5186 | 0.014 |
| GM-8a | 26  | 0.5014 | 292 | 0.5058 | 0.002       | 120 | 0.5222 | 0.014 |
| GM-9  | 1   | 0.5049 | 207 | 0.5060 | 0.010       | 511 | 0.5187 | 0.134 |

## 6  Concluding Remarks

We formulated the problem of finding the best correspondence between two graphs representing a model and an over-segmented image as a combinatorial optimization problem.

A robust randomized construction algorithm was proposed to build feasible solutions for the graph matching problem. We also proposed a local search algorithm based on two neighborhood structures to improve the solutions built by the construction algorithm. Computational results were presented to different test problems. Both algorithms are fast and easily found feasible solutions to realistic problems with up to 250 nodes and 1681 edges in the graph representing the over-segmented image. The local search algorithm consistently improved the solutions found by the construction heuristic. Both algorithms can be easily adapted to handle more complex objective function formulations.

Besides the quality of the solutions found, the randomized algorithm may also be used as a robust generator of initial solutions for population methods such as the genetic algorithm described in [14], replacing the low quality randomly generated solutions originally proposed. The construction and local search algorithms can also be put together into an implementation of the GRASP metaheuristic [16].

# References

1. E. Bengoetxea, P. Larranaga, I. Bloch, A. Perchant, and C. Boeres. Inexact graph matching by means of estimation distribution algorithms. *Pattern Recognition*, 35:2867-2880, 2002.
2. I. Bloch. Fuzzy relative position between objects in image processing: a morphological approach. *IEEE Transactions on Pattern Analysis Machine Intelligence*, 21:657–664, 1999.
3. I. Bloch. On fuzzy distances and their use in image processing under imprecision. *Pattern Recognition*, 32:1873–1895, 1999.
4. I. Bloch, H. Maître, and M. Anvari. Fuzzy adjacency between image objects. *International Journal of Uncertainty, Fuzziness and Knowledge-Based Systems*, 5:615–653, 1997.
5. K.P. Chan and Y.S. Cheung. Fuzzy-attribute graph with application to chinese character recognition. *IEEE Transactions on Systems, Man and Cybernetics*, 22:402–410, 1992.
6. A.D.J. Cross and E.R. Hancock. Relational matching with stochastic optimization. In *International Conference on Computer Vision*, pages 365–370, 1995.
7. A.D.J. Cross, R.C. Wilson, and E.R. Hancock. Inexact graph matching using genetic search. *Pattern Recognition*, 30:953–970, 1997.
8. A.R. Duarte. *New heuristics and an exact integer programming formulation for an inexact graph matching problem* (in Portuguese). M.Sc. Dissertation, Catholic University of Rio de Janeiro, 2004.
9. Y. El-Sonbaty and M. A. Ismail. A new algorithm for subgraph optimal isomorphism. *Pattern Recognition*, 31:205–218, 1998.
10. A. W. Finch, R. C. Wilson, and E. R. Hancock. Symbolic Graph matching with the EM algorithm. *Pattern Recognition*, 31:1777–1790, 1998.
11. H. Moissinac, H. Maître, and I. Bloch. Markov random fields and graphs for uncertainty management and symbolic data fusion in a urban scene interpretation. *EUROPTO Conference on Image and Signal Processing for Remote Sensing*, 2579:298–309, 1995.
12. A. Perchant. *Morphisme de graphes d'attributs flous pour la reconnaissance structurelle de scènes*. Doctorate thesis, École Nationale Supérieure des Télécommunications, 2000.
13. A. Perchant and I. Bloch. A new definition for fuzzy attributed graph homomorphism with application to structural shape recognition in brain imaging. In *Proceedings of the 16th IEEE Conference on Instrumentation and Measurement Technology*, pages 402–410, 1999.
14. A. Perchant, C. Boeres, I. Bloch, M. Roux, and C.C. Ribeiro. Model-based scene recognition using graph fuzzy homomorphism solved by genetic algorithm. In *2nd IAPR-TC-15 Workshop on Graph-based Representations*, pages 61–70, 1999.
15. H.S. Ranganath and L.J. Chipman. Fuzzy relaxaton approach for inexact scene matching. *Image and Vision Computing*, 10:631–640, 1992.
16. M.G.C. Resende and C.C. Ribeiro. "Greedy randomized adaptive search procedures". *Handbook of Metaheuristics* (F. Glover and G. Kochenberger, eds.), pages 219-249, Kluwer, 2002.
17. A. Rosenfeld, R. Hummel, and S. Zucker. Scene labeling by relaxation operations. *IEEE Transactions on Systems, Man and Cybernetics*, 6:420–433, 1976.
18. L. Schrage. A more portable FORTRAN random number generator. *ACM Transactions on Mathematical Software*, 5:132-138, 1979.

19. M. Singh and A. C. S. Chaudhury. Matching structural shape descriptions using genetic algorithms. *Pattern Recognition*, 30:1451–1462, 1997.
20. A.K.C. Wong, M. You, and S.C. Chan. An algorithm for graph optimal monomorphism. *IEEE Transactions on Systems, Man and Cybernetics*, 20:628–636, 1990.
21. E.K. Wong. Model matching in robot vision by subgraph isomorphism. *Pattern Recognition*, 25:287–303, 1992.

# An Efficient Implementation of a Joint Generation Algorithm[*]

E. Boros[1], K. Elbassioni[1], V. Gurvich[1], and L. Khachiyan[2]

[1] RUTCOR, Rutgers University, 640 Bartholomew Road, Piscataway, NJ
08854-8003, USA.
{boros,elbassio,gurvich}@rutcor.rutgers.edu
[2] Department of Computer Science, Rutgers University, 110 Frelinghuysen Road,
Piscataway, NJ 08854-8003, USA.
leonid@cs.rutgers.edu

**Abstract.** Let $\mathcal{C}$ be an $n$-dimensional integral box, and $\pi$ be a monotone property defined over the elements of $\mathcal{C}$. We consider the problems of incrementally generating jointly the families $\mathcal{F}_\pi$ and $\mathcal{G}_\pi$ of all minimal subsets satisfying property $\pi$ and all maximal subsets not satisfying property $\pi$, when $\pi$ is given by a polynomial-time satisfiability oracle. Problems of this type arise in many practical applications. It is known that the above joint generation problem can be solved in incremental quasi-polynomial time. In this paper, we present an efficient implementation of this procedure. We present experimental results to evaluate our implementation for a number of interesting monotone properties $\pi$.

## 1  Introduction

Let $\mathcal{C} = \mathcal{C}_1 \times \cdots \times \mathcal{C}_n$ be an integral box, where $\mathcal{C}_1, \ldots, \mathcal{C}_n$ are finite sets of integers. For a subset $\mathcal{A} \subseteq \mathcal{C}$, let us denote by $\mathcal{A}^+ = \{x \in \mathcal{C} \mid x \geq a, \text{ for some } a \in \mathcal{A}\}$ and $\mathcal{A}^- = \{x \in \mathcal{C} \mid x \leq a, \text{ for some } a \in \mathcal{A}\}$, the ideal and filter generated by $\mathcal{A}$. Any element in $\mathcal{C} \setminus \mathcal{A}^+$ is called *independent of* $\mathcal{A}$, and we let $\mathcal{I}(\mathcal{A})$ denote the set of all *maximal independent* elements for $\mathcal{A}$. Call a family of vectors $\mathcal{A}$ *Sperner* if $\mathcal{A}$ is an antichain, i.e. if no two elements are comparable in $\mathcal{A}$. If $\mathcal{C}$ is the Boolean cube $2^{[n]}$, we get the well-known definitions of a hypergraph $\mathcal{A}$ and its family of maximal independent sets $\mathcal{I}(\mathcal{A})$.

Let $\pi : \mathcal{C} \mapsto \{0, 1\}$ be a *monotone* property defined over the elements of $\mathcal{C}$: if $x \in \mathcal{C}$ satisfies property $\pi$, i.e. $\pi(x) = 1$, then any $y \in \mathcal{C}$ such that $y \geq x$ also satisfies $\pi$. We assume that $\pi$ is described by a polynomial satisfiability oracle $\mathcal{O}_\pi$, i.e. an algorithm that can decide whether a given vector $x \in \mathcal{C}$ satisfies $\pi$, in time polynomial in $n$ and the size $|\pi|$ of the input description of $\pi$. Denote

---

[*] This research was supported by the National Science Foundation (Grant IIS-0118635). The third author is also grateful for the partial support by DIMACS, the National Science Foundation's Center for Discrete Mathematics and Theoretical Computer Science.

C.C. Ribeiro and S.L. Martins (Eds.): WEA 2004, LNCS 3059, pp. 114–128, 2004.
© Springer-Verlag Berlin Heidelberg 2004

respectively by $\mathcal{F}_\pi$ and $\mathcal{G}_\pi$ the families of minimal elements satisfying property $\pi$, and maximal elements not satisfying property $\pi$. Then it is clear that $\mathcal{G}_\pi = \mathcal{I}(\mathcal{F}_\pi)$ for any monotone property $\pi$. Given a monotone property $\pi$, we consider the problem of jointly generating the families $\mathcal{F}_\pi$ and $\mathcal{G}_\pi$:

**GEN($\mathcal{C}, \mathcal{F}_\pi, \mathcal{G}_\pi, \mathcal{X}, \mathcal{Y}$):** *Given a monotone property $\pi$, represented by a satisfiability oracle $\mathcal{O}_\pi$, and two explicitly listed vector families $\mathcal{X} \subseteq \mathcal{F}_\pi \subseteq \mathcal{C}$ and $\mathcal{Y} \subseteq \mathcal{G}_\pi \subseteq \mathcal{C}$, either find a new element in $(\mathcal{F}_\pi \setminus \mathcal{X}) \cup (\mathcal{G}_\pi \setminus \mathcal{Y})$, or prove that these families are complete: $(\mathcal{X}, \mathcal{Y}) = (\mathcal{F}_\pi, \mathcal{G}_\pi)$.*

It is clear that for a given monotone property $\pi$, described by a satisfiability oracle $\mathcal{O}_\pi$, we can generate both $\mathcal{F}_\pi$ and $\mathcal{G}_\pi$ simultaneously by starting with $\mathcal{X} = \mathcal{Y} = \emptyset$ and solving problem GEN($\mathcal{C}, \mathcal{F}_\pi, \mathcal{G}_\pi, \mathcal{X}, \mathcal{Y}$) for a total of $|\mathcal{F}_\pi| + |\mathcal{G}_\pi| + 1$ times, incrementing in each iteration either $\mathcal{X}$ or $\mathcal{Y}$ by the newly found vector $x \in (\mathcal{F}_\pi \setminus \mathcal{X}) \cup (\mathcal{G}_\pi \setminus \mathcal{Y})$, according to the answer of the oracle $\mathcal{O}_\pi$, until we have $(\mathcal{X}, \mathcal{Y}) = (\mathcal{F}_\pi, \mathcal{G}_\pi)$.

In most practical applications, the requirement is to generate either the family $\mathcal{F}_\pi$ or the family $\mathcal{G}_\pi$, i.e. we consider the separate generation problems:

**GEN($\mathcal{C}, \mathcal{F}_\pi, \mathcal{X}$) (GEN($\mathcal{C}, \mathcal{G}_\pi, \mathcal{Y}$)):** *Given a monotone property $\pi$ and a subfamily $\mathcal{X} \subseteq \mathcal{F}_\pi \subseteq \mathcal{C}$ (respectively, $\mathcal{Y} \subseteq \mathcal{G}_\pi \subseteq \mathcal{C}$), either find a new minimal satisfying vector $x \in \mathcal{F}_\pi \setminus \mathcal{X}$ (respectively, maximal non-satisfying vector $x \in \mathcal{G}_\pi \setminus \mathcal{Y}$), or prove that the given partial list is complete: $\mathcal{X} = \mathcal{F}_\pi$ (respectively, $\mathcal{Y} = \mathcal{G}_\pi$).*

Problems GEN($\mathcal{C}, \mathcal{F}_\pi, \mathcal{X}$) and GEN($\mathcal{C}, \mathcal{G}_\pi, \mathcal{Y}$) arise in many practical applications and in a variety of fields, including artificial intelligence [14], game theory [18,19], reliability theory [8,12], database theory [9,14,17], integer programming [4,6,20], learning theory [1], and data mining [2,6,9]. Even though these two problems may be NP-hard in general (see e.g. [20]), it is known that the joint generation problem GEN($\mathcal{C}, \mathcal{F}_\pi, \mathcal{G}_\pi, \mathcal{X}, \mathcal{Y}$) can be solved in incremental quasi-polynomial time $poly(n, \log \|\mathcal{C}\|_\infty) + m^{\text{polylog}\, m}$ for any monotone property $\pi$ described by a satisfiability oracle, where $\|\mathcal{C}\|_\infty = \sum_{i=1}^n |\mathcal{C}_i|$ and $m = |\mathcal{X}| + |\mathcal{Y}|$, see [4,15]. In particular, there is a polynomial-time reduction from the joint generation problem to the following problem, known as dualization on boxes, see [4, 10,16]:

**DUAL($\mathcal{C}, \mathcal{A}, \mathcal{B}$):** *Given a family of vectors $\mathcal{A} \subseteq \mathcal{C}$, and a subset $\mathcal{B} \subseteq \mathcal{I}(\mathcal{A})$ of its maximal independent vectors, either find a new maximal independent vector $x \in \mathcal{I}(\mathcal{A}) \setminus \mathcal{B}$, or prove that no such vector exists, i.e., $\mathcal{B} = \mathcal{I}(\mathcal{A})$.*

The currently best known algorithm for dualization runs in time $poly(n) + m^{o(\log m)}$, see [4,15]. Unfortunately, this joint generation may not be an efficient algorithm for solving either of GEN($\mathcal{C}, \mathcal{F}_\pi, \mathcal{X}$) or GEN($\mathcal{C}, \mathcal{G}_\pi, \mathcal{Y}$) separately for the simple reason that we do not control which of the families $\mathcal{F}_\pi \setminus \mathcal{X}$ and $\mathcal{G}_\pi \setminus \mathcal{Y}$ contains each new vector produced by the algorithm. Suppose we want to generate $\mathcal{F}_\pi$ and the family $\mathcal{G}_\pi$ is exponentially larger than $\mathcal{F}_\pi$. Then, if we are unlucky, we may get elements of $\mathcal{F}_\pi$ with exponential delay, while getting large subfamilies of $\mathcal{G}_\pi$ (which are not needed at all) in between. However, there are two reasons that we are interested in solving the joint generation problem efficiently. First, it is easy to see [17] that no satisfiability oracle based algorithm can generate $\mathcal{F}_\pi$ in fewer than $|\mathcal{F}_\pi| + |\mathcal{G}_\pi|$ steps, in general:

**Proposition 1.** *Consider an arbitrary algorithm A, which generates the family* $\mathcal{F}_\pi$ *by using only a satisfiability oracle* $\mathcal{O}_\pi$ *for the monotone property $\pi$. Then, for the algorithm to generate the whole family $\mathcal{F}_\pi$, it must call the oracle at least* $|\mathcal{F}_\pi| + |\mathcal{G}_\pi|$ *times.*

*Proof.* Clearly, any monotone property $\pi$ can be defined by its value on the boundary $\mathcal{F}_\pi \cup \mathcal{G}_\pi$. For any $y \in \mathcal{F}_\pi \cup \mathcal{G}_\pi$, let us thus define the boundary of the monotone property $\pi_y$ as follows:

$$\pi_y(x) = \begin{cases} \pi(x) & \text{if } x \neq y \\ \overline{\pi}(x) & \text{if } x = y. \end{cases}$$

Then $\pi_y$ is a monotone property different from $\pi$, and algorithm $A$ must be able to distinguish the Sperner families described by $\pi$ and $\pi_y$ for every $y \in \mathcal{F}_\pi \cup \mathcal{G}_\pi$. $\quad\square$

Second, for a wide class of Sperner families (or equivalently, monotone properties), the so-called *uniformly dual-bounded* families, it was realized that the size of the dual family $\mathcal{I}(\mathcal{F}_\pi)$ is uniformly bounded by a (quasi-)polynomial in the size of $\mathcal{F}_\pi$ and the oracle description:

$$|\mathcal{I}(\mathcal{X}) \cap \mathcal{I}(\mathcal{F}_\pi)| \leq (quasi\text{-})poly(|\mathcal{X}|, |\pi|) \tag{1}$$

for any non-empty subfamily $\mathcal{X} \subseteq \mathcal{F}_\pi$. An inequality of the form (1) would imply that joint generation is an incrementally efficient way for generating the family $\mathcal{F}_\pi$ (see Section 2 below and also [5] for several interesting examples).

In [7], we presented an efficient implementation of a quasi-polynomial dualization algorithm on boxes. Direct application of this implementation to solve the joint generation problem may not be very efficient since each call to the dualization code requires the construction of a new recursion tree, wasting therefore information that might have been collected from previous calls. A much more efficient approach, which we implement in this paper, is to use the same recursion tree for generating all elements of the families $\mathcal{F}_\pi$ and $\mathcal{G}_\pi$. The details of this method will be described in Sections 3 and 4. In Section 2, we give three examples of monotone properties, that will be used in our experimental study. Finally, Section 5 presents our experimental findings, and Section 6 provides some conclusions.

## 2    Examples of Monotone Properties

We consider the following three monotone properties in this paper:

*Monotone systems of linear inequalities.* Let $A \in \mathbb{R}^{r \times n}$ be a given non-negative real matrix, $b \in \mathbb{R}^r$ be a given r-vector, $c \in \mathbb{R}_+^n$ be a given non-negative $n$-vector, and consider the system of linear inequalities:

$$Ax \geq b, \quad x \in \mathcal{C} = \{x \in \mathbb{Z}^n \mid 0 \leq x \leq c\}. \tag{2}$$

For $x \in \mathcal{C}$, let $\pi_1(x)$ be the property that $x$ satisfies (2). Then the families $\mathcal{F}_{\pi_1}$ and $\mathcal{G}_{\pi_1}$ correspond respectively to the minimal feasible and maximal infeasible vectors for (2). It is known [4] that the family $\mathcal{F}_{\pi_1}$ is (uniformly) dual bounded:

$$|\mathcal{I}(\mathcal{F}_{\pi_1})| \leq rn|\mathcal{F}_{\pi_1}|. \tag{3}$$

*Minimal infrequent and maximal frequent sets in binary databases.* Let $\mathcal{D} : R \times V \mapsto \{0,1\}$ be a given $r \times n$ binary matrix representing a set $R$ of transactions over a set of attributes $V$. To each subset of columns $X \subseteq V$, let us associate the subset $S(X) = S_{\mathcal{D}}(X) \subseteq R$ of all those rows $i \in R$ for which $\mathcal{D}(i,j) = 1$ in every column $j \in X$. The cardinality of $S(X)$ is called the *support* of $X$. Given an integer $t$, a column set $X \subseteq V$ is called *t-frequent* if $|S(X)| \geq t$ and otherwise, is said to be *t-infrequent*. For each set $X \in \mathcal{C} \overset{\text{def}}{=} 2^V$, let $\pi_2(X)$ be the property that $X$ is $t$-infrequent. Then $\pi_2$ is a monotone property and the families $\mathcal{F}_{\pi_2}$ and $\mathcal{G}_{\pi_2}$ correspond respectively to minimal infrequent and maximal frequent sets for $\mathcal{D}$. It is known [9] that

$$|\mathcal{I}(\mathcal{F}_{\pi_2})| \leq (r - t + 1)|\mathcal{F}_{\pi_2}|. \tag{4}$$

Problems GEN$(\mathcal{C}, \mathcal{F}_{\pi_2}, \mathcal{Y})$ and GEN$(\mathcal{C}, \mathcal{G}_{\pi_2}, \mathcal{Y})$ appear in data mining applications, see e.g. [17].

*Sparse boxes for multi-dimensional data.* Let $S$ be a set of points in $\mathbb{R}^n$, and $t \leq |S|$ be a given integer. A *maximal t-box* is a closed $n$-dimensional interval which contains at most $t$ points of $S$ in its interior, and which is maximal with respect to this property (i.e., cannot be extended in any direction without strictly enclosing more points of $S$). Define $\mathcal{C}_i = \{p_i \mid p \in S\}$ for $i = 1, \ldots, n$ and consider the family of boxes $\mathcal{B} = \{[a,b] \subseteq \mathbb{R}^n \mid a, b \in \mathcal{C}_1 \times \cdots \times \mathcal{C}_n, \ a \leq b\}$. For $i = 1, \ldots, n$, let $u_i = \max \mathcal{C}_i$, and let $\mathcal{C}_{i+n} \overset{\text{def}}{=} \{u_i - p \mid p \in \mathcal{C}_i\}$ be the chain ordered in the direction opposite to $\mathcal{C}_i$. Consider the $2n$-dimensional box $\mathcal{C} = \mathcal{C}_1 \times \cdots \times \mathcal{C}_n \times \mathcal{C}_{n+1} \times \cdots \times \mathcal{C}_{2n}$ and let us represent every $n$-dimensional interval $[a,b] \in \mathcal{B}$ as the $2n$-dimensional vector $(a, u - b) \in \mathcal{C}$, where $u = (u_1, \ldots, u_n)$. This gives a monotone injective mapping $\mathcal{B} \mapsto \mathcal{C}$ (not all elements of $\mathcal{C}$ define a box, since $a_i > b_i$ is possible for $(a, u - b) \in \mathcal{C}$). Let us now define the monotone property $\pi_3$ to be satisfied by an $x \in \mathcal{C}$ if and only if $x$ does not define a box, or the box defined by $x$ contains at most $t$ points of $S$ in its interior. Then the sets

$\mathcal{F}_{\pi_3}$ and $\mathcal{G}_{\pi_3}$ can be identified respectively with the set of maximal $t$-boxes (plus a polynomial number of non-boxes), and the set of minimal boxes of $x \in \mathcal{B} \subseteq \mathcal{C}$ which contain at least $k+1$ points of $\mathcal{S}$ in their interior. It is known [6] that the family of maximal $t$-boxes is (uniformly) dual-bounded:

$$|\mathcal{I}(\mathcal{F}_{\pi_3})| \leq |\mathcal{S}||\mathcal{F}_{\pi_3}|. \tag{5}$$

The problem of generating all elements of $\mathcal{F}_{\pi_3}$ has been studied in the machine learning and computational geometry literatures (see [11,13,23]), and is motivated by the discovery of missing associations or "holes" in data mining applications (see [3,21,22]). [13] gives an algorithm, for solving this problem, whose worst-case time complexity is exponential in the dimension $n$ of the given point set.

## 3   Terminology and Outline of the Algorithm

Throughout the paper, we assume that we are given an integer box $\mathcal{C}^* = \mathcal{C}_1^* \times \ldots \times \mathcal{C}_n^*$, where $\mathcal{C}_i^* = [l_i^* : u_i^*]$, and $l_i^* \leq u_i^*$, are integers, and a monotone property $\pi$, described by a polynomial time satisfiability oracle, for which it is required to generate the families $\mathcal{F}_\pi$ and $\mathcal{G}_\pi$. The generation algorithm, considered in this paper, is based on a reduction to a dualization algorithm of [4], for which an efficient implementation was given in [7]. For completeness, we briefly outline this algorithm. The problem is solved by decomposing it into a number of smaller subproblems and solving each of them recursively. The input to each such subproblem is a sub-box $\mathcal{C}$ of the original box $\mathcal{C}^*$ and two subsets $\mathcal{F} \subseteq \mathcal{F}^*$ and $\mathcal{G} \subseteq \mathcal{G}^*$ of integral vectors, where $\mathcal{F}^* \subseteq \mathcal{F}_\pi$ and $\mathcal{G}^* \subseteq \mathcal{G}_\pi$ denote respectively the subfamilies of minimal satisfying and maximal non-satisfying vectors that have been generated so far. Note that, by definition, the following condition holds for the original problem and all subsequent subproblems:

$$a \not\leq b, \quad \text{for all } a \in \mathcal{F}, b \in \mathcal{G}. \tag{6}$$

Given an element $a \in \mathcal{F}$ ($b \in \mathcal{G}$), we say that a coordinate $i \in [n] \overset{\text{def}}{=} \{1, \ldots, n\}$ is *essential* for $a$ (respectively, $b$), in the box $\mathcal{C} = [l_1 : u_1] \times \cdots \times [l_n : u_n]$, if $a_i > l_i$ (respectively, if $b_i < u_i$). Let us denote by $\text{Ess}(x)$ the set of essential coordinates of an element $x \in \mathcal{F} \cup \mathcal{G}$. Finally, given a sub-box $\mathcal{C} \subseteq \mathcal{C}^*$, and two subsets $\mathcal{F} \subseteq \mathcal{F}^*$ and $\mathcal{G} \subseteq \mathcal{G}^*$, we shall say that $\mathcal{F}$ is *dual to* $\mathcal{G}$ in $\mathcal{C}$ if $\mathcal{F}^+ \cup \mathcal{G}^- \supseteq \mathcal{C}$.

A key lemma, on which the algorithm in [4] is based, is that either (i) there is an element $x \in \mathcal{F} \cup \mathcal{G}$ with at most $1/\epsilon$ essential coordinates, where $\epsilon \overset{\text{def}}{=} 1/(1 + \log m)$ and $m \overset{\text{def}}{=} |\mathcal{F}| + |\mathcal{G}|$, or (ii) one can easily find a new element $z \in \mathcal{C} \setminus (\mathcal{F}^+ \cup \mathcal{G}^-)$, by picking each element $z_i$ independently at random from $\{l_i, u_i\}$ for $i = 1, \ldots, n$; see subroutine Random solution() in the next section. In case (i), one can decompose the problem into two strictly smaller subproblems as follows. Assume, without loss of generality, that $x \in \mathcal{F}$ has at most $1/\epsilon$ essential coordinates. Then, by (6), there is an $i \in [n]$ such that $|\{b \in \mathcal{G} : b_i < x_i\}| \geq \epsilon|\mathcal{G}|$. This allows us to decompose the original problem into two subproblems $\text{GEN}(\mathcal{C}', \pi, \mathcal{F}, \mathcal{G}')$ and $\text{GEN}(\mathcal{C}'', \pi, \mathcal{F}'', \mathcal{G})$, where $\mathcal{C}' = \mathcal{C}_1 \times \cdots \times \mathcal{C}_{i-1} \times [x_i : u_i] \times \mathcal{C}_{i+1} \times \cdots \times \mathcal{C}_n$, $\mathcal{G}' = \mathcal{G} \cap \mathcal{C}^+$, $\mathcal{C}'' = \mathcal{C}_1 \times \cdots \times \mathcal{C}_{i-1} \times [l_i : x_i - 1] \times \mathcal{C}_{i+1} \times \cdots \times \mathcal{C}_n$, and $\mathcal{F}'' = \mathcal{F} \cap \mathcal{C}^-$. This way, the algorithm is guaranteed to reduce the cardinality of one of the sets $\mathcal{F}$ or $\mathcal{G}$ by a factor of at least $1 - \epsilon$ at each recursive step. For efficiency reasons, we do two modifications to this basic approach. First, we use sampling to estimate the sizes of the sets $\mathcal{G}', \mathcal{F}''$ (see subroutine Est() below). Second, once we have determined the new sub-boxes $\mathcal{C}', \mathcal{C}''$ above, we do not compute the *active* families $\mathcal{G}'$ and $\mathcal{F}''$ at each recursion step (this is called the Cleanup step in the next section). Instead, we perform the cleanup step only when the number of vectors reduces by a certain factor $f$, say $1/2$, for two reasons: First, this improves the running time since the elimination of vectors is done less frequently. Second, the expected total memory required by all the nodes of the path from the root of the recursion tree to a leaf is at most $O(nm + m/(1 - f))$, which is linear in $m$ for constant $f < 1$.

## 4 The Algorithm

We use the following data structures in our implementation:

- Two arrays of vectors, $F$ and $G$ containing the elements of $\mathcal{F}^*$ and $\mathcal{G}^*$ respectively.
- Two (dynamic) arrays of indices, index($\mathcal{F}$) and index($\mathcal{G}$), containing the indices of vectors from $\mathcal{F}^*$ and $\mathcal{G}^*$ (i.e. containing pointers to elements of the arrays $F$ and $G$), that appear in the current subproblem. These arrays are used to enable sampling from the sets $\mathcal{F}$ and $\mathcal{G}$, and also to keep track of which vectors are currently active, i.e, intersect the current box.
- Two balanced binary search trees $\mathbf{T}(\mathcal{F}^*)$ and $\mathbf{T}(\mathcal{G}^*)$, built on the elements of $\mathcal{F}^*$ and $\mathcal{G}^*$ respectively using lexicographic ordering. Each node of the tree $\mathbf{T}(\mathcal{F}^*)$ $(\mathbf{T}(\mathcal{G}^*))$ contains an index of an element in the array $F$ $(G)$. This way, checking whether a given vector $x \in C$ belongs to $\mathcal{F}^*$ $(\mathcal{G}^*)$ or not, takes only $O(n \log |\mathcal{F}^*|)$ $(O(n \log |\mathcal{G}^*|))$ time.

In the sequel, we let $m = |\mathcal{F}| + |\mathcal{G}|$ and $\epsilon = 1/(1 + \log m)$. We use the following subroutines in our implementation:

**Minimization** $\min_{\mathcal{F}}(z)$. It takes as input a vector $z \in \mathcal{F}^+$ and returns a minimal vector $z^*$ in $\mathcal{F}^+ \cap \{z\}^-$. Such a vector $z^* = \min_{\mathcal{F}}(z)$ can, for instance, be computed by coordinate descent:

$$z_1^* \leftarrow \min\{y_1 \mid (y_1, y_2, \ldots, y_{n-1}, y_n) \in \mathcal{F}^+ \cap \{z\}^-\},$$
$$z_2^* \leftarrow \min\{y_2 \mid (z_1^*, y_2, \ldots, y_{n-1}, y_n) \in \mathcal{F}^+ \cap \{z\}^-\},$$
$$\ldots$$
$$z_n^* \leftarrow \min\{y_n \mid (z_1^*, z_2^*, \ldots, z_{n-1}^*, y_n) \in \mathcal{F}^+ \cap \{z\}^-\}.$$

Note that each of the $n$ coordinate steps in the above procedure can be reduced via binary search to at most $\log(\|\mathcal{C}\|_\infty + 1)$ satisfiability oracle calls for the monotone property $\pi$.

More efficient procedures can be obtained if we specialize this routine to the specific monotone property under consideration. For instance, for property $\pi_1$ this operation can be performed in $O(nr)$ steps as follows. For $j = 1, \ldots, r$, let $a_j x \geq b_j$ be the $j$th inequality of the system. We initialize $z^* \leftarrow z$ and $w_j = a_j z^* - b_j$ for $j = 1, \ldots, r$. For the $i$th step of the coordinate descend operation, we let $z_i^* \leftarrow z_i^* - \lfloor \lambda \rfloor$ and $w_j \leftarrow w_j - \lfloor \lambda \rfloor a_{ji}$ for $j = 1, \ldots, r$, where

$$\lambda = \min_{1 \leq j \leq r} \left\{ \frac{w_j}{a_{ji}} : a_{ji} > 0 \right\}.$$

Now consider property $\pi_2$. Given a set $Z \in \mathcal{F}_{\pi_2}^+$, the operation $\min_{\mathcal{F}_{\pi_2}}(Z)$ can be done in $O(nr)$ by initializing $Z^* \leftarrow Z$, $s \leftarrow |S(Z)|$ and $c(Y) \leftarrow |Z \setminus Y|$ for all $Y \in \mathcal{D}$, and repeating, for $i \in Z$, the following two steps: (i) $\mathcal{Y} \leftarrow \{Y \in \mathcal{D} : c(Y) = 1, Y \not\ni i\}$; (ii) if $|\mathcal{Y}| + s \leq t - 1$ then 1. $Z^* \leftarrow Z^* \setminus \{i\}$, 2. $s \leftarrow s + |\mathcal{Y}|$, and 3. $c(Y) \leftarrow c(Y) - 1$ for each $Y \in \mathcal{D}$ such that $Y \not\ni i$.

For the monotone property $\pi_3$ and $z \in \mathcal{C}$, the operation $\min_{\mathcal{F}_{\pi_3}}(z)$ can be done in $O(n|\mathcal{S}|)$ as follows. For each point $p \in \mathcal{S}$, let $p' \in \mathbb{R}^{2n}$ be the point with components $p_i' = p_i$ for $i = 1, \ldots, n$, and $p_i' = u_{i-n} - p_{i-n}$ for $i = n+1, \ldots, 2n$. Initialize $s(z) \leftarrow |\{p \in \mathcal{S} : p \text{ is in the interior of } z\}|$ and $c(p) \leftarrow |\{i \in [n] : p_i' \leq z_i\}|$ for all $p \in \mathcal{S}$. Repeat, for $i = 1, \ldots, 2n$, the following steps: (i) $z_i^* \leftarrow \min\{p_i' : p \in \mathcal{S}, c(p) = 1, p_i' \leq z_i \text{ and } |\{q \in \mathcal{S} : c(q) = 1, p_i' < q_i \leq z_i\}| \leq t - s(z)\}$; (ii) $c(p) \leftarrow c(p) - 1$ for each $p \in \mathcal{S}$ such that $z_i^* < p_i' \leq z_i$. Note that (i) can be performed in $O(|\mathcal{S}|)$ steps assuming that we know the sorted order for the points along each coordinate.

**Maximization** $\max_{\mathcal{G}}(z)$. It computes, for a given vector $z \in \mathcal{G}^-$, a maximal vector $z^* \in \mathcal{G}^- \cap z^+$. Similar to $\min_{\mathcal{F}}(z)$, this problem can be done, in general, by coordinate descent. For $\mathcal{G}_{\pi_1}$, $\mathcal{G}_{\pi_2}$ and $\mathcal{G}_{\pi_3}$, this operation can be done in $O(nr)$, $O(nr)$, and $O(n|\mathcal{S}|)$ respectively.

Below, we denote respectively by $T_{min}$ and $T_{max}$ the maximum time taken by the routines $\min_{\mathcal{F}_\pi}(z)$ and $\max_{\mathcal{G}_\pi}(z)$ on any point $z \in \mathcal{C}$.

**Exhaustive duality**$(\mathcal{C}, \pi, \mathcal{F}, \mathcal{G})$. Assuming $|\mathcal{F}||\mathcal{G}| \leq 1$, check if there are no other vectors in $\mathcal{C} \setminus (\mathcal{F}^+ \cup \mathcal{G}^-)$ as follows. First, if $|\mathcal{F}| = |\mathcal{G}| = 1$ then find an $i \in [n]$ such that $a_i > b_i$, where $\mathcal{F} = \{a\}$ and $\mathcal{G} = \{b\}$: (Such a coordinate is guaranteed to exist by (6))
1. If there is a $j \neq i$ such that $b_j < u_j$ then let $z = (u_1, \ldots, u_{i-1}, b_i, u_{i+1}, \ldots, u_n)$.
2. Otherwise, if there is a $j \neq i$ such that $a_j > l_j$ then let $z = (u_1, \ldots, u_{j-1}, a_j - 1, u_{j+1}, \ldots, u_n)$.
3. If $b_i < a_i - 1$ then let $z = (u_1, \ldots, u_{i-1}, a_i - 1, u_{i+1}, \ldots, u_n)$.

In cases 1, 2 and 3, return either $\min_{\mathcal{F}_\pi}(z)$ or $\max_{\mathcal{G}_\pi}(z)$ depending on whether $\pi(z) = 1$ or $\pi(z) = 0$, respectively.

4. Otherwise return $FALSE$ (meaning that $\mathcal{F}$ and $\mathcal{G}$ are dual in $\mathcal{C}$).

Second, if $|\mathcal{F}| = 0$ then check satisfiability of $u$. If $\pi(u) = 1$ then return $\min_{\mathcal{F}_\pi}(u)$. Else, if $\pi(u) = 0$ then let $z = \max_{\mathcal{G}_\pi}(u)$, and return either $FALSE$ or $z$ depending on whether $z \in \mathcal{G}^*$ or not (this check can be done in $O(n \log |\mathcal{G}^*|)$ using the search tree $\mathbf{T}(\mathcal{G}^*)$). Finally, if $|\mathcal{G}| = 0$ then check satisfiability of $l$. If $\pi(l) = 0$ then return $\max_{\mathcal{G}_\pi}(l)$. Else, if $\pi(l) = 1$ then let $z = \min_{\mathcal{F}_\pi}(l)$, and return either $FALSE$ or $z$ depending on whether $z \in \mathcal{F}^*$ or not. This step takes $O(\max\{n \log |\mathcal{F}^*| + T_{min}, n \log |\mathcal{G}^*| + T_{max}\})$ time.

**Random solution**$(\mathcal{C}, \pi, \mathcal{F}^*, \mathcal{G}^*)$. Repeat the following for $k = 1, \dots, t_1$ times, where $t_1$ is a constant (say 10): Find a random point $z^k \in \mathcal{C}$, by picking each coordinate $z_i^k$ randomly from $\{l_i, u_i\}$, $i = 1, \dots, n$. If $\pi(z^k) = 1$ then let $(z^k)^* \leftarrow \min_{\mathcal{F}_\pi}(z^k)$, and if $(z^k)^* \notin \mathcal{F}^*$ then return $(z^k)^* \in \mathcal{F}_\pi \setminus \mathcal{F}^*$. If $\pi(z^k) = 0$ then let $(z^k)^* \leftarrow \max_{\mathcal{G}_\pi}(z^k)$, and if $(z^k)^* \notin \mathcal{G}^*$ then return $(z^k)^* \in \mathcal{G}_\pi \setminus \mathcal{G}^*$. If $\{(z^1)^*, \dots, (z^{t_1})^*\} \subseteq \mathcal{F}^* \cup \mathcal{G}^*$ then return $FALSE$. This step takes $O(\max\{n \log |\mathcal{F}^*| + T_{min}, n \log |\mathcal{G}^*| + T_{max}\})$ time, and is used to check whether $\mathcal{F}^+ \cup \mathcal{G}^-$ covers a large portion of $\mathcal{C}$.

**Count estimation.** For a subset $\mathcal{X} \subseteq \mathcal{F}$ (or $\mathcal{X} \subseteq \mathcal{G}$), use sampling to estimate the number $\mathrm{Est}(\mathcal{X}, \mathcal{C})$ of elements of $\mathcal{X} \subseteq \mathcal{F}$ (or $\mathcal{X} \subseteq \mathcal{G}$) that are active with respect to the current box $\mathcal{C}$, i.e. the elements of the set $\mathcal{X}' \overset{\text{def}}{=} \{a \in \mathcal{X} \mid a^+ \cap \mathcal{C} \neq \emptyset\}$ ($\mathcal{X}' \overset{\text{def}}{=} \{b \in \mathcal{X} \mid b^- \cap \mathcal{C} \neq \emptyset\}$). This can be done as follows. For $t_2 = O(\log(|\mathcal{F}| + |\mathcal{G}|)/\epsilon)$, pick elements $x^1, \dots, x^{t_2} \in \mathcal{F}$ at random, and let the random variable $Y = \frac{|\mathcal{F}|}{t_2} * |\{x^i \in \mathcal{X}' : i = 1, \dots, t_2\}|$. Repeat this step independently for a total of $t_3 = O(\log(|\mathcal{F}| + |\mathcal{G}|))$ times to obtain $t_3$ estimates $Y^1, \dots, Y^{t_3}$, and let $\mathrm{Est}(\mathcal{X}, \mathcal{C}) = \min\{Y^1, \dots, Y^{t_3}\}$. This step requires $O(n \log^3 m)$ time.

**Cleanup**$(\mathcal{F}, \mathcal{C})$ (**Cleanup**$(\mathcal{G}, \mathcal{C})$). Set $\mathcal{F}' \leftarrow \{a \in \mathcal{F} \mid a^+ \cap \mathcal{C} \neq \emptyset\}$ (respectively, $\mathcal{G}' \leftarrow \{b \in \mathcal{G} \mid b^- \cap \mathcal{C} \neq \emptyset\}$), and return $\mathcal{F}'$ (respectively, $\mathcal{G}'$). This step takes $O(n|\mathcal{F}|)$ (respectively, $O(n|\mathcal{G}|)$).

Now, we describe the implementation of procedure $\mathrm{GEN}(\mathcal{C}, \pi, \mathcal{F}, \mathcal{G})$ which is called initially using $\mathcal{C} \leftarrow \mathcal{C}^*$, $\mathcal{F} \leftarrow \emptyset$ and $\mathcal{G} \leftarrow \emptyset$. At the return of this call, the families $\mathcal{F}^*$ and $\mathcal{G}^*$, which are initially empty, are extended respectively by the elements in $\mathcal{F}_\pi$ and $\mathcal{G}_\pi$. Below we assume that $f \in (0, 1)$ is a constant, say $1/2$. The families $\mathcal{F}^o$ and $\mathcal{G}^o$ represent respectively the subfamilies of $\mathcal{F}_\pi$ and $\mathcal{G}_\pi$ that are generated at each recursion tree node.

**Procedure GEN**$(\mathcal{C}, \pi, \mathcal{F}, \mathcal{G})$:
Input: A box $\mathcal{C} = \mathcal{C}_1 \times \cdots \times \mathcal{C}_n$, a monotone property $\pi$, and subsets $\mathcal{F} \subseteq \mathcal{F}_\pi$, and $\mathcal{G} \subseteq \mathcal{G}_\pi$.
Output: Subsets $\mathcal{F}^\circ \subseteq \mathcal{F}_\pi \setminus \mathcal{F}$ and $\mathcal{G}^\circ \subseteq \mathcal{G}_\pi \setminus \mathcal{G}$.

   1.     $\mathcal{F}^\circ \leftarrow \emptyset, \mathcal{G}^\circ \leftarrow \emptyset$.
   2.     While $|\mathcal{F}||\mathcal{G}| \leq 1$
   3.        $z \leftarrow$ Exhaustive duality$(\mathcal{C}, \pi, \mathcal{F}, \mathcal{G})$.
   4.        If $z = FALSE$ then return$(\mathcal{F}^\circ, \mathcal{G}^\circ)$.
   5.        else if $\pi(z) = 1$ then $\mathcal{F} \leftarrow \mathcal{F} \cup \{z\}, \mathcal{F}^\circ \leftarrow \mathcal{F}^\circ \cup \{z\}, \mathcal{F}^* \leftarrow \mathcal{F}^* \cup \{z\}$.
   6.        else $\mathcal{G} \leftarrow \mathcal{G} \cup \{z\}, \mathcal{G}^\circ \leftarrow \mathcal{G}^\circ \cup \{z\}, \mathcal{G}^* \leftarrow \mathcal{G}^* \cup \{z\}$.
   7.     end while
   8.     $z \leftarrow$ Random Solution$(\mathcal{C}, \pi, \mathcal{F}^*, \mathcal{G}^*)$.
   9.     While $(z \neq FALSE)$ do
  10.     if $\pi(z) = 1$ then $\mathcal{F} \leftarrow \mathcal{F} \cup \{z\}, \mathcal{F}^\circ \leftarrow \mathcal{F}^\circ \cup \{z\}, \mathcal{F}^* \leftarrow \mathcal{F}^* \cup \{z\}$.
  11.     else $\mathcal{G} \leftarrow \mathcal{G} \cup \{z\}, \mathcal{G}^\circ \leftarrow \mathcal{G}^\circ \cup \{z\}, \mathcal{G}^* \leftarrow \mathcal{G}^* \cup \{z\}$.
  12.     $z \leftarrow$ Random Solution$(\mathcal{C}, \pi, \mathcal{F}^*, \mathcal{G}^*)$.
  13.     end while
  14.     $x^* \leftarrow \operatorname{argmin}\{|\operatorname{Ess}(y)| \ : \ y \in (\mathcal{F} \cap \mathcal{C}^-) \cup (\mathcal{G} \cap \mathcal{C}^+)\}$.
  15.     If $x^* \in \mathcal{F}$ then
  16.     $i \leftarrow \operatorname{argmax}\{\operatorname{Est}(\{b \in \mathcal{G} \ : \ b_j < x_j^*\}, \mathcal{C}) \ : \ j \in \operatorname{Ess}(x^*)\}$.
  17.     $\mathcal{C}' = \mathcal{C}_1 \times \cdots \times \mathcal{C}_{i-1} \times [x_i^* : u_i] \times \mathcal{C}_{i+1} \times \cdots \times \mathcal{C}_n$.
  18.     If $\operatorname{Est}(\mathcal{G}, \mathcal{C}') \leq f * |\mathcal{G}|$ then
  19.        $\mathcal{G}' \leftarrow$ Cleanup$(\mathcal{G}, \mathcal{C}')$.
  20.     else
  21.        $\mathcal{G}' \leftarrow \mathcal{G}$.
  22.     $(\mathcal{F}_l, \mathcal{G}_l) \leftarrow$ GEN$(\mathcal{C}', \pi, \mathcal{F}, \mathcal{G}')$.
  23.     $\mathcal{F}^\circ \leftarrow \mathcal{F}^\circ \cup \mathcal{F}_l, \mathcal{F} \leftarrow \mathcal{F} \cup \mathcal{F}_l, \mathcal{F}^* \leftarrow \mathcal{F}^* \cup \mathcal{F}_l$.
  24.     $\mathcal{G}^\circ \leftarrow \mathcal{G}^\circ \cup \mathcal{G}_l, \mathcal{G} \leftarrow \mathcal{G} \cup \mathcal{G}_l, \mathcal{G}^* \leftarrow \mathcal{G}^* \cup \mathcal{G}_l$.
  25.     $\mathcal{C}'' = \mathcal{C}_1 \times \cdots \times \mathcal{C}_{i-1} \times [l_i : x_i^* - 1] \times \mathcal{C}_{i+1} \times \cdots \times \mathcal{C}_n$.
  26.     If $\operatorname{Est}(\mathcal{F}, \mathcal{C}'') \leq f * |\mathcal{F}|$ then
  27.        $\mathcal{F}'' \leftarrow$ Cleanup$(\mathcal{F}, \mathcal{C}'')$.
  28.     else
  29.        $\mathcal{F}'' \leftarrow \mathcal{F}$.
  30.     $(\mathcal{F}_r, \mathcal{G}_r) \leftarrow$ GEN$(\mathcal{C}'', \pi, \mathcal{F}'', \mathcal{G})$.
  31.     $\mathcal{F}^\circ \leftarrow \mathcal{F}^\circ \cup \mathcal{F}_r, \mathcal{F}^* \leftarrow \mathcal{F}^* \cup \mathcal{F}_r, \mathcal{G}^\circ \leftarrow \mathcal{G}^\circ \cup \mathcal{G}_r, \mathcal{G}^* \leftarrow \mathcal{G}^* \cup \mathcal{G}_r$.
  32.     else
33-48.     Symmetric versions for Steps 16-31 above (details omitted).
  49.     end if
  50.     Return $(\mathcal{F}^\circ, \mathcal{G}^\circ)$.

The following result, regarding the expected running time of the algorithm, is inherent from [7].

**Proposition 2.** *The expected number of recursive calls until a new element in $(\mathcal{F}_\pi \setminus \mathcal{F}^*) \cup (\mathcal{G}_\pi \setminus \mathcal{G}^*)$ is output, or procedure GEN$(\mathcal{C}, \pi, \mathcal{F}, \mathcal{G})$ terminates is $nm^{O(\log^2 m)}$.*

However, as we shall see from the experiments, the algorithm seems to practically behave much more efficiently than indicated by Proposition 2. In fact, in most of the experiments we performed, we got an almost everage linear delay (in $m$) for generating a new point in $(\mathcal{F}_\pi \setminus \mathcal{F}^*) \cup (\mathcal{G}_\pi \setminus \mathcal{G}^*)$.

## 5   Experimental Results

We performed a number of experiments to evaluate our implementation on random instances of the three monotone properties described in Section 2. The experiments were performed on a *Pentium 4* processor with 2.2 GHz of speed and 512M bytes of memory. For each monotone property $\pi$, we have limited the corresponding parameters defining the property to reasonable values such that the algorithm completes generation of the sets $\mathcal{F}_\pi$ and $\mathcal{G}_\pi$ in reasonable time. Using larger values of the parameters increases the output size, resulting in large total time, although the time per output remains almost constant. For each case, the experiments were performed 5 times, and the numbers shown in the tables below represent averages.

Tables 1 and 2 show our results for linear systems with $n$ variables and $r$ inequalities. Each element of the constraint matrix $A$ and the right-hand side vector $b$ is generated at random from 1 to 15. In the tables we show the output size, the total time taken to generate the output and the average time per each output vector. The parameter $c$ denotes the maximum value that a variable can take. The last row of the table gives the ratio of the size of $\mathcal{F}_{\pi_1}$ to the size of $\mathcal{G}_{\pi_1}$ for comparison with the worst case bound of (3). Note that this ratio is relatively close to 1, making joint generation an efficient method for generating both families $\mathcal{F}_{\pi_1}$ and $\mathcal{G}_{\pi_1}$.

**Table 1.** Performance of the algorithm for property $\pi_1$, where $r = 5$ and $c = 2$.

| $n$ | 10 | | 20 | | 30 | | 40 | | 50 | |
|---|---|---|---|---|---|---|---|---|---|---|
| | $\mathcal{F}_{\pi_1}$ | $\mathcal{G}_{\pi_1}$ | $\mathcal{F}_{\pi_1}$ | $\mathcal{G}_{\pi_1}$ | $\mathcal{F}_{\pi_1}$ | $\mathcal{G}_{\pi_1}$ | $\mathcal{F}_{\pi_1}$ | $\mathcal{G}_{\pi_1}$ | $\mathcal{F}_{\pi_1}$ | $\mathcal{G}_{\pi_1}$ |
| Output size (thousands) | 0.31 | 0.19 | 9.9 | 5.7 | 49.6 | 20.0 | 127.3 | 59.5 | 195.3 | 74.7 |
| Total Time (sec) | 4.7 | 4.7 | 297 | 297 | 1627 | 1625 | 5844 | 5753 | 10703 | 10700 |
| Time/output. (msec) | 13 | 24 | 27 | 62 | 29 | 78 | 40 | 103 | 50 | 133 |
| Ratio $|\mathcal{G}_{\pi_1}|/|\mathcal{F}_{\pi_1}|$ | 0.60 | | 0.57 | | 0.40 | | 0.47 | | 0.38 | |

**Table 2.** Performance of the algorithm for property $\pi_1$, where $n = 30$ and $c = 2$.

| $r$ | 5 | | 15 | | 25 | | 35 | | 45 | |
|---|---|---|---|---|---|---|---|---|---|---|
| | $\mathcal{F}_{\pi_1}$ | $\mathcal{G}_{\pi_1}$ | $\mathcal{F}_{\pi_1}$ | $\mathcal{G}_{\pi_1}$ | $\mathcal{F}_{\pi_1}$ | $\mathcal{G}_{\pi_1}$ | $\mathcal{F}_{\pi_1}$ | $\mathcal{G}_{\pi_1}$ | $\mathcal{F}_{\pi_1}$ | $\mathcal{G}_{\pi_1}$ |
| Output size (thousands) | 20.4 | 11.6 | 68.6 | 27.8 | 122.7 | 43.3 | 196.6 | 61.7 | 317.5 | 115.5 |
| Total Time (sec) | 408 | 408 | 2244 | 2242 | 6495 | 6482 | 15857 | 15856 | 30170 | 30156 |
| Time/output. (msec) | 20 | 50 | 32 | 90 | 50 | 158 | 76 | 258 | 75 | 260 |
| Ratio $|\mathcal{G}_{\pi_1}|/|\mathcal{F}_{\pi_1}|$ | 0.57 | | 0.41 | | 0.35 | | 0.31 | | 0.36 | |

Tables 3 and 4 show the results for minimal infrequent/maximal frequent sets. In the tables, $n$, $r$ and $t$ denote respectively the number of columns, the number of rows of the matrix, and the threshold. Each row of the matrix was generated uniformly at random. As seen from Table 3, for $t = 1, 2$, the bias between the numbers of maximal frequent sets and minimal infrequent sets, for the shown random examples, seem to be large. This makes joint generation an efficient method for generating minimal infrequent sets, but inefficient for generating maximal frequent sets for these examples. However, we observed that this bias in numbers decreases as the threshold $t$ becomes larger. Table 4 illustrates this on a number of examples in which larger values of the threshold were used.

**Table 3.** Performance of the algorithm for property $\pi_2$ for threshold $t = 1, 2$.

| $n, r, t$ | 20,100,1 | | 30,100,1 | | 40,100,1 | | 30,100,2 | | 30,300,2 | | 30,500,2 | |
|---|---|---|---|---|---|---|---|---|---|---|---|---|
| | $\mathcal{F}_{\pi_2}$ | $\mathcal{G}_{\pi_2}$ | $\mathcal{F}_{\pi_2}$ | $\mathcal{G}_{\pi_2}$ | $\mathcal{F}_{\pi_2}$ | $\mathcal{G}_{\pi_2}$ | $\mathcal{F}_{\pi_2}$ | $\mathcal{G}_{\pi_2}$ | $\mathcal{F}_{\pi_2}$ | $\mathcal{G}_{\pi_2}$ | $\mathcal{F}_{\pi_2}$ | $\mathcal{G}_{\pi_2}$ |
| Output size (thousands) | 2.9 | 0.08 | 51.7 | 0.1 | 337.1 | 0.1 | 75.4 | 2.5 | 386.7 | 13.5 | 718.3 | 27.1 |
| Total Time (sec) | 60 | 55 | 1520 | 769 | 22820 | 3413 | 1962 | 1942 | 13269 | 13214 | 28824 | 28737 |
| Time/output. (msec) | 20 | 690 | 30 | 7742 | 68 | 34184 | 28 | 770 | 33 | 979 | 40 | 1062 |
| Ratio $|\mathcal{G}_{\pi_2}|/|\mathcal{F}_{\pi_2}|$ | 0.0280 | | 0.0019 | | 0.0002 | | 0.0335 | | 0.0350 | | 0.0377 | |

**Table 4.** Performance of the algorithm for property $\pi_2$ for large threshold values.

| $n, r, t$ | 30,300,3 | | 30,300,5 | | 30,300,7 | | 30,300,9 | | 30,1000,20 | | 30,1000,25 | |
|---|---|---|---|---|---|---|---|---|---|---|---|---|
| | $\mathcal{F}_{\pi_2}$ | $\mathcal{G}_{\pi_2}$ | $\mathcal{F}_{\pi_2}$ | $\mathcal{G}_{\pi_2}$ | $\mathcal{F}_{\pi_2}$ | $\mathcal{G}_{\pi_2}$ | $\mathcal{F}_{\pi_2}$ | $\mathcal{G}_{\pi_2}$ | $\mathcal{F}_{\pi_2}$ | $\mathcal{G}_{\pi_2}$ | $\mathcal{F}_{\pi_2}$ | $\mathcal{G}_{\pi_2}$ |
| Output size (thousands) | 403.3 | 73.6 | 362.6 | 134.7 | 269.0 | 100.1 | 199.1 | 74.0 | 491.3 | 145.7 | 398.1 | 114.5 |
| Total Time (sec) | 7534 | 7523 | 6511 | 6508 | 4349 | 4346 | 3031 | 3029 | 13895 | 13890 | 9896 | 9890 |
| Time/output. (msec) | 19 | 102 | 18 | 48 | 17 | 43 | 15 | 41 | 28 | 95 | 25 | 86 |
| Ratio $|\mathcal{G}_{\pi_2}|/|\mathcal{F}_{\pi_2}|$ | 0.1826 | | 0.3715 | | 0.3719 | | 0.3716 | | 0.2965 | | 0.2877 | |

Figures 1 and 2 show how the output rate changes for minimal feasible/maximal infeasible solutions of linear systems and for minimal infrequent/maximal frequent sets, respectively. For minimal feasible solutions, we can see that the output rate changes almost linearly as the number of outputs increases. This is not the case for the maximal infeasible solutions, where the algorithm efficiecy decreases (the generation problem for maximal infeasible solutions is NP-hard). For minimal infrequent and maximal frequent sets, Figure 2 shows that the output rate increases very slowly. This illustrates somehow that the algorithm practically behaves much better than the quasi-polynomial bound stated in Proposition 2.

Table 5 shows the results for maximal sparse/minimal non-sparse boxes with dimension $n$, for a set of $r$ random points, threshold $t$, and upper bound $c$ on the coordinate of each point. As in the case of frequent sets, the bias between the numbers $\mathcal{F}_{\pi_3}$ and $\mathcal{G}_{\pi_3}$ is large for $t = 0$ but seems to decrease with larger values of the threshold. In fact, the table shows two examples in which the number of minimal non-sparse boxes is larger than the number of maximal sparse boxes.

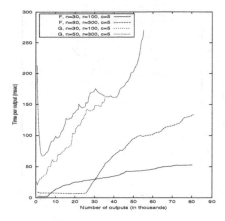

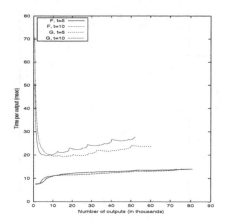

**Fig. 1.** Average time per output, as a function of the number of outputs for minimal feasible/maximal infeasible solutions of linear systems, with $c = 5$ and $(n, r) = (30, 100), (50, 300)$.

**Fig. 2.** Average time per output, as a function of the number of outputs for minimal infrequent/maximal frequent sets, with $n = 30$, $r = 1000$ and $t = 5, 10$.

We are not aware of any implementation of an algorithm for generating maximal sparse boxes except for [13] which presents some experiments for $n = 2$ and $t = 0$. Experiments in [13] indicated that the algorithm suggested there is almost linear in the the number of points $r$. Figure 3 illustrates a similar behaviour exhibited by our algorithm. In the figure, we show the total time required to generate all the 2-dimensional maximal empty boxes, as the number of points is increased from 10,000 to 60,000, for two different values of the upper bound $c$.

**Table 5.** Performance of the algorithm for property $\pi_3$ with $n = 7$ and upper bound $c = 5$.

| $r, t$ | 100,0 | | 300,0 | | 500,0 | | 300,2 | | 300,6 | | 300,10 | |
|---|---|---|---|---|---|---|---|---|---|---|---|---|
| | $\mathcal{F}_{\pi_3}$ | $\mathcal{G}_{\pi_3}$ | $\mathcal{F}_{\pi_3}$ | $\mathcal{G}_{\pi_3}$ | $\mathcal{F}_{\pi_3}$ | $\mathcal{G}_{\pi_3}$ | $\mathcal{F}_{\pi_3}$ | $\mathcal{G}_{\pi_3}$ | $\mathcal{F}_{\pi_3}$ | $\mathcal{G}_{\pi_3}$ | $\mathcal{F}_{\pi_3}$ | $\mathcal{G}_{\pi_3}$ |
| Output size (thousands) | 16.1 | 0.1 | 49.1 | 0.3 | 72.9 | 0.5 | 228.5 | 88.7 | 373.4 | 466.3 | 330.4 | 456.5 |
| Total Time (sec) | 932 | 623 | 2658 | 1456 | 3924 | 2933 | 8731 | 8724 | 17408 | 17404 | 16156 | 16156 |
| Time/output. (msec) | 29 | 6237 | 27 | 4866 | 27 | 5889 | 19 | 98 | 23 | 37 | 24 | 35 |
| Ratio $|\mathcal{G}_{\pi_3}|/|\mathcal{F}_{\pi_3}|$ | 0.0062 | | 0.0061 | | 0.0068 | | 0.3881 | | 1.2488 | | 1.3818 | |

As mentioned in Section 4, it is possible in general to implement the procedures $\min_{\mathcal{F}}(z)$ and $\max_{\mathcal{G}}(z)$ using the coordinate decent method, but more efficient implementations can be obtained if we specialize these procedures to the monotone property under consideration. Figure 4 compares the two different implementations for the property $\pi_3$. Clearly, the gain in performance increases as the upper bound $c$ increases.

Let us finally point out that we have observed that the algorithm tends to run more efficiently when the sets $\mathcal{F}_\pi$ and $\mathcal{G}_\pi$ become closer in size. This observation

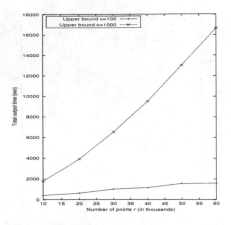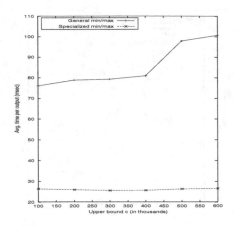

**Fig. 3.** Total generation time as a function of the number of points $r$ for maximal boxes with $n = 2$, $t = 0$, and $c = 100, 1000$.

**Fig. 4.** Comparing general versus specialized minimization for property $\pi_3$. Each plot shows the average CPU time/maximal empty box generated versus the upper bound $c$, for $n = 5$ and $r = 500$.

is illustrated in Figure 5 which plots the average time per output (i.e. total time to output all the elements of $\mathcal{F}_\pi \cup \mathcal{G}_\pi$ divided by $|\mathcal{F}_\pi \cup \mathcal{G}_\pi|$) versus the ratio $|\mathcal{G}_\pi|/|\mathcal{F}_\pi|$. This indicates that, when the elements of the sets $\mathcal{F}_\pi$ and $\mathcal{G}_\pi$ are more uniformly distributed along the space, it becomes easier for the joint generation algorithm to find a new vector not in the already generated sets $\mathcal{F}^* \subseteq \mathcal{F}_\pi$ and $\mathcal{G}^* \subseteq \mathcal{G}_\pi$.

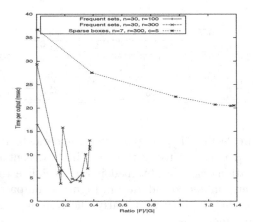

**Fig. 5.** Average generation time as a function of the ratio $|\mathcal{G}_\pi|/|\mathcal{F}_\pi|$, for properties $\pi_2$ and $\pi_3$.

# 6   Conclusion

We have presented an efficient implementation for a quasi-polynomial algorithm for jointly generating the families $\mathcal{F}_\pi$ and $\mathcal{G}_\pi$ of minimal satisfying and maximal non-satisfying vectors for a given monotone property $\pi$. We provided experimental evaluation of the algorithm on three different monotone properties. Our experiments indicate that the algorithm behaves much more efficiently than its worst-case time complexity indicates. The algorithm seems to run faster on instances where the families $\mathcal{F}_\pi$ and $\mathcal{G}_\pi$ are not very biased in size. Finally, our experiments also indicate that such non-bias in size is not a rare situation (for random instances), despite the fact that inequalities of the form (3)-(5) may hold in general.

# References

1. M. Anthony and N. Biggs, *Computational Learning Theory*, Cambridge Univ. Press, 1992.
2. R. Agrawal, T. Imielinski and A. Swami, Mining associations between sets of items in massive databases, *Proc. 1993 ACM-SIGMOD Int. Conf.*, pp. 207-216.
3. R. Agrawal, H. Mannila, R. Srikant, H. Toivonen and A. I. Verkamo, Fast discovery of association rules, in *Advances in Knowledge Discovery and Data Mining* (U. M. Fayyad, G. Piatetsky-Shapiro, P. Smyth and R. Uthurusamy, eds.), pp. 307-328, AAAI Press, Menlo Park, California, 1996.
4. E. Boros, K. Elbassioni, V. Gurvich, L. Khachiyan and K.Makino, Dual-bounded generating problems: All minimal integer solutions for a monotone system of linear inequalities, *SIAM Journal on Computing*, **31** (5) (2002) pp. 1624-1643.
5. E. Boros, K. Elbassioni, V. Gurvich and L. Khachiyan, Generating Dual-Bounded Hypergraphs, *Optimization Methods and Software*, (OMS) **17** (5), Part I (2002), pp. 749–781.
6. E. Boros, K. Elbassioni, V. Gurvich, L. Khachiyan and K. Makino, An Intersection Inequality for Discrete Distributions and Related Generation Problems, in *Automata, Languages and Programming, 30-th International Colloquium, ICALP 2003, Lecture Notes in Computer Science (LNCS)* 2719 (2003) pp. 543–555.
7. E. Boros, K. Elbassioni, V. Gurvich and L. Khachiyan, An Efficient Implementation of a Quasi-Polynomial Algorithm for Generating Hypergraph Transversals, in *the Proceedings of the 11th Annual European Symposium on Algorithms (ESA 2003)*, LNCS 2832, pp. 556–567, Budapest, Hungary, September, 2003.
8. E. Boros, K. Elbassioni, V. Gurvich and L. Khachiyan, On enumerating minimal dicuts and strongly connected subgraphs, to appear in *the 10th Conference on Integer Programming and Combinatorial Optimization (IPCO X)*, DIMACS Technical Report 2003-35, Rutgers University, http://dimacs.rutgers.edu/TechnicalReports/2003.html.
9. E. Boros, V. Gurvich, L. Khachiyan and K. Makino, On the complexity of generating maximal frequent and minimal infrequent sets, in *19th Int. Symp. on Theoretical Aspects of Computer Science, (STACS)*, March 2002, LNCS 2285, pp. 133–141.
10. J. C. Bioch and T. Ibaraki (1995). Complexity of identification and dualization of positive Boolean functions. *Information and Computation*, **123**, pp. 50–63.
11. B. Chazelle, R. L. (Scot) Drysdale III and D. T. Lee, Computing the largest empty rectangle, *SIAM Journal on Computing*, 15(1) (1986) 550-555.

12. C. J. Colbourn, *The combinatorics of network reliability*, Oxford Univ. Press, 1987.
13. J. Edmonds, J. Gryz, D. Liang and R. J. Miller, Mining for empty rectangles in large data sets, in *Proc. 8th Int. Conf. on Database Theory (ICDT)*, Jan. 2001, *Lecture Notes in Computer Science* 1973, pp. 174–188.
14. T. Eiter and G. Gottlob, Identifying the minimal transversals of a hypergraph and related problems. *SIAM Journal on Computing*, 24 (1995) pp. 1278-1304.
15. M. L. Fredman and L. Khachiyan, On the complexity of dualization of monotone disjunctive normal forms, *Journal of Algorithms*, 21 (1996) pp. 618–628.
16. V. Gurvich and L. Khachiyan, On generating the irredundant conjunctive and disjunctive normal forms of monotone Boolean functions, *Discrete Applied Mathematics*, 96-97 (1999) pp. 363-373.
17. D. Gunopulos, R. Khardon, H. Mannila and H. Toivonen, Data mining, hypergraph transversals and machine learning, in *Proc. 16th ACM-PODS Conf.*, (1997) pp. 209–216.
18. V. Gurvich, To theory of multistep games, *USSR Comput. Math. and Math Phys.* **13-6** (1973), pp. 1485–1500.
19. V. Gurvich, Nash-solvability of games in pure strategies, *USSR Comput. Math. and Math. Phys.*, **15** (1975), pp. 357–371.
20. E. Lawler, J. K. Lenstra and A. H. G. Rinnooy Kan, Generating all maximal independent sets: NP-hardness and polynomial-time algorithms, *SIAM Journal on Computing*, 9 (1980), pp. 558–565.
21. B. Liu, L.-P. Ku and W. Hsu, Discovering interesting holes in data, in *Proc. IJCAI*, pp. 930–935, Nagoya, Japan, 1997.
22. B. Liu, K. Wang, L.-F. Mun and X.-Z. Qi, Using decision tree induction for discovering holes in data, in *Proc. 5th Pacific Rim Int. Conf. on Artificial Intelligence*, pp. 182-193, 1998.
23. M. Orlowski, A new algorithm for the large empty rectangle problem, *Algorithmica* 5(1) (1990) 65-73.

# Lempel, Even, and Cederbaum Planarity Method

John M. Boyer[1], Cristina G. Fernandes[2*], Alexandre Noma[2**], and
José C. de Pina[2*]

[1] PureEdge Solutions Inc., 4396 West Saanich Rd., Victoria, BC V8Z 3E9, Canada.
jboyer@acm.org
[2] University of São Paulo, Brazil.
{cris,noma,coelho}@ime.usp.br

**Abstract.** We present a simple pedagogical graph theoretical description of Lempel, Even, and Cederbaum (LEC) planarity method based on
concepts due to Thomas. A linear-time implementation of LEC method
using the PC-tree data structure of Shih and Hsu is provided and described in details. We report on an experimental study involving this
implementation and other available linear-time implementations of planarity algorithms.

## 1 Introduction

The first linear-time planarity testing algorithm is due to Hopcroft and Tarjan [10]. Their algorithm is an ingenious implementation of the method of Auslander and Parter [1] and Goldstein [9]. Some notes to the algorithm were made
by Deo [7], and significant additional details were presented by Williamson [20,
21] and Reingold, Nievergelt, and Deo [16].

The second method of planarity testing proven to achieve linear time is due
to Lempel, Even, and Cederbaum (LEC) [13]. This method was optimized to
linear time thanks to the *st*-numbering algorithm of Even and Tarjan [8] and
the PQ-tree data structure of Booth and Lueker (BL) [2]. Chiba, Nishizeki, Abe,
and Ozawa [6] augmented the PQ-tree operations so that a planar embedding is
also computed in linear time.

All these algorithms are widely regarded as being quite complex [6,12,18].
Recent research efforts have resulted in simpler linear-time algorithms proposed
by Shih and Hsu (SH) [11,18,17] and by Boyer and Myrvold (BM) [4,5]. These
algorithms implement LEC method and present similar and very interesting
ideas. Each algorithm uses its own data structure to efficiently maintain relevant
information on the (planar) already examined portion of the graph.

The description of SH algorithm made by Thomas [19] provided us with the
key concepts to give a simple graph theoretical description of LEC method. This

---

* Research partially supported by PRONEX/CNPq 664107/1997-4.
** Supported by FAPESP 00/03969-2.

C.C. Ribeiro and S.L. Martins (Eds.): WEA 2004, LNCS 3059, pp. 129–144, 2004.
© Springer-Verlag Berlin Heidelberg 2004

description increases the understanding of BL, SH, and BM algorithms, all based on LEC method.

Section 2 contains definitions of the key ingredients used by LEC method. In Section 3, an auxiliary algorithm is considered. LEC method is presented in Section 4 and an implementation of SH algorithm is described in Section 5. This implementation is available at http://www.ime.usp.br/~coelho/sh/ and, as far as we know, is the unique available implementation of SH algorithm, even though the algorithm was proposed about 10 years ago. Finally, Section 6 reports on an experimental study.

## 2    Frames, $XY$-Paths, $XY$-Obstructions and Planarity

This section contains the definitions of some concepts introduced by Thomas [19] in his presentation of SH algorithm. We use these concepts in the coming sections to present both LEC method and our implementation of SH algorithm.

Let $H$ be a planar graph. A subgraph $F$ of $H$ is a *frame of $H$* if $F$ is induced by the edges incident to the external face of a planar embedding of $H$ (Figs. 1(a) and 1(b)).

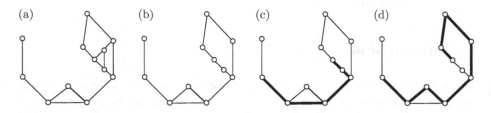

**Fig. 1.** (a) A graph $H$. (b) A frame of $H$. (c) A path $P$ in a frame. (d) The complement of $P$.

If $G$ is a connected graph, $H$ is a planar induced subgraph of $G$ and $F$ is a frame of $H$, then we say that $F$ is a *frame of $H$ in $G$* if it contains all vertices of $H$ that have a neighbor in $V_G \setminus V_H$. Neither every planar induced subgraph of a graph $G$ has a frame in $G$ (Fig. 2(a)) nor every induced subgraph of a planar graph $G$ has a frame in $G$ (Fig. 2(b)). The connection between frames and planarity is given by the following lemma.

**Lemma 1 (Thomas [19]).** *If $H$ is an induced subgraph of a planar graph $G$ such that $G - V_H$ is connected, then $H$ has a frame in $G$.* ∎

Let $F$ be a frame of $H$ and $P$ be a path in $F$. The *basis of $P$* is the subgraph of $F$ formed by all blocks of $F$ which contain at least one edge of $P$. Let $C_1, C_2, \ldots, C_k$ be the blocks in the basis of $P$. For $i = 1, 2, \ldots, k$, let $P_i := P \cap C_i$ and, if $C_i$ is a cycle, let $\bar{P}_i := C_i \setminus P_i$, otherwise let $\bar{P}_i := P_i$. The *complement of $P$ in $F$* is the path $\bar{P}_1 \cup \bar{P}_2 \cup \ldots \cup \bar{P}_k$, which is denoted by $\bar{P}$. If $E_P = \emptyset$ then $\bar{P} := P$ (Figs. 1(c) and 1(d)).

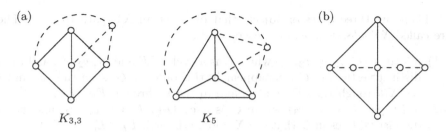

**Fig. 2.** (a) Subgraphs of $K_{3,3}$ and $K_5$ induced by the solid edges have no frames. (b) Subgraph induced by the solid edges has no frame in the graph.

Let $W$ be a set of vertices in $H$ and $Z$ be a set of edges in $H$. A vertex $v$ in $H$ *sees* $W$ *through* $Z$ if there is a path in $H$ from $v$ to a vertex in $W$ with all edges in $Z$. Let $X$ and $Y$ be sets of vertices of a frame $F$ of $H$. A path $P$ in $F$ with basis $S$ is an $XY$-*path* (Fig. 3) if

(p1)  the endpoints of $P$ are in $X$;
(p2)  each vertex of $S$ that sees $X$ through $E_F \setminus E_S$ is in $P$;
(p3)  each vertex of $S$ that sees $Y$ through $E_F \setminus E_S$ is in $\bar{P}$;
(p4)  no component of $F - V_S$ contains vertices both in $X$ and in $Y$.

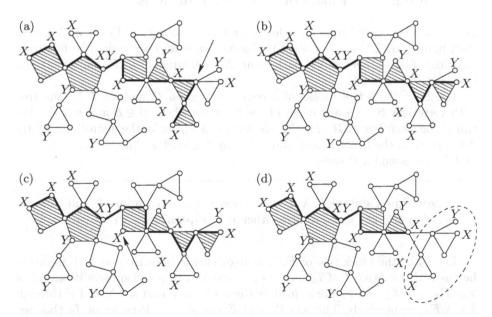

**Fig. 3.** In (a), (b), (c), and (d), let $P$ denote the thick path; its basis is shadowed. (a) $P$ is not an $XY$-path since it violates (p3). (b) $P$ is an $XY$-path. (c) $P$ is not an $XY$-path since it violates (p2). (d) $P$ is not an $XY$-path since it violates (p4).

There are three types of objects that obstruct an $XY$-path to exist. They are called *XY-obstructions* and are defined as

(o1) a 5-tuple $(C, v_1, v_2, v_3, v_4)$ where $C$ is a cycle of $F$ and $v_1$, $v_2$, $v_3$, and $v_4$ are distinct vertices in $C$ that appear in this order in $C$, such that $v_1$ and $v_3$ see $X$ through $E_F \setminus E_C$ and $v_2$ and $v_4$ see $Y$ through $E_F \setminus E_C$;

(o2) a 4-tuple $(C, v_1, v_2, v_3)$ where $C$ is a cycle of $F$ and $v_1$, $v_2$, and $v_3$ are distinct vertices in $C$ that see $X$ and $Y$ through $E_F \setminus E_C$;

(o3) a 4-tuple $(v, K_1, K_2, K_3)$ where $v \in V_F$ and $K_1$, $K_2$, and $K_3$ are distinct components of $F - v$ such that $K_i$ contains vertices in $X$ and in $Y$.

The existence of an $XY$-obstruction is related to non-planarity as follows.

**Lemma 2 (Thomas [19]).** *Let $H$ be a planar connected subgraph of a graph $G$ and $w$ be a vertex in $V_G \setminus V_H$ such that $G - V_H$ and $G - (V_H \cup \{w\})$ is connected. Let $F$ be a frame of $H$ in $G$, let $X$ be the set of neighbors of $w$ in $V_F$ and let $Y$ be the set of neighbors of $V_G \setminus (V_H \cup \{w\})$ in $V_F$. If $F$ has an $XY$-obstruction then $G$ has a subdivision of $K_{3,3}$ or $K_5$.*

**Sketch of the proof:** An $XY$-obstruction of type (o1) or (o3) indicates a $K_{3,3}$-subdivision. An $XY$-obstruction of type (o2) indicates either a $K_5$-subdivision or a $K_{3,3}$-subdivision (Fig. 4). ∎

## 3   Finding $XY$-Paths or $XY$-Obstructions

Let $F$ be a connected frame and let $X$ and $Y$ be subsets of $V_F$. If $F$ is a tree, then finding an $XY$-path or an $XY$-obstruction is an easy task. The following algorithm finds either an $XY$-path or an $XY$-obstruction in $F$ manipulating a tree that represents $F$.

Let $\mathcal{B}$ be the set of blocks of a connected graph $H$ and let $T$ be the tree with vertex set $\mathcal{B} \cup V_H$ and edges of the form $Bv$ where $B \in \mathcal{B}$ and $v \in V_B$. We call $T$ the *block tree* of $H$ (Fig. 5) (the leaves in $V_H$ make the definition slightly different than the usual). Each node of $T$ in $\mathcal{B}$ is said a *C-node* and each node of $T$ in $V_H$ is said a *P-node*.

---

**Algorithm Central$(F, X, Y)$.** Receives a connected frame $F$ and subsets $X$ and $Y$ of $V_F$ and returns either an $XY$-path or an $XY$-obstruction in $F$.

Let $T_0$ be the block tree of $F$. The algorithm is iterative and each iteration begins with a subtree $T$ of $T_0$, subsets $X_T$ and $Y_T$ of $V_T$ and subsets $W$ and $Z$ of $V_F$. The sets $X_T$ and $Y_T$ are formed by the nodes of $T$ that see $X$ and $Y$ through $E_{T_0} \setminus E_T$, respectively. The sets $W$ and $Z$ contain the P-nodes of $T_0$ that see $X$ and $Y$ through $E_{T_0} \setminus E_T$, respectively. At the beginning of the first iteration, $T = T_0$, $X_T = X$, $Y_T = Y$, $W = X$, and $Z = Y$. Each iteration consists of the following:

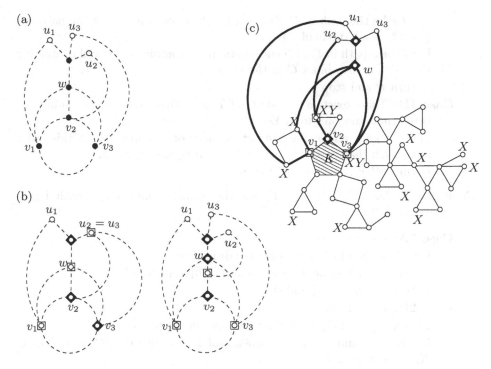

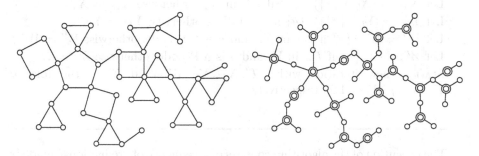

**Fig. 4.** Some situations with an $XY$-obstruction $(C, v_1, v_2, v_3)$ of type (o2). The dashed lines indicate paths. In $u_iv_i$-path, $u_i$ is the only vertex in $(V_H \setminus (V_H \cup \{w\}))$, for $i = 1, 2, 3$. (a) Subdivision of $K_5$ coming from an $XY$-obstruction. (b) Subdivisions of $K_{3,3}$ coming from an $XY$-obstruction. (c) Concrete example of an $XY$-obstruction leading to a $K_{3,3}$-subdivision.

**Fig. 5.** A graph and its block tree.

**Case 1:** Each leaf of $T$ is in $X_T \cap Y_T$ and $T$ is a path.
  Let $R$ be the set of P-nodes of $T$.
  For each C-node $C$ of $T$, let $X_C := V_C \cap (W \cup R)$ and $Y_C := V_C \cap (Z \cup R)$.

  **Case 1A:** Each C-node $C$ of $T$ has a path $P_C$ containing $X_C$ and internally disjoint from $Y_C$.

Let $P_T$ be the path in $F$ obtained by the concatenation of the paths in $\{P_C : C \text{ is a C-node of } T\}$.

Let $P$ be a path in $F$ with endpoints in $X$, containing $P_T$ and containing $V_C \cap W$ for each block $C$ in the basis of $P$.

Return $P$ and stop.

**Case 1B:** There exists a C-node $C$ of $T$ such that no path containing $X_C$ is internally disjoint from $Y_C$.

Let $v_1, v_2, v_3$, and $v_4$ be distinct vertices of $C$ appearing in this order in $C$, such that $v_1$ and $v_3$ are in $X_C$ and $v_2$ and $v_4$ are in $Y_C$.

Return $(C, v_1, v_2, v_3, v_4)$ and stop.

**Case 2:** Each leaf of $T$ is in $X_T \cap Y_T$ and there exists a node $v$ of $T$ with degree greater than 2.

**Case 2A:** $v$ is a C-node.

Let $C$ be the block of $F$ corresponding to $v$.

Let $v_1, v_2$, and $v_3$ be distinct P-nodes adjacent to $v$ in $T$.

Return $(C, v_1, v_2, v_3)$ and stop.

**Case 2B:** $v$ is a P-node.

Let $C_1, C_2$, and $C_3$ be distinct C-nodes adjacent to $v$ in $T$.

Let $K_1, K_2$, and $K_3$ be components of $F - v$ such that $C_i$ is a block of $K_i + v$ $(i = 1, 2, 3)$.

Return $(v, K_1, K_2, K_3)$ and stop.

**Case 3:** There exists a leaf $f$ of $T$ not in $X_T \cap Y_T$.

Let $u$ be the node of $T$ adjacent to $f$.

Let $T' := T - f$.

Let $X_{T'} := (X_T \setminus \{f\}) \cup \{u\}$ if $f$ is in $X_T$; otherwise $X_{T'} := X_T$.

Let $Y_{T'} := (Y_T \setminus \{f\}) \cup \{u\}$ if $f$ is in $Y_T$; otherwise $Y_{T'} := Y_T$.

Let $W' := W \cup \{u\}$ if $f$ is in $X_T$ and $u$ is a P-node; otherwise $W' := W$.

Let $Z' := Z \cup \{u\}$ if $f$ is in $Y_T$ and $u$ is a P-node; otherwise $Z' := Z$.

Start a new iteration with $T', X_{T'}, Y_{T'}, W'$, and $Z'$ in the roles of $T, X_T, Y_T, W$, and $Z$, respectively.

---

The execution of the algorithm consists of a sequence of "reductions" made by Case 3 followed by an occurrence of either Case 1 or Case 2. At the beginning of the last iteration, the leaves of $T$ are called *terminals*. The concept of a terminal node is used in a fundamental way by SH algorithm. The following theorem follows from the correctness of the algorithm.

**Theorem 1 (Thomas [19]).** *If $F$ is a frame of a connected graph and $X$ and $Y$ are subsets of $V_F$, then either there exists an $XY$-path or an $XY$-obstruction in $F$.* ∎

## 4   LEC Planarity Testing Method

One of the ingredients of LEC method is a certain ordering $v_1, v_2, \ldots, v_n$ of the vertices of the given graph $G$ such that, for $i = 1, \ldots, n$, the induced subgraphs $G[\{v_1, \ldots, v_i\}]$ and $G[\{v_{i+1}, \ldots, v_n\}]$ are connected. Equivalently, $G$ is connected and, for $i = 2, \ldots, n - 1$, vertex $v_i$ is adjacent to $v_j$ and $v_k$ for some $j$ and $k$ such that $1 \leq j < i < k \leq n$. A numbering of the vertices according to such an ordering is called a *LEC-numbering* of $G$. If the ordering is such that $v_1 v_n$ is an edge of the graph, the numbering is called an *st-numbering* [13]. One can show that every biconnected graph has a LEC-numbering.

LEC method examines the vertices of a given biconnected graph, one by one, according to a LEC-numbering. In each iteration, the method tries to extend a frame of the subgraph induced by the already examined vertices. If this is not possible, the method declares the graph is non-planar and stops.

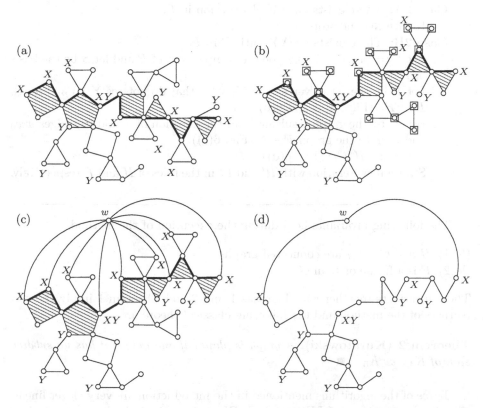

**Fig. 6.** (a) A frame $F$ and an $XY$-path $P$ in thick edges. (b) $F$ after moving the elements of $X$ to one side and the elements of $Y$ to the other side of $P$. Squares mark vertices in $V_F \setminus V_{\bar{P}}$ that do not see $Y \setminus V_{\bar{P}}$ through $E_F \setminus E_S$, where $\bar{P}$ is the complement and $S$ is the basis of $P$. (c) $F$ together with the edges with one endpoint in $F$ and the other in $w$. (d) A frame of $H + w$.

**Method LEC**($G$). Receives a biconnected graph $G$ and returns YES if $G$ is planar, and NO otherwise.

Number the vertices of $G$ according to a LEC-numbering. Each iteration starts with an induced subgraph $H$ of $G$ and a frame $F$ of $H$ in $G$. At the beginning of the first iteration, $H$ and $F$ are empty. Each iteration consists of the following:

**Case 1:** $H = G$.
  Return YES e stop.
**Case 2:** $H \neq G$.
  Let $w$ be the smallest numbered vertex in $G - V_H$.
  Let $X := \{u \in V_F : uw \in E_G\}$.
  Let $Y := \{u \in V_F : \text{ there exists } v \in V_G \setminus (V_H \cup \{w\}) \text{ such that } uv \in E_G\}$.

  **Case 2A:** There exists an $XY$-obstruction in $F$.
    Return NO and stop.
  **Case 2B:** There exists an $XY$-path $P$ in $F$.
    Let $\bar{P} := \langle w_0, w_1, \ldots, w_k \rangle$ be the complement of $P$ and let $S$ be the basis of $P$.
    Let $R$ be the set of vertices in $V_F \setminus V_{\bar{P}}$ that do not see $Y \setminus V_{\bar{P}}$ through $E_F \setminus E_S$ (Figs. 6(a) and 6(b)).
    Let $F'$ be the graph resulting from the addition of $w$ and the edges $ww_0$ and $ww_k$ to the graph $F - R$ (Fig. 6(c)).
    Let $H' := H + w$ (Fig. 6(d)).
    Start a new iteration with $H'$ and $F'$ in the roles of $H$ and $F$ respectively.

The following invariants hold during the execution of the method.

(lec1)  $H$ and $G - V_H$ are connected graphs;
(lec2)  $F$ is a frame of $H$ in $G$.

These invariants together with Lemmas 1 and 2 and Theorem 1 imply the correctness of the method and the following classical theorem.

**Theorem 2 (Kuratowski).** *A graph is planar if and only if it has no subdivision of $K_{3,3}$ or $K_5$.* ∎

Three of the algorithms mentioned in the introduction are very clever linear-time implementations of LEC method. BL use an $st$-numbering instead of an arbitrary LEC-numbering of the vertices and use a PQ-tree to store $F$. SH use a DFS-numbering and a PC-tree to store $F$. BM also use a DFS-numbering and use still another data structure to store $F$. One can use the previous description easily to design a quadratic implementation of LEC method.

# 5   Implementation of SH Algorithm

SH algorithm, as all other linear-time planarity algorithms, is quite complex to implement. The goal of this section is to share our experience in implementing it.

Let $G$ be a connected graph. A *DFS-numbering* is a numbering of the vertices of $G$ obtained from searching a DFS-tree of $G$ in post-order. SH algorithm uses a DFS-numbering instead of a LEC-numbering. If the vertices of $G$ are ordered according to a DFS-numbering, then the graph $G[\{i+1, \ldots, n\}]$ is connected, for $i = 1, \ldots, n$. As a DFS-numbering does not guarantee that $H := G[\{1, \ldots, i-1\}]$ is connected, if there exists a frame $F$ of $H$ and $H$ is not connected, then $F$ is also not connected. Besides, to compute (if it exists) a frame of $H + i$, it is necessary to compute an $XY$-path for each component of $F$ that contains a neighbor of $i$.

Let $v$ be a vertex of $F$ and $C$ be a block of $F$ containing $v$ and, if possible, a higher numbered vertex. We say $v$ is *active* if $v$ sees $X \cup Y$ through $E_F \setminus E_C$.

**PC-Tree**

The data structure proposed by SH to store $F$ is called a *PC-tree* and is here denoted by $T$. Conceptually, a PC-tree is an arborescence representing the relevant information of the block forest of $F$. It consists of *P-nodes* and *C-nodes*. There is a P-node for each active vertex of $F$ and a C-node for each cycle of $F$. We refer to a P-node by the corresponding vertex of $F$. There is an arc from a P-node $u$ to a P-node $v$ in $T$ if and only if $uv$ is a block of $F$. Each C-node $c$ has a circular list, denoted $RBC(c)$, with all P-nodes in its corresponding cycle of $F$, in the order they appear in this cycle. This list starts by the largest numbered P-node in it, which is called its *head*. The head of the list has a pointer to $c$. Each P-node appears in at most one $RBC$ in a non-head cell. It might appear in the head cell of several $RBC$s. Each P-node $v$ has a pointer $nonhead\_RBC\_cell(v)$ to the non-head cell in which it appears in an $RBC$. This pointer is NULL if there is no such cell. The name $RBC$ extends for *representative bounding cycle* (Figs. 7(a)-(c)).

Let $T'$ be the rooted forest whose node set coincides with the node set of $T$ and the arc set is defined as follows. Every arc of $T$ is an arc of $T'$. Besides these arcs, there are some *virtual* arcs: for every C-node $c$, there is an arc in $T'$ from $c$ to the P-node which is the head of $RBC(c)$ and there is an arc to $c$ from all the other P-nodes in $RBC(c)$ (Fig. 7(d)). In the exposition ahead, we use on nodes of $T$ concepts such as *parent, child, leaf, ancestral, descendant* and so on. By these, we mean their counterparts in $T'$.

Forest $T'$ is not really kept by the implementation. However, during each iteration, some of the virtual arcs are determined and temporarily stored to avoid traversing parts of the PC-tree more than once. So, each non-head cell in an $RBC$ and each C-node has a pointer to keep its virtual arc, when it is determined. The pointer is NULL while the virtual arc is not known.

**Values $h(u)$ and $b(v)$**

For each vertex $u$ of $G$, denote by $h(u)$ the largest numbered neighbor of $u$ in $G$. This value can be computed together with a DFS-numbering, and can be stored in an array at the beginning of the algorithm.

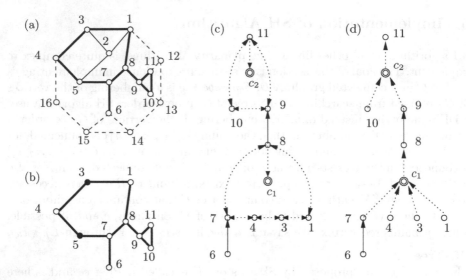

**Fig. 7.** (a) A graph $G$, a DFS-numbering of its vertices and, in thick edges, a frame $F$ of $G[1..11]$ in $G$. (b) Black vertices in frame $F$ are inactive. (c) The PC-tree $T$ for $F$, with $RBC$s indicated in dotted. (d) Rooted tree $T'$ corresponding to $T$; virtual arcs are dashed.

For each node $v$ of $T$, let $b(v) := \max\{h(u) : u$ is a descendant of $v$ in $T\}$. For a C-node of $T$, this number does not change during the execution of the algorithm. On the other hand, for a P-node of $T$, this number might decrease because its set of descendants might shrink when $T$ is modified. So, in the implementation, the value of $b(c)$ for a C-node $c$ is computed and stored when $c$ is created. It is the maximum over $b(u)$ for all $u$ in the path in $T$ corresponding to the $XY$-path in $F$ that originated $c$. One way to keep $b(v)$ for a P-node $v$ is, at the beginning of the algorithm, to build an adjacency list for $G$ sorted by the values of $h$, and to keep, during the algorithm, for each P-node of $T$, a pointer to the last traversed vertex in its sorted adjacency list. Each time the algorithm needs to access $b(v)$ for a P-node $v$, it moves this pointer ahead on the adjacency list (if necessary) until (1) it reaches a vertex $u$ which has $v$ as its parent, in which case $b(v)$ is the maximum between $h(v)$ and $b(u)$, or (2) it reaches the end of the list, in which case $b(v) = h(v)$.

**Traversal of the PC-tree**

The traversal of the PC-tree $T$, inspired by Boyer and Myrvold [4,5], is done as follows. To go from a P-node $u$ to a node $v$ which is an ancestral of $u$ in $T$, one starts with $x = u$ and repeats the following procedure until $x = v$. If $x$ is a P-node and *nonhead_RBC_cell*$(x)$ is NULL, move $x$ up to its parent. If $x$ is a P-node and *nonhead_RBC_cell*$(x)$ is non-NULL, either its virtual arc is NULL or not. If it is non-NULL, move $x$ to the C-node pointed by the virtual arc. Otherwise, starting at *nonhead_RBC_cell*$(x)$, search the $RBC$ in an arbitrary direction until either (1) the head of the $RBC$ is reached or (2) a cell in the $RBC$ with its virtual arc non-NULL is reached or (3) a P-node $y$ such that $b(y) > w$ is reached. If (3) happens before (1), search the $RBC$, restarting at $nonhead\_RBC\_cell(x)$, but in the other direction, until either (1) or (2) happens. If (1) happens, move $x$ to the C-node pointed by the head. If (2) happens, move $x$ to the C-node pointed by the virtual arc. In any case, search all visited cells in the $RBC$ again, setting their virtual arcs to $x$. Also, set the virtual arc from $x$ to the head of its $RBC$.

In a series of moments, the implementation traverses parts of $T$. For each node of $T$, there is a mark to tell whether it was already visited in this iteration or not. By visited, we mean a node which was assigned to $x$ in the traversal process described above. Every time a new node is included in $T$, it is marked as unvisited. Also, during each phase of the algorithm where nodes are marked as visited, the algorithm stacks each visited node and, at the end of the phase, unstacks them all, undoing the marks. This way, at the beginning of each iteration, all nodes of $T$ are marked as unvisited.

The same trick with a stack is done to unset the virtual arcs. When a virtual arc for a node $v$ is set in the traversal, $v$ is included in a second stack and, at the end of the iteration, this stack is emptied and all corresponding virtual arcs are set back to NULL.

## Terminals

The next concept, introduced by SH, is the key on how to search efficiently for an $XY$-obstruction. A node $t$ of $T$ is a *terminal* if

(t1) $b(t) > w$;
(t2) $t$ has a descendant in $T$ that is a neighbor of $w$ in $G$;
(t3) no proper descendant of $t$ satisfies properties (t1) and (t2) simultaneously.

Because of the orientation of the PC-tree, one of the terminals from Section 4 might not be a terminal here. This happens when one of the terminals from Section 4 is an ancestor in the PC-tree of all others. An extra effort in the implementation is necessary to detect and deal with this possible extra terminal.

The first phase of an iteration of the implementation is the search for the terminals. This phase consists of, for each neighbor $v$ of $w$ such that $v < w$, traversing $T$ starting at $v$ until a visited node $z$ is met. (Mark all nodes visited in the traversal; this will be left implicit from now on.) On the way, if a node $u$ such that $b(u) > w$ is seen, mark the first such node as a *candidate-terminal* and, if $z$ is marked as such, unmark it. The result from this phase is the list of terminals for each component of $F$.

## Search for $XY$-Obstructions

The second phase is the search for an $XY$-obstruction. First, if there are three or more terminals for some component of $F$, then there is an $XY$-obstruction of type either (o2) or (o3) in $F$ (Case 2 of Central algorithm). We omit the details on how to effectively find it because this is a terminal case of the algorithm. Second, if there are at most two terminals for each component of $F$, then, for each component of $F$ with at least one terminal, do the following. If it has two terminals, call them $t_1$ and $t_2$. If it has only one terminal, call it $t_1$ and let $t_2$

be the highest numbered vertex in this component. Test each C-node $c$ on the path in $T$ between $t_1$ and $t_2$ for an $XY$-obstruction of type (o1) (Case 1B of Central algorithm). The test decides if the cycle in $F$ corresponding to $c$ plays or not the role of $C$ in (o1). Besides these tests, the implementation performs one more test in the case of two terminals. The least common ancestor $m$ of $t_1$ and $t_2$ in $T$ is tested for an $XY$-obstruction of type (o2), if $m$ is a C-node, or an $XY$-obstruction of type (o3), if $m$ is a P-node. This extra test arises from the possible undetected terminal.

To perform each of these tests, the implementation keeps one more piece of information for each C-node $c$. Namely, it computes, in each iteration, the number of P-nodes in $RBC(c)$ that see $X$ through $E_F \setminus E_C$, where $C$ is the cycle in $F$ corresponding to $c$. This number is computed in the first phase. Each C-node has a counter that, at the beginning of each iteration, values 1 (to account for the head of its $RBC$). During the first phase, every time an $RBC$ is entered through a P-node which was unvisited, the counter of the corresponding C-node is incremented by 1. As a result, at the end of the first phase, each (relevant) C-node knows its number.

For the test of a C-node $c$, the implementation searches $RBC(c)$, starting at the head of $RBC(c)$. It moves in an arbitrary direction, stopping only when it finds a P-node $u$ (distinct from the head) such that $b(u) > w$. On the way, the implementation counts the number of P-nodes traversed. If only one step is given, it starts again at the head of $RBC(c)$ and moves to the other direction until it finds a P-node $u$ such that $b(u) > w$, counting the P-nodes, as before. If the counter obtained matches the number computed for that C-node in the first phase, it passed the test, otherwise, except for two cases, there in an $XY$-obstruction of type (o1). The first of the two cases missing happens when there are exactly two terminals and $c$ is the lower common ancestor of them. The second of the two cases happens when there is exactly one terminal and $c$ is (potentially) the upper block in which the $XY$-path ends. The test required in these two cases is slightly different, but similar, and might give raise to an $XY$-obstruction of type (o1) or (o2). We omit the details.

**PC-Tree update**

The last phase refers to Case 2B in LEC method. It consists of the modification of $T$ according to the new frame. First, one has to add to $T$ a P-node for $w$. Then, parts of $T$ referring to a component with no neighbor of $w$ remain the same. Parts of $T$ referring to a component with exactly one neighbor of $w$ are easily adjusted. So we concentrate on the parts of $T$ referring to components with two or more neighbors of $w$. Each of these originates a new C-node. For each of them, the second phase determined the basis of an $XY$-path, which is given by a path $Q$ in $T$. Path $Q$ consists basically of the nodes visited during the second phase. Let us describe the process in the case where there is only one terminal. The case of two terminals is basically a double application of this one.

Call $c$ the new C-node being created. Start $RBC(c)$ with its head cell, which refers to $w$, and points back to $c$. Traverse $Q$ once again, going up in $T$. For each P-node $u$ in $Q$ such that $nonhead\_RBC\_cell(u)$ is NULL, if $b(u) > w$ (here we refer to the possibly new value of $b(u)$, as $u$ might have lost a child in the traversal), then an $RBC$ cell is created, referring to $u$. It is included in $RBC(c)$ and $nonhead\_RBC\_cell(u)$ is set to point to it. For each P-node $u$ such that $nonhead\_RBC\_cell(u)$ is non-NULL, let $c'$ be its parent in $T$. Concatenate to $RBC(c)$ a part of $RBC(c')$, namely, the part of $RBC(c')$ that was not used to get to $c'$ in any traversal in the second phase. To be able to concatenate without traversing this part, one can use a simple data structure proposed by Boyer and Myrvold [5,4] to keep a doubled linked list. (The data structure consists of the cells with two indistinct pointers, one for each direction. To move in a certain direction, one starts making the first move in that direction, then, to keep moving in the same direction, it is enough to choose always the pointer that does not lead back to the previous cell.)

During the traversal of $Q$, one can compute the value of $b(c)$. Its value is simply the maximum of $b(u)$ over all node $u$ traversed. This completes the description of the core of the implementation.

### Certificate

To be able to produce a certificate for its answer, the implementation carries still more information. Namely, it carries the DFS-tree that originated the DFS-numbering of the vertices and, for each C-node, a combinatorial description of a planar embedding of the corresponding biconnected component where the P-nodes in its $RBC$ appear all on the boundary of the same face. We omit the details, but one can find at http://www.ime.usp.br/~coelho/sh/ the complete implementation, that also certificates its answer.

## 6  Experimental Study

The main purpose of this study was to confirm the linear-time behavior of our implementation and to acquire a deeper understanding of SH algorithm. Boyer et al. [3] made a similar experimental study that does not include SH algorithm.

The LEDA platform has a planarity library that includes implementations of Hopcroft and Tarjan's (HT) and BL algorithms and an experimental study comparing them. The library includes the following planar graph generator routines: maximal_planar_map and random_planar_map. Neither of them generates plane maps according to the uniform distribution [14], but they are well-known and widely used. The following classes of graphs obtained through these routines are used in the LEDA experimental study:

(G1) random planar graphs;
(G2) graphs with a $K_{3,3}$: six vertices from a random planar graph are randomly chosen and edges among them are added to form a $K_{3,3}$;
(G3) graphs with a $K_5$: five random vertices from a random planar graph are chosen and all edges among them are added to form a $K_5$;
(G4) random maximal planar graphs;
(G5) random maximal planar graphs plus a random edge connecting two non-adjacent vertices.

Our experimental study extends the one presented in LEDA including our implementation of SH algorithm made on the LEDA platform and an implementation of BM algorithm developed in C. We performed all empirical tests used in LEDA to compare HT and BL implementations [15]. The experimental environment was a PC running GNU/Linux (RedHat 7.1) on a Celeron 700MHz with 256MB of RAM. The compiler was the gcc 2.96 with options -DLEDA_CHECKING_OFF -O.

In the experiments [15, p. 123], BL performs the planarity test 4 to 5 times faster than our SH implementation in all five classes of graphs above. For the planar classes (G1) and (G4), it runs 10 times faster than our SH to do the planarity test and build the embedding. On (G2) and (G3), it is worse than our SH, requiring 10% to 20% more time for testing and finding an obstruction. On (G5), it runs within 65% of our SH time for testing and finding an obstruction. For the planarity test only, HT runs within 70% of our SH time for the planar classes (G1) and (G4), but performs slightly worse than our SH on (G2) and (G3). On (G5), it outperforms our SH, running in 40% of its time. For the planar classes (G1) and (G4), HT is around 4 times faster when testing and building the embedding. (The HT implementation in question has no option to produce an obstruction when the input graph is non-planar; indeed, there is no linear-time implementation known for finding the obstruction for it [22].) BM performs better than all, but, remember, it is the only one implemented in C and not in the LEDA platform. It runs in around 4% of the time spent by our SH for testing and building the embedding and, for the non-planar classes, when building the obstruction, it runs in about 15% of our SH time on (G2) and (G3) and in about 10% of our SH time on (G5). (There is no implementation of BM available that only does the planarity testing.) The time execution used on these comparisons is the average CPU time on a set of 10 graphs from each class.

Figure 8 shows the average CPU time of each implementation on (a) (G1) for only testing planarity (against BM with testing and building an embedding, as there is no testing only available), (b) (G2) for testing and finding an obstruction (HT is not included in this table, by the reason mentioned above), (c) (G4) for testing and building an embedding, and (d) for testing and finding an obstruction (again, HT excluded).

We believe the results discussed above and shown in the table are initial and still not conclusive because our implementation is yet a prototype. (Also, in our opinion, it is not fair to compare LEDA implementations with C implementations.)

Our current understanding of SH algorithm makes us believe that we can design a new implementation which would run considerably faster. Our belief comes, first, from the fact that our current code was developed to solve the planarity testing only, and was later on modified to also produce a certificate for its answer to the planarity test. Building an implementation from the start thinking about the test and the certificate would be the right way, we believe, to have a more efficient code. Second, during the adaptation to build the certificate (specially the embedding when the input is planar) made us notice several details

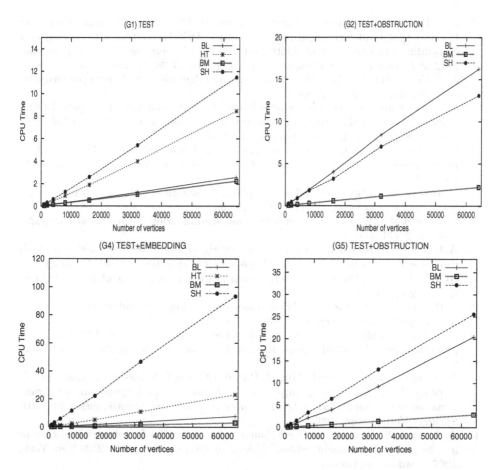

**Fig. 8.** Empirical results comparing SH, HT, BL, and BM implementations.

in the way the implementation of the test was done that could be improved. Even though, we decide to go forward with the implementation of the complete algorithm (test plus certificate) so that we could understand it completely before rewriting it from scratch. The description made on Section 5 already incorporates some of the simplifications we thought of for our new implementation. It is our intention to reimplement SH algorithm from scratch.

# References

1. L. Auslander and S.V. Parter. On imbedding graphs in the plane. *Journal of Mathematics and Mechanics*, 10:517–523, 1961.
2. K.S. Booth and G.S. Lueker. Testing for the consecutive ones property, interval graphs, and graph planarity using PQ–tree algorithms. *Journal of Computer and Systems Sciences*, 13:335–379, 1976.

3. J.M. Boyer, P.F. Cortese, M. Patrignani, and G. Di Battista. Stop minding your P's and Q's: Implementing a fast and simple DFS-based planarity testing and embedding algorithm. In G. Liotta, editor, *Graph Drawing (GD 2003)*, volume 2912 of *Lecture Notes in Computer Science*, pages 25–36. Springer, 2004.
4. J.M. Boyer and W. Myrvold. On the cutting edge: Simplified $O(n)$ planarity by edge addition. Preprint, 29pp.
5. J.M. Boyer and W. Myrvold. Stop minding your P's and Q's: A simplified $O(n)$ planar embedding algorithm. *Proceedings of the Tenth Annual ACM-SIAM Symposium on Discrete Algorithms*, pages 140–146, 1999.
6. N. Chiba, T. Nishizeki, A. Abe, and T. Ozawa. A linear algorithm for embedding planar graphs using PQ–trees. *Journal of Computer and Systems Sciences*, 30:54–76, 1985.
7. N. Deo. Note on Hopcroft and Tarjan planarity algorithm. *Journal of the Association for Computing Machinery*, 23:74–75, 1976.
8. S. Even and R.E. Tarjan. Computing an $st$-numbering. *Theoretical Computer Science*, 2:339–344, 1976.
9. A.J. Goldstein. An efficient and constructive algorithm for testing whether a graph can be embedded in a plane. In *Graph and Combinatorics Conf.* Contract No. NONR 1858-(21), Office of Naval Research Logistics Proj., Dep. of Math., Princeton U., 2 pp., 1963.
10. J. Hopcroft and R. Tarjan. Efficient planarity testing. *Journal of the Association for Computing Machinery*, 21(4):549–568, 1974.
11. W.L. Hsu. An efficient implementation of the PC-tree algorithm of Shih & Hsu's planarity test. Technical report, Inst. of Information Science, Academia Sinica, 2003.
12. M. Jünger, S. Leipert, and P. Mutzel. Pitfalls of using PQ-trees in automatic graph drawing. In G. Di Battista, editor, *Proc. 5th International Symposium on Graph Drawing '97*, volume 1353 of *Lecture Notes in Computer Science*, pages 193–204. Springer Verlag, Sept. 1997.
13. A. Lempel, S. Even, and I. Cederbaum. An algorithm for planarity testing of graphs. In P. Rosenstiehl, editor, *Theory of Graphs*, pages 215–232, New York, 1967. Gordon and Breach.
14. K. Mehlhorn and St. Näher. *The LEDA Platform of Combinatorial and Geometric Computing*. Cambridge Press, 1997.
15. A. Noma. Análise experimental de algoritmos de planaridade (in Portuguese). Master's thesis, Universidade de São Paulo, 2003. http://www.ime.usp.br/dcc/posgrad/teses/noma/dissertation.ps.gz.
16. E.M. Reingold, J. Nievergelt, and N. Deo. *Combinatorial Algorithms: Theory and Practice*. Prentice-Hall, Inc., Englewood Cliffs, New Jersey, 1977.
17. W.K. Shih and W.L. Hsu. A simple test for planar graphs. In *Proceedings of the International Workshop on Discrete Math. and Algorithms*, pages 110–122. University of Hong Kong, 1993.
18. W.K. Shih and W.L. Hsu. A new planarity test. *Theoretical Computer Science*, 223:179–191, 1999.
19. R. Thomas. Planarity in linear time. http://www.math.gatech.edu/~thomas/, 1997.
20. S.G. Williamson. Embedding graphs in the plane – algorithmic aspects. *Ann. Disc. Math.*, 6:349–384, 1980.
21. S.G. Williamson. *Combinatorics for Computer Science*. Computer Science Press, Maryland, 1985.
22. S.G. Williamson. Personal Communication, August 2001.

# A Greedy Approximation Algorithm for the Uniform Labeling Problem Analyzed by a Primal-Dual Technique*

Evandro C. Bracht, Luis A. A. Meira, and Flávio K. Miyazawa

Instituto de Computação, Universidade Estadual de Campinas, Caixa Postal 6176, 13084-971 Campinas, Brazil.
{evandro.bracht,meira,fkm}@ic.unicamp.br

**Abstract.** In this paper we present a new fast approximation algorithm for the Uniform Metric Labeling Problem. This is an important classification problem that occur in many applications which consider the assignment of objects into labels, in a way that is consistent with some observed data that includes the relationship between the objects.
The known approximation algorithms are based on solutions of large linear programs and are impractical for moderated and large size instances. We present an $8 \log n$-approximation algorithm analyzed by a primal-dual technique which, although has factor greater than the previous algorithms, can be applied to large sized instances. We obtained experimental results on computational generated and image processing instances with the new algorithm and two others LP-based approximation algorithms. For these instances our algorithm present a considerable gain of computational time and the error ratio, when possible to compare, was less than 2% from the optimum.

## 1 Introduction

In a traditional classification problem, we wish to assign each of $n$ objects to one of $m$ labels (or classes). This assignment must be consistent with some observed data that includes pairwise relationships among the objects to be classified. More precisely, the classification problem can be defined as follows: Let $P$ be a set of objects, $L$ a set of labels, $w : P \times P \to \mathbb{R}^+$ a weight function, $d : L \times L \to \mathbb{R}^+$ a distance function and $c : P \times L \to \mathbb{R}^+$ an assignment cost function. A *labeling* of $P$ over $L$ is a function $\phi : P \to L$. The *assignment cost* of a labeling $\phi$ is the sum $\sum_{i \in P} c(i, \phi(i))$ and the *separation cost* of a labeling $\phi$ is the sum $\sum_{i,j \in P} d(\phi(i), \phi(j)) w(i, j)$. The function $w$ indicates the strength of the relation between two objects, and the function $d$ indicates the similarity between two labels. The cost of a labeling $\phi$ is the sum of the assignment cost and the separation cost. The Metric Labeling Problem (MLP) consists of finding

---

* This work has been partially supported by MCT/CNPq Project ProNEx grant 664107/97-4, FAPESP grants 01/12166-3, 02/05715-3, and CNPq grants 300301/98-7, 470608/01-3, 464114/00-4, and 478818/03-3.

C.C. Ribeiro and S.L. Martins (Eds.): WEA 2004, LNCS 3059, pp. 145–158, 2004.

a labeling of the objects into labels with minimum total cost. Throughout this paper, we denote the size of the sets $P$ and $L$ by $n$ and $m$, respectively.

In this paper we consider the *Uniform Labeling Problem (ULP)*, a special case of MLP, where the distance $d(i, j)$ has a constant value if $i \neq j$, and 0 otherwise, for all $i, j \in L$.

The MLP has several applications, many listed by Kleinberg and Tardos [12]. Some applications occur in image processing [6,2,8], biometrics [1], text categorization [3], etc. An example of application in image processing is the restoration of images degenerated by noise. In this case, an image can be seen as a grid of pixels, each pixel is an object that must be classified with a color. The assignment cost is given by the similarity between the new and old coloring, and the separation cost is given by the color of a pixel and the color of its neighbors.

The uniform and the metric labeling problems generalizes the Multiway Cut Problem, a known NP-hard problem [7]. These labeling problems were first introduced by Kleinberg and Tardos [12] that present an $O(\log m \log \log m)$-approximation algorithm for the MLP, and a 2-approximation algorithm for the ULP using the probabilistic rounding technique over a solution of a linear program. The associated LP has $O(n^2 m)$ constraints and $O(n^2 m)$ variables.

Chekuri et al. [4] developed a new linear programming formulation that is better for the non-uniform case. However, for the uniform case it maintains a 2-approximation factor and has bigger time complexity. This time complexity is a consequence of solving a linear program with $O(n^2 m^2)$ constraints and $O(n^2 m^2)$ variables.

Gupta and Tardos [10] present a formulation for the Truncated Labeling Problem, a case of the MLP where the labels are positive integers and the metric distance between labels $i$ and $j$ is given by the truncated linear norm, $d(i, j) = \min\{M, |i - j|\}$, where $M$ is the maximum value allowed. They present an algorithm that is a 4-approximation algorithm for the Truncated Labeling Problem and a 2-approximation for the ULP. The algorithm generates a network flow problem instance where the weights of edges come from the assignment and separation costs of the original problem. The resulting graph has $O(n^2 m)$ edges where the *Min Cut* algorithm is applied $O((m/M)(\log Q_0 + \log \varepsilon^{-1}))$ times in order to obtain $(4 + \varepsilon)$-approximation, where $Q_0$ is the cost of the initial solution.

The LP-based and the Flow-based algorithms have a large time complexity, which turns in impractical algorithms for moderate and large sized instances, as in most applications cited above.

In this paper, we present a fast approximation algorithm for the Uniform Metric Labeling Problem and prove that it is an $8 \log n$-approximation algorithm. We also compare the practical performance of this algorithm with the LP-based algorithms of Chekuri et al. and Kleinberg and Tardos. Although this algorithm has higher approximation factor, the solutions obtained have error ratio that was less than 2% from the optimum solution, when it was possible to compare, and the time improvement over the linear programming based algorithms is considerable.

In section 2, we present a proof for a greedy algorithm for the Set Cover Problem via a primal dual analysis. Then we analyze the case when the greedy choice is relaxed to an approximated one. In section 3, we present a general algorithm for the Uniform Labeling Problem using an algorithm for the Set Cover Problem. This last problem is solved approximately via an algorithm for the Quotient Cut Problem[13]. In section 4, we compare the presented algorithm with the LP-based algorithms. Finally, in section 5, we present the concluding remarks.

## 2   A Primal-Dual Analysis for a Greedy Algorithm for the Set Cover Problem

In this section, we present a primal-dual version for a greedy approximation algorithm for the Set Cover Problem, including a proof of its approximation factor. This proof is further generalized for the case when the greedy choice is relaxed to an approximated one.

The Set Cover Problem is a well known optimization problem, that generalizes many others. This problem consists of: Given a set $E = \{e_1, e_2, \ldots, e_n\}$ of elements, a family of subsets $\mathcal{S} = \{S_1, S_2, \ldots, S_m\}$, where each $S_j \subseteq E$, has a cost $w_j$, $\forall j \in \{1, \ldots, m\}$. The goal of the problem is to find a set $Sol \subseteq \{1, \ldots, m\}$ that minimizes $\sum_{j \in Sol} w_j$ and $\cup_{j \in Sol} S_j = E$.

In [5], Chvátal present a greedy $H_g$-approximation algorithm for the Set Cover Problem, where $g$ is the number of elements in the largest set in $\mathcal{S}$ and $H_g$ is the value of the harmonic function of degree $g$. This algorithm iteratively chooses the set with minimum amortized cost, that is the cost of the set divided by the number of non-covered elements. Once a set is chosen to be in the solution, all the elements in this set are considered as covered. In what follows we describe this algorithm more precisely.

GREEDY ALGORITHM FOR THE SET COVER PROBLEM $(E, \mathcal{S}, w)$

**1.**    $Sol \leftarrow \emptyset; \quad U \leftarrow E.$
**2.**    While $U \neq \emptyset$ do
**3.**        $j' \leftarrow \arg\min_{j:S_j \cap U \neq \emptyset} \frac{w_j}{|S_j \cap U|}$
**4.**        $Sol \leftarrow Sol \cup \{j'\}$
**5.**        $U \leftarrow U \setminus S_{j'}$
**6.**    return $Sol$.

To show the primal dual algorithm for the Set Cover Problem, we first present a formulation using binary variables $x_j$ for each set $S_j$, where $x_j = 1$ if and only if $S_j$ is chosen to enter in the solution. The formulation consists of finding $x$ that

$$
\begin{aligned}
\text{minimize} \quad & \sum_j w_j x_j \\
\text{s.t} \quad & \sum_{j:e \in S_j} x_j \geq 1 \qquad \forall e \in E, \\
& x_j \in \{0,1\} \qquad \forall j \in \{1, \ldots, n\},
\end{aligned}
$$

and the dual of the relaxed version consists of finding $\alpha$ that

$$\begin{aligned} \text{maximize} \quad & \sum_{e \in E} \alpha_e \\ \text{s.t} \quad & \sum_{e \in S_j} \alpha_e \leq w_j \quad \forall j \in \{1, \ldots, n\}, \\ & \alpha_e \geq 0 \quad \forall e \in E. \end{aligned}$$

The greedy algorithm can be rewritten as a primal dual algorithm, with similar set of events. The algorithm uses a set $U$ containing the elements not covered in each iteration, initially the set $E$, and a variable $T$ with a notion of time associated with each event. The algorithm also uses (dual) variables $\alpha_e$ for each element $e$, starting at zero, and increasing in each iteration for the elements in $U$.

---

PRIMAL-DUAL ALGORITHM FOR THE SET COVER PROBLEM $(E, \mathcal{S}, w)$

1.    $U \leftarrow E$;    $T \leftarrow 0$ ;    $\alpha_e \leftarrow 0, \forall e \in E$;    $Sol \leftarrow \emptyset$.
2.    while $U \neq \emptyset$ do
3.        grow uniformly the time $T$ and, at the same rate, grow $\alpha_e : e \in U$
          until there exists an index $j$ such that $\sum_{e \in S_j \cap U} \alpha_e$ is equal to $w_j$.
4.        $Sol \leftarrow Sol \cup \{j\}$.
5.        $U \leftarrow U \setminus S_j$.
6.    return $Sol$.

---

**Lemma 1.** *The sequence of events executed by the Greedy algorithm and by the Primal-Dual algorithm is the same.*

*Proof.* Note that the value of $\alpha_e$ when $\sum_{e \in S_j \cap U} \alpha_e = w_j$ is equal to the amortized cost. Since $\alpha_e$ grows uniformly, it is clear that the algorithm, in each iteration, choose a set with minimum amortized cost.    □

This lemma implies that all solutions obtained by the Greedy algorithm can be analyzed by the primal dual techniques.

**Lemma 2.** *Let $S_j = \{e_1, e_2, \ldots, e_k\}$ and $\alpha_i$ the time variable associated to the item $e_i$, generated by the Primal-Dual algorithm. If $\alpha_1 \leq \alpha_2 \leq \ldots \leq \alpha_k$, then $\sum_{i=l}^{k} \alpha_l \leq w_j$.*

*Proof.* In the moment just before the time $\alpha_l$, all variables $\alpha_i$, $l \leq i \leq k$ have the same value, $\alpha_l$, and they are all associated with uncovered elements. Suppose the lemma is false. In this case, $\sum_{i=l}^{k} \alpha_l > w_j$ and, in an instant strictly before $\alpha_l$, the set $S_j$ would enter in the solution and all of its elements would have $\alpha < \alpha_l$, that is a contradiction, since $S_j$ has at least one element greater or equal to $\alpha_l$. Therefore, the lemma is valid.    □

The Primal-Dual algorithm returns a primal solution $Sol$, with value $val(Sol) := \sum_{j \in Sol} w_j$, such that $val(Sol) = \sum_{e \in E} \alpha_e$. Note that the variable $\alpha$ may be dual infeasible. If there exists a value $\gamma$ such that $\alpha/\gamma$ is dual feasible, i.e. $\sum_{e \in S_j} \alpha_e/\gamma \leq w_j$ for each $j \in \{1, \ldots, m\}$, then, by the weak duality theorem, $val(Sol) \leq \gamma OPT$. The idea to prove the approximation factor is to find a value $\gamma$ for which $\alpha/\gamma$ is dual feasible.

**Theorem 3.** *The Primal-Dual algorithm for the Set Cover Problem is an $H_g$-approximation algorithm, where $g$ is the size of the largest set in $S$.*

*Proof.* Consider an arbitrary set $S = \{e_1, \ldots, e_k\}$ with $k$ elements and cost $w$ and let $\alpha_i$ the time variable associated with element $e_i$, $i = 1, \ldots, k$. Without loss of generality, assume that $\alpha_1 \leq \ldots \leq \alpha_k$.

If $\gamma$ is a value such that $\alpha/\gamma$ is dual feasible then

$$\sum_{i=1}^{k} \alpha_i/\gamma \leq w, \tag{1}$$

thus a necessary condition is

$$\gamma \geq \frac{\sum_{i=1}^{k} \alpha_i}{w}. \tag{2}$$

Applying Lemma 2 for each value of $l$, we have

$$\begin{cases} l = 1, & k\alpha_1 \leq w, & \Rightarrow & \alpha_1/w \leq 1/k, \\ l = 2, & (k-1)\alpha_2 \leq w, & \Rightarrow & \alpha_2/w \leq 1/(k-1), \\ & & \vdots & \\ l = k, & \alpha_k \leq w, & \Rightarrow & \alpha_k/w \leq 1. \end{cases}$$

Adding the inequalities above we have

$$\frac{\sum_{i=1}^{k} \alpha_i}{w} \leq H_k. \tag{3}$$

Therefore, when $\gamma = H_g$ we obtain that $\alpha/\gamma$ is dual feasible. □

Now, let us assume a small modification in the previous algorithm. Instead of choosing, in step 3 of the greedy algorithm, the set with minimum amortized cost, we choose a set $S_j$ with amortized cost at most $f$ times greater than the minimum. That is, if $S_{j*}$ is a set with minimum amortized cost then the following inequality is valid

$$\frac{w_j}{|S_j \cap U|} \leq f \frac{w_{j*}}{|S_{j*} \cap U|}.$$

This modification can be understood, in the primal-dual version of the algorithm, as a permission that the sum $\sum_{e \in S_j \cap U} \alpha_e$ can pass the value of $w_j$ by at most a factor of $f$. We denote by $\mathcal{A}_f$ the algorithms with this modification.

**Lemma 4.** *Let $S_j = \{e_1, e_2, \ldots, e_k\}$ and $\alpha_i$ the time variable associated with the item $e_i$ generated by an $\mathcal{A}_f$ algorithm. If $\alpha_1 \leq \alpha_2 \leq \ldots \leq \alpha_k$ then $\sum_{i=l}^{k} \alpha_l \leq f w_j$.*

*Proof.* Suppose the lemma is false. In this case, there exists an execution where $\sum_{i=l}^{k} \alpha_l > f w_j$ and, in an instant $T < \alpha_l$, the set $S_j$ would enter in the solution, which is a contradiction. □

The following theorem can be proved analogously to Theorem 3.

**Theorem 5.** *If $g$ is the size of the largest set in $S$ then any $\mathcal{A}_f$ algorithm is an $f H_g$-approximation algorithm.*

## 3  A New Algorithm for the Uniform Labeling Problem

The algorithm for the ULP uses similar ideas presented by Jain et al. [11] for a facility location problem. To present this idea, we use the notion of a star. A *star* is a connected graph where only one vertex, denoted as *center* of the star, can have degree greater than one. Jain et al. [11] present an algorithm that iteratively select a star with minimum amortized cost, where the center of each star is a facility.

In the Uniform Labeling Problem, we can consider a labeling $\phi : P \to L$ as a set of stars, each one with a label in the center. The algorithm for the ULP iteratively select stars of reduced amortized cost, until all objects have been covered.

Given a star $S = \{l, U_S\}$ for the ULP, where $l \in L$ and $U_S \subseteq P$, we denote by $c_S$ the cost of the star $S$, which is defined as

$$c_S = \sum_{u \in U_S} c_{ul} + \frac{1}{2} \sum_{u \in U_S, v \in (P \backslash U_S)} w_{uv},$$

that is, the cost to assign each element of $U_S$ to $l$ plus the cost to separate each element of $U_S$ with each element of $P \backslash U_S$. We pay just the half of the separation cost, because the other half will appear when we label the elements in $P \backslash U_S$.

In the description of the main algorithm for the Uniform Labeling Problem, we denote by $\mathcal{S}_{\text{ULP}}$ the set of all possible stars of an instance, $U$ the set of unclassified objects, $(E_{\text{SC}}, \mathcal{S}_{\text{SC}}, w')$ an instance for the Set Cover Problem, $\mathcal{U}$ a collection and $\phi$ a labeling. In the following, we describe the algorithm, called GUL, using an approximation algorithm for the Set Cover Problem as parameter.

---

ALGORITHM GREEDY UNIFORM LABELING (GUL) $(L, P, c, w, \mathcal{A}_{\text{SC}})$

    $L :=$ set of labels;    $P :=$ set of objects.

    $c_{ui} :=$ assignment cost between a label $i$ and an object $u$.

    $w_{uv} :=$ separation cost between an object $u$ and an object $v$.

    $\mathcal{A}_{\text{SC}} :=$ A $\beta$-approximation algorithm for the Set Cover Problem.

1.   $E_{\text{SC}} \leftarrow P$
2.   $\mathcal{S}_{\text{SC}} \leftarrow \{U_S : S = \{l, U_S\} \in \mathcal{S}_{\text{ULP}}\}$
3.   $w'_S \leftarrow c_S + \frac{1}{2} \sum_{u \in U_S, v \in P \backslash U_S} w_{uv}, \quad \forall S = \{l, U_S\} \in \mathcal{S}_{\text{ULP}}$
4.   $\mathcal{U} \leftarrow \mathcal{A}_{\text{SC}}(E_{\text{SC}}, \mathcal{S}_{\text{SC}}, w'), \quad$ let $\{U_{S1}, U_{S2}, \ldots, U_{St}\} = \mathcal{U}$
5.   $U \leftarrow P$
6.   for $k \leftarrow 1$ to $t$ do
7.      $\phi(i) \leftarrow l, \forall i \in U_{Sk} \cap U, l = S_k \cap L$
8.      $U \leftarrow U \backslash U_{Sk}$
9.   return $\phi$.

---

## 3.1   Analysis of the Algorithm

To analyze the algorithm we use the following notation:

- $val_{ULP}(\phi)$: value, in the ULP, of the labeling $\phi$.
- $val_{SC}(\mathcal{U})$: value, in the Set Cover Problem, of a collection $\mathcal{U}$.
- $\phi_{OPT}$: an optimum labeling for the ULP.
- $\mathcal{U}_{OPT}$: an optimum solution for the Set Cover Problem.
- $SC(\phi)$: A collection $\{U_{S1}, U_{S2}, \ldots, U_{Sk}\}$ related with a labeling $\phi = \{S_1, S_2, \ldots, S_k\}$ where $S_i$ is the star $\{l, U_{Si}\}$.

**Lemma 6.** *If $\phi$ is a solution returned by the algorithm* GUL *and $\mathcal{U}$ the solution returned by algorithm $\mathcal{A}_{SC}$ for the Set Cover Problem (at step 4) then*

$$val_{ULP}(\phi) \leq val_{SC}(\mathcal{U}).$$

*Proof.*

$$
\begin{aligned}
val_{SC}(\mathcal{U}) &= \sum_{S=\{l,U_S\}:U_S \in \mathcal{U}} w'_S \\
&= \sum_{S=\{l,U_S\}:U_S \in \mathcal{U}} \left( \sum_{u \in U_S} c_{ul} + \sum_{u \in U_S, v \in P \setminus U_S} w_{uv} \right) \\
&\geq \sum_{u \in P} c_{u\phi(u)} + \frac{1}{2} \sum_{u,v \in P:\phi(u) \neq \phi(v)} w_{uv} \\
&= val_{ULP}(\phi).
\end{aligned}
$$

The following inequalities are valid

$$
\sum_{u \in P} c_{u\phi(u)} \leq \sum_{S=\{l,U_S\}:U_S \in \mathcal{U}} \sum_{u \in U_S} c_{ul} \tag{4}
$$

and

$$
\frac{1}{2} \sum_{u,v \in P:\phi(u) \neq \phi(v)} w_{uv} \leq \sum_{S:U_S \in \mathcal{U}} \sum_{u \in U_S, v \in P \setminus U_S} w_{uv}. \tag{5}
$$

The inequality (4) is valid, since the algorithm assign $\phi(u)$ to $u$, if there exists a set $U_S$ in $\mathcal{U}$ such that $u \in U_S$ and $\phi(u) = S \cap L$. Thus, the cost $c_{u\phi(u)}$ also appears in the right hand side of the inequality.

The argument for the inequality (5) is similar. If $\phi(u) \neq \phi(v)$ then, by the execution of the algorithm, there must exist two sets $U_{Sx} \in \mathcal{U}$ and $U_{Sy} \in \mathcal{U}$ such that $\phi(u) = S_x \cap L$ and $\phi(v) = S_y \cap L$. It is not allowed to occur both $\{u, v\} \subseteq U_{Sx}$ and $\{u, v\} \subseteq U_{Sy}$. Therefore, if the cost $w_{uv}$ and $w_{vu}$ appears, once or twice, in the left hand side of the inequality, it must appear, at least once, in right hand side. $\qquad\square$

**Lemma 7.** $val_{SC}(\mathcal{U}_{OPT}) \leq 2 val_{ULP}(\phi_{OPT})$.

*Proof.*

$$2\, val_{ULP}(\phi_{OPT}) = 2 \sum_{S \in \phi_{OPT}} c_S$$

$$\geq \sum_{S=\{l,U_S\} \in \phi_{OPT}} \left( c_S + \frac{1}{2} \sum_{u \in U_S, v \in P \setminus U_S} w_{uv} \right)$$

$$= \sum_{t \in SC(\phi_{OPT})} w'_t$$

$$\geq \sum_{t \in \mathcal{U}_{OPT}} w'_t \tag{6}$$

$$= val_{SC}(\mathcal{U}_{OPT}),$$

where inequality (6) is valid since $SC(\phi_{OPT})$ is a solution for the Set Cover Problem, but not necessarily with optimum value.     □

**Theorem 8.** *If $I$ is an instance for the ULP, $\phi$ is the labeling generated by the algorithm GUL and $\phi_{OPT}$ is an optimum labeling for $I$ then*

$$val_{ULP}(\phi) \leq 2\beta\, val_{ULP}(\phi_{OPT}),$$

*where $\beta$ is the approximation factor of the algorithm $\mathcal{A}_{SC}$ given as a parameter.*

*Proof.* Let $\mathcal{U}$ be the solution returned by the algorithm $\mathcal{A}_{SC}$ (step 4 of the algorithm GUL) and $\mathcal{U}_{OPT}$ an optimal solution for the corresponding Set Cover Instance. In this case, we have

$$val_{ULP}(\phi) \leq val_{SC}(\mathcal{U}) \tag{7}$$

$$\leq \beta\, val_{SC}(\mathcal{U}_{OPT}) \tag{8}$$

$$\leq 2\beta\, val_{ULP}(\phi_{OPT}), \tag{9}$$

where the inequality (7) is valid by Lemma 6, the inequality (8) is valid since $\mathcal{U}$ is found by a $\beta$-approximation algorithm, and the inequality (9) is valid by Lemma 7.     □

To obtain an $8 \log n$-approximation algorithm for the ULP we need to present an algorithm $\mathcal{A}_{SC}$ that is a $4 \log n$-approximation for the Set Cover Problem. The algorithm is based on an approximation algorithm for the Quotient Cut Problem, which we describe in the following subsection.

## 3.2   Showing $\mathcal{A}_{SC}$

In this section we show how to obtain a greedy algorithm for the Set Cover Problem without the explicit generation of all possible sets. The algorithm basically generate a graph $G_l$ for each possible label $l \in L$ and obtain a set $U_S$ with reduced amortized cost.

Consider a label $l \in L$. We wish to find a set $U_S$ that minimize $w_S/|U_S \cap U|$, where $U$ is the set of unclassified objects in the iteration. Denote by $G_l$ the complete graph with vertex set $V(G_l) = P \cup \{l\}$ and edge costs $w_{G_l}$ defined as follows:

$$w_{G_l}(u, v) := w_{uv} \quad \forall u, v \in P, u \neq v,$$
$$w_{G_l}(u, l) := c_{ul} \quad \forall u \in P.$$

In other words, the cost of an edge between the label and an object is the assignment cost and the cost of an edge between two objects is the separation cost.

**Lemma 9.** *If* $C \subseteq P$ *is a cut of* $G_l$, $l \notin C$, *the cost of* $C$, $c(C) = \sum_{u \in C, v \notin C} w_{G_l}(u, v)$, *is equal to the cost of the set* $U_S$, $S = \{l, C\}$, *to the Set Cover Problem.*

*Proof.* The lemma can be proved by counting. In the Set Cover Problem, $w_S$ is equal to

$$\sum_{u \in U_S} c_{ul} + \sum_{u \in U_S, v \in P \setminus U_S} w_{uv}, \quad \text{where } S = \{l, U_S\},$$

that is equal to the cost of the cut $C$. See Figure 1.  □

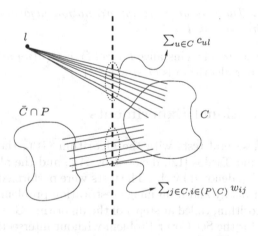

**Fig. 1.** Example of a cut $C$ that has the same cost $w'_S$ for $U_S = C$.

The problem to find a set with smallest amortized cost $c(C)/|C|$ in $G_l$ is a specification of the Quotient Cut Problem, which can be defined as follows.

QUOTIENT CUT PROBLEM (QCP): Given a graph $G$, weights $w_e \in R^+$ for each edge $e \in E(G)$, and weights $c_v \in \mathbb{Z}^+$ for each vertex $v \in V(G)$, find a cut $C$ that minimizes $w(C)/\min\{\pi(C), \pi(\bar{C})\}$, where $w(C) := \sum_{u \in C, v \notin C} w_{uv}$ and $\pi(C) := \sum_{v \in C} c_v$.

If we define a weight function $c_i$ for each vertex of $G_i$ as

$$c_i(v) = \begin{cases} 1 \text{ if } v \in V(G_i) \cap P \cap U, \\ 0 \text{ if } v \in V(G_i) \cap (P \setminus U), \\ n \text{ if } v \in V(G_i) \cap L, \end{cases}$$

then a cut $C := \min_{i \in L}\{QC(G_i, w_i, c_i)\}$, returned by an $f$-approximation algorithm $QC$ for the QCP, corresponds to a set $U_S$, $S = \{l, C\}$ that is at most $f$ times bigger than the set with minimum amortized cost. An $\mathcal{A}_f$ algorithm can be implemented choosing this set. Thus, the following result follows from Theorem 5.

**Theorem 10.** *If there exists an $f$-approximation algorithm for the Quotient Cut Problem with time complexity $T(n)$, then there exists a $2f\,H_n$-approximation algorithm for the ULP, with time complexity $O(n\,m\,T(n))$.*

The Quotient Cut Problem is NP-Hard and the best approximation algorithm has an approximation factor of 4 due to Freivalds [9]. This algorithm has experimental time complexity $O(n^{1.6})$ when the degree of each vertex is a small constant and $O(n^{2.6})$ for dense graphs. Although this algorithm has polynomial time complexity estimated experimentally, it is not proved to be of polynomial time in the worst case.

**Theorem 11.** *There is an $8\log$-approximation algorithm for the ULP, with time complexity estimated in $O(m\,n^{3.6})$.*

Note that the size of an instance $I$, $size(I)$, is $O(n(n+m))$, thus, the estimated complexity of our algorithm is $O(size(I)^{2.3})$.

## 4    Computational Experiments

We performed several tests with the algorithm GUL, the algorithm presented by Kleinberg and Tardos [12], denoted by $A_K$, and the algorithm presented by Chekuri et al. [4], denoted by $A_C$. The tests were performed over instances generated computationally and from image restoration problem. The implementation of the $\mathcal{A}_{SC}$ algorithm, called at step 4 of the algorithm GUL, is such that it generates solutions for the Set Cover Problem without intersections between the sets. We observe that the computational resources needed to solve an instance with the presented algorithm are very small compared with the previous approximation algorithms cited above. All the implemented algorithms and the instances are available under request. The tests were performed in an Athlon XP with 1.2 Ghz, 700 MB of RAM, and the linear programs were solved by the Xpress-MP solver [14].

We started our tests setting $(n, m) \leftarrow (\lfloor \frac{2k}{3} \rfloor, \lceil \frac{k}{3} \rceil)$ for a series of values of $k$ and creating random instances as follows: $c_{ui} \leftarrow random(1000), \forall u \in P, \forall i \in L$ e $w_{uv} = random(10), \forall u, v \in P$ where $random(t)$ returns a random value between 0 and $t$. It must be clear that these are not necessary hard instances for the problem or for the algorithms. Unfortunately, we could not find hard instances for the uniform labeling problem publically available.

The largest instance executed by the algorithm $A_C$ has 46 objects and 24 labels and the time required was about 16722 seconds. When applying the algorithm GUL in the same instance, we obtained a result that is 0.25% worst in 7 seconds.

The largest instance solved by the algorithm $A_K$ has 80 objects and 40 labels and the time required to solve it was about 15560 seconds. When applying the GUL in the same instance, we obtained a result that is 1.01% worst in 73 seconds.

The major limitation to the use of the algorithm $A_K$ was the time complexity, while the major limitation to the use of the algorithm $A_C$ was the memory and time complexity. The times spent by the algorithms $A_K$ and $A_C$ are basically the times to solve the corresponding linear programs. See Figure 2 to compare the time of each algorithm.

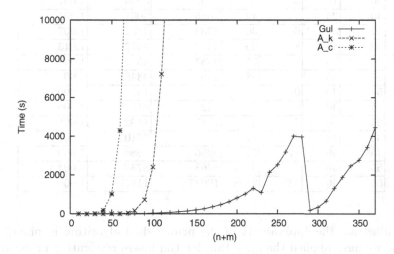

**Fig. 2.** Comparing the experimental time of the implemented algorithms.

One can observe in Figure 2 that there is a strong reduction in the function time associated with the GUL algorithm when $(n + m)$ achieves 290.

This behavior results from the following observation: when the separation costs are relatively bigger than the assignment cost, there are better possibilities that the size of the star become bigger, resulting in a less convergence time. The expectation of the separation costs in our instances, for stars with a small number of objects, is proportional to $n$, while the expectation of the connection costs is kept constant. Thus, growing $n + m$ also grows the separation cost. The average number of objects assigned by iteration just before the time reduction is 1.2, that is near the worst case, while for the instant just after the decreasing, the average number of objects assigned by interaction is 32.3.

For all these instances, for which we could solve the linear program relaxation, the maximum error ratio of the solutions obtained by the algorithm GUL and the solution of the linear relaxation is 0.0127 (or 1.27% worse).

We also perform some tests with the algorithms GUL and $A_K$ for instances generated by fixing $n$ or $m$ and varying the other. The results are presented in Table 1. The maximum error ratio obtained is 1.3% and the time gained is considerable. As expected, when the number of labels is fixed, the time gained is more significant.

**Table 1.** Classification times and costs using Kleinberg and Tardos formulation and the primal dual algorithm.

| Instance | | Times | | Costs | | |
|---|---|---|---|---|---|---|
| Objects | Labels | GUL | $A_K$ | $val(\text{GUL})$ | $val(A_K)$ | $\dfrac{val(\text{GUL})}{val(A_K)}$ |
| 40 | 5 | 0 | 1 | 9388 | 9371 | 1.001 |
| 40 | 15 | 3 | 6 | 5962 | 5923 | 1.007 |
| 40 | 25 | 5 | 12 | 5085 | 5061 | 1.005 |
| 40 | 35 | 6 | 30 | 4763 | 4732 | 1.007 |
| 40 | 45 | 8 | 49 | 4533 | 4477 | 1.013 |
| 40 | 55 | 10 | 75 | 4587 | 4552 | 1.008 |
| 40 | 65 | 11 | 68 | 4364 | 4348 | 1.004 |
| 5 | 40 | 0 | 1 | 137 | 137 | 1.000 |
| 15 | 40 | 0 | 1 | 812 | 812 | 1.000 |
| 25 | 40 | 2 | 3 | 1817 | 1802 | 1.008 |
| 35 | 40 | 4 | 17 | 3736 | 3710 | 1.007 |
| 45 | 40 | 11 | 77 | 5986 | 5986 | 1.000 |
| 55 | 40 | 22 | 400 | 8768 | 8735 | 1.004 |
| 65 | 40 | 33 | 2198 | 12047 | 11975 | 1.006 |

To illustrate the applicability of the primal dual algorithm in practical instances, we have applied the algorithm for the image restoration problem with an image degenerated by noise. The image has pixels in gray scale and dimension 60x60 with a total of 3600 pixels (objects) to be classified in black and white colors. To define the assignment and separation cost, we consider that each color is an integer between 0 and 255. The assignment cost of an object $u$ to a label $i$ is given by $c_{ui} = |clr(u) - clr(i)|$, where $clr(u)$ is the actual color of $u$ and $clr(i)$ is the color assigned to the label $i$. The separation cost of an object $u$ and an object $v$ is given by $w_{uv} = 255 - |clr(u) - clr(v)|$ if $u$ is one of the nine direct neighbors of $v$, $w_{uv} = 0$ in the other case.

The following images, presented in figures 3–6, present the results obtained applying the primal dual algorithm. In each figure, the image in the left is obtained inserting some noise and the image in the right is the image obtained after the classification of the objects.

The time difference between images classification is because the separation cost in images with less noise is bigger than in images with more noise. Thus

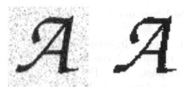 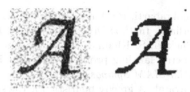

**Fig. 3.** Image with 25% of noise. Time needed is 288 seconds

**Fig. 4.** Image with 50% of noise. Time needed is 319 seconds

**Fig. 5.** Image with 75% of noise. Time needed is 511 seconds

**Fig. 6.** Image with 100% of noise. Time need is 973 seconds

the average size of the stars chosen by the algorithm in images with less noise is bigger, resulting in a minor convergence time.

Although these times are large for image processing applications, it illustrates its performance and solution quality. In addiction, in real instances, it is possible to define a small windows over greater images, and the processing time will decrease. Clearly, the classification problem is more general and the primal dual algorithm is appropriate for moderate and large size instances.

## 5    Concluding Remarks

We have presented a primal dual algorithm with approximation factor $8 \log n$ for the Uniform Labeling Problem. We compared the primal dual algorithm with LP-based approximation algorithms. The previous approximation algorithms for this problem have high time complexity and are adequate only for small and moderate size instances. The presented algorithm could obtain high quality solutions. The average error ratio of the presented algorithm was less than 2% of the optimum solution, when it was possible to compare, and it could obtain solutions for moderate and large size instances.

We would like to thank K. Freivalds to made available his code for the Quotient Cut Problem.

## References

1. J. Besag. Spatial intteraction and the statistical analysis of lattice systems. *J. Royal Statistical Society B*, 36, 1974.
2. J. Besag. On the statistical analysis of dirty pictures. *J. Royal Statistical Society B*, 48, 1986.

3. S. Chakrabarti, B. Dom, and P. Indyk. Enhanced hypertext categorization using hyperlinks. In *Proc. ACM SIGMOD*, 1998.
4. C. Chekuri, S. Khanna, J.S. Naor, and L. Zosin. Approximation algorithms for the metric labeling problem via a new linear programming formulation. In *Proc. of ACM-SIAM Symposium on Discrete Algorithms*, pages 109–118, 2001.
5. V. Chvátal. A greedy heuristic for the set-covering problem. *Math. of Oper. Res.*, 4(3):233–235, 1979.
6. F.S. Cohen. Markov random fields for image modeling and analysis. *Modeling and Application of Stochastic Processes*, 1986.
7. E. Dahlhaus, D.S. Johnson, C.H. Papadimitriou, P.D. Seymour, and M. Yannakakis. The complexity of multiterminal cuts. *SIAM Journal on Computing*, 23(4):864–894, 1994.
8. R. Dubes and A. Jain. Random field models in image analysis. *J. Applied Statistics*, 16, 1989.
9. K. Freivalds. A nondifferentiable optimization approach to ratio-cut partitioning. In *Proc. 2nd Workshop on Efficient and Experimental Algorithms*, Lectures Notes on Computer Science, LNCS 2647. Springer-Verlag, 2003.
10. A. Gupta and E. Tardos. Constant factor approximation algorithms for a class of classification problems. In *Proceedings of the 32nd Annual ACM Symposium on the Theory of Computing*, pages 125–131, 1998.
11. K. Jain, M. Mahdian, E. Markakis, A. Saberi, and V. Vazirani. Greedy facility location algorithms analyzed using dual fitting with factor-revealing lp. *Journal of ACM*, pages 795–824, 2003.
12. J. Kleinberg and E. Tardos. Approximation algorithms for classification problems with pairwise relationships: Metric labeling and markov random fields. In *Proceedings of the 40th Annuall IEEE Symposium on Foundations of Computer Science*, pages 14–23, 1999.
13. T. Leighton and S. Rao. Multicommodity max-flow min-cut theorems and their use in designing approximation algorithms. *Journal of the ACM*, 46 , No. 6:787–832, Nov. 1999.
14. Dash Optimization. *Xpress-MP Manual. Release 13*. 2002.

# Distributed Circle Formation for Anonymous Oblivious Robots*

Ioannis Chatzigiannakis[1], Michael Markou[2], and Sotiris Nikoletseas[1,2]

[1] Computer Technology Institute, P.O. Box 1122, 26110 Patras, Greece.
{ichatz,nikole}@cti.gr
[2] Department of Computer Engineering and Informatics, University of Patras, 26500
Patras, Greece.
markou@ceid.upatras.gr

**Abstract.** This paper deals with systems of multiple mobile robots each of which observes the positions of the other robots and moves to a new position so that eventually the robots form a circle. In the model we study, the robots are anonymous and oblivious, in the sense that they cannot be distinguished by their appearance and do not have a common x-y coordinate system, while they are unable to remember past actions. We propose a new distributed algorithm for circle formation on the plane. We prove that our algorithm is correct and provide an upper bound for its performance. In addition, we conduct an extensive and detailed comparative simulation experimental study with the DK algorithm described in [7]. The results show that our algorithm is very simple and takes considerably less time to execute than algorithm DK.

## 1   Introduction, Our Results, and Related Work

Lately, the field of cooperative mobile robotics has received a lot of attention from various research institutes and industries. A focus of these research and development activities is that of distributed motion coordination, since it allows the robots to form certain patterns and move in formation towards cooperating for the achievement of certain tasks. Motion planning algorithms for robotic systems made up from robots that change their position in order to form a given pattern is very important and may become challenging in the case of severe limitations, such as in communication between the robots, hardware constraints, obstacles etc.

The significance of positioning the robots based on some given patterns may be useful for various tasks, such as in bridge building, in forming adjustable buttresses to support collapsing buildings, satellite recovery, or tumor excision [12].

* This work has been partially supported by the IST Programme of the European Union under contract numbers IST-2001-33116 (FLAGS) and IST-2001-33135 (CRESCCO).

C.C. Ribeiro and S.L. Martins (Eds.): WEA 2004, LNCS 3059, pp. 159–174, 2004.

Also, distributed motion planning algorithms for robotic systems are potentially useful in environments that are inhospitable to humans or are hard to directly control and observe (e.g. space, undersea).

In this paper, we consider a system of multiple robots that move on the plane. The robots are anonymous and oblivious, i.e. they cannot be distinguished by a unique id or by their appearance, do not have a common x-y coordinate system and are unable to remember past actions. Furthermore, the robots are unable to communicate directly (i.e. via a wireless transmission interface) and can only interact by observing each other's position. Based on this model we study the problem of the robots positioning themselves to form a circle.

Remark that the formation of a circle provides a way for robots to agree on a common origin point and a common unit distance [15]. Such an agreement allows a group of robots to move in formation [9]. In addition, formation of patterns and flocking of a group of mobile robots is also useful for providing communication in ad-hoc mobile networks [5,4,3].

**Related Work.** The problem of forming a circle having a given diameter by identical mobile robots was first discussed by Sugihara and Suzuki [14]; they proposed a simple heuristic distributed algorithm, which however forms an approximation of a circle (that reminds a Reuleaux triangle). In an attempt to overcome this problem, Suzuki and Yamasihita [16] propose an algorithm under which the robots eventually reach a configuration where they are arranged at regular intervals on the boundary of a circle. However, to succeed in forming the pattern, the robots must be able to remember all past actions. Lately, Défago and Konogaya [7] designed an algorithm that manages to form a proper circle.

Under a similar model, in which robots have a limited vision, Ando et al. [1] propose an algorithm under which the robots converge to a single point. Flochini et al. [8] study the same problem by assuming that the robots have a common sense of direction (i.e. through a compass) and without considering instantaneous computation and movement.

We here note that there exist other models and problems for the development of motion planning algorithms for robotic systems, e.g. [6,17]. In that model most of the existing motion planning strategies rely on centralized algorithms to plan and supervise the motion of the system components [6], while recently efficient distributed algorithms have been proposed [17]. Our work is also inspired by problems of coordinating pebble motion in a graph, introduced in [10].

**Our Contribution.** In this work we use and extend the system model stated by Défago and Konogaya [7]. Under this model we present a new distributed algorithm that moves a team of anonymous mobile robots in such a way that a (non degenerate) circle is formed. The new algorithm presented here is based on the observation that the Défago - Konogaya algorithm (DK algorithm) [7] is using some very complex computational procedures. In particular, the use of (computationally intensive) Voronoi diagrams in the DK algorithm is necessary to avoid the very specific possibility in which at least two robots share at some time the same position and also have a common coordinate system. We remark that in many cases (e.g. when the system is comprised of not too many robots and/or it covers a large area) this possibility is not very probable. Based on this remark, we provide a new algorithm, which avoids the expensive Voronoi diagram calculations. Instead our algorithm just moves the robots towards the closest point on the circumference of the smallest enclosing circle. We prove that our algorithm is correct and that the robots travel shorter distance than when executing the DK algorithm, i.e. the performance of the DK algorithm is an upper bound for the performance our algorithm.

Furthermore we conduct a very detailed comparative simulation experimental study of our algorithm and the DK algorithm, in order to validate the theoretical results and further investigate the performance of the two algorithms. We remark that this is the first work on implementing and simulating the DK algorithm. The experiments show that the execution of our algorithm is very simple and takes considerably less time to complete than the DK algorithm. Furthermore, our algorithm seems to be more efficient (both with respect to number of moves and distance travelled) in systems that are made up from a large number of mobile robots.

We now provide some definitions that we will use in the following sections.

**Smallest Enclosing Circle.** The *smallest enclosing circle* of a set of points $P$ is denoted by $\mathsf{SEC}(P)$. It can be defined by either two opposite points, or by at least three points. The smallest enclosing circle is unique and can be computed in $\mathcal{O}(n \log n)$ time [13].

**Voronoi Diagram.** The *Voronoi diagram* of a set of points $P = \{p_1, p_2, \ldots, p_n\}$, denoted by $\mathsf{Voronoi}(P)$, is a subdivision of the plane into $n$ cells, one for each point in $P$. The cells have the property that a point $q$ in the plane belongs to the Voronoi cell of point $p_i$, denoted $Vcell_{p_i}(P)$, if and only if, for any other point $p_j \in P, \mathrm{dist}(q, p_i) \leq \mathrm{dist}(p_j, q)$, where $\mathrm{dist}(p, q)$ is the Euclidean distance between two points $p$ and $q$. In particular, the strict inequality means that points located on the boundary of the Voronoi diagram do not belong to any Voronoi cell. More details on Voronoi diagrams can be found in [2].

## 2   The Model

Let $r_1, r_2, \ldots, r_n$ be a set of extremely small robots, modelled as mobile processors with infinite memory and a sensor to detect the instantaneous position of all robots (i.e. a radar). Movement is accomplished with very high precision in an unbounded two dimensional space devoid of any landmark. Each robot $r_i$ uses its own local x-y coordinate system (origin, orientation, distance) and has no particular knowledge of the local coordinate system of other robots, nor of a global coordinate system. It is assumed that initially all robots occupy different positions, although, due to their small size, two or more robots may occupy the same position at some time. We furthermore assume that no two robots are located on the same radius of the Smallest Enclosing Circle, an assumption needed to guarantee the correctness of our protocol. Note that due to the small size of the robots, the probability of failure of our algorithm is very small (as can be shown by a simple balls-and-bins argument).

Robots are anonymous in the sense that they are unable to uniquely identify themselves. All robots execute the same deterministic algorithm, and thus have no way to generate a unique identity for themselves; more generally, no randomization can be used and any two independent executions of the algorithm with identical input values always yield the same output.

Time is represented as an infinite sequence of time instants $t_0, t_1, t_2, \ldots$ and at each time instant $t$, every robot $r_i$ is either *active* or *inactive*. Without loss of generality we assume that at least one robot is active at every time instance. Each time instant during which a robot becomes active, it computes a new position using a given algorithm and moves towards that position. Conversely, when a robot is inactive, it stays still and does not perform any local computation. We use $A_t$ to denote the set of active robots at $t$ and call the sequence $\mathcal{A} = A_0, A_1, \ldots$ an *activation schedule*. We assume that every robot becomes active at infinite many time instants, but no additional assumptions are made on the timing with which the robots become active. Thus $\mathcal{A}$ needs satisfy only the condition that every robot appears in infinitely many $\mathcal{A}'s$.

Given a robot $r_i$, $p_i(t)$ denotes its position at time $t$, according to some global x-y coordinate system (which is not known to the robots) and $p_i(0)$ is its initial position. $P(t) = \{p_i(t) \,\|\, 1 \leq i \leq n\}$ denotes the multiset of the position of all robots at time $t$.

The algorithm that each robot $r_i$ uses is a function $\phi$ that is executed each time $r_i$ becomes active and determines the new position of $r_i$, which must be within one distance unit of the previous position, as measured by $r_i$'s own coordinate system. The arguments to $\phi$ consists of the current position of $r_i$ and the multiset of points containing the observed positions of all robots at the corresponding time instant, expressed in terms of the local coordinate system of $r_i$. It is assumed that obtaining the information about the system, computing the new position and moving towards it is instantaneous. Remark that in this paper we consider oblivious algorithms and thus $\phi$ is not capable of storing any information on past actions or previous observations of the system. The model we use is similar to that of by Défago and Konogaya [7], which in turn is based on the model of Suzuki and Yamashita [16].

## 3    The Problem

In this paper we consider the problem of positioning a set of mobile robots in such a way so that a circle is formed, with finite radius greater than zero. We call such a circle *non degenerate*. We also consider the more difficult problem in which the robots are arranged at regular intervals on the boundary of the circle.

**Problem 1 (Circle Formation).** Given a group of $n$ robots $r_1, r_2, \ldots, r_n$ with distinct positions, located arbitrarily on the plane, arrange them to eventually form a non degenerate circle.

**Problem 2 (Uniform Circle Formation).** Given a group of $n$ robots $r_1, r_2, \ldots, r_n$ with distinct positions, located arbitrarily on the plane, eventually arrange them to eventually at regular intervals on the boundary of a non degenerate circle.

A possible way to solve the problem of forming a uniform circle is to form a "simple" circle and then *transform* the robot configuration as follows.

**Problem 3 (Uniform Transformation).** Given a group of $n$ robots $r_1, r_2, \ldots, r_n$ with distinct positions, located on the boundaries of a non degenerate circle, eventually arrange them at regular intervals on the boundary of the circle.

# 4    A New Circle Formation Algorithm

## 4.1    The Défago-Konogaya (DK) Algorithm

We first briefly describe the Défago-Konogaya circle formation algorithm (the DK algorithm) that is described in [7]. The algorithm relies on two facts: (i) the environment observed by all robots is the same, in spite of the difference in local coordinate system and (ii) the smallest enclosing circle is unique and depends only on the relative positions of the robots. Based on these two facts, the algorithm makes sure that the smallest enclosing circle remains invariant and uses it as a common reference.

Initially, given an arbitrary configuration in which all robots have distinct positions, a sub-algorithm ($\phi_{circle}$) brings the system towards a configuration in which all robots are located on the boundary of the circle (i.e. solves prob. 1). Then, a second sub-algorithm ($\phi_{uniform}$) converges towards a homogeneous distribution of the robots along that circumference, but it does not terminate (i.e. solves prob. 3). Clearly, the combination of the above two sub-algorithms solves the problem of Uniform Circle Formation (prob. 2).

**Circle Formation Algorithm $\phi_{circle}$.** The main idea of the algorithm is very simple: robots that are already on the boundary of the circle do not move and robots that are in the interior of the circle are made to move towards the boundary of the circle. When a robot that is located in the interior of the circle is activated, it observes the positions of the other robots and computes the Voronoi diagram. Given the boundaries of $\mathsf{SEC}(P)$ and the Voronoi cell where the robot

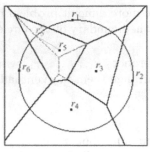 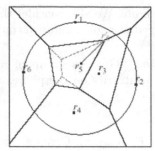 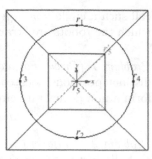

(a) The smallest enclosing    (b) The smallest enclosing    (c) Symmetry is broken by
circle is reachable           circle is unreachable         using the local coordinate
                                                            system

**Fig. 1.** Illustration of algorithm $\phi_{circle}$, executed by robot $r_i$ (in each case, $r_i$ moves towards $r_i'$)

is located, it can find itself in either one of three types of situations:

**Case 1:** When the circle intersects the Voronoi cell of the robot (see Fig. 1a), the robot moves towards the intersection of the circle and the Voronoi cell.

**Case 2:** When the Voronoi cell of the robot does not intersect with the circle (see Fig. 1b), the robot selects the point in its Voronoi cell which is nearest to the boundary of the circle (or farthest from its center).

**Case 3:** Due to symmetry, there exist several points (see Fig. 1c). In this case, all solutions being the same, one is selected arbitrarily. This is for instance done by keeping the solution with the highest x-coordinate (and then y-coordinate) according to the local coordinate system of the robot.

**Uniform Transformation Algorithm** $\phi_{uniform}$. Given that all robots are located on the circumference of a circle, the algorithm $\phi_{uniform}$ converges toward a homogeneous distribution of robots, but does not terminate deterministically. The algorithm $\phi_{uniform}$ works as follows: when a robot $r_i$ becomes active, it considers its two direct neighbors $prev(r_i)$ and $next(r_i)$ and computes the midpoint between them and moves halfway towards it.

The reason for moving halfway toward the midpoint rather than toward the midpoint itself is to prevent situations where the system oscillates endlessly between two different configurations when robots are perfectly synchronized. The system would get stuck into an infinite cycle and hence be unable to progress toward an acceptable solution.

The authors prove that algorithm $\phi_{circle}$ is correct (*Theorem 2*, [7]) and solves the problem of Circle Formation (prob. 1) and that algorithm $\phi_{uniform}$ converges toward a configuration wherein all robots are arranged at regular intervals on the boundary of a circle (*Theorem 3*, [7]) and thus solves the problem of Uniform Transformation (prob. 3).

## 4.2   Our Algorithm ("direct")

We now present a new distributed algorithm (which we call direct) that moves a set of oblivious mobile robots in such a way that a (non degenerate) circle is formed . The new algorithm is motivated by the observation that the DK algorithm is using some very complex (i.e. computationally intensive) procedures. Indeed, computing the Voronoi diagram based on the locations of the robots and then moving towards the smallest enclosing circle, involves computationally complex procedures. Instead, our algorithm is much simple since it just moves the robots towards the closest point on the circumference of the smallest enclosing circle. In Fig. 2, we provide a pseudo-code description. In more details, under algorithm direct a robot can find itself in either one of the following two situations:

**Case 1:** Robot $r$ is located on the boundary of $SEC(P)$; $r$ stays still.
**Case 2:** Robot $r$ is located inside $SEC(P)$; $r$ selects the closest point of the boundary of $SEC(P)$ to its current position and moves towards this point.

```
function φ'_circle(P, p_i)
 1:  begin
 2:     if (p_i ∈ SEC(P) == true)
 3:        begin
 4:          stay still
 5:        end
 6:     else
 7:        begin
 8:          target := point on boundary of SEC(P) closest to p_i
 9:          move toward target
10:        end
11:  end
```

**Fig. 2.** Formation of an (arbitrary) circle (code executed by robot $r_i$)

In order to prove the correctness of algorithm direct we work in a similar way as in [7]. We first prove that when all robots are positioned on the boundary of the smallest enclosing circle, the algorithm terminates (Lemma 1). We then show that the robots that are not positioned on the boundary of $SEC(P)$ will always move towards the boundary (and not further away from $SEC(P)$, Lemma 2). Finally we show that a finite period of time is needed for those robots that are positioned inside $SEC(P)$ to move towards the boundary of $SEC(P)$ (Lemma 4).

**Lemma 1.** Under Algorithm direct, all configurations in which the smallest enclosing circle passes through all robots are stable.

*Proof.* In such configurations, it is $p_i \in SEC(P), \forall i \in \{1, 2, \ldots, n\}$. Thus condition 2 in the pseudo-code description is satisfied for all robots, thus every robot stays still (line 4) and the total configuration is stable.    □

**Lemma 2.** No robot moves to a position further away from $SEC(P)$ (than its current position).

*Proof.* Under algorithm direct, robots move only at lines 4 and 9. Let us consider each case for some arbitrary robot $r$.

1. At line 4, $r$ stays still, so it obviously does not move towards the center.
2. At line 9, $r$ moves from the interior of the circle toward a point located on the boundary of the circle. So, $r$ actually progresses *away* from the center.

Thus, the robot $r$ is unable to move further away (i.e. towards the circle center) from the boundary of the smallest enclosing circle in any of the two cases.    □

**Lemma 3.** There exists at least one robot in the interior of $SEC(P)$ that can progress a non null distance towards $SEC(P)$.

*Proof.* Under algorithm direct the motion of the robots that are located in the interior of the smallest enclosing circle, is not blocked by other robots, since each robot's movement in our algorithm is independent of the positions of other robots. Therefore, at any given time instance $t$, robot $r_i$ that is active and is located in the interior of the smallest enclosing circle, will move towards the boundary of the circle.

Thus, if a set of robots is located in the interior of $SEC(P)$ and at least one robot is active, there exists at least one robot (an active one) in the interior of $SEC(P)$ that can progress a non null distance towards $SEC(P)$. This hence proves the lemma.    □

**Lemma 4.** All robots located in the interior of $SEC(P)$ reach its boundary after a finite number of activation steps.

*Proof.* By Lemma 2 no robot ever moves backwards. Let us denote by $P_{in}$ the set of all robots located in the interior of $SEC(P)$. By Lemma 3, there is at least one robot $r \in P_{in}$ that can reach $SEC(P)$ in a finite number of steps. Once $r$ has reached $SEC(P)$, it does not belong to $P_{in}$ anymore. So by Lemma 3, there must be another robot in $P_{in}$ that can reach $SEC(P)$ in a finite number of steps. Since there is a finite number of robots in $P$, there exists some finite time after which all robots are located on the boundary of the smallest enclosing circle $SEC(P)$.    □

**Theorem 1.** *Algorithm* direct *solves the problem of Circle Formation (prob. 1).*

*Proof.* By Lemma 4, there is a finite time after which all robots are located on the smallest enclosing circle, and this configuration is stable (by Lemma 1). Consequently, algorithm direct solves the Circle Formation problem.     □

**Corollary 1.** The combination of functions $\phi'_{circle}$ and $\phi_{uniform}$ solves the problem of Uniform Circle Formation (prob. 2).

*Remark 1.* Corollary 1 assumes that no two robots end-up at the same circle position after executing $\phi'_{circle}$. This is in fact guaranteed in our model, since no two robots are initially located exactly on the same radius. Note however that this modelling assumption is not very restrictive. Indeed the probability of such an event is upper bounded by

$$m \cdot \left[ 1 - \left( 1 - \frac{1}{m} \right)^n - n \cdot \frac{1}{m} \left( 1 - \frac{1}{m} \right)^{n-1} \right]$$

where $m$ is the number of "different" radiuses and $n$ is the number of robots. Clearly, when $m$ tends to infinity (due to the small size and high precision of robots) then this upper bound tends to 0.

We now provide a comparison of the performance of our algorithm to that of the DK algorithm. More specifically, we show that the distance covered by the robots when executing algorithm direct is at most the distance covered when executing algorithm DK. This is the initial intuition that led us to the design of the new algorithm.

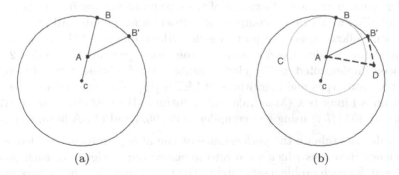

(a)                                              (b)

**Fig. 3.** Distance comparison of next position under the two algorithms

**Theorem 2.** A robot $r$ that executes algorithm direct covers distance less or equal to the distance covered by $r$ when executing function $\phi_{circle}$ of algorithm DK, given that the initial conditions of the system are identical. In other words, the distance covered by $r$ when executing $\phi_{circle}$ is an *upper bound* for the distance covered by $r$ when executing algorithm direct.

*Proof.* Let $r$ be a robot positioned at point $A$ in the plane which is inside the boundaries of $SEC(P)$ (see Fig. 3a) and $r$ is executing algorithm direct. When $r$ is activated, it will move on the line that connects the center of $SEC(P)$ (point $c$) and $A$. Regardless of the number of steps that will be required in order to reach the boundary of the circle, $r$ will keep moving along the line $cA$. Let $B$ the point where $cA$ intersects $SEC(P)$. Therefore the distance covered by $r$ is equal to $|AB|$.

Now let assume that $r$ is executing function $\phi_{circle}$ of algorithm DK. When $r$ is activated, it will start moving away from $A$ towards a point $B'$ located on the boundary of $SEC(P)$, i.e. not necessarily on the line $AB$.

Let $C$ be a circle centered on $A$ having radius equal to $|AB|$. We observe that any point $B'$ (other than $B$) that lies on the boundary of $SEC(P)$ is positioned outside the boundary of $C$ and thus, based on the definition of the circle, further from the center of $C$ (i.e. $A$). Therefore $|AB| < |AB'|$.

In the case when $r$ is moving towards $B'$ after first passing through a point $D$ outside the $C$ circle, its route will clearly greater length than $AB'$ (by the triangle inequality) and thus greater than $AB$. For instance, considering the example of Fig. 3b, it holds that: $|AD| + |DB'| > |AB'| > |AB|$.

Thus, if $r$ follows any other path than $AB$ in order to reach the boundary of $SEC(P)$, it will cover grater distance, which proves the theorem. □

## 5   Experimental Evaluation

In order to evaluate the performance of the two protocols and further investigate their behavior we carried a comparative experimental study. All of our implementations follow closely the protocols described above and the simulation environment that we developed is based on the model presented in section 2. They have been implemented as C++ classes using several advanced two-dimensional geometry data types and algorithms of LEDA [11]. The experiments were conducted on a Linux box (Mandrake 8.2, Pentium III at $933Mhz$, with $512MB$ memory at $133Mhz$) using g++ compiler $ver.2.95.3$ and LEDA library $ver.4.1$.

In order to evaluate the performance of the above protocols we define a set of efficiency measures which we use to measure the efficiency of each protocol considered, for each problem separately. We start by defining the average number of steps performed by each robot, i.e. the total number of steps performed by robot $r_i$ when executing function $\phi$ and then calculate the average number of steps dividing by the number of robots. More formally this is defined as follows:

**Definition 1.** Let $M_\phi^i$ be the *total number of steps* performed by robot $r_i$ when executing function $\phi$. Let $\mathrm{avg}M_\phi = \frac{\sum_{i=1}^{n} M_\phi^i}{n}$ be the *average number of steps* required for function $\phi$ to form a given pattern, where $n$ is the number of robots in the system.

Although $\mathrm{avg}M_\phi$ provides a way to compare the performance of the two protocols, we cannot use the total number of steps for measuring the time complexity of a formation algorithm since a robot may remain inactive for an unpredictable period of time. An alternative measure, also proposed in [16], is the total distance that a robot must move to form a given pattern.

**Definition 2.** Let $D_\phi^i$ be the *total distance* covered by robot $r_i$ when executing function $\phi$. Let $\mathrm{avg}D_\phi = \frac{\sum_{i=1}^{n} D_\phi^i}{n}$ be the *average distance* required for function $\phi$ to form a given pattern, where $n$ is the number of robots in the system.

Finally, another important efficiency measure is the total time of execution required by function $\phi$ to form a given pattern. The computational complexity is a very important efficiency measure since formation algorithms are executed in real time and the computational power of the mobile modules can be a limiting factor. Although the function $\phi$ is executed by each robot in a distributed fashion, we consider the overall time required by the robots until the given pattern is formed. Remark that when calculating the execution of function $\phi$ we assume that the motion of the robots takes negligible time (i.e. the motion of the robots is instantaneous).

**Definition 3.** Let $T_\phi^i$ be the *execution time* required by robot $r_i$ to execute function $\phi$. Let $\mathrm{tot}T = \sum_{i=1}^{n} T_\phi^i$ be the *total execution time* required for function $\phi$ to form a given pattern, where $n$ is the number of robots in the system.

Based on the above, we start our experimentation by considering the effect of the dimensions of the Euclidean space where the mobile robots move on the performance of the two algorithms. More specifically we use a fixed number of mobile robots (100) on an Euclidian plane of dimensions $X \in [10, 50]$ and $X = Y$. The robots are positioned using a random uniform distribution. Each experiment is repeated for at least 100 times in order to get good average results.

Figure 4 depicts the average number of steps and average distance travelled under the two algorithms for the *circle formation problem* (i.e. the first phase). We observe that the performance of the two algorithms is more or less linear in the dimensions of the plane. Actually, it seems that the gradient of the lines is similar for both algorithms. It is evident that algorithm direct manages to form a circle requiring slightly fewer number of steps than the DK algorithm and the robots move in a way such that less distance is covered. This result verifies the theoretical upper bound of Theorem 2.

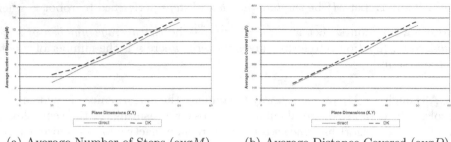

(a) Average Number of Steps (avg$M$)        (b) Average Distance Covered (avg$D$)

**Fig. 4.** Circle Formation: Evaluating the performance for various Plane Dimensions

The results of the second phase of the algorithm, i.e. for the *uniform transformation problem*, are shown in Fig. 5 for the same efficiency measures. In this phase, the DK algorithm is performing substantially better regarding the average number of steps while the robots cover slightly less distance than in algorithm direct. Again we observe that the performance of the two algorithms is more or less linear in the dimensions of the plane.

In Fig. 6 we get the combination of the results for the first and second phase, i.e. when considering the *uniform circle formation problem*. First, we observe that the average number of steps performed by the robots (see Fig. 6a) in the second phase ($\phi_{uniform}$) is dominating the overall performance of the two algorithms. Thus, the DK algorithm manages to form a uniform circle with a much smaller average number of steps per robot. However, regardless of the number of steps, under both algorithms the robots seem to cover similar distances (see Fig. 6b).

Figure 10 depicts the total time required to execute each algorithm for both phases. Clearly, algorithm direct executes in significantly less time and actually the graph suggests that it takes about 60 times less than the DK algorithm. It is evident that algorithm direct is very simple to execute making the total execution time almost independent from the plane dimensions while the execution time of the DK algorithm seems to be linear to the plane dimensions.

In the second set of experiments we investigate the performance of the two algorithms as the number of mobile robots increases ($m \in [50, 1000]$) while keeping the dimensions of the Euclidean plane fixed ($X = Y = 20$); that is, we investigate the effect of the density of robots on the performance of the two algorithms.

Figure 7a depicts the average number of steps as the density of robots increases for the first phase of the two algorithms (i.e. the *circle formation problem*). The graph shows that the performance of the DK algorithm, in terms of average number of steps, is linear in the total number of mobile robots. On the other hand, algorithm direct seems to be unaffected by this system parameter, especially when $m \geq 200$. This threshold behavior of the two algorithms is also observed in Fig. 7b. For low densities of robots, the average distance increases

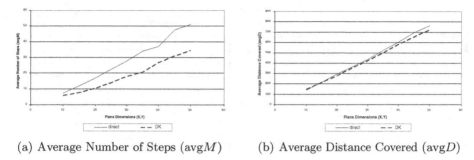

(a) Average Number of Steps (avg$M$)    (b) Average Distance Covered (avg$D$)

**Fig. 5.** Uniform Transformation: Evaluating the performance for various Plane Dimensions

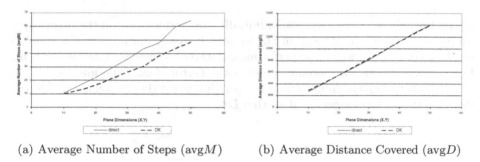

(a) Average Number of Steps (avg$M$)    (b) Average Distance Covered (avg$D$)

**Fig. 6.** Uniform Circle Formation: Evaluating the performance for various Plane Dimensions

slowly with $m$ until a certain number of robots is reached ($m = 200$) when no further increase is observed. Again, we observe that algorithm direct manages to form a circle requiring the robots to move in a way such that less distance is covered which verifies the theoretical upper bound of Theorem 2.

Regarding the second phase, in figure 8a we observe that both algorithms require a high number of steps, which, interestingly, decreases as the density of robots increases. However, for the DK algorithm, when the number of robots crosses the value $m = 200$, the average number of steps starts to increase linearly to the number of robots that make up the system. On the other hand, when $m \geq 200$, the performance of algorithm direct seems to remain unaffected by the density of the robots. As a result, when $m = 300$ the two algorithms achieve the same performance, while for $m \geq 300$ algorithm direct outperforms the DK algorithm.

The average distance covered by the robots under the two different algorithms, as the density of the system increases is shown in figure 8b. In this graph we observe that the two algorithms have a different behavior than that

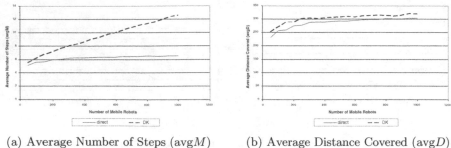

(a) Average Number of Steps (avg$M$)      (b) Average Distance Covered (avg$D$)

**Fig. 7.** Circle Formation: Evaluating the performance as the number of Mobile Robots increases

in the first phase. More specifically, initially the performance of the algorithms drops very fast as the density of the system increases until a certain value is reached ($m = 200$) when the performance starts to increase linearly with $m$. Furthermore, taking into account the statistical error of our experiments, Fig. 8b suggests that when the robots execute algorithm direct they travel about 5% less distance than when executing algorithm DK.

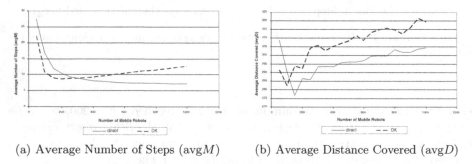

(a) Average Number of Steps (avg$M$)      (b) Average Distance Covered (avg$D$)

**Fig. 8.** Uniform Transformation: Evaluating the performance as the number of Mobile Robots increases

When considering the *uniform circle formation* problem (i.e. for both phases of the algorithms) it is clear that the overall behavior of the two algorithm is dominated by the second phase (see Fig. 9a). It seems that algorithm direct requires a constant number of steps when $m \geq 300$ while the DK algorithm's performance is linear to the number of mobile robots that make up the system. As for the average distance covered, again, the behavior of the algorithms is dominated by the second phase. Fig. 9b follows the same pattern as in Fig. 8b.

Finally, Fig. 11 depicts the overall time required by the robots to execute each algorithm. In this figure we clearly see that the DK algorithm has a very poor performance as the density of the robots in the system increases. On the other hand, the performance of algorithm direct is independent from the number of robots and is clearly more suitable for systems comprised of very large number of robots.

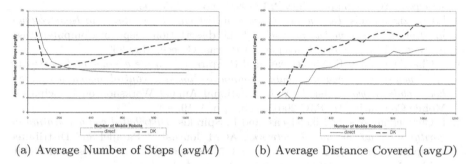

(a) Average Number of Steps (avg$M$)    (b) Average Distance Covered (avg$D$)

**Fig. 9.** Uniform Circle Formation: Evaluating the performance as the number of Mobile Robots increases

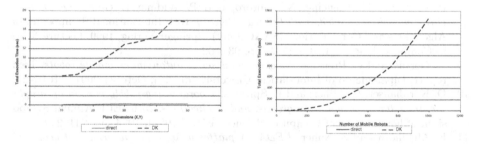

**Fig. 10.** Uniform Circle Formation: Total Execution Time (avg$T$) vs. Plane Dimensions

**Fig. 11.** Uniform Circle Formation: Total Execution Time (avg$T$) as the number of Mobile Robots increases

## 6   Closing Remarks

We presented a new algorithm for the problem of circle formation for systems made up from anonymous mobile robots that cannot remember past actions. We provided a proof of correctness and provided an upper bound on its performance. The experiments show that the execution our algorithm is very simple and takes considerably less time to complete than the DK algorithm and in systems that are made up from a large number of mobile robots, our algorithm is more efficient than DK (with respect to number of moves and distance travelled by the robots).

# References

1. H. Ando, I. Suzuki, and M. Yamashita, *Distributed memoryless point convergence algorithm for mobile robots with limited visibility*, IEEE Trans. on Robotics and Automation **15** (1999), no. 5, 818–828.
2. F. Aurenhammer, *Voronoi diagrams - a survey of a fundamental geometric data structure*, ACM Comput. Surv. **23** (1991), no. 3, 345–405.
3. I. Chatzigiannakis, *Design and analysis of distributed algorithms for basic communication in ad-hoc mobile networks*, Ph.D. dissertation, Dept. of Computer Engineering and Informatics, University of Patras, Greece, May 2003.
4. I. Chatzigiannakis, S. Nikoletseas, and P. Spirakis, *On the average and worst-case efficiency of some new distributed communication and control algorithms for ad-hoc mobile networks*, 1st ACM International Annual Workshop on Principles of Mobile Computing (POMC 2001), 2001, pp. 1–19.
5. I. Chatzigiannakis, S. Nikoletseas, and P. Spirakis, *Distributed communication algorithms for ad hoc mobile networks*, ACM Journal of Parallel and Distributed Computing (JPDC) **63** (2003), no. 1, 58–74, Special issue on Wireless and Mobile Ad-hoc Networking and Computing, edited by A. Boukerche.
6. G. Chirikjian, *Kinematics of a metamorphic robotic system*, IEEE International Conference on Robotics and Automation, 1994, pp. 449–455.
7. X. Défago and A. Konagaya, *Circle formation for oblivious anonymous mobile robots with no common sense of orientation*, 2nd ACM International Annual Workshop on Principles of Mobile Computing (POMC 2002), 2002.
8. P. Flocchini, G. Prencipe, N. Santoro, and P. Widmayer, *Gathering of asynchronous oblivious robots with limited visibility*, 10th International Symposium on Algorithms and Computation (ISAAC 1999), Springer-Verlag, 1999, Lecture Notes in Computer Science, LNCS 1741, pp. 93–102.
9. V. Gervasi and G. Prencipe, *Flocking by a set of autonomous mobile robots*, Tech. report, Dipartmento di Informatica, Università di Pisa, October 2001, TR-01-24.
10. D. Kornhauser, G. Miller, and P. Spirakis, *Coordinating pebble motion on graphs, the diameter of permutation groups, and applications*, 25th IEEE Annual Symposium of Foundations of Computer Science (FOCS 1984), 1984, pp. 241–250.
11. K. Mehlhorn and S. Näher, *LEDA: A platform for combinatorial and geometric computing*, Cambridge University Press, 1999.
12. A. Pamacha, I. Ebert-Uphoff, and G. Chirikjian, *Useful metrics for modular robot motion planning*, Transactions on Robotics and Automation **13** (1997), no. 4, 531–545.
13. S. Skyum, *A simple algorithm for computing the smallest enclosing circle*, Information Processing Letters **37** (1991), no. 3, 121–125.
14. K. Sugihara and I. Suzuki, *Distributed motion coordination of multiple mobile robots*, IEEE International Symposium on Intelligence Control, 1990, pp. 138,143.
15. I. Suzuki and M. Yamashita, *Agreement on a common x-y coordinate system by a group of mobile robots*, Dagstuhl Seminar on Modeling and Planning for Sensor-Based Intelligent Robots (Dagstuhl, Germany), September 1996.
16. I. Suzuki and M. Yamashita, *Distributed anonymous mobile robots: Formation of geometric patterns*, SIAM Journal of Computer Science **28** (1999), no. 4, 1347–1363.
17. J.E. Walter, J.L. Welch, and N.M. Amato, *Distributed reconfiguration of metamorphic robot chains*, 19th ACM Annual Symposium on Principles of Distributed Computing (PODC 2000), 2000, pp. 171–180.

# Dynamic Programming and Column Generation Based Approaches for Two-Dimensional Guillotine Cutting Problems

Glauber Cintra* and Yoshiko Wakabayashi**

Instituto de Matemática e Estatística, Universidade de São Paulo, Rua do Matão
1010, São Paulo 05508-090, Brazil.
{glauber,yw}@ime.usp.br

**Abstract.** We investigate two cutting problems and their variants in
which orthogonal rotations are allowed. We present a dynamic programming based algorithm for the *Two-dimensional Guillotine Cutting Problem with Value* (GCV) that uses the recurrence formula proposed by
Beasley and the discretization points defined by Herz. We show that if
the items are not so small compared to the dimension of the bin, this
algorithm requires polynomial time. Using this algorithm we solved all
instances of GCV found at the OR–LIBRARY, including one for which
no optimal solution was known. We also investigate the *Two-dimensional
Guillotine Cutting Problem with Demands* (GCD). We present a column
generation based algorithm for GCD that uses the algorithm above mentioned to generate the columns. We propose two strategies to tackle the
residual instances. We report on some computational experiments with
the various algorithms we propose in this paper. The results indicate that
these algorithms seem to be suitable for solving real-world instances.

## 1 Introduction

Many industries face the challenge of finding solutions that are the most econom-
ical for the problem of cutting large objects to produce specified smaller objects.
Very often, the large objects (bins) and the small objects (items) have only two
relevant dimensions and have rectangular shape. Besides that, a usual restriction
for cutting problems is that in each object we may use only *guillotine cuts*, that
is, cuts that are parallel to one of the sides of the object and go from one side
to the opposite one; problems of this type are called two-dimensional guillotine
cutting problems. This paper focuses on algorithms for such problems. They are
classical $\mathcal{NP}$-hard optimization problems and are of great interest, both from
theoretical as well as practical point-of-view.

This paper is organized as follows. In Section 2, we present some definitions
and establish the notation. In Section 3, we focus on the Two-dimensional Guil-

* Supported by CNPq grant 141072/1999-7.
** Partially supported by MCT/CNPq Project ProNEx 664107/97-4 and CNPq grants
304527/89-0 and 470608/01-3.

lotine Cutting Problem with Value (GCV) and also a variant of it in which the items are allowed to be rotated orthogonally.

Section 4 is devoted to the Two-dimensional Guillotine Cutting Problem with Demands (GCD). We describe two algorithms for it, both based on the column generation approach. One of them uses a perturbation strategy we propose to deal with the residual instances. We also consider the variant of GCD in which orthogonal rotations are allowed. Finally, in Section 5 we report on the computational results we have obtained with the proposed algorithms. In the last section we present some final remarks.

Owing to space limitations we do not prove some of the claims and we do not describe one of the approximation algorithms we have designed. For more details on these results we refer to [12].

## 2    Preliminaries

The *Two-dimensional Guillotine Cutting Problem with Value* (GCV) is the following: given a two-dimensional bin (a large rectangle), $B = (W, H)$, with width $W$ and height $H$, and a list of $m$ items (small rectangles), each item $i$ with width $w_i$, height $h_i$, and value $v_i$ $(i = 1, \ldots, m)$, determine how to cut the bin, using only guillotine cuts, so as to maximize the sum of the value of the items that are produced. We assume that many copies of the same item can be produced.

The *Two-dimensional Guillotine Cutting Problem with Demands* (GCD) is defined as follows. Given an unlimited quantity of two-dimensional bins $B = (W, H)$, with width $W$ and height $H$, and a list of $m$ items (small rectangles) each item $i$ with dimensions $(w_i, h_i)$ and demand $d_i$ $(i = 1, \ldots, m)$, determine how to produce $d_i$ unities of each item $i$, using the smallest number of bins $B$.

In both problems GCV and GCD we assume that the items are oriented (that is, rotations of the items are not allowed); moreover, $w_i \leq W$, $h_i \leq H$ for $i = 1, \ldots, m$. The variants of these problems in which the items may be rotated orthogonally are denoted by GCV$^r$ and GCD$^r$.

Our main interest in the problem GCV lies in its use as a routine in the column generation based algorithm for the problem GCD. While the first problem was first investigated in the sixties [18], we did not find in the literature results on the problem GCD. We observe that any instance of GCD can be reduced to an instance of the two-dimensional cutting stock problem (without demands), by substituting each item $i$ for $d_i$ copies of this item; but this reduction is not appropriate as the size of the new instance may become exponential in $m$.

We call each possible way of cutting a bin a *cutting pattern* (or simply *pattern*). To represent the patterns (and the cuts to be performed) we use the following convention. We consider the Euclidean plane $\mathbb{R}^2$, with the $xy$ coordinate system, and assume that the width of a rectangle is represented in the $x$-axis, and the height is represented in the $y$-axis. We also assume that the position $(0, 0)$ of this coordinate system represents the bottom left corner of the bin. Thus a bin of width $W$ and height $H$ corresponds to the region defined by the rectangle whose bottom left corner is at the position $(0, 0)$ and the top right corner is at the position $(W, H)$. To specify the position of an item $i$ in the bin, we specify the coordinates of its bottom left corner. Using these conventions, it is not difficult to define more formally what is a pattern and how we can represent one.

A *guillotine pattern* is a pattern that can be obtained by a sequence of guillotine cuts applied to the original bin and to the subsequent small rectangles that are obtained after each cut (see Figure 1).

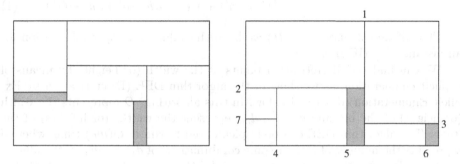

**Fig. 1.** (a) Non-guillotine pattern; (b) Guillotine pattern.

## 3  The Problem GCV

In this section we focus on the *Two-dimensional Guillotine Cutting Problem with Value* (GCV). We present first some concepts and results needed to describe the algorithm.

Let $I = (W, H, w, h, v)$, with $w = (w_1, \ldots, w_m)$, $h = (h_1, \ldots, h_m)$ and $v = (v_1, \ldots, v_m)$, be an instance of the problem GCV. We consider that $W$, $H$, and the entries of $w$ and $h$ are all integer numbers. If this is not the case, we can obtain an equivalent integral instance simply by multiplying the widths and/or the heights of the bin and of the items by appropriate numbers.

The first dynamic programming based algorithm for GCV was proposed by Gilmore and Gomory [18]. (It did not solve GCV in its general form.) We use here a dynamic programming approach for GCV that was proposed by Beasley [4] combined with the concept of discretization points defined by Herz [19]. A *discretization point of the width* (respectively of the *height*) is a value $i \leq W$ (respectively $j \leq H$) that can be obtained by an integer conic combination of $w_1, \ldots, w_m$ (respectively $h_1, \ldots, h_m$). We denote by $P$ (respectively $Q$) the set of all discretization points of the width (respectively height).

Following Herz, we say a *canonical pattern* is a pattern for which all cuts are made at discretization points (*e.g.*, the pattern indicated in Figure 1(b)).

It is immediate that it suffices to consider only canonical patterns (for every pattern that is not canonical there is an equivalent one that is canonical). To refer to them, the following functions will be useful. For a rational $x \leq W$, let $p(x) := \max(i \mid i \in P, \ i \leq x)$ and for a rational $y \leq H$, let $q(y) := \max(j \mid j \in Q, \ j \leq y)$.

Using these functions, it is not difficult to verify that the recurrence formula below, proposed by Beasley [4], can be used to calculate the value $V(w, h)$ of an optimal canonical guillotine pattern of a rectangle of dimensions $(w, h)$. In this formula, $v(w, h)$ denotes the value of the most valuable item that can be cut in a rectangle of dimensions $(w, h)$, or 0 if no item can be cut in the rectangle.

$$V(w, h) = \max \left( v(w, h), \{V(w', h) + V(p(w - w'), h) \mid w' \in P\}, \right.$$
$$\left. \{V(w, h') + V(w, q(h - h')) \mid h' \in Q\}\right). \quad (1)$$

Thus, if we calculate $V(W, H)$ we have the value of an optimal solution for an instance $I = (W, H, w, h, v)$.

We can find the discretization points of the width (or height) by means of explicit enumeration, as we show in the algorithm DEE (Discretization by Explicit Enumeration) described below. In this algorithm, $D$ represents the width (or height) of the bin and $d_1, \ldots, d_m$ represent the widths (or heights) of the items. The algorithm DEE can be implemented to run in $O(m\delta)$ time, where $\delta$ represents the number of integer conic combinations of $d_1, \ldots, d_m$ with value at most $D$. This means that when we multiply $D$, $d_1, \ldots, d_m$ by a constant, the time required by DEE is not affected.

It is easy to construct instances such that $\delta \geq 2^{\sqrt{D}}$. Thus, an explicit enumeration may take exponential time. But if we can guarantee that $d_i > \frac{D}{k}$ $(i = 1, \ldots, m)$, the sum of the $m$ coefficients of any integer conic combination of $d_1, \ldots, d_m$ with value at most $D$ is not greater than $k$. Thus, $\delta$ is at most the number of $k$-combinations of $m$ objects with repetition. Therefore, for fixed $k$, $\delta$ is polynomial in $m$ and consequently the algorithm DEE is polynomial in $m$.

---

**Algorithm 3.1** DEE

---

*Input*: $D$ (width or height), $d_1, \ldots, d_m$.
*Output*: a set $\mathcal{P}$ of discretization points (of the width or height).

$\mathcal{P} = \emptyset$, $k = 0$.
While $k \geq 0$ do
  For $i = k + 1$ to $m$ do $z_i = \lfloor (D - \sum_{j=1}^{i-1} d_j z_j)/d_i \rfloor$.
  $\mathcal{P} = \mathcal{P} \cup \{\sum_{j=1}^{m} z_j d_j\}$.
  $k = \max(\{i \mid z_i > 0, 1 \leq i \leq m\} \cup \{-1\})$.
  If $k > 0$ then $z_k = z_k - 1$ and $\mathcal{P} = \mathcal{P} \cup \{\sum_{j=1}^{k} z_j d_j\}$.
Return $\mathcal{P}$.

---

We can also use dynamic programming to find the discretization points. The basic idea is to solve a knapsack problem in which every item $i$ has weight and value $d_i$ $(i = 1, \ldots, m)$, and the knapsack has capacity $D$. The well-known dynamic programming technique for the knapsack problem (see [13]) gives the optimal value of knapsacks with (integer) capacities taking values from 1 to $D$.

It is easy to see that $j$ is a discretization point if and only if the knapsack with capacity $j$ has optimal value $j$. We have then an algorithm, which we call DDP (Discretization using Dynamical Programming), described in the sequel.

---

**Algorithm 3.2** DDP

---

*Input*: $D$, $d_1$, ..., $d_m$.
*Output*: a set $\mathcal{P}$ of discretization points.

$\mathcal{P} = \{0\}$.
For $j = 0$ to $D$ do $c_j = 0$.
For $i = 1$ to $m$ do
   For $j = d_i$ to $D$
      If $c_j < c_{j-d_i} + d_i$ then $c_j = c_{j-d_i} + d_i$
For $j = 1$ to $D$
   If $c_j = j$ then $\mathcal{P} = \mathcal{P} \cup \{j\}$.
Return $\mathcal{P}$.

---

We note that the algorithm DDP requires time $O(mD)$. Thus, the scaling (if needed) to obtain an integral instance may render the use of DDP unsuitable in practice. On the other hand, the algorithm DDP is suited for instances in which $D$ is small. If $D$ is large but the dimensions of the items are not so small compared to the dimension of the bin, the algorithm DDP has a satisfactory performance. In the computational tests, presented in Section 5, we used the algorithm DDP.

We describe now the algorithm $\mathcal{DP}$ that solves the recurrence formula (1). We have designed this algorithm in such a way that a pattern corresponding to an optimal solution can be easily obtained. For that, the algorithm stores in a matrix, for every rectangle of width $w' \in P$ and height $h' \in Q$, which is the direction and the position of the first guillotine cut that has to be made in this rectangle. In case no cut should made in the rectangle, the algorithm stores which is the item that corresponds to this rectangle.

When the algorithm $\mathcal{DP}$ halts, for each rectangle with dimensions $(p_i, q_j)$, we have that $V(i, j)$ contains the optimal value that can be obtained in this rectangle, $guillotine(i, j)$ indicates the direction of the first guillotine cut, and $position(i, j)$ is the position (in the $x$-axis or in the $y$-axis) where the first guillotine cut has to be made. If $guillotine(i, j) = nil$, then no cut has to be made in this rectangle. In this case, $item(i, j)$ (if nonzero) indicates which item corresponds to this rectangle. The value of the optimal solution will be in $V(r, s)$.

Note that each attribution of value to the variable $t$ can be done in $O(\log r + \log s)$ time by using binary search in the set of the discretization points. If we use the algorithm DEE to calculate the discretization points, the algorithm $\mathcal{DP}$ can be implemented to have time complexity $O(\delta_1 + \delta_2 + r^2 s \log r + r s^2 \log s)$, where $\delta_1$ and $\delta_2$ represent the number of integer conic combinations that produce the discretization points of the width and of the height, respectively.

## Algorithm 3.3 $\mathcal{DP}$

*Input*: An instance $I = (W, H, w, h, v)$ of GCV, where $w = (w_1, \ldots, w_m)$,
$\quad\quad h = (h_1, \ldots, h_m)$ and $v = (v_1, \ldots, v_m)$.

*Output*: An optimal solution for $I$.

Find $p_1 < \ldots < p_r$, the discretization points of the width $W$.
Find $q_1 < \ldots < q_s$, the discretization points of the height $H$.
For $i = 1$ to $r$
$\quad$ For $j = 1$ to $s$
$\quad\quad V(i,j) = \max(\{v_k \mid 1 \leq k \leq m,\ w_k \leq p_i \text{ and } h_k \leq q_j\} \cup \{0\})$.
$\quad\quad item(i,j) = \max(\{k \mid 1 \leq k \leq m,\ w_k \leq p_i,\ h_k \leq q_j \text{ and } v_k = V(i,j)\} \cup \{0\})$.
$\quad\quad guillotine(i,j) = nil$.
For $i = 2$ to $r$
$\quad$ For $j = 2$ to $s$
$\quad\quad n = \max(k \mid 1 \leq k \leq r \text{ and } p_k \leq \lfloor \frac{p_i}{2} \rfloor)$.
$\quad\quad$ For $x = 2$ to $n$
$\quad\quad\quad t = \max(k \mid 1 \leq k \leq r \text{ and } p_k \leq p_i - p_x)$.
$\quad\quad\quad$ If $V(i,j) < V(x,j) + V(t,j)$ then
$\quad\quad\quad\quad V(i,j) = V(x,j) + V(t,j)$, $position(i,j) = p_x$ and $guillotine(i,j) = \text{'V'}$.
$\quad\quad n = \max(k \mid 1 \leq k \leq s \text{ and } q_k \leq \lfloor \frac{q_j}{2} \rfloor)$.
$\quad\quad$ For $y = 2$ to $n$
$\quad\quad\quad t = \max(k \mid 1 \leq k \leq s \text{ and } q_k \leq q_j - q_y)$.
$\quad\quad\quad$ If $V(i,j) < V(i,y) + V(i,t)$ then
$\quad\quad\quad\quad V(i,j) = V(i,y) + V(i,t)$, $position(i,j) = q_y$ and $guillotine(i,j) = \text{'H'}$.

For the instances of GCV with $w_i > \frac{W}{k}$ and $h_i > \frac{H}{k}$ ($k$ fixed and $i = 1, \ldots, m$), we have that $\delta_1$, $\delta_2$, $r$ and $s$ are polynomial in $m$. For such instances the algorithm $\mathcal{DP}$ requires time polynomial in $m$.

We can use a vector $X$ (resp. $Y$), of size $W$ (resp. $H$), and let $X_i$ (resp. $Y_j$) contain $p(X_i)$ (resp. $q(Y_j)$). Once the discretization points are calculated, it requires time $O(W + H)$ to determine the values in the vectors $X$ and $Y$. Using these vectors, each attribution to the variable $t$ can be done in constant time. In this case, an implementation of the algorithm $\mathcal{DP}$, using DEE (resp. DDP) as a subroutine, would have time complexity $O(\delta_1 + \delta_2 + W + H + r^2 s + r s^2)$ (resp. $O(mW + mH + r^2 s + r s^2)$). In any case, the amount of memory required by the algorithm $\mathcal{DP}$ is $O(rs)$.

We can use the algorithm $\mathcal{DP}$ to solve the variant of GCV, denoted by GCV$^r$, in which orthogonal rotations of the items are allowed. For that, given an instance $I$ of GCV$^r$, we construct another instance (for GCV) as follows. For each item $i$ in $I$, of width $w_i$, height $h_i$ and value $v_i$, we add another item of width $h_i$, height $w_i$ and value $v_i$, whenever $w_i \neq h_i$, $w_i \leq H$ and $h_i \leq W$.

## 4 The Problem GCD and the Column Generation Method

We focus now on the *Two-dimensional Guillotine Cut Problem with Demands* (GCD). First, let us formulate GCD as an ILP (Integer Linear Program).

Let $I = (W, H, w, h, d)$ be an instance of GCD. Represent each pattern $j$ of the instance $I$ as a vector $p_j$, whose $i$-th entry indicates the number of times item $i$ occurs in this pattern. The problem GCD consists then in deciding how many times each pattern has to be used to meet the demands and minimize the total number of bins that are used. Let $n$ be the number of all possible patterns for $I$, and let $P$ denote an $m \times n$ matrix whose columns are the patterns $p_1, \ldots, p_n$. If we denote by $d$ the vector of the demands, then the following is an ILP formulation for GCD: minimize $\sum_{j=1}^{n} x_j$ subject to $Px = d$ and $x_j \geq 0$ and $x_j$ integer for $j = 1, \ldots, n$. (The variable $x_j$ indicates how many times the pattern $j$ is selected.)

Gilmore and Gomory [17] proposed a column generation method to solve the relaxation of the above ILP, shown below. The idea is to start with a few columns and then generate new columns of $P$, only when they are needed.

$$\begin{aligned} \text{minimize} \quad & x_1 + \ldots + x_n \\ \text{subject to} \quad & Px = d \\ & x_j \geq 0 \quad j = 1, \ldots, n. \end{aligned} \tag{2}$$

The algorithm $\mathcal{DP}$ given in Section 3 finds guillotine patterns. Moreover, if each item $i$ has value $y_i$ and occurs $z_i$ times in a pattern produced by $\mathcal{DP}$, then $\sum_{i=1}^{m} y_i z_i$ is maximum. This is exactly what we need to generate new columns. We describe below the algorithm SimplexCG$_2$ that solves (2).

The computational tests indicated that on the average the number of columns generated by SimplexCG$_2$ was $O(m^2)$. This is in accordance with the theoretical results that are known with respect to the average behavior of the Simplex method [1,7].

We describe now a procedure to find an optimal integer solution from the solutions obtained by SimplexCG$_2$. The procedure is iterative. Each iteration starts with an instance $I$ of GCD and consists basically in solving (2) with SimplexCG$_2$ obtaining $B$ and $x$. If $x$ is integral, we return $B$ and $x$ and halt. Otherwise, we calculate $x^* = (x_1^*, \ldots, x_m^*)$, where $x_i^* = \lfloor x_i \rfloor$ ($i = 1, \ldots, m$). For this new solution, possibly part of the demand of the items is not fulfilled. More precisely, the demand of each item $i$ that is not fulfilled is $d_i^* = d_i - \sum_{j=1}^{m} B_{i,j} x_j^*$. Thus, if we take $d^* = (d_1^*, \ldots, d_m^*)$, we have a residual instance $I^* = (W, H, l, c, d^*)$ (we may eliminate from $I^*$ the items with no demand).

If some $x_i^* > 0$ ($i = 1, \ldots, m$), part of the demand is fulfilled by the solution $x^*$. In this case, we return $B$ and $x$, we let $I = I^*$ and start a new iteration. If $x_i^* = 0$ ($i = 1, \ldots, m$), no part of the demand is fulfilled by $x^*$. We solve then the instance $I^*$ with the algorithm *Hybrid First Fit* (HFF) [10]. We present in

---

**Algorithm 4.1** SimplexCG$_2$

---

*Input*: An instance $I = (W, H, w, h, d)$ of GCD, where $w = (w_1, \ldots, w_m)$,
$\quad h = (h_1, \ldots, h_m)$ and $d = (d_1, \ldots, d_m)$.

*Output*: An optimal solution for (2), where the columns of $P$ are the patterns for $I$.

1 Let $x = d$ and $B$ be the identity matrix of order $m$.

2 Solve $y^T B = \mathbb{1}^T$.

3 Generate a new column $z$ executing the algorithm $\mathcal{DP}$ with parameters $W, H, l, a, y$.

4 If $y^T z \leq 1$, return $B$ and $x$ and halt ($x$ corresponds to the columns of $B$).

5 Otherwise, solve $Bw = z$.

6 Let $t = min(\frac{x_j}{w_j} \mid 1 \leq j \leq m, \ w_j > 0)$.

7 Let $s = min(j \mid 1 \leq j \leq m, \ \frac{x_j}{w_j} = t)$.

8 For $i = 1$ to $m$ do

$\quad$ **8.1** $B_{i,s} = z_i$.

$\quad$ **8.2** If $i = s$ then $x_i = t$; otherwise, $x_i = x_i - w_i t$.

9 Go to step 2.

---

what follows the algorithm $\mathcal{CG}$ that implements the iterative procedure we have described.

$\quad$ Note that in each iteration either part of the demand is fulfilled or we go to step 4. Thus, after a finite number of iterations the demand will be met (part of it eventually in step 4). In fact, one can show that step 3.6 of the algorithm $\mathcal{CG}$ is executed at most $m$ times. It should be noted however, that step 5 of the algorithm $\mathcal{CG}$ may require exponential time. This step is necessary to transform the representation of the last residual instance in an input for the algorithm HFF, called in the next step. Moreover, HFF may also take exponential time to solve this last instance.

$\quad$ We designed an approximation algorithm for GCD, called $\mathcal{SH}$, that makes use of the concept of semi-homogeneous patterns and has absolute performance bound 4 (see [12]). The reasons for not using $\mathcal{SH}$, instead of HFF, to solve the last residual instance are the following: first, to generate a new column with the algorithm $\mathcal{DP}$ requires time that can be exponential in $m$. Thus, $\mathcal{CG}$ is already exponential, even on the average case. Besides that, the algorithm HFF has asymptotic approximation bound 2.125. Thus, we may expect that using HFF we may produce solutions of better quality.

$\quad$ On the other hand, if the items are not too small with respect to the bin[1], the algorithm $\mathcal{DP}$ can be implemented to require polynomial time (as we mentioned in Section 3). In this case, we could eliminate steps 4 and 5 of $\mathcal{CG}$ and use $\mathcal{SH}$ instead of HFF to solve the last residual instance. The solutions may have worse quality, but at least, theoretically, the time required by such an algorithm would be polynomial in $m$, on the average.

$\quad$ We note that the algorithm $\mathcal{CG}$ can be used to solve the variant of GCD, called GCD$^r$, in which orthogonal rotations of the items are allowed. For that,

---

[1] More precisely, for fixed $k$, $w_i > \frac{W}{k}$ and $h_i > \frac{H}{k}$ $(i = 1, \ldots, m)$.

---

**Algorithm 4.2 $\mathcal{CG}$**

---

*Input*: An instance $I = (W, H, w, h, d)$ of GCD, where $w = (w_1, \ldots, w_m)$,
$\quad$ $h = (h_1, \ldots, h_m)$ and $d = (d_1, \ldots, d_m)$.

*Output*: A solution for $I$

**1** Execute the algorithm SimplexCG$_2$ with parameters $W, H, l, a, d$ obtaining $B$ and $x$.
**2** For $i = 1$ to $m$ do $x_i^* = \lfloor x_i \rfloor$.
**3** If $x_i^* > 0$ for some $i$, $1 \le i \le m$, then
$\quad$ **3.1** Return $B$ and $x_1^*, \ldots, x_m^*$ (but do not halt).
$\quad$ **3.2** For $i = 1$ to $m$ do
$\quad\quad$ **3.2.1** For $j = 1$ to $m$ do $d_i = d_i - B_{i,j} x_j^*$.
$\quad$ **3.3** Let $m' = 0$.
$\quad$ **3.4** For $i = 1$ to $m$ do
$\quad\quad$ **3.4.1** If $d_i > 0$ then $m' = m' + 1$, $w_{m'} = w_i$, $h_{m'} = h_i$ and $d_{m'} = d_i$.
$\quad$ **3.5** If $m' = 0$ then halt.
$\quad$ **3.6** Let $m = m'$ and go to step 1.
**4** $w' = \emptyset$, $h' = \emptyset$.
**5** For $i = 1$ to $m$ do
$\quad$ **5.1** For $j = 1$ to $d_i$ do
$\quad\quad$ **5.1.1** $w' = w' \cup \{w_i\}$, $h' = h' \cup \{h_i\}$.
**6** Return the solution of algorithm HFF executed with parameters $W, H, w', h'$.

---

before we call the algorithm $\mathcal{DP}$, in step 3 of SimplexCG$_2$, it suffices to make the transformation explained at the end of Section 3. We will call SimplexCG$_2^r$ the variant of SimplexCG$_2$ with this transformation. It should be noted however that the algorithm HFF, called in step 6 of $\mathcal{CG}$, does not use the fact that the items can be rotated.

We designed a simple algorithm for the variant of GCD$^r$ in which all items have demand 1, called here GCDB$^r$. This algorithm, called *First Fit Decreasing Height using Rotations* (FFDHR), has asymptotic approximation bound at most 4, as we have shown in [12]. Substituting the call to HFF with a call to FFDHR, we obtain the algorithm $\mathcal{CGR}$, that is a specialized version of $\mathcal{CG}$ for the problem GCD$^r$.

We also tested another modification of the algorithm $\mathcal{CG}$ (and of $\mathcal{CGR}$). This is the following: when we solve an instance, and the solution returned by SimplexCG$_2$ rounded down is equal to zero, instead of submitting this instance to HFF (or FFDHR), we use HFF (or FFDHR) to obtain a *good* pattern, updating the demands, and if there is some item for which the demand is not fulfilled, we go to step 1.

Note that, the basic idea is to *perturb* the residual instances whose relaxed LP solution, rounded down, is equal to zero. With this procedure, it is expected that the solution obtained by SimplexCG$_2$ for the residual instance has more variables with value greater than 1. The algorithm $\mathcal{CG}^p$, described below, incorporates this modification.

It should be noted that with this modification we cannot guarantee anymore that we have to make at most $m + 1$ calls to SimplexCG$_2$. It is however, easy

---

**Algorithm 4.3 $\mathcal{GC}^p$**

---

*Input*: An instance $I = (W, H, w, h, d)$ of GCD, where $w = (w_1, \ldots, w_m)$,
$\qquad h = (h_1, \ldots, h_m)$ and $d = (d_1, \ldots, d_m)$.

*Output*: A solution for $I$

**1** Execute the algorithm SimplexCG$_2$ with parameters $W, H, w, h, d$ obtaining $B$ and $x$.

**2** For $i = 1$ to $m$ do $x_i^* = \lfloor x_i \rfloor$.

**3** If $x_i^* > 0$ for some $i$, $1 \le i \le m$, then

    **3.1** Return $B$ and $x_1^*, \ldots, x_m^*$ (but do not halt).

    **3.2** For $i = 1$ to $m$ do

        **3.2.1** For $j = 1$ to $m$ do $d_i = d_i - B_{i,j} x_j^*$.

    **3.3** Let $m' = 0$.

    **3.4** For $i = 1$ to $m$ do

        **3.4.1** If $d_i > 0$ then $m' = m' + 1$, $w_{m'} = w_i$, $h_{m'} = h_i$ and $d_{m'} = d_i$.

    **3.5** If $m' = 0$ then halt.

    **3.6** Let $m = m'$ and go to step 1.

**4** $w' = \emptyset$, $h' = \emptyset$.

**5** For $i = 1$ to $m$ do

    **5.1** For $j = 1$ to $d_i$ do

        **5.1.1** $w' = w' \cup \{w_i\}$, $h' = h' \cup \{h_i\}$.

**6** Return a pattern generated by the algorithm HFF, executed with parameters $W, H, w', h'$, that has the smallest waste of area, and update the demands.

**7** If there are demands to be fulfilled, go to step 1.

---

to see that the algorithm $\mathcal{CG}^p$ in fact halts, as each time step 1 is executed, the demand decreases strictly. After a finite number of iterations the demand will be fulfilled and the algorithm halts (in step 3.5 or step 7).

# 5   Computational Results

The algorithms described in sections 3 and 4 were implemented in C language, using *Xpress-MP* [27] as the LP solver. The tests were run on a computer with two processors *AMD Athlon MP 1800+*, clock of 1.5 Ghz, memory of 3.5 GB and operating system *Linux* (distribution *Debian GNU/Linux 3.0*).

The performance of the algorithm $\mathcal{DP}$ was tested with the instances of GCV available in the OR-LIBRARY[2] (see Beasley [6] for a brief description of this library). We considered the 13 instances of GCV, called $gcut1, \ldots, gcut13$ available in this library. For all these instances, with exception of instance $gcut13$, optimal solutions had already been found [4]. We solved to optimality this instance as well. In these instances the bins are squares, with dimensions between 250 and 3000, and number of items ($m$) between 10 and 50. The value of each item is precisely its area. We show in Figure 2 the optimal solution for $gcut13$ found by the algorithm $\mathcal{DP}$.

---

[2] http://mscmga.ms.ic.ac.uk/info.html

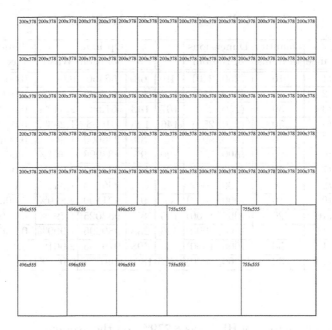

**Fig. 2.** The optimal solution for $gcut13$ found by the algorithm $\mathcal{DP}$.

In Table 1 we show the characteristics of the instances solved and the computational results. The column "Waste" gives —for each solution found— the percentage of the area of the bin that does not correspond to any item. Each instance was solved 100 times; the column "Time" shows the average CPU time in seconds, considering all these 100 resolutions.

We have also tested the algorithm $\mathcal{DP}$ for the instances $gcut1, \ldots, gcut13$, this time allowing rotations (we called these instances $gcut1r, \ldots, gcut13r$). Owing to space limitations, we omit the table showing the computational results. It can be found at http://www.ime.usp.br/$\sim$glauber/gcut. We only remark that for some instances the time increased (it did not doubled) but the waste decreased, as one would expect.

We did not find instances for GCD in the OR-LIBRARY. We have then tested the algorithms $\mathcal{CG}$ and $\mathcal{CG}^p$ with the instances $gcut1, \ldots, gcut12$, associating with each item $i$ a randomly generated demand $d_i$ between 1 and 100. We called these instances $gcut1d, \ldots, gcut12d$.

We show in Table 2 the computational results obtained with the algorithm $\mathcal{CG}$. In this table, LB denotes the lower bound (given by the solution of (2)) for the value of an optimal integer solution. Each instance was solved 10 times; the column "Average Time" shows the average time considering these 10 experiments.

The algorithm $\mathcal{CG}$ found optimal or quasi-optimal solutions for all these instances. On the average, the difference between the solution found by $\mathcal{CG}$ and the lower bound (LB) was only 0,161%. We note also that the time spent to solve these instances was satisfactory. Moreover, the gain of the solution of $\mathcal{CG}$

**Table 1.** Performance of the algorithm $\mathcal{DP}$.

| Instance | Quantity of items | Dimensions of the bin | $r$ | $s$ | Optimal Solution | Waste | Time (sec) |
|---|---|---|---|---|---|---|---|
| gcut1 | 10 | (250, 250) | 19 | 68 | 56460 | 9,664% | 0,003 |
| gcut2 | 20 | (250, 250) | 112 | 95 | 60536 | 3,142% | 0,010 |
| gcut3 | 30 | (250, 250) | 107 | 143 | 61036 | 2,342% | 0,012 |
| gcut4 | 50 | (250, 250) | 146 | 146 | 61698 | 1,283% | 0,022 |
| gcut5 | 10 | (500, 500) | 76 | 39 | 246000 | 1,600% | 0,004 |
| gcut6 | 20 | (500, 500) | 120 | 95 | 238998 | 4,401% | 0,008 |
| gcut7 | 30 | (500, 500) | 126 | 179 | 242567 | 2,973% | 0,017 |
| gcut8 | 50 | (500, 500) | 262 | 225 | 246633 | 1,347% | 0,062 |
| gcut9 | 10 | (1000, 1000) | 41 | 91 | 971100 | 2,890% | 0,006 |
| gcut10 | 20 | (1000, 1000) | 155 | 89 | 982025 | 1,798% | 0,009 |
| gcut11 | 30 | (1000, 1000) | 326 | 238 | 980096 | 1,990% | 0,066 |
| gcut12 | 50 | (1000, 1000) | 363 | 398 | 979986 | 2,001% | 0,140 |
| gcut13 | 32 | (3000, 3000) | 2425 | 1891 | 8997780 | 0,025% | 144,915 |

compared to the solution of HFF was 8,779%, on the average, a very significant improvement.

We have also used the algorithm $\mathcal{CG}^p$ to solve the instances gcut1d, …, gcut12d. The results are shown in Table 3. We note that the number of columns generated increased approximately 40%, on the average, and the time spent increased approximately 15%, on the average. On the other hand, an optimal solution was found for the instance gcut4d.

We also considered the instances gcut1d, …, gcut12d as being for the problem GCD$^r$ (rotations are allowed), and called them gcut1dr, …, gcut12dr. We omit the table with the computational results we have obtained (the reader can find it at the URL we mentioned before). We only remark that the algorithm $\mathcal{CGR}$ found optimal or quasi-optimal solutions for all instances. The difference between the value found by $\mathcal{CGR}$ and the lower bound (LB) was only 0,408%, on the average.

Comparing the value of the solutions obtained by $\mathcal{CGR}$ with the solutions obtained by FFDHR, we note that there was an improvement of 12,147%, on the average. This improvement would be of 16,168% if compared with the solution obtained by HFF.

The instances gcut1dr, …, gcut12dr were also tested with the algorithm $\mathcal{CGR}^p$. The computational results are shown in Table 4. We remark that the performance of the algorithm $\mathcal{CGR}^p$ was a little better than the performance of $\mathcal{CGR}$, with respect to the quality of the solutions. The difference between the value of the solution obtained by $\mathcal{CGR}^p$ and the lower bound decreased to 0,237%, on the average. The gain on the quality was obtained at the cost of an increase of approximately 97% (on the average) of the number of columns generated and of an increase of approximately 44% of time.

Table 2. Performance of the algorithm $\mathcal{CG}$.

| Instance | Solution of $\mathcal{CG}$ | LB | Difference from LB | Average Time (sec) | Columns Generated | Solution of HFF | Improvement over HFF |
|---|---|---|---|---|---|---|---|
| gcut1d | 294 | 294 | 0,000% | 0,059 | 9 | 295 | 0,339% |
| gcut2d | 345 | 345 | 0,000% | 0,585 | 68 | 402 | 14,179% |
| gcut3d | 333 | 332 | 0,301% | 2,340 | 274 | 393 | 14,834% |
| gcut4d | 838 | 836 | 0,239% | 11,693 | 820 | 977 | 11,323% |
| gcut5d | 198 | 197 | 0,507% | 0,088 | 18 | 198 | 0,000% |
| gcut6d | 344 | 343 | 0,291% | 0,362 | 101 | 418 | 17,308% |
| gcut7d | 591 | 591 | 0,000% | 1,184 | 136 | 615 | 4,523% |
| gcut8d | 691 | 690 | 0,145% | 30,361 | 952 | 764 | 9,555% |
| gcut9d | 131 | 131 | 0,000% | 0,068 | 11 | 143 | 7,092% |
| gcut10d | 293 | 293 | 0,000% | 0,172 | 20 | 335 | 12,537% |
| gcut11d | 331 | 330 | 0,303% | 8,570 | 222 | 353 | 6,232% |
| gcut12d | 673 | 672 | 0,149% | 39,032 | 485 | 727 | 7,428% |

# 6   Concluding Remarks

In this paper we presented algorithms for the problems GCV and GCD and their variants GCV$^r$ and GCD$^r$. For the problem GCV we presented the (exact) pseudo-polynomial algorithm $\mathcal{DP}$. This algorithm can either use the algorithm DDE or DDP to generate the discretization points. Both of these algorithms were described. We have also shown that these algorithms can be implemented to run in polynomial time when the items are not so small compared to the size of the bin. In this case the algorithm $\mathcal{DP}$ also runs in polynomial time. We have also mentioned how to use $\mathcal{DP}$ to solve the problem GCV$^r$.

We presented two column generation based algorithms to solve GCD: $\mathcal{CG}$ and $\mathcal{CG}^p$. Both use the algorithm $\mathcal{DP}$ to generate the columns: the first uses the algorithm HFF to solve the last residual instance and the second uses a perturbation strategy. The algorithm $\mathcal{CG}$ combines different techniques: Simplex method with column generation, an exact algorithm for the discretization points, and an approximation algorithm (HFF) for the last residual instance. An approach of this nature has shown to be promising, and has been used in the one-dimensional cutting problem with demands [26,11].

The algorithm $\mathcal{CG}^p$ is a variant of $\mathcal{CG}$, in which we use an idea that consists in perturbing the residual instances. We have also designed the algorithms $\mathcal{CGR}$ and $\mathcal{CGR}^p$, for the problem GCD$^r$, a variation of GCD in which orthogonal rotations are allowed. The algorithm $\mathcal{CGR}$ uses as a subroutine the algorithm FFDHR, we have designed.

We noted that the algorithm $\mathcal{CG}$ and $\mathcal{CGR}$ are polynomial, on the average, when the items are not so small compared to the size of the bin. The computational results with these algorithms were very satisfactory: optimal or quasi-optimal solutions were found for the instances we have considered. As expected,

**Table 3.** Performance of the algorithm $C\mathcal{G}^p$.

| Instance | Solution of $C\mathcal{G}^p$ | LB | Difference from LB | Average Time (sec) | Columns Generated | Solution of HFF | Improvement over HFF |
|---|---|---|---|---|---|---|---|
| gcut1d | 294 | 294 | 0,000% | 0,034 | 9 | 295 | 0,339% |
| gcut2d | 345 | 345 | 0,000% | 0,552 | 68 | 402 | 14,179% |
| gcut3d | 333 | 332 | 0,301% | 3,814 | 492 | 393 | 14,834% |
| gcut4d | 837 | 836 | 0,120% | 16,691 | 1271 | 977 | 11,429% |
| gcut5d | 198 | 197 | 0,507% | 0,086 | 25 | 198 | 0,000% |
| gcut6d | 344 | 343 | 0,291% | 0,400 | 121 | 418 | 17,308% |
| gcut7d | 591 | 591 | 0,000% | 1,202 | 136 | 615 | 4,523% |
| gcut8d | 691 | 690 | 0,145% | 32,757 | 1106 | 764 | 9,555% |
| gcut9d | 131 | 131 | 0,000% | 0,042 | 11 | 143 | 7,092% |
| gcut10d | 293 | 293 | 0,000% | 0,153 | 20 | 335 | 12,537% |
| gcut11d | 331 | 330 | 0,303% | 10,875 | 416 | 353 | 6,232% |
| gcut12d | 673 | 672 | 0,149% | 42,616 | 692 | 727 | 7,428% |

$C\mathcal{G}^p$ (respectively $C\mathcal{GR}^p$) found solutions of a little better quality than $C\mathcal{G}$ (respectively $C\mathcal{GR}$) at the cost of a slight increase in the running time.

We exhibit in Table 5 the list of the algorithms we proposed in this paper.

A natural development of our work would be to adapt the approach used in the algorithm $C\mathcal{G}$ to the *Two-dimensional cutting stock problem with demands* (CSD), a variant of GCD in which the cuts need not be guillotine. One can find an initial solution using homogeneous patterns; the columns can be generated using any of the algorithms that have appeared in the literature for the two-dimensional cutting stock problem with value [5,2]. To solve the last residual instance one can use approximation algorithms [10,8,20].

One can also use column generation for the variant of CSD in which the quantity of items in each bin is bounded. This variant, proposed by Christofides and Whitlock [9], is called *restricted two-dimensional cutting stock problem*. Each new column can be generated with any of the known algorithms for the restricted two-dimensional cutting stock problem with value [9,24], and the last residual instance can be solved with the algorithm HFF. This restricted version with guillotine cut requirement can also be solved using the ideas we have just described: the homogeneous patterns and the patterns produced by HFF can be obtained with guillotine cuts, and the columns can be generated with the algorithm of Cung, Hifi and Le Cun [16].

A more audacious step would be to adapt the column generation approach for the *three-dimensional cutting stock problem with demands*. For the initial solutions one can use homogeneous patterns. The last residual instances can be dealt with some approximation algorithms for the three-dimensional bin packing problem [3,14,15,22,23,21,25]. We do no know however, exact algorithms to generate columns. If we require the cuts to be guillotine, one can adapt the algorithm $\mathcal{DP}$ to generate new columns.

**Table 4.** Performance of the algorithm $C\mathcal{G}\mathcal{R}^p$.

| Instance | Solution of $C\mathcal{G}\mathcal{R}^p$ | LB | Difference from LB | Average Time (sec) | Columns Generated | Solution of FFDHR | Improvement over FFDHR |
|---|---|---|---|---|---|---|---|
| gcut1dr | 291 | 291 | 0,000% | 0,070 | 5 | 291 | 0,000% |
| gcut2dr | 283 | 282 | 0,355% | 5,041 | 252 | 330 | 14,242% |
| gcut3dr | 314 | 313 | 0,319% | 10,006 | 740 | 355 | 11,549% |
| gcut4dr | 837 | 836 | 0,120% | 25,042 | 1232 | 945 | 11,429% |
| gcut5dr | 175 | 174 | 0,575% | 0,537 | 58 | 200 | 12,500% |
| gcut6dr | 301 | 301 | 0,000% | 2,884 | 175 | 405 | 25,679% |
| gcut7dr | 542 | 542 | 0,000% | 4,098 | 121 | 599 | 9,516% |
| gcut8dr | 651 | 650 | 0,153% | 68,410 | 1154 | 735 | 11,429% |
| gcut9dr | 123 | 122 | 0,820% | 0,494 | 42 | 140 | 12,143% |
| gcut10dr | 270 | 270 | 0,000% | 1,546 | 33 | 330 | 18,182% |
| gcut11dr | 299 | 298 | 0,336% | 86,285 | 686 | 329 | 9,119% |
| gcut12dr | 602 | 601 | 0,166% | 181,056 | 945 | 682 | 11,730% |

**Table 5.** Algorithms proposed in this paper.

| Algorithm | Problems | Comments |
|---|---|---|
| $\mathcal{DP}$ | GCV and GCV$^r$ | Polynomial for large items |
| DEE | Discretization points | Polynomial for large items |
| DDP | Discretization points | |
| $\mathcal{CG}$ | GCD | Polynomial, on the average, for large items |
| $\mathcal{CG}^p$ | GCD | |
| $\mathcal{CG}\mathcal{R}$ | GCD$^r$ | Polynomial, on the average, for large items |
| $\mathcal{CG}\mathcal{R}^p$ | GCD$^r$ | |

# References

1. Ilan Adler, Nimrod Megiddo, and Michael J. Todd. New results on the average behavior of simplex algorithms. *Bull. Amer. Math. Soc. (N.S.)*, 11(2):378–382, 1984.
2. M. Arenales and R. Morábito. An and/or-graph approach to the solution of two-dimensional non-guillotine cutting problems. *European Journal of Operations Research*, 84:599–617, 1995.
3. N. Bansal and M. Srividenko. New approximability and inapproximability results for 2-dimensional bin packing. In *Proceedings of 15th ACM-SIAM Symposium on Discrete Algorithms*, pages 189–196, New York, 2004. ACM.
4. J. E. Beasley. Algorithms for unconstrained two-dimensional guillotine cutting. *Journal of the Operational Research Society*, 36(4):297–306, 1985.
5. J. E. Beasley. An exact two-dimensional nonguillotine cutting tree search procedure. *Oper. Res.*, 33(1):49–64, 1985.
6. J. E. Beasley. Or-library: distributing test problems by electronic mail. *Journal of the Operational Research Society*, 41(11):1069–1072, 1990.
7. Karl-Heinz Borgwardt. Probabilistic analysis of the simplex method. In *Mathematical developments arising from linear programming (Brunswick, ME, 1988)*, volume 114 of *Contemp. Math.*, pages 21–34. Amer. Math. Soc., Providence, RI, 1990.

8. A. Caprara. Packing 2-dimensional bins in harmony. In *Proceedings of the 43-rd Annual IEEE Symposium on Foundations of Computer Science*, pages 490–499. IEEE Computer Society, 2002.
9. N. Christofides and C. Whitlock. An algorithm for two dimensional cutting problems. *Operations Research*, 25:30–44, 1977.
10. F. R. K. Chung, M. R. Garey, and D. S. Johnson. On packing two-dimensional bins. *SIAM J. Algebraic Discrete Methods*, 3:66–76, 1982.
11. G. F. Cintra. Algoritmos híbridos para o problema de corte unidimensional. In *XXV Conferência Latinoamericana de Informática*, Assunção, 1999.
12. G. F. Cintra. Algoritmos para problemas de corte de guilhotina bidimensional (PhD thesis in preparation). Instituto de Matemática e Estatística, 2004.
13. Thomas H. Cormen, Charles E. Leiserson, Ronald L. Rivest, and Clifford Stein. *Introduction to algorithms*. MIT Press, Cambridge, MA, second edition, 2001.
14. J.R. Correa and C. Kenyon. Approximation schemes for multidimensional packing. In *Proceedings of 15th ACM-SIAM Symposium on Discrete Algorithms*, pages 179–188, New York, 2004. ACM.
15. J. Csirik and A. Van Vliet. An on-line algorithm for multidimensional bin packing. *Operations Research Letters*, 13:149–158, 1993.
16. Van-Dat Cung, Mhand Hifi, and Bertrand Le Cun. Constrained two-dimensional cutting stock problems a best-first branch-and-bound algorithm. *Int. Trans. Oper. Res.*, 7(3):185–210, 2000.
17. P. Gilmore and R. Gomory. A linear programming approach to the cutting stock problem. *Operations Research*, 9:849–859, 1961.
18. P. Gilmore and R. Gomory. Multistage cutting stock problems of two and more dimensions. *Operations Research*, 13:94–120, 1965.
19. J. C. Herz. A recursive computational procedure for two-dimensional stock-cutting. *IBM Journal of Research Development*, pages 462–469, 1972.
20. Claire Kenyon and Eric Rémila. A near-optimal solution to a two-dimensional cutting stock problem. *Math. Oper. Res.*, 25(4):645–656, 2000.
21. Y. Kohayakawa, F.K. Miyazawa, P. Raghavan, and Y. Wakabayashi. Multidimensional cube packing. In *Brazilian Symposium on Graphs and Combinatorics. Electronic Notes of Discrete Mathematics (GRACO'2001)*. Elsevier Science, 2001. (to appear in Algorithmica).
22. Keqin Li and Kam-Hoi Cheng. Generalized first-fit algorithms in two and three dimensions. *Internat. J. Found. Comput. Sci.*, 1(2):131–150, 1990.
23. F. K. Miyazawa and Y. Wakabayashi. Parametric on-line algorithms for packing rectangles and boxes. *European J. Oper. Res.*, 150(2):281–292, 2003.
24. J. F. Oliveira and J. S. Ferreira. An improved version of Wang's algorithm for two-dimensional cutting problems. *European Journal of Operations Research*, 44:256–266, 1990.
25. S. Seiden and R. van Stee. New bounds for multidimensional packing. *Algorithmica*, 36:261–293, 2003.
26. G. Wäscher and T. Gau. Heuristics for the integer one-dimensional cutting stock problem: a computational study. *OR Spektrum*, 18:131–144, 1996.
27. Xpress. *Xpress Optimizer Reference Manual*. DASH Optimization, Inc, 2002.

# Engineering Shortest Path Algorithms*

Camil Demetrescu[1] and Giuseppe F. Italiano[2]

[1] Dipartimento di Informatica e Sistemistica, Università di Roma "La Sapienza", via Salaria 113, 00198 Roma, Italy.
demetres@dis.uniroma1.it
[2] Dipartimento di Informatica, Sistemi e Produzione, Università di Roma "Tor Vergata", via del Politecnico 1, 00133 Roma, Italy.
italiano@disp.uniroma2.it

**Abstract.** In this paper, we report on our own experience in studying a fundamental problem on graphs: all pairs shortest paths. In particular, we discuss the interplay between theory and practice in engineering a simple variant of Dijkstra's shortest path algorithm. In this context, we show that studying heuristics that are efficient in practice can yield interesting clues to the combinatorial properties of the problem, and eventually lead to new theoretically efficient algorithms.

## 1 Introduction

The quest for efficient computer programs for solving real world problems has led in recent years to a growing interest in experimental studies of algorithms. Producing efficient implementations requires taking into account issues such as memory hierarchy effects, hidden constant factors in the performance bounds, implications of communication complexity, numerical precision, and use of heuristics, which are sometimes overlooked in classical analysis models. On the other hand, developing and assessing heuristics and programming techniques for producing codes that are efficient in practice is a difficult task that requires a deep understanding of the mathematical structure and the combinatorial properties of the problem at hand. In this context, experiments can raise new conjectures and theoretical questions, opening unexplored research directions that may lead to further theoretical improvements and eventually to more practical algorithms. The whole process of designing, analyzing, implementing, tuning, debugging and experimentally evaluating algorithms is usually referred to as *Algorithm Engineering*. As shown in Figure 1, algorithm engineering is a cyclic process: designing algorithmic techniques and analyzing their performance according to theoretical models provides a sound foundation to writing efficient computer programs. On

---

* Work partially supported by the Sixth Framework Programme of the EU under contract number 507613 (Network of Excellence "EuroNGI: Designing and Engineering of the Next Generation Internet"), by the IST Programme of the EU under contract number IST-2001-33555 ("COSIN: COevolution and Self-organization In dynamical Networks"), and by the Italian Ministry of University and Scientific Research (Project "ALINWEB: Algorithmics for Internet and the Web").

C.C. Ribeiro and S.L. Martins (Eds.): WEA 2004, LNCS 3059, pp. 191–198, 2004.

the other hand, analyzing the practical performance of a program can be helpful in spotting bottlenecks in the code, designing heuristics, refining the theoretical analysis, and even devising more realistic cost models in order to get a deeper insight into the problem at hand. We refer the interested reader to [4] for a broader survey of algorithm engineering issues.

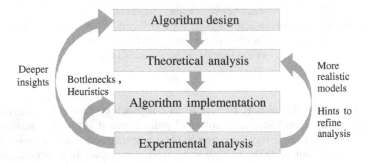

**Fig. 1.** The algorithm engineering cycle.

In this paper, we report on our own experience in studying a fundamental problem on graphs: the all pairs shortest paths problem. We discuss the interplay between theory and practice in engineering a variant of Dijkstra's shortest path algorithm. In particular, we present a simple heuristic that can improve substantially the practical performances of the algorithm on many typical instances. We then show that a deeper study of this heuristic can reveal interesting combinatorial properties of paths in a graph. Surprisingly, exploiting such properties can lead to new theoretically efficient methods for updating shortest paths in a graph subject to dynamic edge weight changes.

## 2   From Theory to Experiments: Engineering Dijkstra's Algorithm

In 1959, Edsger Wybe Dijkstra devised a simple algorithm for computing shortest paths in a graph [6]. Although more recent advances in data structures led to faster implementations (see, e.g., the Fibonacci heaps of Fredman and Tarjan [7]), Dijkstra's algorithmic invention is still in place after 45 years, providing an ubiquitous standard framework for designing efficient shortest path algorithms. If priority queues with constant amortized time per decrease are used, the basic method computes the shortest paths from a given source vertex in $O(m + n \log n)$ worst-case time in a graph with $n$ vertices and $m$ edges. To compute all pairs shortest paths, we get an $O(mn + n^2 \log n)$ bound in the worst case by simply repeating the single-source algorithm from each vertex.

In Figure 2, we show a simple variant of Dijkstra's algorithm where all pairs shortest paths are computed in an interleaved fashion, rather than one source

**algorithm** DijkstraAPSP(graph $G = (V, E, w)) \to$ matrix
1.    let $d$ be an $n \times n$ matrix and let $H$ be an empty priority queue
2.    **for each** pair $(x, y) \in V \times V$ **do**                    {initialization}
3.       **if** $(x, y) \in E$ **then** $d[x, y] \leftarrow w(x, y)$ **else** $d[x, y] \leftarrow +\infty$
4.       add edge $(x, y)$ to $H$ with priority $d[x, y]$
5.    **while** $H$ is not empty **do**                    {main loop}
6.       extract from $H$ a pair $(x, y)$ with minimum priority
7.       **for each** edge $(y, z) \in E$ leaving $y$ **do**             {forward extension}
8.          **if** $d[x, y] + w(y, z) < d[x, z]$ **then**             {relaxation}
9.             $d[x, z] \leftarrow d[x, y] + w(y, z)$
10.             decrease the priority of $(x, z)$ in $H$ to $d[x, z]$
11.       **for each** edge $(z, x) \in E$ entering $x$ **do**        {backward extension}
12.          **if** $w(z, x) + d[x, y] < d[z, y]$ **then**             {relaxation}
13.             $d[z, y] \leftarrow w(z, x) + d[x, y]$
14.             decrease the priority of $(z, y)$ in $H$ to $d[z, y]$
15.    **return** $d$

**Fig. 2.** All pairs variant of Dijkstra's algorithm. $w(x, y)$ denotes the weight of edge $(x, y)$ in $G$.

at a time. The algorithm maintains a matrix $d$ that contains at any time an upper bound to the distances in the graph. The upper bound $d[x, y]$ for any pair of vertices $(x, y)$ is initially equal to the edge weight $w(x, y)$ if there is an edge between $x$ and $y$, and $+\infty$ otherwise. The algorithm also maintains in a priority queue $H$ each pair $(x, y)$ with priority $d[x, y]$. The main loop of the algorithm repeatedly extracts from $H$ a pair $(x, y)$ with minimum priority, and tries to extend at each iteration the corresponding shortest path by exactly one edge in every possible direction. This requires scanning all edges leaving $y$ and entering $x$, performing the classical *relaxation* step to decrease the distance upper bounds $d$. It is not difficult to prove that at the end of the procedure, if edge weights are non-negative, $d$ contains the exact distances. The time required for loading and unloading the priority queue is $O(n^2 \log n)$ in the worst case. Each edge is scanned at most $O(n)$ times and for each scanned edge we spend constant amortized time if $H$ is, e.g., a Fibonacci heap. This yields $O(mn + n^2 \log n)$ worst-case time.

While faster algorithms exist for very sparse graphs [8,9,10], Dijkstra's algorithm appears to be still a good practical choice in many real world settings. For this reason, the quest for fast implementations has motivated researchers to study methods for speeding up Dijkstra's basic method based on priority queues and relaxation. For instance, it is now well understood that efficient data structures play a crucial role in the case of sparse graphs, while edge scanning is the typical bottleneck in dense graphs [1]. In the rest of this section, we focus on the algorithm of Figure 2, and we investigate heuristic methods for reducing the number of scanned edges.

Consider line 7 of Figure 2 (or, similarly, line 11): the loop scans all edges $(y, z)$, seeking for tentative shortest paths of the form $x \rightsquigarrow y \to z$ with weight

$d[x, y] + w(y, z)$. Do we really need to scan all edges $(y, z)$ leaving $y$? More to the point, is there any way to avoid scanning an edge $(y, z)$ if it cannot belong to a shortest path from $x$ to $z$? Let $\pi_{xy}$ be a shortest path from $x$ to $y$ and let $\pi_{xz} = \pi_{xy} \cdot (y, z)$ be the path obtained by going from $x$ to $y$ via $\pi_{xy}$ and then from $y$ to $z$ via edge $(y, z)$ (see Figure 3). Consider now the well-known optimal-substructure property of shortest paths (see e.g., [2]):

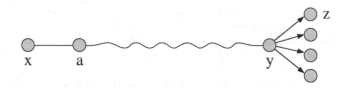

**Fig. 3.** Extending the shortest path from $x$ to $y$ by one edge.

**Lemma 1.** *Every subpath of a shortest path is a shortest path.*

This property implies that, if $\pi_{xz}$ is a shortest path and we remove either one of its endpoints, we still get a shortest path. Thus, edge $(y, z)$ can be the last edge of a shortest path $\pi_{xz}$ *only if* both $\pi_{xy} \subset \pi_{xz}$ and $\pi_{az} \subset \pi_{xz}$ are shortest paths, where $(x, a)$ is the first edge in $\pi_{xz}$. Let us now exploit this property in our shortest path algorithm in order to avoid scanning "unnecessary" edges. In the following, we assume without loss of generality that shortest paths in the graph are unique. We can just maintain for each pair of vertices $(x, y)$ a list of edges $R_{xy} = \{\ (y, z)$ s.t. $\pi_{xz} = \pi_{xy} \cdot (y, z)$ is a shortest path $\}$, and a list of edges $L_{xy} = \{\ (z, x)$ s.t. $\pi_{zy} = (z, x) \cdot \pi_{xy}$ is a shortest path $\}$, where $\pi_{xy}$ is the shortest path from $x$ to $y$. Such lists can be easily constructed incrementally as the algorithm runs at each pair extraction from the priority queue $H$. Suppose now to modify line 7 and line 11 of Figure 2 to scan just edges in $R_{ay}$ and $L_{xb}$, respectively, where $(x, a)$ is the first edge and $(b, y)$ is the last edge in $\pi_{xy}$. It is not difficult to see that in this way we consider in the relaxation step only paths whose proper subpaths are shortest paths. We call such paths *locally shortest* [5]:

**Definition 1.** *A path $\pi_{xy}$ is* locally shortest *in $G$ if either:*

*(i) $\pi_{xy}$ consists of a single vertex, or*
*(ii) every proper subpath of $\pi_{xy}$ is a shortest path in $G$.*

With the modifications above, the algorithm requires $O(|LSP| + n^2 \log n)$ worst-case time, where $|LSP|$ is the number of locally shortest paths in the graph. A natural question seems to be: how many locally shortest paths can we have in a graph? The following lemma is from [5]:

**Lemma 2.** *If shortest paths are unique in $G$, then there can be at most $mn$ locally shortest paths in $G$. This bound is tight.*

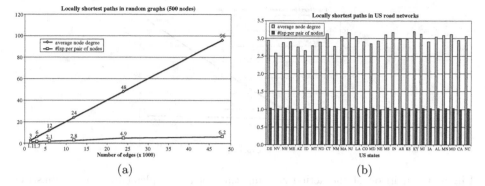

**Fig. 4.** Average number of locally shortest paths connecting a pair of vertices in: (a) a family of random graphs with increasing density; (b) a suite of US road networks obtained from `ftp://edcftp.cr.usgs.gov`.

This implies that our modification of Dijkstra's algorithm does not produce an asymptotically faster method. But what about typical instances? In [3], we have performed some counting experiments on both random and real-world graphs (including road networks and Internet Autonomous Systems subgraphs), and we have discovered that in these graphs $|LSP|$ tends to be surprisingly very close to $n^2$. In Figure 4, we show the average number of locally shortest paths connecting a pair of vertices in a family of random graphs with 500 vertices and increasing number of edges, and in a suite of US road networks. According to these experiments, the computational savings that we might expect using locally shortest paths in Dijkstra's algorithm increase as the edge density grows. In Figure 5, we compare the actual running time of a C implementation the algorithm given in Figure 2 (S-DIJ), and the same algorithm with the locally shortest path heuristic described above (S-LSP). Our implementations are described in [3] and are available at: `http://www.dis.uniroma1.it/~demetres/experim/dsp/`. Notice that in a random graph with density 20% S-LSP can be 16 times faster than S-DIJ. This confirms our expectations based on the counting results given in Figure 4. On very sparse graphs, however, S-LSP appears to be slightly slower than S-DIJ due to the data structure overhead of maintaining lists $L_{xy}$ and $R_{xy}$ for each pair of vertices $x$ and $y$.

## 3    Back from Engineering to Theory: A New Dynamic Algorithm

Let us now consider a scenario in which the input graph changes over time, subject to a sequence of edge weight updates. The goal of a dynamic shortest paths algorithm is to update the distances in the graph more efficiently than recomputing the whole solution from scratch after each change of the weight of an edge. In this section, we show that the idea of using locally shortest paths, which appears to be a useful heuristic for computing all pairs shortest paths as

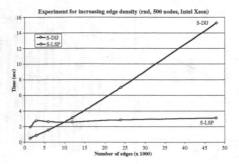

**Fig. 5.** Comparison of the actual running time of a C implementation of Dijkstra's algorithm with (S-LSP) and without (S-DIJ) the locally shortest paths heuristic in a family of random graphs with 500 vertices, increasing edge density, and integer edge weights in $[1, 1000]$. Experiments were performed on an Intel Xeon 500MHz, 512KB L2 cache, 512MB RAM.

we have seen in Section 2, can play a crucial role in designing asymptotically fast update algorithms for the dynamic version of the problem. Let us consider the case where edge weights can only be increased (the case where edge weights can only be decreased in analogous). Notice that, after increasing the weight of an edge, some of the shortest paths containing it may stop being shortest, while other paths may become shortest, replacing the old ones. The goal of a dynamic update algorithm is find efficiently such replacement paths. Intuitively, a locally shortest path is either shortest itself, or it just falls short of being shortest. Locally shortest paths are therefore natural candidates for being replacement paths after an edge weight increase. A possible approach could be to keep in a data structure all the locally shortest paths of a graph, so that a replacement path can be found quickly after an edge update. On the other hand, keeping such a data structure up to date should not be too expensive. To understand if this is possible at all, we first need to answer the following question: how many paths can start being locally shortest and how many paths can stop being locally shortest after an edge weight increase? The following theorem from [5] answers this question:

**Theorem 1.** *Let $G$ be a graph subject to a sequence of edge weight increases. If shortest paths are unique in $G$, then during each update:*

*(1) $O(n^2)$ paths can stop being locally shortest;*
*(2) $O(n^2)$ paths can start being locally shortest, amortized over $\Omega(n)$ operations.*

According to Theorem 1, we might hope to maintain explicitly the set of locally shortest paths in a graph in quadratic amortized time per operation. Since shortest paths in a graph are locally shortest, then maintaining such a set would allow us to keep information also about shortest paths. As a matter of fact, there exists a dynamic variant of the algorithm given in Figure 2 that is able to update the locally shortest paths (and thus the shortest paths) of a graph in $O(n^2 \log n)$ amortized time per edge weight increase; details can be found in [5]. For graphs with $m = \Omega(n \log n)$ edges, this is asymptotically faster than recomputing the solution from scratch using Dijkstra's algorithm. Furthermore, if the distance matrix has to be maintained explicitly, this is only a polylogarithmic factor away from the best possible bound. Surprisingly, no previous result was known for this problem until recently despite three decades of research in this topic.

To solve the dynamic all pairs shortest paths problem in its generality, where the sequence of updates can contain both increases and decreases, locally shortest paths can no longer be used directly: indeed, if increases and decreases can be intermixed, we may have worst-case situations with $\Omega(mn)$ changes in the set of locally shortest paths during each update. However, using a generalization of locally shortest paths, which encompasses the history of the update sequence to cope with pathological instances, we devised a method for updating shortest paths in $O(n^2 \log^3 n)$ amortized time per update [5]. This bound has been recently improved to $O(n^2 \cdot (\log n + \log^2(m/n)))$ by Thorup [11].

## 4    Conclusions

In this paper, we have discussed our own experience in engineering different all pairs shortest paths algorithms based on Dijkstra's method. The interplay between theory and practice yielded significant results. We have shown that the novel notion of locally shortest paths, which allowed us to design a useful heuristic for improving the practical performances of Dijkstra's algorithm on dense graphs, led in turn to the first general efficient dynamic algorithms for maintaining the all pairs shortest paths in a graph.

Despite decades of research, many aspects of the shortest paths problem are still far from being fully understood. For instance, can we compute all pairs shortest paths in $o(mn)$ in dense graphs? As another interesting open problem, can we update a shortest paths tree asymptotically faster than recomputing it from scratch after an edge weight change?

## References

1. B.V. Cherkassky, A.V. Goldberg, and T. Radzik. Shortest paths algorithms: Theory and experimental evaluation. *Mathematical Programming*, 73:129–174, 1996.
2. T.H. Cormen, C.E. Leiserson, R.L. Rivest, and C. Stein. *Introduction to Algorithms*. McGraw-Hill, 2001.
3. C. Demetrescu, S. Emiliozzi, and G.F. Italiano. Experimental analysis of dynamic all pairs shortest path algorithms. In *Proceedings of the 15th Annual ACM-SIAM Symposium on Discrete Algorithms (SODA'04)*, 2004.
4. C. Demetrescu, I. Finocchi, and G.F. Italiano. Algorithm engineering. *The Algorithmics Column (J. Diaz), Bulletin of the EATCS*, 79:48–63, 2003.
5. C. Demetrescu and G.F. Italiano. A new approach to dynamic all pairs shortest paths. In *Proceedings of the 35th Annual ACM Symposium on Theory of Computing (STOC'03), San Diego, CA*, pages 159–166, 2003.

6. E.W. Dijkstra. A note on two problems in connexion with graphs. *Numerische Mathematik*, 1:269–271, 1959.
7. M.L. Fredman and R.E. Tarjan. Fibonacci heaps and their use in improved network optimization algorithms. *Journal of the ACM*, 34:596–615, 1987.
8. S. Pettie. A faster all-pairs shortest path algorithm for real-weighted sparse graphs. In *Proceedings of 29th International Colloquium on Automata, Languages, and Programming (ICALP'02), LNCS Vol. 2380*, pages 85–97, 2002.
9. S. Pettie and V. Ramachandran. Computing shortest paths with comparisons and additions. In *Proceedings of the 13th Annual ACM-SIAM Symposium on Discrete Algorithms (SODA'02)*, pages 267–276. SIAM, January 6–8 2002.
10. S. Pettie, V. Ramachandran, and S. Sridhar. Experimental evaluation of a new shortest path algorithm. In *4th Workshop on Algorithm Engineering and Experiments (ALENEX'02), LNCS Vol. 2409*, pages 126–142, 2002.
11. M. Thorup. Tighter fully-dynamic all pairs shortest paths, 2003. Unpublished manuscript.

# How to Tell a Good Neighborhood from a Bad One: Satisfiability of Boolean Formulas

Tassos Dimitriou[1] and Paul Spirakis[2]

[1] Athens Information Technology, Greece.
**tassos@ait.gr**
[2] Computer Technology Institute, Greece.
**spirakis@cti.gr**

**Abstract.** One of the major problems algorithm designers usually face is to know in advance whether a proposed optimization algorithm is going to behave as planned, and if not, what changes are to be made to the way new solutions are examined so that the algorithm performs nicely. In this work we develop a methodology for differentiating good neighborhoods from bad ones. As a case study we consider the structure of the space of assignments for random 3-SAT formulas and we compare two neighborhoods, a simple and a more refined one that we already know the corresponding algorithm behaves extremely well. We give evidence that it is possible to tell in advance what neighborhood structure will give rise to a good search algorithm and we show how our methodology could have been used to discover some recent results on the structure of the SAT space of solutions. We use as a tool "Go with the winners", an optimization heuristic that uses many particles that independently search the space of all possible solutions. By gathering statistics, we compare the combinatorial characteristics of the different neighborhoods and we show that there are certain features that make a neighborhood better than another, thus giving rise to good search algorithms.

## 1 Introduction

Satisfiability (SAT) is the problem of determining, given a Boolean formula $f$ in conjunctive normal form, whether there exists a truth assignment to the variables that makes the formula true. If all clauses consist of exactly $k$ literals then the formula is said to be an instance of $k$-SAT. While 2-SAT is solvable in polynomial time, $k$-SAT, $k \geq 3$ is known to be $NP$-complete, so we cannot expect to have good performance in the worst case. In practice, however, one is willing to trade "completeness" for "soundness". Incomplete algorithms may fail to find a satisfying assignment even if one exists but they usually perform very well in practice. A well studied algorithm of this sort is the WalkSat heuristic [SKC93]. The distinguishing characteristic of this algorithm is the way new assignments are examined. The algorithm chooses to flip only variables that appear in *unsatisfied* clauses as opposed to flipping any variable and testing the new assignment. Moreover, flips are made even if occasionally they increase the number of unsatisfied clauses.

C.C. Ribeiro and S.L. Martins (Eds.): WEA 2004, LNCS 3059, pp. 199–212, 2004.
© Springer-Verlag Berlin Heidelberg 2004

In this work we make an attempt to differentiate "good" neighborhoods from "bad" ones by discovering the combinatorial characteristics a neighborhood must have in order to produce a good algorithm. We study the WALKSAT neighborhood mentioned above and a simpler one that we call GREEDY in which neighboring assignments differ by flipping any variable. Our work was motivated by a challenge reported by D. S. Johnson [J00] which we quote below:

> **Understanding Metaheuristics**
> "...Currently the only way to tell whether an algorithm of this sort [optimization heuristic] will be effective for a given problem is to implement and run it. We currently do not have an adequate theoretical understanding of the design of search neighborhoods, rules for selecting starting solutions, and the effectiveness of various search strategies. Initial work has given insight in narrow special cases, but no useful general theory has yet been developed. We need one."

Here, we compare the two neighborhoods by examining how the space of neighboring assignments, the search graph as we call it, decomposes into smaller regions of related solutions by imposing a quality threshold to them. Our main tool is the "Go with the winners" (GWW) strategy which uses many particles that independently search the search graph for a solution of large value. Dimitriou and Impagliazzo [DI98] were able to relate the performance of GWW with the existence of a combinatorial property of the search graph, the so called "local expansion". Intuitively, if local expansion holds then particles remain uniformly distributed and *sampling* can be used to deduce properties of the search space.

Although the process of collecting statistics using GWW may not be a cheap one since many particles are used to search the space for good solutions, it is a more accurate one than collecting statistics by actually running the heuristic under investigation. The reason is that heuristic results may be biased towards certain regions of the space and any conclusions we draw may not be true. This cannot happen with GWW since when the local expansion property is true, particles remain uniformly distributed inside the space of solutions and one may use this information to unravel the combinatorial characteristics of the search space. These properties can then help optimize heuristic performance and design heuristics that take advantage of this information.

Our goal in this work is to provide algorithm designers with a method of testing whether a proposed neighborhood structure will give rise to a good search algorithm. We do this following a two-step approach. Initially, we verify experimentally that the search graph has good local expansion. Then by gathering statistics, we compare the *combinatorial characteristics* of the neighborhood and we show that there are certain features that make a neighborhood better than another, thus giving rise to good search algorithms. Our results are in some sense complementary to the work of Schuurmans and Southey [SS01] for SAT where emphasis is given on the characteristics of successful search algorithms. Here, instead, the emphasis is on the properties of the search space itself.

# 2  Go with the Winners Strategy (GWW)

Most optimization heuristics (Simulated Annealing, Genetic algorithms, Greedy, WalkSat, etc.) can be viewed as *strategies*, possibly probabilistic, of moving a particle in a search graph, where the goal is to find a solution of optimal value. The placement of the particle in a node of the search graph corresponds to examining a potential solution.

An immediate generalization of this idea is to use many particles to explore the space of solutions. This generalization together with the extra feature of *interaction* between the particles is essentially "Go with the Winners" [DI96]. The algorithm uses many particles that independently search the space of all solutions. The particles however interact with each other in the following way: each "dead" particle (a particle that ended its search at a local optimum) is moved to the position of a randomly chosen particle that is still "alive". The goal of course is to find a solution of optimal value. Progress is made by imposing a quality threshold to the solutions found, thus improving at each stage the average number of satisfied clauses.

In Figure 1, a "generic" version of GWW is presented without any reference to the underlying neighborhood of solutions (see Section 3). Three conventions are made throughout this work: i) creating a particle means placing the particle in a (random or predefined) node of the search graph, ii) the value of a particle is the value of the solution it represents and iii) more than one particle may be in the same node of the search graph, thus examining the same solution.

The algorithm uses two parameters: the number of particles $B$ and the length $S$ of each particle's random walk. The intuition is that in the $i$-th stage and beyond we are eliminating from consideration assignments which satisfy less than

---

**Generic GWW**

Let $\mathcal{N}_i$ be the *subset* of all solutions during stage $i$ of the algorithm, that is assignments satisfying *at least i* clauses.

- Stage 0 (*Initialization*): Generate $B$ random solutions and place one particle in each one of them.
- Stage $i$ (*Main loop*): Proceed in two phases.
  1. In the *randomization* phase, for each particle, perform an $S$ steps *random walk* in $\mathcal{N}_i$, where a step is defined as follows: Select a neighbor of the current solution. If its value is $\geq i$, move the particle to this neighbor, otherwise leave the particle in its current position.
  2. In the *redistribution* phase, if all particles have value $i$, stop. Otherwise, replace each particle of value $i$ with a copy of a randomly chosen particle whose value is $> i$.
  3. Go to the next stage, by raising the threshold $i$.
- Output the best solution found.

---

**Fig. 1.** Description of the algorithm

$i$ clauses. This divides the search space into *components* of assignments. The redistribution phase will make sure that particles in locally optimal solutions will be distributed among non-local solutions, while the randomization phase will ensure that particles remain *uniformly distributed* inside these emerging components. This is shown abstractly in Figure 2 for a part of the search space and two thresholds $T_A$, $T_B$. When the algorithm is in stage $T_A$, all solutions of value larger than $T_A$ form a connected component and particles are uniformly distributed inside this large component. However, when the threshold is increased to $T_B$, the search graph decomposes into smaller components and the goal is to have particles in all of them in *proportion* to their sizes.

Dimitriou and Impagliazzo [DI98] characterized the search spaces where GWW works well in terms of a combinatorial parameter, the *local expansion* of the search space. Intuitively this property suggests that if a particle starts a random walk, then the resulting solution will be uncorrelated to its start and it is unlikely that particles will be trapped into small regions of the search space.

Studying the behavior of the algorithm with respect to its two important parameters will offer a tradeoff between using larger populations or longer walks. This in turn may give rise to implementations of the algorithm which produce high quality solutions, even when some of these parameters are kept small. In particular, our goal will be to show that for certain neighborhoods only *a few* particles suffice to search the space of solutions provided the space has good expansion properties and the length of the random walk is sufficiently long. When this is true, having a few particles is statistically the same as having one, therefore *this particle together with the underlying neighborhood can be thought of as defining an optimization heuristic.*

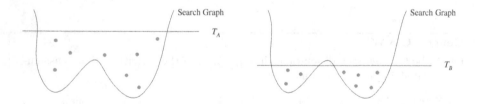

**Fig. 2.** Decomposition of the search graph.

# 3    Distribution of SAT Instances and Search Graphs

It is known that random 3-SAT formulas exhibit a *threshold* behavior. What this means is that there is a constant $r$ so that formulas with more than $rn$ clauses are likely to be unsatisfiable, while formulas with less than $rn$ clauses are likely to be satisfiable. In all experiments we generated formulas with ratio of clauses to variables $r = 4.26$, since as was found experimentally [MSL92,GW94] the hardest to solve instances occur at this ratio.

We proceed now to define the neighborhood structure of solutions. GWW is an optimization heuristic that tries to locate optimal solutions in a *search graph* whose nodes represent all feasible solutions for the given problem. Two nodes in the search graph are connected by an edge, if one solution results from the other by making a local change. The set of neighbors of a given node defines the *neighborhood* of solutions. Such a search graph is implicitly defined by the problem at hand and doesn't have to be computed explicitly. The only operations required by the algorithm are i) generating a random node (solution) in the search graph, ii) compute the value of a given node, and iii) list efficiently all the neighboring solutions. The two search graphs we consider are:

**GREEDY**

Nodes correspond to assignments. Two such nodes are connected by an edge if one results from the other by flipping the truth value of a single variable. Thus any node in this search graph has exactly $n$ neighbors, where $n$ is the number of variables in the formula. The value of a node is simply the number of clauses satisfied by the corresponding assignment.

**WALKSAT**

Same as above except that two such nodes are considered neighboring if one results from the other by flipping the truth value of a variable that belongs to an *unsatisfied* clause.

Our motivation of studying these search graphs is to explain why the WALK-SAT heuristic performs so well in practice as opposed to the simple GREEDY heuristic (which essentially corresponds to GSAT). Does this second search graph has some interesting combinatorial properties that explain the apparent success of this heuristic? We will try to answer some of these questions in the sections that follow.

# 4    Implementation Details

We used a highly optimized version of GWW in which we avoided the explicit enumeration of neighboring solutions in order to perform the random walks for each particle.

In particular we followed a "randomized approach" to picking the right neighbor. To perform one step of the random walk, instead of just enumerating all possible neighbors we simply pick one of the $n$ potential neighbors at random and check its value. If it is larger than the current threshold we place the particle there, otherwise we repeat the experiment. How many times do we have to do this in order to perform one step of the random walk? It can be shown (see for example Figure 4 illustrating the number of neighbors at each threshold) that in the earlier stages of the algorithm we only need a few tries. At later stages, when neighbors become scarce, we simply have to remember which flips we have tried and in any case never perform more than $n$ flips.

The savings are great because as can be seen in the same figure it is only in the last stages that the number of neighbors drops below $n$, thus requiring more than one try on average. In all other stages each random step takes constant time.

# 5   Experiments

The two important parameters of GWW are population size $B$ and random walk length $S$. Large, interacting populations allow the algorithm to reach deeper levels of the search space while long random walks allow GWW to escape from local optima that are encountered when one is using only greedy moves. The use of these two ingredients has the effect of maintaining a *uniform distribution* of particles throughout the part of the search graph being explored.

The goal of the experiments is to understand the relative importance of these parameters and the structure of the search graph. In the first type of experiments (Section 5.1) we will try to reveal the expansion characteristics of the search graph and validate the implementation choices of the algorithm. If the search graph has good expansion, this can be shown by a series of tests that indicate that sufficiently long random walks uniformly distribute the particles.

The purpose of the second type of experiments (Sections 5.2 and 5.3) is to study the quality of the solutions found as a function of $B$ and $S$. In particular, we would like to know what is more beneficial to the algorithm: *to invest in more particles or longer random walks?* Any variation in the quality of solutions returned by the algorithm as a function of these two parameters will illustrate different characteristics of the two neighborhoods. Then, hopefully, all these observations can be used to tell what makes a neighborhood better than another and how to design heuristics that take advantage of the structure of the search space.

In the following experiments we tested formulas with 200 variables and 857 clauses and each sample point on the figures was based on averaging over 500 such random formulas. To support our findings we repeated the experiments with formulas with many more variables and again the same patterns of behavior were observed.

## 5.1   Optimal Random Walk Length

Before we proceed with the core of the experiments we need to know whether our "randomized" version of GWW returns valid statistics. This can be tested with a series of experiments that compare actual data collected by the algorithm with data expected to be true for random formulas. For these comparisons to be valid we need to be sure that particles remain well distributed in the space of solutions. If particles are biased towards certain regions of the space this will be reflected on the statistics and any conclusions we draw may not be true. So, how can we measure the right walk length so that particles remain uniformly distributed?

One nice such test is to compute for each particle the Hamming distance of the assignments *before* and *after* the random walk, and compare with the

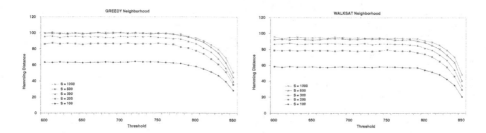

**Fig. 3.** Choosing the right walk length that achieves uniformity for GREEDY (*left*) and WALKSAT (*right*) neighborhoods: Average Hamming distance between start and end of the random walk for $S = 100, 200, 300, 500, 1000$.

distance expected in the random formula model. If the length $S$ of the walk is sufficiently large, these quantities should match.

In Figure 3 we show the results of this experiment for a medium population size of $B = 100$ particles and various walk lengths ($S = 100, 200, 300, 500, 1000$). For each stage (shown are stages where the number of satisfied formulas is $\geq 600$) we computed the average Hamming distance between solutions (assignments) at start and end of the random walk, and we plotted the average over all particles. As we can see the required length to achieve uniformity in the GREEDY neighborhood is about 1000 steps. For this number of steps the average Hamming distance matches the one expected when working with random formulas, which is exactly $n/2$. It is only in the last stages of the algorithm that the Hamming distance begins to drop below the $n/2$ value as not all flips give rise to good neighbors (compare with Figure 4). The same is true for the WALKSAT neighborhood. The required length is again about 1000 steps. Here however, the Hamming distance is slightly smaller as flips are confined only to variables found in unsatisfied clauses. Thus we conclude that 1000 steps seems to be a sufficient walk length so that particles remain uniformly distributed in the space of solutions.

A second test we performed to compute the appropriate walk length was to calculate the average number of neighbors of each particle at each threshold and compare it with the expected values in the random formulas model. The results are shown in Figure 4(left) for a population size $B = 100$ and the optimal random walk length ($S = 1000$). For the GREEDY neighborhood, at least in the early stages of the algorithm, all the flips should lead to valid neighbors, so their number should be equal to $n$, the number of variables. For the WALKSAT neighborhood the number of neighbors should be smaller than $n$ as flips are confined to variables in unsatisfied clauses. As it can be seen, the collected averages match with the values expected to be true for random formulas, proving the validity of the randomized implementation and the uniform distribution of particles.

Finally in Figure 4(right), we performed yet another test to examine the hypothesis that 1000 steps are sufficient to uniformly distribute the particles. In this test we examined the average number of satisfied clauses of the GWW pop-

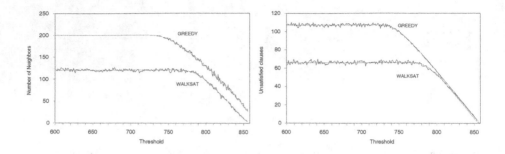

**Fig. 4.** Comparing the number of neighbors (*left*) and unsatisfied clauses (*right*) at various thresholds with those *expected* for random formulas, for both neighborhoods. The collected data matched the expectations.

ulation at each threshold and compared them with those expected for random formulas. Again the two curves matched showing that particles remain well distributed. In this particular plot it is instructive to observe a qualitative difference between the two neighborhoods. In the GREEDY one, the algorithm starts with 1/8 of the clauses unsatisfied and it is only in the last stages that this number begins to drop. In the WALKSAT neighborhood however, this number is much smaller. It seems that the algorithm already has an advantage over the GREEDY implementation as it has to satisfy fewer unsatisfied clauses (approximately 5/64 of the clauses against 8/64 of the GREEDY neighborhood). Does this also mean that the algorithm will need fewer "resources" to explore the space of solutions? We will try to answer this next.

## 5.2   Characterizing the Two Neighborhoods

In this part we would like to study the quality of the solutions found by GWW as a function of its two mechanisms, population size and random walk length. Studying the question of whether to invest in more particles or longer random walks will illustrate different characteristics of the two neighborhoods. In particular, correlation of the quality of the solutions with population size will provide information about the *connectedness* of the search graphs, while correlation with random walk length will tell us more about the *expansion characteristics* of these search spaces.

We first studied the effect of varying the population size $B$ while keeping the number of random steps $S$ constant, for various walk lengths ($S = 4, 8, 16, 32$). The number of satisfied clauses increased as a function of $B$ and the curves looked asymptotic, like those of Figure 5. In general, the plots corresponding to the two neighborhoods were similar with the exception that results were slightly better for the WALKSAT neighborhood.

In Figure 5 we studied the effect of varying the number of steps while keeping the population size constant, for various values of $B$ ($B = 2, 4, 8, 16, 32$). As it

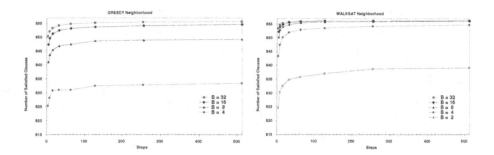

**Fig. 5.** Average number of clauses satisfied vs random walk length for various population sizes. Results shown are for GREEDY (*left*) and WALKSAT neighborhoods (*right*).

can be seen, increasing the population size resulted in better solutions found. But there is a distinctive difference between the two neighborhoods. While in the first increasing the particles had a clear impact on the quality of solutions found, in the second, having 8 particles was essentially the same as having 32 particles. Furthermore, 2 particles searching the WALKSAT neighborhood obtained far better results than 4 particles searching the GREEDY one, and in general the GREEDY neighborhood required twice as many particles to achieve results comparable to those obtained in the WALKSAT one. Thus we note immediately that the second neighborhood is *more suited for search with fewer particles*, reaching the same quality on the solutions found provided the walks are kept sufficiently long. This agrees with results found by Hoos [Ho99] where a convincing explanation is given for this phenomenon: Greedy (i.e. GSAT) is essentially incomplete, so numerous restarts are necessary to remedy this situation. Since a restart is like introducing a new search particle, although without the benefit of interaction between particles as in GWW, we see that our findings come to verify this result.

It is evident from these experiments that increasing the computational resources (particles or walk steps) improves the quality of the solutions. We also have an indication that the number of particles is not so important in the WALK-SAT neighborhood. So, to make this difference even more striking, we proceeded to examine the effect on the solutions found when the product $B \times S$, and hence the running time, was kept *constant*[1]. The results are shown in Figure 6.

It is clear that as the number of particles increases the average value also increases, but there is a point where this behavior stops and the value of solutions starts to decline. This is well understood. As the population becomes large, the number of random steps decreases (under the constant time constraint) and this has the effect of not allowing the particles to escape from bad regions.

---

[1] The running time of GWW is $O(BSm)$, where $m$ is the maximum number of stages (formula clauses)

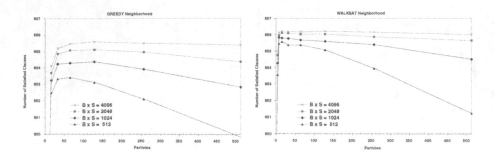

**Fig. 6.** Average solution value vs number of particles for different values of $B \times S$. Results shown are for GREEDY (*left*) and WALKSAT neighborhoods (*right*).

Perhaps what is more instructive to observe is the point in the two neighborhoods when this happens. In the GREEDY neighborhood this behavior is more balanced between particles and length of random walk, as one resource does not seem more important than the other. In the WALKSAT neighborhood however, things change drastically as the quality of the solutions found degrades when fewer steps and more particles are used! Thus *having longer random walks is more beneficial to the algorithm than having larger populations*. This comes to validate the observation we made in the beginning of this section that the second neighborhood is more suited for search with fewer particles. When taken to the extreme this explains the apparent success of WALKSAT type of algorithms in searching for satisfying assignments as they can be viewed as strategies for moving just one particle around the WALKSAT neighborhood.

To illustrate the difference in the two neighborhoods we performed 300 runs of GWW using $B = 2$ and $S = 1000$ and we presented the results in Figure 7 as a histogram of solutions. The histograms of the two neighborhoods were displayed on the same axis as the overlap of values was very small. There was such strong separation between the two neighborhoods that even when we normalized for running time the overlap was still very small. Indeed the best assignment found in the GREEDY neighborhood satisfied only 827 clauses, which is essentially the worst case for the WALKSAT neighborhood. One can thus conclude that the second implementation is intrinsically more powerful and more suited for local optimization than the first one, even when running time is taken into account.

### 5.3 Further Properties of the Search Graphs

The results of the previous section suggest that the WALKSAT neighborhood is more suited for local search as only a few particles can locate the optimal solutions. So, we may ask what is the reason behind this discrepancy between the two neighborhoods?

We believe the answer must lie in the structure of the search graphs. As the GREEDY space decomposes into components by imposing the quality threshold to solutions, good solutions must reside in pits with high barriers that render

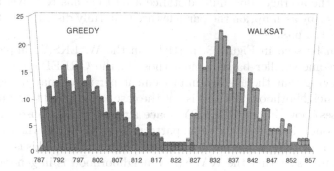

**Fig. 7.** Histogram of satisfied clauses for 300 runs of GWW with $B = 2$ and $S = 1000$. Results shown are for GREEDY and WALKSAT neighborhoods.

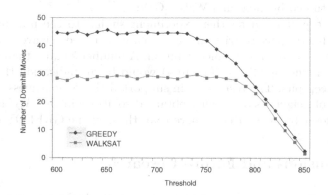

**Fig. 8.** Results of performing simple greedy optimization starting from particles at various thresholds. The plots show the average number of downhill moves required before getting trapped into a local optimum in both neighborhoods.

long random walks useless. So the algorithm requires more particles as smaller populations usually get trapped. In the case of the WALKSAT neighborhood, the search graph must be "smooth" in the sense that good solutions must not be hidden in such deep pits. Thus the algorithm needs only a few particles to hit these regions.

We tried to test this hypothesis with the following experiment: Once the particles were uniformly distributed after the randomization phase, we performed simple *greedy optimization* starting from that particle's position and recorded the number of downhill moves (we use this term even if this is a maximization problem) required before the particle gets trapped into a local optimum. Then we averaged over all particles and proceeded with the next stage. So, essentially we counted the average (downhill) distance a particle has to travel before gets trapped. Since by assumption the particles are uniformly distributed, this reveals the depth of the pits.

As it can be seen in Figure 8, particles in the WALKSAT implementation had to overcome smaller barriers than those in the GREEDY one. (One may object by saying that the average improvement per move may be larger in the WALKSAT neighborhood. But this is not the case as we found that improvement is $\approx 2$ clauses/move in the GREEDY space against $\approx 1.6$ clauses/move in the WALKSAT one). Moreover, the same patterns were observed when we repeated the experiment with formulas having 500 and 1000 variables. This is a remarkable result. It simply says that the WALKSAT neighborhood is in general smoother than the GREEDY one and easier to be searched with only a few particles, which perhaps also explains why GSAT doesn't meet the performance of WalkSat. Again our findings come to verify some old results: Schuurmans and Southey [SS01] identify three measures of local search effectiveness, one of which is *depth* corresponding to number of unsatisfied clauses. Similarly, the depth of GSAT is intensively studied by Gent and Walsh [GW93].

Finally we performed another experiment similar to the previous one (not shown here due to space restrictions) in which we counted the average number of neighbors of each *greedily* obtained solution. A number smaller than the number of neighbors found by GWW (Figure 4) would be an indication that solutions lied inside deep pits. The results again supported the "smoothness" conjecture; the number of neighbors of greedily obtained solutions in the WALKSAT space was much closer to the expected curve than those of the GREEDY one.

## 6    Conclusions and Future Research

Many optimizations algorithms can be viewed as strategies for searching a space of potential solutions in order to find a solution of optimal value. The success of these algorithms depends on the way the underlying search graph is implicitly defined and in particular on the way a new potential solution is generated by making local changes to the current one. As was mentioned in [J00], currently the only way to tell whether an algorithm of this sort will be effective for a given problem is simply to implement and run it. So, one of the challenges algorithm designers face is to design the right search neighborhood so that the corresponding optimization heuristic behaves as expected.

In this work we considered the problem of differentiating good neighborhoods from bad ones. In particular, we studied two search neighborhoods for the 3-SAT problem, a simple one which we called GREEDY and a more refined one that we called WALKSAT. In the first one, neighboring assignments were generated by flipping the value of *any* of the $n$ variables, while in the second one only variables that belong to *unsatisfied* formulas were flipped. Our motivation for this work was inspired by the challenge mentioned above and the need to explain the apparent success of WALKSAT type of algorithms in finding good satisfying assignments since all of them are based on the WALKSAT neighborhood.

We gave evidence that it is possible to tell in advance what neighborhood structure will give rise to a good search algorithm by comparing the *combinatorial characteristics* of the two neighborhoods. We used as a platform for testing neighborhoods "Go with the winners", an algorithm that uses many particles that independently search the space of solutions. By gathering statistics we showed that there are certain features that make one neighborhood better than another, thus giving rise to a good search algorithm.

In particular, we noticed that the WALKSAT neighborhood was more suited for search with fewer particles, statistically the same as one. We expected this to be true since we knew that the WalkSat heuristic performs extremely well, but we were surprised by the extend to which this was true. We thought that having a more balanced setting between particles and walk lengths would be more beneficial to GWW but this was the case only for the GREEDY neighborhood.

Although we studied only one type of problem (SAT), we believe that search spaces for which good heuristics exist must have similar "characteristics" as the WALKSAT space and can be verified using our approach. Specifically, *to test if a neighborhood will give rise to a good search algorithm run GWW and study the tradeoff between particles and random walk steps. If GWW can discover good solutions with just a few particles and long enough walks, then this space is a candidate for a good search heuristic.* These observations lead to some interesting research directions:

- Provide more evidence for the previous conjecture by trying to analyze the search spaces of other optimization problems for which good algorithms exist. Do the corresponding search graphs have similar properties to the WALK-SAT one? This line of research would further validate GWW as a tool for collecting valid statistics.
- What are the properties of search spaces for which *no good* algorithms exist? Are in any sense complementary to those defined here?
- Understand how the WALKSAT space decomposes into components by imposing the quality threshold to solutions. We believe that the WALKSAT space decomposes into only *a few* components (which also explains why one doesn't need many particles) but this remains to be seen.
- Can we find a simple neighborhood that behaves even better than WALK-SAT? If this neighborhood has similar characteristics to WALKSAT's (sufficiency of a few particles to locate good solutions, small barriers, etc.) it will probably give rise to even better satisfiability algorithms. (Introducing weights [F97,WW97] smooths out the space but does not meet our definition of a neighborhood.)
- More ambitiously, try to analyze WalkSat and its variants and prove that they work in polynomial time (for certain ratios of clauses to variables, of course). This ambitious plan is supported by the fact that similar findings for graph bisection [DI98,Ca01]) ultimately led to a polynomial time, local search algorithm for finding good bisections [CI01].

**Acknowledgements:** The authors wish to thank Christos Papadimitriou and the reviewers for their useful comments.

# References

[Ca01]    T. Carson. *Empirical and Analytic Approaches to understanding Local Search Heuristics.* PhD thesis, University of California, San Diego, 2001.

[CI01]    T. Carson and R. Impagliazzo. Hill climbing finds random planted bisections. In *Proc. 12th Annual ACM Symposium on Discrete Algorithms (SODA)*, 2001.

[DI96]    T. Dimitriou and R. Impagliazzo. Towards an analysis of local optimization algorithms. In *Proc. 28th Annual ACM Symposium on Theory of Computing (STOC)*, 1996.

[DI98]    T. Dimitriou and R. Impagliazzo. Go with the Winners for Graph Bisection. In *Proc. 9th Annual ACM Symposium on Discrete Algorithms (SODA)*, 510–520, 1998.

[F97]    Frank, J. Learning weights for GSAT. In *Proc. IJCAI-97*, 384-391.

[GJ79]    M. R. Garey and D. S. Johnson. *Computers and Intractability.* Freeman, San Francisco, CA, 1979.

[GW93]    I. Gent and T. Walsh. An empirical analysis of search in GSAT. *Journal of Artificial Intelligence Research*, 1:23-57, 1993.

[GW94]    I. Gent and T. Walsh. The SAT Phase Transition. In *Proc. of ECAI-94*, 105-109, 1994.

[Ho99]    Hoos, H. On the run-time behavior of stochastic local search algorithms for SAT. In *Proc. AAAI-99*, 661–666, 1999.

[J00]    D. S. Johnson. Report on Challenges for Theoretical Computer Science. Workshop on Challenges for Theoretical Computer Science, Portland, 2000, http://www.research.att.com/~dsj/

[MSL92]    Mitchell, D., Selman, B., and Levesque, H.J. Hard and easy distributions of SAT problems. In *Proc. AAAI-92*, pp. 459–465, San Jose, CA, 1992.

[SKC93]    B. Selman, H. A. Kautz and B. Cohen. Local search strategies for satisfiability testing. In *Second DIMACS Challenge on Cliques, Coloring and Satisfiability*, October 1993.

[SLM92]    Selman, B., Levesque, H.J. and Mitchell, D. A new method for solving hard statisfiability problems. In *Proc. AAAI-92*, San Jose, CA, 1992.

[SS01]    D. Schuurmans, F. Southey. Local search characteristics of incomplete SAT procedures. *Artif. Intelligence*, 132(2), 121-150, 2001.

[WW97]    Wu, Z. and Wah, W. Trap escaping strategies in discrete Langrangian methods for solving hard satisfiability and maximum satisfiability problems. In *Proc. AAAI-99*, 673-678.

# Implementing Approximation Algorithms for the Single-Source Unsplittable Flow Problem

Jingde Du[1]* and Stavros G. Kolliopoulos[2]**

[1] Department of Mathematical Sciences, University of New Brunswick, Saint John, Canada.
jdu@unbsj.ca
[2] Department of Computing and Software, McMaster University, Canada.
stavros@mcmaster.ca

**Abstract.** In the *single-source unsplittable flow* problem, commodities must be routed simultaneously from a common source vertex to certain sinks in a given graph with edge capacities. The demand of each commodity must be routed along a single path so that the total flow through any edge is at most its capacity. This problem was introduced by Kleinberg [12] and generalizes several NP-complete problems. A cost value per unit of flow may also be defined for every edge. In this paper, we implement the 2-approximation algorithm of Dinitz, Garg, and Goemans [6] for congestion, which is the best known, and the $(3, 1)$-approximation algorithm of Skutella [19] for congestion and cost, which is the best known bicriteria approximation. We study experimentally the quality of approximation achieved by the algorithms and the effect of heuristics on their performance. We also compare these algorithms against the previous best ones by Kolliopoulos and Stein [15].

## 1 Introduction

In the *single-source unsplittable flow* problem (UFP), we are given a directed graph $G = (V, E)$ with edge capacities $u : E \to \mathbb{R}^+$, a designated source vertex $s \in V$, and $k$ commodities each with a terminal (sink) vertex $t_i \in V$ and associated demand $d_i \in \mathbb{R}^+$, $1 \leq i \leq k$. For each $i$, we have to route $d_i$ units of commodity $i$ along a single path from $s$ to $t_i$ so that the total flow through an edge $e$ is at most its capacity $u(e)$. As is standard in the relevant literature we assume that no edge can be a bottleneck, i.e., the minimum edge capacity is assumed to have value at least $\max_i d_i$. We will refer to instances which satisfy this assumption as *balanced*, and ones which violate it as *unbalanced*. Instances in which the maximum demand is $\rho$ times the minimum capacity, for $\rho > 1$, are $\rho$-*unbalanced*. A relaxation of UFP is obtained by allowing the demands of commodities to be split along more than one path; this yields a standard maximum

* Part of this work was done while at the Department of Computing and Software, McMaster University. Partially supported by NSERC Grant 227809-00.
** Partially supported by NSERC Grant 227809-00.

C.C. Ribeiro and S.L. Martins (Eds.): WEA 2004, LNCS 3059, pp. 213–227, 2004.
© Springer-Verlag Berlin Heidelberg 2004

flow problem. We will call a solution to this relaxation, a *fractional* or *splittable* flow.

In this paper we use the following terminology. A flow $f$ is called *feasible* if it satisfies all the demands and respects the capacity constraints, i.e., $f(e) \leq u(e)$ for all $e \in E$. An *unsplittable flow* $f$ can be specified as a flow function on the edges or equivalently by a set of paths $\{P_1, \cdots P_k\}$, where $P_i$ starts at the source $s$ and ends at $t_i$, such that $f(e) = \sum_{i:e \in P_i} d_i$ for all edges $e \in E$. Hence in the UFP a feasible solution means a feasible unsplittable flow. If a cost function $c : E \to \mathbb{R}^+$ on the edges is given, then the cost $c(f)$ of flow $f$ is given by $c(f) = \sum_{e \in E} f(e) \cdot c(e)$. The cost $c(P_i)$ of an path $P_i$ is defined as $c(P_i) = \sum_{e \in P_i} c(e)$ so that the cost of an unsplittable flow $f$ given by paths $P_1, \cdots, P_k$ can also be written as $c(f) = \sum_{i=1}^{k} d_i \cdot c(P_i)$. In the version of the UFP *with costs*, apart from the cost function $c : E \to \mathbb{R}^+$ we are also given a budget $B \geq 0$. We seek a feasible unsplittable flow whose total cost does not exceed the budget. Finally, we set $d_{\max} = \max_{1 \leq i \leq k} d_i$, $d_{\min} = \min_{1 \leq i \leq k} d_i$, and $u_{\min} = \min_{e \in E} u_e$. For $a, b \in \mathbb{R}^+$ we write $a \,|\, b$ and say that b is *a-integral* if and only if $b \in a \cdot \mathbb{N}$.

The feasibility question for UFP (without costs) is strongly $NP$-complete [12]. Various optimization versions can be defined for the problem. In this study we focus on *minimizing congestion:* Find the smallest $\alpha \geq 1$ such that there exists a feasible unsplittable flow if all capacities are multiplied by $\alpha$. Among the different optimization versions of UFP the congestion metric admits the currently best approximation ratios. Moreover congestion has been studied extensively in several settings for its connections to multicommodity flow and cuts.

*Previous work.* UFP was introduced by Kleinberg [12] and contains several well-known NP-complete problems as special cases: Partition, Bin Packing, scheduling on parallel machines to minimize makespan [12]. In addition UFP generalizes single-source edge-disjoint paths and models aspects of virtual circuit routing. The first constant-factor approximations were given in [13]. Kolliopoulos and Stein [14,15] gave a 3-approximation algorithm for congestion with a simultaneous performance guarantee 2 for cost, which we denote as a $(3, 2)$-approximation. Dinitz, Garg, and Goemans [6] improved the congestion bound to 2. To be more precise, their basic result is: any splittable flow satisfying all demands can be turned into an unsplittable flow while increasing the total flow through any edge by less than the maximum demand and this is tight [6]. It is known that when the congestion of the fractional flow is used as a lower bound the factor 2 increase in congestion is unavoidable. Skutella [19] improved the $(3, 2)$-approximation algorithm for congestion and cost [14] to a $(3, 1)$-approximation algorithm.

In terms of negative results, Lenstra, Shmoys, and Tardos [16] show that the minimum congestion problem cannot be approximated within less than $3/2$, unless $P = NP$. Skutella [19] shows that, unless $P = NP$, congestion cannot be approximated within less than $(1 + \sqrt{5})/2 \approx 1.618$ for the case of $((1 + \sqrt{5})/2)$-unbalanced instances. Erlebach and Hall [7] prove that for arbitrary $\varepsilon > 0$ there is no $(2-\varepsilon,1)$-approximation algorithm for congestion and cost unless $P = NP$. Matching this bicriteria lower bound is a major open question.

*This work.* As a continuation of the experimental study initiated by Kolliopoulos and Stein [15], we present an evaluation of the current state-of-the-art algorithms from the literature. We implement the two currently best approximation algorithms for minimizing congestion: (i) the 2-approximation algorithm of Dinitz, Garg, and Goemans [6] (denoted DGGA) and (ii) the $(3,1)$-approximation algorithm of Skutella [19] (denoted SA) which simultaneously mininimizes the cost. We study experimentally the quality of approximation achieved by the algorithms, and the effect of heuristics on approximation and running time. We also compare these algorithms against two implementations of the Kolliopoulos and Stein [14,15] 3-approximation algorithm (denoted KSA). Extensive experiments on the latter algorithm and its variants were reported in [15].

The goal of our work is to examine primarily the quality of approximation. We also consider the time efficiency of the approximation algorithms we implement. Since our main focus is on the performance guarantee we have not extensively optimized our codes for speed and we use a granularity of seconds to indicate the running time. Our input data comes from four different generators introduced in [15]. The performance guarantee is compared against the congestion achieved by the fractional solution, which is always taken to be 1. This comparison between the unsplittable and the fractional solution mirrors the analyses of the algorithms we consider. Moreover it has the benefit of providing information on the "integrality" gap between the two solutions. In general terms, our experimental study shows that the approximation quality of the DGGA is typically better, by a small absolute amount, than that of the KSA. Both algorithms behave consistently better than the SA. However the latter remains competitive for minimum congestion even though it is constrained by having to meet the budget requirement. All three algorithms achieve approximation ratios which are typically well below the theoretical ones. After reviewing the algorithms and the experimental setting we present the results in detail in Section 5.

## 2    The 2-Approximation Algorithm for Minimum Congestion

In this section we briefly present the DGGA [6] and give a quick overview of the analysis as given in [6]. The skeleton of the algorithm is given in Fig. 1.

We explain the steps of the main loop. Certain edges, labeled as *singular*, play a special role. These are the edges $(u,v)$ such that $v$ and all the vertices reachable from $v$ have out-degree at most 1. To construct an *alternating cycle* $C$ we begin from an arbitrary vertex $v$. From $v$ we follow outgoing edges as long as possible, thereby constructing a *forward* path. Since the graph is acyclic, this procedure stops, and we can only stop at a terminal, $t_i$. We then construct a *backward* path by beginning from any edge entering $t_i$ distinct from the edge that was used to reach $t_i$ and following singular incoming edges as far as possible. We thus stop at the first vertex, say $v'$, which has another edge leaving it. We now continue by constructing a forward path from $v'$. We proceed in this manner till we reach a vertex, say $w$, that was already visited. This creates a cycle. If the two paths containing $w$ in the cycle are of the same type then they both have to be forward paths, and we glue them into one forward path. Thus the cycle consists of alternating forward and backward paths.

DGG-ALGORITHM:

> **Input:** A directed graph $G = (V, E)$ with a source vertex $s \in V$, $k$ commodi-
> ties $i = 1, \cdots, k$ with terminals $t_i \in V \setminus \{s\}$ and positive demands $d_i$, and
> a (splittable) flow on $G$ satisfying all demands.
> **Output:** An unsplittable flow given by a path $P_i$ from $s$ to each terminal $t_i$,
> $1 \leq i \leq k$.
>
> remove all edges with zero flow and all flow cycles from $G$;
> preliminary phase:
> $i := 1$;
> **while** $i \leq k$ **do**
> > **while** there is an incoming edge $e = (v, t_i)$ with flow $\geq d_i$ **do**
> > > move $t_i$ to $v$;
> > > add $e$ to $P_i$;
> > > decrease the flow on $e$ by $d_i$;
> > > remove $e$ from $G$ if the flow on $e$ vanishes;
> > $i := i + 1$;
> main loop:
> **while** outdegree$(s) > 0$ **do**
> > construct an alternating cycle $C$;
> > augment flow along $C$;
> > move terminals as in the preliminary phase giving preference to singular
> > edges;
> **return** $P_1, \cdots, P_k$;

**Fig. 1.** Algorithm DGGA.

We augment the flow along $C$ by decreasing the flow along the forward paths and increasing the flow along the backward paths by the same amount equal to $\min\{\varepsilon_1, \varepsilon_2\}$. The quantity $\varepsilon_1 > 0$, is the minimum flow along an edge on a forward path of the cycle. The second quantity, $\varepsilon_2$, is equal to $\min(d_j - f(e))$ where the minimum is taken over all edges $e = (u, v)$ lying on backward paths of the cycle and over all terminals $t_j$ at $v$ for which $d_j > f(e)$. If the minimum is achieved for an edge on a forward path then after the augmentation the flow on this edge vanishes and so the edge disappears from the graph. If the minimum is achieved for an edge $(u, t_j)$ on a backward path, then after the augmentation the flow on $(u, t_j)$ is equal to $d_j$.

*Analysis Overview.* The correctness of the algorithm is based on the following two facts: the first is that at the beginning of any iteration, the in-degree of any vertex containing one or more terminals is at least 2; the second, which is a consequence of the first fact, is that as long as all terminals have not reached the source, the algorithm always finds an alternating cycle.

At each iteration, after augmentation either the flow on some forward edge vanishes and so the edge disappears from the graph or the flow on a backward edge $(u, t_j)$ is equal to $d_j$ and so the edge disappears from the graph after moving the terminal $t_j$ to $u$, decreasing the flow on the edge $(u, t_j)$ to zero and removing this edge from the graph. So, as a result of each iteration, at least one edge is eliminated and the algorithm makes progress. Before an edge becomes a singular edge, the flow on it does not increase. After the edge becomes a singular edge we move at most one terminal along this edge and then this edge vanishes. Thus the total unsplittable flow through this edge is less than the sum of its initial flow and the maximum demand and the performance guarantee is at most 2. We refer the reader to [6] for more details.

*Running Time.* Since every augmentation removes at least one edge, the number of augmentations is at most $m$, where $m = |E|$. An augmenting cycle can be found in $O(n)$ time, where $n = |V|$. The time for moving terminals is $O(kn)$, where $k$ denotes the number of terminals. Since there are $k$ terminals, computing $\varepsilon_2$ requires $O(k)$ time in each iteration. Therefore the running time of the algorithm is $O(nm + km)$.

*Heuristic Improvement.* We have a second implementation with an added heuristic. The purpose of the heuristic is to try to reduce the congestion. The heuristic is designed so that it does not affect the theoretical performance guarantee of the original algorithm, but as a sacrifice, the running time is increased. In our second implementation, we use the heuristic only when we determine an alternating cycle. We always pick an outgoing edge with the smallest flow to go forward and choose an incoming edge with the largest flow to go backward. For most of the cases we tested, as we show in the experimental results in Section 5, the congestion is reduced somewhat. For some cases, it is reduced a lot. The running time for the new implementation with the heuristic is $O(dnm + km)$, where $d$ is the maximum value of incoming and outgoing degrees among all vertices, since the time for finding an alternating cycle is now $O(dn)$.

## 3    The $(3, 1)$-Approximation Algorithm for Congestion and Cost

The KSA [14,15] iteratively improves a solution by doubling the flow amount on each path utilized by a commodity which has not yet been routed unsplittably. This scheme can be implemented to give a $(2, 1)$-approximation if all demands are powers of 2. Skutella's algorithm [19] combines this idea with a clever initialization which rounds down the demands to powers of 2 by removing the most costly paths from the solution. In Fig. 2 we give the algorithm for the case of powers of 2 [14,15] in the more efficient implementation of Skutella [19]. The main idea behind the analysis of the congestion guarantee is that the total increase on an edge capacity across all iterations is bounded by $\sum_{l < \log(d_{max}/d_{min})} d_{min} 2^l \le d_{max}$.

POWER-ALGORITHM:

**Input:** A directed graph $G = (V, E)$ with non-negative costs on the edges, a
source vertex $s \in V$, $k$ commodities $i = 1, \cdots, k$ with terminals $t_i \in V \setminus \{s\}$
and positive demands $d_i = d_{\min} \cdot 2^{q_i}$, $q_i \in \mathbb{N}$, $q_1 \leq q_2 \leq \cdots \leq q_k$, and a
(splittable) flow $f_0$ on $G$ satisfying all demands.
**Output:** An unsplittable flow given by a path $P_i$ from $s$ to each terminal $t_i$,
$1 \leq i \leq k$.

$i := 1; j := 0;$
**while** $d_{\min} \cdot 2^j \leq d_{\max}$ **do**
    $j := j + 1; \delta_j := d_{\min} \cdot 2^{j-1};$
    for every edge $e \in E$, set its capacity $u_e^j$ to $f_{j-1}(e)$ rounded up to the
    nearest multiple of $\delta_j$;
    compute a feasible $\delta_j$-integral flow $f_j$ satisfying all demands with $c(f_j) \leq$
    $c(f_{j-1})$;
    remove all edges $e$ with $f_j(e) = 0$ from $G$;
    **while** $i \leq k$ and $d_i = \delta_j$ **do**
        determine an arbitrary path $P_i$ from $s$ to $t_i$ in $G$;
        decrease $f_j$ along $P_i$ by $d_i$;
        remove all edges $e$ with $f_j(e) = 0$ from $G$;
        $i := i + 1;$
    **return** $P_1, \cdots, P_k;$

**Fig. 2.** The SA after all demands have been rounded to powers of 2

*Running time.* The running time of the POWER-ALGORITHM is dominated by
the time to compute a $\delta_j$-integral flow $f_j$ in each while-loop-iteration $j$. Given the
flow $f_{j-1}$, this can be done in the following way [19]. We consider the subgraph
of the current graph $G$ which is induced by all edges $e$ whose flow value $f_{j-1}(e)$
is not $\delta_j$-integral. Starting at an arbitrary vertex of this subgraph and ignoring
directions of edges, we greedily determine a cycle $C$; this is possible since, due
to flow conservation, the degree of every vertex is at least two. Then, we choose
the *orientation of the augmentation* on $C$ so that the cost of the flow is not
increased. We augment flow on the edges of $C$ whose direction is identical to
the augmentation orientation and decrease flow by the same amount on the
other edges of $C$ until the flow value on one of the edges becomes $\delta_j$-integral. We
delete all $\delta_j$-integral edges and continue iteratively. This process terminates after
at most $m$ iterations and has thus running time $O(nm)$. The number of while-
loop-iterations is $1 + \log(d_{\max}/d_{\min})$. The running time of the first iteration
is $O(nm)$ as discussed above. However, since $f_{j-1}$ is $(d_{\min} \cdot 2^{j-2})$-integral in
each further iteration $j \geq 2$, the amount of augmented flow along a cycle $C$ is
$d_{\min} \cdot 2^{j-2}$ and after the augmentation the flow on each edge of $C$ is $(d_{\min} \cdot 2^{j-1})$-
integral and thus all edges of $C$ will not be involved in the remaining cycle
augmentation steps of this iteration. So the computation of $f_j$ from $f_{j-1}$ takes
only $O(m)$ time. Moreover the path $P_i$ can be determined in $O(n)$ time for
each commodity $i$ and the total running time of the POWER-ALGORITHM is
$O(kn + m \log(d_{\max}/d_{\min}) + nm)$.

We now present the GENERAL-ALGORITHM [19] which works for arbitrary demand values. In the remainder of the paper when we refer to the SA we mean the GENERAL-ALGORITHM. It constructs an unsplittable flow by rounding down the demand values such that the rounded demands satisfy the condition for using the POWER-ALGORITHM. Then, the latter algorithm is called to compute paths $P_1, \cdots, P_k$. Finally, the original demand of commodity $i$, $1 \leq i \leq k$, is routed across path $P_i$. In contrast the KSA rounds demands *up* to the closest power of 2 before invoking the analogue of the POWER-ALGORITHM.

We may assume that the graph is acyclic, which can be achieved by removing all edges with flow value 0 and iteratively reducing flow along directed cycles. This can be implemented in $O(nm)$ time using standard methods.

In the first step of the GENERAL-ALGORITHM, we round down all demands $d_i$ to

$$\bar{d}_i := d_{\min} \cdot 2^{\lfloor \log(d_i/d_{\min}) \rfloor}.$$

Then, in a second step, we modify the flow $f$ such that it only satisfies the rounded demands $\bar{d}_i$, $1 \leq i \leq k$. The algorithm deals with the commodities $i$ one after another and iteratively reduces the flow $f$ along the most expensive $s$-$t_i$-paths within $f$ (ignoring or removing edges with flow value zero) until the inflow in node $t_i$ has been decreased by $d_i - \bar{d}_i$. So, when we reroute this amount of reduced flow along any $s$-$t_i$-paths within the updated $f$, the cost of this part of the flow will not increase. Since the underlying graph has no directed cycles, a most expensive $s$-$t_i$-path can be computed in polynomial time. Notice that the resulting flow $\bar{f}$ satisfies all rounded demands. Thus, the POWER-ALGORITHM can be used to turn $\bar{f}$ into an unsplittable flow $\tilde{f}$ for the rounded instance with $c(\tilde{f}) \leq c(\bar{f})$. The GENERAL-ALGORITHM constructs an unsplittable flow $\hat{f}$ for the original instance by routing, for each commodity $i$, the total demand $d_i$ (instead of only $\bar{d}_i$) along the path $P_i$ returned by the POWER-ALGORITHM and the cost of $\hat{f}$ is bounded by $c(\hat{f}) = c(\tilde{f}) + \sum_{i=1}^k (d_i - \bar{d}_i) c(P_i) \leq c(\bar{f}) + \sum_{i=1}^k (d_i - \bar{d}_i) c(P_i) \leq c(f)$.

Skutella [19] shows that the GENERAL-ALGORITHM finds an unsplittable flow whose cost is bounded by the cost of the initial flow $f$ and the flow value on any edge $e$ is less than $2f(e) + d_{\max}$. Therefore, if the instance is balanced, i.e., the assumption that $d_{\max} \leq u_{\min}$ is satisfied, an unsplittable flow whose cost is bounded by the cost of the initial flow and whose congestion is less than 3 can be obtained. Furthermore, if we use a minimum-cost flow algorithm to find a feasible splittable flow of minimum cost for the initial flow, the cost of an unsplittable flow obtained by the GENERAL-ALGORITHM is bounded by this minimum cost.

*Running time.* The procedure for obtaining $\bar{f}$ from $f$ can be implemented to run in $O(m^2)$ time; in each iteration of the procedure, computing the most expensive paths from $s$ to all vertices in the current acyclic network takes $O(m)$ time, and the number of iterations can be bounded by $O(m)$. Thus, the running time of the GENERAL-ALGORITHM is $O(m^2)$ plus the running time of the POWER-ALGORITHM, i.e., $O(m^2 + kn + m \log(d_{\max}/d_{\min}))$. The first term can be usually improved using a suitable min-cost flow algorithm [19]. We examine this further in Section 5.

In our implementation, the variable $\delta_j$ adopts only the distinct rounded demand values. We have two reasons for doing that. The first is that it is not necessary for $\delta_j$ to adopt a value of the form $d_{\min} \cdot 2^i$ when it is not a rounded demand value and as a result of this we could have fewer iterations. The second reason is because of the following heuristic we intend to use.

*Heuristic improvement.* We have a second implementation of the SA in which we try to select augmenting cycles in a more sophisticated manner. When we look for an augmenting cycle in iteration $j$, at the current vertex we always pick an outgoing or incoming edge on which the flow value is not $\delta_j$-integral and the difference between $\delta_j$ and the remainder of the flow value with respect to $\delta_j$ is minimal. Unfortunately, the benefit of this heuristic seems to be very limited. We give details in Section 5. As mentioned above, in our implementation the variable $\delta_j$ adopts only the different rounded demand values. Since the time for finding an augmenting cycle in the implementation with the heuristic is $O(dn)$, where $d$ is the maximum value of in- and outdegrees among all vertices, the worst-case running time for the implementation with the heuristic is $O(m^2 + dknm)$.

## 4 Experimental Framework

*Software and hardware resources.* We conducted our experiments on a sun4u sparc SUNW Ultra-5_10 workstation with 640 MB of RAM and 979 MB swap space. The operating system was SunOS, release 5.8 Generic_108528-14. Our programs were written in C and compiled using gcc, version 2.95, with the -O3 optimization option.

*Codes tested.* The fastest maximum flow algorithm to date is due to Goldberg and Rao [8] with a running time of $O(\min\{n^{2/3}, m^{1/2}\}m\log(n^2/m)\log U)$, where $U$ is an upper bound on the edge capacities which are assumed to be integral. However in practice preflow-push [10] algorithms are the fastest. We use the preflow-push Cherkassky-Goldberg code kit [5] to find a maximum flow as an initial fractional flow. We assume integral capacities and demands in the unsplittable flow input. We implement and test the following codes:

2alg:  this is the DGGA without any heuristic.
2alg_h:  version of 2alg with the heuristic described in Section 2.
3skut:  this is the SA without any heuristic.
3skut_h:  version of 3skut with the heuristic described at the end of Section 3.

In addition we compare against the programs 3al and 3al2 used in [15], where 3al is an implementation of the KSA. The program 3al2 is an implementation of the same algorithm, where to improve the running time the edges carrying zero flow in the initial fractional solution are discarded. Note that both the DGGA and the SA discard these edges as well.

*Input classes.* We generated data from the same four input classes designed by Kolliopoulos and Stein [15]. For each class we generated a variety of instances varying different parameters. The generators use randomness to produce different instances for the same parameter values. To make our experiments repeatable the seed of the pseudorandom generator is an input parameter for all generators. If no seed is given, a fixed default value is chosen. We used the default seed in generating all inputs. The four classes used are defined next. Whenever the term "randomly" is used in the following, we mean *uniformly at random*. For the inputs to 3skut and 3skut_h, we also generate randomly a cost value on each edge using the default seed.

genrmf. This is adapted from the GENRMF generator of Goldfarb and Grigoriadis [11,2]. The input parameters are a b c1 c2 k d. The generated network has $b$ frames (grids) of size $a \times a$, for a total of $a * a * b$ vertices. In each frame each vertex is connected with its neighbors in all four directions. In addition, the vertices of a frame are connected one-to-one with the vertices of the next frame via a random permutation of those vertices. The source is the lower left vertex of the first frame. Vertices become sinks with probability $1/k$ and their demand is chosen uniformly at random from the interval $[1, d]$. The capacities are randomly chosen integers from $(c1, c2)$ in the case of interframe edges, and $(1, c2 * a * a)$ for the in-frame edges.

noigen. This is adapted from the noigen generator used in [3,17] for minimum cut experimentation. The input parameters are n d t p k. The network has $n$ nodes and $\lfloor n(n-1)d/200 \rfloor$ edges. Vertices are randomly distributed among $t$ components. Capacities are chosen uniformly from a prespecified range $[l, 2l]$ in the case of intercomponent edges and from $[pl, 2pl]$ for intracomponent edges, $p$ being a positive integer. Only vertices belonging to one of the $t-1$ components not containing the source can become sinks, each with probability $1/k$. The desired effect of the construction is for commodities to contend for the light intercomponent cuts. Demand for commodities is chosen uniformly form the range $[1, 2l]$.

rangen. This generates a random graph $G(n, p)$ with input parameters n p c1 c2 k d, where $n$ is the number of nodes, $p$ is the edge probability, capacities are in the range $(c1, c2)$, $k$ is the number of commodities and demands are in the range $(0, d)$.

Table 1. family sat_density: satgen -a 1000 -b i -c1 8 -c2 16 -k 10000 -d 8; 9967, 20076, 59977, 120081, 250379, 500828 edges; 22, 61, 138, 281, 682, 1350 commodities; i is the expected percentage of pairs of vertices joined by an edge.

| | 1 | | 2 | | 6 | | 12 | | 25 | | 50 | |
|---|---|---|---|---|---|---|---|---|---|---|---|---|
| program | cong. | time | cong. | time | cong. | time | cong. | time | cong. | time | cong. | time |
| 2alg | 1.40 | 0 | 1.50 | 0 | 1.63 | 0 | 1.63 | 0 | 1.75 | 0 | 1.75 | 1 |
| 2alg_h | 1.33 | 0 | 1.50 | 0 | 1.42 | 0 | 1.60 | 0 | 1.75 | 0 | 1.64 | 0 |
| 3al | 1.55 | 1 | 1.56 | 1 | 1.67 | 4 | 1.64 | 9 | 1.64 | 23 | 1.70 | 54 |
| 3al2 | 1.45 | 0 | 1.67 | 1 | 1.78 | 2 | 1.57 | 5 | 1.67 | 13 | 1.75 | 32 |
| 3skut | 1.70 | 0 | 1.64 | 1 | 2.22 | 1 | 2.11 | 1 | 2.11 | 2 | 2.33 | 6 |
| 3skut_h | 1.70 | 1 | 1.64 | 0 | 2.22 | 1 | 2.11 | 1 | 2.11 | 2 | 2.33 | 6 |

**Table 2.** family noi_commodities: noigen 1000 1 2 10 i; 7975 edges; 2-unbalanced family; i is the expected percentage of sinks in the non-source component.

| | 2 | | 6 | | 12 | | 25 | | 50 | | 100 | |
|---|---|---|---|---|---|---|---|---|---|---|---|---|
| program | cong. | time | cong. | time | cong. | time | cong. | time | cong. | time | cong. | time |
| 2alg | 1.30 | 0 | 2.37 | 0 | 2.14 | 0 | 1.43 | 0 | 1.26 | 0 | 1.09 | 0 |
| 2alg_h | 1.30 | 0 | 1.73 | 0 | 2.14 | 0 | 1.49 | 0 | 1.17 | 0 | 1.09 | 0 |
| 3al | 2.21 | 0 | 1.73 | 0 | 2.41 | 1 | 1.70 | 1 | 1.26 | 1 | 1.14 | 1 |
| 3al2 | 2.21 | 0 | 1.88 | 0 | 2.29 | 0 | 1.52 | 1 | 1.27 | 1 | 1.18 | 1 |
| 3skut | 1.40 | 0 | 1.65 | 0 | 2.15 | 0 | 1.85 | 0 | 1.45 | 0 | 1.49 | 1 |
| 3skut_h | 1.40 | 0 | 1.65 | 0 | 2.15 | 0 | 1.85 | 0 | 1.45 | 0 | 1.49 | 0 |

**Table 3.** family rmf_commDem: genrmf -a 10 -b 64 -c1 64 -c2 128 -k i -d 128; 29340 edges; 2-unbalanced family; i is the expected percentage of sinks among the vertices.

| | 2 | | 5 | | 10 | | 20 | | 50 | | 70 | |
|---|---|---|---|---|---|---|---|---|---|---|---|---|
| program | cong. | time | cong. | time | cong. | time | cong. | time | cong. | time | cong. | time |
| 2alg | 2.71 | 8 | 1.77 | 14 | 1.41 | 15 | 1.21 | 21 | 1.09 | 33 | 1.06 | 42 |
| 2alg_h | 2.47 | 8 | 1.75 | 14 | 1.36 | 15 | 1.19 | 19 | 1.08 | 31 | 1.06 | 42 |
| 3al | 2.76 | 6 | 1.89 | 13 | 1.57 | 23 | 1.37 | 49 | 1.19 | 138 | 1.14 | 196 |
| 3al2 | 2.79 | 4 | 1.89 | 7 | 1.54 | 14 | 1.30 | 33 | 1.16 | 91 | 1.15 | 132 |
| 3skut | 3.35 | 1 | 2.39 | 1 | 2.07 | 2 | 1.76 | 1 | 1.63 | 2 | 1.62 | 3 |
| 3skut_h | 3.35 | 1 | 2.39 | 1 | 2.07 | 1 | 1.76 | 2 | 1.63 | 2 | 1.62 | 3 |

**satgen.** It first generates a random graph $G(n,p)$ as in rangen and then uses the following procedure to designate commodities. Two vertices $s$ and $t$ are picked from $G$ and maximum flow is computed from $s$ to $t$. Let $v$ be the value of the flow. New nodes corresponding to sinks are incrementally added each connected only to $t$ and with a randomly chosen demand value. The process stops when the total demand reaches $v$, the value of the minimum $s$-$t$ cut or when the number of added commodities reaches the input parameter $k$, typically given as a crude upper bound.

**Table 4.** family ran_dense: rangen -a i*2 -b 30 -c1 1 -c2 16 -k i -d 16; 1182, 4807, 19416, 78156, 313660, 1258786 edges; 16-unbalanced family; i*2 is the number of vertices.

| | 32 | | 64 | | 128 | | 256 | | 512 | | 1024 | |
|---|---|---|---|---|---|---|---|---|---|---|---|---|
| program | cong. | time | cong. | time | cong. | time | cong. | time | cong. | time | cong. | time |
| 2alg | 4.67 | 0 | 3.75 | 0 | 7.50 | 0 | 7.00 | 0 | 7.50 | 0 | 8.00 | 0 |
| 2alg_h | 2.50 | 0 | 3.25 | 0 | 6.50 | 0 | 7.00 | 0 | 7.50 | 1 | 7.50 | 0 |
| 3al | 4.67 | 0 | 8.00 | 0 | 8.00 | 1 | 8.00 | 6 | 7.50 | 30 | 7.50 | 136 |
| 3al2 | 4.67 | 0 | 6.50 | 0 | 7.50 | 0 | 8.50 | 3 | 7.50 | 17 | 8.00 | 82 |
| 3skut | 5.00 | 0 | 7.00 | 0 | 7.50 | 0 | 7.00 | 1 | 7.50 | 4 | 7.50 | 19 |
| 3skut_h | 5.00 | 0 | 7.00 | 0 | 7.50 | 0 | 7.00 | 1 | 7.50 | 4 | 7.50 | 18 |

# 5   Experimental Results

In this section we give an overview of the experimental results. In all algorithms we study, starting with a different fractional flow may give different unsplittable solutions. Hence in order make a meaningful comparison of the experimental results of the SA against the results of the DGGA and KSA, we use the same initial fractional flow for all three. If the SA was used in isolation, one could use, as mentioned in Section 3, a min-cost flow algorithm to find the initial fractional flow and therefore obtain a best possible budget.

The implementations follow the algorithm descriptions as given earlier. In the case of the SA, after finding the initial fractional flow $f$, one has to to iteratively reduce flow, for each commodity $i$, along the most expensive $s$-$t_i$-paths used by $f$ until the inflow in terminal $t_i$ has been decreased by $d_i - \bar{d}_i$, where $\bar{d}_i$ stands for the rounded demand. Instead of doing this explicitly, as Skutella [19] suggests, we set the capacity of each edge $e$ to $f(e)$ and use an arbitrary min-cost flow algorithm to find a min-cost flow that satisfies the rounded demands. Because of this, the term $O(m^2)$ in the running time of Algorithm 3 in Section 3 can be replaced by the running time of an arbitrary min-cost flow algorithm. The running times of the currently best known min-cost flow algorithms are $O(nm \log(n^2/m) \log(nC))$ [9], $O(nm(\log \log U) \log(nC))$ [1], and $O((m \log n)(m + n \log n))$ [18]. The code we use is again due to Cherkassky and Goldberg [4]. The experimental results for all the implementations are given in Tables 1–7. The wall-clock running time is given in seconds, with a running time of 0 denoting a time less than a second. We gave earlier the theoretical running times for the algorithms we implement but one should bear in mind that the real running time depends also on other factors such as the data structures used. Apart from standard linked and adjacency lists no other data structures were used in our codes. As mentioned in the introduction, speeding up the codes was not our primary focus. This aspect could be pursued further in future work.

*The DGGA vs. the KSA.* We first compare the results of the 2- and the 3-approximation algorithms since they are both algorithms for congestion without costs. On a balanced input (see Table 1), the congestion achieved by the DGGA, with or without heuristics, was typically less than or equal to 1.75. The congestion achieved by the KSA was almost in the same range. For each balanced input, the difference in the congestion achieved by these two algorithms was not obvious, but the DGGA's congestion was typically somewhat better. The obvious difference occurred in running time. Before starting measuring running time, we use the Cherkassky-Goldberg code kit to find a feasible splittable flow (if necessary we use other subroutines to scale up the capacities by the minimum amount needed to achieve a fractional congestion of 1), and then we create an array of nodes to represent this input graph (discarding all zero flow edges) and delete all flow cycles. After that, we start measuring the running time and applying the DGGA. The starting point for measuring the running time in the implementation of the KSA is also set after the termination of the Cherkassky-Goldberg code.

To test the robustness of the approximation guarantee we relaxed, on several instances, the balance assumption allowing the maximum demand to be twice the minimum capacity or more, see Tables 2–6. The approximation guarantee of each individual algorithm was not significantly affected. Even in the extreme case when the balance assumption was violated by a factor of 16, as in Table 4, the code 2alg achieved 8 and the code 2alg_h achieved 7.5. Relatively speaking though the difference in the congestion achieved between 2alg, 2alg_h and 3al, 3al2 is much more pronounced compared to the inputs with small imbalance. See the big differences in first two columns of Table 4 (in Column 2, 2alg: 3.75 and 3al2: 6.5). Hence the DGGA is more robust against highly unbalanced inputs. This is consistent with the behavior of the KSA which keeps increasing the edge capacities by a fixed amount in each iteration before routing any new flow. In contrast, the DGGA increases congestion only when some flow is actually rerouted through an edge. As shown in Table 4, the SA which behaves similarly to the KSA exhibits the same lack of robustness for highly unbalanced inputs.

We also observed that the benefit of the heuristic used in our 2alg_h implementation showed up in our experimental results. For most of the inputs, the congestion was improved, although rarely by more than 5%. Some significant improvements were observed when the input was very unbalanced, see Table 4. Theoretically, the running time for the program with the heuristic should increase by a certain amount. But in our experiments, the running time stayed virtually the same. This phenomenon was beyond what we expected.

In summary, the DGGA performs typically better than the KSA for congestion. The average improvement for Tables 1–4 is 6.5%, 6.3%, 9.4%, and 53% with the occasional exception where the KSA outperformed the DGGA. This behavior is consistent with the fact that the DGGA is a theoretically better algorithm. Moreover the theoretical advantage typically translates to a much smaller advantage in practice.

The difference in the running time for these two approximation algorithms was fairly significant in our experiments especially for dense graphs with a large number of commodities. The DGGA runs much faster than the implementation of the KSA we used. We proceed to give two possible reasons for this phenomenon.

The first reason is the difference in complexity for these two implementations. Recall that the running time of the DGGA is $O(mn + km)$ and the running time of the implementation of the KSA that we used is $O(k(d_{max}/d_{min})m)$ [14,15]. We emphasize that a polynomial-time implementation is possible (see [14,15]). In fact Skutella's POWER-ALGORITHM can be seen as a much more efficient implementation of essentially the same algorithm. A second reason is that the DGGA processes the graph in a localized manner, i.e., finding an alternating cycle locally and increasing a certain amount of flow on it, while the 3al, 3al2 codes repeatedly compute maximum flows on the full graph.

*The Skutella algorithm.* We now examine the congestion achieved by the SA. On a balanced input (see Table 1), the congestion achieved by the SA was typically greater than 1.64 and less than or equal to 2.33. This range of congestion is bigger than the range [1.33, 1.75] achieved by the DGGA and the range [1.45, 1.78] by

**Table 5.** family noi_components: noigen 1000 1 i 10 50; 7975 edges; 2-unbalanced family; i is the number of components.

| program | 3 cong. | time | 6 cong. | time | 12 cong. | time | 24 cong. | time | 48 cong. | time |
|---------|---------|------|---------|------|----------|------|----------|------|----------|------|
| 2alg    | 1.10    | 0    | 1.07    | 0    | 1.04     | 0    | 1.01     | 0    | 1.01     | 0    |
| 2alg_h  | 1.12    | 0    | 1.05    | 0    | 1.02     | 0    | 1.01     | 0    | 1.01     | 0    |
| 3al     | 1.22    | 2    | 1.13    | 1    | 1.12     | 1    | 1.07     | 1    | 1.08     | 1    |
| 3al2    | 1.22    | 1    | 1.13    | 0    | 1.11     | 1    | 1.04     | 1    | 1.09     | 1    |
| 3skut   | 1.46    | 0    | 1.43    | 0    | 1.37     | 0    | 1.37     | 0    | 1.38     | 0    |
| 3skut_h | 1.46    | 0    | 1.43    | 0    | 1.37     | 0    | 1.37     | 0    | 1.38     | 0    |

**Table 6.** family rmf_depthDem: genrmf -a 10 -b i -c1 64 -c2 128 -k 40 -d 128; 820, 1740, 7260, 29340 edges; 2-unbalanced family; i is the number of frames.

| program | 2 cong. | time | 4 cong. | time | 16 cong. | time | 64 cong. | time |
|---------|---------|------|---------|------|----------|------|----------|------|
| 2alg    | 2.30    | 0    | 2.47    | 0    | 1.36     | 1    | 1.11     | 28   |
| 2alg_h  | 1.89    | 0    | 2.64    | 0    | 1.37     | 1    | 1.09     | 27   |
| 3al     | 2.26    | 0    | 2.18    | 0    | 1.48     | 5    | 1.21     | 103  |
| 3al2    | 1.82    | 0    | 2.72    | 0    | 1.46     | 3    | 1.20     | 77   |
| 3skut   | 2.72    | 0    | 2.71    | 0    | 1.87     | 1    | 1.72     | 1    |
| 3skut_h | 2.72    | 0    | 2.71    | 0    | 1.87     | 0    | 1.72     | 2    |

the KSA. More precisely, the absolute difference in congestion between the SA and the DGGA or KSA is on average around 0.4. We think that this nontrivial difference in congestion is partially or mainly caused by the involvement of costs on the edges and the simultaneous performance guarantee of 1 for cost of the SA. The constraint that the flow found at each step should not increase the cost limits the routing options.

In the implementation of the $(3, 1)$-approximation algorithm we start measuring the running time just before applying a min-cost flow algorithm [4] to find a min-cost flow for the rounded demands. Before that starting point of running time, we use the Cherkassky-Goldberg code kit to find a feasible splittable flow (if necessary, as we did before, we use other subroutines to scale up the capacities by the minimum amount to get the optimal fractional congestion), and then we create the input data for the min-cost flow subroutine, i.e., setting the capacity of each edge to its flow value and the demand of each commodity $i$ to $\bar{d}_i$. For balanced input instances in Table 1, the running time of the SA is much better than that of the KSA but slightly more than that of the DGGA. Actually, as we observed in testing, most of the running time for the SA is spent in finding the initial min-cost flow.

To test the robustness of the approximation guarantee achieved by the SA we used the instances with the relaxed balanced assumption. Even in the extreme case when the balance assumption was violated by a factor of 16, as in Table 4, the code 3skut achieved 7.50. The absolute difference in congestion achieved by the codes 2alg, 3al and 3skut is typically small. The only big difference occurred

**Table 7.** Effect of our heuristic on the SA. Here the input instances in Columns 1 to 6 are the modified input instances in the last columns of Tables 1 to 6 whose original demands, denoted by $d$, are modified as follows to the value $d'$: $d' = 1$ if $d = 2$; $d' = 2^2$ if $d = 3$; $d' = 2^4$ if $8 \leq d < 16$; $d' = 2^6$ if $32 \leq d < 64$; for all other cases, $d$ are not changed. Note that the maximum demand value in our input instances is equal to $128 = 2^7$.

| | 1 | | 2 | | 3 | | 4 | | 5 | | 6 | |
|---|---|---|---|---|---|---|---|---|---|---|---|---|
| program | cong. | time | cong. | time | cong. | time | cong. | time | cong. | time | cong. | time |
| 3skut | 2.22 | 6 | 1.26 | 0 | 1.50 | 3 | 5.33 | 18 | 1.19 | 0 | 1.61 | 2 |
| 3skut_h | 2.22 | 6 | 1.28 | 0 | 1.48 | 3 | 5.33 | 18 | 1.20 | 0 | 1.58 | 3 |

in the second output column in Table 4 (2alg: 3.75, 3al: 8.00 and 3skut: 7.00). However, similar to the output in Table 1, the congestion achieved by the codes 2alg and 3al for an unbalanced input was typically better, see Tables 2–6. Given the similarities between the KSA and SA the reason is, as mentioned above, the involvement of costs on the edges and a simultaneous performance guarantee of 1 for cost in the $(3, 1)$-approximation algorithm. For the running time, things are different. We can see from Tables 3 and 6 that the code 3skut runs much faster than 2alg and 3al when the size of the input is large. This is probably because after the rounding stage the number of the distinct rounded demand values, which is the number of iterations in our implementation, is small (equal to 7 in Tables 3 and 6) and the number of augmenting cycles (to be chosen iteratively) in most of the iterations is not very large. If this is the case, the execution of these iterations could be finished in a very short period of time and the total running time is thus short also.

*Effect of the heuristic on the* SA. No benefit of the heuristic used in our 3skut_h implementation showed in Tables 1–6. This is because in each iteration (except the stage of finding a min-cost flow) the non-zero remainder of flow value on each edge with respect to the rounded demand value of the current iteration is exactly the same in our input instances. More precisely, in our input instances, the variable $\delta_j$ adopts all values $d_{\min} \cdot 2^i$ between $d_{\min}$ and $d_{\max}$, and in this case, in iteration $j$ the remainder of flow value on any edge with respect to $\delta_j = d_{\min} \cdot 2^{j-1}$ is either $d_{\min} \cdot 2^{j-2}$ or 0. So the amount of augmented flow along an augmenting cycle $C$ is $d_{\min} \cdot 2^{j-2}$ and after the augmentation the flow on each edge of $C$ is $\delta_j$-integral and thus all edges of $C$ will not be involved in the remaining augmentation procedure of this iteration. This is also probably the reason why sometimes the SA runs faster than the DGGA. When the variable $\delta_j$ would not adopt all values $d_{\min} \cdot 2^i$ between $d_{\min}$ and $d_{\max}$, the heuristic proved to be of some marginal usefulness. This can be seen from Table 7. The congestion was improved in Columns 3 and 6 by 0.02 and 0.03, respectively, but in Columns 2 and 5 the congestion increased by 0.02 and 0.01, respectively.

In summary, in most of our experiments the DGGA and KSA achieved lower congestion than the SA. Relative gains of the order of 35% or more are common especially for Tables 1, 3 and 4. This is mainly because the SA has a simultaneous performance guarantee for the cost. The SA remains competitive and typically achieved approximation ratios well below the theoretical guarantee. The 3skut code runs much faster than 3al and occasionally faster than the 2alg code.

# References

1. R. K. Ahuja, A. V. Goldberg, J. B. Orlin, and R. E. Tarjan. Finding minimum-cost flows by double scaling. *Mathematical Programming* 53, 243–266, 1992.
2. T. Badics. GENRMF. ftp://dimacs.rutgers.edu/pub/netflow/generators/network/genrmf/, 1991.
3. C. Chekuri, A. V. Goldberg, D. Karger, M. Levine, and C. Stein. Experimental study of minimum cut algorithms. In *Proc. 8th ACM-SIAM SODA*, 324–333, 1997.
4. B. V. Cherkassky and A. V. Goldberg. An efficient implementation of a minimum-cost flow algorithm. *J. Algorithms,* 22:1-29, 1997.
5. B. V. Cherkassky and A. V. Goldberg. On implementing the push-relabel method for the maximum flow problem. *Algorithmica*, 19:390–410, 1997.
6. Y. Dinitz, N. Garg, and M. X. Goemans. On the single-source unsplittable flow problem. *Combinatorica*, 19:1–25, 1999.
7. T. Erlebach and Λ. Hall. NP-hardness of broadcast scheduling and inapproximability of single-source unsplittable min-cost flow. *Proc. 13th ACM-SIAM SODA* , 2002, 194-202.
8. A. V. Goldberg and S. Rao. Beyond the flow decomposition barrier. *J. ACM*, 45:783–797, 1998.
9. A. V. Goldberg and R. E. Tarjan. Solving minimum cost flow problems by successive approximation. *Mathematics of Operations Research* 15, 430–466, 1990.
10. A. V. Goldberg and R. E. Tarjan. A new approach to the maximum flow problem. *J. ACM*, Vol. 35, No. 4, pages 921–940, October 1988.
11. D. Goldfarb and M. Grigoriadis. A computational comparison of the Dinic and Network Simplex methods for maximum flow. *Annals of Operations Research,* 13:83–123, 1988.
12. J. M. Kleinberg. *Approximation algorithms for disjoint paths problems.* Ph.D. thesis, MIT, Cambridge, MA, May 1996.
13. J. M. Kleinberg. Single-source unsplittable flow. In *Proceedings of the 37th Annual Symposium on Foundations of Computer Science*, pages 68–77, October 1996.
14. S. G. Kolliopoulos and C. Stein. Approximation algorithms for single-source unsplittable flow. *SIAM J. Computing*, 31(3): 919–946, 2002.
15. S. G. Kolliopoulos and C. Stein. Experimental evaluation of approximation algorithms for single-source unsplittable flow. *Proc. 7th IPCO, Springer-Verlag LNCS vol. 1610*, 153–168, 1999.
16. J. K. Lenstra, D. B. Shmoys, and E. Tardos. Approximation algorithms for scheduling unrelated parallel machines. *Mathematical Programming 46*, 259–271, 1990.
17. Hiroshi Nagamochi, Tadashi Ono, and Toshihide Ibaraki. Implementing an efficient minimum capacity cut algorithm. *Mathematical Programming*, 67:325–241, 1994.
18. J. B. Orlin. A faster strongly polynomial minimum cost flow algorithm. *Operations Research 41*, 338–350, 1993.
19. M. Skutella. Approximating the single source unsplittable min-cost flow problem. *Mathematical Programming*, Ser. B 91(3), 493-514, 2002.

# Fingered Multidimensional Search Trees*

Amalia Duch and Conrado Martínez

Departament de Llenguatges i Sistemes Informàtics, Universitat Politècnica de
Catalunya, E-08034 Barcelona, Spain.
{duch,conrado}@lsi.upc.es

**Abstract.** In this work, we propose two variants of $K$-d trees where
*fingers* are used to improve the performance of orthogonal range search
and nearest neighbor queries when they exhibit locality of reference. The
experiments show that the second alternative yields significant savings.
Although it yields more modest improvements, the first variant does it
with much less memory requirements and great simplicity, which makes
it more attractive on practical grounds.

## 1   Introduction

The well-known, time-honored aphorism says that "80% of computer time in
spent on 20% of the data". The actual percentages are unimportant but the
moral is that we can achieve significant improvements in performance if we
are able to exploit this fact. In on-line settings, where requests arrive one at
a time and they must be attended as soon as they arrive (or after some small
delay), we frequently encounter *locality of reference*, that is, for any time frame
only a small number of different requests among the possible ones are made or
consecutive requests are close to each other in some sense. Locality of reference is
systematically exploited in the design of memory hierarchies (disk and memory
caches) and it is the rationale for many other techniques like buffering and self-
adjustment [2,11].

The performance of searches and updates in data structures can be improved
by augmenting the data structure with *fingers*, pointers to the hot spots in the
data structure where most activity is going to be performed for a while (see for
instance [9,3]). Thus, successive searches and updates do not start from scratch
but use the clues provided by the finger(s), so that when the request affects some
item in the "vicinity" of the finger(s) the request can be attended very efficiently.

To the best of the authors' knowledge, fingering techniques haven't been
applied so far to multidimensional data structures. In this paper, we will specifically concentrate in two variants of $K$-dimensional trees, namely *standard $K$-d
trees* [1] and *relaxed $K$-d trees* [6], but the techniques can easily be applied to
other multidimensional search trees and data structures. In general, multidimensional data structures maintain a collection of items or records, each holding a

* This research was partially supported by the Future and Emergent Technologies
programme of the EU under contract IST-1999-14186 (ALCOM-FT) and the Spanish
Min. of Science and Technology project TIC2002-00190 (AEDRI II).

C.C. Ribeiro and S.L. Martins (Eds.): WEA 2004, LNCS 3059, pp. 228–242, 2004.

distinct $K$-dimensional key (which we may assume w.l.o.g. is a point in $[0,1]^K$).
Also, we will identify each item with its key and use both terms interchangeably.
Besides usual insertions, deletions and (exact) search, we will be interested in
providing efficient answers to questions like which records do fall within a given
hyper-rectangle $Q$ (orthogonal range search) or which is the closest record to
some given point $q$ (nearest neighbor search) [10,8]. After a brief summary of
basic concepts, definitions and previous results in Section 2, we propose two alternative designs that augment $K$-d trees[1] with fingers to improve the efficiency of
orthogonal range and nearest neighbor searches (Section 3). We thereafter study
their performance under reasonable models which exhibit locality of reference
(Section 4). While it seems difficult to improve the performance of multidimensional data structures using self-adjusting techniques (as reorganizations in this
type of data structures is too expensive), fingering yields significant savings and
it is easy to implement. Our experiments show that the second, more complex
scheme of $m$-finger $K$-d trees exploits better the locality of reference than the
simpler 1-finger $K$-d trees; however these gains probably do not compensate for
the amount memory that it needs, so that 1-finger $K$-d trees are more attractive
on a practical ground.

## 2    Preliminaries and Basic Definitions

A *standard $K$-d tree* [1] for a set $F$ of $K$-dimensional data points is a binary
tree in which: (a) each node contains a $K$-dimensional data point and has an
associated discriminant $j \in \{0, 1, \ldots, K-1\}$ cyclically assigned starting with
$j = 0$. Thus the root of the tree discriminates with respect to the first coordinate
$(j = 0)$, its sons at level 1 discriminate w.r.t. the second coordinate $(j = 1)$, and
in general, all nodes at level $m$ discriminate w.r.t. coordinate $j = m \mod K$; (b)
for each node $x = (x_0, x_1, \ldots, x_{K-1})$ with discriminant $j$, the following invariant
is true: any data point $y$ in the left subtree satisfies $y_j < x_j$ and any data point
$z$ in the right subtree satisfies $z_j \geq x_j$ (see Fig. 1).

   *Randomized relaxed $K$-d trees* [6] (relaxed $K$-d trees, for short) are $K$-d trees
where the sequence of discriminants in a path from the root to any leaf is random
instead of cyclic. Hence, discriminants must be explicitly stored in the nodes.
Other variants of $K$-d tree use different alternatives to assign discriminants (for
instance, the *squarish $K$-d trees* of Devroye *et al.* [5]) or combine different methods to partition the space (not just axis-parallel hyper-planes passing through
data coordinates as standard and relaxed $K$-d trees do).

   We say that a $K$-d tree of size $n$ is *random* if it is built by $n$ insertions where
the points are independently drawn from a continuous distribution in $[0,1]^K$.
In the case of random relaxed $K$-d trees, the discriminants associated to the
internal nodes are uniformly and independently drawn from $\{0, \ldots, K-1\}$.

---

[1] They actually apply to any variant of $K$-d trees, not just the two mentioned; some
   additional but minor modifications would be necessary to adapt them to quad trees,
   $K$-d tries, etc.

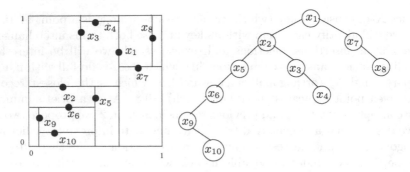

**Fig. 1.** A standard 2-d tree and the corresponding induced partition of $[0,1]^2$

An *(orthogonal) range query* is a $K$-dimensional hyper-rectangle $Q$. We shall write $Q = [\ell_0, u_0] \times \cdots \times [\ell_{K-1}, u_{K-1}]$, with $\ell_i \leq u_i$, for $0 \leq i < K$. Alternatively, a range query can be specified given its center $z$ and the length of its edges $\Delta_0, \Delta_1, \ldots \Delta_{K-1}$, with $0 \leq \Delta_i \leq 1/2$, for $0 \leq i < K$. We assume that $-\Delta_i/2 \leq z_i \leq 1 + \Delta_i/2$, for $0 \leq i < K$, so that a range query $Q$ may fall partially outside of $[0,1]^K$.

Range searching in any variant of $K$-d trees is straightforward. When visiting a node $x$ that discriminates w.r.t. the $j$-th coordinate, we must compare $x_j$ with the $j$-th range $[\ell_j, u_j]$ of the query. If the query range is totally above (or below) that value, we must search only the right subtree (respectively, left) of that node. If, on the contrary, $\ell_j \leq x_j \leq u_j$ then both subtrees must be searched; additionally, we must check whether $x$ falls or not inside the query hyper-rectangle. This procedure continues recursively until empty subtrees are reached.

One important concept related to orthogonal range searches is that of *bounding box* or *bounding hyper-rectangle* of a data point. Given an item $x$ in a $K$-d tree $T$, its bounding hyper-rectangle $B(x) = [\ell_0(x), u_0(x)] \times \ldots \times [\ell_{K-1}(x), u_{K-1}(x)]$ is the region of $[0,1]^K$ corresponding to the leaf that $x$ replaced when it was inserted into $T$. Formally, it is defined as follows: a) if $x$ is the root of $T$ then $B(x) = [0,1]^K$; b) if $y = (y_0, \ldots, y_{K-1})$ is the father of $x$ and it discriminates w.r.t. the $j$-th coordinate and $x_j < y_j$ then $B(x) = [\ell_0(y), u_0(y)] \times \ldots \times [\ell_j(y), y_j] \times \ldots \times [\ell_{K-1}(y), u_{K-1}(y)]$; c) if $y = (y_0, \ldots, y_{K-1})$ is the father of $x$ and it discriminates w.r.t. the $j$-th coordinate and $x_j > y_j$ then $B(x) = [\ell_0(y), u_0(y)] \times \ldots \times [y_j, u_j(y)] \times \ldots \times [\ell_{K-1}(y), u_{K-1}(y)]$. The relation of bounding hyper-rectangles with range search is established by the following lemma.

**Lemma 1 ([4,5]).** *A point $x$ with bounding hyper-rectangle $B(x)$ is visited by a range search with query hyper-rectangle $Q$ if and only if $B(x)$ intersects $Q$.*

The cost of orthogonal range queries is usually measured as the number of nodes of the $K$-d tree visited during the search. It has two main parts: a term corresponding to the unavoidable cost of reporting the result plus an *overwork*.

**Theorem 1** ([7]). *Given a random $K$-d tree storing $n$ uniformly and independently drawn data points in $[0,1]^K$, the expected overwork $\mathbb{E}[W_n]$, of an orthogonal range search with edge lengths $\Delta_0, \ldots, \Delta_{K-1}$ such that $\Delta_i \to 0$ as $n \to \infty$, $0 \le i < K$, and with center uniformly and independently drawn from $Z_\Delta = [-\Delta_0/2, 1 + \Delta_0/2] \times \cdots \times [-\Delta_{K-1}/2, 1 + \Delta_{K-1}/2]$ is given by*

$$\mathbb{E}[W_n] = \sum_{1 \le j \le K} c_j \cdot n^{\alpha(j/K)} + 2 \cdot (1 - \Delta_0) \cdots (1 - \Delta_{K-1}) \cdot \log n + \mathcal{O}(1),$$

*where $\alpha(x)$ and the $c_j$'s depend on the particular type of $K$-d trees.*

In the case of standard $K$-d trees $\alpha(x) = 1 - x + \phi(x)$, where $\phi(x)$ is the unique real solution of $(\phi(x) + 3 - x)^x (\phi(x) + 2 - x)^{(1-x)} - 2 = 0$; for any $x \in [0,1]$, we have $\phi(x) < 0.07$. For relaxed $K$-d trees, $\alpha(x) = 1 - x + \phi(x)$ where $\phi(x) = (\sqrt{9 - 8x} - 3)/2 + x$. Of particular interest is the case where $\Delta_i = \Theta(n^{-1/K})$ corresponding to the situation where each orthogonal range search reports a constant number of points on the average. Then the cost is dominated by the overwork and we have

$$\mathbb{E}[R_n] = \mathbb{E}[W_n] = \Theta(n^{\phi(1/K)} + n^{\phi(2/K)} + \cdots + n^{\phi(1-1/K)} + \log n).$$

A *nearest neighbor query* is a multidimensional point $q = (q_0, q_1, \ldots, q_{K-1})$ lying in $[0,1]^K$. The goal of the search is to find the point in the data structure which is closest to $q$ under a predefined distance measure.

There are several variants for nearest neighbor searching in $K$-d trees. One of the simplest, which we will use for the rest of this paper works as follows. The initial closest point is the root of the tree. Then we traverse the tree as if we were inserting $q$. When visiting a node $x$ that discriminates w.r.t. the $j$-th coordinate, we must compare $q_j$ with $x_j$. If $q_j$ is smaller than $x_j$ we follow the left subtree, otherwise we follow the right one. At each step we must check whether $x$ is closer or not to $q$ than the closest point seen so far and update the candidate nearest neighbor accordingly. The procedure continues recursively until empty subtrees are reached. If the hyper-sphere, say $B_q$, defined by the query $q$ and the candidate closest point is totally enclosed within the bounding boxes of the visited nodes then the search is finished. Otherwise, we must visit recursively the subtrees corresponding to nodes whose bounding box intersects but does not enclose $B_q$.

The performance of nearest neighbor search is similar to the overwork in range search. Given a random $K$-d tree, the expected cost $\mathbb{E}[NN_n]$ of a nearest neighbor query $q$ uniformly drawn from $[0,1]^K$ is given by

$$\mathbb{E}[NN_n] = \Theta(n^\rho + \log n),$$

where $\rho = \max_{0 \le s \le K}(\alpha(s/K) - 1 + s/K)$. In the case of standard $K$-d trees $\rho \in (0.0615, 0.064)$. More precisely, for $K = 2$ we have $\rho = (\sqrt{17} - 4)/2 \approx 0.0615536$ and for $K = 3$ we have $\rho \approx 0.0615254$, which is minimal. For relaxed $K$-d trees $\rho \in (0.118, 0.125)$. When $K = 2$ we have $\rho = \frac{\sqrt{5}}{2} - 1 \approx 0.118$, which is minimal, whereas for $K = 8$ we have $\rho = \frac{1}{8}$, which is maximal.

## 3   Finger $K$-d Trees

In this section we introduce two different schemes of fingered $K$-d trees. We call the first and simpler scheme 1-finger $K$-d trees; we augment the data structure with one finger pointer. The second scheme is called multiple finger $K$-d tree (or $m$-finger $K$-d tree, for short). Each node of the new data structure is equipped with two additional pointers or fingers, each pointing to descendent nodes in the left and right subtrees, respectively. The search in this case proceeds by recursively using the fingers whenever possible.

### 3.1   One-Finger $K$-d Trees

A *one-finger $K$-d tree* (1-finger $K$-d tree) for a set $F$ of $K$-dimensional data points is a $K$-d tree in which: (a) each node contains its bounding box and a pointer to its father; (b) there is a pointer called *finger* that points to an arbitrary node of the tree.

The finger is initially pointing to the root but it is updated after each individual search. Consider first orthogonal range searches. The orthogonal range search algorithm starts the search at some node $x$ pointed to by the finger $f$. Let $B(x)$ be the bounding box of node $x$ and $Q$ the range query. If $Q \subset B(x)$ then all the points to be reported must necessarily be in the subtree rooted at $x$. Thus, the search algorithm proceeds from $x$ down following the classical range search algorithm. Otherwise, some of the points that are inside the query $Q$ can be stored in nodes which are not descendants of $x$. Hence, in this situation the algorithm backtracks until it finds the first ancestor $y$ of $x$ such that $B(y)$ completely contains $Q$. Once $y$ has been found the search proceeds as in the previous case. The finger is updated to point to the first node where the range search must follow recursively into both subtrees (or to the last visited node if no such node exists). In other terms, $f$ is updated to point to the node whose bounding box completely contains $Q$ and none of the bounding boxes of its descendants does. The idea is that if consecutive queries $Q$ and $Q'$ are close in geometric terms then either the bounding box $B(x)$ that contains $Q$ does also contain $Q'$ or only a limited amount of backtrack suffices to find the appropriate ancestor $y$ to go on with the usual range searching procedure. Of course, the finger is initialized to point to the tree's root before the first search is made. Algorithm 1 describes the orthogonal range search in 1-finger $K$-d trees. It invokes the standard `range_search` algorithm once the appropriate starting point has been found. We use the notation $p \to field$ to refer to the field *field* in the node pointed to by $p$. For simplicity, the algorithm assumes that each node stores its bounding box; however, it is not difficult to modify the algorithm so that only the nodes in the path from the root to $f$ contain this information or to use an auxiliary stack to store the bounding boxes of the nodes in the path from the root to the finger. Additionally, the explicit pointers to the father can be avoid using pointer reversal plus a pointer to finger's father.

The single finger is exploited for nearest neighbor searches much in the same vein. Let be $q$ the query and let $x$ be the node pointed to by the finger $f$. Initially

---

**Algorithm 1** The orthogonal range search algorithm for 1-finger $K$-d trees.

▷   Initial call: $f := \text{one\_finger\_range\_search}(f, Q)$
**function** one_finger_range_search$(f, Q)$
    **if** $f = $ **nil then return** $f$
    $B := f \rightarrow bounding\_box$
    **if** $Q \not\subseteq B$ **then** ▷   Backtrack
       **return** one_finger_range_search$(f \rightarrow father, Q)$
    $x := f \rightarrow info; j := f \rightarrow discr$
    **if** $Q.u[j] < x[j]$ **then**
       **return** one_finger_range_search$(f \rightarrow left, Q)$
    **if** $Q.l[j] \geq x[j]$ **then**
       **return** one_finger_range_search$(f \rightarrow right, Q)$
    **if** $x \in Q$ **then** Add $x$ to result
    range_search$(f \rightarrow left, Q)$
    range_search$(f \rightarrow right, Q)$
    **return** $f$
**end**

---

$f$ will point to the root of the tree, but on successive searches it will point to the last closest point reported. The first step of the algorithm is then to calculate the distance $d$ between $x$ and $q$ and to determine the ball with center $q$ and radius $d$. If this ball is completely included in the bounding box of $x$ then nearest neighbor search algorithm proceeds down the tree exactly in the same way as the standard nearest neighbor search algorithm. If, on the contrary, the ball is not included in $B(x)$, the algorithm backtracks until it finds the least ancestor $y$ whose bounding box completely contains the ball. Then the algorithm continues as the standard nearest neighbor search. Algorithm 2 describes the nearest neighbor algorithm for 1-finger $K$-d trees; notice that it behaves just as the standard nearest neighbor search once the appropriate node where to start has been found.

## 3.2    Multiple Finger $K$-d Trees

A *multiple-finger $K$-d tree* ($m$-finger $K$-d tree) for a set $F$ of $K$-dimensional data points is a $K$-d tree in which each node contains its bounding box, a pointer to its father and two pointers, `fleft` and `fright`, pointing to two nodes in its left and right subtrees, respectively.

Given a $m$-finger $K$-d tree $T$ and an orthogonal range query $Q$ the orthogonal range search in $T$ returns the points in $T$ which fall inside $Q$ as usual, but it also modifies the finger pointers of the nodes in $T$ to improve the response time of future orthogonal range searches. The algorithm for $m$-finger search trees follows by recursively applying the 1-finger $K$-d tree scheme at each stage of the orthogonal range search trees. The fingers of visited nodes are updated as the search proceeds; we have considered that if a search continues in just one subtree of the current node the finger corresponding to the non-visited subtree should

---

**Algorithm 2** The nearest neighbor search algorithm for 1-finger $K$-d trees.

▷  Initial call: $nn := T \to root; md := \text{dist}(nn \to info, q);$
▷            $f := \text{one\_finger\_NN}(f, q, md, nn)$
**function** one_finger_NN$(f, q, min\_dist, nn)$
  **if** $f = $ **nil then return** $f$
  $x := f \to info$
  $d := \text{dist}(q, x)$
  $B := f \to bounding\_box$
  **if** $d < min\_dist$ **then**
    $min\_dist := d$
    $nn := f$
  **if** BALL$(q, min\_dist) \not\subset B$ **then** ▷    Backtrack
    **return** one_finger_NN$(f \to father, q, min\_dist, nn)$
  $j := f \to discr$
  **if** $q[j] < x[j]$ **then**
    $nn := $ one_finger_NN$(f \to left, q, min\_dist, nn)$
    $other := f \to right$
  **else**
    $nn := $ one_finger_NN$(f \to right, q, min\_dist, nn)$
    $other := f \to left$
  **if** $q[j] - min\_dist \le x[j]$ **and** $q[j] + min\_dist \ge x[j]$ **then**
    $nn := $ one_finger_NN$(other, q, min\_dist, nn)$
  **return** $nn$
**end**

---

be reset, for it was not providing useful information. The pseudo-code for this algorithm is given as Algorithm 3.

It is important to emphasize that while it is not too difficult to code 1-finger search trees using $\mathcal{O}(\log n)$ additional memory[2], the implementation of $m$-finger search trees does require $\Theta(n)$ additional memory for the father pointer, finger pointers and bounding boxes, that is, a total of $3n$ additional pointers and $2n$ $K$-dimensional points. This could be a high price for the improvement in search performance which, perhaps, might not be worth paying.

## 4    Locality Models and Experimental Results

Both 1-finger and $m$-finger try to exploit locality of reference in long sequences of queries, so one of the main aspects of this work was to devise meaningful models on which we could carry out the experiments. The programs used in the experiments described in this section have been written in C, using the GNU compiler `gcc-2.95.4`. The experiments themselves have been run in a computer with Intel Pentium 4 CPU at 2.8 GHz with 1 Gb of RAM and 512 Kb of cache memory.

---

[2] Actually, the necessary additional space is proportional to the height of the $K$-d tree, which on average is $\Theta(\log n)$ but can be as much $\Theta(n)$ in the worst-case.

---

**Algorithm 3** The orthogonal range search algorithm in a $m$-finger $K$-d tree.

▷    Initial call: multiple_finger_range_search($T \to root, Q$)

function multiple_finger_range_search($f, Q$)

    if $f =$ nil then return $f$

    $B := f \to bounding\_box$

    if $Q \not\subset B$ then  ▷    Backtrack

        $f \to fleft := f \to left$

        $f \to fright := f \to right$

        return multiple_finger_range_search($f \to father, Q$)

    $x := f \to info$

    if $Q.u[j] < x[j]$ then

        $f \to fright := f \to right$

        $f \to fleft :=$ multiple_finger_range_search($f \to fleft, Q$)

        if $f \to fleft =$ nil then return $f$

        if $f \to fleft = f$ then $f \to fleft := f \to left$

        return $T \to fleft$

    if $Q.l[j] \geq x[j]$ then

        $f \to fleft := f \to left$

        $f \to fright :=$ multiple_finger_range_search($f \to fright, Q$)

        if $f \to fright =$ nil then return $f$

        if $f \to fright = f$ then $f \to fright := f \to right$

        return $T \to fright$

    if $x \in Q$ then Add $x$ to result

    $f \to fleft :=$ multiple_finger_range_search($f \to fleft, Q'$)

    if $f \to fleft =$ nil then return $f$

    if $f \to fleft = f$ then $f \to fleft := f \to left$

    $f \to fright :=$ multiple_finger_range_search($f \to fright, Q''$)

    if $f \to fright =$ nil then return $f$

    if $f \to fright = f$ then $f \to fright := f \to right$

    return $f$

end

---

## 4.1    The Models

In the case of orthogonal range search, given a size $n$, and a dimension $K$, we generate $T = 1000$ sets of $n$ $K$-dimensional points drawn uniformly and independently at random in $[0, 1]^K$. Each point of each set is inserted into two initially empty trees, so that we get a random standard $K$-d tree $T_s$ and a random relaxed $K$-d tree $T_r$ of size $n$ which contain the same information. For each pair $(T_s, T_r)$, we generate $S = 300$ sequences of $Q = 100$ orthogonal range queries and make the corresponding search with the standard and the fingered variants of the algorithm, collecting the basic statistics on the performance of the search.

We have used in all experiments fixed size queries: the length of the $K$ edges of each query was $\Delta = 0.01$. Since we have run experiments with up to $n = 50000$ elements per tree, the number of reported points by any range search is typically small (from 1 to 5 reported points). To modelize locality, we introduced the notion of $\delta$-close queries: given to queries $Q$ and $Q'$ with identical edge lengths $\Delta_0, \Delta_1, \ldots, \Delta_{K-1}$, we say that $Q$ and $Q'$ are $\delta$-close if their respective centers $z$ and $z'$ satisfy $z - z' = (d_0, d_1, \ldots, d_{K-1})$ and $|d_j| \leq \delta \cdot \Delta_j$, for any $0 \leq j < K$. The sequences of $\delta$-close queries were easily generated at random by choosing the initial center $z_0$ at random and setting each successive center $z_{m+1} = z_m + d_m$ for some randomly generated vector $d_m$; in particular, the $i$-th coordinate of $d_m$ is generated uniformly at random in $[-\delta \cdot \Delta_j, \delta \cdot \Delta_j]$.

The real-valued parameter $\delta$ is a simple way to capture into a single number the degree of locality of reference. If $\delta < 1$ then $\delta$-close queries must overlap at least a fraction $(1 - \delta)^K$ of their volume. When $\delta \to \infty$ (in fact it suffices to set $\delta = \max\{\Delta_i^{-1}\}$) we have no locality.

For nearest neighbor searches, the experimental setup was pretty much the same as for orthogonal search; for each pair $(T_s, T_r)$ of randomly built $K$-d trees, we perform nearest neighbor search on each of the $Q = 100$ queries of each of the $S = 300$ generated sequences.

Successive queries $q$ and $q'$ are said to be $\delta$-close if $q - q' = (d_0, d_1, \ldots, d_{K-1})$ and $|d_j| \leq \delta$, for any $0 \leq j < K$. It is interesting to note that the locality of reference parameter $\delta$ now bounds the distance between queries in absolute terms, whereas it is used in relative terms in the case of orthogonal range queries. As such, only values in the range $[0, \sqrt{K}]$ are meaningful, although we find convenient to say $\delta \to \infty$ to indicate that there is no locality of reference.

## 4.2    The Experiments

**Range Queries.** Due to the space limitations, we show here only the results corresponding to relaxed $K$-d trees; the results for standard $K$-d trees are qualitatively identical. To facilitate the comparison between the standard algorithms and their fingered counterparts we use the ratio of the respective overworks; namely, if $W_n^{(1)}$ denotes the overwork of 1-finger search, $W_n^{(m)}$ denotes the overwork of $m$-finger search and $W_n^{(0)}$ denotes the overwork of standard search (no fingers), we will use the ratios $\tau_n^{(1)} = W_n^{(1)}/W_n^{(0)}$ and $\tau_n^{(m)} = W_n^{(m)}/W_n^{(0)}$. Recall that the overwork is the number of visited nodes during a search minus the number of nodes (points) which satisfied the range query. The graphs of Figures 2, 3 and 4 depict $\tau_n^{(1)}$ and $\tau_n^{(m)}$ and $\delta = 0.25, 0.75, 2$, respectively.

All the plots confirm that significant savings can be achieved thanks to the use of fingers; in particular, $m$-finger $K$-d trees do much better than 1-finger $K$-d trees for all values of $K$ and $\delta$. As $\delta$ increases the savings w.r.t. non-fingered search decrease, but even for $\delta = 2$ the overwork of 1-finger search is only about 60% of the overwork of the standard search.

As we already expected, the performance of both 1-finger $K$-d trees and $m$-finger $K$-d trees heavily depends on the locality parameter $\delta$, a fact that is well illustrated by Figures 5 and 6. The first one shows the plot of the overwork $W_n^{(m)}$ of standard $m$-finger $K$-d trees for various values of $\delta$ and dimensions. In particular, when the dimension increases we shall expect big differences in the savings that fingered search yield as $\delta$ varies; for lower dimensions, the variability of $W_n^{(m)}$ with $\delta$ is not so "steep". Similar phenomena can be observed for relaxed $m$-finger $K$-d trees and standard and relaxed 1-finger $K$-d trees. On the other hand, Figure 6 shows the variation of $\tau_{50000}^{(1)}$ and $\tau_{50000}^{(m)}$ as functions of $\delta$ for relaxed 1-finger and $m$-finger $K$-d trees.

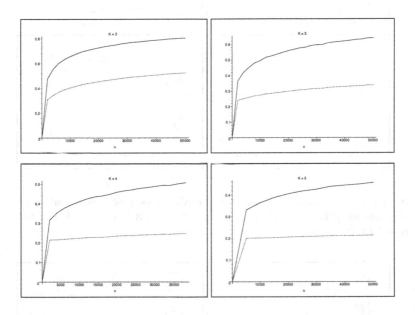

**Fig. 2.** Overwork ratios for relaxed 1-finger $K$-d trees (*solid line*) and $m$-finger $K$-d trees (*dashed line*), for $\delta = 0.25$, $K = 2$ (*up left*), $K = 3$ (*up right*), $K = 4$ (*down left*), and $K = 5$ (*down right*)

Taking into account the way the algorithm works, we conjecture that 1-finger search reduces by a constant factor the logarithmic term in the overwork. Thus, if standard search has overwork $W_n^{(0)} \approx \sum_{0<j<K} \beta_j n^{\phi(j/K)} + \gamma \log n$ then

$$W_n^{(1)} \approx \sum_{0<j<K} \beta_j n^{\phi(j/K)} + \gamma' \log n, \tag{1}$$

with $\gamma' = \gamma'(\delta)$. However, since the $\phi$'s and $\beta$'s are quite small it is rather difficult to disprove this hypothesis on an experimental basis; besides it is fully consistent with the results that we have obtained.

On the other hand, and again, following our intuitions on its *modus operandi*, we conjecture that the overwork of $m$-finger search is equivalent to skipping the initial logarithmic path and then performing a standard range search on a random tree whose size is a fraction of the total size, say $n/x$, for some $x > 1$

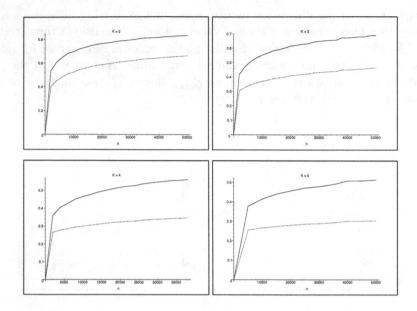

**Fig. 3.** Overwork ratios for relaxed 1-finger $K$-d trees (*solid line*) and $m$-finger $K$-d trees (*dashed line*), for $\delta = 0.75$, $K = 2$ (*up left*), $K = 3$ (*up right*), $K = 4$ (*down left*), and $K = 5$ (*down right*)

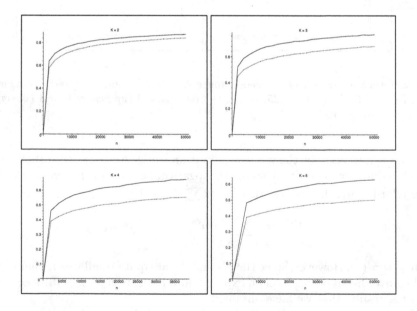

**Fig. 4.** Overwork ratios for relaxed 1-finger $K$-d trees (*solid line*) and $m$-finger $K$-d trees (*dashed line*), for $\delta = 2$, $K = 2$ (*up left*), $K = 3$ (*up right*), $K = 4$ (*down left*), and $K = 5$ (*down right*)

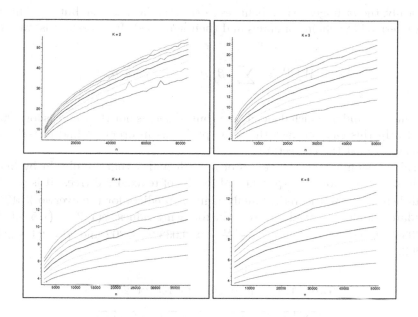

**Fig. 5.** Overwork in relaxed $m$-finger $K$-d trees for several values of the locality parameter $\delta$

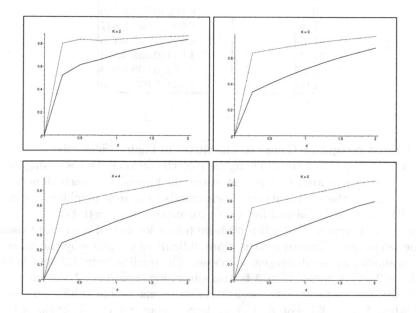

**Fig. 6.** Overwork ratios for $n = 50000$ for relaxed 1-finger $K$-d trees (*solid line*) and $m$-finger $K$-d trees (*dashed line*), $K = 2$ (*up left*), $K = 3$ (*up right*), $K = 4$ (*down left*), and $K = 5$ (*down right*)

(basically, the $m$-finger search behaves as the standard search, but skips more or long intermediate chains of nodes and their subtrees). In other words, we would have

$$W_n^{(m)} \approx \sum_{0<j<K} \beta_j' n^{\phi(j/K)} + \gamma' \log n, \qquad (2)$$

for some $\beta_j'$'s and $\gamma'$ which depend on $\delta$ (but $\gamma'$ here is not the same as for 1-finger search). In this case we face the same problems in order to find experimental evidence against the conjecture.

Table 1 summarizes the values of $\beta \equiv \beta_1$ and $\gamma$ that we obtain by finding the best-fit curve for the experimental results of relaxed 2-d trees. It is worth to recall here that the theoretical analysis in [7] predicts for the overwork $W_n^{(0)}$ of standard search in relaxed 2-d trees the following values: $\phi(1/2) = (\sqrt{5}-1)/2 \approx$ .618033989, $\beta = 4\Delta(1-\Delta)\frac{\Gamma(2\alpha+1)}{(\alpha+1)\alpha^3\Gamma^3(\alpha)} \approx 0.03828681802$ and $\gamma = 2(1-\Delta)^2 = 1.9602$.

**Table 1.** Best-fit $\beta$ and $\gamma$ for relaxed 2-d trees

| $\delta$ | no finger | | 1-finger | | $m$-finger | |
|---|---|---|---|---|---|---|
| | $\beta$ | $\gamma$ | $\beta$ | $\gamma$ | $\beta$ | $\gamma$ |
| 0.25 | 0.037 | 1.912 | 0.040 | 0.746 | 0.028 | 0.338 |
| 0.50 | " | " | " | 0.0836 | 0.031 | 0.450 |
| 0.75 | " | " | " | 0.902 | 0.034 | 0.550 |
| 1.00 | " | " | " | 0.955 | 0.035 | 0.647 |
| 1.25 | " | " | " | 0.999 | 0.037 | 0.733 |
| 1.50 | " | " | " | 1.054 | 0.038 | 0.824 |
| 1.75 | " | " | " | 1.103 | 0.039 | 0.908 |
| 2.00 | " | " | " | 1.151 | 0.039 | 0.986 |

**Nearest Neighbor Queries.** The curves in Figure 7 show the performance of relaxed 1-finger $K$-d trees. There, we plot the ratio of the cost using 1-finger nearest neighbor search to the cost using no fingers. For each dimension $K$ ($K = 2, 3, 4, 5$), the solid line curve corresponds to nearest neighbor search with $\delta = 0.01$, whereas the dashed line curve corresponds to $\delta = 0.005$.

It is not a surprise that when we have better locality of reference (a smaller $\delta$) the performance improves. It is more difficult to explain why the variability on $\delta$ is smaller as the dimension increases. The qualitatively different behavior for $K = 2$, $K = 3$ and $K > 3$ is also surprising. For $K = 2$ the ratio of the costs increases as $n$ increases until it reaches some stable value (e.g., roughly 90% when $\delta = 0.005$). For $K = 3$ we have rather different behavior when we pass from $\delta = 0.005$ to $\delta = 0.01$. For $K = 4$, $K = 5$ and $K = 6$ we have the same qualitative behavior in all cases[3]: a decrease of the ratio as $n$ grows until the

---

[3] The plot for $K = 6$ is not shown in the figure.

ratio reaches a limit value. A similar phenomenon occurs for $K = 2$ and $K = 3$ provided that $\delta$ is even smaller than 0.005.

We did not find significant improvements of 1-finger search with respect to standard search in none of our experiments, in particular, the cost of 1-finger nearest neighbor search was not below 90% of the standard cost even for large dimensions and small (but not unrealistic) $\delta$'s.

A preliminary explanation for the observed behavior is that unless the locality of reference is high then the NN search will have to backtrack a significant fraction of the path that it had skipped thanks to the finger. On the other hand, we define the locality parameter $\delta$ in an absolute manner, but the actual degree of locality that it expresses depends on the dimension. For instance, take $n = 10000$ and $\delta = 0.01$. If $K = 2$ then we would expect to find one point in a $L_\infty$-sphere of radius $\delta$; but we would find only 0.01 points in a $L_\infty$-sphere of radius $\delta$ if $K = 3$, etc. Hence, many NN searches in a large dimension and with some reasonably small value of $\delta$ will have the same result as the immediately preceding query and 1-finger search does pretty well under that (easy) situation.

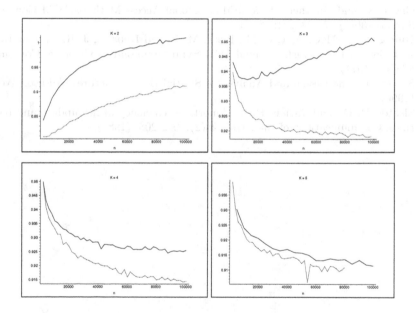

**Fig. 7.** Nearest neighbor queries in relaxed 1-finger $K$-d trees for $\delta = 0.01$ (*solid line*) and $\delta = 0.005$ (*dashed line*), $K = 2$ (*up left*), $K = 3$ (*up right*), $K = 4$ (*down left*), and $K = 5$ (*down right*)

# References

1. Bentley, J.L., Multidimensional binary search trees used for associative retrieval, Communications of the ACM, 18(9):509-517, (1975).
2. Borodin, A. and El-Yaniv, R., Online Computation and Competitive Analysis, Cambridge University Press, (1998).
3. Brown, M. R. and Tarjan, R. E., Design and Analysis of a Data Structure for Representing Sorted Lists, SIAM Journal of Computing, 9(3):594-614, (1980).
4. Chanzy, P. and Devroye, L. and Zamora-Cura, C., Analysis of range search for random $k$-d trees, Acta Informatica, 37:355-383, (2001).
5. Devroye, L. and Jabbour, J. and Zamora-Cura, C., Squarish $k$-d trees, SIAM Journal of Computing, 30:1678-1700, (2000).
6. Duch, A. and Estivill-Castro, V. and Martínez, C., Randomized $K$-dimensional binary search trees, Int. Symposium on Algorithms and Computation (ISAAC'98), Eds. K.-Y. Chwa and O. H. Ibarra, Lecture Notes of Computer Science Vol. 1533: 199-208, Springer-Verlag, (1998).
7. Duch, A. and Martínez, C., On the Average Performance of Orthogonal Range Search in Multidimensional Data Structures, Journal of Algorithms, 44(1):226-245, (2002).
8. Gaede, V. and Günther, O., Multidimensional Access Methods,ACM Computing Surveys, 30(2):170-231, (1998).
9. Guibas, L. J., McCreight, E. M., Plass, M. F. and Roberts, J. R., A New Representation for Linear Lists, Annual ACM Symposium on the Theory of Computing, STOC, (1977).
10. Samet, H., The Design and Analysis of Spatial Data Structures, Addison-Wesley, (1990).
11. Sleator, D. D., and Tarjan, R. E., Amortized efficiency of list update and paging rules, Communications of the ACM, 28(2):2002-208, (1985).

# Faster Deterministic and Randomized Algorithms on the Homogeneous Set Sandwich Problem

Celina M.H. de Figueiredo[1], Guilherme D. da Fonseca[1], Vinícius G.P. de Sá[1], and Jeremy Spinrad[2]

[1] Instituto de Matemática and COPPE, Universidade Federal do Rio de Janeiro, Brazil.
celina@cos.ufrj.br, fonseca@cs.umd.edu, vigusmao@uol.com.br
[2] Computer Science Department, Vanderbilt University, USA.
spin@vuse.vanderbilt.edu

**Abstract.** A homogeneous set is a non-trivial, proper subset of a graph's vertices such that all its elements present exactly the same outer neighborhood. Given two graphs, $G_1(V, E_1)$, $G_2(V, E_2)$, we consider the problem of finding a sandwich graph $G_s(V, E_S)$, with $E_1 \subseteq E_S \subseteq E_2$, which contains a homogeneous set, in case such a graph exists. This is called the Homogeneous Set Sandwich Problem (HSSP). We give an $O(n^{3.5})$ deterministic algorithm, which updates the known upper bounds for this problem, and an $O(n^3)$ Monte Carlo algorithm as well. Both algorithms, which share the same underlying idea, are quite easy to be implemented on the computer.

## 1 Introduction

Given two graphs $G_1(V, E_1)$, $G_2(V, E_2)$ such that $E_1 \subseteq E_2$, a sandwich problem with input pair $(G_1, G_2)$ consists in finding a *sandwich graph* $G_s(V, E_S)$, with $E_1 \subseteq E_S \subseteq E_2$, which has a desired property $\Pi$ [6]. In this paper, the property we are interested in is the ownership of a homogeneous set. A *homogeneous set* $H$, in a graph $G(V, E)$, is a subset of $V$ such that *(i)* $2 \leq |H| \leq |V| - 1$ and *(ii)* for all $v \in V \setminus H$, either $(v, v') \in E$ is true for all $v' \in H$ or $(v, v') \notin E$ is true for all $v' \in H$. In other words, a homogeneous set $H$ is a subset of $V$ such that the outside-$H$ neighborhood of all its vertices is the same and which also satisfies the necessary, above mentioned size constraints. A *sandwich homogeneous set* of a pair $(G_1, G_2)$ is a homogeneous set for at least one among all possible sandwich graphs for $(G_1, G_2)$.

Graph sandwich problems have attracted much attention lately arising from many applications and as a natural generalization of recognition problems [1,3, 5,6,7,8,9].

The importance of homogeneous sets in the conetxt of graph decomposition has been well acknowledged, specially in the context of perfect graphs [10]. There are many algorithms which find homogeneous sets quickly in a single graph. The

C.C. Ribeiro and S.L. Martins (Eds.): WEA 2004, LNCS 3059, pp. 243–252, 2004.
© Springer-Verlag Berlin Heidelberg 2004

most efficient one is due to McConnell and Spinrad [11] and has $O(|E|)$ time complexity.

On the other hand, the known algorithms for the homogeneous set sandwich problem are far less efficient. The first polynomial time algorithm was presented by Cerioli *et al.* [1] and has $O(n^4)$ time complexity (where $n = |V|$, as throughout the whole text). We refer to it as the *Exhaustive Bias Envelopment algorithm* (EBE algorithm, for short), as in [2]. An $O(\Delta n^2)$ algorithm (where $\Delta$ stands for the maximum vertex degree in $G_1$) has been found by Tang *et al.* [12], but in [4, 2] it is proved incorrect. Although all efforts to correct Tang *et al.*'s algorithm (referred to as the *Bias Graph Components algorithm*, in [2]) have been in vain, some of its ideas were used, in [4,2], to build a hybrid algorithm, inspired by both [1] and [12]. This one has been called the *Two-Phase algorithm* (2-P algorithm, for short) and currently sets the HSSP's upper bounds at its $O(m_1 \overline{m_2})$ time complexity, where $m_1$ and $\overline{m_2}$ respectively refer to the number of edges in $G_1$ and the number of edges *not* in $G_2$.

After defining some concepts and auxiliary procedures in Section 2, we present, in Section 3, a new $O(n^{3.5})$ deterministic algorithm for the HSSP. It offers a good alternative to the *2-P* algorithm (whose time complexity is not better than $O(n^4)$ if we express it only as a function of $n$) when dealing with dense input graphs, whereas the *2-P* would remain the best choice when sparse graphs are dealt with. Besides, Section 4 is devoted to a fast, randomized Monte Carlo algorithm, which solves this problem in $O(n^3)$ time with whatever desired error ratio.

## 2   Bias Envelopment

We define the *bias set* $B(H)$ of a vertex subset $H$ as the set of vertices $b \notin H$ such that $(b, v_i) \in E_1$ and $(b, v_j) \notin E_2$, for some $v_i, v_j \in H$. Such vertices $b$ are called *bias vertices* of the set $H$ [12]. It is easy to see that $H$, with $2 \leq |H| \leq n - 1$, is a sandwich homogeneous set if and only if $B(H) = \varnothing$.

It is proved in [1] that any sandwich homogeneous set containing the set of vertices $H$ should also contain $B(H)$. This result, along with the fact that no homogeneous sets are allowed with less than two vertices, gave birth in that same paper to a procedure which we call *Bias Envelopment* [2]. The Bias Envelopment procedure answers whether a given pair of vertices is contained in any sandwich homogeneous sets of the input instance. The procedure starts from an initial sandwich homogeneous set candidate $H_1 = \{v_1, v_2\}$ and successively computes $H_{t+1} = H_t \cup B(H_t)$ until either *(i)* $B(H_t) = \varnothing$, whereby $H_t$ is a sandwich homogeneous set and it answers *yes*; or *(ii)* $|H_t| + |B(H_t)| = n$, when it answers *no*, meaning that there is no sandwich homogeneous set containing $\{v_1, v_2\}$. The Bias Envelopment procedure runs in $O(n^2)$ time, granted some appropriate data structures are used, as described in [1].

The EBE algorithm, presented in [1], tries to find a sandwich homogeneous set exhaustively, running the Bias Envelopment procedure on all $n(n-1)/2$ pairs

**IncompleteBiasEnvelopment** $(G_1(V, E_1), G_2(V, E_2), v_1, v_2, k)$

1.       $H \leftarrow \{v_1, v_2\}$
2.       while $|H| \leq k$
2.1.         if $B(H) = \varnothing$ and $|H| < |V|$
2.1.1.           return $H$ and *yes*   // a sandwich homogeneous set was found.
2.2.         else
2.2.1.           $H \leftarrow H \cup B(H)$
3.       return *no*   // there are no sandwich homogeneous sets with $k$ vertices
                 // or less which contain $\{v_1, v_2\}$.

Fig. 1. The Incomplete Bias Envelopment procedure.

of the input graphs' vertices, in the worst case. Thus, the time complexity of the EBE algorithm is $O(n^4)$.

Both algorithms we introduce in this paper are based on a variation of the Bias Envelopment procedure, which we call the *Incomplete Bias Envelopment*. The input of the Incomplete Bias Envelopment is a pair of vertices $\{v_1, v_2\}$ and a *stop parameter* $k < n$. The only change in this incomplete version is that, whenever $|H_t| > k$, the envelopment stops prematurely, answering *no* and rejecting $\{v_1, v_2\}$. Notice that a *no* answer from the Incomplete Bias Envelopment with parameter $k$ means that $\{v_1, v_2\}$ is not contained in any homogeneous sets *of size at most* $k$. Using the same data structures as in [1], the Incomplete Bias Envelopment runs in $O(nk)$ time.

The Incomplete Bias Envelopment generalizes its complete version, as a normal Bias Envelopment is equivalent to an Incomplete Bias Envelopment with parameter $k = n$.

The pseudo-code for the Incomplete Bias Envelopment is in Figure 1.

## 3   The Balanced Subsets Algorithm

The algorithm we propose in this section is quite similar to the EBE algorithm, in the sense that it submits each of the input vertices' pairs to the process of Bias Envelopment. The only difference is that this algorithm establishes a particular order in which the vertex pairs are chosen, in such a way that it can benefit, at a certain point, from unsuccessful envelopments that have already taken place. After some unsuccessful envelopments, a number of vertex pairs have been found not to be contained in any sandwich homogeneous sets. This knowledge is then made useful by the algorithm, which will stop further envelopments earlier by means of calling *Incomplete* Bias Envelopments instead of complete ones, saving relevant time without loss of completeness.

When the algorithm starts, it partitions all $n$ vertices of the input graphs into $O(\sqrt{n})$ disjoint subsets $C_i$ of size $O(\sqrt{n})$, each. Then all pairs of vertices will be submitted to Bias Envelopment in two distinct phases: in the first phase, all

**BalancedSubsetsHSSP** $(G_1(V, E_1), G_2(V, E_2))$

**1.**     label all vertices in $V$ from $v_1$ to $v_n$.
**2.**     create $\lceil \sqrt{n} \, \rceil$ empty sets $C_i$.
**3.**     for each vertex $v_j \in V$, do $C_j \ \mathbf{modulo}_{\lceil \sqrt{n} \, \rceil} = C_j \ \mathbf{modulo}_{\lceil \sqrt{n} \, \rceil} \cup \{v_j\}$.
**4.**     for each pair of vertices $\{x, y\}$ in the same subset $C_i$
**4.1.**       if BiasEnvelopment$(G_1, G_2, x, y) = yes$
**4.1.1.**         return $yes$.
**5.**     for each pair of vertices $\{x, y\}$ not in the same subset $C_i$
**5.1.**       if IncompleteBiasEnvelopment$(G_1, G_2, x, y, \lceil \sqrt{n} \, \rceil) = yes$
**5.1.1.**         return $yes$.
**6.**     return $no$.

---

**Fig. 2.** The Balanced Subsets algorithm for the HSSP.

pairs consisting of vertices from the same subset $C_i$ are bias enveloped (and only those); in the second phase, all remaining pairs (i.e. those comprising vertices that are not from the same subset $C_i$) are then bias-enveloped. In the end, all possible vertex pairs will have been checked out as to belong or not to some sandwich homogeneous set from the input instance, just like in the EBE algorithm. The point is: if all Bias Envelopments in the first phase fail to find a sandwich homogeneous set, then the input instance does not admit any sandwich homogeneous sets which contain two vertices from the same subset $C_i$. Thence, the maximum size of any possibly existing homogeneous set is $O(\sqrt{n})$ (the number of subsets into which the vertices had been dispersed), which grants that all Bias Envelopments of the second phase need not search for large homogeneous sets! That is why an Incomplete Bias Envelopment with stop parameter $k = O(\sqrt{n})$ can be used instead.

Figure 2 illustrates the pseudo-code for the Balanced Subsets algorithm.

**Theorem 1** *The Balanced Subsets algorithm correctly answers whether there exists a sandwich homogeneous set in the input graphs.*

*Proof.* If the algorithm returns $yes$, then it has successfully found a set $H \subset V$, with $2 \leq |H| \leq |V| - 1$, such that the bias set of $H$ is empty. Thus, $H$ is indeed a valid sandwich homogeneous set.

Now, suppose the input has a sandwich homogeneous set $H$. If $|H| > \lceil \sqrt{n} \, \rceil$ then there are more vertices in $H$ than subsets into which all input vertices were spread, in the beginning of the algorithm (line 3). Thus, by the pigeon hole principle, there must be two vertices $x, y \in H$ which were assigned to the same subset $C_i$. So, whenever $\{x, y\}$ is submitted to Bias Envelopment (line 4), the algorithm is doomed to find a sandwich homogeneous set. On the other hand, if $|H| \leq \lceil \sqrt{n} \, \rceil$, then it is possible that $H$ does not contain any two vertices from the same subset $C_i$, which would cause all Bias Envelopments of the first phase (line 4.1) to fail. In this case, however, when a pair $\{x, y\} \subseteq H$ happens to be bias enveloped in line 5, the Incomplete Bias Envelopment is meant to be successful, for the size of $H$ is, by hypothesis, less than or equal its stop parameter $k = \lceil \sqrt{n} \rceil$.                                                                □

### 3.1  Complexity Analysis

As each subset $C_i$ has $O(n)$ pairs of vertices and there are $O(\sqrt{n})$ such subsets, the number of pairs that are bias enveloped in the first phase of the algorithm (line 4) is $O(n\sqrt{n})$. All Bias Envelopments, in this phase, are complete and take $O(n^2)$ time to be executed, which yields a subtotal of $O(n^{3.5})$ time in the whole first phase.

The number of pairs that are only submitted to Bias Envelopment in the second phase (line 5) is $O(n^2) - O(n\sqrt{n}) = O(n^2)$ pairs. Each Bias Envelopment is, now, an incomplete one with parameter $k = \lceil \sqrt{n} \rceil = O(\sqrt{n})$. Because the time complexity of each Incomplete Bias Envelopment with parameter $k$ is $O(nk)$, then the total time complexity of the whole second phase of the algorithm is $O(n^3 k) = O(n^{3.5})$.

Thus, the overall time complexity of the Balanced Subsets algorithm is $O(n^{3.5})$.

## 4  The Monte Carlo Algorithm

An yes-biased Monte Carlo algorithm for a decision problem is one which always answers *no* when the correct answer is *no* and which answers *yes* with probability $p$ whenever the correct answer is *yes*.

In order to gather some intuition, let us suppose the input has a sandwich homogeneous set $H$ with $h$ vertices or more.

What would be, in this case, the probability $\overline{p_1}$ that a random pair of vertices $\{x, y\} \in V$ is *not* contained in $H$? Clearly,

$$\overline{p_1} \leq 1 - \frac{h(h-1)}{n(n-1)}.$$

What about the probability $\overline{p_t}$ that $t$ random pairs of vertices fail to be contained in $H$? It is easy to see that

$$\overline{p_t} \leq \left(1 - \frac{h(h-1)}{n(n-1)}\right)^t.$$

Now, what is the probability $p_t$ that, after $t$ Bias Envelopment procedures have been run (starting from $t$ randomly chosen pairs of vertices), a sandwich homogeneous set have been found? Again, it is quite simple to reach the following expression, which will be vital to the forthcoming reasoning.

$$p_t \geq 1 - \left(1 - \frac{h(h-1)}{n(n-1)}\right)^t. \tag{1}$$

If, instead of obtaining the probability $p_t$ from the expression above, we fix $p_t$ at some desired value $p = 1 - \epsilon$, we will be able to calculate the minimum integer value of $h_t$ (which will denote $h$ as a function of $t$) that satisfies the inequality 1. This value $h_t$ is such that the execution of $t$ independent Bias Envelopment procedures (on $t$ random pairs) *is sufficient to find a sandwich homogeneous set of the input instance with probability at least $p$, in case there exists any with $h_t$ vertices or more* (see equation 2):

$$h_t = \left\lfloor \frac{1 + \sqrt{1 + 4(n^2 - n)(1 - (1 - p)^{1/t})}}{2} \right\rfloor. \tag{2}$$

However, we want an algorithm that finds a sandwich homogeneous set with some fixed probability $p$ *in case there exists any, no matter its size*. But as $h_t$ decreases with the growth of $t$, the following question arises: how many random pairs do we need to submit to Bias Envelopment in order to achieve that? The answer is rather simple: the minimum integer $t'$ such that $h_{t'} = 2$, for 2 is the shortest possible size of a sandwich homogeneous set!

Determining $t'$ comes straightforwardly from equation 2 (please refer to Section 4.1 for the detailed calculations):

$$t' = \frac{\ln(1 - p)}{\ln\left(1 - \frac{2}{n(n-1)}\right)} = \Theta(n^2). \tag{3}$$

Once the number $t'$ of Bias Envelopment procedures that need to be undertaken on randomly chosen pairs of vertices is $\Theta(n^2)$ and the time complexity of each Bias Envelopment is $O(n^2)$, so far we seem to have been lead to an $O(n^4)$ randomized algorithm, which is totally undesirable, for we could already solve the problem *deterministically* with less asymptotical effort (see Section 3)!

Now we have come to a point where the *incomplete* version of the Bias Envelopment procedure will play an essential role as far as time saving goes. We show that, by the time the $t$-th Bias Envelopment is run, its incomplete version with stop parameter $k = h_{t-1}$ serves exactly the same purpose as its complete version would do.

**Lemma 2** *In order to find a sandwich homogeneous set, with probability $p$, in case there exists any with $h_t$ vertices or more, the $t$-th Bias Envelopment need not go further when the size of the candidate set has exceeded $h_{t-1}$.*

*Proof.* Two are the possibilities regarding the input: *(i)* there is a sandwich homogeneous set with more than $h_{t-1}$ vertices; or *(ii)* there are no sandwich homogeneous sets with more than $h_{t-1}$ vertices.

If *(i)* is true, then no more than $t - 1$ Bias Envelopments would even be required to achieve that. Hence the $t$-th Bias Envelopment can stop as early as it pleases.

If *(ii)* is the case, then an Incomplete Bias Envelopment with stop parameter $k = h_{t-1}$ is meant to give the exact same answer as the complete Bias Envelop-

**MonteCarloHSSP** $(G_1(V, E_1), G_2(V, E_2), p)$

**1.**    $h \leftarrow |V|$
**2.**    $t \leftarrow 1$
**3.**    while $h \geq 2$
**3.1.**        $(v_1, v_2) \leftarrow$ random pair of distinct vertices of $V$
**3.2.**        if IncompleteBiasEnvelopment$(G_1, G_2, v_1, v_2, h) = yes$
**3.2.1.**            return $yes$
**3.3.**        $h \leftarrow \lfloor (1 + \sqrt{1 + 4(|V|^2 - |V|)(1 - (1 - p)^{1/t})})/2 \rfloor$
**3.4.**        $t \leftarrow t + 1$
**4.**    return $no$

---

**Fig. 3.** The Monte Carlo algorithm for the HSSP.

ment would, for there are no sandwich homogeneous sets with more than $h_{t-1}$ vertices to be found.

Whichever the case, thus, such an Incomplete Bias Envelopment is perfectly sufficient.                                                                                      □

Now we can describe an efficient Monte Carlo algorithm which gives the correct answer to the HSSP with probability at least $p$.

The algorithm's idea is to run several Incomplete Bias Envelopment procedures on randomly chosen initial candidate sets (pairs of vertices). At each iteration $t$ of the algorithm we run an Incomplete Bias Envelopment with stop parameter $k = h_{t-1}$ and either it succeeds in finding a sandwich homogeneous set (and the algorithm stops with an $yes$ answer) or else it aborts the current envelopment whenever the number of vertices in the sandwich homogeneous set candidate exceeds the $h_{t-1}$ threshold. (In this case, Lemma 2 grants its applicability.) For the first iteration, the stop parameter $k$ is set to $h_0 = n$, as the first iteration corresponds to a complete Bias Envelopment. At the end of each iteration, the value of $h_t$ is then updated (see equation 2), which makes it progressively decrease throughout the iterations until it reaches 2 (the minimum size allowed for a homogeneous set), which necessarily happens after $\Theta(n^2)$ iterations (see equation 3).

The pseudo-code for this algorithm is in Figure 3.

**Theorem 3** *The Monte Carlo HSSP algorithm correctly answers whether there exists a sandwich homogeneous set in the input graphs with probability at least $p$.*

*Proof.* If the algorithm returns $yes$, then it is the consequence of having found a set $H \subset V$, with $2 \leq |H| \leq |V| - 1$, such that the bias set of $H$ is empty, which makes a valid sandwich homogeneous set out of $H$. In other words, if the correct answer is $no$ then the algorithm gives a correct $no$ answer with probability 1.

If the correct answer is $yes$, we want to show that it gives a correct $yes$ answer with probability $p$. Let $h^*$ be the size of the largest sandwich homogeneous set of the input instance. As $h_0 = n$ and the algorithm only answers *no* after $h_t$ has lowered down to 2, there must exist an index $d$ such that $h_d \leq h^* < h_{d-1}$. From the definition of $h_t$ we know that, on the hypothesis that the input has a sandwich homogeneous set with $h_t$ vertices or more, $t$ Bias Envelopments are sufficient to find one, with probability at least $p$. As, by hypothesis, there is a sandwich homogeneous set with $h^* \geq h_d$ vertices, then $d$ independent Bias Envelopments are sufficient to find a sandwich homogeneous set with probability $p$. So, it is enough to show that this quota of $d$ Bias Envelopments is achieved. It is true that Incomplete Bias Envelopments that stop *before* the candidate set has reached the size of $h^*$ cannot find a sandwich homogeneous set with $h^*$ vertices. Nevertheless, the first $d$ iterations alone perform this minimum quota of Bias Envelopments. Because $h^*$ is the size of the largest sandwich homogeneous set, the fact of being *incomplete* simply does not matter for these first $d$ Bias Envelopments, none among which being allowed to stop before the size of the candidate set has become larger than $h_{d-1} > h^*$.                           □

## 4.1   Complexity Analysis

The first iteration of the algorithm runs the complete Bias Envelopment in $O(n^2)$ time [1]. (Actually, a more precise bound is given by $O(m_1 + \overline{m_2})$ [2], but, as the complexities of the Incomplete Bias Envelopment procedures do not benefit at all from having edge quantities in their analysis, we prefer to write time bounds only as functions of $n$, however.) The remaining iterations take $O(nh_t)$ time each. To analyze the time complexity of the algorithm, we have to calculate

$$\sum_{t=1}^{t'} O\left(nh_{t-1}\right),$$

where $t'$ is the number of iterations in the worst case.

The value of $h_t$, obtained at the end of iteration $t$, is defined by equation 2. To calculate $t'$, we replace $h_{t'}$ for 2 and have

$$\left(1 - \frac{2}{n(n-1)}\right)^{t'} = 1 - p, \text{ and finally}$$

$$t' = \frac{\ln(1-p)}{\ln\left(1 - \frac{2}{n(n-1)}\right)}.$$

For $0 < x < 1$, it is known that

$$\ln(1-x) = -x - \frac{x^2}{2} - \frac{x^3}{3} - \cdots.$$

Consequently,

$$t' = \frac{\ln(1-p)}{-\frac{2}{n(n-1)} - \frac{1}{\Theta(n^4)}} = \ln\frac{1}{\epsilon}\Theta(n^2) = \Theta(n^2).$$

Now, we will show that $q = h(h-1)/n(n-1) \geq h^2/2n^2$. This result is useful to simplify some calculations. We have

$$\frac{n}{n-1} \cdot \frac{h-1}{h} \cdot \frac{h^2}{n^2} = \frac{h(h-1)}{n(n-1)}, \text{ and}$$

$$\frac{h-1}{h} \cdot \frac{h^2}{n^2} \leq \frac{h(h-1)}{n(n-1)}.$$

Since $h \geq 2$,

$$\frac{h^2}{2n^2} \leq \frac{h(h-1)}{n(n-1)} = q.$$

To calculate the total time complexity, we replace $h(h-1)/n(n-1)$ for $h^2/2n^2$ and $p_t$ for the fixed value $p$ in equation 1, and have

$$\left(1 - \frac{h^2}{2n^2}\right)^t \geq 1 - p,$$

$$\frac{h^2}{2n^2} \leq 1 - (1-p)^{1/t}, \text{ and}$$

$$h \leq \Theta(n)\sqrt{1 - (1-p)^{1/t}}.$$

It is well known that

$$e^x = 1 + x + \frac{x^2}{2!} + \frac{x^3}{3!} + \cdots.$$

Consequently, for $x > 1$,

$$e^{1/x} = 1 + 1/\Theta(x).$$

Using this approximation, we have

$$h \leq \Theta(n)\sqrt{1 - (1 + 1/\Theta(t))} = \Theta(n)/\Theta(\sqrt{t}).$$

The total time complexity of the algorithm is

$$\sum_{t=1}^{\Theta(n^2)} O(nh_t) = \sum_{t=1}^{\Theta(n^2)} \frac{O(n^2)}{O(\sqrt{t})} = O(n^2) \sum_{t=1}^{\Theta(n^2)} 1/O(\sqrt{t}).$$

Using elementary calculus, we have

$$\sum_{t=1}^{\Theta(n^2)} 1/O(\sqrt{t}) = O(n).$$

Consequently, the total time complexity of the algorithm is $O(n^3)$.

## 5  Conclusion

In this article, we presented two efficient algorithms for the Homogeneous Set Sandwich Problem: the first was an $O(n^{3.5})$ deterministic algorithm and the other, an $O(n^3)$ Monte Carlo one. The best results so far had been $O(n^4)$, if only functions of $n$ are used to express time complexities.

A natural step, after having developed such a Monte Carlo algorithm, is often the development of a related Las Vegas algorithm, i.e. an algorithm which *always* gives the right answer in some expected polynomial time. Unfortunately, we do not know of any short certificate for the *non-existence* of sandwich homogeneous sets in some given HSSP instance, which surely complicates matters and suggests a little more research on this issue.

## References

1. M. R. Cerioli, H. Everett, C. M. H. de Figueiredo, and S. Klein. The homogeneous set sandwich problem. *Information Processing Letters*, 67:31–35, 1998.
2. C. M. H. de Figueiredo and V. G. P. de Sá. A new upper bound for the homogeneous set sandwich problem. Technical report, COPPE/Sistemas, Universidade Federal do Rio de Janeiro, 2004.
3. C. M. H. de Figueiredo, S. Klein, and K. Vušković. The graph sandwich problem for 1-join composition is NP-complete. *Discrete Appl. Math.*, 121:73–82, 2002.
4. V. G. P. de Sá. The sandwich problem for homogeneous sets in graphs. Master's thesis, COPPE / Universidade Federal do Rio de Janeiro, May 2003. In Portuguese.
5. M. C. Golumbic. Matrix sandwich problems. *Linear Algebra Appl.*, 277:239–251, 1998.
6. M. C. Golumbic, H. Kaplan, and R. Shamir. Graph sandwich problems. *Journal of Algorithms*, 19:449–473, 1995.
7. M. C. Golumbic and A. Wassermann. Complexity and algorithms for graph and hypergraph sandwich problems. *Graphs Combin.*, 14:223–239, 1998.
8. H. Kaplan and R. Shamir. Pathwidth, bandwidth, and completion problems to proper interval graphs with small cliques. *SIAM J. Comput.*, 25:540–561, 1996.
9. H. Kaplan and R. Shamir. Bounded degree interval sandwich problems. *Algorithmica*, 24:96–104, 1999.
10. L. Lovász. Normal hypergraphs and the perfect graph conjecture. *Discrete Math.*, 2:253–267, 1972.
11. R. M. McConnell and J. Spinrad. Modular decomposition and transitive orientations. *Discrete Math.*, 201:189–241, 1999.
12. S. Tang, F. Yeh, and Y. Wang. An efficient algorithm for solving the homogeneous set sandwich problem. *Information Processing Letters*, 77:17–22, 2001.

# Efficient Implementation of the BSP/CGM Parallel Vertex Cover FPT Algorithm*

Erik J. Hanashiro[1], Henrique Mongelli[1], and Siang W. Song[2]

[1] Universidade Federal de Mato Grosso do Sul, Campo Grande, MS, Brazil.
{erik,mongelli}@dct.ufms.br
[2] Universidade de São Paulo, São Paulo, Brazil.
song@ime.usp.br

**Abstract.** In many applications NP-complete problems need to be solved exactly. One promising method to treat some intractable problems is by considering the so-called *Parameterized Complexity* that divides the problem input into a main part and a parameter. The main part of the input contributes polynomially on the total complexity of the problem, while the parameter is responsible for the combinatorial explosion. We consider the parallel FPT algorithm of Cheetham *et al.* to solve the $k$-Vertex Cover problem, using the CGM model. Our contribution is to present a refined and improved implementation. In our parallel experiments, we obtained better results and obtained smaller cover sizes for some input data. The key idea for these results was the choice of good data structures and use of the backtracking technique. We used 5 graphs that represent conflict graphs of amino acids, the same graphs used also by Cheetham *et al.* in their experiments. For two of these graphs, the times we obtained were approximately 115 times better, for one of them 16 times better, and, for the remaining graphs, the obtained times were slightly better. We must also emphasize that we used a computational environment that is inferior than that used in the experiments of Cheetham *et al.*. Furthermore, for three graphs, we obtained smaller sizes for the cover.

## 1 Introduction

In many applications, we need to solve NP-complete problems exactly. This means we need a new approach in addition to solutions such as approximating algorithms, randomization or heuristics.

One promising method to treat some intractable problems is by considering the so-called *Parameterized Complexity* [1]. The input problem is divided into two parts: the main part containing the data set and a parameter. For example, in the parameterized version of the Vertex Cover problem for a graph $G = (V, E)$, also known as the $k$-Vertex Cover, we want to determine if there is a subset in $V$ of size smaller than $k$, whose edges are incident with the vertices of this subset.

* Partially supported by FAPESP grant 1997/10982-0, CNPq grants 30.5218/03-4, 30.0482/02-7, 55.2028/02-9 and DS-CAPES.

C.C. Ribeiro and S.L. Martins (Eds.): WEA 2004, LNCS 3059, pp. 253–268, 2004.
© Springer-Verlag Berlin Heidelberg 2004

In this problem, the input is a graph $G$ (the main part) and a non-negative integer $k$ (the parameter). For simplicity, a problem whose input can be divided like this is said to be *parameterized*.

A parameterized problem is said to be *fixed-parameter tractable*, or *FPT* for short, if there is an algorithm that solves the problem in $O(f(k)n^\alpha)$ time, where $\alpha$ is a constant and $f$ is an arbitrary function [1]. If we exchange the multiplicative connective between these two contributions by an additive connective ($f(k) + n^\alpha$), the definition of FPT problems remains unchanged. The main part of the input contributes polynomially on the total complexity of the problem, while the parameter is responsible for the combinatorial explosion. This approach is feasible if the constant $\alpha$ is small and the parameter $k$ is within a tight interval. The $k$-Vertex Cover problem is one of the first problems proved to be FPT and is the focus of this work. One of the well-known FPT algorithms for this problem is the algorithm of Balasubramanian *et al.* [2], of time complexity $O(kn + 1.324718^k k^2)$, where $n$ is the size of the graph and $k$ is the maximum size of the cover. This problem is very important from the practical point of view. For example, in Bioinformatics we can use it in the analysis of multiple sequences alignment.

Two techniques are usually applied in the FPT algorithms design: the reduction to problem kernel and the bounded search tree. These techniques can be combined to solve the problem.

FPT algorithms have been implemented and they constitute a promising approach to solve problems to get the exact solution. Nevertheless, the exponential complexity on the parameter can still result in a prohibitive cost. In this article, we show how we can solve larger instances of the $k$-Vertex Cover, using the CGM parallel model.

A CGM (Coarse-Grained Multicomputer) [3] consists of $p$ processors connected by some interconnection network. Each processor has local memory of size $O(N/p)$, where $N$ is the problem size. A CGM algorithm alternates between computation and communication rounds. In a communication round each processor can send and receive a total of $O(N/p)$ data.

The CGM algorithm presented in this paper has been designed by Cheetham *et al.* [4] and requires $O(\log p)$ communication rounds. It has two phases: in the first phase a reduction to problem kernel is applied; the second phase consists of building a bounded search tree that is distributed among the processors.

Cheetham *et al.* implemented the algorithm and the results are presented in [4]. Our contribution is to present a refined and improved implementation. In our parallel experiments, we obtained better results and obtained better cover sizes for some input data. The key idea for these results was the choice of good data structures and use of the backtracking technique. We used 5 graphs that represent conflict graphs of amino acids, and these same graphs were used also by Cheetham *et al.* [4] in their experiments. For two of these graphs, the times we obtained were approximately 115 times better, for one of them 16 times better, and, for the remaining graphs, the obtained times were slightly better. We must also emphasize that we used a computational environment that is inferior than that used in the experiments of Cheetham *et al.* [4]. Furthermore, for three graphs, we obtained smaller sizes for the cover.

In the next section we introduce some important concepts. In Section 3 we present the main FPT sequential algorithms for the problem and the CGM version. In Section 4 we present the data structures and discuss the implementation and in Section 5 we show the experimental results. In Section 6 we present some conclusions.

# 2    Parameterized Complexity and $k$-Vertex Cover Problem

We present some fundamental concepts for sequential and CGM versions of the FPT algorithm for the $k$-Vertex Cover problem.

Parameterized complexity [1,5,6,7,8] is another way of dealing with the intractability of some problems. This method has been successfully used to solve problems of instance sizes that otherwise cannot be solved by other means [7].

The input of the problem is divided into two parts: the main part and the parameter. There exist some computational problems that can be naturally specified in this manner [5].

In classical computational complexity, the entire input of the problems is considered to be responsible for the combinatorial explosion of the intractable problem. In parameterized complexity, we try to understand how the different parts of the input contribute in the total complexity of the problem, and we wish to identify those input parts that cause the apparently inevitable combinatorial explosion. The main input part contributes in a polynomial way in the total complexity of the problem, while the parameter part probably contributes exponentially in the total complexity. Thus, in cases where we manage to do this, NP-complete problems can be solved by algorithms of exponential time with respect to the parameter and polynomial time with respect to the main input part. Even then we need to confine the parameter to a small, but useful, interval. In many applications, the parameter can be considered "very small" when compared to the main input part.

A *parameterizable problem* is a set $L \subseteq \Sigma^* \times \Sigma^*$, where $\Sigma$ is a fixed alphabet. If the pair $(x, y) \in L$, we call $x$ the main input part (or instance) and $y$ the parameter.

According to Downey and Fellows [1], a *parameterizable problem* $L \subseteq \Sigma^* \times \mathbb{N}^*$ is fixed parameter tractable if there exists an algorithm that, given an input $(x, y) \in L$, solves it in $O(f(k)n^\alpha)$ time, where $n$ is the size of the main input part $x$, $|x| = n$, $k$ is the size of parameter $y$, $|y| = k$, $\alpha$ is a constant independent of $k$, and $f$ is an arbitrary function.

The arbitrary function $f(k)$ of the definition is the contribution of the parameter $y$ to the total complexity of the problem. Probably this contribution is exponential. However, the main input part contributes polynomially to the total complexity of the problem. The basic assumption is that $k \ll n$ [8]. The polynomial contribution is acceptable if the constant $\alpha$ is small. However, the definition of fixed parameter tractable problem remains unchanged if we exchange the multiplicative connective between the two contributions, $f(k)n^\alpha$, by an additive connective $f(k) + n^\alpha$ [1].

The fixed parameter tractable problems form a class of problems called *FPT* (*Fixed-Parameter Tractability*). There are NP-complete problems that has been proven not to be in FPT class.

An important issue to compare the performance of FPT algorithms is the maximum size for the parameter $k$, without affecting the desired efficiency of the algorithm. This value is called *klam* and is defined as the largest value of $k$ such that $f(k) \leq U$, where $U$ is some absolute limit on the number of computational steps. Downey and Fellows [1] suggest $U = 10^{20}$. A challenge in the fixed parameter tractable problems is the design of FPT algorithms with increasingly larger values of *klam*.

Two elementary methods are used to design algorithms for fixed parameter tractable problems: reduction to problem kernel and bounded search tree. The application of these methods, in this order, as an algorithm of two phases, is the basis of several FPT algorithms. In spite of being simple algorithmic strategies, these techniques do not come into mind immediately, since they involve exponential costs relative to the parameter [6].

- **Reduction to problem kernel**: The goal is to reduce, in polynomial time, an instance $I$ of the parameterizable problem into another equivalent instance $I'$, whose size is limited by a function of the parameter $k$. If a solution of $I'$ is found, probably after an exhaustive analysis of the generated instance, this solution can be transformed into a solution of $I$. The use of this technique always results in an additive connective between the contributions $n^\alpha$ and $f(k)$ on the total complexity.
- **Bounded search tree**: This technique attempts to solve the problem through an exhaustive tree search, whose size is to be bounded by a function of the parameter $k$. Therefore, we use the instance generated by the reduction to problem kernel method in the search tree, which must be traversed until we find a node with the solution of the instance. In the worst case, we have to traverse all the tree. However, it is important to emphasize that the tree size depends only on the parameter, limiting the search space by a function of $k$.

In the parameterized version of the Vertex Cover problem, also known as $k$-Vertex Cover problem, we must have a graph $G = (V, E)$ (the instance) and a non-negative integer $k$ (the parameter). We want to answer the following question: "Is there a set $V' \subseteq V$ of vertices, whose maximum size is $k$, so that for every edge $(u, v) \in E$, $u \in V'$ or $v \in V'$?". Many other graph problems can be parameterized similarly.

The set $V'$ is not unique. An application of the vertex cover problem is the analysis of multiple sequences alignment [4]. A solution to resolve the conflicts among sequences is to exclude some of them from the sample. A conflict exists when two sequences have a score below a certain threshold. We can construct a graph, called the conflict graph, where each sequence is a vertex and an edge links two conflict sequences. Our goal is to remove the least number of sequences so that the conflict will be deleted. We thus want to find a minimum vertex cover for the conflict graph.

A trivial exact algorithm for this problem is to use brute force. In this case all the possible subsets whose size is smaller or equal to $k$ are verified to be a cover [1], where $k$ is the maximum size desired for the cover and $n$ is the number of vertices in the graph ($k \leq n$). The number of subsets with $k$ elements is $C_{n,k}$, so the algorithm to find all these subsets has time complexity of $O(n^k)$. The costly brute force approach is usually not feasible in practice.

## 3    FPT Algorithms for the $k$-Vertex Cover Problem

In this section we present FPT algorithms that solve the vertex cover problem and are used in our implementation. Initially we show the algorithm of Buss [9], responsible for the phase of reduction to problem kernel. Then we show two algorithms of Balasubramanian *et al.* [2] that present two forms to construct the bounded search tree. Finally we present the CGM algorithm of Cheetham *et al.* [4]. In all these algorithms, the input is formed by a graph $G$ and the size of the vertex cover desired (parameter $k$).

### 3.1    Algorithm of Buss

The algorithm of Buss [9] is based on the idea that all the vertices of degree greater than $k$ belong to any vertex cover for graph $G$ of size smaller or equal to $k$. Therefore, such vertices must be added to the partial cover and removed from the graph. If there are more than $k$ vertices in this situation, there is no vertex cover of size smaller or equal to $k$ for the graph $G$.

The edges incident with the vertices of degree greater than $k$ can also be removed since they are joined to at least one vertex of the cover, and the isolated vertices are removed once there are no vertices to cover. The graph produced is denominated $G'$.

From now on, our goal is to find a vertex cover of size smaller or equal to $k'$ for the graph $G'$, where $k'$ is the difference between $k$ and the number of elements of the partial vertex cover. This is only possible if there do no exist more than $kk'$ edges in $G'$. This is because $k'$ vertices can cover at most $kk'$ edges in the graph, since the vertices of $G'$ have degree bounded by $k$. Furthermore, if we do not have more than $kk'$ edges in $G'$, nor isolated vertices, we can conclude that there are at most $2kk'$ vertices in $G'$. As $k'$ is at most $k$, the size of the graph $G'$ is $O(k^2)$.

Given the adjacency list of the graph, the steps described until here spend $O(kn)$ time and form the basis for the reduction to problem kernel phase. Observe that graph $G$ is reduced, in polynomial time, to an equivalent graph $G'$, whose size is bounded by a function of the parameter $k$. The kernellization phase as described is used in the algorithms presented in the next subsection.

To determine finally if there exists or not a vertex cover for $G'$ of size smaller or equal to $k'$, the algorithm of Buss [9] executes a brute force algorithm. If a vertex cover for $G'$ of size smaller or equal to $k'$ exists, these vertices and the vertices of degree greater than $k$ form a vertex cover for $G$ of size smaller or equal to $k$. The algorithm of Buss [9] spends a total time of $O(kn + (2k^2)^k k^2)$.

## 3.2    Algorithms of Balasubramanian *et al.*

The algorithms of Balasubramanian *et al.* [2] execute initially the phase of reduction to problem kernel based on the algorithm of Buss [9]. In the second phase, a bounded search tree is generated. The two options to generate the bounded search tree are shown in Balasubramanian *et al.* [2] and described below as Algorithm B1 and Algorithm B2. In both cases, we search the tree nodes exhaustively for a solution of the vertex cover problem, by depth first tree traversal. The difference between the two algorithms is the form we choose the vertices to be added to the partial cover and, consequently, the format of such a tree.

Each node of the search tree stores a partial vertex cover and a reduced instance of the graph. This partial cover is composed of the vertices that belong to the cover. The reduced instance is formed by the graph resulting from the removal of the vertices of $G$ that are in the partial cover, as well as the edges incident with them and any isolated vertex. We call this graph $G''$ and an integer $k''$ that is the maximum desired size for the vertex cover of $G''$. The root of the search tree, for example, represents the situation after the method of reduction to problem kernel. In other words, in the partial cover we have the vertices of degree greater than $k$ and the instance $\langle G', k' \rangle$.

The edges of the search tree represents the several possibilities of adding vertices to the existing partial cover. Notice that the son of a tree node has more elements in the partial vertex cover and a graph with less nodes and edges than its parent, since every time a vertex is added to the partial cover, we remove it from the graph, together with the incident edges and any isolated vertices. We actually do not generate all the nodes before the depth first tree traversal. We only generate a node of the bounded search tree when this node is visited.

The search tree has the following property: for each existing vertex cover for graph $G$ of size smaller or equal to $k$, there exists a corresponding tree node with a resulting empty graph and a vertex cover (not necessarily the same) of size smaller or equal to $k$. However, if there is no vertex cover of size smaller or equal to $k$ for graph $G$, then no tree node possesses a resulting empty graph. Actually the growth of the search tree is interrupted when the node has a partial vertex cover of size smaller or equal to $k$ or a resulting empty graph (case in which we find a valid vertex cover for graph $G$). Notice that this bounds the size of the tree in terms of the parameter $k$. Therefore, in the worst case, we have to traverse all the search tree to determine if there exists or not a vertex cover of size smaller or equal to $k$ for graph $G$.

Given the adjacency list of the graph, we spend $O(m)$ time in each node, where $m$ is the number of vertices of the current graph. Therefore, if $C(k)$ is the number of nodes of the search tree, then the time spent to traverse all the tree is

$O(mC(k))$. Recall that the root node of the search tree, whose size is bounded by $O(k^2)$, stores the resulting graph of the phase of reduction to problem kernel.

**Algorithm B1.** In this algorithm, the choice of the vertices of $G''$ to be added to the partial cover in any tree node is done according to a path generated from any vertex $v$ of $G''$ that passes through at least three edges.

If this path has size one or two, then we add the neighbor of the node of degree one to the partial cover, remove their incident edges and any isolated vertices. This new graph instance with the new partial cover is kept in the same node of the bounded search tree and the Algorithm B1 is applied again in this node.

If this path is a simple path of size three, passing by vertices $v$, $v_1$, $v_2$ and $v_3$, any vertex cover must contain $\{v, v_2\}$ or $\{v_1, v_2\}$ or $\{v_1, v_3\}$. If the path is a simple cycle of size three, passing by vertices $v$, $v_1$, $v_2$ and $v$, any vertex cover must contain $\{v, v_1\}$ or $\{v_1, v_2\}$ or $\{v, v_2\}$. In both cases, the tree node is ramified into three three sons to add one of the three pairs of suggested vertices. We can then go to the next node of the tree, recalling the depth first traversal.

Notice that this algorithm generates a tertiary search tree and that at each tree level the partial cover increases by at least two vertices. The Algorithm B1 spends $O(kn + (\sqrt{3})^k k^2)$ time to solve the $k$-Vertex Cover problem.

**Algorithm B2.** In this algorithm, the choice of vertices of $G''$ to be added to the partial cover in any node of the tree is done according to five cases by considering the degree of the vertices of the resulting graph. We deal first with the vertices of degree 1 (Case 1), then with vertices of degree 2 (Case 2), then with vertices of degree 5 or more (Case 3), then with vertices of degree 3 (Case 4) and, finally, with vertices of degree 4 (Case 5).

We use the following notation. $N(v)$ represents the set of vertices that are neighbors of vertices $v$ and $N(S)$ represents the set $\bigcup_{v \in S} N(v)$.

In Case 1, if there exists a vertex $v$ of degree 1 in the graph, then we create a new son to add $N(v)$ to the partial cover.

In Case 2, if there exists a vertex $v$ of degree 2 in the graph, then we can have three subcases, to be tested in the following order. Let $x$ and $y$ be the neighbors of $v$. In Subcase 1, if there exists an edge between $x$ and $y$, then we create a new son to add $N(v)$ to the partial cover. In Subcase 2, if $x$ and $y$ have at least two neighbors different from $v$, then we ramify the node of the tree into two sons to add $N(v)$ and $N(\{x, y\})$ to the partial cover. In Subcase 3, if $x$ and $y$ share an only neighbor $a$ different from $v$, then we create a new son to add $\{v, a\}$.

In Case 3, if there exists a vertex of degree 5 or more in the graph, then we ramify the node of the tree into two sons to add $v$ and $N(v)$ to the partial cover.

If none of the three previous cases occurs, then we have a 3 or 4-regular graph. In case 4, if there exists a vertex $v$ of degree 3, then we can have four subcases, to be treated in the following order. Let $x$, $y$ and $z$ be the neighbors of $v$. In Subcase 1, if there exists an edge between two neighbors of $v$, say $x$ and $y$, the we ramify the node of the tree into two sons to add $N(v)$ and $N(z)$ to the partial cover. In Subcase 2, if a pair of neighbors of $v$, say $x$ and $y$, share another common neighbor $a$ (but different from $v$), then we ramify the node of the tree into two sons to add $N(v)$ and $\{v, a\}$ to the partial cover. In Subcase 3, if a neighbor of $v$, say $x$, has at least three neighbors different from $v$, then we ramify the node of the tree into three sons to add $N(v)$, $N(x)$ and $x \bigcup N(\{y, z\})$ to the partial cover. In Subcase 4, the neighbors of $v$ have exactly two private neighbors, not considering vertex $v$ proper. Let $x$ be a neighbor of $v$ and let $a$ and $b$ be the neighbors of $x$, then we ramify the node of the tree into three sons to add $N(v)$, $\{v, a, b\}$ and $N(\{y, z, a, b\})$ to the partial cover.

In Case 5, we have a 4-regular graph and we can have three subcases, to be tested in the following order. Let $v$ be a vertex of the graph and $x$, $y$, $z$ and $w$ its neighbor vertices. In Subcase 1, if there exists an edge between two neighbors of $v$, say $x$ and $y$, then we ramify the node of the tree into three sons to add $N(v)$, $N(z)$ and $z \bigcup N(w)$ to the partial cover. In Subcase 2, if three neighbors of $v$, say $x$, $y$ and $z$, share common neighbor $a$, then we ramify the node of the tree into two sons to add $N(v)$ and $(v, a)$ to the partial cover. In Subcase 3, if each of the neighbors of $v$ has three neighbors different from $v$, then we ramify the node of the tree into four sons to add $N(v)$, $N(y)$, $y \bigcup N(w)$ and $\{y, w\} \bigcup N(\{x, z\})$ to the partial cover.

Contrary to Algorithm B1, a node in the search tree can be ramified into two, three or four sons, and the partial cover can increase up to 8 vertices, depending on the selected case. Algorithm B2 spends $O(kn + 1.324718^k k^2)$ time to solve the $k$-Vertex Cover problem.

## 3.3    Algorithm of Cheetham *et al.*

The CGM algorithm proposed by Cheetham *et al.* [4] to solve the $k$-Vertex Cover problem parallelizes both phases of an FPT algorithm, reduction to problem kernel and bounded search tree. Previous works designed for the PRAM model parallelize only the method of reduction to problem kernel [4]. However, as the implementations of FPT algorithms usually spends minutes in the reduction to problem kernel and hours, or maybe even days in the bounded search tree, the parallelization of the bounded search tree designed in the CGM algorithm is an important contribution.

The CGM algorithm of Cheetham *et al.* [4] solves even larger instances of the $k$-Vertex Cover problem than those solved by sequential FPT algorithms. The implementation of this algorithm can solve instances with $k \geq 400$ in less than 75 minutes of processing time. It is important to emphasize that the $k$-Vertex Cover is considered well solved for instances of $k \leq 200$ (sequential FPT algorithms) [7]. Not only there is a considerable increase in the parameter $k$, it is important to recall that the time of a FPT algorithm grows exponentially in relation to $k$.

The phase of reduction to problem kernel is parallelized through a parallel integer sorting. The $p$ processors that participate in the parallel sort are identified as $P_i$, $0 \leq i \leq p - 1$. To identify vertices of the graph with degree larger than $k$, the edges are sorted by the label of the vertex they are incident with through deterministic sample sort [10], that require $O(1)$ parallel integer sorts, i.e. in constant time. The partial vertex cover (vertices with degree larger than $k$) and the instance $\langle G', k' \rangle$ is sent to all the processors.

The basic idea of the parallelization of the phase of bounded search tree is to generate a complete tertiary tree $T$ with $O(\log_3 p)$ tree levels and $p$ leaf nodes $(\gamma_0...\gamma_{p-1})$. Each one of these $p$ leaf nodes is then assigned to one of the $p$ processors, that search locally for a solution in the subtree generated from the leaf node $\gamma_i$, as shown in Fig. 1. A detailed description of this phase is presented in the following.

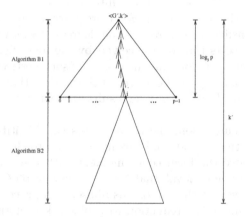

**Fig. 1.** A processor $P_i$ computes the unique path in $T$ from the root to leaf $\gamma_i$, using the Algorithm B1. Then, $P_i$ computes the entire subtree below $\gamma_i$, using the Algorithm B2.

- Consider the tertiary search tree $T$. Each processor $P_i$, $0 \le i < p$, starts this phase with the instance obtained at the previous phase ($\langle G', k' \rangle$), and uses Algorithm B1 to compute the unique path in $T$ from the tree root to the leaf node $\gamma_i$. Let $\langle G_i'', k_i'' \rangle$, be the instance computed at the leaf node $\gamma_i$.
- Each processor $P_i$, $0 \le i < p$, searches locally for a solution in the subtree generated from $\langle G_i'', k_i'' \rangle$, based on Algorithm B2. Processor $P_i$ chooses a son of the node at random and expands it until a solution is found or the partial cover is larger than $k$. If a solution is not found, return to the subtree to get a still not explored son, until all the subtree is traversed. If a solution is encountered, the other processors are notified to interrupt.

In the algorithm of Cheetham et al. [4], the major part of the computational work occurs when each processor $P_i$, $0 \le i < p$, computes locally the search tree from $\langle G_i'', k_i'' \rangle$, where Algorithm B2 is used. As all the $p$ subtrees are traversed simultaneously, it is possible that the parallel algorithm visits nodes that the sequential algorithm would not visit.

# 4    Implementation Details

In this section we present some implementation details of the parallel FPT algorithm and discuss the data structures utilized in our implementation. We use C/C++ and the MPI communication library.

The program receives as input a text file describing a graph $G$ by its adjacency list and an integer $k$ that determines the maximum size for the vertex cover desired. Let $n$ be the number of vertices and $m$ the number of edges of graph $G$ and $p$ the number of processors to run the program.

At the beginning of the reduction to problem kernel phase, the input adjacency list of graph $G$ is transformed into a list of corresponding edges and distributed among the $p$ processors. Each processor $P_i$, $0 \leq i < p$, receives $m/p$ edges and is responsible for controlling the degrees of $n/p$ vertices.

Each processor sorts the edges received by the identifier of the first vertex they are incident with, and obtains the degree of such vertices. Notice it is possible for a processor to compute the degree of the vertices that are of responsibility for another processor. In this case, the results are sent to the corresponding processor.

After this communication, the $p$ processors can identify the local vertices with degree larger than $k$ and send this information to the others, so that each processor can remove the local edges incident with these vertices. All the remaining edges after the removal, that form the new graph $G'$, are sent to all the processors. In this way, at the end of this phase, each processor has the instance generated by the method of reduction to problem kernel and the partial cover (vertices of degree smaller than $k$), that is, the root of the bounded search tree. The $p$ processors transform the list of edges corresponding to graph $G'$ again into an adjacency list, that will be used in the next phase.

The resulting adjacency list from the reduction to problem kernel is implemented as a doubly linked list of vertices. Each node $x$ of this list of vertices contains a pointer to a doubly linked list of pointers, whose elements represent all the edges incident with $x$, that we denote, for simplicity, by the list of edges of $x$. Each node of the list of edges of $x$ points to the node of the list of vertices that contains the other extreme of the edge. In spite of the fact that graph is not a directed graph, each edge is represented twice in distinct lists of edges. Thus each node of the list of edges contains also a pointer to its other representation. In Fig. 2 we present an example of a graph and the data structure to store it.

The insertion of a new element in the list of vertices takes $O(n)$ time, since it is necessary to check if such elements already exist. In the list of edges, the insertion of a new element, in case it does not yet exist, results in the insertion of elements in the two lists of edges incident with its two extremes and also takes $O(n)$ time.

The removal of a vertex or an edge is a rearrangement of the pointers of previous and next elements of the list. They are not effectively deallocated from memory, they are only removed from the list. Notice that the edges incident with it are removed automatically with the vertex. However, we still have to remove the other representation. As each edge has a pointer to its other representation,

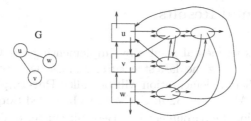

**Fig. 2.** The data structure used to store the graph $G$.

we spend $O(1)$ time to remove it from the list of edges of the other vertex. Therefore, we spend $O(k)$ time to remove a vertex from the list, since the vertices of the graph have degree bounded by $k$. In our implementation, we store in memory only the data relative to the node of the bounded search tree being worked on.

Since we use depth first traversal in the bounded search tree, we need to store some information that enables us to go up the tree and recover a previous instance of the graph. Thus our program uses the *backtracking* technique. Such information is stored in a stack of pointers to removed vertices and edges. Adding an element in the stack takes $O(1)$ time. Removing an element from the stack and put it back in the graph also takes $O(1)$ time, since the removed vertex or edge has pointers to the previous and next elements in the list.

The partial vertex cover is also a stack of pointers to vertices known to be part of the cover. To add or remove an element from the cover takes $O(1)$ time.

At the beginning of the bounded search tree phase, all the $p$ processors contain the instance $(\langle G', k' \rangle)$ and the partial vertex cover resulting from the phase of reduction to problem kernel. As seen in Section 3.3, there exists a bounded tertiary complete search tree $T$ with $p$ leaf nodes. Each processor $P_i$, $0 \leq i < p$, uses Algorithm B1, generates the unique path in tree $T$ from the root to the leaf node $\gamma_i$ of tree $T$. Then, each processor $P_i$ applies Algorithm B2 in the subtree whose root is the leaf node $\gamma_i$, until finding a solution or finishing the traversal.

In Algorithm B1, we search a path that starts at a vertex and passes through at most three edges. In our implementation, this initial vertex is always the first vertex of the list and, therefore, the same tree $T$ is generated in all the executions of the program.

In the implementation of Algorithm B2, to obtain constant time for the selection of a vertex for the cases of this algorithm, we use 6 auxiliary lists of pointers to organize the vertices of the graph according to its degree (0, 1, 2, 3, 4 and 5 or more). Furthermore, each vertex of the graph also has a pointer to its representative in the list of degrees, therefore in any change of degree of a vertex implies $O(1)$ time to change it in the list of degrees.

## 5   Experimental Results

We present the experimental results by implementing the CGM algorithm of Cheetham et al. [4], using the data structures and the description of the previous section. Our parallel implementation will be called **Par-Impl**. Furthermore, we also implemented Algorithm B2 in C/C++, to be called **Seq-Impl**.

The computational environment is a Beowulf cluster of 64 Pentium III processors, with 500 MHz and 256 MB RAM memory each processor. All the nodes are interconnected by a Gigabit Ethernet switch. We used Linux Red Hat 7.3 with g++ 2.96 and MPI/LAM 6.5.6.

The sequential times were measured as wall clock times in seconds, including reading input data, data structures deallocation and writing output data. The parallel times were also measured as wall clock time between the start of the first processor and termination of the last process, including I/O operations and data structures deallocation.

In our experiments we used conflict graphs that were kindly provided by Professor Frank Dehne (Carleton University). These graphs represent sequences of amino acid collected from the NCBI database. They are Somatostatin, WW, Kinase, SH2 (src-homology domain 2) and PHD (pleckstrin homology domain). The Table 1 shows a summary of the characteristics of these graphs (name, number of vertices, number of edges, size of desired cover and size of the cover to search for after the reduction to problem kernel).

**Table 1.** Sequences and corresponding graphs and cover sizes used in experiments.

| Graph | $|V|$ | $|E|$ | k | k' |
|---|---|---|---|---|
| Kinase | 647 | 113122 | 495 | 391 |
| PHD | 670 | 147054 | 601 | 600 |
| SH2 | 730 | 95463 | 461 | 397 |
| Somatostatin | 559 | 33652 | 272 | 254 |
| WW | 425 | 40182 | 322 | 318 |

In Fig. 3 we compare the times obtained by executing Seq-Impl and Par-Impl in a single processor (3 virtual processors) and Par-Impl in 27 processors. To run Par-Impl in a single processor we used MPI/LAM simulation mode, that simulates $p$ virtual processors as independent processes on the same physical processor. The time obtained by Par-Impl in a single processor is the sum of the wall clock times of the individual processes plus the overhead created by their communication. The tests were carried out for the graphs PHD, Somatostatin and WW. These input data were chosen because their sequential times are reasonable. To obtain the averages, we ran Seq-Impl 10 times for each data set and Par-Impl 30 times for each data set. In spite of the fact we are using a single processor to run the parallel implementation, the time was significantly much smaller. This is justified by the fact of having more initial distinct points in the

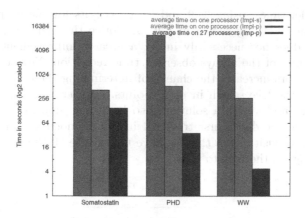

**Fig. 3.** Comparison of sequential and parallel times.

bounded search tree, such that from one of them we can find a path that takes to the cover more quickly.

In Fig. 4 we show the average of the parallel times obtained in 27 processors. Our parallel implementation can solve problem instances of size $k \geq 400$ in less than 3 minutes. For example, graph PHD ($k = 601$) can be solved in less than 1 minute. Notice that $k$-Vertex Cover problem is considered well solved for instances of $k \leq 200$ by sequential FPT algorithms [7]. It is important to emphasize that the time of FPT algorithm grows exponentially in relation to $k$. Again we use 30 time samples to get the average time. Observe the times obtained and the Table 1. We see that the parallel wall clock times do not strictly increase with either $k$ or $k'$. This makes us conclude that the graph structure is the responsible for the running time.

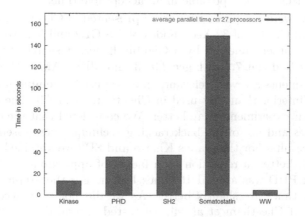

**Fig. 4.** Average wall clock times for the data sets on 27 real processors.

The parallel times, using 3, 9 and 27 processors for the graphs PHD, Somatostatin and WW are shown in Fig. 5. Notice the increase in the number of processors does not necessarily imply a greater improvement on the average time, in spite of the always observed time reduction. Nevertheless, the use of more processors increases the chance of determining the cover more quickly, since we start the tree search in more points. Furthermore, it seems that the number of tree nodes with a solution also has some influence on the running times. As we do the depth first traversal in the bounded search tree, a wrong choice of a son to visit means that we have to traverse all the subtree of the son before choosing another son to visit.

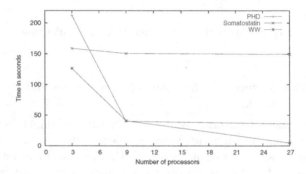

**Fig. 5.** Average wall clock times on 3, 9 and 27 processors for PHD, Somatostatin and WW.

For the graphs PHD, SH2, Somatostatin and WW we could guarantee, in less than 75 minutes, the non existence of covers smaller than that determined by the parallel algorithm, confirming the minimality of the values obtained. For this, all the possible nodes of the bounded search tree were generated. For the graph Kinase this was not possible in an acceptable time.

Our results were compared with those presented in Cheetham et al. [4], who used a Beowulf Cluster of 32 Xeon nodes of 1.8 GHz and 512 MB of RAM. All the nodes were interconnected by a Gigabit Ethernet switch. Every node was running Linux Red Hat 7.2 with gcc 2.95.3 and MPI/LAM 6.5.6.

Our experiments are very relevant, since we used a computational platform that is much inferior than that used in Cheetham et al. [4]. The parallel times obtained in our experiments were better. We considered that the choice of good data structures and use of the backtracking technique were essential to obtain our relevant results. For the graphs Kinase and SH2 we obtained parallel times that are much better, a reduction by a factor of approximately 115. The time for the graph PHD was around 16 times better. For the graphs Somatostatin and WW the times are slightly better. As we did not have access to the implementation of Cheetham et al. [4], we tested several data structures in our implementation. In the final version we used that implementation that gave the best performance, together with the backtracking technique. More details can be found in Hanashiro [11].

Furthermore, the size of the covers obtained were smaller for the following graphs: Kinase (from 497 to 495), PHD (from 603 to 601) and Somatostatin (from 273 to 272). It is important to emphasize that the reduction in the size of the cover implies the reduction on the universe of existing solutions in the bounded search tree, which in turn gives rise to an increase in the running time.

# 6   Conclusion

FPT algorithms constitute an alternative approach to solve NP-complete problems for which it is possible to fix a parameter that is responsible for the combinatorial explosion. The use of parallelism improve significantly the running time of the FPT algorithms, as in the case of the $k$-Vertex Cover problem.

In the implementation of the presented CGM algorithm, the choice of the data structures and the use of the backtracking technique were essential to obtain the relevant experimental results. During the program design, we utilized several alternative data structures and their results were compared with those of Cheetham et al. [4]. Then we chose the design that obtained the best performance. Unfortunately we did not have access to the implementation of Cheetham et al. to compare it with our code.

We obtained great improvements on the running times as compared to those of Cheetham et al. [4]. This is more significant if we take into account the fact that we used an inferior computational environment. Furthermore, we improved the values for the minimum cover and guaranteed the minimality for some of the graphs.

The speedups of our implementation with that of Cheetham et al. [4] vary very much. The probable cause of this may lie in the structures of the input graphs, and also in the number of solutions and how these solutions are distributed among the nodes of the bounded search tree.

For the input used, only for the Thrombin graph we did not obtain better average times, as compared to those of Cheetham et al. [4]. To improve the results, we experimented two other implementations, by introducing randomness in some of the choices. With these modifications, in more experiments we get lower times for the Thrombin graph, though we did not improve the average. For some of the graphs, the modification increases the times obtained, and does not justify its usage.

**Acknowledgments.** The authors would like to thank Prof. Frank Dehne (Carleton University) who kindly provided us the conflict graphs, Prof. Edson N. Cáceres (UFMS) for his assistance, the Institute of Computing/UNICAMP for giving the permission to use the machines, and finally the referees for their helpful comments.

# References

1. Downey, R.G., Fellows, M.R.: Parameterized Complexity. Springer-Verlag (1998)
2. Balasubramanian, R., Fellows, M.R., Raman, V.: An improved fixed-parameter algorithm for vertex cover. Information Processing Letters **65** (1998) 163–168
3. Dehne, F., Fabri, A., Rau-Chaplin, A.: Scalable parallel computational geometry for coarse grained multicomputers. In: Proceedings of the ACM 9th Annual Computational Geometry. (1993) 298–307
4. Cheetham, J., Dehne, F., Rau-Chaplin, A., Stege, U., Taillon, P.: Solving large FPT problems on coarse grained parallel machines. Journal of Computer and System Sciences **4** (2003) 691–706
5. Downey, R.G., Fellows, M.R.: Fixed-parameter tractability and completeness I: Basic results. SIAM J. Comput. **24** (1995) 873–921
6. Downey, R.G., Fellows, M.R.: Parameterized complexity after (almost) 10 years: Review and open questions. In: Combinatorics, Computation & Logic, DMTCS'99 and CATS'99. Volume 21, número 3., Australian Comput. Sc. Comm., Springer-Verlag (1999) 1–33
7. Downey, R.G., Fellows, M.R., Stege, U.: Parameterized complexity: A framework for systematically confronting computational intractability. In: Contemporary Trends in Discrete Mathematics: From DIMACS and DIMATIA to the Future. Volume 49 of AMS-DIMACS Proceedings Series. (1999) 49–99
8. Niedermeier, R.: Some prospects for efficient fixed parameter algorithms. In: Conf. on Current Trends in Theory and Practice of Informatics. (1998) 168–185
9. Buss, J.F., Goldsmith, J.: Nondeterminism within P. SIAM J. Comput. **22** (1993) 560–572
10. Chan, A., Dehne, F.: A note on course grained parallel integer sorting. In: 13th Annual Int. Symposium on High Performance Computers. (1999) 261–267
11. Hanashiro, E.J.: O problema da $k$-Cobertura por Vértices: uma implementação FPT no modelo CGM. Master's thesis, Universidade Federal de Mato Grosso do Sul (2004)

# Combining Speed-Up Techniques for Shortest-Path Computations*

Martin Holzer, Frank Schulz, and Thomas Willhalm

Universität Karlsruhe, Fakultät für Informatik, Postfach 6980, 76128 Karlsruhe,
Germany.
{mholzer,fschulz,willhalm}@ira.uka.de

**Abstract.** Computing a shortest path from one node to another in a
directed graph is a very common task in practice. This problem is classi-
cally solved by Dijkstra's algorithm. Many techniques are known to speed
up this algorithm heuristically, while optimality of the solution can still
be guaranteed. In most studies, such techniques are considered individually. The focus of our work is the *combination* of speed-up techniques
for Dijkstra's algorithm. We consider all possible combinations of four
known techniques, namely *goal-directed search, bi-directed search, multi-
level approach*, and *shortest-path bounding boxes*, and show how these
can be implemented. In an extensive experimental study we compare
the performance of different combinations and analyze how the techniques harmonize when applied jointly. Several real-world graphs from
road maps and public transport and two types of generated random
graphs are taken into account.

## 1 Introduction

We consider the problem of (repetitively) finding single-source single-target
shortest paths in large, sparse graphs. Typical applications of this problem include route planning systems for cars, bikes, and hikers [1,2] or scheduled vehicles like trains and buses [3,4], spatial databases [5], and web searching [6].
Besides the classical algorithm by Dijkstra [7], with a worst-case running time of
$\mathcal{O}(m + n \log n)$ using Fibonacci heaps [8], there are many recent algorithms that
solve variants and special cases of the shortest-path problem with better running
time (worst-case or average-case; see [9] for an experimental comparison, [10] for
a survey and some more recent work [11,12,13]).

It is common practice to improve the running time of Dijkstra's algorithm
heuristically while correctness of the solution is still provable, i.e., it is guaranteed
that a shortest path is returned but not that the modified algorithm is faster.
In particular, we consider the following four speed-up techniques:

---

* This work was partially supported by the Human Potential Programme of the European Union under contract no. HPRN-CT-1999-00104 (AMORE) and by the DFG
under grant WA 654/12-1.

C.C. Ribeiro and S.L. Martins (Eds.): WEA 2004, LNCS 3059, pp. 269–284, 2004.

**Goal-Directed Search** modifies the given edge weights to favor edges leading towards the target node [14,15]. With graphs from timetable information, a speed-up in running time of a factor of roughly 1.5 is reported in [16].

**Bi-Directed Search** starts a second search backwards, from the target to the source (see [17], Section 4.5). Both searches stop when their search horizons meet. Experiments in [18] showed that the search space can be reduced by a factor of 2, and in [19] it was shown that combinations with the goal-directed search can be beneficial.

**Multi-Level Approach** takes advantage of hierarchical coarsenings of the given graph, where additional edges have to be computed. They can be regarded as distributed to multiple levels. Depending on the given query, only a small fraction of these edges has to be considered to find a shortest path. Using this technique, speed-up factors of more than 3.5 have been observed for road map and public transport graphs [20]. Timetable information queries could be improved by a factor of 11 (see [21]), and also in [22] good improvements for road maps are reported.

**Shortest-Path Bounding Boxes** provide a necessary condition for each edge, if it has to be respected in the search. More precisely, the bounding box of all nodes that can be reached on a shortest path using this edge is given. Speed-up factors in the range between 10 and 20 can be achieved [23].

Goal-directed search and shortest-path bounding boxes are only applicable if a layout of the graph is provided. Multi-level approach and shortest-path bounding boxes both require a preprocessing, calculating additional edges and bounding boxes, respectively. All these four techniques are tailored to Dijkstra's algorithm. They crucially depend on the fact that Dijkstra's algorithm is label-setting and that it can be terminated when the destination node is settled.

The focus of this paper is the *combination* of the four speed-up techniques. We first show that, with more or less effort, all $2^4 = 16$ combinations can be implemented. Then, an extensive experimental study of their performance is provided. Benchmarks were run on several real-world and generated graphs, where operation counts as well as CPU time were measured.

The next section contains, after some definitions, a description of the speed-up techniques and shows how to combine them. Section 3 presents the experimental setup and data sets for our statistics, and the belonging results are given in Section 4. Section 5, finally, gives some conclusions.

## 2    Definitions and Problem Description

### 2.1    Definitions

A directed simple *graph* $G$ is a pair $(V, E)$, where $V$ is the set of nodes and $E \subseteq V \times V$ the set of edges in $G$. Throughout this paper, the number of nodes, $|V|$, is denoted by $n$ and the number of edges, $|E|$, by $m$.

A *path* in $G$ is a sequence of nodes $u_1, \ldots, u_k$ such that $(u_i, u_{i+1}) \in E$ for all $1 \le i < k$. Given non-negative edge lengths $l : E \to \mathbb{R}_0^+$, the *length* of a path

$u_1, \ldots, u_k$ is the sum of weights of its edges, $\sum_{i=1}^{k-1} l(u_i, u_{i+1})$. The *(single-source single-target) shortest-path problem* consists of finding a path of minimum length from a given source $s \in V$ to a target $t \in V$.

A graph *layout* is a mapping $L : V \to \mathbb{R}^2$ of the graph's nodes to the Euclidean plane. For ease of notation, we will identify a node $v \in V$ with its location $L(v)$ in the plane. The Euclidean distance between two nodes $u, v \in V$ is then denoted by $d(u, v)$.

## 2.2   Speed-Up Techniques

Our base algorithm is Dijkstra's algorithm using Fibonacci heaps as priority queue. In this section, we provide a short description of the four speed-up techniques, whose combinations are discussed in the next section.

**Goal-Directed Search.** This technique uses a potential function on the node set. The edge lengths are modified in order to direct the graph search towards the target. Let $\lambda$ be such a potential function and $l(e)$ be the length of $e$. The new length of an edge $(v, w)$ is defined to be $\bar{l}(v, w) := l(v, w) - \lambda(v) + \lambda(w)$. The potential must fulfill the condition that for each edge $e$, its new edge length $\bar{l}(e)$ is non-negative, in order to guarantee optimal solutions.

In case edge lengths are Euclidean distances, the Euclidean distance $d(u, t)$ of a node $u$ to the target $t$ is a valid potential, due to the triangular inequality. Otherwise, a potential function can be defined as follows: let $v_{max}$ denote the maximum "edge-speed" $d(u, v)/l(e)$, over all edges $e = (u, v)$. The potential of a node $u$ can now be defined as $\lambda(u) = d(u, t)/v_{max}$.

**Bi-Directed Search.** The bi-directed search simultaneously applies the "normal", or forward, variant of the algorithm, starting at the source node, and a so-called reverse, or backward, variant of Dijkstra's algorithm, starting at the destination node. With the reverse variant, the algorithm is applied to the reverse graph, i.e., a graph with the same node set $V$ as that of the original graph, and the reverse edge set $\overline{E} = \{(u, v) \mid (v, u) \in E\}$. Let $d_f(u)$ be the distance labels of the forward search and $d_b(u)$ the labels of the backward search, respectively. The algorithm can be terminated when one node has been designated to be permanent by both the forward and the reverse algorithm. Then the shortest path is determined by the node $u$ with minimum value $d_f(u) + d_b(u)$ and can be composed of the one from the start node to $u$, found by the forward search, and the edges reverted again on the path from the destination to $u$, found by the reverse search.

**Multi-Level Approach.** This speed-up technique requires a preprocessing step at which the input graph $G = (V, E)$ is decomposed into $l + 1$ ($l \geq 1$) levels and enriched with additional edges representing shortest paths between certain nodes. This decomposition depends on subsets $S_i$ of the graph's node set for each level, called selected nodes at level $i$: $S_0 := V \supseteq S_1 \supseteq \ldots \supseteq S_l$. These node sets can be determined on diverse criteria; with our implementation, they consist of the desired numbers of nodes with highest degree in the graph, which has turned out to be an appropriate criterion [20].

There are three different types of edges being added to the graph: *upward edges*, going from a node that is not selected at one level to a node selected at that level, *downward edges*, going from selected to non-selected nodes, and *level edges*, passing between selected nodes at one level. The weight of such an edge is assigned the length of a shortest path between the end-nodes.

To find a shortest path between two nodes, then, it suffices for Dijkstra's algorithm to consider a relatively small subgraph of the "multi-level graph" (a certain set of upward and of downward edges and a set of level edges passing at a maximal level that has to be taken into account for the given source and target nodes).

**Shortest-Path Bounding Boxes.** This speed-up technique requires a preprocessing computing all shortest path trees. For each edge $e \in E$, we compute the set $S(e)$ of those nodes to which a shortest path starts with edge $e$. Using a given layout, we then store for each edge $e \in E$ the bounding box of $S(e)$ in an associative array $BB$ with index set $E$.

It is then sufficient to perform Dijkstra's algorithm on the subgraph induced by the edges $e \in E$ with the target node included in $BB[e]$. This subgraph can be determined on the fly, by excluding all other edges in the search. (One can think of bounding boxes as traffic signs which characterize the region that they lead to.)

A variation of this technique has been introduced in [16], where as geometric objects angular sectors instead of bounding boxes were used, for application to a timetable information system. An extensive study in [23] showed that bounding boxes are the fastest geometric objects in terms of running time, and competitive with much more complex geometric objects in terms of visited nodes.

## 2.3   Combining the Speed-Up Techniques

In this section, we enlist for every pair of speed-up techniques how we combined them. The extension to a combination of three or four techniques is straight forward, once the problem of combining two of them is solved.

**Goal-Directed Search and Bi-Directed Search.** Combining goal-directed and bi-directed search is not as obvious as it may seem at first glance. [18] provides a counter-example for the fact that simple application of a goal-directed search forward and backward yields a wrong termination condition. However, the alternative condition proposed there has been shown in [19] to be quite inefficient, as the search in each direction almost reaches the source of the other direction. This often results in a slower algorithm.

To overcome these deficiencies, we simply use the very same edge weights $\bar{l}(v,w) := l(v,w) - \lambda(v) + \lambda(w)$ for both the forward and the backward search. With these weights, the forward search is directed to the target $t$ and the backward search has no preferred direction, but favors edges that are directed towards $t$. This should be (and indeed is) faster than each of the two speed-up techniques. This combination computes a shortest path, because a shortest $s$-$t$-path is the same for given edge weights $l$ and edge weights modified according to goal-directed search, $\bar{l}$.

**Goal-Directed Search and Multi-Level Approach.** As described in Section 2.2, the multi-level approach basically determines for each query a subgraph of the multi-level graph, on which Dijkstra's algorithm is run to compute a shortest path. The computation of this subgraph does not involve edge lengths and thus goal-directed search can be simply performed on it.

**Goal-Directed Search and Shortest-Path Bounding Boxes.** Similar to the multi-level approach, the shortest-path bounding boxes approach determines for a given query a subgraph of the original graph. Again, edge lengths are irrelevant for the computation of the subgraph and goal-directed search can be applied offhand.

**Bi-Directed Search and Multi-Level Approach.** Basically, bi-directed search can be applied to the subgraph defined by the multi-level approach. In our implementation, that subgraph is computed on the fly during Dijkstra's algorithm: for each node considered, the set of necessary outgoing edges is determined. If applying bi-directed search to the multi-level subgraph, a symmetric, backward version of the subgraph computation has to be implemented: for each node considered in the backward search, the incoming edges that are part of the subgraph have to be determined.

**Bi-Directed Search and Shortest-Path Bounding Boxes.** In order to take advantage of shortest-path bounding boxes in both directions of a bi-directional search, a second set of bounding boxes is needed. For each edge $e \in E$, we compute the set $S_b(e)$ of those nodes from which a shortest path ending with $e$ exists. We store for each edge $e \in E$ the bounding box of $S_b(e)$ in an associative array $BB_b$ with index set $E$. The forward search checks whether the target is contained $BB(e)$, the backward search, whether the source is in $BB_b(e)$.

**Multi-Level Approach and Shortest-Path Bounding Boxes.** The multi-level approach enriches a given graph with additional edges. Each new edge $(u_1, u_k)$ represents a shortest path $(u_1, u_2, \ldots, u_k)$ in $G$. We annotate such a new edge $(u_1, u_k)$ with $BB(u_1, u_2)$, the associated bounding box of the first edge on this path.

**Table 1.** Number of nodes and edges for all test graphs

| | street | | | | | | | | | |
|---|---|---|---|---|---|---|---|---|---|---|
| n | 1444 | 3045 | 16471 | 20466 | 25982 | 38823 | 45852 | 45073 | 51510 | 79456 |
| m | 3060 | 7310 | 34530 | 42288 | 57620 | 79988 | 98098 | 91314 | 110676 | 172374 |

| | public transport | | | | | | | | | |
|---|---|---|---|---|---|---|---|---|---|---|
| n | 409 | 705 | 1660 | 2279 | 2399 | 4598 | 6884 | 10815 | 12070 | 14335 |
| m | 1215 | 1681 | 4327 | 6015 | 8008 | 14937 | 18601 | 29351 | 33728 | 39887 |

| | planar | | | | | | | | | |
|---|---|---|---|---|---|---|---|---|---|---|
| n | 1000 | 2000 | 3000 | 4000 | 5000 | 6000 | 7000 | 8000 | 9000 | 10000 |
| m | 5000 | 10000 | 15000 | 20000 | 25000 | 30000 | 35000 | 40000 | 45000 | 50000 |

| | waxman | | | | | | | | | |
|---|---|---|---|---|---|---|---|---|---|---|
| n | 938 | 1974 | 2951 | 3938 | 4949 | 5946 | 6943 | 7917 | 8882 | 9906 |
| m | 4070 | 9504 | 14506 | 19658 | 24474 | 29648 | 34764 | 39138 | 44208 | 48730 |

# 3  Experimental Setup

In this section, we provide details on the input data used, consisting of real-world and randomly generated graphs, and on the execution of the experiments.

## 3.1  Data

**Real-World Graphs.** In our experiments we included a set of graphs that stem from real applications. As in other experimental work, it turned out that using realistic data is quite important as the performance of the algorithms strongly depends on the characteristics of the data.

**Street Graphs.** Our street graphs are street networks of US cities and their surroundings. These graphs are bi-directed, and edge lengths are Euclidean distances. The graphs are fairly large and very sparse because bends are represented by polygonal lines. (With such a representation of a street network, it is possible to efficiently find the nearest point in a street by a point-to-point search.)

**Public Transport Graphs.** A public transport graph represents a network of trains, buses, and other scheduled vehicles. The nodes of such a graph correspond to stations or stops, and there exists an edge between two nodes if there is a non-stop connection between the respective stations. The weight of an edge is the average travel time of all vehicles that contribute to this edge. In particular, the edge lengths are not Euclidean distances in this set of graphs.

**Random Graphs.** We generated two sets of random graphs that have an estimated average out-degree of 2.5 (which corresponds to the average degree in the real-world graphs). Each set consists of ten connected, bi-directed graphs with (approximately) $1000 \cdot i$ nodes ($i = 1, \ldots, 10$).

**Random Planar Graphs.** For the construction of random planar graphs, we used a generator provided by LEDA [24]. A given number of $n$ nodes are uniformly distributed in a square with a lateral length of 1, and a triangulation of the nodes is computed. This yields a complete undirected planar graph. Finally, edges are deleted at random until the graph contains $2.5 \cdot n$ edges, and each of these is replaced by two directed edges, one in either direction.

**Random Waxman Graphs.** The construction of these graphs is based on a random graph model introduced by Waxman [25]. Input parameters are the number of nodes $n$ and two positive rational numbers $\alpha$ and $\beta$. The nodes are again uniformly distributed in a square of a lateral length of 1, and the probability that an edge $(u, v)$ exists is $\beta \cdot \exp(-d(u, v)/(\sqrt{2}\alpha))$. Higher $\beta$ values increase the edge density, while smaller $\alpha$ values increase the density of short edges in relation to long edges. To ensure connectedness and bi-directedness of the graphs, all nodes that do not belong to the largest connected component are deleted (thus, slightly less than $n$ nodes remain) and the graph is bi-directed by insertion of missing reverse edges. We set $\alpha = 0.01$ and empirically determined that setting $\beta = 2.5 \cdot 1620/n$ yields an average degree of 2.5, as wished.

## 3.2 Experiments

We have implemented all combinations of speed-up techniques as described in Sections 2.2 and 2.3 in C++, using the graph and Fibonacci heap data structures of the LEDA library [24] (version 4.4). The code was compiled with the GNU compiler (version 3.3), and experiments were run on an Intel Xeon machine with 2.6 GHz and 2 GB of memory, running Linux (kernel version 2.4).

For each graph and combination, we computed for a set of queries shortest paths, measuring two types of *performance*: the mean values of the *running times* (CPU time in seconds) and the *number of nodes* inserted in the priority queue. The queries were chosen at random and the amount of them was determined such that statistical relevance can be guaranteed (see also [23]).

# 4    Experimental Results

The outcome of the experimental study is shown in Figures 1–4. Further diagrams that we used for our analysis are depicted in Figures 5–10. Each combination is referred to by a 4-tuple of shortcuts: go (goal-directed), bi (bi-directed), ml (multi-level), bb (bounding box), and xx if the respective technique is not used (e.g., go-bi-xx-bb). In all figures, the graphs are ordered by size, as listed in Table 1.

We calculated two different values denoting relative *speed-up*: on the one hand, Figures 1–4 show the speed-up that we achieved compared to plain Dijkstra, i.e., for each combination of techniques the ratio of the performance of plain Dijkstra and the performance of Dijkstra with the specific combination of

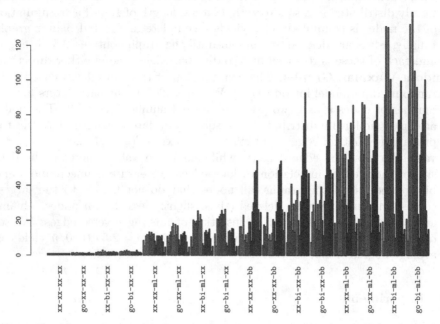

**Fig. 1.** Speed-up relative to Dijkstra's algorithm in terms of visited nodes for real-world graphs (in this order: street graphs in red and public transport graphs in blue)

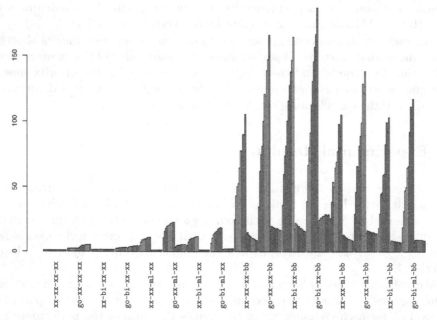

**Fig. 2.** Speed-up relative to Dijkstra's algorithm in terms of visited nodes for generated graphs (in this order: random planar graphs in yellow and random Waxman graphs in green)

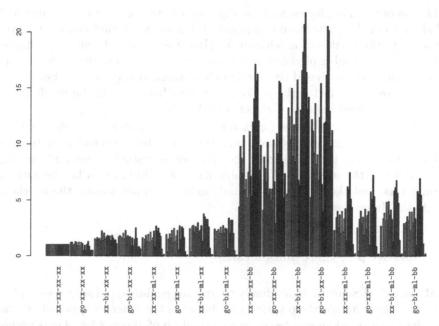

**Fig. 3.** Speed-up relative to Dijkstra's algorithm in terms of running time for real-world graphs (in this order: street graphs in red and public transport graphs in blue)

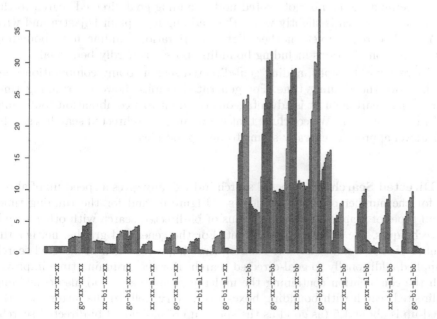

**Fig. 4.** Speed-up relative to Dijkstra's algorithm in terms of running time for generated graphs (in this order: random planar graphs in yellow and random Waxman graphs in green)

techniques applied. There are separate figures for real-world and random graphs, for the number of nodes and running time, respectively.

On the other hand, for each of the Figures 5–8, we focus on one technique $\mathcal{T}$ and show for each combination containing $\mathcal{T}$ the speed-up that can be achieved compared to the combination without $\mathcal{T}$. (Because of lack of space only figures dealing with the number of visited nodes are depicted.) For example, when focusing on bi-directed search and considering the combination `go-bi-xx-bb`, say, we investigate by which factor the performance gets better when the combination `go-bi-xx-bb` is used instead of `go-xx-xx-bb` only.

In the following, we discuss, for each technique separately, how combinations with the specific technique behave, and then turn to the relation of the two performance parameters measured, the number of visited nodes and running time: we define the *overhead* of a combination of techniques to be the ratio of running time and the number of visited nodes. In other words, the overhead reflects the time spent per node.

## 4.1    Speed-Up of the Combinations

**Goal-Directed Search.** Individually comparing goal-directed search with plain Dijkstra (Figure 5), speed-up varies a lot between the different types of graphs: Considering the random graphs, we get a speed-up of about 2 for planar graphs but of up to 5 for the Waxman graphs, which is quite surprising. Only little speed-up, of less than 2, can be observed for the real-world graphs.

Concerning the number of visited nodes, adding goal-directed search to the multi-level approach is slightly worse than adding it to plain Dijkstra and with bi-directed search, we get another slight deterioration. Adding it to bounding boxes (and combinations including bounding boxes) is hardly beneficial.

For real-world graphs, adding goal-directed search to any combination does not improve the running time. For generated graphs, however, running time decreases. In particular, it is advantageous to add it to a combination containing multi-level approach. We conclude that combining goal-directed search with the multi-level approach generally seems to be a good idea.

**Bi-Directed Search.** Bi-directed search individually gives a speed-up of about 1.5 for the number of visited nodes (see Figure 6) and for the running time, for all types of graphs. For combinations of bi-directed search with other speed-up techniques, the situation is different: For the generated graphs, neither the number of visited nodes nor the running time improves when bi-directed search is applied additionally to goal-directed search. However, running time improves with the combination containing the multi-level approach, and also combining bi-directed search with bounding boxes works very well. In the latter case, the speed-up is about 1.5 (as good as the speed-up of individual bi-directed search) for all types of graphs.

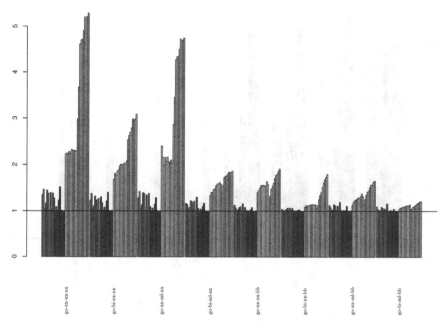

**Fig. 5.** Speed-up relative to the combination without goal-directed search in terms of visited nodes (in this order: street graphs in red, public transport graphs in blue, random planar graphs in yellow, and random Waxman graphs in green)

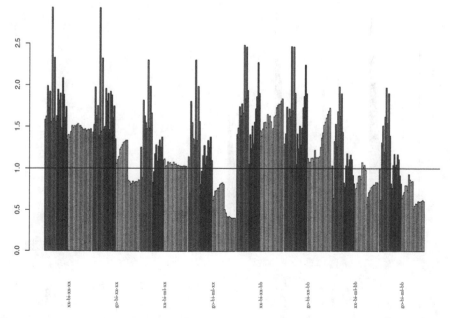

**Fig. 6.** Speed-up relative to the combination without bi-directed search in terms of visited nodes (in this order: street graphs in red, public transport graphs in blue, random planar graphs in yellow, and random Waxman graphs in green)

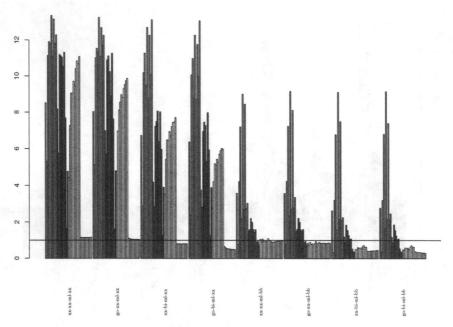

**Fig. 7.** Speed-up relative to the combination without multi-level approach in terms of visited nodes (in this order: street graphs in red, public transport graphs in blue, random planar graphs in yellow, and random Waxman graphs in green)

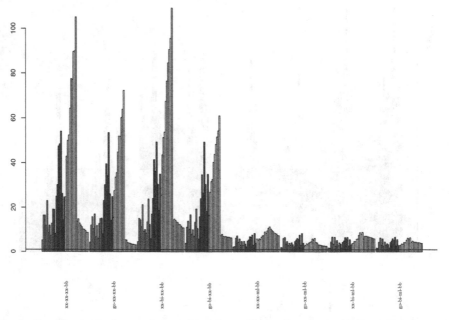

**Fig. 8.** Speed-up relative to the combination without shortest-path bounding boxes in terms of visited nodes (in this order: street graphs in red, public transport graphs in blue, random planar graphs in yellow, and random Waxman graphs in green)

**Multi-Level Approach.** The multi-level approach crucially depends on the decomposition of the graph. The Waxman graphs could not be decomposed properly by the multi-level approach, and therefore all combinations containing the latter yield speed-up factors of less than 1, which means a slowing down. Thus we consider only the remaining graph classes.

Adding multi-levels to goal-directed and bi-directed search and their combination gives a good improvement in the range between 5 and 12 for the number of nodes (see Figure 7). Caused by the big overhead of the multi-level approach, however, we get a considerable improvement in running time only for the real-world graphs. In combination with bounding boxes, the multi-level approach is beneficial only for the number of visited nodes in the case of street graphs.

The multi-level approach allows tuning of several parameters, such as the number of levels and the choice of the selected nodes. The tuning crucially depends on the input graph [20]. Hence, we believe that considerable improvements of the presented results are possible if specific parameters are chosen for every single graph.

**Shortest-Path Bounding Boxes.** Shortest-path bounding boxes work especially well when applied to planar graphs, actually speed-up even increases with the size of the graph (see Figure 8). For Waxman graphs, the situation is completely different: with the graph size the speed-up gets smaller. This can be explained by the fact that large Waxman graphs have, due to construction, more long-distance edges than small ones. Because of this, shortest paths become more tortuous and the bounding boxes contain more "wrong" nodes.

Throughout the different types of graphs, bounding boxes individually as well as in combination with goal-directed and bi-directed search yield exceptionally high speed-ups. Only the combinations that include the multi-level approach cannot be improved that much.

## 4.2  Overhead

For goal-directed and bi-directed search, the overhead (time per visited node) is quite small, while for bounding boxes it is a factor of about 2 compared to plain Dijkstra (see Figures 9 and 10). The overhead caused by the multi-level approach is generally high and quite different, depending on the type of graph. As Waxman graphs do not decompose well, the overhead for the multi-level approach is large and becomes even larger when the size of the graph increases. For very large street graphs, the multi-level approach overhead increases dramatically. We assume that it would be necessary to add a third level for graphs of this size.

It is also interesting to note that the relative overhead of the combination goal-directed, bi-directed, and multi-level is smaller than just multi-level —especially for the generated graphs.

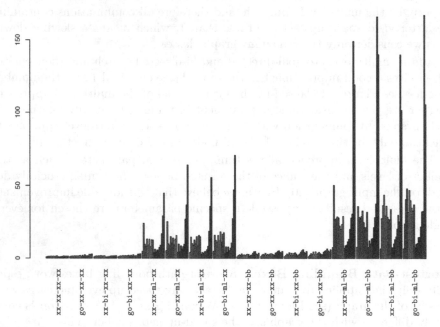

**Fig. 9.** Average running time per visited node in $\mu s$ for real-world graphs (in this order: street graphs in red and public transport graphs in blue)

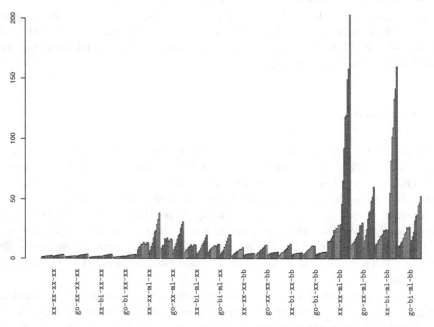

**Fig. 10.** Average running time per visited node in $\mu s$ for generated graphs (in this order: random planar graphs in yellow and random Waxman graphs in green)

# 5   Conclusion and Outlook

To summarize, we conclude that there are speed-up techniques that combine well and others where speed-up does not scale. Our result is that goal-directed search and multi-level approach is a good combination and bi-directed search with shortest-path bounding boxes complement each other.

For real-world graphs, a combination including bi-directed search, multi-level, and bounding boxes is the best choice as to the number of visited nodes. In terms of running time, the winner is bi-directed search in combination with bounding boxes. For generated graphs, the best combination is goal-directed, bi-directed, and bounding boxes for both the number of nodes and running time.

Without an expensive preprocessing, the combination of goal-directed and bi-directed search is generally the fastest algorithm with smallest search space— except for Waxman graphs. For these graphs, pure goal-directed is better than the combination with bi-directed search. Actually, goal-directed search is the only speed-up technique that works comparatively well for Waxman graphs. Because of this different behaviour, we conclude that planar graphs are a better approximation of the real-world graphs than Waxman graphs (although the public transport graphs are not planar).

Except bi-directed search, the speed-up techniques define a modified graph in which a shortest path is searched. From this shortest path one can easily determine a shortest path in the original graph. It is an interesting question whether the techniques can be applied directly, or modified, to improve also the running time of other shortest-path algorithms.

Furthermore, specialized priority queues used in Dijkstra's algorithm have been shown to be fast in practice [26,27]. Using such queues would provide the same results for the number of visited nodes. Running times, however, would be different and therefore interesting to evaluate.

# References

1. Zhan, F.B., Noon, C.E.: A comparison between label-setting and label-correcting algorithms for computing one-to-one shortest paths. Journal of Geographic Information and Decision Analysis **4** (2000)
2. Barrett, C., Bisset, K., Jacob, R., Konjevod, G., Marathe, M.: Classical and contemporary shortest path problems in road networks: Implementation and experimental analysis of the TRANSIMS router. In: Proc. 10th European Symposium on Algorithms (ESA). Volume 2461 of LNCS., Springer (2002) 126–138
3. Nachtigall, K.: Time depending shortest-path problems with applications to railway networks. European Journal of Operational Research **83** (1995) 154–166
4. Preuss, T., Syrbe, J.H.: An integrated traffic information system. In: Proc. 6th Int. Conf. Appl. Computer Networking in Architecture, Construction, Design, Civil Eng., and Urban Planning (europIA '97). (1997)
5. Shekhar, S., Fetterer, A., Goyal, B.: Materialization trade-offs in hierarchical shortest path algorithms. In: Proc. Symp. on Large Spatial Databases. (1997) 94–111
6. Barrett, C., Jacob, R., Marathe, M.: Formal-language-constrained path problems. SIAM Journal on Computing **30** (2000) 809–837

7. Dijkstra, E.W.: A note on two problems in connexion with graphs. Numerische Mathematik **1** (1959) 269–271
8. Fredman, M.L., Tarjan, R.E.: Fibonacci heaps and their uses in improved network optimization algorithms. Journal of the ACM **34** (1987) 596–615
9. Cherkassky, B.V., Goldberg, A.V., Radzik, T.: Shortest paths algorithms: Theory and experimental evaluation. Mathematical Programming **73** (1996) 129–174
10. Zwick, U.: Exact and approximate distances in graphs - a survey. In: Proc. 9th European Symposium on Algorithms (ESA). LNCS, Springer (2001) 33–48
11. Goldberg, A.V.: A simple shortest path algorithm with linear average time. In: Proc. 9th European Symposium on Algorithms (ESA). Volume 2161 of LNCS., Springer (2001) 230–241
12. Meyer, U.: Single-source shortest-paths on arbitrary directed graphs in linear average-case time. In: Proc. 12th Symp. on Discrete Algorithms. (2001) 797–806
13. Pettie, S., Ramachandran, V., Sridhar, S.: Experimental evaluation of a new shortest path algorithm. In: Proc. Algorithm Engineering and Experiments (ALENEX). Volume 2409 of LNCS., Springer (2002) 126–142
14. Hart, P., Nilsson, N.J., Raphael, B.A.: A formal basis for the heuristic determination of minimum cost paths. IEEE Trans. Sys. Sci. Cybernet. **2** (1968)
15. Shekhar, S., Kohli, A., Coyle, M.: Path computation algorithms for advanced traveler information system (ATIS). In: Proc. 9th IEEE Int. Conf. Data Eng. (1993) 31–39
16. Schulz, F., Wagner, D., Weihe, K.: Dijkstra's algorithm on-line: An empirical case study from public railroad transport. ACM Journal of Exp. Algorithmics **5** (2000)
17. Ahuja, R., Magnanti, T., Orlin, J.: Network Flows. Prentice–Hall (1993)
18. Pohl, I.: Bi-directional and heuristic search in path problems. Technical Report 104, Stanford Linear Accelerator Center, Stanford, California (1969)
19. Kaindl, H., Kainz, G.: Bidirectional heuristic search reconsidered. Journal of Artificial Intelligence Research **7** (1997) 283–317
20. Holzer, M.: Hierarchical speed-up techniques for shortest-path algorithms. Technical report, Dept. of Informatics, University of Konstanz, Germany (2003) http://www.ub.uni-konstanz.de/kops/volltexte/2003/1038/.
21. Schulz, F., Wagner, D., Zaroliagis, C.: Using multi-level graphs for timetable information in railway systems. In: Proc. 4th Workshop on Algorithm Engineering and Experiments (ALENEX). Volume 2409 of LNCS., Springer (2002) 43–59
22. Jung, S., Pramanik, S.: An efficient path computation model for hierarchically structured topographical road maps. IEEE Transactions on Knowledge and Data Engineering **14** (2002) 1029–1046
23. Wagner, D., Willhalm, T.: Geometric speed-up techniques for finding shortest paths in large sparse graphs. In: Proc. 11th European Symposium on Algorithms (ESA). Volume 2832 of LNCS., Springer (2003) 776–787
24. Näher, S., Mehlhorn, K.: The LEDA Platform of Combinatorial and Geometric Computing. Cambridge University Press (1999) (http://www.algorithmic-solutions.com).
25. Waxman, B.M.: Routing of multipoint connections. IEEE Journal on Selected Areas in Communications **6** (1988)
26. Dial, R.: Algorithm 360: Shortest path forest with topological ordering. Communications of ACM **12** (1969) 632–633
27. Goldberg, A.V.: Shortest path algorithms: Engineering aspects. In: Proc. International Symposium on Algorithms and Computation (ISAAC). Volume 2223 of LNCS., Springer (2001) 502–513

# Increased Bit-Parallelism for Approximate String Matching

Heikki Hyyrö[1,2]*, Kimmo Fredriksson[3]**, and Gonzalo Navarro[4]***

[1] PRESTO, Japan Science and Technology Agency, Japan.
[2] Department of Computer Sciences, University of Tampere, Finland.
Heikki.Hyyro@cs.uta.fi
[3] Department of Computer Science, University of Joensuu, Finland.
Kimmo.Fredriksson@cs.joensuu.fi
[4] Department of Computer Science, University of Chile, Chile.
gnavarro@dcc.uchile.cl

**Abstract.** Bit-parallelism permits executing several operations simultaneously over a set of bits or numbers stored in a single computer word. This technique permits searching for the approximate occurrences of a pattern of length $m$ in a text of length $n$ in time $O(\lceil m/w \rceil n)$, where $w$ is the number of bits in the computer word. Although this is asymptotically the optimal speedup over the basic $O(mn)$ time algorithm, it wastes bit-parallelism's power in the common case where $m$ is much smaller than $w$, since $w - m$ bits in the computer words get unused.

In this paper we explore different ways to increase the bit-parallelism when the search pattern is short. First, we show how multiple patterns can be packed in a single computer word so as to search for multiple patterns simultaneously. Instead of paying $O(rn)$ time to search for $r$ patterns of length $m < w$, we obtain $O(\lceil r/\lfloor w/m \rfloor \rceil n)$ time. Second, we show how the mechanism permits boosting the search for a single pattern of length $m < w$, which can be searched for in time $O(n/\lfloor w/m \rfloor)$ instead of $O(n)$. Finally, we show how to extend these algorithms so that the time bounds essentially depend on $k$ instead of $m$, where $k$ is the maximum number of differences permitted.

Our experimental results show that that the algorithms work well in practice, and are the fastest alternatives for a wide range of search parameters.

## 1 Introduction

Approximate string matching is an old problem, with applications for example in spelling correction, bioinformatics and signal processing [7]. It refers in general to searching for substrings of a text that are within a predefined edit distance

---

* Supported by the Academy of Finland and Tampere Graduate School in Information Science and Engineering.
** Supported by the Academy of Finland.
*** Supported in part by Fondecyt grant 1-020831.

threshold from a given pattern. Let $T = T_{1...n}$ be a text of length $n$ and $P = P_{1...m}$ a pattern of length $m$. Here $A_{a...b}$ denotes the substring of $A$ that begins at its $a$th character and ends at its $b$th character, for $a \leq b$. Let $ed(A, B)$ denote the edit distance between the strings $A$ and $B$, and $k$ be the maximum allowed distance. Then the task of approximate string matching is to find all text indices $j$ for which $ed(P, T_{h...j}) \leq k$ for some $h \leq j$.

The most common form of edit distance is Levenshtein distance [5]. It is defined as the minimum number of single-character insertions, deletions and substitutions needed in order to make $A$ and $B$ equal. In this paper $ed(A, B)$ will denote Levenshtein distance. We also use $w$ to denote the computer word size in bits, $\sigma$ to denote the size of the alphabet $\Sigma$ and $|A|$ to denote the length of the string $A$.

Bit-parallelism is the technique of packing several values in a single computer word and updating them all in a single operation. This technique has yielded the fastest approximate string matching algorithms if we exclude filtration algorithms (which need anyway to be coupled with a non-filtration one). In particular, the $O(\lceil m/w \rceil kn)$ algorithm of Wu and Manber [13], the $O(\lceil km/w \rceil n)$ algorithm of Baeza-Yates and Navarro [1], and the $O(\lceil m/w \rceil n)$ algorithm of Myers [6] dominate for almost every value of $m$, $k$ and $\sigma$.

In complexity terms, Myers' algorithm is superior to the others. In practice, however, Wu & Manber's algorithm is faster for $k = 1$ and Baeza-Yates and Navarro's is faster when $(k + 2)(m - k) \leq w$ or $k/m$ is low. The reason is that, despite that Myers' algorithm packs better the state of the search (needing to update less computer words), it needs slightly more operations than its competitors. Except when $m$ and $k$ are small, the need to update less computer words makes Myers' algorithm faster than the others. However, when $m$ is much smaller than $w$, Myers' advantage disappears because all the three algorithms need to update just one (or very few) computer words. In this case, Myers' representation wastes many bits of the computer word and is unable to take advantage of its more compact representation.

The case where $m$ is much smaller than $w$ is very common in several applications. Typically $w$ is 32 or 64 in a modern computer, and for example the Pentium 4 processor allows one to use even words of size 128. Myers' representation uses $m$ bits out of those $w$. In spelling, for example, it is usual to search for words, whose average length is 6. In computational biology one can search for short DNA or amino acid sequences, of length as small as 4. In signal processing applications one can search for sequences composed of a few audio, MIDI or video samples.

In this paper we concentrate on reducing the number of wasted bits in Myers' algorithm, so as to take advantage of its better packing of the search state even when $m \leq w$. This has been attempted previously [2], where $O(m \lceil n/w \rceil)$ time was obtained. Our technique is different. We first show how to search for several patterns simultaneously by packing them all in the same computer word. We can search for $r$ patterns of length $m \leq w$ in $O(\lceil r/\lfloor m/w \rfloor \rceil n + occ)$ rather than $O(rn)$ time, where $occ \leq rn$ is the total number of occurrences of all the patterns. We then show how this idea can be pushed further to boost the search for a single pattern, so as to obtain $O(n/\lfloor w/m \rfloor)$ time instead of $O(n)$ for $m \leq w$.

Our experimental results show that the presented schemes work well in practice.

## 2    Dynamic Programming

In the following $\epsilon$ denotes the empty string. To compute Levenshtein distance $ed(A, B)$, the dynamic programming algorithm fills an $(|A|+1) \times (|B|+1)$ table $D$, in which each cell $D[i, j]$ will eventually hold the value $ed(A_{1..i}, B_{1..j})$. Initially the trivially known *boundary values* $D[i, 0] = ed(A_{1..i}, \epsilon) = i$ and $D[0, j] = ed(\epsilon, B_{1..j}) = j$ are filled. Then the cells $D[i, j]$ are computed for $i = 1 \ldots |A|$ and $j = 1 \ldots |B|$ until the desired solution $D[|A|, |B|] = ed(A_{1\ldots|A|}, B_{1\ldots|B|}) = ed(A, B)$ is known. When the values $D[i - 1, j - 1]$, $D[i, j - 1]$ and $D[i - 1, j]$ are known, the value $D[i, j]$ can be computed by using the following well-known recurrence.

$$D[i, 0] = i, \quad D[0, j] = j.$$
$$D[i, j] = \begin{cases} D[i - 1, j - 1], \text{ if } A_i = B_j. \\ 1 + \min(D[i - 1, j - 1], D[i - 1, j], D[i, j - 1]), \text{ otherwise.} \end{cases}$$

This distance computation algorithm is easily modified to find approximate occurrences of $A$ somewhere inside $B$ [9]. This is done simply by changing the boundary condition $D[0, j] = j$ into $D[0, j] = 0$. In this case $D[i, j] = \min(ed(A_{1\ldots i}, B_{h\ldots j}), h \leq j)$, which corresponds to the earlier definition of approximate string matching if we replace $A$ with $P$ and $B$ with $T$.

The values of $D$ are usually computed by filling it in a column-wise manner for increasing $j$. This corresponds to scanning the string $B$ (or the text $T$) one character at a time from left to right. At each character the corresponding column is completely filled in order of increasing $i$. This order makes it possible to save space by storing only one column at a time, since then the values in column $j$ depend only on already computed values in it or values in column $j - 1$.

Some properties of matrix $D$ are relevant to our paper [11]:

-The diagonal property:   $D[i, j] - D[i - 1, j - 1] = 0$ or $1$.
-The adjacency property:  $D[i, j] - D[i, j - 1] = -1, 0,$ or $1$, and
$$D[i, j] - D[i - 1, j] = -1, 0, \text{ or } 1.$$

## 3    Myers' Bit-Parallel Algorithm

In what follows we will use the following notation in describing bit-operations: '&' denotes bitwise "and", '|' denotes bitwise "or", '^' denotes bitwise "xor", '~' denotes bit complementation, and '<<' and '>>' denote shifting the bit-vector left and right, respectively, using zero filling in both directions. The $i$th bit of the bit vector $V$ is referred to as $V[i]$ and bit positions are assumed to grow from right to left. In addition we use superscripts to denote repetition. As an example let $V = 1011010$ be a bit vector. Then $V[1] = V[3] = V[6] = 0$, $V[2] = V[4] = V[5] = V[7] = 1$, and we could also write $V = 101^2010$ or $V = 101(10)^2$.

We describe here a version of the algorithm [3,8] that is slightly simpler than the original by Myers [6]. The algorithm is based on representing the dynamic programming table $D$ with vertical, horizontal and diagonal differences and pre-computing the matching positions of the pattern into an array of size $\sigma$. This is done by using the following length-$m$ bit-vectors:

-Vertical positive delta: $VP[i] = 1$ at text position $j$ if and only if $D[i,j] - D[i-1,j] = 1$.
-Vertical negative delta: $VN[i] = 1$ at text position $j$ if and only if $D[i,j] - D[i-1,j] = -1$.
-Horizontal positive delta: $HP[i] = 1$ at text position $j$ if and only if $D[i,j] - D[i,j-1] = 1$.
-Horizontal negative delta: $HN[i] = 1$ at text position $j$ if and only if $D[i,j] - D[i,j-1] = -1$.
-Diagonal zero delta: $D0[i] = 1$ at text position $j$ if and only if $D[i,j] = D[i-1,j-1]$.
-Pattern match vector $PM_\lambda$ for each $\lambda \in \Sigma$: $PM_\lambda[i] = 1$ if and only if $P_i = \lambda$.

Initially $VP = 1^m$ and $VN = 0^m$ to enforce the boundary condition $D[i,0] = i$. At text position $j$ the algorithm first computes vector $D0$ by using the old values $VP$ and $VN$ and the pattern match vector $PM_{T_j}$. Then the new $HP$ and $HN$ are computed by using $D0$ and the old $VP$ and $VN$. Finally, vectors $VP$ and $VN$ are updated by using the new $D0$, $HN$ and $HP$. Fig. 1 shows the complete formula for updating the vectors, and Fig. 2 shows the preprocessing of table $PM$ and the higher-level search scheme. We refer the reader to [3,6] for a more detailed explanation of the formula in Fig. 1.

| Step($j$) |
|---|
| 1.     $D0 \leftarrow (((PM_{T_j} \& VP) + VP) \wedge VP) \mid PM_{T_j} \mid VN$ |
| 2.     $HP \leftarrow VN \mid \sim (D0 \mid VP)$ |
| 3.     $HN \leftarrow VP \& D0$ |
| 4.     $VP \leftarrow (HN << 1) \mid \sim (D0 \mid (HP << 1))$ |
| 5.     $VN \leftarrow (HP << 1) \& D0$ |

Fig. 1. Updating the delta vectors at column $j$.

The algorithm in Fig. 2 computes the value $D[m,j]$ explicitly in the $currDist$ variable by using the horizontal delta vectors (the initial value of $currDist$ is $D[m,0] = m$). A pattern occurrence with at most $k$ errors is found at text position $j$ whenever $D[m,j] \le k$.

We point out that the boundary condition $D[0,j] = 0$ is enforced on lines 4 and 5 in Fig. 1. After the horizontal delta vectors $HP$ and $HN$ are shifted

```
ComputePM(P)
  1.      For λ ∈ Σ Do PM_λ ← 0^m
  2.      For i ∈ 1...m Do PM_{P_i} ← PM_{P_i} | 0^{m-i}10^{i-1}

Search(P, T, k)
  1.      ComputePM(P)
  2.      VN ← 0^m, VP ← 1^m, currDist ← m
  3.      For j ∈ 1...n Do
  4.          Step(j)
  5.          If HP & 10^{m-1} = 10^{m-1} Then
  6.              currDist ← currDist + 1
  7.          Else If HN & 10^{m-1} = 10^{m-1} Then
  8.              currDist ← currDist − 1
  9.          If currDist ≤ k Then
 10.              Report occurrence at j
```

Fig. 2. Preprocessing the $PM$-table and conducting the search.

left, their first bits correspond to the difference $D[0,j] - D[0, j-1]$. This is the only phase in the algorithm where the values from row 0 are relevant. And as we assume zero filling, the left shifts correctly set $HP[1] = HN[1] = 0$ to encode the difference $D[0,j] - D[0, j-1] = 0$.

The running time of the algorithm is $O(n)$ when $m \leq w$, as there are only a constant number of operations per text character. The general running time is $O(\lceil m/w \rceil n)$ as a vector of length $m$ may be simulated in $O(\lceil m/w \rceil)$ time using $O(\lceil m/w \rceil)$ bit-vectors of length $w$.

## 4   Searching for Several Patterns Simultaneously

We show how Myers' algorithm can be used to search for $r$ patterns of length $m$ simultaneously. For simplicity we will assume $rm \leq w$; otherwise the search patterns must be split into groups of at most $\lfloor w/m \rfloor$ patterns each, and each group searched for separately. Hence our search time will be $O(\lceil r/\lfloor w/m \rfloor \rceil n + occ)$, as opposed to the $O(rn)$ time that would be achieved by searching for each pattern separately. Here $occ \leq rn$ stands for the total number of occurrences of all the patterns. When $w/m \geq 2$, our complexity can be written as $O(\lceil rm/w \rceil n + occ)$.

Consider the situation where $w/m \geq 2$ and Myers' algorithm is used. Fig. 3a shows how the algorithm fails to take full advantage of bit-parallelism in that situation as at least one half of the bits in the bit vectors is not used. Fig. 3b depicts our proposal: encode several patterns into the bit vectors and search for them in parallel. There are several obstacles in achieving this goal correctly, which will be discussed next.

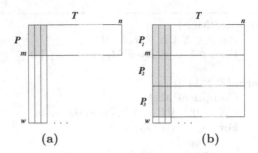

**Fig. 3.** For short patterns ($m < w$) Myers' algorithm (a) wastes $w - m$ bits. Our proposal (b) packs several pattern into the same computer word, and wastes only $w - rm$ bits.

## 4.1   Updating the Delta Vectors

A natural starting point is the problem of encoding and updating several patterns in the delta vectors. Let us denote a parallel version of a delta vector with the superscript $p$. We encode the patterns repeatedly into the vectors without leaving any space between them. For example $D0^p[i]$ corresponds to the bit $D0[((i-1) \bmod m) + 1]$ in the $D0$-vector of the $\lceil i/m \rceil$th pattern. The pattern match vectors $PM$ are computed in normal fashion for the concatenation of the patterns. This correctly aligns the patterns with their positions in the bit vectors.

When the parallel vectors are updated, we need to ensure that the values for different patterns do not interfere with each other and that the boundary values $D[0, j] = 0$ are used correctly. From the update formula in Fig. 1 it is obvious that only the addition ("+") on line 2 and the left shifts on lines 4 and 5 can cause incorrect interference.

The addition operation may be handled by temporarily setting off the bits in $VP^p$ that correspond to the last characters of the patterns. When this is done before the addition, there cannot be an incorrect overflow, and on the other hand the correct behaviour of the algorithm is not affected: The value $VP^p[i]$ can affect only the values $D0^p[i+h]$ for some $h > 0$. It turns out that a similar modification works also with the left shifts. If the bits that correspond to the last characters of the patterns are temporarily set off in $HP^p$ and $HN^p$ then, after shifting left, the positions in $HP^p$ and $HN^p$ that correspond to the first characters of the patterns will correctly have a zero bit. The first pattern gets the zero bits from zero filling of the shift. Therefore, this second modification both removes possible interference and enforces the boundary condition $D[0, j] - D[0, j - 1] = 0$.

Both modifications are implemented by *and*ing the corresponding vectors with the bit mask $ZM = (01^{m-1})^r$. Figure 4 gives the code for a step.

## 4.2   Keeping the Scores

A second problem is computing the value $D[m, j]$ explicitly for each of the $r$ patterns. We handle this by using bit-parallel counters in a somewhat similar

```
MStep(j)
  1.    XP ← VP & ZM
  2.    D0 ← (((PM_{T_j} & XP) + XP) ^ XP) | PM_{T_j} | VN
  3.    HP ← VN | ~(D0 | VP)
  4.    HN ← VP & D0
  5.    XP ← HP & ZM, XN ← HN & ZM
  6.    VP ← ((XN << 1) | ~(D0 | (XP << 1)))
  7.    VN ← (XP << 1) & D0
```

**Fig. 4.** Updating the delta vectors at column $j$, when searching for multiple patterns.

fashion to [4]. Let $MC$ be a length-$w$ bit-parallel counter vector. We set up into $MC$ an $m$-bit counter for each pattern. Let $MC(i)$ be the value of the $i$th counter. The counters are aligned with the patterns so that $MC(1)$ occupies the first $m$ bits, $MC(2)$ the next $m$ bits, and so on. We will represent value zero in each counter as $b = 2^{m-1}+k$, and the value $MC(i)$ will be translated to actually mean $b - MC(i)$. This gives each counter $MC(i)$ the following properties: (1) $b < 2^m$. (2) $b - m \geq 0$. (3) The $m$th bit of $MC(i)$ is set if and only if $b - MC(i) \leq k$. (4) In terms of updating the translated value of $MC(i)$, the roles of adding and subtracting from it are reversed.

The significance of properties (1) and (2) is that they ensure that the values of the counters will not overflow outside their regions. Their correctness depends on the assumption $k < m$. This is not a true restriction as it excludes only the case of trivial matching ($k = m$).

We use a length-$w$ bit-mask $EM = (10^{m-1})^r$ to update $MC$. The bits set in $HP^p$ & $EM$ and $HN^p$ & $EM$ correspond to the last bits of the counters that need to be incremented and decremented, respectively. Thus, remembering to reverse addition and subtraction, $MC$ may be updated by setting $MC \leftarrow MC + ((HN^p \& EM) >> (m-1)) - ((HP^p \& EM) >> (m-1))$.

Property (3) means that the last bit of $MC(i)$ signals whether the $i$th pattern matches at the current position. Hence, whenever $MC$ & $EM \neq 0^{rm}$ we have an occurrence of some of the patterns in $T$. At this point we can examine the bit positions of $EM$ one by one to determine which patterns have matched and report their occurrences. This, however, adds $O(r \min(n, occ))$ time in the worst case to report the $occ$ occurrences of all the patterns. We show next how to reduce this to $O(occ)$.

Fig. 5 gives the code to search for the patterns $P^1 \ldots P^r$.

### 4.3   Reporting the Occurrences

Let us assume that we want to identify which bits in mask $OM = MC$ & $EM$ are set, in time proportional to the number of bits set. If we achieve this, the total time to report all the $occ$ occurrences of all the patterns will be $O(occ)$. One choice is to precompute a table $F$ that, for any value of $OM$, gives the position of the first bit set in $OM$. That is, if $F[OM] = s$, then we report an occurrence

**MComputePM**$(P^1 \ldots P^r)$
1.     **For** $\lambda \in \Sigma$ **Do** $PM_\lambda \leftarrow 0^{mr}$
2.     **For** $s \in 1 \ldots r$ **Do**
3.         **For** $i \in 1 \ldots m$ **Do** $PM_{P_i^s} \leftarrow PM_{P_i^s} \mid 0^{m(r-s+1)-i}10^{m(s-1)+i-1}$

**MSearch**$(P^1 \ldots P^r, T, k)$
1.     **MComputePM**$(P)$
2.     $ZM \leftarrow (01^{m-1})^r$, $EM \leftarrow (10^{m-1})^r$
3.     $VN \leftarrow 0^{mr}$, $VP \leftarrow 1^{mr}$
4.     $MC \leftarrow (2^{m-1} + k) \times (0^{m-1}1)^r$
5.     **For** $j \in 1 \ldots n$ **Do**
6.         **MStep**$(j)$
7.         $MC \leftarrow MC + ((HN \ \& \ EM) >> (m-1)) - ((HP \ \& \ EM) >> (m-1))$
8.         **If** $MC \ \& \ EM \neq 0^{rm}$ **Then MReport**$(j, MC \ \& \ EM)$

**Fig. 5.** Preprocessing the $PM$-table and conducting the search for multiple patterns.

of the $(s/m)$th pattern at the current text position $j$, clear the $s$th bit in $OM$ by doing $OM \leftarrow OM \ \& \ \sim (1 << (s-1))$, and repeat until $OM$ becomes zero.

The only problem of this approach is that table $F$ has $2^{rm}$ entries, which is too much. Fortunately, we can compute the $s$ values efficiently without resorting to look-up tables. The key observation is that the position of the highest bit set in $OM$ is effectively the function $\lfloor \log_2(OM) \rfloor + 1$ (we number the bits from 1 to $w$), i.e. it holds that

$$2^{\lfloor \log_2(x) \rfloor} \ \leq \ x \ < \ 2^{\lfloor \log_2(x) \rfloor + 1},$$
$$1 << \lfloor \log_2(x) \rfloor \ \leq \ x \ < \ 1 << (\lfloor \log_2(x) \rfloor + 1).$$

The function $\lfloor \log_2(x) \rfloor$ for an integer $x$ can be computed in $O(1)$ time in modern computer architectures by converting $x$ into a floating point number, and extracting the exponent, which requires only two additions and a shift. This assumes that the floating point number is represented in a certain way, in particular that the radix is 2, and that the number is normalized. The "industry standard" IEEE floating point representation meets these requirements. For the details and other solutions for the integer logarithm of base 2, refer e.g. to [12]. ISO C99 standard conforming C compilers also provide a function to extract the exponent directly, and many CPUs even have a dedicated machine instruction for $\lfloor \log_2(x) \rfloor$ function. Fig. 6 gives the code.

For architectures where $\lfloor \log_2(x) \rfloor$ is hard to compute, we can still manage to obtain $O(\min(n, occ) \log r)$ time as follows. To detect the bits set in $OM$, we check its two halves. If some half is zero, we can finish there. Otherwise, we recursively check its two halves. We continue the process until we have isolated each individual bit set in $OM$. In the worst case, each such bit has cost us $O(\log r)$ halving steps.

```
MReport(j, OM)
  1.      While OM ≠ 0^r Do
  2.          s ← ⌊log_2(OM)⌋
  3.          Report occurrence of P^{(s+1)/m} at text position j
  4.          OM ← OM & ~ (1 << s)
```

**Fig. 6.** Reporting occurrences at current text position.

## 4.4 Handling Different Lengths and Thresholds

For simplicity we have assumed that all the patterns are of the same length and are all searched with the same $k$. The method, however, can be adapted with little problems to different $m$ and $k$ for each pattern.

If the lengths are $m_1 \ldots m_r$ and the thresholds are $k_1 \ldots k_r$, we have to *and* the vertical and horizontal vectors with $ZM = 01^{m_r-1} 01^{m_{r-1}-1} \ldots 01^{m_1-1}$, and this fixes the problem of updating the delta vectors. With respect to the counters, the $i$th counter must be represented as $b_i - MC(i)$, where $b_i = 2^{m_i-1} + k_i$.

One delicacy is the update of $MC$, since the formula we gave to align all the $HP^p$ bits at the beginning of the counters involved ">> $(m - 1)$", and this works only when all the patterns are of the same length. If they are not, we could align the counters so that they start at the end of their areas, hence removing the need for the shift at all. To avoid overflows, we should sort the patterns in increasing length order prior to packing them in the computer word. The price is that we will need $m_r$ extra bits at the end of the bit mask to hold the largest counter. An alternative solution would be to handle the last counter separately. This would avoid the shifts, and effectively adds only a few operations.

Finally, reporting the occurrences works just as before, except that the pattern number we report is no longer $(s + 1)/m$ (Fig. 6). The correct pattern number can be computed efficiently e.g. using a look-up table indexed with $s$. The size of the table is only $O(w)$, as $s \leq w - 1$.

# 5   Boosting the Search for One Pattern

Up to now we have shown how to take advantage of wasted bits by searching for several patterns simultaneously. Yet, if we only want to search for a single pattern, we still waste the bits. In this section we show how the technique developed for multiple patterns can be adapted to boost the search for a single pattern.

The main idea is to search for multiple copies of the same pattern $P$ and parallelize the access to the text. Say that $r = \lfloor w/m \rfloor$. Then we search for $r$ copies of $P$ using a single computer word, with the same technique developed for multiple patterns.

Of course this is of little interest in principle, as all the copies of the pattern will report the same occurrences. However, the key idea here will be to search a *different* text segment for each pattern copy. We divide the text $T$ into $r$ equal-sized subtexts $T = T^1T^2 \ldots T^r$. Text $T^s$, of length $\ell = \lceil n/r \rceil$, will be searched for the $s$th copy of $P$, and therefore all the occurrences of $P$ in $T$ will be found.

Our search will perform $\lceil n/r \rceil$ steps, where step $j$ will access $r$ text characters $T_j$, $T_{j+\ell}$, $T_{j+2\ell}, \ldots, T_{j+(r-1)\ell}$. With those $r$ characters $c_1 \ldots c_r$ we should build the corresponding $PM$ mask to execute a single step. This is easily done by using

$$PM \;\leftarrow\; PM_{c_1} \mid (PM_{c_2} << m) \mid (PM_{c_3} << 2m) \mid \; \ldots \; \mid (PM_{c_r} << (r-1)m)$$

We must exercise some care at the boundaries between consecutive text segments. On the one hand, processing of text segment $T_s$ $(1 \le s < r)$ should continue up to $m + k - 1$ characters in $T_{s+1}$ in order to provide the adequate context for the possible occurrences in the beginning of $T_{s+1}$. On the other hand, the processing of $T_{s+1}$ must avoid reporting occurrences at the first $m + k - 1$ positions to avoid reporting them twice. Finally, occurrences may be reported out of order if printed immediately, so it is necessary to store them in $r$ buffer arrays in order to report them ordered at the end.

Adding up the $\lceil n/r \rceil = \lceil n/\lfloor w/m \rfloor \rceil$ bit-parallel steps required plus the $n$ character accesses to compute $PM$, we obtain $O(\lceil n/\lfloor w/m \rfloor \rceil)$ complexity for $m \le w$.

## 6    Long Patterns and $k$-Differences Problem

We have shown how to utilize the bits in computer word economically, but our methods assume that $m \le w$. We now sketch a method that can handle longer patterns, and can pack more patterns in the same computer word. The basic assumption here is that we are only interested in pattern occurrences that have at most $k$ differences. This is the situation that is most interesting in practice, and usually we can assume that $k$ is much smaller than $m$. Our goal is to obtain similar time bounds as above, but replace $m$ with $k$ in the complexities. The difference will be that these become average case complexities now.

The method is similar to our basic algorithms, but now we use an adaptation of Ukkonen's well-known "cut-off" algorithm [10]. That algorithm fills the table $D$ in column-wise order, and computes the values $D[i, j]$ in column $j$ for only $i \le \ell_j$, where

$$\ell_j = 1 + \max\{i \mid D[i, j-1] \le k\}.$$

The cut-off heuristic is based on the fact that the search result does not depend on cells whose value is larger than $k$. From the diagonal property it follows that once $D[i, j] > k$, then $D[i + h, j + h] > k$ for all $h \ge 0$ (within the bounds of $D$). And a consequence of this is that $D[i, j] > k$ for $i > \ell_j$.

After evaluating the current column of the matrix up to the row $\ell_j$, the value $\ell_{j+1}$ is computed, and the algorithm continues with the next column $j+1$. The evaluation of $\ell_j$ takes $O(1)$ amortized time, and its expected value $L(k)$ is $O(k)$, and hence the whole algorithm takes only $O(nk)$ time.

Myers adapted his $O(n\lceil m/w \rceil)$ algorithm to use the cut-off heuristic as well. In principle the idea is very simple; since on average the search ends at row $L(k)$, it is enough to use only $L(k)$ bits of the computer word on average (actually he used $w\lceil L(k)/w \rceil$ bits), and only in some text positions (e.g. when the pattern matches) one has to use more bits. Only two modifications to the basic method are needed. We must be able to decide which is the last active row in order to compute the number of bits required for each text position, and we must be able to handle the communication between the boundaries of the consecutive computer words. Both problems are easy to solve, for details refer to [6]. With these modifications Myers was able to obtain his $O(n\lceil L(k)/w \rceil)$ average time algorithm.

We can do exactly the same here. We use only $b = \max\{L(k), \lceil \log(m+k) \rceil + 1\}$ bits for each pattern and pack them into the same computer word just like in our basic method. We need $L(k)$ bits as $L(k)$ is the row number where the search is expected to end, and at least $\lceil \log(m+k) \rceil + 1$ bits to avoid overflowing the counters. Therefore we are going to search for $\lfloor w/b \rfloor$ patterns in parallel.

If for some text positions $b$ bits are not enough, we use as many computer words as needed, each having $b$ bits allocated for each pattern. Therefore, the $b$-bit blocks in the first computer word correspond to the first $b$ characters of the corresponding patterns, and the $b$-bit blocks in the second word correspond to the next $b$ characters of the patterns, and so on. In total we need $\lceil m/b \rceil$ computer words, but on average use only one for each text position.

The counters for each pattern have only $b$ bits now, which means that the maximum pattern length is limited to $2^{b-1} - k$. The previous counters limited the pattern length to $2^{m-1} - k$, but at the same time assumed that the pattern length was $\leq w/2$. Using the cut-off method, we have less bits for the counters, but in effect we can use longer patterns, the upper bound being $m = 2^{w/2-1} - k$.

The tools we have developed for the basic method can be applied to modify Myers' cut-off algorithm to search for $\lfloor w/b \rfloor$ patterns simultaneously. The only additional modification we need is that we must add a new computer word whenever *any* of the pattern counters has accumulated $k$ differences, and this is trivial to detect with our counters model. On the other hand, this modification means that $L(k)$ must grow as the function of $r$. It has been shown in [7] that $L(k) = k/(1 - e/\sqrt{\sigma}) + O(1)$ for $r = 1$. For reasonably small $r$ this bound should not be affected much, as the probability of a match is exponentially decreasing for $m > L(k)$.

The result is that we can search for $r$ patterns with at most $k$ differences in $O(n\lceil r/\lfloor w/b \rfloor \rceil)$ expected time. Finally, it is possible to apply the same scheme for single pattern search as well, resulting in $O(\lceil n/\lfloor w/b \rfloor \rceil)$ expected time. The method is useful even for short patterns (where we could apply our basic method also), because we can use tighter packing when $b < m$.

## 7    Experimental Results

We have implemented all the algorithms in C and compiled them using GCC 3.3.1 with full optimizations. The experiments were run on a Sparc Ultra 2 with 128 MB RAM that was dedicated solely for our tests. The word size of the machine is 64 bits.

In the experiments we used DNA from baker's yeast and natural language English text from the TREC collection. Each text was cut into 4 million characters for testing. The patterns were selected randomly from the texts. We compared the performance of our algorithms against previous work. The algorithms included in the experiments were:

**Parallel BPM:** Our parallelized single-pattern search algorithm (Section 5). We used $r = 3$ for $m = 8$ and $m = 16$, $r = 2$ for $m = 32$, and $r = 2$ and cut-off (Section 6) for $m = 64$.

**Our multi-pattern algorithm:** The basic multipattern algorithm (Section 4) or its cut-off version (Section 6). We determined which version to use by using experimentally computed estimates for $L(k)$.

**BPM:** Myers' original algorithm [6], whose complexity is $O(\lceil m/w \rceil n)$. We used our implementation, which was roughly 20 % faster than the original code of Myers on the test computer.

**BPD:** Non-deterministic finite state automaton bit-parallelized by diagonals [1]. The complexity is $O(\lceil km/w \rceil n)$. Implemented by its original authors.

**BPR:** Non-deterministic finite state automaton bit-parallelized by rows [13]. The complexity is $O(\lceil m/w \rceil kn)$. We used our implementation, with hand optimized special code for $k = 1 \ldots 7$.

For each tested combination of $m$ and $k$, we measured the average time per pattern when searching for 50 patterns. The set of patterns was the same for each algorithm. The results are shown in Fig. 7. Our algorithms are clear winners in most of the cases.

Our single-pattern parallel search algorithm is beaten only when $k = 1$, as **BPR** needs to do very little work in that case, or when the probability of finding occurrences becomes so high that our more complicated scheme for occurrence checking becomes very costly. At this point we would like to note, that the occurrence checking part of our single-pattern algorithm has not yet been fully optimized in practice.

Our multi-pattern algorithm is also shown to be very fast: in these tests it is worse than a single-pattern algorithm only when $w/2 < m$ and $k$ is moderately high with relation to the alphabet size.

## 8    Conclusions

Bit-parallel algorithms are currently the fastest approximate string matching algorithms when Levenshtein distance is used. In particular, the algorithm of Myers [6] dominates the field when the pattern is long enough, thanks to its

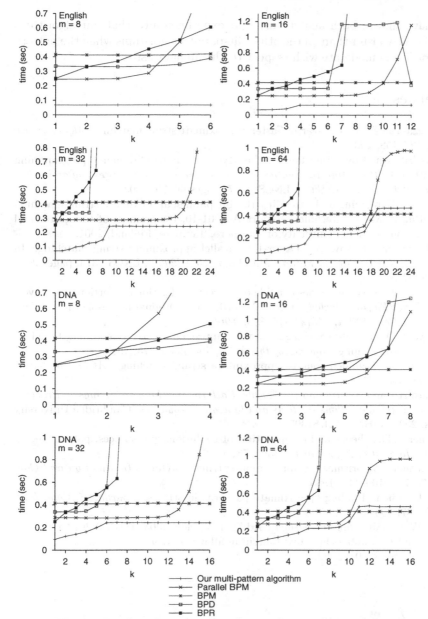

**Fig. 7.** The plots show the average time for searching a single pattern.

better packing of the search state in the bits of the computer word. In this paper we showed how this algorithm can be modified to take advantage of the wasted bits when the pattern is short. We have shown two ways to do this. The first one permits searching for several patterns simultaneously. The second one boosts the search for a single pattern by processing several text positions simultaneously.

We have shown, both analytically and experimentally, that our algorithms are significantly faster than all the other bit-parallel algorithms when the pattern is short or if $k$ is moderate with respect to the alphabet size.

# References

1. R. Baeza-Yates and G. Navarro. Faster approximate string matching. *Algorithmica*, 23(2):127–158, 1999.
2. K. Fredriksson. Row-wise tiling for the Myers' bit-parallel dynamic programming algorithm. In *Proc. 10th International Symposium on String Processing and Information Retrieval (SPIRE'03)*, LNCS 2857, pages 66–79, 2003.
3. H. Hyyrö. Explaining and extending the bit-parallel approximate string matching algorithm of Myers. Technical Report A-2001-10, Department of Computer and Information Sciences, University of Tampere, Tampere, Finland, 2001.
4. H. Hyyrö and G. Navarro. Faster bit-parallel approximate string matching. In *Proc. 13th Combinatorial Pattern Matching (CPM'2002)*, LNCS 2373, pages 203–224, 2002.
5. V. Levenshtein. Binary codes capable of correcting deletions, insertions and reversals. *Soviet Physics Doklady*, 10(8):707–710, 1966. Original in Russian in *Doklady Akademii Nauk SSSR, 163(4):845–848, 1965*.
6. G. Myers. A fast bit-vector algorithm for approximate string matching based on dynamic progamming. *Journal of the ACM*, 46(3):395–415, 1999.
7. G. Navarro. A guided tour to approximate string matching. *ACM Computing Surveys*, 33(1):31–88, 2001.
8. G. Navarro and M. Raffinot. *Flexible Pattern Matching in Strings – Practical on-line search algorithms for texts and biological sequences*. Cambridge University Press, 2002. ISBN 0-521-81307-7.
9. P. Sellers. The theory and computation of evolutionary distances: pattern recognition. *Journal of Algorithms*, 1:359–373, 1980.
10. E. Ukkonen. Algorithms for approximate string matching. *Information and Control*, 64(1–3):100–118, 1985.
11. Esko Ukkonen. Finding approximate patterns in strings. *Journal of Algorithms*, 6:132–137, 1985.
12. H. S. Warren Jr. *Hacker's Delight*. Addison Wesley, 2003. ISBN 0-201-91465-4.
13. S. Wu and U. Manber. Fast text searching allowing errors. *Communications of the ACM*, 35(10):83–91, 1992.

# The Role of Experimental Algorithms in Genomics

Richard M. Karp

Department of Electrical Engineering and Computer Sciences, University of California, 378 Soda Hall, Berkeley, CA 94720, USA.
karp@icsi.berkeley.edu

Biology has become a computational science. In their efforts to understand the functioning of cells at a molecular level, biologists make use of a growing array of databases that codify knowledge about genomes, the genes within them, the structure and function of the proteins encoded by the genes, and the interactions among genes, RNA molecules, proteins, molecular machines and other chemical components of the cell. Biologists have access to high-throughput measurement technologies such as DNA microarrays, which can measure the expression levels of tens of thousands of genes in a single experiment.

Most computational problems in genomics do not fit the standard computer science paradigms in which a well-defined function is to be computed exactly or approximately. Rather, the goal is to determine nature's ground truth. The object to be determined may be well-defined - a genomic sequence, an evolutionary tree, or a classification of biological samples, for example - but the criterion used to evaluate the result of the computation may be ill-defined and subjective. In such cases several different computational methods may be tried, in the hope that a consensus solution will emerge. Often, the goal may be simply to explore a body of data for significant patterns, with no predefined objective. Often great effort goes into understanding the particular characteristics of a single important data set, such as the human genome, rather than devising an algorithm that works for all possible data sets. Sometimes there is an iterative process of computation and data acquisition, in which the computation suggests patterns in data and experimentation generates further data to confirm or disconfirm the suggested patterns. Sometimes the computational method used to extract patterns from data is less important than the statistical method used to evaluate the validity of the discovered patterns.

All of these characteristics hold not only in genomics but throughout the natural sciences, but they have not received sufficient consideration within computer science.

Many problems in genomics are attacked by devising a stochastic model of the generation of the data. The model includes both observed variables, which are available as experimental data, and hidden variables, which illuminate the structure of the data and need to be inferred from the observed variables. For example, the observed variables may be genomic sequences, and the hidden variables may be the positions of genes within those sequences. Machine learning theory provides general methods based on maximum likelihood or maximum a

C.C. Ribeiro and S.L. Martins (Eds.): WEA 2004, LNCS 3059, pp. 299–300, 2004.

posteriori probability for estimating the hidden variables. A useful goal for experimental algorithmics would be to characterize the performance of these general
methods.

Often a genomics problem can be viewed as piecing together a puzzle using
diverse types of evidence. Commitments are made to those features of the solution that are supported by the largest weight of evidence. Those commitments
provide a kind of scaffold for the solution, and commitments are successively
made to further features that are supported by less direct evidence but are consistent with the commitments already made. This was the approach used by
the Celera group in sequencing the human genome and other large, complex
genomes.

The speaker will illustrate these themes in connection with three specific
problems: finding the sites at which proteins bind to the genome to regulate
the transcription of genes, finding large-scale patterns of protein-protein interaction that are conserved in several species, and determinining the variations
among individuals that occur at so-called polymorphic sites in their genomes.
No knowledge of molecular biology will be assumed.

# A Fast Algorithm for Constructing Suffix Arrays for Fixed-Size Alphabets*

Dong K. Kim[1], Junha Jo[1], and Heejin Park[2]

[1] School of Electrical and Computer Engineering, Pusan National University, Busan
609-735, South Korea.
dkkim1@pusan.ac.kr, jhjo@islab.ce.pusan.ac.kr
[2] College of Information and Communications, Hanyang University, South Korea.
hjpark@hanyang.ac.kr

**Abstract.** The suffix array of a string $T$ is basically a sorted list of all
the suffixes of $T$. Suffix arrays have been fundamental index data structures in computational biology. If we are to search a DNA sequence in a
genome sequence, we construct the suffix array for the genome sequence
and then search the DNA sequence in the suffix array. In this paper,
we consider the construction of the suffix array of $T$ of length $n$ where
the size of the alphabet is fixed. It has been well-known that one can
construct the suffix array of $T$ in $O(n)$ time by constructing suffix tree
of $T$ and traversing the suffix tree. Although this approach takes $O(n)$
time, it is not appropriate for practical use because it uses a lot of spaces
and it is complicated to implement. Recently, almost at the same time,
several algorithms have been developed to directly construct suffix arrays in $O(n)$ time. However, these algorithms are developed for integer
alphabets and thus do not exploit the properties given when the size of
the alphabet is fixed. We present a fast algorithm for constructing suffix
arrays for the fixed-size alphabet. Our algorithm constructs suffix arrays
faster than any other algorithms developed for integer or general alphabets when the size of the alphabet is fixed. For example, we reduced the
time required for constructing suffix arrays for DNA sequences by 25%-
38%. In addition, we do not sacrifice the space to improve the running
time. The space required by our algorithm is almost equal to or even less
than those required by previous fast algorithms.

## 1 Introduction

The string searching problem is finding a pattern string $P$ of length $m$ in a
text string $T$ of length $n$. It occurs in many practical applications and has long
been studied [7]. Recently, searching DNA sequences in full genome sequences
is becoming one of the primary operations in bioinformatics areas. The studies
for efficient pattern search are divided into two approaches: One approach is to
preprocess the pattern. Preprocessing takes $O(m)$ time and then searching takes
$O(n)$ time. The other is to build a full-text index data structure for the text.

---

* This work is supported by Korea Research Foundation grant KRF-2003-03-D00343.

C.C. Ribeiro and S.L. Martins (Eds.): WEA 2004, LNCS 3059, pp. 301–314, 2004.
© Springer-Verlag Berlin Heidelberg 2004

Building the index data structure takes $O(n)$ time and searching takes $O(m)$ time. The latter approach is more appropriate than the former when we are to search DNA sequences in full genome sequences because the text is much longer than the pattern and we have to search many patterns in the text.

Two well-known such index data structures are suffix trees and suffix arrays. The suffix tree due to McCreight [17] is a compacted trie of all suffixes of the text. It was designed as a simplified version of Weiner's position tree [22]. The suffix array due to Manber and Myers [16] and independently due to Gonnet et al. [6] is basically a sorted list of all the suffixes of the text. Since suffix arrays consume less space than suffix trees, suffix arrays are preferred to suffix trees.

When we consider the complexity of index data structures, there are three types of alphabets from which text $T$ of length $n$ is drawn: (i) a fixed-size alphabet, (ii) an integer alphabet where symbols are integers in the range $[0, n^c]$ for a constant $c$, and (iii) a general alphabet in which the only operations on string $T$ are symbol comparisons.

We only consider fixed-size alphabets in this paper. The suffix tree of $T$ can be constructed in $O(n)$ time due to McCreight [17], Ukkonen [21], Farach et al. [2,3] and so on. The suffix array of $T$ can be constructed in $O(n)$ time by constructing the suffix tree of $T$ and then traversing it. Although this algorithm constructs the suffix array in $O(n)$ time, it is not appropriate for practical use because it uses a lot of spaces and is complicated to implement.

Manber and Myers [16] and Gusfield [8] proposed $O(n \log n)$-time algorithms for constructing suffix arrays without using suffix trees. Recently, almost at the same time, several algorithms have been developed to directly construct suffix arrays in $O(n)$ time. They are Kim et al.'s [14] algorithm, Ko and Aluru's [15] algorithm, Kärkkäinen and Sanders' [13] algorithm, and Hon et al.'s [11] algorithm. They are are based on similar recursive divide-and-conquer scheme. The recursive divide-and-conquer scheme is as follows.

1. Partition the suffixes of $T$ into two groups $A$ and $B$, and generate a string $T'$ such that the suffixes in $T'$ corresponds to the suffixes in $A$. This step requires encoding several symbols in $T$ into a new symbol in $T'$.
2. Construct the suffix array of $T'$ recursively.
3. Construct the suffix array for $A$ directly from the suffix array of $T'$.
4. Construct the suffix array for $B$ using the suffix array for $A$.
5. Merge the two suffix arrays for $A$ and $B$ to get the suffix array of $T$.

Kim et al. [14] followed Farach et al.'s [3] odd-even scheme, that is, divided the suffixes of $T$ into odd suffixes and even suffixes to get an algorithm running in $O(n)$ time. Kärkkäinen and Sanders [13] used skew scheme, that is, divided the suffixes of $T$ into suffixes beginning at positions $i \bmod 3 \neq 0$ (group $A$) and the other suffixes beginning at positions $i \bmod 3 = 0$ (group $B$). Ko and Aluru [15] divided the suffixes of $T$ into $S$-type and $L$-type suffixes. This algorithm does not require a string $T'$ and performs steps 3-5 in somewhat different way. Hon et al. [11] followed the odd-even scheme. They seems to have focused on reducing space rather than enhancing the running time. They used the backward search to merge the succinctly represented odd and even arrays.

In practice, Ko and Aluru's algorithm and Kärkkäinen and Sanders' skew algorithm run fast but Kim et al.'s odd-even algorithm runs slower than the two algorithms above. It is quite counter-intuitive that an algorithm based on the odd-even scheme is slower than an algorithm based on the skew scheme because the odd-even scheme has some advantages over the skew scheme such as less recursive calls and fast encoding. Although the odd-even scheme has some advantages, the merging step presented in Kim et al.'s algorithm is quite complicated and too slow and thus Kim et al.'s algorithm is slow overall. Therefore, it is natural to ask if there is a faster algorithm using the odd-even scheme by adopting a fast odd-even merging algorithm.

We got an affirmative answer to this question when the size of the alphabet is fixed. We present a fast odd-even algorithm constructing suffix arrays for the fixed-size alphabet by developing a fast merging algorithm. Our merging algorithm uses the backward search and thus requires a data structure for the backward search in suffix arrays. The data structure for the backward search in suffix arrays is quite different from the backward search in succinctly represented suffix arrays suggested by Hon et al. [11]. Our algorithm runs in $O(n \log \log n)$ time asymptotically. However, the experiments show that our algorithm is faster than any other previous algorithms running in $O(n)$ time. The reason for this is that $\log \log n$ is a small number ($\log \log n = 6$ if $n = 2^{64}$) in practical situation and thus $\log \log n$ can be considered to be a constant.

We describe the construction algorithm in Section 2. In Section 3, we measure the running time of this algorithm and compare it with those of previous algorithms. In Section 4, we further analyze our construction algorithm to explain why our algorithm runs so fast. In Section 5, we conclude with some remarks.

# 2   Construction Algorithm

We first introduce some notations and then we describe our construction algorithm. In describing our construction algorithm, we first describe the odd-even scheme, then describe the merging algorithm, and finally analyze the time complexity.

Consider a string $T$ of length $n$ over an alphabet $\Sigma$. Let $T[i]$ for $1 \leq i \leq n$ denote the $i$th symbol of string $T$. We assume that $T[n]$ is a special symbol # which is lexicographically smaller than any other symbol in $\Sigma$. The suffix array $SA_T$ of $T$ is basically a sorted list of all the suffixes of $T$. However, suffixes themselves are too heavy to be stored and thus only the starting positions of the suffixes are stored in $SA_T$. Figure 1 shows an example of a suffix array of $aaaabbbbaaabbbaabbb\#$. An odd suffix of $T$ is a suffix of $T$ starting at an odd position and an even suffix is a suffix starting at an even position. The odd array $SA_o$ and the even array $SA_e$ of $T$ are sorted lists of all odd suffixes and all even suffixes, respectively.

$T$

| 1 | 2 | 3 | 4 | 5 | 6 | 7 | 8 | 9 | 10 | 11 | 12 | 13 | 14 | 15 | 16 | 17 | 18 | 19 | 20 |
|---|---|---|---|---|---|---|---|---|----|----|----|----|----|----|----|----|----|----|----|
| a | a | a | a | b | b | b | b | a | a  | a  | b  | b  | b  | a  | a  | b  | b  | b  | #  |

$SA_T$

| 1 | 2 | 3 | 4 | 5 | 6 | 7 | 8 | 9 | 10 | 11 | 12 | 13 | 14 | 15 | 16 | 17 | 18 | 19 | 20 |
|---|---|---|---|---|---|---|---|---|----|----|----|----|----|----|----|----|----|----|----|
| 20 | 1 | 9 | 2 | 15 | 10 | 3 | 16 | 11 | 4 | 19 | 8 | 14 | 18 | 7 | 13 | 17 | 6 | 12 | 5 |

| # | a | a | a | a | a | a | a | a | a | b | b | b | b | b | b | b | b | b | b |
|---|---|---|---|---|---|---|---|---|---|---|---|---|---|---|---|---|---|---|---|
|   | a | a | a | a | a | a | b | b | b | # | a | a | b | b | b | b | b | b | b |
|   | a | a | a | b | b | b | b | b | b |   | a | a | # | a | a | b | b | b | b |
|   | a | b | b | b | b | b | b | b | b |   | a | b |   | a | a | # | a | a | b |
|   | b | b | b | b | b | b | # | a | b |   | b | b |   | a | b |   | a | a | a |
|   | b | b | b | # | a | b |   | a | a |   | b | b |   | b | b |   | a | b | a |
|   | b | a | b |   | a | a |   | b | a |   | b |   |   | b | b |   | b | b | a |

**Fig. 1.** The suffix array of $aaaabbbbaaabbbaabbb\#$. We depicted corresponding suffix to each entry of the suffix array. If the corresponding suffix is too long, we depicted a prefix of it rather than the whole suffix.

## 2.1 Odd-Even Scheme

We describe the odd-even scheme by elaborating the recursive divide-and-conquer scheme presented in the previous section.

1. Encode the given string $T$ into a half-sized string $T'$: We encode $T$ into $T'$ by replacing each pair of adjacent symbols $(T[2i-1], T[2i])$, $1 \le i \le n/2$, with a new symbol. How to encode $T$ into $T'$ is as follows.
   - Sort the pairs of adjacent symbols $(T[2i-1], T[2i])$ lexicographically and then remove duplicates: We use radix-sort to sort the pairs and we perform a scan on the sorted pairs to remove duplicates. Both the radix-sort and the scan take $O(n)$ time.
   - Map the $i$th lexicographically smallest pair of adjacent symbols into integer $i$: The integer $i$ is in the range $[1, n/2]$ because the number of pairs is at most $n/2$.
   - Replace $(T[2i-1], T[2i])$ with the integer it is mapped into.

   Fig. 2 shows how to encode $T = aaaabbbbaaabbbaabbb\#$ of length 20. After we sort the pairs $(T[1], T[2]), (T[3], T[4]), ..., (T[19], T[20])$ and remove duplicates, we are left with 4 distinct pairs which are $aa$, $ab$, $b\#$, and $bb$. We map $aa$, $ab$, $b\#$, and $bb$ into 1, 2, 3, and 4, respectively. Then, we get $T_o = 1144124143$ of length 10.
2. Construct the suffix array $SA_{T'}$ of $T'$ recursively.
3. Construct the odd array $SA_o$ of $T$ from $SA_{T'}$: Since the $i$th suffix of $T'$ corresponds to the $(2i-1)$st suffix of $T$, we get $SA_o[k]$ by computing $2SA_{T'}[k] - 1$ for all $k$. For example, $SA_o[2] = 2SA_{T'}[2] - 1 = 9$ in Fig. 2.
4. Construct the even array $SA_e$ of $T$ from the odd array $SA_o$: An even suffix is one symbol followed by an odd suffix. For example, the 8th suffix of $T$ is

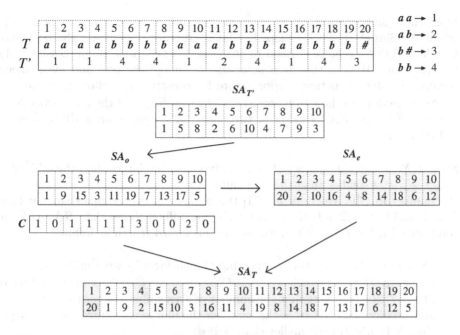

**Fig. 2.** An example of constructing the suffix array for $aaaabbbbaaabbbaabbb\#$.

$T[8]$ followed by the 9th suffix of $T$. We make tuples for even suffixes: the first element of a tuple is $T[2i]$ and the second element is the $(2i+1)$st suffix of $T$. First, we sort the tuples by the second elements (this result is given in $SA_o$). Then we stably sort the tuples by the first elements and we get $SA_e$.

5. Merge the odd array $SA_o$ and even array $SA_e$.

The odd-even scheme takes $O(n)$ time except the merging step [14].

## 2.2  Merging

We first describe the backward search on which our merging algorithm based and then describe our merging algorithm.

The backward search is finding the location of a pattern $P = p_1 p_2 \cdots p_m$ in the suffix array $SA_T$ by scanning $P$ by one symbol in reverse order: We first find the location of $p_m$ in $SA_T$, then the location of $p_{m-1}p_m$, the location of $p_{m-2}p_{m-1}p_m, \cdots$, and etc. We repeat this procedure until we find the location of $p_1 p_2 \cdots p_m$. One advantage of the backward search is that the locations of all suffixes of $P$ are found while we are finding the location of $P$.

The backward search was introduced by Ferragina and Manzini [4,5]. They used it to develop an opportunistic data structures that uses Burrow-Wheeler transformation [1]. The backward search has been used to search patterns in the succinctly represented suffix arrays [9,18,19] which is suggested by Grossi and Vitter [10]. Hon et al. [11] used the backward search to merge two succinctly represented suffix arrays. When it comes to suffix arrays (not succinctly represented), Sim et al. [20] developed a data structure that makes the backward search of $P$ in $SA_T$ be performed in $O(m \log |\Sigma|)$ time with preprocessing $SA_T$ in $O(n)$ time. We use this data structure to merge the odd and even arrays because this data structure is appropriate for constructing suffix arrays fast.

Now, we describe how to merge the odd array $SA_o$ and the even array $SA_e$. Merging $SA_o$ and $SA_e$ consists of two steps and requires an additional array $C[1..n/2+1]$.

**Step 1.** We count the number of even suffixes that are larger than the odd suffix $SA_o[i-1]$ and smaller than the odd suffix $SA_o[i]$ for all $2 \leq i \leq n/2$ and store the number in $C[i]$. We store in $C[1]$ the number of even suffixes smaller than $SA_o[1]$ and in $C[n/2+1]$ the number of even suffixes larger than $SA_o[n/2]$. To compute $C[i]$, $1 \leq i \leq n/2+1$, we use the backward search as follows.

1. Generate a data structure supporting the backward search in $SA_{T'}$.
2. Generate $T''$ where each symbol $T''[i]$, $1 \leq i \leq n/2$, is an encoding of $(T[2i], T[2i+1])$. (Note that the number of even suffixes larger than $SA_o[i-1]$ and smaller than $SA_o[i]$ is the same as the number of suffixes of $T''$ larger than $SA_{T'}[i-1]$ and smaller than $SA_{T'}[i]$.)
3. Initialize all the entries of array $C[1..n/2+1]$ to zero.
4. Perform the backward search of $T''$ in $SA_{T'}$: During the backward search, we increment $C[i]$ if a suffix of $T''$ is larger than $SA_{T'}[i-1]$ and smaller than $SA_{T'}[i]$.

**Step 2.** We store the suffixes in $SA_o$ and $SA_e$ into $SA_T$ using array $C$. Let $p_i$, $1 \leq i \leq n/2+1$, denote prefix sum $C[1] + \cdots + C[i]$. We should store the odd suffix $SA_o[i]$ into $SA_T[i+p_i]$ because $p_i$ even suffixes are smaller than $SA_o[i]$. We should store the even suffixes $SA_e[p_i + 1..p_{i+1}]$ into $SA_T[i+p_i+1..i+p_{i+1}]$ because $i$ odd suffixes are smaller than the even suffix $SA_e[j]$, $p_i + 1 \leq j \leq p_{i+1}$. To store the odd and even suffixes into $SA_T$ in $O(n)$ time, we do as follows: We store the suffixes into $SA_T$ from the smallest to the largest. We first store $C[1]$ smallest even suffixes into $SA_T$ and then we store the odd suffix $SA_o[1]$ into $SA_T$. Then, we store next $C[2]$ smallest even suffixes then store the odd suffix $SA_o[2]$. We repeat this procedure until all the odd and even suffixes are stored in $SA_T$. Consider the example in Fig 2. Since $C[1] = 1$, we store the smallest even suffix $SA_e[1]$ into $SA_T[1]$ and store the odd suffix $SA_o[1]$ into $SA_T[2]$. Since $C[2] = 0$, we store no even suffixes and then store the odd suffix $SA_o[2]$ into $SA_T[3]$.

We consider the time complexity of this merging algorithm. Since the backward search in step 1 takes $O(n \log |\Sigma'|)$ time where $\Sigma'$ is the set of the alphabet of $T'$ and the other parts of the merging step take $O(n)$ time, we get the following lemma.

**Lemma 1.** *This merging algorithm takes $O(n \log |\Sigma'|)$ time.*

## 2.3   Time Complexity

We consider the time complexity of our algorithm. Since the odd-even scheme takes $O(n)$ time except the merging [14], we only consider the time required for merging in all recursive calls. We first compute the time required for merging in each recursive call. Let $T_i$ denote the text and $\Sigma_i$ denote the set of the alphabet in the $i$th recursive call. We generalize Lemma 1 as follows.

**Corollary 1.** *The merging step in the ith recursive call takes* $O(|T_i| \log |\Sigma_{i+1}|)$ *time.*

Now, we compute $|T_i|$ and the upper bound of $|\Sigma_i|$. Since the length of text in the $i$th recursive call is the half length of text in the $(i-1)$st recursive call,

$$|T_i| = n/2^{i-1}. \tag{1}$$

Since the size of the alphabet in the $i$th call is at most the square of the alphabet size in the $(i-1)$st call,

$$|\Sigma_i| \le |\Sigma|^{2^{i-1}}. \tag{2}$$

Since $|\Sigma_i|$ is cannot be larger than $|T_i|$ and by equation (1),

$$|\Sigma_i| \le n/2^{i-1}. \tag{3}$$

We first compute the time required for merging in the first $\log \log n$ recursive calls and then the time required in the other recursive calls.

**Lemma 2.** *The merging steps in the first* $\log \log n$ *recursive calls take* $O(n \log \log n)$ *time.*

*Proof.* We first show that the merging step in each recursive call takes $O(n)$ time. The time required for the merging step in the $i$th, $1 \le i \le \log \log n$, recursive call is $O(|T_i| \log |\Sigma_{i+1}|)$ by Corollary 1. Since $|T_i| = n/2^{i-1}$ and $|\Sigma_{i+1}| \le |\Sigma|^{2^i}$ by equations 1 and 2, $O(|T_i| \log |\Sigma_{i+1}|) = O(n/2^{i-1} \cdot \log |\Sigma|^{2^i}) = O(n \cdot \log |\Sigma|)$. Since we only consider the case that $|\Sigma|$ is fixed, it is $O(n)$. Thus, the total time required for merging in the first $\log \log n$ recursive calls is $O(n \log \log n)$.

**Lemma 3.** *The merging steps in all the ith, $i > \log \log n$, recursive calls take* $O(n)$ *time.*

*Proof.* In the $i$th, $i > \log \log n$, recursive call, it takes $O(n/2^{i-1} \cdot \log(n/2^i))$ time by Corollary 1 and equations 1 and 3. If we replace $i$ by $\log \log n + j$ for $j > 0$, $O(n/2^{i-1} \cdot \log(n/2^i)) = O(n/(2^{j-1} \cdot \log n) \cdot \log(n/2^{\log \log n + j}))$. Since $\log n \ge \log(n/2^{\log \log n + j})$, $O(n/(2^{j-1} \cdot \log n) \cdot \log(n/2^{\log \log n + j})) = O(n/2^{j-1})$. Thus, the total time required for the merging steps in all the $i$th, $i > \log \log n$, recursive calls is $O(n + n/2 + n/4 + \cdots) = O(n)$.

By Lemma 2 and 3, we get the following theorem.

**Theorem 1.** *Our construction algorithm runs in* $O(n \log \log n)$ *time.*

# 3  Experimental Results

We measure the running time of our construction algorithm and compare it with those of previous algorithms due to Manber and Myers' (MM), Ko and Aluru's (KA), and Kärkkäinen and Sanders' (KS).

We made experiments on both random strings and DNA sequences. We generated different kinds of random strings which are differ in lengths (1 million, 5 million, 10 million, 30 million, and 50 million) and in the sizes of alphabets (2, 4, 64, and 128) from which they are drawn. For each pair of text length and alphabet size, we generated 100 random strings and made experiments on them. We also selected six DNA sequences of lengths 3.2M, 3.6M, 4,7M, 12.2M, 16.9M, and 31.0M, respectively. The data obtained from experiments on random strings are given in Table 1 and those on DNA sequences are given in Table 2. We measured the running time in mili-second on the 2.8Ghz Pentium VI with 2GB main memory.

**Table 1.** The data obtained from experiments on random strings. We used the C language to implement the algorithms. The code implementing algorithm MM is received from Myers and the code implementing algorithm KS is obtained from Sander's homepage (http://www.mpi-sb.mpg.de/~sanders/ programs/suffix/). The codes for algorithm KA and our algorithm were implemented by us. We tried to implement algorithm KA running as fast as possible.

| Algorithm Length($n$) | MM | KA | KS | Ours | MM | KA | KS | Ours |
|---|---|---|---|---|---|---|---|---|
| | $\mid\Sigma\mid = 2$ | | | | $\mid\Sigma\mid = 4$ | | | |
| 1M | 11,964 | 1,808 | 2,045 | 1,717 | 13,145 | 2,350 | 3,347 | 1,681 |
| 5M | 70,916 | 10,983 | 12,417 | 8,736 | 75,897 | 12,058 | 15,033 | 9,419 |
| 10M | 150,923 | 22,555 | 25,314 | 18,475 | 106,492 | 24,780 | 38,413 | 19,630 |
| 30M | 573,659 | 72,010 | 79,789 | 59,824 | 576,364 | 79,205 | 94,888 | 62,664 |
| 50M | 1,007,094 | N/A | 137,908 | 103,717 | 1,106,789 | N/A | 162,591 | 108,419 |
| | $\mid\Sigma\mid = 64$ | | | | $\mid\Sigma\mid = 128$ | | | |
| 1M | 1,439 | 2,933 | 1,850 | 2,149 | 2,305 | 3,183 | 1,950 | 2,241 |
| 5M | 83,683 | 19,403 | 20,200 | 13,636 | 13,853 | 15,680 | 10,411 | 12,367 |
| 10M | 175,174 | 38,503 | 40,416 | 26,413 | 25,347 | 42,677 | 42,845 | 29,347 |
| 30M | 595,766 | 118,114 | 126,206 | 81,706 | 86,215 | 127,984 | 134,980 | 86,227 |
| 50M | 1,094,356 | N/A | 211,792 | 141,276 | 158,491 | N/A | 226,425 | 152,249 |

Table 1 shows that our algorithm runs faster than the other algorithms we've tested in most cases of random strings. Our algorithm is slower than the other algorithms when $\mid\Sigma\mid = 128$ and $n \leq 10M$ and when $\mid\Sigma\mid = 64$ and $n = 1M$, i.e., when the size of the alphabet is large and the length of text is rather small. Thus, our algorithm runs faster than the other algorithms when the size of the alphabet is small or the length of text is large. This implies that our algorithm

**Table 2.** The data obtained from experiments on six DNA sequences

| DNA string (length) | *mito.nt* (3.2M) | *vector* (3.6M) | *ecoli.nt* (4.7M) | *yeast.nt* (12.2M) | *month.est human* (16.9M) | *month.est mouse* (31.0M) |
|---|---|---|---|---|---|---|
| KA | 8,234 | 7,781 | 11,750 | 32,609 | 41,094 | 87,656 |
| KS | 9,531 | 10,062 | 14,640 | 40,734 | 53,125 | 101,756 |
| Ours | 5,500 | 6,828 | 8,453 | 23,875 | 32,969 | 66,031 |

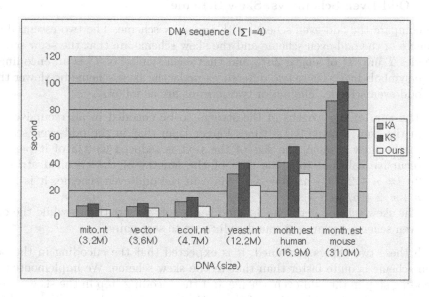

**Fig. 3.** A graphic representation of the data in Table 2.

is appropriate for constructing the suffix arrays of DNA sequences. Table 2 and Fig. 3 show the results of experiments made on DNA sequences. On average, our algorithm requires less time than algorithm KS by 38% and algorithm KA by 25%.

We consider the space required by the algorithms. Our algorithm, algorithm KA, and algorithm KS require $O(n)$ space asymptotically. To compare the hidden constants in asymptotic notations, we estimate the constant of each algorithm by increasing the length of text until the algorithm uses the virtual memory in the secondary storage. Our algorithm and algorithm KS start to use the virtual memory when $n$ is about 70M and algorithm KA start to use when $n$ is about 40M. (Thus, we got no data for algorithm KA when $n = 50$M in Table 1.) Hence, the space required by our algorithm is almost equal to or even less than those required by previous fast algorithms. This implies we do not sacrifice the space for the running time.

## 4   Discussion

We explain why our algorithm runs so fast. First, the odd-even scheme on which our algorithm is based is quite efficient. We show this by comparing the odd-even scheme with the skew scheme. Second, the part of our algorithm whose time complexity is $O(n \log \log n)$, i.e., the backward search in the merging step, does not dominate the total running time of our algorithm.

### 4.1   Odd-Even Scheme vs. Skew Scheme

We compare the odd-even scheme with the skew scheme. The two essential differences of the odd-even scheme and the skew scheme are that the skew scheme encodes $T$ into $T'$ of length $2n/3$ and that each symbol of $T'$ is an encoding of three symbols in $T$. These two differences make the skew scheme be slower than the odd-even scheme. The major two reasons are as follows.

- The sum of the lengths of the strings to be encoded in all recursive calls in the skew scheme is 1.5 times longer than that in the odd-even scheme. In the skew scheme, the size of the text is reduced to $2/3$ of it per each recursive call and thus the sum of the lengths of the encoded strings is $3n$ $(= n + 2n/3 + 4n/9 + \cdots)$, while in the odd-even scheme, it is $2n$ $(= n + n/2 + n/4 + \cdots)$.
- The skew scheme performs 3-round radix sort to sort triples while the odd-even scheme performs 2-round radix sort to sort pairs.

With these two things combined, it is expected that the encoding in the odd-even scheme is quite faster than that in the skew scheme. We implemented the encoding step in the odd-even scheme and the encoding step in the skew scheme and measured the running time of them. We summarized the results in Table 3.

**Table 3.** Comparion of the encoding time. The first table is the result when the size of the alphabet is 4 and the second table is the result when the size of the alphabet is 128.

| | $|\Sigma| = 4$ | | | | |
|---|---|---|---|---|---|
| $n$ | 1M | 5M | 10M | 30M | 50M |
| odd-even | 1,078 | 4,110 | 8,719 | 28,703 | 47,202 |
| skew | 1,860 | 10,672 | 21,233 | 64,956 | 110,438 |

| | $|\Sigma| = 128$ | | | | |
|---|---|---|---|---|---|
| $n$ | 1M | 5M | 10M | 30M | 50M |
| odd-even | 1,406 | 6,249 | 16,390 | 45,034 | 79,563 |
| skew | 1,608 | 7,985 | 35,094 | 107,578 | 179,326 |

Table 3 shows that the encoding in the odd-even scheme is about 2 times faster than that in the skew scheme. We carefully implemented the two schemes such that anything that is not inherent to the schemes such as code tuning, cannot affect the running time. We attached the C codes used to implement these two schemes in the appendix. (The code for encoding in the skew scheme is the same as that presented by Kärkkäinen and Sanders [15].)

## 4.2 The Ratio of $O(n \log \log n)$-Time Backward Search

We compute the ratio of the running time of the backward search which runs in $O(n \log \log n)$ time to the total running time of our algorithm. Table 4 shows that the backward search consumes 34% of the total running time when $|\Sigma| = 4$ and 27% when $|\Sigma| = 128$. Hence, the running time of the backward search is not crucial to the total running time.

**Table 4.** The ratio of the running time of the backward search to the total running time. The first table is the result when the size of the alphabet is 4 and the second table is the result when the size of the alphabet is 128.

| | $\|\Sigma\| = 4$ | | | | |
|---|---|---|---|---|---|
| $n$ | 1M | 5M | 10M | 30M | 50M |
| backward search | 829 | 4,999 | 5,874 | 20,157 | 35,390 |
| total | 2,376 | 11,859 | 19,220 | 63,767 | 106,984 |
| rate | 34.89% | 42.15% | 30.56% | 31.61% | 33.08% |

| | $\|\Sigma\| = 128$ | | | | |
|---|---|---|---|---|---|
| $n$ | 1M | 5M | 10M | 30M | 50M |
| backward search | 1,078 | 5,876 | 5,501 | 17,687 | 31,283 |
| total | 3,126 | 14,438 | 29,892 | 85,953 | 154,469 |
| rate | 34.48% | 40.70% | 18,40% | 20.58% | 20.25% |

## 5  Concluding Remarks

We presented a fast algorithm for constructing suffix arrays for the fixed-size alphabet. Our algorithm constructs suffix arrays faster than any other algorithms developed for integer or general alphabets when the alphabet is fixed-size. Our algorithm runs 1.33 - 1.6 times faster than the previous fast algorithms without sacrificing the space in constructing suffix arrays of DNA sequences.

In this paper, we considered the suffix array of $T$ as the lexicographically sorted list of the suffixes of $T$. Sometimes, the suffix array is defined as a pair of two arrays, which are the sorted array and the lcp array. The sorted array stores the sorted list of the suffixes and the lcp array stores the lcp(longest common prefix)'s of the suffixes. If the lcp array is needed, it can be computed from the sorted array in $O(n)$ time due to Kasai et al. [12].

Besides, there's another line of research related to reducing the space of suffix arrays. They use only $O(n)$-bit space representing a suffix array. Since we only focused on fast construction of suffix arrays in this paper, we used $O(n \log n)$-bit space. It will be interesting to develop a practically fast construction algorithm that uses only $O(n)$-bit space.

# References

1. M. Burrows and D. Wheeler, A block sorting lossless data compression algorithm, Technical Report 124 (1994), Digital Equipment Corporation, 1994.
2. M. Farach, Optimal suffix tree construction with large alphabets, *IEEE Symp. Found. Computer Science* (1997), 137–143.
3. M. Farach-Colton, P. Ferragina and S. Muthukrishnan, On the sorting-complexity of suffix tree construction, *J. Assoc. Comput. Mach.* 47 (2000), 987-1011.
4. P. Ferragina and G. Manzini, Opportunistic data structures with applications, *IEEE Symp. Found. Computer Science* (2001), 390–398.
5. P. Ferragina and G. Manzini, An experimental study of an opportunistic index, *ACM-SIAM Symp. on Discrete Algorithms* (2001), 269–278.
6. G. Gonnet, R. Baeza-Yates, and T. Snider, New indices for text: Pat trees and pat arrays. In W. B. Frakes and R. A. Baeza-Yates, editors, Information Retrieval: Data Structures & Algorithms, *Prentice Hall* (1992), 66–82.
7. D. Gusfield, Algorithms on Strings, Trees, and Sequences, *Cambridge Univ. Press* 1997.
8. D. Gusfield, An "Increment-by-one" approach to suffix arrays and trees, *manuscript* (1990).
9. R. Grossi, A. Gupta and J.S. Vitter, When indexing equals compression: Experiments with compressing suffix arrays and applications, *ACM-SIAM Symp. on Discrete Algorithms* (2004).
10. R. Grossi and J.S. Vitter, Compressed suffix arrays and suffix trees with applications to text indexing and string matching, *ACM Symp. Theory of Computing* (2000), 397–406.
11. W.K. Hon, K. Sadakane and W.K. Sung, Breaking a time-and-space barrier in constructing full-text indices, *IEEE Symp. Found. Computer Science* (2003), 251–260.
12. T. Kasai, G. Lee, H. Arimura, S. Arikawa, and K. Park, Linear-time longest-common-prefix computation in suffix arrays and its applications, *Symp. Combinatorial Pattern Matching* (2001), 181–192.
13. J. Kärkkäinen and P. Sanders, Simpler linear work suffix array construction, *Int. Colloq. Automata Languages and Programming* (2003), 943–955.
14. D.K. Kim, J.S. Sim, H. Park and K. Park, Linear-time construction of suffix arrays, *Symp. Combinatorial Pattern Matching* (2003), 186–199.
15. P. Ko and S. Aluru, Space-efficient linear time construction of suffix arrays, *Symp. Combinatorial Pattern Matching* (2003), 200–210.
16. U. Manber and G. Myers, Suffix arrays: A new method for on-line string searches, *SIAM J. Comput.* 22 (1993), 935–938.
17. E.M. McCreight, A space-economical suffix tree construction algorithm, *J. Assoc. Comput. Mach.* 23 (1976), 262–272.
18. K. Sadakane, Compressed text databases with efficient query algorithms based on the compressed suffix array, *Int. Symp. Algorithms and Computation* (2000), 410–421.

19. K. Sadakane, Succinct representations of lcp Information and improvements in the compressed suffix arrays, *ACM-SIAM Symp. on Discrete Algorithms* (2002), 225–232.
20. J.S. Sim, D.K. Kim, H. Park and K. Park, Linear-time search in suffix arrays, *Australasian Workshop on Combinatorial Algorithms* (2003), 139–146.
21. E. Ukkonen, On-line construction of suffix trees, *Algorithmica* 14 (1995), 249–260.
22. P. Weiner, Linear pattern matching algorithms, *Proc. 14th IEEE Symp. Switching and Automata Theory* (1973), 1–11.

# Appendix

```
-------------- Radix Sort(shared by both schemes) ----------------
static void radixPass(long* a, long* b, long* r, long n, long K)
{  // count occurrences
    long sum;
    long* c = new long[K + 1];              // counter array
    for (long i=0; i <= K; i++) c[i] = 0; // reset counters
    for (i=0; i < n; i++) c[r[a[i]]]++;   // count occurrences
    for (i=0, sum=0; i <= K; i++) {  // exclusive prefix sums
        long t = c[i];  c[i] = sum;  sum += t; }
    for (i=0; i < n; i++)                   // sort
        b[c[r[a[i]]]++] = a[i];
    delete [] c;
}
----------- Codes for encoding in the skew scheme ---------------
// lsb radix sort the mod 1 and mod 2 triples
radixPass(s12 , SA12, s+2, n02, K);
radixPass(SA12, s12 , s+1, n02, K);
radixPass(s12 , SA12, s  , n02, K);

// find lexicographic names of triples
int name = 0;
int c0 = -1, c1 = -1, c2 = -1;
for (i = 0;  i < n02;  i++)
{   if (s[SA12[i]]!=c0 || s[SA12[i]+1]!=c1 || s[SA12[i]+2]!=c2) {
        name++; c0=s[SA12[i]]; c1=s[SA12[i]+1]; c2=s[SA12[i]+2];
    }
    if (SA12[i] % 3 == 1)
            s12[SA12[i]/3] = name;          // left half
        else    s12[SA12[i]/3 + n0] = name;  // right half
}
----------- Codes for encoding in the odd-even scheme ------------
// radix sort
radixPass(SA, encodedText, s+1, n+1, K);
radixPass(encodedText, SA, s, n+1, K);
```

```
// find lexicographic names of couples
for name = 0, c0 = -1, c1 = -1;
for (i = 0; i <= n; i++)
{   if (SA[i] % 2 != 0) continue;
    if (s[SA[i]] != c0 || s[SA[i]+1] != c1) {
        name++; c0 = s[SA[i]]; c1 = s[SA[i]+1];
    }
    textEven[SA[i]/2] = name;
} // We construct the even array recursively rather than the odd
    array
// because 0 is the starting index of an array in the C language.
```

# Pre-processing and Linear-Decomposition Algorithm to Solve the k-Colorability Problem*

Corinne Lucet[1], Florence Mendes[1], and Aziz Moukrim[2]

[1] LaRIA EA 2083, 5 rue du Moulin Neuf 80000 Amiens, France.
{Corinne.Lucet,Florence.Mendes}@laria.u-picardie.fr
[2] HeuDiaSyC UMR CNRS 6599 UTC, BP 20529 60205 Compiègne, France.
Aziz.Moukrim@hds.utc.fr

**Abstract.** We are interested in the graph coloring problem. We studied the effectiveness of some pre-processings that are specific to the k-colorability problem and that promise to reduce the size or the difficulty of the instances. We propose to apply on the reduced graph an exact method based on a linear-decomposition of the graph. We present some experiments performed on literature instances, among which DIMACS library instances.

## 1 Introduction

The Graph Coloring Problem constitutes a central problem in a lot of applications such as school timetabling, scheduling, or frequency assignment [5,6]. This problem belongs to the class of NP-hard problems [10]. Various heuristics approaches have been proposed to solve it (see for instance [2,8,9,11,13,17,19,21]). Efficient exact methods are less numerous: implicit enumeration strategies [14, 20,22], column generation and linear programming [18], branch-and-bound [3], branch-and-cut [7], without forgetting the well-known exact version of Brelaz's DSATUR [2].

A *coloring* of a graph $G = (V, E)$ is an assignment of a color $c(x)$ to each vertex such that $c(x) \neq c(y)$ for all edges $(x, y) \in E$. If the number of colors used is $k$, the coloring of $G$ is called a *k-coloring*. The minimum value of $k$ for which a k-coloring is possible is called the *chromatic number* of $G$ and is denoted $\chi(G)$. The *graph coloring problem* consists in finding the chromatic number of a graph. Our approach to solve this problem is to solve for different values of $k$ the *k-colorability problem*: "does there exist a k-coloring of $G$ ?".

We propose to experiment the effectiveness of some pre-processings that are directly related to the k-colorability problem. The aim of these processings is to reduce the size of the graph by deleting vertices and to constrain it by adding edges. Then we apply a linear-decomposition algorithm on the reduced graph in order to solve the graph coloring problem. This method is strongly related to notions of tree-decomposition and path-decomposition, well studied by Bodlaender [1]. Linear-decomposition has been implemented efficiently by Carlier,

---

* With the support of Conseil Régional de Picardie and FSE.

C.C. Ribeiro and S.L. Martins (Eds.): WEA 2004, LNCS 3059, pp. 315–325, 2004.
© Springer-Verlag Berlin Heidelberg 2004

Lucet and Manouvrier to solve various NP-hard problems [4,15,16] and has for main advantage that the exponential factor of its complexity depends on the linearwidth of the graph but not on its size.

Our paper is organized as follows. We present in Sect. 2 some pre-processings related to the k-colorability problem and test their effectiveness on various benchmark instances. In Sect. 3, we describe our linear-decomposition algorithm. We report the results of our experiments in Sect. 4. Finally, we conclude and discuss about the perspectives of this work.

## 2    Pre-processings

In this section, we present several pre-processings to reduce the difficulty of a k-colorability problem. These pre-processings are iterated until the graph remains unchanged or the whole graph is reduced.

### 2.1    Definitions

An *undirected graph* $G$ is a pair $(V, E)$ made up of a vertex set $V$ and an edge set $E \subset V \times V$. Let $N = |V|$ and $M = |E|$. A graph $G$ is *connected* if for all vertices $w, v \in V (w \neq v)$, there exists a path from $w$ to $v$. Without loss of generality, the graphs we will consider in the following of this paper will be only undirected and connected graphs. Given a graph $G = (V, E)$ and a vertex $x \in V$, let $\vartheta(x) = \{y \in V/(x, y) \in E\}$. $\vartheta(x)$ represents the neighborhood of $x$ in $G$. The *subgraph* of $G = (V, E)$ induced by $I \subseteq V$, is the graph $G(I) = (I, E_I)$ such that $E_I = E \cap (I \times I)$. A *clique* of $G = (V, E)$ is a subset $C \subseteq V$ such that every two vertices in $C$ are joined by an edge in $E$. Let $\overline{E} = (V \times V) \setminus E$ be the set made up of all pairs of vertices that are not neighbors in $G = (V, E)$. Let $d$ be the degree of $G$, i.e. the maximal vertex degree among all vertices of $G$.

### 2.2    Reduction 1

A vertex reduction using the following property of the neighborhood of the vertices can be applied to the representative graph before any other computation with time complexity $O(|\overline{E}| * d)$, upper bounded by $O(N^3)$. Given a graph $G$, for each pair of vertices $x, y \in V$ such that $(x, y) \notin E$, if $\vartheta(y) \subseteq \vartheta(x)$ then $y$ and its adjacent edges can be erased from the graph. Indeed, suppose that $k - 1$ colors are needed to color the neighbors of $x$. The vertex $x$ can take the $k^{th}$ color. Vertices $x$ and $y$ are not neighbors. Moreover, the neighbors of $y$ are already colored with at most $k - 1$ colors. So, if $G \setminus \{y\}$ is k-colorable then $G$ is k-colorable as well and we can delete $y$ from the graph. This principle can be applied recursively as long as vertices are removed from the graph.

### 2.3    Reduction 2

Suppose that we are searching for a k-coloring of a graph $G = (V, E)$. Then we can use the following property: for each vertex $x$, if the degree of $x$ is strictly lower than $k$, $x$ and its edges can be erased from the graph [11]. Assume $x$ has $k - 1$ neighbors. In the worst case, those neighbors must have different colors. Then the vertex $x$ can take the $k^{th}$ color. It does not interfere in the coloring of the remaining vertices because all its neighbors have already been colored. Therefore we can consider from the beginning that it will take a color unused by its neighbors and delete it from the graph before the coloring. The time complexity of this reduction is $O(N)$. We apply this principle recursively by examining the remaining vertices until having totally reduced the graph or being enable to delete any other vertex.

## 2.4   Vertex Fusion

Suppose that we are searching for a k-coloring of $G = (V, E)$ and that a clique $C$ of size $k$ has been previously determined. For each couple of non-adjacent vertices $x, y \in V$ such that $x \notin C$ and $y \in C$, if $x$ is adjacent to all vertices of $C \setminus y$ then $x$ and $y$ can be merged by the following way: each neighbor of $x$ becomes a neighbor of $y$, then $x$ and its adjacent edges are erased from the graph. Indeed, since we are searching for a k-coloring, $x$ and $y$ must have the same color. Then $\forall z \in \vartheta(x)$ $c(y) \neq c(z)$ and the edge $(y, z)$ can be added to $G$. Then $\vartheta(x) \subseteq \vartheta(y)$ and $x$ can be erased from the graph (cf Sect. 2.2). The time complexity of this pre-processing is $O(N * k)$.

## 2.5   Edge Addition

Suppose that we are searching for a k-coloring of $G = (V, E)$ and that a clique $C$ of size $k$ has been previously determined. For each couple of non-adjacent vertices $x, y \in V$, if $\forall z \in C$ we have $(x, z) \in E$ or $(y, z) \in E$, then the edge $(x, y)$ can be added to the graph. Necessarily, $x$ must take a color from the colors of $C \setminus \vartheta(x)$. Since $\vartheta(y) \supseteq C \setminus \vartheta(x)$, $c(x) \neq c(y)$. This constraint can be represented by an edge between $x$ and $y$. The time complexity of this pre-processing is $O(|\overline{E}| * k)$, upper bounded by $O(N^2 * k)$.

---

**Algorithm 1** Pre-processings

---

**Input:** a graph $G$ and an integer $k$
**Output:** a graph $G'$ k-colorable if and only if $G$ is k-colorable
   **repeat**
      reduction 1
      reduction 2
      **if** $\exists$ at least 1 clique of size $k$ **then**
         apply vertex fusion and edge addition on $G$
      **end if**
   **until** there is no more change in $G$
   $G' = G$

---

## 2.6   Pre-processing Experiments

Our algorithms have been implemented on a PC AMD Athlon Xp 2000+ in C language. The method used is as follows. To start with, we apply on the entry graph $G$ a fast clique search algorithm: as long as the graph is not triangulated, we remove a vertex of smallest degree, and then we color the remaining triangulated graph by determining a perfect elimination order [12] on the vertices of $G$. The size of the clique provided by this algorithm, denoted $LB$, constitutes a lower bound of the chromatic number of $G$. Then we apply on $G$ the pre-processings described in Algorithm 1, supposing that we are searching for a k-coloring of the graph with $k = LB$. We performed tests on benchmark instances used at the computational symposium COLOR02, including well-known DIMACS instances (see description of the instances at http://mat.gsia.cmu.edu/COLOR02). Results are reported in Table 1. For each graph, we indicate the initial number of vertices $N$ and the number of edges $M$. The column $LB$ contains the size of the maximal clique found. The percentage of vertices deleted by the pre-processings is reported in column $Del$. The number of remaining vertices after the pre-processing step is reported in column $new\_N$. Remark that some of the instances are totally reduced by the pre-processings when $k = LB$, and that some of them are not reduced at all.

**Table 1.** Pre-processings results

| Graph | N | M | LB | new_N | Del | Graph | N | M | LB | new_N | Del |
|---|---|---|---|---|---|---|---|---|---|---|---|
| 1-FullIns3 | 30 | 100 | 3 | 15 | 50% | 1-FullIns4 | 93 | 593 | 3 | 35 | 62% |
| 1-FullIns5 | 282 | 3247 | 3 | 75 | 73% | 2-FullIns3 | 52 | 201 | 4 | 9 | 81% |
| 2-FullIns4 | 212 | 1621 | 4 | 41 | 81% | 2-FullIns5 | 852 | 12201 | 4 | 89 | 90% |
| 3-FullIns3 | 80 | 346 | 5 | 11 | 86% | 3-FullIns4 | 405 | 3524 | 2 | 51 | 87% |
| 3-FullIns5 | 2030 | 33751 | 2 | 107 | 95% | 4-FullIns3 | 114 | 541 | 6 | 13 | 89% |
| 4-FullIns4 | 690 | 6650 | 2 | 58 | 92% | 5-FullIns3 | 154 | 792 | 7 | 15 | 90% |
| 5-FullIns4 | 1085 | 11395 | 2 | 65 | 94% | fpsol2.i.1 | 496 | 11654 | 65 | 228 | 54% |
| fpsol2.i.2 | 451 | 8691 | 30 | 175 | 61% | fpsol2.i.3 | 425 | 8688 | 30 | 149 | 65% |
| inithx.i.1 | 864 | 18707 | 54 | 443 | 49% | inithx.i.2 | 645 | 13979 | 31 | 215 | 67% |
| inithx.i.3 | 621 | 13969 | 31 | 190 | 69% | mulsol.i.1 | 197 | 3925 | 49 | 60 | 70% |
| mulsol.i.2 | 188 | 3885 | 31 | 88 | 53% | mulsol.i.3 | 184 | 3916 | 31 | 83 | 55% |
| mulsol.i.4 | 185 | 3946 | 31 | 85 | 54% | mulsol.i.5 | 186 | 3973 | 31 | 84 | 55% |
| school1 | 385 | 19095 | 14 | 360 | 6% | school1_nsh | 352 | 14612 | 14 | 331 | 6% |
| 3-Inser_3 | 56 | 110 | 2 | 56 | 0% | 4-Inser_3 | 79 | 156 | 2 | 79 | 0% |
| le450_25a | 450 | 8260 | 20 | 297 | 34% | le450_25b | 450 | 8263 | 25 | 294 | 35% |
| anna | 138 | 493 | 11 | 0 | 100% | david | 87 | 812 | 11 | 0 | 100% |
| homer | 561 | 1629 | 13 | 0 | 100% | jean | 80 | 508 | 10 | 0 | 100% |
| mug100-1 | 100 | 166 | 3 | 100 | 0% | mug100-25 | 100 | 166 | 3 | 100 | 0% |
| mug88-1 | 88 | 146 | 3 | 88 | 0% | mug88-25 | 88 | 146 | 3 | 88 | 0% |
| miles250 | 128 | 387 | 7 | 34 | 73% | miles500 | 128 | 2340 | 20 | 0 | 100% |
| miles750 | 128 | 4226 | 31 | 0 | 100% | miles1000 | 128 | 6432 | 42 | 0 | 100% |
| miles1500 | 128 | 10396 | 73 | 0 | 100% | DSJR500_1 | 500 | 3555 | 12 | 28 | 94% |
| zeroin.i.1 | 211 | 4100 | 49 | 86 | 59% | zeroin.i.2 | 211 | 3541 | 30 | 55 | 74% |
| zeroin.i.3 | 206 | 3540 | 30 | 50 | 76% | games120 | 120 | 638 | 9 | 0 | 100% |

# 3    Linear-Decomposition Applied to the k-Colorability Problem

In this section, we propose a method which uses linear-decomposition mixed with Dsatur heuristic in order to solve the k-colorability problem.

## 3.1    Definitions

We will consider a graph $G = (V, E)$. Let $N = |V|$ and $M = |E|$. A *vertex linear ordering* of $G$ is a bijection $\mathcal{N} : V \to \{1, \ldots, N\}$. For more clarity, we denote $i$ the vertex $\mathcal{N}^{-1}(i)$. Let $V_i$ be subset of $V$ made of the vertices numbered from 1 to $i$. Let $H_i = (V_i, E_i)$ be the subgraph of $G$ induced by $V_i$. Let $F_i = \{j \in V / \exists (j, l) \in E \ j \leq i < l\} \ \forall i \in \{1, \ldots, |V|\}$. $F_i$ is the *boundary set* of $H_i$. Let $H_i' = (V_i', E_i')$ be the subgraph of $G$ such that $V_i' = (V \setminus V_i) \cup F_i$ and $E_i' = E \cap (V_i' \times V_i')$. The boundary set $F_i$ corresponds to the set of vertices joining $H_i$ to $H_i'$ (see Fig. 1).

The *linearwidth* of a vertex linear ordering $\mathcal{N}$ is $F_{max}(\mathcal{N}) = max_{i \in V}(|F_i|)$. We use a vertex linear ordering of the graph to resolve the k-colorability problem with a linear-decomposition. The resolution method is based on a sequential insertion of the vertices, using a vertex linear ordering previously determined. This will be developed in the following section.

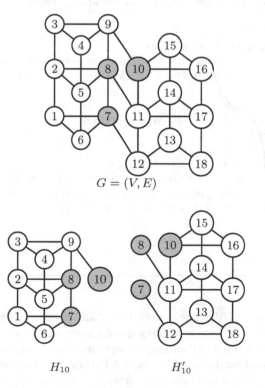

$$G = (V, E)$$

$$H_{10} \qquad\qquad H_{10}'$$

**Fig. 1.** A subgraph $H_{10}$ of $G$ and its boundary set $F_{10} = \{7, 8, 10\}$

## 3.2 Linear-Decomposition Algorithm

The details of the implementation of the linear-decomposition method are reported in Algorithm 2. The vertices of $G$ are numbered according to a linear ordering $\mathcal{N} : V \to \{1, \ldots, N\}$. Then, during the coloring, we will consider $N$ subgraphs $H_1, \ldots, H_N$ and the $N$ corresponding boundary sets $F_1, \ldots, F_N$, as defined in Sect. 3.1.

---

**Algorithm 2** k-colorability

---

**Input:** a graph $G$ and an integer $k$
**Output:** $Result$ : $True$ if and only if $G$ is k-coloriable
$\quad H_1 = (\{1\}, \emptyset)$
$\quad F_1 = \{1\}$
$\quad C(H_1, 1) = [1]$
$\quad i = 2$
$\quad Result = True$
$\quad$ **while** $i \leq N$ and $Result$ **do**
$\quad\quad Result = False$
$\quad\quad$ Build $H_i$ and $F_i$
$\quad\quad$ **for** each configuration $C(H_{i-1}, x)$ of $F_{i-1}$ **do**
$\quad\quad\quad$ **for** $j = 1$ to number of blocks of $C(H_{i-1}, x)$ **do**
$\quad\quad\quad\quad$ **if** $i$ does not have any neighbor in the block $j$ **then**
$\quad\quad\quad\quad\quad Result = True$
$\quad\quad\quad\quad\quad part = C(H_{i-1}, x)$
$\quad\quad\quad\quad\quad$ insert $i$ in the block $j$ of $part$
$\quad\quad\quad\quad\quad$ generate the configuration $C(H_i, y)$ corresponding to $part$
$\quad\quad\quad\quad\quad val(C(H_i, y)) = min(val(C(H_i, y)), val(C(H_{i-1}, x)))$
$\quad\quad\quad\quad$ **end if**
$\quad\quad\quad$ **end for**
$\quad\quad\quad$ **if** number of blocks of $C(H_{i-1}, x) < k$ **then**
$\quad\quad\quad\quad Result = True$
$\quad\quad\quad\quad part = C(H_{i-1}, x)$
$\quad\quad\quad\quad$ add to $part$ a new block containing $i$
$\quad\quad\quad\quad val(part) = max(val(C(H_{i-1}, x)),$ number of blocks of $part)$
$\quad\quad\quad\quad$ generate the configuration $C(H_i, y)$ corresponding to $part$
$\quad\quad\quad\quad val(C(H_i, y)) = min(val(C(H_i, y)), val(part))$
$\quad\quad\quad$ **end if**
$\quad\quad$ **end for**
$\quad\quad i = i + 1$
$\quad$ **end while**

---

The complexity of the linear-decomposition is exponential with respect to $F_{max}(\mathcal{N})$, so it is necessary to make a good choice when numbering the vertices of the graph. Unfortunately, finding an optimal vertex linear ordering in order to obtain the smallest linearwidth is a NP-complete problem [1]. After some experiments on various heuristics of vertex numbering, we choose to begin the numbering from the biggest clique provided by our clique search heuristic (cf

Sect. 2.6). Then we order the vertices by decreasing number of already numbered neighbors.

Starting from a vertex linear ordering, we build at first iteration a subgraph $H_1$ which contains only the vertex 1, then at each step the next vertex and its corresponding edges are added, until $H_N$. To each subgraph $H_i$ corresponds a boundary set $F_i$ containing the vertices of $H_i$ which have at least one neighbor in $H_i'$. The boundary set $F_i$ is built from $F_{i-1}$ by adding the vertex $i$ and removing the vertices whose neighbors have all been numbered with at most $i$. Several colorings of $H_i$ may correspond to the same coloring of $F_i$. Moreover, the colors used by the vertices $V_i \setminus F_i$ do not interfere with the coloring of the vertices which have an ordering number greater than $i$, since no edge exists between them. So, only the partial solutions corresponding to different colorings of $F_i$ have to be stored in memory. This way, several partial solutions on $H_i$ may be summarized by a unique partial solution on $F_i$, called *configuration* of $F_i$.

A configuration of the boundary set $F_i$ is a given coloring of the vertices of $F_i$. This can be represented by a partition of $F_i$, denoted $B_1, \ldots, B_j$, such that two vertices $u, v$ of $F_i$ are in the same block $B_c$ if they have the same color. The number of configurations of $F_i$ depends obviously on the number of edges between the vertices of $F_i$. The minimum number of configurations is 1. If the vertices of $F_i$ form a clique, only one configuration is possible: $B_1, \ldots, B_{|F_i|}$, with exactly one vertex in each block. The maximal number of configurations of $F_i$ equals the number of possible partitions of a set with $|F_i|$ elements. When no edge exists between the boundary set vertices, all the partitions are to be considered. Their number $T(F_i)$ grows exponentially according to the size of $F_i$. Their ordering number $x$, included between 1 and $T(F_i)$, is computed by an algorithm according to their number of blocks and their number of elements. This algorithm uses the recursive principle of Stirling numbers of the second kind. The partitions of sets with at most four elements and their ordering number are reported in Table 2. Let $C(H_i, x)$ be the $x^{th}$ configuration of $F_i$ for the subgraph $H_i$. Its value, denoted $val(C(H_i, x))$ equals the minimum number of colors necessary to color $H_i$ for this configuration.

At step $i$, fortunately we do not examine all the possible configurations of the step $i - 1$, but only those which have been created at precedent step, it means those for which there is no edge between two vertices of the same block. For each configuration of $F_{i-1}$, we introduce the vertex $i$ in each block successively. Each time the introduction is possible without breaking the coloring rules, the corresponding configuration of $F_i$ is generated. Moreover, for each configuration of $F_{i-1}$ with value strictly lower than $k - 1$, we generate also the configuration obtained by adding a new block containing the vertex $i$.

In order to improve the linear-decomposition, we apply the Dsatur heuristic evenly on the remaining graph $H_i'$, for different configurations of $F_i$. If Dsatur finds a k-coloring then the process ends and the result of the k-coloring is yes. Otherwise the linear-decomposition continues until a configuration is generated at step N, in this case the graph is k-colorable, or no configuration can be generated from the precedent step, in this case the graph is not k-colorable. The

**Table 2.** Classification of the partitions of sets containing from 1 to 4 elements

| | j=1 | j=2 | j=3 | j=4 | $T(i)$ |
|---|---|---|---|---|---|
| i=1 | **1** [1] | | | | $T(1) = 1$ |
| i=2 | **1** [12] | **2** [1][2] | | | $T(2) = 2$ |
| i=3 | **1** [123] | **2** [13][2] | **5** [1][2][3] | | |
| | | **3** [1][23] | | | |
| | | **4** [12][3] | | | $T(3) = 5$ |
| i=4 | **1** [1234] | **2** [134][2] | **9** [14][2][3] | **15** [1][2][3][4] | |
| | | **3** [13][24] | **10** [1][24][3] | | |
| | | **4** [14][23] | **11** [1][2][34] | | |
| | | **5** [1][234] | **12** [13][2][4] | | |
| | | **6** [124][3] | **13** [1][23][4] | | |
| | | **7** [12][34] | **14** [12][3][4] | | |
| | | **8** [123][4] | | | $T(4) = 15$ |

complexity of the linear-decomposition algorithm, upper bounded by $N * 2^{F_{max}}$, is exponential according to the linearwidth of the graph, but linear according to its number of vertices.

## 3.3    Example of Configuration Computing

Assume that we are searching for a 3-coloring of the graph $G$ of Fig. 2. Suppose that at step $i - 1$ we had $F_{i-1} = \{u, v\}$. The configurations of $F_{i-1}$ were $C(H_{i-1}, 1) = [uv]$ of value $\alpha$ and $C(H_{i-1}, 2) = [u][v]$ of value $\beta$. The value of $\beta$ is 2 or 3, since the corresponding configuration has 2 blocks and $k = 3$.

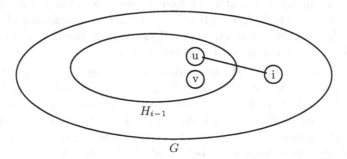

**Fig. 2.** Construction of $H_i = (V_{i-1} \cup \{i\}, E_{i-1} \cup \{(u, i)\})$

Suppose that at step $i$, vertex $u$ is deleted from the boundary set (we suppose that it has no neighbor in $H_i'$), so $F_i = \{v, i\}$. We want to generate the configurations of $F_i$ from the configurations of $F_{i-1}$. The insertion of $i$ in the unique block of $C(H_{i-1}, 1)$ is impossible, since $u$ and $i$ are neighbors. It is possible to add a new block, it provides the partition [uv][i] of 2 blocks, corresponding to the configuration $C(H_i, 2) = [v][i]$ with $val(C(H_i, 2)) = max(\alpha, 2)$. Vertex $i$ can be introduced in the second block of $C(H_{i-1}, 2)$. It provides the partition $[u][vi]$ corresponding to the configuration $C(H_i, 1) = [vi]$ with value $\beta$. It is also possible to add a new block to $C(H_{i-1}, 2)$, it provides the partition $[u][v][i]$ of 3 blocks corresponding to the configuration $C(H_i, 2) = [v][i]$. This configuration already exists, so $val(C(H_i, 2)) = min(val(C(H_i, 2)), max(\beta, 3))$. Thus two configurations are provided at step $i$, they are used to determine the configurations of the following step, and so on until the whole graph is colored.

## 4    k-Colorability Experiments

We performed experiments on the reduced instances of Table 1. Obviously, we did not test instances that were already solved by pre-processings. Results of these experiments are reported in Table 3. For each instance, we tested successive k-colorings, $k$ starting from $LB$ and increasing by step 1 until a coloring exists. We report the result and computing time of our linear-decomposition algorithm $kColor$, for one or two relevant values of $k$. We give also in column $F_{max}$ the linearwidth of the vertex linear ordering chosen $F_{max}(\mathcal{N})$. Most of these instances are easily solved. Configurations generated by instance 2-FullIns5 for a 6-coloring exceeded the memory capacity of our computer, so we give for this instance the results for a 5-coloring and for a 7-coloring. Instances 2-FullIns4, 3-FullIns4, 4-FullIns4, 5-FullIns4 and 4-Inser3 are solved exactly, whereas no exact method had been able to solve them at the COLOR02 computational symposium (see http://mat.gsia.cmu.edu/COLOR02/ summary.htm for all results).

**Table 3.** k-colorability results

| Problem | $F_{max}$ | k | kColor | Time | k | kColor | Time |
|---------|-----------|---|--------|------|---|--------|------|
| 1-FullIns3 | 17 | 3 | no | 0.00 | 4 | yes | 0.00 |
| 1-FullIns4 | 46 | 4 | no | 0.02 | 5 | yes | 0.02 |
| 1-FullIns5 | 132 | 5 | no | 436.17 | 6 | yes | 0.05 |
| 2-FullIns3 | 22 | 4 | no | 0.00 | 5 | yes | 0.00 |
| 2-FullIns4 | 93 | 5 | no | 0.02 | 6 | yes | 0.02 |
| 2-FullIns5 | 359 | 5 | no | 29.50 | 7 | yes | 0.15 |
| 3-FullIns3 | 36 | 5 | no | 0.00 | 6 | yes | 0.00 |
| 3-FullIns4 | 200 | 6 | no | 0.03 | 7 | yes | 0.03 |
| 4-FullIns3 | 49 | 6 | no | 0.00 | 7 | yes | 0.00 |
| 4-FullIns4 | 302 | 7 | no | 0.08 | 8 | yes | 0.13 |
| 5-FullIns3 | 64 | 7 | no | 0.00 | 8 | yes | 0.00 |
| 5-FullIns4 | 447 | 8 | no | 0.22 | 9 | yes | 0.20 |
| 3-Inser3 | 16 | 3 | no | 29.27 | 4 | yes | 0.00 |
| 4-Inser3 | 20 | 3 | no | 1772.95 | 4 | yes | 0.00 |
| mug88-1 | 8 | 3 | no | 0.00 | 4 | yes | 0.00 |
| mug88-25 | 8 | 3 | no | 0.00 | 4 | yes | 0.00 |
| | | | | | | | *continued on next page* |

| continued from previous page | | | | | | | |
|:---:|:---:|:---:|:---:|:---:|:---:|:---:|:---:|
| Problem | $F_{max}$ | k | kColor | Time | k | kColor | Time |
| mug100-1 | 7 | 3 | no | 0.02 | 4 | yes | 0.00 |
| mug100-25 | 8 | 3 | no | 0.00 | 4 | yes | 0.00 |
| miles250 | 16 | 7 | no | 0.00 | 8 | yes | 0.00 |
| le450-25a | 293 | 24 | no | 0.00 | 25 | yes | 0.98 |
| le450-25b | 303 | 25 | yes | 0.00 | | | |
| fpsol2.i.1 | 82 | 65 | yes | 0.00 | | | |
| fpsol2.i.2 | 50 | 30 | yes | 0.00 | | | |
| fpsol2.i.3 | 50 | 30 | yes | 0.00 | | | |
| inithx.i.1 | 69 | 54 | yes | 0.00 | | | |
| inithx.i.2 | 42 | 31 | yes | 0.00 | | | |
| inithx.i.3 | 42 | 31 | yes | 0.00 | | | |
| mulsol.i.1 | 61 | 49 | yes | 0.00 | | | |
| mulsol.i.2 | 45 | 31 | yes | 0.00 | | | |
| mulsol.i.3 | 46 | 31 | yes | 0.00 | | | |
| mulsol.i.4 | 45 | 31 | yes | 0.00 | | | |
| mulsol.i.5 | 47 | 31 | yes | 0.00 | | | |
| school1 | 291 | 14 | yes | 0.00 | | | |
| school1_nsh | 258 | 14 | yes | 0.00 | | | |
| zeroin.i.1 | 62 | 49 | yes | 0.00 | | | |
| zeroin.i.2 | 45 | 30 | yes | 0.00 | | | |
| zeroin.i.3 | 45 | 30 | yes | 0.00 | | | |
| DSJR500_1 | 68 | 12 | yes | 0.02 | | | |

## 5   Conclusions

In this paper, we have presented some pre-processings that are effective to reduce the size of some of difficult coloring instances. We presented also an original
method to solve the graph coloring problem by an exact way. This method has
the advantage of solving easily large instances which have a bounded linearwidth.
The computational results obtained on literature instances are very satisfactory.
We consider using the linear-decomposition mixed with heuristics approach to
deal with unbounded linearwidth instances. We are also looking for more reduction techniques to reduce the size or the difficulty of these instances.

## References

1. H. L. Bodlaender. A tourist guide through treewidth. *Acta Cybernetica*, 11:1–21, 1993.
2. D. Brelaz. New methods to color the vertices of a graph. *Communications of the ACM*, 22(4):251–256, april 1979.
3. M. Caramia and P. Dell'Olmo. Vertex coloring by multistage branch-and-bound. In *Computational Symposium on Graph Coloring and its Generalizations*, Corneil University, september 2002.

4. J. Carlier and C. Lucet. A decomposition algorithm for network reliability evaluation. *Discrete Appl. Math.*, 65:141–156, 1996.
5. D. de Werra. An introduction to timetabling. *European Journal of Operation Research*, 19:151–162, 1985.
6. D. de Werra. On a multiconstrained model for chromatic scheduling. *Discrete Appl. Math.*, 94:171–180, 1999.
7. I. Mendez Diaz and P. Zabala. A branch-and-cut algorithm for graph coloring. In *Computational Symposium on Graph Coloring and its Generalizations*, Corneil University, september 2002.
8. N. Funabiki and T. Higashino. A minimal-state processing search algorithm for graph coloring problems. *IEICE Transactions on Fundamentals*, E83-A(7):1420–1430, 2000.
9. P. Galinier and J.K. Hao. Hybrid evolutionary algorithms for graph coloring. *Journal of Combinatorial Optimization*, 3(4):379–397, 1999.
10. M. R. Garey and D. S. Johnson. *Computers and Intractability – A Guide to the Theory of NP-Completeness*. Freeman, San Francisco, 1979.
11. F. Glover, M. Parker, and J. Ryan. Coloring by tabu branch and bound. In Trick and Johnson [23], pages 285–308.
12. M. C. Golumbic. *Algorithmic Graph Theory and Perfect Graphs*. Academic Press, New York, 1980.
13. A. Hertz and D. De Werra. Using tabu search techniques for graph coloring. *Computing*, 39:345–351, 1987.
14. M. Kubale and B. Jackowski. A generalized implicit enumeration algorithm for graph coloring. *Communications of the ACM*, 28(4):412–418, 1985.
15. C. Lucet. *Méthode de décomposition pour l'évaluation de la fiabilité des réseaux*. PhD thesis, Université de Technologie de Compiègne, 1993.
16. J.F. Manouvrier. *Méthode de décomposition pour résoudre des problèmes combinatoires sur les graphes*. PhD thesis, Université de Technologie de Compiègne, 1998.
17. B. Manvel. Extremely greedy coloring algorithms. In F. Harary and J.S. Maybee, editors, *Graphs and applications: Proceedings of the First Colorado Symposium on Graph Theory*, pages 257–270, New York, 1985. John Wiley & Sons.
18. A. Mehrotra and M. A. Trick. A column generation approach for graph coloring. *INFORMS Journal on Computing*, 8(4):344–354, 1996.
19. C. A. Morgenstern. Distributed coloration neighborhood search. In Trick and Johnson [23], pages 335–357.
20. T. J. Sager and S. Lin. A pruning procedure for exact graph coloring. *ORSA Journal on Computing*, 3:226–230, 1991.
21. S. Sen Sarma and S. K. Bandyopadhyay. Some sequential graph colouring algorithms. *International Journal of Electronic*, 67(2):187–199, 1989.
22. E. Sewell. An improved algorithm for exact graph coloring. In Trick and Johnson [23], pages 359–373.
23. Michael A. Trick and David S. Johnson, editors. *Cliques, Coloring, and Satisfiability: Proceedings of the Second DIMACS Implementation Challenge*. American Mathematical Society, 1993.

# An Experimental Study of Unranking Algorithms*

Conrado Martínez and Xavier Molinero

Departament de Llenguatges i Sistemes Informàtics, Universitat Politècnica de
Catalunya, E-08034 Barcelona, Spain.
{conrado,molinero}@lsi.upc.es

**Abstract.** We describe in this extended abstract the experiments that
we have conducted in order to fullfil the following goals: a) to obtain
a working implementation of the unranking algorithms that we have
presented in previous works; b) to assess the validity and range of application of our theoretical analysis of the performance of these algorithms;
c) to provide preliminary figures on the practical performance of these
algorithms under a reasonable environment; and finally, d) to compare
these algorithms with the algorithms for random generation. Additionally, the experiments support our conjecture that the average complexity of *boustrophedonic* unranking is $\Theta(n \log n)$ for many combinatorial
classes (namely, those whose specification requires recursion) and that it
performs only slightly worse than *lexicographic* unranking for iterative
classes (those which do not require recursion to be specified).

## 1 Introduction

The problem of *unranking* asks for the generation of the $i$th combinatorial object
of size $n$ in some combinatorial class $\mathcal{A}$, according to some well defined order
among the objects of size $n$ of the class. Efficient unranking algorithms have been
devised for many different combinatorial classes, like binary and Cayley trees,
Dyck paths, permutations, strings or integer partitions, but most of the work in
this area concentrates in efficient algorithms for particular classes, whereas we
aim at generic algorithms that apply to a broad family of combinatorial classes.
The problem of unranking is intimately related with its converse, the *ranking*
problem, as well as with the problems of random generation and exhaustive
generation of all combinatorial objects of a given size. The interest of this whole
subject is witnessed by the vast number of research papers and books that has
appeared in over four decades (see for instance [1,2,7,8,9,13,14,15]).

In [10,11] we have designed *generic* unranking algorithms for a large family of
combinatorial classes, namely, those which can be inductively built from the basic
$\epsilon$-class (a class which contains only one object of size 0), atomic classes (classes

---

* This research was supported by the Future and Emergent Technologies programme
of the EU under contract IST-1999-14186 (ALCOM-FT) and the Spanish "Ministerio
de Ciencia y Tecnología" programme TIC2002-00190 (AEDRI II).

C.C. Ribeiro and S.L. Martins (Eds.): WEA 2004, LNCS 3059, pp. 326–340, 2004.

that contain only one object of size 1 or *atom*) and a collection of admissible combinatorial operators: disjoint unions, labelled and unlabelled products, sequence, set, etc. We say our algorithms are generic in the sense that they receive a finite description of the combinatorial class $\mathcal{A}$ to which the sought object belongs, besides the rank $i$ and size $n$ of the object to be generated. Our approach provides a considerable flexibility, thus making these algorithms attractive for their inclusion in general purpose combinatorial libraries such as combstruct for MAPLE [3] and MuPAD-combinat for MuPAD (see mupad-combinat.sourceforge.net), as well as for rapid prototyping.

Besides designing the unranking algorithms, we tackled in [10,11] the analysis of their performance. We were able to prove that for classes whose specification does not involve recursion (the so-called *iterative* classes) unranking can be performed in linear time on the size of the object. On the other hand, for those classes whose specification requires the use of recursion—for instance, binary trees—the worst case complexity of unranking is $\mathcal{O}(n^2)$, whereas the average case complexity is typically $\mathcal{O}(n\sqrt{n})$.

The results just mentioned on the average and worst-case complexity of unranking apply when we generate objects according to the *lexicographic* order; the use of the somewhat "extravagant" *boustrophedonic* order substantially improves the complexity, namely to $\mathcal{O}(n \log n)$ in the worst-case. Later on we will precisely define the lexicographic and boustrophedonic orders.

Our analysis showed that the performance of unranking algorithms coincides with that of the random generation algorithms devised by Flajolet *et al.* [6], where the same generic approach for the random generation of combinatorial objects was firstly proposed.Thus if the random generation algorithm has average-case (worst-case) complexity $\Theta(f(n))$ to generate objects of size $n$ in some class $\mathcal{A}$, so does the unranking algorithm for that class.

However, the analysis of the performance of the unranking algorithms—the same applies to the analysis of random generation algorithms—introduced several simplifications, so two important questions were left unanswered:

1. Which is the constant factor in the average-case complexity of the unranking algorithm for a given class $\mathcal{A}$? In particular, for objects unranked using the boustrophedonic order the analysis didn't provide even a rough estimation of such constant.
2. How does unranking compare random generation? Both unranking and random generation have complexities with identical order of magnitude, but the analysis did not settled how the respective constant factors compare to each other.

In order to provide an answer (even if partial) to these questions we have developed working implementations of our unranking algorithms in MAPLE and conducted a series experiments to measure their practical performance. We have instrumented our programs to collect data about the number of arithmetical operations that they perform. In this context this is the usual choice to measure implementation-independent performance, much in the same manner that key comparisons are used to measure the performance of sorting and searching algorithms. On the other hand, we have also collected data on their actual execution times. While these last measurements should have been done in several platforms and for several different implementations in order to extract well-grounded conclusions, they confirm the main trends that the implementation-independent measurements showed up when we compared unranking and random generation. In spite of the theoretical equivalence of the complexity of unranking and random generation, the comparison on practical terms is interesting if we want to produce samples without repetition. We can use the rejection method, generating objects of size $n$ at random until we have collected $m$ distinct objects, or we can generate $m$ distinct ranks at random in $\Theta(m)$ time using Floyd's algorithm and then make the corresponding $m$ calls to the unrank function. While the second is theoretically better, the answer could be different on practical grounds (especially if $m$ is not very big and random generation were more efficient than unranking by a large factor).

Our experiments also show that the unranking algorithms have acceptable running times —there is some penalty that we have to pay for the generality and flexibility— in a relatively common and modest platform when $n \leq 800$. For larger $n$, we face the problem of storing and manipulating huge counters (the number of combinatorial structures typically grows exponentially in $n$ or even as $n!$).

Last but not least, the experiments have been useful to show that our asymptotic analyses provide good estimations for already small values of $n$.

The paper is organized as follows. In Section 2 we briefly review basic definitions and concepts, the unranking algorithms and the theoretical analysis of their performance. Then, in sections 3 and 4 we describe the experimental setup and the results of our experiments.

## 2 Preliminaries

As it will become apparent, all the unranking algorithms in this paper require an efficient algorithm for counting, that is, given a specification of a class and a size, they need to compute the number of objects with the given size. Hence, we will only deal with so-called *admissible combinatorial classes* [4,5]. Those are constructed from *admissible operators*, operations over classes that yield new classes, and such that the number of objects of a given size in the new class can be computed from the number of objects of that size or smaller sizes in the constituent classes. In this paper we consider labelled and unlabelled objects built from these admissible combinatorial operators. *Unlabelled* objects are those whose atoms are indistinguishable. On the contrary, each of the $n$ atoms of a *labelled* object of size $n$ bears a distinct label drawn from the numbers 1 to $n$.

For labelled classes, the finite specifications are built from the $\epsilon$-class (with a single object of size 0 and no labels), labelled atomic classes, and the following combinatorial operators: union ('+'), partitional product ('$\star$'), sequence ('Seq'), set ('Set'), cycle ('Cycle'), substitution ('Subst'), and sequence, set and cycle with restricted cardinality. For unlabelled classes, the finite specifications are generated from the $\epsilon$-class, atomic classes, and combinatorial operators including union ('+'), Cartesian product ('×'), sequence ('Seq'), powerset, ('PowerSet'), set[1] ('Set'), substitution ('Subst'), and sequence, set and powerset with restricted cardinality. Figure 2 gives a few examples of both labelled and unlabelled admissible classes.

For the rest of this paper, we will use calligraphic uppercase letters to denote classes: $\mathcal{A}, \mathcal{B}, \mathcal{C}, \ldots$. Given a class $\mathcal{A}$ and a size $n$, $\mathcal{A}_n$ will denote the subset of objects of size $n$ in $\mathcal{A}$ and $a_n$ the number of such objects. Furthermore, given a class $\mathcal{A}$ we denote $\Upsilon\mathcal{A}_n$ the cumulated cost of unranking all the elements of size $n$ in $\mathcal{A}$. The average cost of unranking will be then given by $\mu_{\mathcal{A},n} = \Upsilon\mathcal{A}_n/a_n$, if we assume all objects of size $n$ to be equally likely.

The unranking algorithms themselves are not too difficult, except perhaps those for unlabelled powersets, multisets, cycles and their variants. Actually, we have not found any efficient algorithm for the unranking of unlabelled cycles, and they will not be considered any further in this paper. One important observation is that any unranking algorithm depends, by definition, on the order that we have imposed on the combinatorial class, and that the order itself will depend on a few basic rules plus the given specification of the class.

For instance, the order $\prec_{\mathcal{C}_n}$ among the objects of size $n$ for a class $\mathcal{C} = \mathcal{A} + \mathcal{B}$ is naturally defined by $\gamma \prec_{\mathcal{C}_n} \gamma'$ if both $\gamma$ and $\gamma'$ belong to the same class (either $\mathcal{A}_n$ or $\mathcal{B}_n$) and $\gamma \prec \gamma'$ within their class, or if $\gamma \in \mathcal{A}_n$ and $\gamma' \in \mathcal{B}_n$. It is then clear that although $\mathcal{A} + \mathcal{B}$ and $\mathcal{B} + \mathcal{A}$ are isomorphic ("the same class"), these two specifications induce quite different orders. The unranking algorithm for disjoint unions compares the given rank with the cardinality of $\mathcal{A}_n$ to decide if the sought object belongs to $\mathcal{A}$ or to $\mathcal{B}$ and then solves the problem by recursively calling the unranking on whatever class ($\mathcal{A}$ or $\mathcal{B}$) is appropriate.

For Cartesian products the order in $\mathcal{C}_n = (\mathcal{A} \times \mathcal{B})_n$ depends on whether $\gamma = (\alpha, \beta)$ and $\gamma' = (\alpha', \beta')$ have first components of the same size. If $|\alpha| = |\alpha'| = j$ then we have $\gamma \prec_{\mathcal{C}_n} \gamma'$ if $\alpha \prec_{\mathcal{A}_j} \alpha'$ or $\alpha = \alpha'$ and $\beta \prec_{\mathcal{B}_{n-j}} \beta'$. But when $|\alpha| \neq |\alpha'|$, we must provide a criterion to order $\gamma$ and $\gamma'$. The *lexicographic* order stems from the specification

$$\mathcal{C}_n = \mathcal{A}_0 \times \mathcal{B}_n + \mathcal{A}_1 \times \mathcal{B}_{n-1} + \ldots + \mathcal{A}_n \times \mathcal{B}_0,$$

in other words, the smaller object is that with smaller first component. On the other hand, the *boustrophedonic* order is induced by the specification

$$\mathcal{C}_n = \mathcal{A}_0 \times \mathcal{B}_n + \mathcal{A}_n \times \mathcal{B}_0 + \mathcal{A}_1 \times \mathcal{B}_{n-1} + \mathcal{A}_{n-1} \times \mathcal{B}_1 + \mathcal{A}_2 \times \mathcal{B}_{n-2} + \ldots,$$

in other words, we consider that the smaller pairs of total size $n$ are those whose $\mathcal{A}$-component has size 0, then those with $\mathcal{A}$-component of size $n$, then those with $\mathcal{A}$-component of size 1, and so on. Figure 1 shows the lists of unlabelled binary trees of size 4 in lexicographic (a) and boustrophedonic order (b).

---

[1] We shall sometimes use the term 'multisets' to refer to these, to emphasize that repetition is allowed.

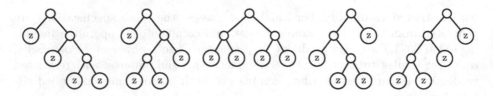

(a) Lexicographic order

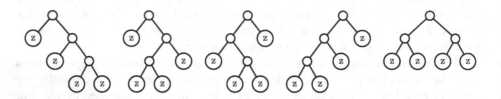

(b) Boustrophedonic order

**Fig. 1.** Binary trees of size 4.

Other orders are of course possible, but they either do not help improving the performance of unranking or they are too complex to be useful or of general applicability. The unranking algorithm for products is also simple: find the least $j$ such the given rank $i$ satisfies

$$\sum_{k=0}^{j-1} a_k b_{n-k} \le i < \sum_{k=0}^{j} a_k b_{n-k},$$

with the provision that all ranks begin at 0 (thus the rank of $\alpha \in \mathcal{A}_n$ is the number of objects in $\mathcal{A}_n$ which are strictly smaller than $\alpha$).

For labelled products we use the same orders (lexicographic, boustrophedonic), but we must also take the labels of the atoms into account. An object $\gamma$ in the labelled product $\mathcal{A} \star \mathcal{B}$ is actually a 3-tuple $(\alpha, \beta, \rho)$ where $\rho$ is a partition of the labels $\{1, \dots, n\}$ into the set of labels attached to $\alpha$'s atoms and the set of labels attached to $\beta$'s atoms. We will assume that if $\alpha = \alpha'$ and $\beta = \beta'$ then $\gamma = (\alpha, \beta, \rho) \prec_{\mathcal{C}_n} \gamma' = (\alpha', \beta', \rho')$ whenever $\rho \prec \rho'$ according to the natural lexicographical criterion. The rest of combinatorial constructs work much in the same way as Cartesian products and we will not describe them here; the distinction between lexicographic and boustrophedonic ordering makes sense for the other combinatorial constructs, and so we shall speak, for instance, of unranking labelled sets in lexicographic order or of unranking unlabelled sequences in boustrophedonic order.

The theoretical performance of the unranking algorithms is given by the following results (see [11], also [6]).

| Labelled class | Specification |
|---|---|
| Cayley trees | $\mathcal{A} = Z \star \mathsf{Set}(\mathcal{A})$ |
| Binary plane trees | $\mathcal{B} = Z + \mathcal{B} \star \mathcal{B}$ |
| Hierarchies | $\mathcal{C} = Z + \mathsf{Set}(\mathcal{C}, \mathrm{card} \geq 2))$ |
| Surjections | $\mathcal{D} = \mathsf{Seq}(\mathsf{Set}(Z, \mathrm{card} \geq 1)))$ |
| Functional graphs | $\mathcal{E} = \mathsf{Set}(\mathsf{Cycle}(\mathcal{A}))$ |

| Unlabelled class | Specification |
|---|---|
| Binary sequences | $\mathcal{A} = \mathsf{Seq}(Z + Z)$ |
| Rooted unlabelled trees | $\mathcal{C} = Z \times \mathsf{Set}(\mathcal{C})$ |
| Non plane ternary trees | $\mathcal{D} = Z + \mathsf{Set}(\mathcal{D}, \mathrm{card} = 3)$ |
| Integer partitions with distinct parts | $\mathcal{E} = \mathsf{PowerSet}(\mathsf{Seq}(Z, \mathrm{card} \geq 1))$ |

**Fig. 2.** Examples of labelled and unlabelled classes and their specifications

**Theorem 1.** *The worst-case time complexity of unranking for objects of size n in any admissible labelled class $\mathcal{A}$ using lexicographic ordering is of $\mathcal{O}(n^2)$ arithmetic operations.*

**Theorem 2.** *The worst-case time complexity of unranking for objects of size n in any admissible labelled class $\mathcal{A}$ using boustrophedonic ordering is of $\mathcal{O}(n \log n)$ arithmetic operations.*

A particular important case which deserves explicit treatment is that of *iterative* combinatorial class. A class is iterative if it is specified without using recursion (in technical terms, if the dependency graph of the specification is acyclic). Examples of iterative classes include surjections ($\mathsf{Seq}(\mathsf{Set}(Z, \mathrm{card} \geq 1)))$) and permutations ($\mathsf{Set}(\mathsf{Cycle}(Z)))$).

**Theorem 3.** *The cost of unranking any object of size n, using either lexicographic or boustrophedonic order, in any iterative class $\mathcal{A}$ is $\Theta(n)$.*

Last but least, the average-case cost of unranking and random generation under the lexicographic order can be obtained by means of a cost algebra that we describe very briefly here (for more details see [6,11]). Let $\alpha$ be the $i$-th object of $\mathcal{A}_n$ and $\mathrm{cu}(\alpha)$ the cost of unranking this object $\alpha$, then the cumulated cost of unranking all the objects in $\mathcal{A}_n$ is defined by $\Upsilon \mathcal{A}_n = \sum_{\alpha \in \mathcal{A}_n} \mathrm{cu}(\alpha)$. Now, if we introduce exponential and ordinary generating functions for the cumulated costs in labelled and unlabelled classes respectively, $\Upsilon \mathcal{A}(z) = \sum_{n \geq 0} \Upsilon \mathcal{A}_n z^n / n! = \sum_{\alpha \in \mathcal{A}} \mathrm{cu}(\alpha) z^{|\alpha|} / |\alpha|!$ and $\Upsilon \mathcal{A}(z) = \sum_{n \geq 0} \Upsilon \mathcal{A}_n z^n = \sum_{\alpha \in \mathcal{A}} \mathrm{cu}(\alpha) z^{|\alpha|}$, then the average cost of unranking all objects in $\mathcal{A}_n$ is given by $\mu_{n,\mathcal{A}} = [z^n] \Upsilon \mathcal{A}(z) / [z^n] A(z)$, where $A(z)$ denotes the generating function of the class $\mathcal{A}$ (exponential GF if $\mathcal{A}$ is labelled, ordinary GF is $\mathcal{A}$ is unlabelled), and $[z^n] f(z)$ denotes the $n$-th coefficient of the generating function $f(z)$.

The computation of $\mu_{n,\mathcal{A}}$ for labelled classes is then possible thanks to the application of the rules stated in the following theorem.

**Theorem 4.** *Let $\mathcal{A}$ be a labelled combinatorial class such that $\epsilon \notin \mathcal{A}$ (equivalently, $A(0) = 0$). Then*

1. $\Upsilon(\epsilon) = \Upsilon(Z) = 0$.
2. $\Upsilon(\mathcal{A} + \mathcal{B}) = \Upsilon\mathcal{A} + \Upsilon\mathcal{B}$.
3. $\Upsilon(\mathcal{A} \star \mathcal{B}) = \Theta A \cdot B + \Upsilon\mathcal{A} \cdot B + A \cdot \Upsilon\mathcal{B}$.
4. $\Upsilon(Seq(\mathcal{A})) = (\Theta A + \Upsilon\mathcal{A})/(1 - A)^2$.
5. $\Upsilon(Set(\mathcal{A})) = \exp(A) \cdot (\Theta A + \Upsilon\mathcal{A})$.
6. $\Upsilon(Cycle(\mathcal{A})) = (\Theta A + \Upsilon\mathcal{A})/(1 - A)$.
7. $\Upsilon(Seq(\mathcal{A}, card = k)) = k A^{k-1}(\Theta A + \Upsilon\mathcal{A})$,
8. $\Upsilon(Set(\mathcal{A}, card = k)) = A^{k-1}/(k - 1)! \, (\Theta A + \Upsilon\mathcal{A})$,
9. $\Upsilon(Subst(\mathcal{A}, \mathcal{B})) = (\Theta A + \Upsilon\mathcal{A})(B(z)) + (\Theta B + \Upsilon\mathcal{B}) \cdot A'(B(z))$,

*where the operator $\Theta$ for generating functions is $\Theta \equiv z\frac{d}{dz}$, $\Upsilon(Seq(\mathcal{A}, card = 0)) = \Upsilon(Seq(\mathcal{A}, card = 0)) = 0$ and $k > 0$.*

Similar rules exist for other variants of restricted cardinality, e.g. Set($\mathcal{A}$, card $\geq k$)) and for the unlabelled combinatorial constructs, although these are somewhat more complex.

Thanks to Theorem 4 we can compute very precise estimates of $\mu_{n\mathcal{A}}$; however, they are based upon a few simplification which are worth recalling here. First of all, we do not count the preprocessing cost of parsing and converting specifications to the so-called standard form. We also disregard the cost of calling the count function, as we assume that this is performed as a preprocessing step and that the values are stored into tables. Finally, we also neglect the cost of converting the object produced by the unranking algorithm back to the non-standard form in which the original specification were given. Last but not least, we assume that the cost of the non-recursive part of the unranking algorithms is exactly the number of iterations made to determine the size of the first component in a product or sequence, the leading component in a cycle or set, etc. For example, if the size of $\alpha$ in the object $(\alpha, \beta)$ of size $n$ which we want to produce is $j$ then we assume that the cost of unranking is $j + cu(\alpha) + cu(\beta)$. While this simplification does not invalidate the computation from the point of view of the order of growth of the average-complexity, a more accurate computation is necessary to determine the constant factors and lower order terms in the average-case complexity of unranking.

## 3     Experiments on the Performance of Unranking

We have implemented all the unranking algorithms for the combinatorial constructions described in the previous section, except for unlabelled cycles (this is a difficult open problem, see for instance [12]). Our programs[2] have been written for MAPLE (Maple V Release 8) and run under Linux in a Pentium 4 at 1.7 GHz with 512 Mb of RAM. The unranking algorithms use the basic facilities

---

[2] They are available on request from the second author; send email to molinero@lsi.upc.es.

for counting and parsing of specifications already provided by the combstruct package. We have also used the function draw in the combstruct package for random generation in order to compare its performance with that of our unrank function, but that is the subject of the next section. The interface to unrank is similar to that of the draw; for instance, we might write

```
bintree:= B = Union(Z, Prod(B, B)) ;
unrank([B, bintree, labelled], size = 10, rank = 3982254681);
```

to obtain the following labelled binary tree of size 10:

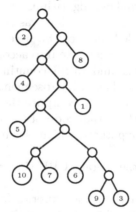

The function unrank also accepts several optional parameters, in particular we can specify which order we want to use: lexicographic (default) or boustrophedonic.

The first piece of our experimental setup was the choice of the combinatorial classes $\mathcal{A}$ to be used. Our aim was to find a representative collection. It was specially interesting to find cases where both the unlabelled and labelled versions made sense. As a counterexample, the labelled class $\mathsf{Seq}(Z)$ is interesting (permutations), but the unlabelled class $\mathsf{Seq}(Z)$ is not, since there is only one element of each size. The selected collection was[3]:

1. Binary trees: $\mathcal{B} = Z + \mathcal{B} \times \mathcal{B}$.
2. Unary-binary trees or Motzkin trees: $\mathcal{M} = Z + Z \times \mathcal{M} + Z \times \mathcal{M} \times \mathcal{M}$.
3. Integer partitions (unlabelled):

$$\mathcal{P} = \mathsf{Set}(\mathsf{Seq}(Z, \mathrm{card} \geq 1)).$$

4. Integer compositions (unlabelled) / Surjections (labelled):

$$\mathcal{C} = \mathsf{Seq}(\mathsf{Set}(Z, \mathrm{card} \geq 1)).$$

5. Non-ordered rooted trees (unlabelled) / Cayley trees (labelled):

$$\mathcal{T} = Z \times \mathsf{Set}(\mathcal{T}).$$

---

[3] For labelled versions of the class we have to use labelled products $\star$ instead of standard Cartesian product $\times$.

6. Functional graphs (labelled):

$$\mathcal{F} = \mathsf{Set}(\mathsf{Cycle}(\mathcal{T})).$$

One goal of the first set of experiments that we have conducted was to empirically measure $\overline{\mu}_{n,\mathcal{A}}$ when using lexicographic ordering and to compare it with the theoretical asymptotic estimate, so that we could determine the range of practical validity of that asymptotic estimate. A second goal was to measure $\overline{\mu}'_{n,\mathcal{A}}$, which is similar to $\overline{\mu}_{n,\mathcal{A}}$ but it accurately counts all the arithmetical operations used by the unranking algorithms; recall that the theorical analysis briefly sketched in the previous section was based upon several simplifying assumptions which disregarded some of the necessary arithmetical operations. Nevertheless $\mu'_{n,\mathcal{A}}$, as $\mu_{n,\mathcal{A}}$, does not take into account the arithmetic operations needed to parse specifications and to fill counting tables, since once this has been done as a preprocessing phase, all subsequent **unrank** or **draw** calls on the corresponding class do not need to recompute the tables or to parse the specification. The inspection of the actual code of the unranking programs suggests that $\mu'_{n,\mathcal{A}} \approx \mu_{n,\mathcal{A}} + 3n$, an hypothesis which is consistent with the data that we have collected.

We have also measured the average CPU time $\tau_{n,\mathcal{A}}$ to unrank objects of size $n$.

We have also performed the same experiments using the unranking algorithms with boustrophedonic order. The theoretical analysis establishes that the worst-case for unranking any class using this order is $\mathcal{O}(n \log n)$; however, there are no specific results for its average-case complexity. The experiments support our conjecture that the average complexity is $\Theta(n \log n)$ whenever the class is not iterative.

Due to the huge size of the numbers involved in the arithmetical computations performed by unranking we have considered in our experiments objects of size up to 800. For instance, initializing the tables of counts to unrank objects of size 300 can take up to 30 seconds of CPU! In each experiment we use $N$ random ranks to gather the statistics; for sizes of the objects up to 300 we have used samples of $N = 10000$ ranks, whereas for large objects we have used smaller samples, with $N = 100$ ranks.

The theoretical cost of unranking binary trees (unlabelled or labelled) in lexicographic order is $\mu_{n,\mathcal{B}} = \frac{\sqrt{\pi}}{2} n\sqrt{n} - \frac{1}{2}n + \frac{1}{16} + o(1) = 0.8862n\sqrt{n} - 0.5n + 0.0625 + o(1)$. The best fit for the measured data is $\overline{\mu}_{n,\mathcal{B}} = 0.897n\sqrt{n} - 0.673n + 2.086$. The collected data shows that the theoretical asymptotic estimate is very accurate even for $n = 50$, the relative error being less than 1.1% (see Figure 3). Looking at the plot itself the difference is almost unnoticeable.

All the samples used in this experiments had $N = 10000$ random ranks. The average CPU times (in seconds) also grow as $n\sqrt{n}$ in the considered range of sizes; for much larger sizes, it is not reasonable to assume that the cost of each single arithmetic operation is constant. The experimental data in the case of unlabelled binary trees does not substantially differ and the conclusions are basically identical, except that average CPU times are noticeably smaller, as the magnitude of the involved counts is very small compared with that of the labelled case.

On the other hand, the experimental data for $\overline{\mu}'_{n,\mathcal{B}}$ supports the hypothesis $\mu'_{n,\mathcal{B}} = \mu_{n,\mathcal{B}} + 3n$; a best fit gives $\overline{\mu}'_{n,\mathcal{B}} = 1.010436960\overline{\mu}_{n,\mathcal{B}} + 2.784136376n + 7.752450064$.

| $n$ | $\overline{\mu}_{n,\mathcal{B}}$ | $\overline{\mu}'_{n,\mathcal{B}}$ | $\tau_{n,\mathcal{B}}$ |
|---|---|---|---|
| 50 | 285.21 | 432.21 | 0.04 |
| 100 | 829.93 | 1130.86 | 0.13 |
| 150 | 1559.29 | 1995.21 | 0.26 |
| 200 | 2401.32 | 3003.24 | 0.43 |
| 250 | 3378.20 | 4100.15 | 0.66 |
| 300 | 4467.72 | 5364.72 | 0.91 |

(a) Experimental data.

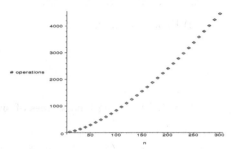

(b) Plot of the experimental $\overline{\mu}_{n,\mathcal{B}}$ vs. the theoretical $\mu_{n,\mathcal{B}}$.

**Fig. 3.** Unranking binary trees of size $n$ in lexicographic order.

Figure 4 gives the experimental data for the unranking of labelled binary trees using boustrophedonic order. The data is consistent with the hypothesis that on average this cost is $\Theta(n \log n)$ and in particular the best fit curve is

$$0.623n \ln n + 1.379n - 3.304.$$

If we compute the best fit for the data corresponding to the boustrophedonic unranking of unlabelled binary trees we get

$$0.653n \ln n + 1.198n + 0.116,$$

which suggests that the performance of the labelled and unlabelled versions of the boustrophedonic unranking for products is essentially the same. We will see later that this is no longer true when we compare the performance of the unranking (either lexicographic or boustrophedonic) of labelled and unlabelled sets.

The collected data for boustrophedonic unranking involves objects of sizes from $n = 50$ to $n = 300$, with samples of $N = 10000$ random ranks. Side by side, we give the corresponding values for lexicographic order to ease the comparison. The graph to the right also plots $\overline{\mu}_{n,\mathcal{B}}$ for boustrophedonic order (solid line) and lexicographic order (dashed line); the $n \log n$ behavior of the first versus the $n\sqrt{n}$ of the second is made quite evident in the plot.

| $n$ | $\overline{\mu}_{n,\mathcal{B}}$ | $\overline{\mu}_{n,\mathcal{B}}^{[\text{lex}]}$ | $\tau_{n,\mathcal{B}}$ | $\tau_{n,\mathcal{B}}^{[\text{lex}]}$ |
|---|---|---|---|---|
| 50 | 187.75 | 285.21 | 0.03 | 0.04 |
| 100 | 421.21 | 829.93 | 0.07 | 0.13 |
| 150 | 672.17 | 1559.29 | 0.11 | 0.26 |
| 200 | 932.90 | 2401.32 | 0.15 | 0.43 |
| 250 | 1202.21 | 3378.20 | 0.21 | 0.66 |
| 300 | 1476.65 | 4467.72 | 0.26 | 0.91 |

(a) Experimental data.

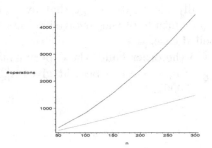

(b) Plot of the experimental data for boustrophedonic order vs. lexicographic order.

**Fig. 4.** Unranking binary trees of size $n$ in boustrophedonic order.

Functional graphs ($\mathcal{F}$) are sets of cycles of Cayley trees. For instance, the graph for the function $f$, with $f(1) = 8, f(2) = 6, f(3) = 2, f(4) = 5, f(5) = 10, f(6) = 5, f(7) = 5, f(8) = 1, f(9) = 10, f(10) = 6$ is

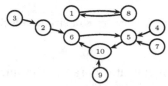

which is the 5000000-th element of size 10 in $\mathcal{F}$. The theoretical analysis of the cost of unranking such objects gives

$$\mu_{n,\mathcal{F}} = \frac{\sqrt{2\pi}}{4} n\sqrt{n} + \frac{7}{3}n + \mathcal{O}(\sqrt{n}) = 0.6266n\sqrt{n} + 2.3333n + \mathcal{O}(\sqrt{n}).$$

Now, the best fit for the measured data is $\overline{\mu}_{n,\mathcal{B}} = 0.609n\sqrt{n} + 2.532n$, in good accordance with the asymptotic estimate; the relative error is less than %1 in the range of sizes we have considered (see Figure 5). Again the plots of $\mu_{n,\mathcal{F}}$ and $\overline{\mu}_{n,\mathcal{F}}$ show almost no difference.

When comparing $\overline{\mu}_{n,\mathcal{F}}$ to $\overline{\mu}'_{n,\mathcal{F}}$ we again find further evidence for the relation $\mu'_n = \mu_n + 3n$; the comparison of the performance of the boustrophedonic order with that of the lexicographic order is qualitatively similar to what we found in the case of binary trees.

Also, we find that the average cost of unranking in boustrophedonic seems to be $\Theta(n \log n)$; the best fit for the experimental data (see Table 1) is

$$\overline{\mu}_{n,\mathcal{F}} = 2.427n \ln n - 6.235n + 77.612.$$

Together with the other examples of non-iterative classes (binary trees, Motzkin trees, Cayley trees, ... ) the data for the boustrophedonic unranking of functional graphs is consistent with our conjecture that the average cost

| $n$ | $\bar{\mu}_{n,\mathcal{F}}$ | $\bar{\tau}_{n,\mathcal{F}}$ |
|-----|------|------|
| 50 | 338.79 | 0.05 |
| 100 | 861.11 | 0.13 |
| 150 | 1499.93 | 0.23 |
| 200 | 2239.00 | 0.36 |
| 250 | 3038.26 | 0.57 |
| 300 | 3928.05 | 0.79 |

(a) Experimental data.

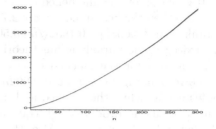

(b) Plot of $\bar{\mu}_{n,\mathcal{F}}$ vs. $\mu_{n,\mathcal{F}}$.

**Fig. 5.** Unranking functional graphs of size $n$ in lexicographic order.

of boustrophedonic unranking is $\Theta(n \log n)$ for non-iterative class (it is linear for iterative classes).

**Table 1.** Unranking functional graphs of size $n$ in boustrophedonic order.

| $n$ | $\bar{\mu}_{n,\mathcal{F}}$ | $\tau_{n,\mathcal{F}}$ |
|-----|------|------|
| 50 | 237.74 | 0.04 |
| 100 | 577.06 | 0.10 |
| 150 | 968.80 | 0.17 |
| 200 | 1398.67 | 0.26 |
| 250 | 1865.22 | 0.36 |
| 300 | 2364.55 | 0.48 |

The results for the remaining considered classes are very similar. We summarize here our main findings.

For the lexicographic unranking of Motzkin trees we have $\mu_{n,\mathcal{M}} = \frac{\sqrt{3\pi}}{6}n\sqrt{n} + \mathcal{O}(n) = 0.5116n\sqrt{n} + \mathcal{O}(n)$, while the best fit for the experimental data gives us $\bar{\mu}_{n,\mathcal{M}} = 0.518n\sqrt{n} - 0.454n + 6.74$, with relative errors less than 1% between the asymptotic estimate and the empirical values already for $n = 50$. For boustrophedonic unranking the best fit is $\bar{\mu}_{n,\mathcal{M}} = 0.642n \ln n + 0.6n + 1.609$. If we repeat the experiments for labelled Motzkin trees (which have the same theoretical performance as in the unlabelled case) we find that the best fit for lexicographic unranking is $0.513n\sqrt{n} - 0.355n + 4.54$ and for boustrophdonic it is $0.633n \ln n + 0.655\sqrt{n} + 0.872$, again supporting the equivalence of labelled and unlabelled unranking when only unions, products and sequences are involved, even for boustrophedonic unranking.

For the unlabelled class $\mathcal{P}$ of integer partitions, the best fit for lexicographic unranking is $\bar{\mu}_{n,\mathcal{P}} = 1.8162n + 19.382$, while the best fit for boustrophedonic unranking is $\bar{\mu}_{n,\mathcal{P}} = 1.95n + 24.301$. The theoretical analysis gives $\mu_{n,\mathcal{P}} = 2n$ for lexicographic unranking.

If we consider the unlabelled class $\mathcal{C} = \mathsf{Seq}(\mathsf{Set}(Z, \text{card} \geq 1))$ of integer compositions then the best fit curves are $2.7566n - 0.145\sqrt{n} + 0.2221$ for lexicographic unranking and $3.0083n - 0.1952\sqrt{n} + 0.5066$ for boustrophedonic unranking. Here, the lexicographic unranking has theorical average cost $\mu_{n,\mathcal{C}} = \frac{5}{2}n + \frac{1}{2}$. Interestingly enough, for the labelled class $\mathcal{C} = \mathsf{Seq}(\mathsf{Set}(Z, \text{card} \geq 1))$ of surjections the corresponding best fits are $2.002n$ and $2.9122n + 0.0584\sqrt{n} - 2.855$. This confirms that while the labelled and unlabelled versions of the unranking algorithms for unions, products and sequences behave identically, it is not the case for labelled and unlabelled sets. Also, these experiments indicate that boustrophedonic unranking could be slightly more inefficient than lexicographic unranking for iterative classes, contrary to what happens with the non-iterative classes, where boustrophedonic unranking clearly outperforms lexicographic unranking, even for small $n$.

## 4    An Empirical Comparison of Unranking and Random Generation

From a theoretical point of view this question has a clear cut answer: they have identical complexity. However, the hidden constant factor and lower order terms may markedly differ; the goal of the experiments described in this short section is to provide evidence about this particular aspect. We have chosen the implementation of random generation already present in the combstruct package for a fair comparison, since the platform, programming language, etc. are identical then. From the point of view of users, the underlying ordering of the class is irrelevant when generating objects at random, hence the default order used by the function draw is the boustrophedonic order, which is never significantly worse than the lexicographic order and it is frequently much faster (see Section 3). Hence, to get a meningful comparison we will also use boustrophedonic unranking in all the experiments of this section.

As in the previous section we will not take into account all the preprocess needed by both unranking and random generation, therefore we will "start the chrono" once the initialization of counting tables and the parsing of specifications have already been finished. Also, in the previous section we have given the data concerning the average CPU time for unranking; for the largest objects ($n = 800$) this time is around 1.2 seconds; for moderately sized objects ($n = 300$) the average time is typically around 0.3 secons. Since we want here to compare unranking and random generation we will systematically work with the ratio $\rho_{n,\mathcal{A}}$ between the average time of unrank and the average time of draw.

Table 2 summarizes this comparative analysis. Here $\mathcal{B}$ denotes unlabelled binary trees, $\mathcal{F}$ are the (labelled) functional graphs, $\mathcal{M}$ denotes unlabelled Motzkin trees and $\mathcal{P}$ the (unlabelled) class of integer partitions. The data varies widely from one case to other, so that no firm conclusions can be drawn. Some preliminary results (with small samples for large sizes) confirm the theoretical prediction that $\rho_{n,\mathcal{A}}$ must tend to a constant $\rho_{\mathcal{A}}$ as $n \to \infty$, but there seems that no easy rule to describe/compute these constants. For instance, the first three classes are non-iterative and the fourth is iterative, but there is no "commonality" in the behavior of the three first classes. Neither there is such commonality between the arborescent classes $\mathcal{B}$ and $\mathcal{M}$, or the unlabelled classes. From the small collection we have worked with, however, it seems that $\rho_{\mathcal{A}}$ will be typically between 1.0 and 3.0. Further tests and analytical work is necessary to confirm this hypothesis, but if it were so, then sampling based on unranking plus Floyd's algorithm should be the alternative of choice as long as the number of elements to be sampled were $\geq 1000$.

**Table 2.** Ratios of average times for unranking and random generation.

| $n$ | $\rho_{n,\mathcal{B}}$ | $\rho_{n,\mathcal{F}}$ | $\rho_{n,\mathcal{M}}$ | $\rho_{n,\mathcal{P}}$ |
|-----|------|------|------|------|
| 50  | 1.00 | 4.00 | 1.50 | 1.00 |
| 100 | 1.16 | 3.33 | 1.75 | 1.00 |
| 150 | 1.22 | 2.83 | 2.20 | 2.00 |
| 200 | 1.07 | 2.88 | 2.12 | 2.00 |
| 250 | 1.10 | 2.76 | 2.10 | 3.00 |
| 300 | 1.08 | 2.66 | 2.45 | 2.66 |

# References

1. J. Nievergelt E.M. Reingold and N. Deo. *Combinatorial Algorithms: Theory and Practice*. Prentice-Hall, Englewood Cliffs, NJ, 1977.
2. S. Even. *Combinatorial Algorithms*. MacMillan, New York, 1973.
3. P. Flajolet and B. Salvy. Computer algebra libraries for combinatorial structures. *J. Symbolic Comp. (also INRIA, num. 2497)*, 20:653–671, 1995.
4. P. Flajolet and R. Sedgewick. The average case analysis of algorithms: Counting and generating functions. Technical Report 1888, INRIA, Apr 1993.
5. P. Flajolet and J.S. Vitter. Average-case Analysis of Algorithms and Data Structures. In J. Van Leeuwen, editor, *Handbook of Theoretical Computer Science*, chapter 9. North-Holland, 1990.
6. P. Flajolet, P. Zimmermann, and B. Van Cutsem. A calculus for the random generation of combinatorial structures. *Theor. Comp. Sci.*, 132(1):1–35, 1994.
7. D.L. Kreher and D.R. Stinson. *Combinatorial Algorithms: Generation, Enumeration and Search*. CRC Press LLC, 1999.
8. J. Liebehenschel. Ranking and unranking of lexicographically ordered words: An average-case analysis. *Journal of Automata, Languages and Combinatorics*, 2(4):227–268, 1997.
9. J. Liebehenschel. Ranking and unranking of a generalized dyck language and the application to the generation of random trees. In *The Fifth International Seminar on the Mathematical Analysis of Algorithms*, Bellaterra (Spain), Jun 1999.
10. C. Martínez and X. Molinero. Unranking of labelled combinatorial structures. In D. Krob, A. A. Mikhalev, and A. V. Mikhalev, editors, *Formal Power Series and Algebraic Combinatorics : 12th International Conference ; Procedings / FPSAC '00*, pages 288–299. Springer-Verlag, Jun 2000.

11. C. Martínez and X. Molinero. A generic approach for the unranking of labeled combinatorial classes. *Random Structures & Algorithms*, 19(3-4):472–497, Oct-Dic 2001.
12. C. Martínez and X. Molinero. Generic algorithms for the generation of combinatorial objects. In Branislav Rovan and Peter Vojtáš, editors, *Mathematical Foundations of Computer Science 2003 (28th International Symposium, MFCS 2003; Bratislava, Slovakia, August 2003; Proceedings)*, volume 2747, pages 572–581. Springer, Aug 2003.
13. A. Nijenhuis and H.S. Wilf. *Combinatorial Algorithms: For Computers and Calculators*. Academic Press, Inc., 1978.
14. J.M. Pallo. Enumerating, ranking and unranking binary trees. *The Computer Journal*, 29(2):171–175, 1986.
15. H.S. Wilf. East side, west side ... an introduction to combinatorial families-with MAPLE programming, Feb 1999.

# An Improved Derandomized Approximation Algorithm for the Max-Controlled Set Problem

Carlos A. Martinhon[1] and Fábio Protti[2]

[1] Instituto de Computação, Universidade Federal Fluminense, Rua Passo da Pátria
156, Bloco E, Sala 303, 24210-230, Niterói, Brazil.
mart@dcc.ic.uff.br
[2] IM and NCE, Universidade Federal do Rio de Janeiro, Caixa Postal 2324,
20001-970, Rio de Janeiro, Brazil.
fabiop@nce.ufrj.br

**Abstract.** A vertex $i$ of a graph $G = (V, E)$ is said to be *controlled*
by $M \subseteq V$ if the majority of the elements of the neighborhood of
$i$ (including itself) belong to $M$. The set $M$ is a *monopoly* in $G$ if
every vertex $i \in V$ is controlled by $M$. Given a set $M \subseteq V$ and two
graphs $G_1 = (V, E_1)$ and $G_2 = (V, E_2)$ where $E_1 \subseteq E_2$, the MONOPOLY
VERIFICATION PROBLEM (MVP) consists of deciding whether there exists
a sandwich graph $G = (V, E)$ (i.e., a graph where $E_1 \subseteq E \subseteq E_2$)
such that $M$ is a monopoly in $G = (V, E)$. If the answer to the MVP
is No, we then consider the MAX-CONTROLLED SET PROBLEM (MCSP),
whose objective is to find a sandwich graph $G = (V, E)$ such that
the number of vertices of $G$ controlled by $M$ is maximized. The MVP
can be solved in polynomial time; the MCSP, however, is NP-hard. In
this work, we present a deterministic polynomial time approximation
algorithm for the MCSP with ratio $\frac{1}{2} + \frac{1+\sqrt{n}}{2n-2}$, where $n = |V| > 4$.
(The case $n \leq 4$ is solved exactly by considering the parameterized
version of the MCSP.) The algoritm is obtained through the use of
randomized rounding and derandomization techniques, namely the
method of conditional expectations. Additionally, we show how to improve this ratio if good estimates of expectation are obtained in advance.

## 1 Preliminaries

Given two graphs $G_1 = (V, E_1)$ and $G_2 = (V, E_2)$ such that $E_1 \subseteq E_2$, we say
that $G = (V, E)$, where $E_1 \subseteq E \subseteq E_2$, is a *sandwich graph* for some property
$\Pi$ if $G = (V, E)$ satisfies $\Pi$. A *sandwich problem* consists of deciding whether
there exists some sandwich graph satisfying $\Pi$. Many different properties may
be considered in this context. In general, the property $\Pi$ is non-hereditary by
(not induced) subgraphs (otherwise $G_1$ would trivially be a solution, if any) and
non-ancestral by supergraphs (otherwise $G_2$ would trivially be a solution, if any.)
As discussed by Golumbic *et al.* [7], sandwich problems generalize recognition
problems arising in various situations (when $G_1 = G_2$, the sandwich problem
becomes simply a recognition problem.)

C.C. Ribeiro and S.L. Martins (Eds.): WEA 2004, LNCS 3059, pp. 341–355, 2004.
© Springer-Verlag Berlin Heidelberg 2004

One of the most known sandwich problems is the CHORDAL SANDWICH PROBLEM, where we require $G$ to be a chordal graph (a graph where every cycle of length at least four possesses a chord - an edge linking two non-consecutive vertices in the cycle). The CHORDAL SANDWICH PROBLEM is closely related to the MINIMUM FILL-IN PROBLEM [19]: given a graph $G$, find the minimum number of edges to be added to $G$ so that the resulting graph is chordal. The MINIMUM FILL-IN POBLEM has applications to areas such as solution of sparse systems of linear equations [15]. Another important sandwich problem is the INTERVAL SANDWICH PROBLEM, where we require the sandwich graph $G$ to be an interval graph (a graph whose vertices are in a one-to-one correspondence with intervals on the real line in such a way that there exists an edge between two vertices if and only if the corresponding intervals intersect.) Kaplan and Shamir [8] describe applications to DNA physical mapping via the INTERVAL SANDWICH PROBLEM. In this work we consider a special kind of sandwich problem, the MAX-CONTROLLED SET PROBLEM (MCSP) [11], which is described in the sequel.

Given an undirected graph $G = (V, E)$ and a set of vertices $M \subseteq V$, a vertex $i \in V$ is said to be *controlled* by $M$ if $|N_G[i] \cap M| \geq |N_G[i]|/2$, where $N_G[i] = \{i\} \cup \{j \in V | (i,j) \in E\}$. The set $M$ defines a *monopoly* in $G$ if every vertex $i \in V$ is controlled by $M$. Following the notation of [11], if $cont(G, M)$ denotes the set of vertices controlled by $M$ in $G$, $M$ will be a monopoly in $G$ if and only if $cont(G, M) = V$.

In order to defined formally the MCSP, we first define the MONOPOLY VERIFICATION PROBLEM (MVP) : given a set $M \subseteq V$ and two graphs $G_1 = (V, E_1)$ and $G_2 = (V, E_2)$, where $E_1 \subseteq E_2$, the question is to decide whether there exists a set $E$ such that $E_1 \subseteq E \subseteq E_2$ and $M$ is a monopoly in $G = (V, E)$. If the answer of the MVP applied to $M$, $G_1$, and $G_2$ is No, we then consider the MCSP, whose goal is to find a set $E$ such that $E_1 \subseteq E \subseteq E_2$ and the number of vertices controlled by $M$ in $G = (V, E)$ is maximized.

The MVP can be solved in polynomial time by formulating it as a network flow problem [11]. If the answer to the MVP is No, then a natural alternative is to solve the MCSP. Unfortunately, the MCSP is NP-hard, even for those instances where $G_1$ is an empty graph and $G_2$ is a complete graph. In [11] a reduction from INDEPENDENT SET to the MCSP is given. In the same work, an approximation algorithm for the MCSP with ratio $\frac{1}{2}$ is presented.

The notion of monopoly has applications to local majority voting in distributed environments and agreement in agent systems [1,6,10,13,16,17]. For instance, suppose that the agents must agree on one industrial standard between two proposed candidate standards. Suppose also that the candidate standard supported by the majority of the agents is to be selected. When every agent knows the opinion of his neighbors, a natural heuristic to obtain a reasonable agreement is: every agent $i$ takes the majority opinions in $N[i]$. This is known as the deterministic local majority polling system. In such a system, securing the support by the members of a monopoly $M$ implies securing unanimous agreement. In this context, the motivation for the MCSP is to find an efficient way of controlling the maximum number of objects by modifying the system's topology.

In this work, we present a linear integer programming formulation and a randomized rounding procedure for the MCSP. As far as we know, our procedure achieves the best polynomial time approximation ratio for the MCSP. If $y^*$ denotes the optimum value of the linear relaxation and $y^* > A(k)$ (for some fixed $k \in (1, 2]$ and some function $A(k) > 4$), the approximation ratio $\frac{1}{k} + \frac{1+\sqrt{y^*}}{k(y^*-1)}$ improves the $\frac{1}{2}$-approximation algorithm presented in [11]. As described later, the case $y^* \leq A(k)$ may be solved exactly by considering a polynomial time algorithm for the parameterized version of the MCSP. This procedure is based on the ideas presented in [11] for the MVP.

This paper is organized as follows. In Section 2 some basic notation and results from [11] are presented. These are fundamental for the development of our algorithm. In Section 3, we introduce the PARAMETERIZED MCSP. For a given parameter $A \geq 0$, we solve exactly the PARAMETERIZED MCSP in time $O(n^A)$. Section 4 gives a detailed description of our MCSP formulation and outlines our randomized rounding procedure. In randomized rounding techniques, we first solve the linear relaxation and "round" the resulting solution to produce feasible solutions. In Section 5 we present an approximation analysis via the probabilistic method (introduced by Erdös and Spencer [5]). In this case, the main objective is to construct probabilistic existence proofs of some particular combinatorial structure for actually exhibiting this structure. This is performed through the use of derandomization techniques. In Section 6 we describe a derandomized procedure via the method of conditional expectations, achieving an improved deterministic approximation algorithm for the MCSP with performance ratio $\frac{1}{2} + \frac{1+\sqrt{n}}{2n-2}$. Finally, in Section 7, we present some conclusions and suggestions for future work.

## 2 The $\frac{1}{2}$-Approximation Algorithm for the MCSP by Makino *et al.*

Consider a maximization optimization problem $P$ and an arbitrary input instance $I$ of $P$. Denote by $\hat{z}(I)$ the optimal objective function value for $I$, and by $z_H(I)$ the value of the objective function delivered by an algorithm $H$. Without loss of generality, it is assumed that each feasible solution for $I$ has a non-negative objective function value. Recall that $H$ is a $\varphi$-approximation algorithm for $P$ if and only if a feasible solution of value $z_H(I) \geq \varphi\hat{z}(I)$ is delivered for all instances $I$ and some $\varphi$ satisfying $0 < \varphi \leq 1$.

From now on, we suppose that the answer for the MVP when applied to $M$, $G_1$ and $G_2$ is No. Let us briefly describe the deterministic $\frac{1}{2}$-approximation algorithm for the MCSP presented in [11]. For $A, B \subseteq V$, define the edge set $D(A, B) = \{(i, j) \in E_2 \backslash E_1 \mid i \in A, j \in B\}$. Let $U = V \backslash M$. Two reduction rules are used: a new edge set $E_1^*$ is obtained by the union of $E_1$ and $D(M, M)$, and a new edge set $E_2^*$ is obtained by removing $D(U, U)$ from $E_2$. Since we are maximizing the total number of vertices controlled by $M$, these reduction rules do not modify the optimal solution. In other words, the edge set $E$ in the sandwich graph $G$ satisfies $E_1 \cup D(M, M) \subseteq E \subseteq E_1 \cup D(M, M) \cup D(U, U)$.

For simplicity, assume from now on $E_1 = E_1^*$ (Reduction Rule 1) and $E_2 = E_2^*$ (Reduction Rule 2). In the MYK algorithm, $W1, W2 \subseteq V$ denote the sets of vertices controlled by $M$ in $G = (V, E)$ for $E = E_1$ and $E = E_2$, respectively.

Algorithm 1: MYK ALGORITHM [11]
1. compute $|W_1|$ by removing from $G_2$ the edge set $D(U, M)$
2. compute $|W_2|$ by adding to $G_1$ the edge set $D(U, M)$
3. $z_H \leftarrow \max\{|W_1|, |W_2|\}$

Formally, they proved the following result:

**Theorem 1.** *The value $z_H$ returned by Algorithm 1 satisfies $z_H(I) \geq \frac{1}{2}\hat{z}(I)$, for all instances $I$ of the* MCSP.

# 3 Parameterizing the MCSP

In this Section we introduce the PARAMETERIZED MCSP. Let $A$ be a fixed non-negative integer. The objective is to find, in polynomial time, a solution for the MCSP with value at least $A$. In other words, we require the parameter $A$ to be a lower bound for the maximum number of vertices that can be controlled by $M$ in a sandwich graph $G = (V, E)$ with $E_1 \subseteq E \subseteq E_2$.

Let us describe an $O(n^A)$ algorithm for the PARAMETERIZED MCSP. We first consider a partition of $V$ into six special subsets (some of them are implicitly described in [11]):

- $M_C$ and $U_C$, consisting of the vertices in $M$ and $U$, respectively, which are controlled by $M$ in any sandwich graph (vertices which are "always controlled");
- $M_N$ and $U_N$, consisting of the vertices in $M$ and $U$, respectively, which are not controlled by $M$ in any sandwich graph (vertices which are "never controlled");
- $M_R$ and $U_R$, defined as $M_R = M \backslash (M_C \cup M_N)$ and $U_R = U \backslash (U_C \cup U_N)$.

Define the binary variables $x_{ij} \in \{0, 1\}$ for $i, j \in V$ and assume that $x_{ij} = x_{ji}$, $\forall i, j \in V$. Define binary constants $a_{ij} \in \{0, 1\}$ such that $a_{ij} = 1$ if and only if $i = j$ or $(i, j) \in E_2$. Consider now the following auxiliary equations:

$$B_i = \sum_{j \in M} a_{ij} x_{ij} - \sum_{j \in U} a_{ij} x_{ij}, \quad \text{for} \quad i = 1, \dots, |V| \tag{1}$$

In these equations, assume that: $x_{ii} = 1$ for every $i \in V$, $x_{ij} = 1$ for every $(i, j) \in E_1$, and $x_{ij} = 0$ for every $(i, j) \notin E_2$. This means that the remaining binary variables are associated to edges in $E_2 \backslash E_1$ (set of optional edges). It is clear that $B_i \geq 0$ for some 0-1 assignment (to variables associated to optional edges) if and only if vertex $i$ can be controlled by $M$ in some sandwich graph. Observe that the subsets $M_C, U_C, M_N, U_N$ can be thus characterized by the following properties:

- $i \in M_C \cup U_C$ if and only if $B_i \geq 0$ for every 0-1 assignment;
- $i \in M_N \cup U_N$ if and only if $B_i < 0$ for every 0-1 assignment.

In fact, it is easy to construct these four sets, since it is sufficient to look at "worst-case" assignments. For instance, if $i \in M$ then $i \in M_C$ if and only if $B_i \geq 0$ for a 0-1 assignment which sets $x_{ij} = 1$ for every $(i, j) \in D(\{i\}, U)$.

Use now the following new reduction rules:

- set $x_{ij} = 1$ for every $(i, j) \in D(M_C \cup M_N, U_R)$ (Reduction Rule 3);
- set $x_{ij} = 0$ for every $(i, j) \in D(M_R, U_C \cup U_N)$ (Reduction Rule 4);
- set arbitrary values to variables $x_{ij}$ for $(i, j) \in D(M_C \cup M_N, U_C \cup U_N)$ (Reduction Rule 5).

It is clear that if $A \leq |M_C| + |U_C|$ then the algorithm for the PARAMETERIZED MCSP answers Yes. Hence, from now on, assume that $A > |M_C| + |U_C|$.

Clearly, $M_R = \emptyset$ if and only if $U_R = \emptyset$. Thus, if $M_R = \emptyset$ or $U_R = \emptyset$, the algorithm must answer No. Add then the assumption $M_R, U_R \neq \emptyset$.

We fix an arbitrary subset $S \subseteq M_R \cup U_R$ of cardinality $|S| = A - |M_C| - |U_C|$, and check whether it is possible to control $S$. Similarly to the reduction rules described above, set $x_{ij} = 1$ for every edge $(i, j) \in D(M_R \backslash S, U_R \cap S)$, and $x_{ij} = 0$ for every edge $(i, j) \in D(M_R \cap S, U_R \backslash S)$. Finally, set $x_{ij} = 0$ for every edge $(i, j) \in D(M_R \cap S, U_R \cap S)$, and calculate the corresponding $B_i$'s according to equations (1).

Following the ideas in [11], we construct a network $\mathcal{N}$ whose vertex set consists of $S$ together with two additional vertices $s, t$. Create an edge $(s, i)$ with capacity $B_i$ for every $i \in S \cap M_R$, an edge $(j, t)$ with capacity $B'_j = \max\{-B_j, 0\}$ for every $j \in S \cap U_R$, and an edge $(i, j)$ with capacity 1 for every $(i, j) \in D(S \cap M_R, S \cap U_R)$. Notice that $\mathcal{N}$ can be constructed in constant time, since its edge set contains at most $O(A^2)$ elements. Now, if the maximum flow in $\mathcal{N}$ is equal to $\sum_{j \in S \cap U_R} B'_j$ then $S$ can be controlled by selecting the edges of the form $(i, j) \in D(S \cap M_R, S \cap U_R)$ with unitary flow value. This maximum flow problem can be solved in constant time, depending on $A$.

By repeating this procedure for every $S \subseteq M_R \cup U_R$ such that $|S| = A - |M_C| - |U_C|$, we obtain an algorithm with complexity $O(n^A)$.

## 4    An Improved Randomized Rounding Procedure for the MCSP

The definition of performance ratio in randomized approximation algorithms is the same as in the deterministic ones. In this case, however, $z_H(I)$ is replaced by $E(z_H(I))$, where the expectation is taken over the random choices made by the algorithm. Then, an algorithm $H$ for a maximization problem is a randomized $\varphi$-approximation algorithm if and only if $E(z_H(I)) \geq \varphi \hat{z}(I)$ is delivered for all instances $I$ and some $0 < \varphi \leq 1$.

In randomized rounding techniques (introduced by Raghavan and Thompson [14]), one usually solves a relaxation of a combinatorial optimization problem (by using linear or semidefinite programming), and uses randomization to return from the relaxation to the original optimization problem. The main idea is to use fractional solutions to define tuned probabilities in the randomized rounding procedure. Additional executions of this randomized procedure arbitrarily reduce the failure probability (Monte Carlo method).

In order to introduce a new integer programming formulation for the MCSP, we define the binary variables $z_i$ for $i \in V$, which determine whether vertex $i$ is controlled or not by $M$. Binary variables $x_{ij}$ are used to decide whether optional edges belonging to $E_2 \backslash E_1$ will be included or not in the sandwich graph. The objective function (2) computes the maximum number of controlled vertices. As defined before, binary constants $a_{ij} \in \{0, 1\}$ are associated to edges $(i, j) \in E_2$ with $a_{ij} = 1$ if and only if $i = j$ or $(i, j) \in E_2$. (Assume that $a_{ij} = a_{ji}, \forall i, j \in V$.) Inequalities (3) guarantee that every time a vertex $i$ is controlled by $M$, the left hand side will be greater than or equal to 1. On the other hand, if the left hand side is less than 1, vertex $i$ will not be controlled by $M$ and $z_i$ will be set to 0. The divisions by $n$ are used to maintain the difference between the two summations always greater than $-1$. Equalities (4) define the set of fixed edges. The linear programming relaxation is obtained by replacing integrality constraints (5) and (6) by $x_{ij} \in [0, 1]$ and $z_i \in [0, 1]$, respectively.

$$\hat{z} = \max \sum_{i \in V} z_i \tag{2}$$

subject to:

$$\sum_{j \in M} (a_{ij}/n) x_{ij} - \sum_{j \in V} (a_{ij}/2n) x_{ij} + 1 \geq z_i, \forall i \in V \tag{3}$$

$$x_{ij} = 1, \forall (i, j) \in E_1 \tag{4}$$

$$x_{ij} \in \{0, 1\}, \forall (i, j) \in E_2 \backslash E_1 \tag{5}$$

$$z_i \in \{0, 1\}, \forall i \in V \tag{6}$$

It is assumed from now on that $(x^*, z^*)$ and $y^*$ will denote, respectively, an optimal solution of the relaxed integer programming formulation and its associated objective function value. The value of the original integer problem will be denoted by $\hat{z}$.

The value of linear programming relaxation may be improved if the reduction rules for the MCSP are used. As will be observed in Section 5, the performance ratio of our randomized algorithm is based on the value of the linear relaxation and an improvement of this ratio is attained if good upper bounds are obtained. Thus, assume without loss of generality that MCSP instances satisfy Reduction Rules 1 to 5.

Algorithm 2, based on randomized rounding techniques, is a Monte Carlo procedure and delivers, in polynomial time and with high probability, a value within a prescribed approximation ratio. In Step 3 of the algorithm we define a function $A(k)$ for a given parameter $k \in (1,2]$ conveniently chosen. The construction of $A(k)$ will be detailed in the next section. For the time being, an "oracle" is used.

Algorithm 2: RANDOMIZED ALGORITHM FOR THE MCSP
1. compute $z_1$ using Algorithm 1
2. solve the linear programming relaxation and return $x^*$ and $y^*$
3. compute $A(k)$ for some $k \in (1,2]$ (conveniently chosen)
4. **if** $y^* \leq A(k)$

> **then** compute $z_H$ by executing the algorithm for the PARAMETERIZED
> MCSP in Section 3 for parameters $A = \lfloor y^* \rfloor, \lfloor y^* \rfloor - 1, \ldots, 1, 0$, until
> obtaining a Yes answer
> **else** **for** each $(i,j) \in E_2 \setminus E_1$ **do**
> $\qquad Pr(\bar{x}_{ij} = 1) = x^*_{ij}$ (constructing the integer feasible soluction)
> $\qquad Pr(\bar{x}_{ij} = 0) = 1 - x^*_{ij}$
> $\qquad$ compute $z_2$ by using the integer feasible function $\bar{x}$
> $\qquad z_H \leftarrow \max\{z_1, z_2\}$

5. return $z_H$

We can use, for example, an interior point method in Step 2 (introduced by Karmarkar [9]) to compute the fractional solution $x^*$, yielding in this way a polynomial time execution for Algorithm 2. Observe that Algorithm 2 always produce a feasible solution, and additional executions of Step 4 (for $y^* > A(k)$) arbitrarily reduce the failure probability, provided that a prescribed approximation ratio is given. Moreover, it is obviously a $\frac{1}{2}$-approximation algorithm since Algorithm 1 was used in Step 1. As will be pointed out in the next section, this will directly help us to build an improved approximation algorithm with ratio $\frac{1}{k} + \frac{1+\sqrt{y^*}}{k(y^*-1)}$ for some $k \in (1,2]$, conveniently chosen. It is straightforward to observe that, even for $k = 2$, this ratio is strictly greater than $\frac{1}{2}$, thus improving the previous result of [11]. In addition, recall that all those instances with $y^* \leq A(k)$ (for some parameter $A = \lfloor y^* \rfloor$) are polinomially solved by the algorithm for the PARAMETERIZED MCSP given in Section 3.

## 5 Approximation Analysis

Before to proceed to the approximation analysis, consider the following auxiliary definitions and lemmas. We first present the notion of negative association.

**Definition 1. (Negative Association)** *Let $X = (X_1, X_2, \ldots, X_n)$ be a vector of random variables. The random variables $X$ are negatively associated if for every two disjoint index sets $I, J \subseteq \{1, 2, \ldots, n\}$, $E(f(X_i, i \in I)g(X_j, j \in J)) \leq E(f(X_i, i \in I))E(g(X_j, j \in J))$ for all functions $f : \Re^{|I|} \to \Re$ and $g : \Re^{|J|} \to \Re$ that are both non-decreasing and both non-increasing.*

For a more detailed study concerning negative dependence see Dubhashi and Ranjan [4].

The next lemma ensures that the lower Chernoff-Hoeffding bound (lower CH bound) may be applied to not necessarily independent random variables. See Motwani and Raghavan [12] and Dubhashi and Ranjan [4] for the proof. An analogous result may be established for the upper CH bound.

**Lemma 1. (Lower Chernoff-Hoeffding Bound and Negative Association)** *Let $X_1, X_2, \ldots, X_n$ be negatively associated Poisson trials such that, for $1 \leq i \leq n$, $Pr(X_i = 1) = p_i$, where $0 < p_i < 1$. Then, for $X = \sum_{i=1}^{n} X_i$, $\mu = E(X) = \sum_{i=1}^{n} p_i$, and any $0 < \delta \leq 1$, we have that $Pr(X < (1 - \delta)\mu) < exp(-\mu\delta^2/2)$.*

Finally, consider the following auxiliary lemma:

**Lemma 2.** *Let $X, Y$ be arbitrary random variables. Then $E(min(X, Y)) \leq min(E(X), E(Y))$.*

Now, in order to describe the approximation analysis of Algorithm 2, we define random variables $Z_i \in \{0, 1\}$ for every $i \in V$. These variables denote the set of vertices controlled by $M$. We also define random variables $X_{ij} \in \{0, 1\}$ for every $i, j \in V$. Assume $X_{ii} = 1$ for every $i \in V$, and $X_{ij} = 1$ for every $(i, j) \in E_1$. Observe that variables $X_{ij}$ for $(i, j) \in E_2 \backslash E_1$ are associated to the set of optional edges. Additionally, let $Z_H$ be the sum of not necessarily independent random variables $Z_i \in \{0, 1\}$ for $i \in V$. Thus, we have the following preliminary result:

**Lemma 3.** *The random variables $Z_i$ for all $i \in V$ are negatively associated.*

**Proof:** Consider two arbitrary disjoint index sets $I, J \subseteq \{1, 2, \ldots, n\}$. Then we want to show that:

$$E(\sum_{i \in I} Z_i \sum_{j \in J} Z_j) - E(\sum_{i \in I} Z_i)E(\sum_{j \in J} Z_j) = \sum_{i \in I, j \in J} (E(Z_i Z_j) - E(Z_i)E(Z_j)) \leq 0$$

In particular, it is easy to observe (from the definition of the MCSP) that $Z_i$ and $Z_j$ (for $i \neq j$) are independent random variables if they simultaneously belong to $M$ (or $U$). However, $Z_i$ and $Z_j$ are negatively associated if they are not in the same set. Generally, for arbitrary index sets $I$ and $J$, we can establish that $Pr(Z_j = 1 \mid Z_i = 1) \leq Pr(Z_j = 1)$ or, equivalently, $Pr(Z_i = 1 \mid Z_j = 1) \leq Pr(Z_i = 1)$. Thus, for every pair $i \in I$ and $j \in J$, we have that $E(Z_i Z_j) = Pr(Z_i Z_j = 1) = Pr(Z_i = 1)Pr(Z_j = 1 \mid Z_i = 1) \leq Pr(Z_i = 1)Pr(Z_j = 1) = E(Z_i)E(Z_j)$, which proves the lemma. $\square$

Now, consider our relaxed integer programming formulation. For $i, j \in V$, assume $X_{ij} = 1$ if $i = j$ or $(i, j) \in E_1$. Assume also we assign, as described in Algorithm 2, arbitrarily values $X_{ij}$ for every $(i, j) \in E_2 \backslash E_1$. If $Z_H$ is the sum of random variables denoting the value of the randomized solution, it follows from constraints (3)-(6) that:

$$Z_H = \sum_{i \in V} Z_i \leq \sum_{i \in V} \min\{1; \sum_{j \in M} (a_{ij}/n)X_{ij} - \sum_{j \in V} (a_{ij}/2n)X_{ij} + 1\} \qquad (7)$$

From Lemma 2 and the linearity of expectation one obtains:

$$E(Z_H) = \sum_{i \in V} E(Z_i) \leq \min\{1; \sum_{j \in M}(a_{ij}/n)x_{ij}^* - \sum_{j \in V}(a_{ij}/2n)x_{ij}^* + 1\},$$

where $E(X_{ij}) = x_{ij}^*$. Therefore:

$$E(Z_H) \leq \sum_{i \in V}\min\{1; z_i^*\} \Rightarrow E(Z_H) \leq \sum_{i \in V} z_i^* = y^* \text{ for } z_i^* \in [0,1] \qquad (8)$$

Recall that Step 1 Algorithm 2 guarantees a performance ratio equal to $\frac{1}{2}$. Therefore, each iteration of Algorithm 2 returns a solution with $Z_H \geq \hat{z}/2$ (where $\hat{z}$ denotes the value of the optimal integer solution). Now, as the optimal solution itself may be generated at random, one may concludes, without loss of generality, that $E(Z_H)$ is strictly greater than $\hat{z}/2$ (otherwise, the solution generated by Algorithm 1 would be optimal). Thus, we assume from (8) that $\hat{z}/2 < E(Z_H) \leq y^*$, where $E(Z_H) = \mu = y^*/\beta$ for some $\beta \in [1,2)$.

Now, for some $\alpha > 1$, to be considered later, define a bad event $B \equiv (Z_H < y^*/\alpha)$. Equivalently to the definition of a randomized approximation algorithm (described in the preceding Section), $Z_H$ defines an $\frac{1}{\alpha}$-approximation solution for the MCSP if $B^c \neq \emptyset$ holds (complementary event).

How small a value for $\alpha$ can we achieve while guaranteeing good events $B^c \neq \emptyset$? Since we expect to obtain an approximation algorithm with a superior performance ratio (greater than $\frac{1}{2}$), it suffices to consider $\alpha \in (\beta, k)$ for some $k \in (\beta, 2]$. The parameter $k$ will be fixed later. This give us an improved $\frac{1}{\alpha}$-approximation $Z_H$ with nonzero probability. As discussed later, this solution will be made deterministic through derandomization techniques, namely, the method of conditional expectations.

Therefore, a bad event $B$ occurs if $Z_H < y^*/\alpha$. Then:

$$Pr(B) = Pr(Z_H < \frac{y^*}{\alpha}) = Pr(Z_H < \frac{\beta\mu}{\alpha}) = Pr(Z_H < (1-\delta)\mu),$$

where $\delta = 1 - \frac{\beta}{\alpha} > 0$.

In order to apply the lower CH bound, in addition to the negative association (Lemma 3), all random variables must assume values in the interval $(0,1)$. In our case, however, as observed in Section 3, $Pr(Z_i = 1) = 1$ for every $i \in M_C \cup U_C$ (set of vertices which are always controlled by $M$) and $Pr(Z_i = 1) = 0$ for every $i \in M_N \cup U_N$ (set of vertices which are never controlled by $M$). Despite of that, CH bounds may be applied, since the linear programming relaxation is being solved by some interior point method (see Wright [18].)

Therefore, from the lower CH bound and assuming $\mu > \hat{z}/2$, it follows that:

$$Pr(B) < \frac{1}{\exp\left((1-\beta/\alpha)^2 \frac{\mu}{2}\right)} < \frac{1}{\exp\left((1-\beta/\alpha)^2 \frac{\hat{z}}{4}\right)} \qquad (9)$$

This implies:

$$Pr(B) < \frac{1}{\exp((1-\alpha/\beta)^2 \frac{\hat{z}}{4}) - \exp(\frac{1}{4})} \qquad (10)$$

We expect that $Pr(B) < 1$ (probability of bad event). Thus, if we impose this last condition, it follows from (10) that:

$$Pr(B) = \frac{1}{\exp((1 - \alpha/\beta)^2 \frac{\hat{z}}{4}) - \exp(\frac{1}{4})} < 1 \text{ for some } \alpha \in (\beta, k) \quad (11)$$

Additional executions of Step 4 in Algorithm 2 for $y > A(k)$ arbitrarily reduce the failure probability (Monte Carlo method). Therefore, without loss of generality, if $Pr(B) = C < 1$ is the probability of a bad event, and $\delta > 0$ is a given error, $\lceil \ | \log \delta / \log C \ | \ \rceil$ iterations are sufficient to ensure a $\frac{1}{\alpha}$-approximation algorithm with probability $1 - \delta > 0$.

Then, we need to determine if there is some value $\alpha \in (\beta, k)$ (where $\beta \geq 1$ and $k \leq 2$) for which inequality (11) makes sense. Equivalently, we expect to obtain $(\hat{z} - 1)\alpha^2 - (2\hat{z}\beta)\alpha + \beta^2\hat{z} > 0$ for some $\alpha \in (\beta, k)$. By solving the quadratic equation, we obtain the roots:

$$\alpha' = \frac{\beta(\hat{z} - \sqrt{\hat{z}})}{\hat{z} - 1} \text{ and } \alpha'' = \frac{\beta(\hat{z} + \sqrt{\hat{z}})}{\hat{z} - 1}.$$

Since $\alpha > \beta$, it is easy to observe that inequality $(\hat{z} \pm \sqrt{\hat{z}})/(\hat{z} - 1) > 1$ holds only for $\alpha''$ with $\hat{z} > 1$. In addition, we expect that $\alpha'' < k$ for some $k \in (\beta, 2]$. Thus, since $\beta = y^*/\mu \geq 1$, it follows that:

$$\alpha'' = \frac{y^*(\hat{z} + \sqrt{\hat{z}})}{\mu(\hat{z} - 1)} < k \Rightarrow \frac{y^*(\hat{z} + \sqrt{\hat{z}})}{k(\hat{z} - 1)} < \mu \leq y^*. \quad (12)$$

Therefore:

$$\frac{y^*(\hat{z} + \sqrt{\hat{z}})}{k(\hat{z} - 1)} < y^* \Rightarrow \hat{z} + \sqrt{\hat{z}} < k(\hat{z} - 1) \quad (13)$$

Now, inequality (13) holds only for $\hat{z} > A(k)$ with:

$$A(k) = \frac{2k(k - 1) + 1 + \sqrt{4k(k - 1) + 1}}{2(k - 1)^2}$$

Notice that constraint $\hat{z} > 1$ above is immediately verified since we have $A(k) \geq 4$ for every $k \in (1, 2]$. Finally, from expression (12), since $E(Z_H) = \mu$, $\hat{z} \leq y^*$ and $y^* > A(k)$, it follows that:

$$E(Z_H) > \left(\frac{\hat{z} + \sqrt{\hat{z}}}{\hat{z} - 1}\right)\left(\frac{y^*}{k}\right) > \left(\frac{y^* + \sqrt{y^*}}{k(y^* - 1)}\right)\hat{z} = \left(\frac{1}{k} + \frac{1 + \sqrt{y^*}}{k(y^* - 1)}\right)\hat{z} \quad (14)$$

Moreover, observe from the above expression that:

$$\frac{1 + \sqrt{y^*}}{k(y^* - 1)} > 0 \quad \text{and} \quad \lim_{y^* \to \infty} \frac{1 + \sqrt{y^*}}{k(y^* - 1)} = 0 \quad \text{for every } y^* > A(k).$$

Thus, inequality (14) gives us a randomized $\frac{1}{k} + \frac{1+\sqrt{y^*}}{k(y^*-1)}$-approximation algorithm for every $y^* > A(k)$ and $k \in (\beta, 2]$. Therefore, with high probability and for a large class of instances, this ratio improves the $\frac{1}{2}$-approximation algorithm in [11]. The case $y^* \leq A(k)$ may be solved exactly in time $O(n^{A(k)})$ through the algorithm for the PARAMETERIZED MCSP in Section 3. Observe for instance that $A(k_1) > A(k_2)$ for every $k_1, k_2 \in (\beta, 2]$ with $k_1 < k_2$. In other words, despite the increase in the computational time of the algorithm for the PARAMETERIZED MCSP, small values of $k$ (for $k > \beta$) guarantee improved approximation ratios for every $y^* > A(k)$. Formally, we proved the following result:

**Theorem 2.** *Consider $y^*$ and $\mu$ as above. Then, for a given parameter $k \in (\beta, 2]$ with $\beta = y^*/\mu$, Algorithm 2 defines a randomized $\frac{1}{k} + \frac{1+\sqrt{y^*}}{k(y^*-1)}$-approximation algorithm for the MCSP.*

Unfortunately, we do not know explicitly the value of $\beta = y^*/\mu$ since the expectation $\mu$ is unknown and hard to compute. Moreover, we cannot guarantee a parameter $k$ strictly less than 2. This problem is minimized if some good estimations of $E(Z_H) = \mu$, and thus of $\beta$, are obtained. By running independent experiments with respect to $Z_H$, the recent work of Dagum *et al.* [2] ensures, for given $\delta$ and $\epsilon$, an estimator $\mu'$ of $\mu$ within a factor $1 + \epsilon$ and probability at least $1 - \delta$. Therefore, if this approximation is performed in advance, and if we assume $k = \min\{2, \frac{y^*}{\mu'(1-\epsilon)}\}$, an improved randomized approximation algorithm (for every instance of the MCSP) may be achieved if $k < 2$. Notice for instance that, given an interval $(\beta, k)$, the proof of Theorem 2 guarantees the existence of $\alpha \in (\beta, k)$, thus improving the performance ratio.

## 6   A Derandomized Algorithm

Derandomization techniques convert a randomized algorithm into a deterministic one. Here, this is performed through the probabilistic method (introduced by Erdös and Spencer [5]). The main idea is to use the existence proof of some combinatorial structure for actually exhibiting this structure.

The purpose of this section is to derandomize Algorithm 2 by using the method of conditional expectations. In this case, the goal is to convert the expected approximation ratio into a guaranteed approximation ratio while increasing the running time by a factor that is polynomial on the input size. Basically, the method of conditional expectations analyzes the behavior of a randomized approximation algorithm as a computation tree, in a such way that each path from the root to a leaf of this tree corresponds to a possible computation generated by the algorithm.

In order to describe our derandomized procedure for the MCSP, consider inequality (7). Then, it follows that:

$$Z_H = \sum_{i \in V} Z_i \leq \sum_{i \in V} Y_i, \quad \text{where } Y_i = \min\{1, \sum_{j \in M} (a_{ij}/n)X_{ij} - \sum_{j \in V} (a_{ij}/2n)X_{ij} + 1\}$$

Recall that $X_{ii} = 1$ for every $i \in V$, and $X_{ij} = 1, \forall (i,j) \in E_1$. In addition, suppose that all optional edges in $E_2 \backslash E_1$ are arbitrarily ordered and indexed by $k = 1, \ldots, |E_2 \backslash E_1|$. In this section, the notation $x_{ij}^{(k)} = 1$ has the following meaning: the $k$-th edge of $E_2 \backslash E_1$, with endpoints $i$ and $j$, belongs to the sandwich graph $G = (V, E)$. Otherwise, $x_{ij}^{(k)} = 0$ means that $(i,j) \notin E$. For simplicity, we will suppress indexes $i$ and $j$ and simply write $x^{(k)}$. Capital letters $X^{(k)}$ mean that a value 0 or 1 was assigned to variable $x^{(k)}$, for some $k \in \{1, \ldots, |E_2 \backslash E_1|\}$. Furthermore, the notation $E(Z_H \mid x^{(k)} = 0$ or $1)$ denotes the average value produced by the randomized algorithm by computations that set $x^{(k)} = 0$ or 1.

Thus, from de definition of conditional expectation and from its linearity property one concludes that:

$$E(Z_H) = E(Z_H \mid x^{(1)} = 1)Pr(x^{(1)} = 1) + E(Z_H \mid x^{(1)} = 0)Pr(x^{(1)} = 0)$$
$$\leq \max\{ \textstyle\sum_{i \in V} E(Z_i \mid x^{(1)} = 1) ; \sum_{i \in V} E(Z_i \mid x^{(1)} = 0) \} = \sum_{i \in V} E(Z_i \mid X^{(1)})$$
$$\leq \max\{ \textstyle\sum_{i \in V} E(Z_i \mid X^{(1)}, x^{(2)} = 1) ; \sum_{i \in V} E(Z_i \mid X^{(1)}, x^{(2)} = 0)\}$$
$$= \textstyle\sum_{i \in V} E(Z_i \mid X^{(1)}, X^{(2)})$$

By repeating this process for every edge in $E_2 \backslash E_1$, one obtains:

$$E(Z_H) \leq \max\{ \sum_{i \in V} E(Z_i \mid X^{(1)}, \ldots, X^{(|E_2 \backslash E_1|)}); \sum_{i \in V} E(Z_i \mid X^{(1)}, \ldots, X^{(|E_2 \backslash E_1|)}) \}$$

Therefore, within this framework, a guaranteed performance ratio is polynomially attained through an expected approximation ratio, gathering, in this way, an improved deterministic approximation solution.

Now, from the definition of conditional expectation,

$$E(Z_i \mid X^{(1)}, \ldots, X^{(k-1)}, x^{(k)} = 0 \text{ or } 1) = Pr(Z_i = 1 \mid X^{(1)}, \ldots, X^{(k-1)}, x^{(k)} = 0 \text{ or } 1),$$

for every $i \in V$ and $k = 1, \ldots, |E_2 \backslash E_1|$. Unfortunately, for the MCSP, these probabilities are hard to compute. Lemma 4 will give us an alternate way to deal with these expectations without explicitly consider conditional probabilities.

**Lemma 4.** *Suppose that $Z_i$ and $Y_i$, for some $i \in V$, are random variables as described above. Then:*

(a) $E(Z_i \mid X^{(1)}, \ldots, X^{(k-1)}, x^{(k)} = 1) \geq E(Z_i \mid X^{(1)}, \ldots, X^{(k-1)}, x^{(k)} = 0) \Leftrightarrow$
    $E(Y_i \mid X^{(1)}, \ldots, X^{(k-1)}, x^{(k)} = 1) \geq E(Y_i \mid X^{(1)}, \ldots, X^{(k-1)}, x^{(k)} = 0)$

(b) $E(Z_i \mid X^{(1)}, \ldots, X^{(k-1)}, x^{(k)} = 0) \geq E(Z_i \mid X^{(1)}, \ldots, X^{(k-1)}, x^{(k)} = 1) \Leftrightarrow$
    $E(Y_i \mid X^{(1)}, \ldots, X^{(k-1)}, x^{(k)} = 0) \geq E(Y_i \mid X^{(1)}, \ldots, X^{(k-1)}, x^{(k)} = 1)$

**Proof:** We will prove item (a), the proof of (b) follows analogously. Consider without loss of generality that $E(Z_i \mid X^{(1)}, \ldots, X^{(k-1)}, x^{(k)} = 1) \geq E(Z_i \mid X^{(1)}, \ldots, X^{(k-1)}, x^{(k)} = 0)$. Thus, since $E(Z_H) = y^*/\beta$ for some $\beta \in [1, 2)$, it follows from inequalities (7)-(8) and from the definition of conditional expectations that:

$$E(Z_i \mid X^{(1)}, \ldots, X^{(k-1)}, x^{(k)} = 0 \text{ or } 1) = (1/\beta) E(Y_i \mid X^{(1)}, \ldots, X^{(k-1)}, x^{(k)} = 0 \text{ or } 1).$$

Thus:

$$(1/\beta)E(Y_i \mid X^{(1)}, \ldots, X^{(k-1)}, x^{(k)} = 1) \geq (1/\beta)E(Y_i \mid X^{(1)}, \ldots, X^{(k-1)}, x^{(k)} = 0).$$

By multiplying both sides by $\beta$, we get the desired inequality

$$E(Y_i \mid X^{(1)}, \ldots, X^{(k-1)}, x^{(k)} = 1) \geq E(Y_i \mid X^{(1)}, \ldots, X^{(k-1)}, x^{(k)} = 0).$$

The converse is obtained in the same way by first multiplying this last inequality by $\frac{1}{\beta}$. $\square$

Now, from Lemma 4, it follows that $E(Z_H)$ is less than or equal to

$$\max\{\sum_{i \in V} E(Y_i \mid X^{(1)}, \ldots, X^{(k-1)}, x^{(k)} = 1); \sum_{i \in V} E(Y_i \mid X^{(1)}, \ldots, X^{(k-1)}, x^{(k)} = 0)\},$$

for $k = 1, \ldots, |E_2 \backslash E_1|$.

We repeat the process above for every optional edge in $E_2 \backslash E_1$. Therefore, the sequence $X^{(1)}, \ldots, X^{(|E_2 \backslash E_1|)}$ is obtained deterministically in polynomial time while improving the approximation ratio.

From the preceding section, we have described a randomized algorithm whose expectation $E(Z_H)$ is greater than or equal to $\frac{1}{k} + \frac{1 + \sqrt{y^*}}{k(y^* - 1)}$ for some $k \in (\beta, 2]$ conveniently chosen. Since we expect to obtain a deterministic procedure, it suffices to consider (in the worst case) $k = 2$ and $y^* = \theta(n)$. Observe, from the preceding section, that by setting $k = 2$ one obtains $A(2) = 4$. This will give us (for an arbitrary instance) an improved deterministic polynomial time approximation algorithm with performance ratio equal to $\frac{1}{2} + \frac{1 + \sqrt{n}}{2n - 2}$.

Algorithm 3: DERANDOMIZED ALGORITHM FOR THE MCSP
1. compute $z_1$ using Algorithm 1
2. solve the linear programming relaxation and return $y^*$
3. if $y^* \leq 4$

    **then** compute $z_H$ by executing the algorithm for the PARAMETERIZED
        MCSP in Section 3 for parameters $A = \lfloor y^* \rfloor, \lfloor y^* \rfloor - 1, \ldots, 1, 0$, until
        obtaining a Yes answer
    **else**  **for** $k = 1, \ldots, |E_2 \backslash E_1|$ **do**
        **if** $E(Y_k \mid X^{(1)}, \ldots, X^{(k-1)}, x^{(k)} = 1) \geq E(Y_k \mid X^{(1)}, \ldots, X^{(k-1)}, x^{(k)} = 0)$
          **then**  $X^{(k)} \leftarrow 1$
          **else**  $X^{(k)} \leftarrow 0$
        compute $z_2$ by using the integer feasible function $X$
        $z_H \leftarrow \max\{z_1, z_2\}$

5. return $z_H$

Observe above that expectations $E(Y_k \mid X^{(1)}, \ldots, X^{(k-1)}, x^{(k)} = 0$ or $1)$ are easily obtained. This may be accomplished in polynomial time by solving a linear programming problem for every optional edge (settled 0 or 1).

If $L$ denotes the length of the input, the linear relaxation has complexity $O(n^3 L)$ [18], and thus the total complexity of Algorithm 3 will be equal to $O(\max\{n^4, |E_2 \backslash E_1| n^3 L\})$. Moreover, from Theorem 2, it is straightforward to observe that an improvement of the approximation ratio may be attained if good upper bounds are obtained via the linear relaxation. This may be accomplished, for example, through the use of new reduction rules and/or through the use of additional cutting planes. Notice for instance that, even in the worst case, when $y^* = \theta(n)$, one obtains an improved approximation ratio. Formally, we can establish the following result:

**Theorem 3.** *Algorithm 3 guarantees in polynomial time an approximation ratio equal to* $\frac{1}{2} + \frac{1+\sqrt{n}}{2n-2}$ *for* $n > 4$.

## 7   Conclusions

We presented an improved deterministic polynomial time approximation algorithm for the Max-Controlled Set Problem through the use of randomized rounding and derandomization techniques. As far as we know, this is the best approximation result for the MCSP. This improves the $\frac{1}{2}$-approximation procedure presented by Makino, Yamashita and Kameda [11]. A new linear integer programming formulation was presented to define tuned probabilities in our randomized procedure. Through the use of the probabilistic method, we converted a probabilistic proof of existence of an approximated solution into an efficient deterministic algorithm for actually constructing this solution. Additionally, we show that if some good estimations of expectation are obtained in advance, some improved approximation ratios may be attained.

As future work, an interesting question is to decide whether the PARAMETER-IZED MCSP is Fixed Parameter Tractable - FPT. (A problem with parameter $A$ is FPT if it admits an $O(f(A)n^\gamma)$ time algorithm, for some function $f$ and some constant $\gamma$ independent of $A$. For details, see [3].) Obtaining non-approximability results for the MCSP and using semidefinite programming relaxation in the randomized rounding procedure are also interesting attempts of research.

**Acknowledgments.** We thank Marcos Kiwi and Prabhakar Raghavan for their valuable comments and pointers to the literature.

## References

1. J.-C. Bermond and D. Peleg, The power of small coalitions in graphs, *Proc. 2nd Structural Information and Communication Complexity*, Olympia, Carleton University Press, Ottawa, pp. 173–184, 1995.
2. P. Dagum, R. Karp, M. Luby, and S. Ross, An optimal algorithm for Monte Carlo estimation, *SIAM Journal on Computing*, **29**(5) (2000) 1484–1496.
3. R. G. Downey and M. R. Fellows, Fixed parameter tractability and completeness I: Basic results, *21st Manitoba Conference on Numerical Mathematics and Computing*, Winnipeg, Canada, 1991.

4. D. Dubashi and D. Ranjan, Balls and bins: A study of negative dependence, *Random Structures and Algorithms* **13**(2) (1998) 99–124.
5. P. Erdös and J. Spencer, "The Probabilistic Method in Combinatorics" , Academic Press, San Diego, 1974.
6. D. Fitoussi and M. Tennenholtz, Minimal social laws, *Proc. AAAI'98*, pp. 26–31, 1998.
7. M. C. Golumbic, H. Kaplan, and R. Shamir, Graph sandwich problems, *Journal of Algorithms*, **19** (1994), 449–473.
8. H. Kaplan and R. Shamir, Physical maps and interval sandwich problems: Bounded degrees help, *Proceedings of the 5th Israeli Symposium on Theory of Computing and Systems - ISTCS*, 1996, pp. 195–201. To appear in *Algorithmica* under the title "Bounded degree interval sandwich problems".
9. N. Karmarkar, A new polynomial time algorithm for linear programming, *Combinatorica*, **4** (1984), 375–395.
10. N. Linial, D. Peleg, Y. Rabinovich, and N. Saks, Sphere packing and local majorities in graphs, *Proc. 2nd Israel Symposium on Theoretical Computer Science*, IEEE Computer Society Press, Rockville, MD, pp. 141–149, 1993.
11. K. Makino, M. Yamashita, and T. Kameda, Max-and min-neighborhood monopolies, *Algorithmica*, **34** (2002), 240–260.
12. R. Motwani and P. Raghavan, "Randomized Algorithms", Cambridge University Press, London, 1995.
13. D. Peleg, Local majority voting, small coalitions and controlling monopolies in graphs: A review, Technical Report CS96-12, Weizmann Institute, Rehovot, 1996.
14. P. Raghavan and C. D. Thompson, Randomized rounding: A technique for provably good algorithms and algorithmic proofs, *Combinatorica*, **7**(4) (1987), 365–374.
15. J. D. Rose, A graph-theoretic study of the numerical solution of sparse positive definite systems of linear equations, in *Graph Theory and Computing* (R. C. Reed, ed.), Academic Press, New York, 1972, pp. 183–217.
16. Y. Shoham and M. Tennenholtz, Emergent conventions in multi-agent systems: Initial experimental results and observations, *Proc. International Conference on Principles of Knowledge Representation and Reasoning*, pp. 225–231, 1992.
17. Y. Shoham and M. Tennenholtz, On the systhesis of useful social laws for artificial agent societies, *Proc. AAAI'92*, pp. 276–281, 1992.
18. S. J. Wright, "Primal-Dual Interior-Point Methods", SIAM, 1997.
19. M. Yannakakis, Computing the minimum fill-in is NP-complete, *SIAM Journal on Algebraic and Discrete Methods* **2** (1981), 77–79.

# GRASP with Path-Relinking for the Quadratic Assignment Problem

Carlos A.S. Oliveira[1], Panos M. Pardalos[1], and Mauricio G.C. Resende[2]

[1] Department of Industrial and Systems Engineering, University of Florida, 303 Weil Hall, Gainesville, FL 32611, USA.
{oliveira,pardalos}@ufl.edu
[2] Algorithms and Optimization Research Department, AT&T Labs Research, Room C241, 180 Park Avenue, Florham Park, NJ 07932, USA.
mgcr@research.att.com

**Abstract.** This paper describes a GRASP with path-relinking heuristic for the quadratic assignment problem. GRASP is a multi-start procedure, where different points in the search space are probed with local search for high-quality solutions. Each iteration of GRASP consists of the construction of a randomized greedy solution, followed by local search, starting from the constructed solution. Path-relinking is an approach to integrate intensification and diversification in search. It consists in exploring trajectories that connect high-quality solutions. The trajectory is generated by introducing in the initial solution, attributes of the guiding solution. Experimental results illustrate the effectiveness of GRASP with path-relinking over pure GRASP on the quadratic assignment problem.

## 1 Introduction

The quadratic assignment problem (QAP) was first proposed by Koopmans and Beckman [10] in the context of the plant location problem. Given $n$ facilities, represented by the set $F = \{f_1, \ldots, f_n\}$, and $n$ locations represented by the set $L = \{l_1, \ldots, l_n\}$, one must determine to which location each facility must be assigned. Let $A^{n \times n} = (a_{i,j})$ be a matrix where $a_{i,j} \in \mathbb{R}^+$ represents the flow between facilities $f_i$ and $f_j$. Let $B^{n \times n} = (b_{i,j})$ be a matrix where entry $b_{i,j} \in \mathbb{R}^+$ represents the distance between locations $l_i$ and $l_j$. Let $p : \{1 \ldots n\} \to \{1 \ldots n\}$ be an assignment and define the cost of this assignment to be

$$c(p) = \sum_{i=1}^{n} \sum_{j=1}^{n} a_{i,j} b_{p(i),p(j)}.$$

In the QAP, we want to find a permutation vector $p \in \Pi_n$ that minimizes the assignment cost, i.e. $\min c(p)$, subject to $p \in \Pi_n$, where $\Pi_n$ is the set of all permutations of $\{1, \ldots, n\}$. The QAP is well known to be strongly NP-hard [18].

GRASP, or greedy randomized adaptive search procedures [5,6,8,17], have been previously applied to the QAP [12,14,15]. For a survey on heuristics and metaheuristics applied to the QAP, see Voß[19]. In this paper, we present a new

C.C. Ribeiro and S.L. Martins (Eds.): WEA 2004, LNCS 3059, pp. 356–368, 2004.
© Springer-Verlag Berlin Heidelberg 2004

GRASP for the QAP, which makes use of path-relinking as an intensification mechanism. In Section 2, we briefly review GRASP and path-relinking, and give a description of how both are combined to find approximate solutions to the QAP. Experimental results with benchmark instances are presented in Section 3. Finally, in Section 4 we draw some concluding remarks.

## 2   GRASP and Path-Relinking

GRASP is a multi-start procedure, where different points in the search space are probed with local search for high-quality solutions. Each iteration of GRASP consists of the construction of a randomized greedy solution, followed by local search, starting from the constructed solution. A high-level description of GRASP for QAP, i.e. solving $\min c(p)$ for $p \in \Pi_n$, is given in Algorithm 1.

---

**Algorithm 1** GRASP for minimization

---

1: $c^* \leftarrow \infty$
2: **while** stopping criterion not satisfied **do**
3:     $p \leftarrow \texttt{GreedyRandomized}()$
4:     $p \leftarrow \texttt{LocalSearch}(p)$
5:     **if** $c(p) < c^*$ **then**
6:         $p^* \leftarrow p$
7:         $c^* \leftarrow c(p)$
8:     **end if**
9: **end while**
10: **return** $p^*$

---

The greedy randomized construction and the local search used in the new algorithm are similar to the ones described in [12]. The construction phase consists of two stages.

In stage 1, two initial assignments are made: facility $F_i$ is assigned to location $L_k$ and facility $F_j$ is assigned to location $L_l$. To make the assignment, elements of the distance matrix are sorted in increasing order:

$$b_{i(1),j(1)} \leq b_{i(2),j(2)} \leq \cdots \leq b_{i(n),j(n)},$$

while the elements of the flow matrix are sorted in increasing order:

$$a_{k(1),l(1)} \geq a_{k(2),l(2)} \geq \cdots \geq a_{k(n),l(n)}.$$

The product elements

$$a_{k(1),l(1)} \cdot b_{i(1),j(1)}, a_{k(2),l(2)} \cdot b_{i(2),j(2)}, ..., a_{k(n),l(n)} \cdot b_{i(n),j(n)}$$

are sorted and the term $a_{k(q),l(q)} \cdot b_{i(q),j(q)}$ is selected at random from among the smallest elements. This product corresponds to the initial assignments: facility $F_{k(q)}$ is assigned to location $L_{i(q)}$ and facility $F_{l(q)}$ is assigned to location $L_{j(q)}$.

In stage 2, the remaining $n-2$ assignments of facilities to locations are made, one facility/location pair at a time. Let $\Omega = \{(i_1, k_1), (i_2, k_2), \ldots, (i_q, k_q)\}$ denote the first $q$ assignments made. Then, the cost assigning facility $F_j$ to location $L_l$ is $c_{j,l} = \sum_{i,k \in \Omega} a_{i,j} b_{k,l}$. To make the $q + 1$-th assignment, select at random an assignment from among the feasible assignments with smallest costs and add the assignment to $\Omega$.

Once a solution is constructed, local search is applied to it to try to improve its cost. For each pair of assignments $(F_i \to L_k; F_j \to L_l)$ in the current solution, check if the swap $(F_i \to L_l; F_j \to L_k)$ improves the cost of the assignment. If so, make the swap, and repeat. A solution is locally optimal, when no swap improves the cost of the solution.

Path-relinking [9] is an approach to integrate intensification and diversification in search. It consists in exploring trajectories that connect high-quality solutions. The trajectory is generated by introducing in the initial solution, attributes of the guiding solution. It was first used in connection with GRASP by Laguna and Martí [11]. A recent survey of GRASP with path-relinking is given in Resende and Ribeiro [16]. The objective of path-relinking is to integrate features of good solutions, found during the iterations of GRASP, into new solutions generated in subsequent iterations. In pure GRASP (i.e. GRASP without path-relinking), all iterations are independent and therefore most good solutions are simply "forgotten." Path-relinking tries to change this, by retaining previous solutions and using them as "guides" to speed up convergence to a good-quality solution.

Path-relinking uses an *elite set* $P$, in which good solutions found by the GRASP are saved to be later combined with other solutions produced by the GRASP. The maximum size of the elite set is an input parameter. During path-relinking, one of the solutions $q \in P$ is selected to be combined with the current GRASP solution $p$. The elements of $q$ are incrementally incorporated into $p$. This relinking process can result in an improved solution, since it explores distinct neighborhoods of high-quality solutions.

Algorithm 2 shows the steps of GRASP with path-relinking. Initially, the elite set $P$ is empty, and solutions are added if they are different from the solutions already in the set. Once the elite set is full, path-relinking is done after each GRASP construction and local search.

A solution $q \in P$ is selected, at random, to be combined, through path-relinking, with the GRASP solution $p$. Since we want to favor long paths, which have a better change of producing good solutions, we would like to choose an elite solution $q$ with a high degree of differentiation with respect to $p$. Each element $q \in P$, let $d(q)$ denote the number of facilities in $q$ and $p$ that have different assignments, and let $D = \sum_{q \in P} d(q)$. A solution $q$ is selected from the elite set with probability $d(q)/D$. The selected solution $q$ is called the *guiding* solution. The output of path-relinking, $r$, is at least as good as solutions $p$ and $q$, that were combined by path-relinking.

If the combined solution $r$ is not already in the elite set and its cost is not greater than cost of the highest-cost elite set solution, then it is inserted into

---

**Algorithm 2** GRASP with path-relinking

---

1: $P \leftarrow \emptyset$
2: **while** stopping criterion not satisfied **do**
3:  $\quad p \leftarrow$ GreedyRandomized$()$
4:  $\quad p \leftarrow$ LocalSearch$(p)$
5:  $\quad$ **if** $P$ is full **then**
6:  $\quad\quad$ Select elite solution $q \in P$ at random
7:  $\quad\quad r \leftarrow$ PathRelinking$(p, q)$
8:  $\quad\quad$ **if** $c(r) \leq \max\{c(q) \mid q \in P\}$ and $r \notin P$ **then**
9:  $\quad\quad\quad$ Let $P' = \{q \in P \mid c(q) \geq c(r)\}$
10: $\quad\quad\quad$ Let $q' \in P'$ be the most similar solution to $r$
11: $\quad\quad\quad P \leftarrow P \cup \{r\}$
12: $\quad\quad\quad P \leftarrow P \setminus \{q'\}$
13: $\quad\quad$ **end if**
14: $\quad$ **else**
15: $\quad\quad$ **if** $p \notin P$ **then**
16: $\quad\quad\quad P \leftarrow P \cup \{p\}$
17: $\quad\quad$ **end if**
18: $\quad$ **end if**
19: **end while**
20: **return** $p^* = \min\{c(p) \mid p \in P\}$

---

the elite set. Among the elite set solutions having cost not smaller than $c(r)$, the one most similar to $r$ is deleted from the set. This scheme keeps the size of the elite set constant and attempts to maintain the set diversified.

We next give details on our implementation of path-relinking for the QAP, shown in Algorithm 3. Let $p$ be the mapping implied by the current solution and $q$ the mapping implied by the guiding solution. For each location $i = 1, \ldots, n$, path-relinking attempts to exchange facility $p(i)$ assigned to location $i$ in the current solution with facility $q(i)$ assigned to $i$ in the guiding solution. To maintain the mapping $p$ feasible, it exchanges $p(i)$ with $p(k)$, where $p(k) = q(i)$.

The change in objective function caused by this swap is found using the function evalij, which is limited to the part of the objective function affected by these elements. If the change is positive, then the algorithm applies local search to the resulting solution. This is done only for positive changes in the objective value function to reduce the total computational time spent in local search. The algorithm also checks if the generated solution is better than the best known solution and, if so, saves it.

The path-relinking procedure described above can be further generalized, by observing that path-relinking can also be done in the reverse direction, from the solution in the elite set to the current solution. This modification of the path-relinking procedure is called *reverse path-relinking*. In our implementation, a reverse path-relinking is also applied at each iteration. As a last step, we use a post-optimization procedure where path-relinking is applied among all solutions

---

**Algorithm 3** Path-relinking

---

**Require:** $p$, the current GRASP solution; $q$, the guiding solution
1: $c^* \leftarrow \infty$
2: **for** $i \leftarrow 1, \ldots, n$ **do**
3:    **if** $p(i) \neq q(i)$ **then**
4:       Let $j$ be such that $p(j) = q(i)$
5:       $\delta \leftarrow \texttt{evalij}(p, i, j)$
6:       $\tau \leftarrow p(i)$
7:       $p(i) \leftarrow p(j)$
8:       $p(j) \leftarrow \tau$
9:       **if** $\delta > 0$ **then**
10:          $r \leftarrow \texttt{LocalSearch}(p)$
11:          **if** $c(r) < c^*$ **then**
12:             $r^* \leftarrow r$
13:          **end if**
14:       **end if**
15:    **end if**
16: **end for**
17: **return** $r^*$

---

of the elite set. This procedure, which can be viewed as an extended local search, is repeated while an improvement in the best solution is possible.

One of the computational burdens associated with path-relinking is the local search done on all new solutions found during path-relinking. To ameliorate this, we modified the local search phase proposed in GRASP [12] by using a non-exhaustive improvement phase. In the local search in [12], each pair of assignments was exchanged until the best one was found. In our implementation, only one of the assignments is verified and exchanged with the one that brings the best improvement. This reduces the complexity of local search by a factor of $n$, leading to a $O(n^2)$ procedure. This scheme is used after the greedy randomized construction and at each iteration during path-relinking.

To enhance the quality of local search outside path-relinking, after the modified local search discussed above is done, the algorithm performs a random 3-exchange step, equivalent to changing, at random, two pair of elements in the solution. The algorithm then continues with the local search, until a local optimum is found. This type of random shaking is similar to what is done in variable neighborhood search [13].

## 3   Computational Experiments

Before we present the results, we first describe a plot used in several of our papers to experimentally compare different randomized algorithms or different versions of the same randomized algorithm [1,3,7]. This plot shows empirical distributions of the random variable *time to target solution value*. To plot the empirical distribution, we fix a solution target value and run each algorithm $T$

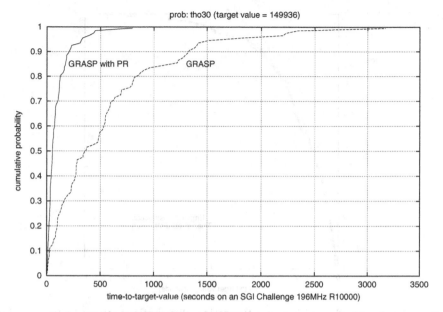

**Fig. 1.** Probability distribution of time-to-target-value on instance tho30 from QAPLIB for GRASP and GRASP with path-relinking.

independent times, recording the running time when a solution with cost at least as good as the target value is found. For each algorithm, we associate with the $i$-th sorted running time $(t_i)$ a probability $p_i = (i - \frac{1}{2})/T$, and plot the points $z_i = (t_i, p_i)$, for $i = 1, \ldots, T$. Figure 1 shows one such plot comparing the pure GRASP with the GRASP with path-relinking for QAPLIB instance tho30 with target (optimal) solution value of 149936. The figure shows clearly that GRASP with path-relinking (GRASP+PR) is much faster than pure GRASP to find a solution with cost 149936. For instance, the probability of finding such a solution in less than 100 seconds is about 55% with GRASP with path-relinking, while it is about 10% with pure GRASP. Similarly, with probability 50% GRASP with path-relinking finds such a target solution in less than 76 seconds, while for pure GRASP, with probability 50% a solution is found in less than 416 seconds.

In [3], Aiex, Resende, and Ribeiro showed experimentally that the distribution of the random variable *time to target solution value* for a GRASP is a shifted exponential. The same result holds for GRASP with path-relinking [2]. Figure 2 illustrates this result, depicting the superimposed empirical and theoretical distributions observed for one of the cases studied in [3].

In this paper, we present extensive experimental results, showing that path-relinking substantially improves the performance of GRASP. We compare an implementation of GRASP with and without path-relinking. The instances are taken from QAPLIB [4], a library of quadratic assignment test problems.

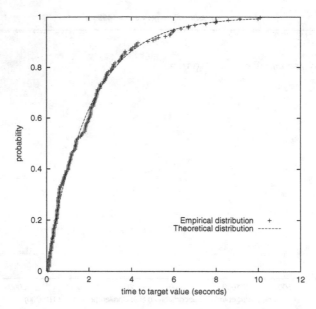

**Fig. 2.** Superimposed empirical and theoretical distributions (times to target values measured in seconds on an SGI Challenge computer with 196 MHz R10000 processors).

For each instance considered in our experiments, we make $T = 100$ independent runs with GRASP with and without path-relinking, recording the time taken for each algorithm to find the best known solution for each instance. (Due to the length of the runs on a few of the instances, fewer than 100 runs were done.) The probability distributions of time-to-target-value for each algorithm are plotted for each instance considered. We consider 91 instances from QAPLIB. Since it is impractical to fit 91 plots in this paper, we show the entire collection of plots at the URL http://www.research.att.com/~mgcr/exp/gqapspr. In this paper, we show only a representative set of plots.

Table 3 summarizes the runs in the representative set. The numbers appearing in the names of the instances indicate the dimension ($n$) of the problem. For each instance, the table lists for each algorithm the number of runs, and the times in seconds for 25%, 50%, and 75% of the runs to find a solution having the target value.

The distributions are depicted in Figures 1 and 3 to 9.

The table and figures illustrate the effect of path-relinking on GRASP. On all instances, path-relinking improved the performance of GRASP. The improvement went from about a factor of two speedup to over a factor of 60.

**Table 1.** Summary of experiments. For each instance, the table lists for each algorithm, the number of independent runs, and the time (in seconds) for 25%, 50%, and 75% of the runs to find the target solution value.

| problem | runs | GRASP 25% | 50% | 75% | runs | GRASP with PR 25% | 50% | 75% |
|---------|------|-------|-----|-----|------|------|-----|-----|
| esc32h | 100 | .5 | 1.4 | 2.5 | 100 | .2 | .5 | 1.0 |
| bur26h | 100 | 2.5 | 1.4 | 2.5 | 100 | .7 | 1.4 | 2.8 |
| kra30a | 100 | 47 | 115 | 241 | 100 | 11 | 26 | 57 |
| tho30 | 100 | 208 | 410 | 944 | 100 | 30 | 76 | 154 |
| nug30 | 100 | 583 | 1334 | 2841 | 100 | 63 | 149 | 283 |
| chr22a | 100 | 723 | 1948 | 4188 | 100 | 234 | 449 | 726 |
| lipa40a | 75 | 12,366 | 23,841 | 39,649 | 100 | 360 | 526 | 708 |
| ste36a | 17 | 27,034 | 91,075 | 135,011 | 100 | 1787 | 4047 | 8503 |

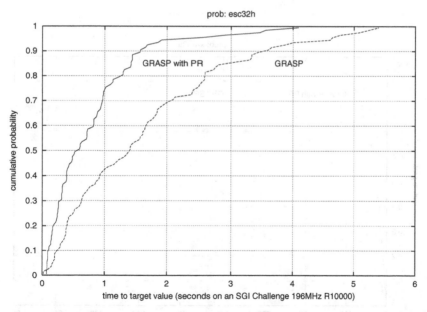

**Fig. 3.** Probability distribution of time-to-target-value on instance esc32h from QAPLIB for GRASP and GRASP with path-relinking.

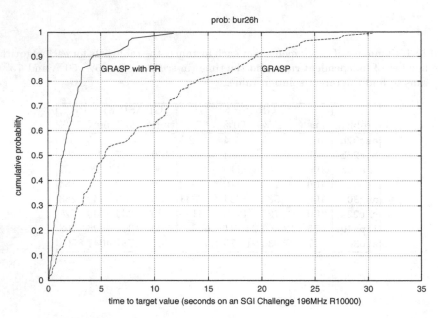

**Fig. 4.** Probability distribution of time-to-target-value on instance **bur26h** from QAPLIB for GRASP and GRASP with path-relinking.

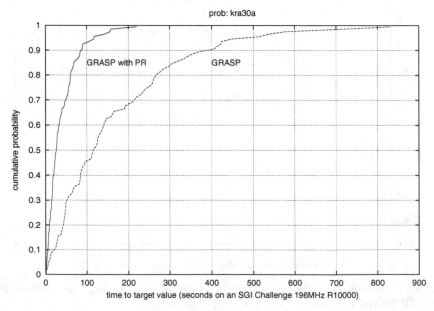

**Fig. 5.** Probability distribution of time-to-target-value on instance **kra30a** from QAPLIB for GRASP and GRASP with path-relinking.

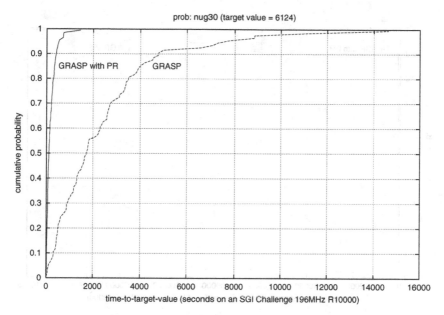

**Fig. 6.** Probability distribution of time-to-target-value on instance nug30, from QAPLIB for GRASP and GRASP with path-relinking.

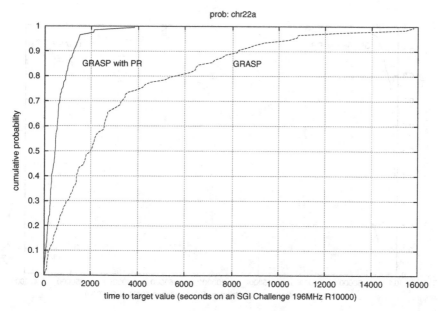

**Fig. 7.** Probability distribution of time-to-target-value on instance chr22a from QAPLIB for GRASP and GRASP with path-relinking.

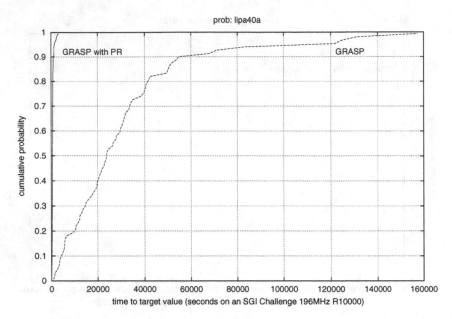

**Fig. 8.** Probability distribution of time-to-target-value on instance `lipa40a` from QAPLIB for GRASP and GRASP with path-relinking.

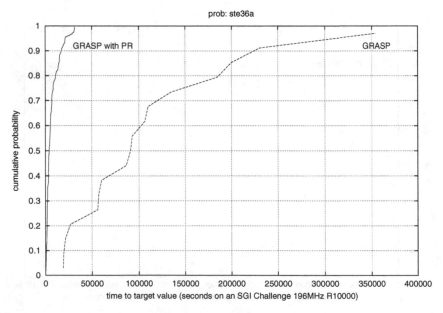

**Fig. 9.** Probability distribution of time-to-target-value on instance `ste36a` from QAPLIB for GRASP and GRASP with path-relinking.

# 4    Concluding Remarks

In this paper, we propose a GRASP with path-relinking for the quadratic assignment problem. The algorithm was implemented in the ANSI-C language and was extensively tested. Computational results show that path-relinking speeds up convergence, sometimes by up to two orders of magnitude. The source code for both GRASP and GRASP with path-relinking, as well as the plots for the extended experiment, can be downloaded from the URL http://www.research.att.com/~mgcr/exp/gqapspr.

# References

1. R.M. Aiex. *Uma investigação experimental da distribuição de probabilidade de tempo de solução em heurísticas GRASP e sua aplicação na análise de implementações paralelas.* PhD thesis, Department of Computer Science, Catholic University of Rio de Janeiro, Rio de Janeiro, Brazil, 2002.
2. R.M. Aiex, S. Binato, and M.G.C. Resende. Parallel GRASP with path-relinking for job shop scheduling. *Parallel Computing*, 29:393–430, 2003.
3. R.M. Aiex, M.G.C. Resende, and C.C. Ribeiro. Probability distribution of solution time in GRASP: An experimental investigation. *Journal of Heuristics*, 8:343–373, 2002.
4. R. Burkhard, S. Karisch, and F. Rendl.  QAPLIB – A quadratic assignment problem library.  *Eur. J. Oper. Res.*, 55:115–119, 1991.  Online version on http://www.opt.math.tu-graz.ac.at/qaplib.
5. T. A. Feo and M. G. C. Resende. Greedy randomized adaptive search procedures. *Journal of Global Optimization*, 6:109–133, 1995.
6. T.A. Feo and M.G.C. Resende.  A probabilistic heuristic for a computationally difficult set covering problem. *Operations Research Letters*, 8:67–71, 1989.
7. P. Festa, P.M. Pardalos, M.G.C. Resende, and C.C. Ribeiro. Randomized heuristics for the max-cut problem. *Optimization Methods and Software*, 7:1033–1058, 2002.
8. P. Festa and M. G. C. Resende. GRASP: An annotated bibliography. In C.C. Ribeiro and P. Hansen, editors, *Essays and Surveys on Metaheuristics*, pages 325–368. Kluwer Academic Publishers, 2002.
9. F. Glover. Tabu search and adaptive memory programing – Advances, applications and challenges. In R.S. Barr, R.V. Helgason, and J.L. Kennington, editors, *Interfaces in Computer Science and Operations Research*, pages 1–75. Kluwer, 1996.
10. T. C. Koopmans and M. J. Berkmann. Assignment problems and the location of economic activities. *Econometrica*, 25:53–76, 1957.
11. M. Laguna and R. Martí.  GRASP and path relinking for 2-layer straight line crossing minimization. *INFORMS Journal on Computing*, 11:44–52, 1999.
12. Y. Li, P.M. Pardalos, and M.G.C. Resende. A greedy randomized adaptive search procedure for the quadratic assignment problem. In P.M. Pardalos and H. Wolkowicz, editors, *Quadratic assignment and related problems*, volume 16 of *DIMACS Series on Discrete Mathematics and Theoretical Computer Science*, pages 237–261. American Mathematical Society, 1994.
13. N. Mladenović and P. Hansen.  Variable neighborhood search. *Computers and Operations Research*, 24:1097–1100, 1997.

14. P.M. Pardalos, L.S. Pitsoulis, and M.G.C. Resende. Algorithm 769: FORTRAN subroutines for approximate solution of sparse quadratic assignment problems using GRASP. *ACM Trans. Math. Software*, 23(2):196–208, 1997.
15. M.G.C. Resende, P.M. Pardalos, and Y. Li. Algorithm 754: FORTRAN subroutines for approximate solution of dense quadratic assignment problems using GRASP. *ACM Transactions on Mathematical Software*, 22:104–118, March 1996.
16. M.G.C. Resende and C.C. Ribeiro. GRASP and path-relinking: Recent advances and applications. Technical Report TD-5TU726, AT&T Labs Research, 2003.
17. M.G.C. Resende and C.C. Ribeiro. Greedy randomized adaptive search procedures. In F. Glover and G. Kochenberger, editors, *Handbook of Metaheuristics*, pages 219–249. Kluwer Academic Publishers, 2003.
18. S. Sahni and T. Gonzalez. P-complete approximation problems. *Journal of the Association of Computing Machinery*, 23:555–565, 1976.
19. S. Voß. Heuristics for nonlinear assignment problems. In P.M. Pardalos and L. Pitsoulis, editors, *Nonlinear Assignment Problems*, pages 175–215. Kluwer, Boston, 2000.

# Finding Minimum Transmission Radii for Preserving Connectivity and Constructing Minimal Spanning Trees in Ad Hoc and Sensor Networks

Francisco Javier Ovalle-Martínez[1], Ivan Stojmenović[1], Fabián García-Nocetti[2], and Julio Solano-González[2]

[1] SITE, University of Ottawa, Ottawa, Ontario K1N 6N5, Canada.
{fovalle,ivan}@site.uottawa.ca
[2] DISCA, IIMAS, UNAM, Ciudad Universitaria, México D.F. 04510, Mexico.
{fabian,julio}@uxdea4.iimas.unam.mx

**Abstract.** The minimum transmission radius $R$ that preserves ad hoc network connectivity is equal to the longest edge in the minimum spanning tree. This article proposes to use the longest *LMST* (local *MST*, recently proposed message free approximation of *MST*) edge to approximate $R$ using a wave propagation quazi-localized algorithm. Despite small number of additional edges in *LMST* with respect to *MST*, they can extend $R$ by about 33% its range on networks with up to 500 nodes. We then prove that *MST* is a subset of *LMST* and describe a quazi-localized scheme for constructing *MST* from *LMST*. The algorithm eliminates *LMST* edges which are not in *MST* by a loop breakage procedure, which iteratively follows dangling edges from leaves to *LMST* loops, and breaks loops by eliminating their longest edges, until the procedure finishes at a single leader node, which then broadcasts $R$ to other nodes.

## 1 Introduction

Due to its potential applications in various situations such as battlefield, emergency relief, environment monitoring, etc., wireless ad hoc networks have recently emerged as a prime research topic. Wireless networks consist of a set of wireless nodes which are spread over a geographical area. These nodes are able to perform processing and are capable communicating with each other by means of a wireless ad hoc network. Wireless nodes cooperate to perform data communication tasks, and the network may function in both urban and remote environments.

Energy conservation is a critical issue in wireless networks for the node and the network life, as the nodes are powered by batteries only. Each wireless node typically has transmission and reception processing capabilities. To transmit a signal from a node $A$ to other node $B$, the consumed power (in the most common power-attenuation model) is proportional to $\|AB\|^\alpha + c$, where $\|AB\|$ is the Euclidean distance between $A$ and $B$; $\alpha$ is a real constant between 2 and 5 which depends on the transmission environment, and constant $c$ represents the minimal power to receive a signal, and power to store and then process that signal. For simplicity, this overhead cost can be inte

C.C. Ribeiro and S.L. Martins (Eds.): WEA 2004, LNCS 3059, pp. 369–382, 2004.

grated into one cost, which is almost the same for all nodes. The expression represents merely the minimal power, assuming that the transmission radius is adjusted to the distance between nodes. While adjusting transmission radius is technologically feasible, medium access layer (e.g. the standard IEEE 802.11) works properly only when all nodes use the same transmission radius. Otherwise hidden terminal problem is more difficult to control and magnifies its already negative impact on the network performance.

Ad hoc networks are normally modeled by unit graphs, where two nodes are connected if and only if their distance is at most $R$, where $R$ is the transmission radius, equal for all nodes. Finding the minimum $R$ that preserves connectivity is an important problem for network functionality. Larger than necessary values of $R$ cause communication interference and consumption of increased energy, while smaller values of $R$ may disable data communication tasks such as routing and broadcasting.

The two objectives that have been mainly considered in literature are: minimizing the maximum power at each node (as in this paper) and minimizing the total power assigned if transmission ranges can be adjusted (for example, [11] shows that MST used as nonuniform power assignment yields a factor of 2 approximation. Instead of transmitting using the maximum possible power, nodes in an ad hoc network may collaboratively determine the optimal common transmission power. It corresponds to the minimal necessary transmission radius to preserve network connectivity. It was recognized [7,8,10] that the minimum value of $R$ that preserves the network connectivity is equal to the longest edge in the minimum spanning tree (MST). However, all existing solutions for finding $R$ rely on algorithms that require global network knowledge or inefficient straightforward distributed adaptations of centralized algorithms. Therefore almost all existing solutions for energy-efficient broadcast communications are globalized, meaning that each node needs global network information.

Global network information requires huge communication overhead for its maintenance when nodes are mobile or have frequent changes between active and sleeping periods. Localized solutions are therefore preferred. In a localized solution to any problem, the node makes (forwarding or structure) decisions solely based on the position of itself and its neighboring nodes. In addition, it may require constant amount of additional information, such as position of destination in case of routing. In case of LMST, both construction and maintenance are fully localized. In some cases (such as MST), however, local changes to a structure may trigger changes in a different part of the network, and therefore could have global impact to a structure. We refer to such protocols are being quazi-localized, if any node in the update process still makes decision based on local information, but a 'wave' of propagation messages may occur.

Nodes in ad hoc network are assumed to either know their geographic position (using, for example, GPS), or are able to determine mutual distances based on signal strength or time delays. This assumption is in accordance with literature.

Li, Hou and Sha [4] recently proposed a localized algorithm to approximate MST. The algorithms constructs local minimal spanning tree, where each node finds MST of the subgraph of its neighbors, and an edge is kept in LMST (localized minimal spanning tree) if and only if both endpoints have it in their respective local trees. In this article, we propose to use the longest edge in LMST edge to approximate $R$ using a wave propagation quazi-localized algorithm. The wave propagation algorithm is adapted from [9], where it was used for leader election. We simply propagate the longest LMST edge instead of propagating the winning leader information.

In order to determine whether the longest *LMST* edge is a reasonably good approximation of desired *R*, we study some characteristics of *LMST* in two dimensions (2D) and three dimensions (3D). We observed that, although the number of additional edges in *LMST* respect to *MST* is very small (under 5% of additional edges), they tend to be relatively large edges, and can extend *R* by about 33% of its range on networks with up to 500 nodes.

In some applications, such as mesh networks for wireless Internet access, or sensor networks for monitoring environment, the nodes are mostly static and network does not change too frequently (which is the case if mobility is involved). The increased transmisison range by 33% may easily double the energy expenditure, depending on constants $\alpha$ and $c$ in the energy consumption model, and increases interference with other traffic. On the other hand, the increased, the larger value for *R* provides redundancy in routing, which is useful especially in a dynamic setting. These observations about the structure of *LMST* and increased power consumption motivated us to design an algorithm for constructing *MST* topology from *LMST* topology, without the aid of any central entity. The proposed algorithm needs less than 7 messages per node on average (on networks up to 500 nodes). It eliminates *LMST* edges which are not in *MST* by a loop breakage procedure, which iteratively follows dangling edges from leaves to *LMST* loops, and breaks loops by eliminating their longest edges, until the procedure finishes at a single node (as a byproduct, this single node can also be considered as an elected leader of the network). This so elected leader also learns longest *MST* edge in the process, and may broadcast it to other nodes.

We made two sets of experiments (using Matlab environment) for the *MST* construction from the *LMST*. In one scenario, nodes are static and begin constructing *MST* from *LMST* more or less simultaneously. In the second set of experiments, we study the maintenance of already constructed *MST* when a new node is added to the network.

This paper is organized as follows: Section 2 presents the related work for the stated problem. In section 3 we present some characteristics of *MST* and *LMST* obtained by experiments, for 2D and 3D. The main characteristics of interest to this study are the lengths of the longest *MST* and *LMST* edges. In section 4, we describe the adaptation of wave propagation protocol [3] for disseminating *R* throughout the network. The algorithm for constructing *MST* from the *LMST* is explained in detail in section 5. Section 6 gives performance evaluation of the algorithm for constructing *MST* from *LMST*. Section 7 describes an algorithm for updating *MST* when a single node is added to the network, and gives results of its performance evaluation. Section 8 concludes this paper and discusses relevant future work.

## 2  Related Work

In [2], Dai and Wu proposed three different algorithms to compute the minimal uniform transmission power of the nodes, using Area-based binary search, Prim's Minimum Spanning Tree, and it's extension with Fibonacci heap implementation. However, all solutions are globalized, where each node is assumed to have full network information (or centralized, assuming a specific station has this information and informs network nodes about *MST*). We are interested in quazi-localized algorithm, where each node uses only local knowledge of its 1-hop neighbors, and the communication propagates throughout the network until *MST* is constructed.

In [6], Narayanaswamy et al. presented a distributed protocol that attempts to determine the minimum common transmitting range needed to ensure network connectivity. Their algorithm runs multiple routing daemons (RDs), one at each power level available. Each RD maintains a separate routing table where the number of entries in the routing table gives the number of reachable nodes at that power level. The node power level is set to be the smallest power level at which the number of reachable nodes is the same as that of the max power level. The kernel routing table is set to be the routing table corresponding to this power level. The protocol apparently requires more messages per each node, and at higher power levels, than the protocol presented here.

Penrose [7] [8] investigated the longest edge of the minimal spanning tree. The critical transmission range for preserving network connectivity is the length of the longest edge of the Euclidean *MST* [7] [8] [10]. The only algorithm these articles offer is to find *MST* and then its longest edge, without even discussing the distributed implementation of the algorithm.

Santi and Blough [10] show that, in two and three dimensions, the transmitting range can be reduced significantly if weaker requirements on connectivity are acceptable. Halving the critical transmission range, the longest connected component contains 90% of nodes, approximately. This means that a considerable amount of energy is spent to connect relatively few nodes.

A localized *MST* based topology control algorithm for ad hoc networks was proposed in [4] by Li, Hou and Sha. Each node $u$ first collects the positions of its one-hop neighbours $NI(u)$. Node $u$ then computes the minimum spanning tree $MST(NI(u))$ of $NI(u)$. Node $u$ keeps a directed edge $uv$ in *LMST* if and only if $uv$ is also an edge in $MST(NI(v))$. If each node already has 2-hop neighbouring information, the construction does not involve any message exchange between neighboring nodes. Otherwise each node contacts neighbors along its *LMST* link candidates, to verify the status at other node. The variant with the union of edge candidates rather than their common intersection is also considered in [4], possibly leading to a directed graph (no message exchange is then needed even with 1-hop neighbour information). In [5], Li et al. showed that *LMST* is a planar graph (no two edges intersect). Then they extended the *LMST* definition to $k$-hop neighbours, that is, the same construction but with each node having more local knowledge. They also prove that *MST* is subset of 2-hop based *LMST*, but not that *MST* is a subset of 1-hop based *LMST* considered in this article. We observed, however, on their diagrams that *LMST* with 2-hop and higher local knowledge was mostly identical to the one constructed with merely 1-hop knowledge, and decided to use only that limited knowledge, therefore conserving the communication overhead needed to maintain $k$-hop knowledge.

In [3], Dulman et al. proposed a wave leader election protocol. Each node is assigned an unique ID from an ordered set. Their algorithm selects as leader the node with minimum ID. In the wave propagation algorithm [3], each node maintains a record of the minimum ID it has seen so far (initially its own). Each node receiving a smaller ID than the one it kept as currently smallest updates it before the next round. In each round, each node that received smaller ID in the previous round will propagate this minimum on all of its outgoing edges. After a number of rounds, a node elects itself as leader if the minimum value seen in this node is the node's own ID; otherwise it is a non-leader.

We will apply *FACE* routing algorithm [1] in our protocol for converting *LMST* into an *MST*. *FACE* routing guarantees delivery and needs a planar graph to be applied. Starting from source node, faces that intersect imaginary line from source to destination are traversed. The traversal of each face is made from the first intersecting edge (with mentioned imaginary line) to the second one. Reader is referred to [1] for more details.

# 3 Comparing Longest Edges of MST and LMST

**Theorem 1.** *MST* is a subset of *LMST*.

**Proof.** The well known Kruskal's algorithm for constructing *MST* sorts all edges in the increasing order, and considers these edges one by one for inclusion in *MST*. *MST* initially has all vertices but no edges. An edge is included into already constructed *MST* if and only if its addition does not create a cycle in the already constructed *MST*. Let *LMST(A)* be the minimal spanning trees constructed from *n(A)*, which is set containing A and its 1-hop neighbors. We will show that if an edge from *MST* has endpoints in *n(A)* then it belongs to *LMST(A)*. Suppose that this is not correct, and let e be the shortest such edge. *LMST(A)* may also be constructed by following the same Kruskal's algorithm. Thus edges from A to its neighbors and between neighbors of A are sorted in the increasing order. They are then considered for inclusion in *LMST(A)*. Thus, when e is considered, since it is not included in *LMST(A)*, it creates a cycle in *LMST(A)*. All other edges in the cycle are shorter than e. Since e is in *MST*, at least one of edges from the cycle cannot be in *MST*, but is in *LMST(A)* since it was already added to it. However, this contradicts the choice of e being the shortest edge of *MST* not being included in *LMST(A)*. Therefore each edge *AB* from *MST* belongs to both *LMST(A)* and *LMST(B)*, and therefore to *LMST*. ♦

We are interested in the viability of using the *LMST* topology for approximating the minimal transmission radius R. Matlab was used to derive some characteristics of the *LMST* topology in 2D and in 3D. We generate unit graph of n nodes ($n = 10, 20, 50, 100, 200$ and $500$), each randomly distributed over an area of 1 x 1 for the 2D case and over a volume of 1 x 1 x 1 for the 3D case. The following characteristics (some of them are presented for possible other applications) of *LMST* and *MST* were compared:

- Average degree (average number of neighbors for each node)
- Average maximal number of neighbors
- Percentage of nodes which have degrees 1, 2, 3, 4, 5, degree > 5
- Highest degree of a node ever found in any of tests
- Average Maximal radius
- Standard deviation of average maximal radius

For each n we ran 200 tests in order to have more confident results. The following tables show the obtained results. It is well known that the average degree (average number of neighbors per node) of an *MST* with n nodes is always 2-2/n, since it has n-1 edges and n nodes. This value is entered in tables bellow. Table 1 shows the results for 500 nodes (similar data are obtained for other values of n). Note that *LMST* has <5% more edges than *MST*.

**Table 1.** MST vs LMST for 500 nodes

|                                   | MST 2D | LMST 2D | MST 3D | LMST 3D |
|-----------------------------------|--------|---------|--------|---------|
| **Average Degree**                | 1.996  | 2.114   | 1.996  | 2.192   |
| **Average Max Number of Neighbors** | 3.900 | 3.900   | 4.400  | 4.500   |
| **Highest Degree**                | 4.000  | 4.000   | 5.000  | 5.000   |
| **Average Maximal Radius**        | 0.081  | 0.103   | 0.177  | 0.214   |
| **Std Deviation Maximal Radius**  | 0.010  | 0.005   | 0.014  | 0.017   |

Table 2 presents the ratios of the longest *MST* and *LMST* edges, for various numbers of nodes. The ratio is always >0.75. This means that, on average, longest *LMST* edge may have about 1/3 longer length than the longest *MST* edge. It may lead to about twice as much additional energy for using the longest transmission radius from *LMST* instead of *MST*. Such discovery motivated us to design a procedure for converting *LMST* into an *MST*.

**Table 2.** Ratio of Average Maximal Radius of MST and LMST

| N | MST/LMST Average Maximal Radius, 2D | MST/LMST Average Maximal Radius, 3D |
|-----|--------|--------|
| 10  | 0.9392 | 0.9421 |
| 20  | 0.8635 | 0.8650 |
| 50  | 0.7923 | 0.8043 |
| 100 | 0.7944 | 0.7978 |
| 200 | 0.7515 | 0.7588 |
| 500 | 0.7864 | 0.8271 |

As we can see from table 1, the maximum degree of any node obtained for *LMST* in all the tests was 5. This means that *LMST* maintains a relatively low degree independently on the size of the network and its density (our study is based on maximal density, or complete graphs). Since the area where nodes were placed remained fixed, and more nodes were placed, the maximal transmission radius was decreasing when number of nodes was increasing.

**Table 3.** Percentage of nodes which have degree 1, 2, 3, 4, 5 and > 5 for the 2d MST

| Degree    | 1       | 2      | 3       | 4      | 5 | >5 |
|-----------|---------|--------|---------|--------|---|----|
| N = 10    | 35.5 %  | 49.5 % | 14.5 %  | 0.5 %  | 0 | 0  |
| N = 20    | 28.75 % | 53 %   | 17.75 % | 0.5 %  | 0 | 0  |
| N = 50    | 23.8 %  | 56.8 % | 19 %    | 0.4 %  | 0 | 0  |
| N = 100   | 22.8 %  | 57.05 %| 19.5 %  | 0.65 % | 0 | 0  |
| N = 200   | 22.48 % | 56.7 % | 20.18 % | 0.65 % | 0 | 0  |
| N = 500   | 22.1 %  | 55.9 % | 21.2 %  | 0.8 %  | 0 | 0  |

Tables 3-6 show the percentages of nodes which have degree 1, degree 2, degree 3, degree 4, degree 5, and degree > 5, for the *MST* and the *LMST*. It can be observed that about half nodes have degree two, and <2% of nodes in 2D have degree >3 and <5% of nodes in 3D have degree >3.

**Table 4.** Percentage of nodes which have degree 1, 2, 3, 4, 5 and > 5 for the 2d LMST

| Degree | 1 | 2 | 3 | 4 | 5 | >5 |
|---|---|---|---|---|---|---|
| N = 10 | 27 % | 55.5 % | 16 % | 1.5 % | 0 | 0 |
| N = 20 | 20.75 % | 54 % | 24.25 % | 1 % | 0 | 0 |
| N = 50 | 16.3 % | 58.8 % | 23.9 % | 1 % | 0 | 0 |
| N = 100 | 16 % | 58.75 % | 24.4 % | 0.85 % | 0 | 0 |
| N = 200 | 15.82 % | 58.7 % | 24.62 % | 0.85 % | 0 | 0 |
| N = 500 | 15.52 % | 58.64 % | 24.82 % | 1.02 % | 0 | 0 |

**Table 5.** Percentage of nodes which have degree 1, 2, 3, 4, 5 and > 5 for the 3d MST

| Degree | 1 | 2 | 3 | 4 | 5 | >5 |
|---|---|---|---|---|---|---|
| N = 10 | 35.5 % | 50 % | 13.5 % | 1 % | 0 | 0 |
| N = 20 | 35.75 % | 40.75 % | 21.25 % | 2.25 % | 0 | 0 |
| N = 50 | 31.2 % | 45.2 % | 20 % | 3.6 % | 0 | 0 |
| N = 100 | 30.65 % | 44.3 % | 21.55 % | 3.4 % | 0.1 % | 0 |
| N = 200 | 29.02 % | 46.05 % | 21.98 % | 2.8 % | 0.15 % | 0 |
| N = 500 | 28.02 % | 46.07 % | 22.02 % | 3.67 % | 0.22 % | 0 |

**Table 6.** Percentage of nodes which have degree 1, 2, 3, 4, 5 and > 5 for the 3d LMST

| Degree | 1 | 2 | 3 | 4 | 5 | >5 |
|---|---|---|---|---|---|---|
| N = 10 | 26 % | 55 % | 17 % | 2 % | 0 % | 0 |
| N = 20 | 23.75 % | 47.25 % | 24.5 % | 4.25 % | 0.25 % | 0 |
| N = 50 | 22 % | 46.8 % | 26 % | 5.2 % | 0 % | 0 |
| N = 100 | 21.55 % | 45.95 % | 27.7 % | 4.65 % | 0.15 % | 0 |
| N = 200 | 18.85 % | 47.67 % | 29.07 % | 4.17 % | 0.23 % | 0 |
| N = 500 | 17.34 % | 48.12 % | 31. 03 % | 3.24 % | 0.27 % | 0 |

## 4 Wave Propagation Quasi-localized Algorithm for Finding Transmission Radius from the Longest LMST Edge

We adapt the wave propagation leader election algorithm [3] for the use in finding the longest *LMST* edge. Our basic idea is to substitute the node ID with the longest edge adjacent to each node in its *LMST* topology. Each node maintains a record of the longest edge it has seen so far (initially its own longest edge in its *LMST*). In each round, each node receiving larger edge in the previous round will broadcasts its new longest edge. At end, all nodes will receive the same longest edge, which will be used as transmission radius. One of drawbacks, or perhaps advantages, of given protocol is that a node does not know when the wave propagation process is finished. It is draw-back in the sense that it may not use the proper transmission radius, same for all nodes, but a smaller one. However, this smaller transmission radius will still preserve network connectivity, since it is not equal to all nodes. It is advantage in the sense that it is a very simple protocol, and can be an ongoing process with dynamic ad hoc networks. It is straightforward to apply it when an *LMST* edge has increased over current transmission radius *R*, in which case this new value can be propagated. However, when an edge that was equal to *R* has decreased, the process of reducing *R* in the net-work is not straightforward, since the length of the second longest edge was not preserved with wave propagation algorithm. To address this issue, $k$ longest *LMST* edges may be maintained, and the message to use the next smaller value is broadcast from the neighbourhood of the event. The alternative is to initiate new wave propagation from a node detecting the problem with edge that decided currently recognized $R$.

We did not implement this protocol, since it does not significantly differ from one in [3]. The reader can see that article about its performance. Most importantly, the number of messages per node was <7 in all measurements, done in somewhat different settings, with denser graphs than *LMSTs*. Therefore we can expect much lower message count per node in our application.

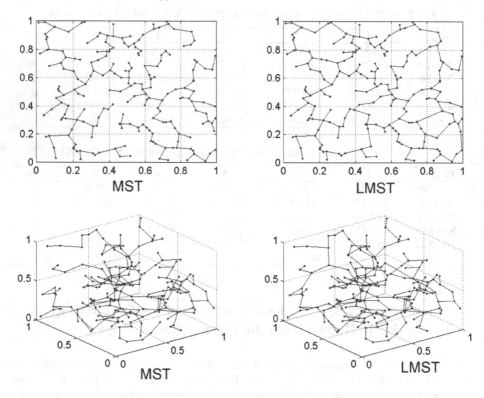

**Fig. 1.** MST and LMST comparison for 200 nodes

Figure 1 illustrate *MST* and *LMST* graphs for $n$= 200 nodes, in 2D and 3D. These figures helped us in gaining an insight on how to construct efficiently *MST* from *LMST*.

## 5   Constructing MST from LMST in Ad Hoc Networks

We can observe (see Fig. 1) that the only differences between the *LMST* and the *MST* are the loops that appear in the *LMST* but not in *MST*, created by edges present in *LMST* but not in *MST*. Our main idea is to somehow 'break' the *LMST* loops in order to obtain the *MST*.

We are now ready to describe our proposed scheme for converting *LMST* into *MST*. It consists of several iterations. Each iteration consists of two steps: traversing or eliminating dangling (tree) edges, and breaking some loops. These two steps repeat until the process ends in a node. That node is, as a byproduct, network leader, and learns longest *MST* edge in the process (it also may learn the longest *LMST* edge). The value of the longest *MST* edge can then be broadcast to other network nodes. Details of this process are as follows.

**Tree step.** Each leaf of *LMST* initiates a broadcast up the tree. Each node receiving such upward messages from all but one neighbor declares itself as dangling node and continues with upward messages. All edges traversed in this way belong to *MST*. Each such traversed subtree at end decides what is its candidate for *R* (maximal radius of *MST*). This tree advance will stop at an *LMST* loop, at a node that will be called the breaking node, because of its role in the sequel. Figure 2 illustrates this tree step. Traversed dangling nodes and edges are shown by arrows. More precisely, breaking node is a node that receives at least one message from dangling node/edge and, after some predefined timeout, remains with two or more neighbors left. Breaking nodes are exactly nodes that initiate the loop step, the second of the two steps that are repeating. Note that, an alternate choice is that all nodes on any loop become breaking nodes. We did not select this option since, on average, it would generate more messages, and our goal is to reduce message count. However, in some special cases, *LMST* may not have any leaf (e.g. *LMST* being a ring). In this case, node(s) that decided to create *MST* may declare themselves as breaking nodes after certain timeout, if no related message is received in the meanwhile. In some special cases, such as sensor network trained from the air or a monitoring station, signals to create *MST* may arrive externally as part of training.

**Loop step.** Each of the breaking nodes from the previous step initiates the loop traversal to find and break the longest edge in the loop. Consider breaking node *A* on one such loop. It has, in general, two neighbors on the loop, and edges *AB* and *AC*. Note that in some special cases breaking node may also be branching node, that is, it could have more than two neighbors; in that case consider clockwise and counterclockwise neighbors with respect to incoming dangling edge. Also, in general, breaking node *A* belongs to two loops (that is, two faces of considered planar graph). Let $\|AB\| > \|AC\|$. Node *A* will select direction *AC* to traverse the cycle, that is, the direction of shorter edge. The message that started at *A* will advance using *FACE* routing algorithm [1]. Dangling edges from previous tree steps are ignored in the loop steps. If a branching node is encountered, the traversal splits into two traversals, one of each face of edges traversed so far. The advance will stop at a node *D* whose following neighbor *E* is such that $\|AB\| < \|DE\|$. This means that a longer edge in the cycle is detected, and *AB* is not eliminated. Node *D* is then declared as the new breaking node, and starts the same longest edge verification algorithm. If the message returns to *A* then the longest edge *DE* from the loop is eliminated. Endpoints *D* and *E* of each such broken *LMST* edges follow then step 1, upward tree climbing, until they reach another *LMST* cycle. This process continues until all upward tree messages meet at a single node, which means that *MST* is constructed.

The described algorithm has several byproducts, in addition to constructing *MST*. The last node in the construction can be selected as the leader in the network, especially because it is expected to be somehow centrally positioned. Next, this leader node may, in the process, learn the longest *LMST* and longest *MST* edges, and may initiate simple broadcasting algorithm about these obtained values, which will be used as transmission radius for the whole network. This is especially needed for *MST*, as part of algorithm to find minimum transmission radius, and inform nodes about it. In case of *LMST*, we already observed that wave propagation algorithm can be used instead immediately.

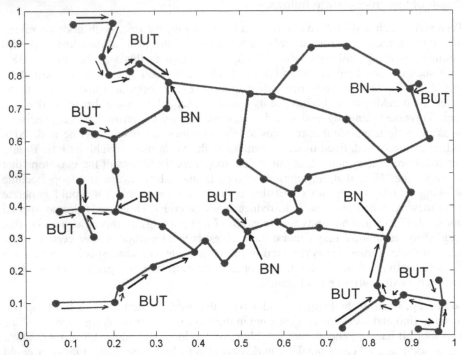

**Fig. 2.** Tree step in MST construction from *LMST* (BUT=Broadcast Up the Tree, BN=Breaking Node)

**Theorem 2.** The tree obtained from *LMST* by applying the described loop breakage algorithm correctly is *MST*.

**Proof.** Planarity of *LMST* [5] shows that it consists of well defined faces and therefore loop breakage process does create tree at end, by 'opening' up all closed faces. Theorem 1 proves that *MST* is a subset of *LMST*. The algorithm described above clearly breaks every loop, by eliminating its longest edge. Suppose that the tree obtained at the end of the process is not *MST*. Since both trees have equal number of edges, let *e* be the shortest *MST* edge that was not included in the tree. Edge *e* was eliminated at some point, since it was the longest edge in a loop of *LMST*. Subsequently, more edges from that loop may be eliminated, thus increasing the length of loop that would have been created if *e* was returned to the graph. However, in all these subsequent loops, including one at the very end of process, *e* remains the longest edge, since subsequent edges, from loops that would contain *e*, are originally shorter than *e*, but always longest among new edges that appears in the loop when they are eliminated. Since *e* is in *MST*, at least one other edge *f* from that final loop is not in *MST*. This edge $f$ is shorter than $e$ as explained. However, if $e$ is replaced by $f$, the graph remains connected and remains a tree, with overall smaller weight than $MST$. This is a contradiction. Therefore the tree that remains at end is indeed an $MST$. ◆

# 6  Performance Evaluation of the Algorithm for Constructing Minimal Spanning Trees

We consider a network of n nodes, with randomly distributed nodes over an area of 1 x 1. We constructed the *LMST* for several values of $n$ ($n$ = 10, 20, 60, 100, 200, 300 and 500), with 200 generated networks for each $n$. The described algorithm was simulated, and we measured the following characteristics:

- Average number of messages in the network, for constructing *MST* from *LMST*
- Maximal number of messages in any of generated networks
- Minimal number of messages in any of generated networks
- Average number of messages per node
- Standard deviation of message counts per node
- Average number of iterations (that is, how many times loop step was applied)

Table 7. Message counts in the algorithm that constructs MST from LMST

| Node s | Average number of messages | Max number of messages | Min number of messages | Mean of messages per node | Std of messages per node | Average Number of Iterations |
|--------|-----------|-----------|-----------|-----------|-----------|-----------|
| 10  | 16.10   | 34   | 9    | 1.61 | 0.73 | 0.120 |
| 20  | 48.64   | 81   | 19   | 2.43 | 1.37 | 0.700 |
| 60  | 257.16  | 1083 | 108  | 4.29 | 3.08 | 1.860 |
| 100 | 518.94  | 1381 | 253  | 5.19 | 4.01 | 4.660 |
| 200 | 1265.25 | 2161 | 717  | 6.33 | 5.11 | 8.120 |
| 300 | 1952.35 | 2695 | 1462 | 6.51 | 5.13 | 8.900 |
| 500 | 3490.75 | 3957 | 2473 | 6.98 | 5.80 | 13.10 |

It appears that the average number of messages per node is approximately $\log_2 n - 2$. Therefore it increases with the network size, but does it very slowly. It is not surprising since *MST* is a global structure, where change in one part of the network has impact on the decision made in other part of the network, and this happens at various levels of hierarchy.

# 7  Updating MST after Adding One Node

Mobility of nodes, or changes in node activity status, will cause changes in *MST* topology. We will now design a simple algorithm for updating *MST* when a new node is added to the network. The added node first constructs its own *LMST*, that is, *MST* of itself and its 1-hop neighbors. There are two cases to consider. Simpler case is when such *LMST* contains only one edge. In this case, the edge is added to *MST*, and no further updates are needed.

The case when the *LMST* at given node contains more than one edge is non-trivial, and requires a procedure for loop breaking to find and eliminate longest edges in newly created cycles. For example, in Fig. 3, *LMST* of new node has three edges. The update procedure assumes that *MST* is organized as a tree, rooted at the leader found in the construction process. This tree can be constructed during *MST* construction algorithm, or leader can additionally construct or complete the tree while informing about the length of longest edge in *MST*. All edges in *MST* are oriented toward the leader. In this way, branching is completely avoided in the traversal. The decision to use leader in the update process is made in order to avoid traversing long open *LMST* face (in *MST*, this open face contains all nodes and edges of *MST;* in fact each edge is found twice on that open face) unnecessarily and with each non-trivial node addition.

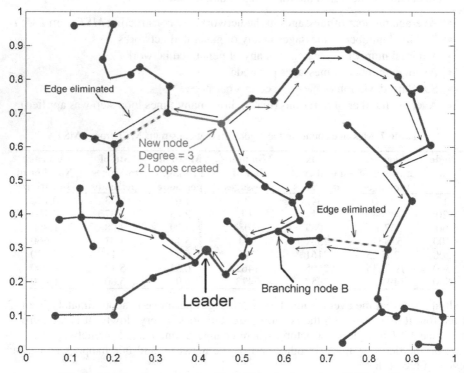

**Fig. 3.** Added node, traversal, and eliminated edges when *MST* is updated

Thus, if *LMST* at added node contains $k > 1$ neighbors, $k$ traversals toward the leader are initiated. Each of traversals records the longest edge along the path traversed. Each of the traversals stops at the leader node. However, some branching nodes may receive two copies of such traversal messages. These nodes recognize completion of a loop, and can make decision about longest loop edge to be eliminated. In example in Fig. 3, the traversal that starts to the left of new node will end at the leader node, with each intermediate node being visited once only. The other two will 'meet' at indicated branching node *B*. Such node *B* may forward only one of traversals toward the leader, and may stop the second incoming traversal. Starting with $k$ traversals, $k$-$1$ loops will be recognized, each at leader node or an interim branching node. These nodes may learn the longest edge in the process, and may send backward messages toward it to ask for breaking the edge.

The described algorithm is implemented. After 100 tests we measured that the average number of messages in the network with 200 nodes for updating *MST* when one node is added was 34.8. Therefore it appears that the *MST* construction with synchronous start from *LMST*, requiring less than 7 messages per node, leads to significant communication savings.

# 8   Conclusions and Future Work

*LMST* is a message free localized structure in ad hoc networks, which contains *MST* as a subset and which has less that 5% additional edges not already contained in *MST*. We proposed to use the longest *LMST* edge to approximate the minimal transmission radius $R$, whose exact value is the longest edge in the *MST*. We compared some characteristics of *LMST* and *MST*. The average degree of the *MST* is $\approx 1.94$ compared to the $\approx 2.06$ obtained with the *LMST* in the 2D case. In the 3D case the degree for the *MST* is $\approx 1.94$ and the degree for the *LMST* is $\approx 2.12$. Also we noted that most of the nodes had degree two or three. The longest *LMST* edge can be spread throughout the network by applying a wave propagation algorithm, previously proposed to be used as an leader election algorithm.

Existing *MST* construction algorithms were based on global knowledge of the network, or on some operations that, in distributed implementations, were not performed between neighboring nodes. The main novelty of our proposed scheme is that all the communication was restricted between neighboring nodes; therefore the message count is realistic one. To design the new *MST* construction algorithm, we observed that the difference between *LMST* and *MST* is in some loops present in *LMST*, and that the number of these loops was not large. The construction was based on 'breaking' these loops in iterations, with *MST* edges being recognized between iterations. The proposed algorithm appears to have logarithmic (in number of nodes in the network) number of messages per node.

Constructing *MST* from *LMST*, following the procedure, is beneficial when the network considered is not very dynamic. Such scenarios include mesh networks for wireless Internet access, with antennas placed on the roofs of buildings. Sensor networks normally are static, but the usefulness of the construction depends on the frequency of sleep period operations. We assume that sensors are divided into groups, and that changing between active and passive states in sensors occurs inside groups, while at the same time *MST* is constructed between groups, not between individual sensors. This treatment of sensor networks justifies the application of our *MST* construction, with certain limitations regarding accuracies involved coming from changes in active participating sensors from each group. In particular, reduced transmission range in sensor networks leads toward energy savings and prolonged network life. Reduced energy expenditure may also allow less frequent topological changes.

We described also a very simple algorithm for updating *MST* when a new node is added to the network. If an existing node is deleted from the network, its neighbors may similarly construct their new *LMSTs*, and similar procedure (with somewhat more details) can be applied.

Our proposed construction of *MST* from *LMST* works only for 2D case. It cannot be applied in 3D since the *FACE* algorithm [1] does not work in 3D. It is therefore an open problem to design an algorithm for constructing *MST* from *LMST* in 3D. Similarly, generalizing *FACE* routing with guaranteed delivery to 3D [1] remains an outstanding open problem.

**Acknowledgments.** This research is supported by CONACYT project 37017-A, CONACYT Scholarship for the first author, and NSERC grant of the second author.

# References

1.  Prosenjit Bose, Pat Morin, Ivan Stojmenovic and Jorge Urrutia, Routing with guaranteed delivery in ad hoc wireless networks, ACM Wireless Networks, 7, 6, November 2001, 609-616.
2.  Qing Dai and Jie Wu, Computation of Minimal Uniform Transmission Power in Ad Hoc Wireless Networks, 23 International Conference on Distributed Computing Systems Worksshops (ICDCSW'03), May 19–22, 2003, Providence, Rhode Island.
3.  Stefan Dulman, Paul Havinga, and Johann Jurink, Wave Leader Election Protocol for Wireless Sensor Networks, MMSA Workshop Delft The Netherlands, December 2002.
4.  Ning Li, Jennifer C. Hou, and Lui Sha, Design and Analysis of an MST-Based Topology Control Algorithm, Proc. INFOCOM 2003, San Francisco, USA, 2003.
5.  Xiang-Yang Li, Yu Wang, Peng-Jun Wan, and Ophir Frieder, Localized low weight graph and its applications in wireless ad hoc networks, INFOCOM, 2004.
6.  S. Narayanaswamy, S. Kawadia, V. Sreenivas, P. Kumar, Power control in ad hoc networks: Theory, architecture, algorithm and implementation of compow protocol, Proc. European Wireless, 2002, 156-162.
7.  M. Penrose, The longest edge of the random minimal spanning tree, The Annals of Applied Probability, 7, 2, 340-361, 1997.
8.  M. Penrose, A strong law for the longest edge of the minimal spanning tree, The Annals of Probability, 27, 1, 246-260, 1999.
9.  M. Sanchez, P. Manzoni, Z. Haas, Determination of critical transmission range in ad hoc networks, Proc. IEEE Multiaccess, Mobility and Teletraffic for Wireless Communication Conf., Venice, Italy, Oct. 1999.
10. P. Santi and D. Blough, The critical transmitting range for connectivity in sparse wireless ad hoc networks, IEEE Transactions on Mobile Computing, 2, 1, 1-15, 2003.
11. E. Althaus, G. Calinescu, I. Mandoiu, S. Prasad, N. Tchervenski, and A. Zelikovsky, Power Efficient Range Assignment in Ad-hoc Wireless Networks, submitted for publication.

# A Dynamic Algorithm for Topologically Sorting Directed Acyclic Graphs

David J. Pearce and Paul H.J. Kelly

Department of Computing, Imperial College, London SW7 2BZ, UK.
{djp1,phjk}@doc.ic.ac.uk

**Abstract.** We consider how to maintain the topological order of a directed acyclic graph (DAG) in the presence of edge insertions and deletions. We present a new algorithm and, although this has marginally inferior time complexity compared with the best previously known result, we find that its simplicity leads to better performance in practice. In addition, we provide an empirical comparison against three alternatives over a large number of random DAG's. The results show our algorithm is the best for sparse graphs and, surprisingly, that an alternative with poor theoretical complexity performs marginally better on dense graphs.

## 1 Introduction

A topological ordering, *ord*, of a directed acyclic graph $G = (V, E)$ maps each vertex to a priority value such that, for all edges $x \rightarrow y \in E$, it is the case that $ord(x) < ord(y)$. There exist well known linear time algorithms for computing the topological order of a DAG (see e.g. [4]). However, these solutions are considered *offline* as they compute the solution from scratch.

In this paper we examine *online* algorithms, which only perform work necessary to update the solution after a graph change. We say that an online algorithm is *fully dynamic* if it supports both edge insertions and deletions. The main contributions of this paper are as follows:

1. A new fully dynamic algorithm for maintaining the topological order of a directed acyclic graph.

2. The first experimental study of algorithms for this problem. We compare against two online algorithms [15,1] and a simple offline solution.

We show that, compared with [1], our algorithm has marginally inferior time complexity, but its simplicity leads to better overall performance in practice. This is mainly because our algorithm does not need the Dietz and Sleator ordered list structure [7]. We also find that, although [15] has the worst theoretical complexity overall, it outperforms the others when the graphs are dense.

**Organisation:** Section 2 covers related work; Section 3 begins with the presentation of our new algorithm, followed by a detailed discussion of the two

C.C. Ribeiro and S.L. Martins (Eds.): WEA 2004, LNCS 3059, pp. 383–398, 2004.

previous solutions [1,15]. Section 4 details our experimental work. This includes a comparison of the three algorithms and the standard offline topological sort; finally, we summarise our findings and discuss future work in Section 5.

## 2    Related Work

At this point, it is necessary to clarify some notation used throughout the remainder. Note, in the following we assume $G = (V, E)$ is a digraph:

**Definition 1.** *The path relation,* $\rightsquigarrow$, *holds if* $\forall x, y \in V.[x \rightsquigarrow y \iff x \rightarrow y \in E_T]$, *where* $G_T = (V, E_T)$ *is the transitive closure of* $G$. *If* $x \rightsquigarrow y$, *we say that* $x$ *reaches* $y$ *and that* $y$ *is reachable from* $x$.

**Definition 2.** *The set of edges involving vertices from a set,* $S \subseteq V$, *is* $E(S) = \{x \rightarrow y \mid x \rightarrow y \in E \wedge (x \in S \vee y \in S)\}$.

**Definition 3.** *The extended size of a set of vertices,* $K \subseteq V$, *is denoted* $\|K\| = |K| + |E(K)|$. *This definition originates from [1].*

The offline topologicreferences.bibal sorting problem has been widely studied and optimal algorithms with $\Theta(\|V\|)$ (i.e. $\Theta(|V| + |E|)$) time are known (see e.g. [4]). However, the problem of maintaining a topological ordering online appears to have received little attention. Indeed, there are only two existing algorithms which, henceforth, we refer to as AHRSZ [1] and MNR [15]. We have implemented both and will detail their working in Section 3. For now, we wish merely to examine their theoretical complexity. We begin with results previously obtained:

- AHRSZ - Achieves $O(\|\delta\| log \|\delta\|)$ time complexity per edge insertion, where $\delta$ is the minimal number of nodes that must be reprioritised [1,19].
- MNR - Here, an amortised time complexity of $O(|V|)$ over $\Theta(|E|)$ insertions has been shown [15].

There is some difficulty in relating these results as they are expressed differently. However, they both suggest that each algorithm has something of a difference between best and worst cases. This, in turn, indicates that a standard worse-case comparison would be of limited value. Determining average-case performance might be better, but is a difficult undertaking.

In an effort to find a simple way of comparing online algorithms the notion of *bounded complexity analysis* has been proposed [20,1,3,19,18]. Here, cost is measured in terms of a parameter $\delta$, which captures in some way the minimal amount of work needed to update the solution after some incremental change. For example, an algorithm for the online topological order problem will update *ord*, after some edge insertion, to produce a valid ordering *ord'*. Here, $\delta$ is viewed as the smallest set of nodes whose priority must change between *ord* and *ord'*. Under this system, an algorithm is described as *bounded* if its worse-case complexity can be expressed purely in terms of $\delta$.

Ramalingam and Reps have also shown that any solution to the online topological ordering problem cannot have a constant competitive ratio [19]. This suggests that competitive analysis may be unsatisfactory in comparing algorithms for this problem.

In general, online algorithms for directed graphs have received scant attention, of which the majority has focused on shortest paths and transitive closure (see e.g. [14,6,8,5,9,2]). For undirected graphs, there has been substantially more work and a survey of this area can be found in [11].

# 3    Online Topological Order

We now examine the three algorithms for the online topological order problem: PK, MNR and AHRSZ. The first being our contribution. Before doing this however, we must first present and discuss our complexity parameter $\delta_{xy}$.

**Definition 4.** *Let $G = (V, E)$ be a directed acyclic graph and ord a valid topological order. For an edge insertion $x \to y$, the affected region is denoted $AR_{xy}$ and defined as $\{k \in V \mid ord(y) \leq ord(k) \leq ord(x)\}$.*

**Definition 5.** *Let $G = (V, E)$ be a directed acyclic graph and ord a valid topological order. For an edge insertion $x \to y$, the complexity parameter $\delta_{xy}$ is defined as $\{k \in AR_{xy} \mid y = k \vee x = k \vee y \leadsto k \vee k \leadsto x\}$.*

Notice that $\delta_{xy}$ will be empty when $x$ and $y$ are already prioritised correctly (i.e. when $ord(x) < ord(y)$). We say that *invalidating* edge insertions are those which cause $|\delta_{xy}| > 0$. To understand how $\delta_{xy}$ compares with $\delta$ and the idea of minimal work, we must consider the *minimal cover* (from [1]):

**Definition 6.** *For a directed acyclic graph $G = (V, E)$ and an invalidated topological order ord, the set $K$ of vertices is a cover if $\forall x, y \in V.[x \leadsto y \wedge ord(y) < ord(x) \Rightarrow x \in K \vee y \in K]$.*

This states that, for any connected $x$ and $y$ which are incorrectly prioritised, a cover $K$ must include $x$ or $y$ or both. We say that $K$ is minimal, written $K_{min}$, if it is not larger than any valid cover. Furthermore, we now show that $K_{min} \subseteq \delta_{xy}$:

**Lemma 1.** *Let $G = (V, E)$ be a directed acyclic graph and ord a valid topological order. For an edge insertion $x \to y$, it holds that $K_{min} \subseteq \delta_{xy}$.*

*Proof.* Suppose this were nreferences.bibot the case. Then a node $a \in K_{min}$, where $a \notin \delta_{xy}$ must be possible. By Definition 6, $a$ is incorrectly prioritised with respect to some node $b$. Let us assume (for now) that $b \leadsto a$ and, hence, $ord(a) < ord(b)$. Since $ord$ is valid $\forall e \in E$, except $x \to y$, any path from $b$ to $a$ must cross $x \to y$. Therefore, $y \leadsto a$ and $b \leadsto x$ and we have $a \in AR_{xy}$ as $ord(y) \leq ord(a) \leq ord(b) \leq ord(x)$. A contradiction follows as, by Definition 5, $a \in \delta_{xy}$. The case when $a \leadsto b$ is similar.

In fact, $K_{min} = \delta_{xy}$ only when they are both empty. Now, the complexity of AHRSZ is defined in terms $K_{min}$ only and, thus, we know that $\delta_{xy}$ is not strictly a measure of minimal work for this problem. Nevertheless, we choose $\delta_{xy}$ as it facilitates a meaningful comparison between the algorithms being studied.

## 3.1   The PK Algorithm

We now present our algorithm for maintaining the topological order of a graph online. As we will see in the coming Sections, it is similar in design to MNR, but achieves a much tighter complexity bound on execution time. For a DAG $G$, the algorithm implements the topological ordering, *ord*, using an array of size $|V|$, called the *node-to-index* map or *n2i* for short. This maps each vertex to a unique integer in $\{1 \ldots |V|\}$ such that, for any edge $x \rightarrow y$ in $G$, $n2i[x] < n2i[y]$. Thus, when an invalidating edge insertion $x \rightarrow y$ is made, the algorithm must update *n2i* to preserve the topological order properreferences.bibty. The key insight is that we can do this by simply reorganising nodes in $\delta_{xy}$. That is, in the new ordering, $n2i'$, nodes in $\delta_{xy}$ are repositioned to ensure a valid ordering, *using only positions previously held by members of $\delta_{xy}$*. All other nodes remain unaffected. Consider the following caused by invalidating edge $x \rightarrow y$:

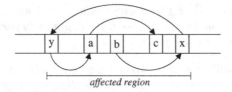

Here, nodes are laid out in topological order (i.e. increasing in *n2i* value from left to right) with members of $\delta_{xy}$ shown. As *n2i* is a total and contiguous ordering, the gaps must contain nodes, omitted to simplify the discussion. The affected region contains all nodes (including those not shown) between $y$ and $x$. Now, let us partition the nodes of $\delta_{xy}$ into two sets:

**Definition 7.** *Assume $G = (V, E)$ is a DAG and ord a valid topological order. Let $x \rightarrow y$ be an invalidating edge insertion, which does not introduce a cycle. The sets $R_F$ and $R_B$ are defined as $R_F = \{z \in AR_{xy} \mid z = y \lor y \rightsquigarrow z\}$ and $R_B = \{z \in AR_{xy} \mid z = x \lor z \rightsquigarrow x\}$.*

Note, there can be no edge from a member of $R_F$ to any in $R_B$, otherwise $x \rightarrow y$ would a introduce a cycle. Thus, for the above graph, we have $R_F = \{y, a, c\}$ and $R_B = \{b, x\}$. Now, we can obtain a correct ordering by repositioning nodes to ensure all of $R_B$ are left of $R_F$, giving:

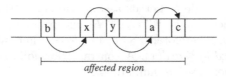

*affected region*

---

**procedure** add_edge($x, y$)
  $lb = n2i[y]$; $ub = n2i[x]$; **if** $lb < ub$ **then** dfs-f($y$); dfs-b($x$); reassign();

**procedure** dfs-f($n$)
  mark $n$ as *visited*; $R_F \cup= \{n\}$;
  **forall** $n \to w \in E$ **do**
    **if** $n2i[w] = ub$ **then** abort; //*cycle*
    **if** $w$ not *visited* $\wedge$ $n2i[w] < ub$ **then** dfs-f($w$);

**procedure** reassign()
  sort($R_B$); sort($R_F$); $L = \emptyset$;
  **for** $i = 0$ to $|R_B|-1$ **do** $w = R_B[i]$; $R_B[i] = n2i[w]$; unmark $w$; push($w, L$);
  **for** $i = 0$ to $|R_F|-1$ **do** $w = R_F[i]$; $R_F[i] = n2i[w]$; unmark $w$; push($w, L$);
  merge($R_B, R_F, R$);
  **for** $i = 0$ to $|L|-1$ **do** $n2i[L[i]] = R[i]$;

---

**Fig. 1.** The PK algorithm. The "sort" function sorts an array such that $x$ comes before $y$ iff $n2i[x] < n2i[y]$. "merge" combines two arrays into one whilst maintaining sortedness. "dfs-b" is similar to "dfs-f" except it traverses in the reverse direction, loads into $R_B$ and compares against $lb$. Note, $L$ is a temporary.

In doing this, the original order of nodes in $R_F$ must be preserved and likewise for $R_B$. The reason is that a subtle invariant is being maintained:

$$\forall x \in R_F. \left[ \, n2i[x] \leq n2i'[x] \, \right] \wedge \forall y \in R_B. \left[ \, n2i'[y] \leq n2i[y] \, \right]$$

This states that members of $R_F$ cannot be given lower priorities than they already have, whilst those in $R_B$ cannot get higher ones. This is because, for any node in $R_F$, we have identified all in the affected region which must be higher (i.e. right) than it. However, we have not determined all those which must come lower and, hence, cannot safely move them in this direction. A similar argument holds for $R_B$. Thus, we begin to see how the algorithm works: it first identifies $R_B$ and $R_F$. Then, it pools the indices occupied by their nodes and, starting with the lowest, allocates increasing indices first to members of $R_B$ and then $R_F$. So, in the above example, the algorithm proceeds by allocating $b$ the lowest available index, like so:

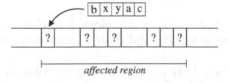

after this, it will allocate $x$ the next lowest index, then $y$ and so on. The algorithm is presented in Figure 1 and the following summarises the two stages:

**Discovery:** The set $\delta_{xy}$ is identified using a forward depth-first search from $y$ and a backward depth-first search from $x$. Nodes outside the affected region are not explored. Those visited by the forward and backward search are placed into $R_F$ and $R_B$ respectively. The total time required for this stage is $\Theta(||\delta_{xy}||)$.

**Reassignment:** The two sets are now sorted separately into increasing topological order (i.e. according to $n2i$), which we assume takes $\Theta(|\delta_{xy}|log\,|\delta_{xy}|)$ time. We then load $R_B$ into array $L$ followed by $R_F$. In addition, the pool of available indices, $R$, is constructed by merging indices used by elements of $R_B$ and $R_F$ together. Finally, we allocate by giving index $R[i]$ to node $L[i]$. This whole procedure takes $\Theta(|\delta_{xy}|log\,|\delta_{xy}|)$ time.

Therefore, algorithm PK has time complexity $\Theta((|\delta_{xy}|log\,|\delta_{xy}|) + ||\delta_{xy}||)$. As we will see, this is a good improvement over MNR, but remains marginally inferior to that for AHRSZ and we return to consider this in Section 3.3. Finally, we provide the correctness proof:

**Lemma 2.** *Assume $D = (V, E)$ is a DAG and $n2i$ an array, mapping vertices to unique values in $\{1 \dots |V|\}$, which is a valid topological order. If an edge insertion $x \to y$ does not introduce a cycle, then algorithm PK obtains a correct topological ordering.*

*Proof.* Let $n2i'$ be the new ordering found by the algorithm. To show this is a correct topological order we must show, for any two vertices $a, b$ where $a \to b$, that $n2i'[a] < n2i'[b]$ holds. An important fact to remember is that the algorithm only uses indices of those in $\delta_{xy}$ for allocation. Therefore, $z \in \delta_{xy} \Rightarrow n2i[y] \leq n2i'[z] \leq n2i[x]$. There are six cases to consider:

Case 1: $a, b \notin AR_{xy}$. Here neither $a$ or $b$ have been moved as they lie outside affected region. Thus, $n2i[a] = n2i'[a]$ and $n2i[b] = n2i'[b]$ which (by defn of $n2i$) implies $n2i'[a] < n2i'[b]$.

case 2: $(a \in AR_{xy} \wedge b \notin AR_{xy}) \vee (a \notin AR_{xy} \wedge b \in AR_{xy})$. When $a \in AR_{xy}$ we know $n2i[a] \leq n2i[x] < n2i[b]$. If $a \in \delta_{xy}$ then $n2i'[a] \leq n2i[x]$. Otherwise, $n2i'[a] = n2i[a]$. A similar argument holds when $b \in AR_{xy}$.

Case 3: $a, b \in AR_{xy} \wedge a, b \notin \delta_{xy}$. Similar to case 1 as neither $a$ or $b$ have been moved.

Case 4: $a, b \in \delta_{xy} \wedge x \leadsto a \wedge x \neq a$. Here, $a$ reachable from $x$ only along $x \to y$, which means $y \leadsto a \wedge y \leadsto b$. Thus, $a, b \in R_F$ and their relative order is preserved in $n2i'$ by sorting.

Case 5: $a, b \in \delta_{xy} \wedge b \leadsto y \wedge y \neq b$. Here, $b$ reaches $y$ along $x \to y$, so $b \leadsto x$ and $a \leadsto x$. Therefore, $a, b \in R_B$ and their relative order is preserved in $n2i'$ by sorting.

Case 6: $x = a \wedge y = b$. Here, we have $a \in R_B \wedge b \in R_F$ and $n2i'[a] < n2i'[b]$ follows because all elements of $R_B$ are allocated lower indices than those of $R_F$.

## 3.2   The MNR Algorithm

The algorithm of Marchetti-Spaccamela *et al.* operates in a similar way to PK by using a total ordering of vertices. This time two arrays, $n2i$ and $i2n$, of size

$|V|$ are used with $n2i$ as before. The second array $i2n$, is the reverse mapping of $n2i$, such that $i2n[n2i[x]] = x$ holds and its purpose is to bound the cost of updating $n2i$. The difference between PK is that only the set $R_F$ is identified, using a forward depth-first search. Thus, for the example we used previously only $y, a, c$ would be visited:

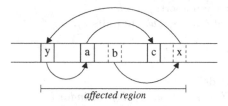

*affected region*

To obtain a correct ordering the algorithm shifts nodes in $R_F$ up the order so that they hold the highest positions within the affected region, like so:

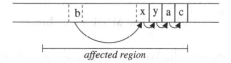

*affected region*

Notice that these nodes always end up alongside $x$ and that, unlike PK, each node in the affected region receives a new position. We can see that this has achieved a similar effect to PK as every node in $R_B$ now has a lower index than any in $R_F$. For completeness, the algorithm is presented in Figure 2 and the two stages are summarised in the following, assuming an invalidating edge $x \to y$:

**Discovery:** A depth-first search starting from $y$ and limited to $AR_{xy}$ marks those visited. This requires $O(||\delta_{xy}||)$ time.

**Reassignment:** Marked nodes are shifted up into the positions immediately after $x$ in $i2n$, with $n2i$ being updated accordingly. This requires $\Theta(|AR_{xy}|)$ time as each node between $y$ and $x$ in $i2n$ is visited.

Thus we obtain, for the first time, the following complexity result for algorithm MNR: $O(||\delta_{xy}|| + |AR_{xy}|)$. This highlights an important difference in the expected behaviour between PK and MNR as the affected region ($AR_{xy}$) can contain many more nodes than $\delta_{xy}$. Thus, we would expect MNR to perform badly when this is so.

### 3.3    The AHRSZ Algorithm

The algorithm of Alpern *et al.* employs a special data structure, due to Dietz and Sleator [7], to implement a priority space which permits new priorities to be created between existing ones in $O(1)$ worse-case time. This is a significant departure from the other two algorithms. Like PK, the algorithm employs a

```
procedure add_edge(x, y)
  lb = n2i[y]; ub = n2i[x]; if lb < ub then dfs(y);shift();

procedure dfs(n)
  mark n as visited;
  forall n → s ∈ E do
    if n2i[s] = ub then abort; // cycle
    if s not visited ∧ n2i[s] < ub then dfs(s);

procedure shift()
  L = ∅;
  for i = lb to ub do
    w = i2n[i]; // w is node at topological index i
    if w marked visited then unmark w; push(w, L); shift=shift+1;
    else n2i[L[w]] = i−shift; i2n[i−shift] = w;
    for j = 0 to |L|−1 do
      n2i[L[j]] = i−shift; i2n[i−shift] = L[j]; i=i+1;
```

**Fig. 2.** The MNR algorithm.

forward and backward search: We now examine each stage in detail, assuming an invalidating edge insertion $x \rightarrow y$:

**Discovery:** The set of nodes, $K$, to be reprioritised is determined by simultaneously searching forward from $y$ and backward from $x$. During this, nodes queued for visitation by the forward (backward) search are said to be on the forward (backward) frontier. At each step the algorithm extends the frontiers toward each other. The forward (backward) frontier is extending by visiting a member with the lowest (largest) priority. Consider the following:

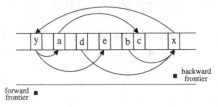

Initially, the frontiers consists of a single starting node, determined by the invalidating edge and the algorithm proceeds by extending each:

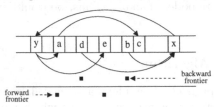

Here, members of each frontier are marked with a dot. We see the forward frontier has been extended by visiting $y$ and this results in $a, e$ being added and

$y$ removed. In the next step, $a$ will be visited as it has the lowest priority of any on the frontier. Likewise, the backward frontier will be extended next time by visiting $b$ as it has the *largest* priority. Thus, we see that the two frontiers are moving toward each other and the search stops either when one frontier is empty or they "meet" — when each node on the forward frontier has a priority greater than any on the backward frontier. An interesting point here is that the frontiers may meet before $R_B$ and $R_F$ have been fully identified. Thus, the discovery stage may identify fewer nodes than that of algorithm PK. In fact, it can be shown that at most $O(|K_{min}|)$ nodes are visited [1], giving an $O(||K_{min}||log||K_{min}||)$ bound on discovery. The *log* factor arises from the use of priority queues to implement the frontiers, which we assume are heaps.

The algorithm also uses another strategy to further minimise work. Consider

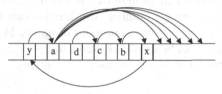

where node $a$ has high outdegree (which can be imagined as much larger). Thus, visiting node $a$ is expensive as its outedges must be iterated. Instead, we could visit $d, c, b$ in potentially much less time. Therefore, AHRSZ maintains a counter, $C(n)$, for each node $n$, initialised by outdegree. Now, let $x$ and $y$ be the nodes to be chosen next on the forward and backward frontiers respectively. Then, the algorithm subtracts $min(C(x), C(y))$ from $C(x)$ and $C(y)$ , extending the forward frontier if $C(x) = 0$ and the backward if $C(y) = 0$.

**Reassignment:** The reassignment process also operates in two stages. The first is a depth-first search of $K$, those visited during discovery, and computes a ceiling on the new priority for each node, where:

$$ceiling(x) = min(\{ord(y) \mid y \notin K \ \land \ x \rightarrow y\} \cup$$
$$\{ceiling(y) \mid y \in K \ \land \ x \rightarrow y\} \cup \{+\infty\})$$

In a similar fashion, the second stage of reassignment computes the floor:

$$floor(y) = max(\{ord'(x) \mid x \rightarrow y\} \cup \{-\infty\})$$

Note that, $ord'(x)$ is the topological ordering being generated. Once the floor has been computed the algorithm assigns a new priority, $ord'(k)$, such that $floor(k) < ord'(k) < ceiling(k)$. An $O(|K_{min}|log|K_{min}|) + |E(K_{min})|)$ bound on the time for reassignment is obtained. Again, the log factor arises from the use of a priority queue. The bound is slightly better than for discovery as only nodes in $K$ are placed onto this queue.

Therefore, we arrive at a $O(||K_{min}||log||K_{min}||)$ time bound on AHRSZ [1,19]. Finally, there has been a minor improvement on the storage requirements of AHRSZ [21], although this does not affect our discussion.

---

**procedure** add_edges($B$) // $B$ is a batch of updates
   **if** $\exists x \to y \in B.[ord(y) < ord(x)]$ **then** perform standard topological sort

---

**Fig. 3.** Algorithm DFS. Note that $ord$ is implemented as an array of size $|V|$.

## 3.4    Comparing PK and AHRSZ

We can now see the difference between PK and AHRSZ is that the latter has a tighter complexity bound. However, there are some intriguing differences between them which may offset this. In particular, AHRSZ relies on the Dietz and Sleator ordered list structure [7] and this does not come for free: firstly, it is difficult to implement and suffers high overheads in practice (both in time and space); secondly, only a certain number of priorities can be created for a given word size, thus limiting the maximum number of nodes. For example, only 32768 priorities (hence nodes) can be created if 32bit integers are being used, although with 64bit integers the limit is a more useful $2^{31}$ nodes.

## 4    Experimental Study

To experimentally compare the three algorithms, we measured their performance over a large number of randomly generated DAGs. Specifically, we investigated how insertion cost varies with $|V|$, $|E|$ and batch size. The latter relates to the processing of multiple edges and, although none of the algorithms discussed offer an advantage from this, the standard offline topological sort does. Thus, it is interesting to consider when it becomes economical to use and we have implemented a simple algorithm for this purpose, shown in Figure 3.

To generate a random DAG, we select uniformly from the probability space $G_{dag}(n, p)$, a variation on $G(n, p)$ [12], first suggested in [10]:

**Definition 8.** The model $G_{dag}(n, p)$ is a probability space containing all graphs having a vertex set $V = \{1, 2, \ldots, n\}$ and an edge set $E \subseteq \{(i, j) \mid i < j\}$. Each edge of such a graph exists with a probability $p$ independently of the others.

For a DAG in $G_{dag}(n, p)$, we know that there are at most $\frac{n(n-1)}{2}$ possible edges. Thus, we can select uniformly from $G_{dag}(n, p)$ by enumerating each possible edge and inserting with probability $p$. In our experiments, we used $p = \frac{2x}{n-1}$ to generate a DAG with $n$ nodes and expected average outdegree $x$.

Our procedure for generating a data point was to construct a random DAG and measure the time taken to insert 5000 edges. We present the exact method in Figure 4 and, for each data point, this was repeated 25 times with the average taken. Note that, we wanted the number of insertions measured over to increase proportionally with $|V|$, but this was too expensive in practice. Also, for the batch experiments, we always measured over a multiple of the batch size and chose the least value over 5000. We also recorded the values of our complexity parameters $|\delta_{xy}|, ||\delta_{xy}||$ and $|AR_{xy}|$, in an effort to correlate our theoretical

---

**procedure** measure_acpi($V, E, B, O$)
    // measure, in B sized batches, O insertions over a DAG with V
    // nodes and E edges and we assume O = cB, for some c.
    $edgeS$ = gen_random_acyclic_edgeset($V, E + O$);
    $overS$ = randomly select $O$ edges from $edgeS$;

    $G = (\{1 \ldots V\}, edgeS - overS)$;
    $startT$ = timestamp(); // *start timing now*

    **while** $overS \neq \emptyset$
        $T$ = randomly select $B$ edges from $overS$;
        $overS = overS - T$;
        add_edges($T, G$);
        randomly erase $B$ edges from $G$;

    **return** (timestamp() $- startT)/O$;

---

**Fig. 4.** Our procedure for measuring insertion cost over a random DAG. The algorithm maintains a constant number of edges in $G$ in an effort to eliminate interference caused by varying $V$, whilst keeping $O$ fixed. Note that, through careful implementation, we have minimised the cost of the other operations in the loop, which might have otherwise interfered. In particular, erasing edges is fast (unlike adding them) and independent of the algorithm being investigated.

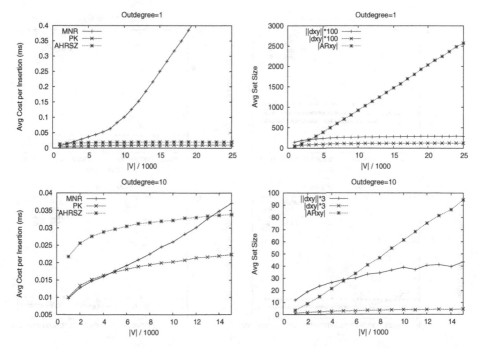

**Fig. 5.** Experimental data on random graphs with varying $|V|$.

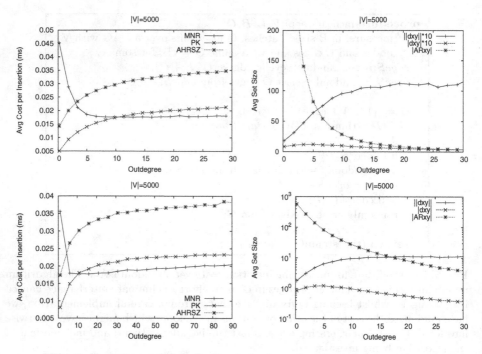

**Fig. 6.** Experimental data for fixed sized graphs with varying outdegree.

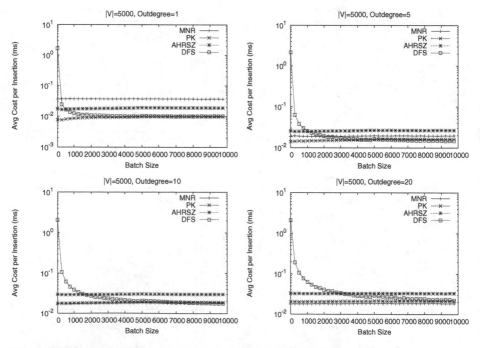

**Fig. 7.** Experimental data for varying batch sizes comparing the three algorithms against a DFS based offline topological sort

analysis. This was done using the same procedure as before, but instead of measuring time, we traversed the graph on each insertion to determine their values. These were averaged over the total number of edges inserted for 25 runs of the procedure from Figure 4.

Non-invalidating edges were included in all measurements and this dilutes the execution time and parameter counts, since all three algorithms do no work for these cases. Our purpose, however, was to determine what performance can be expected in practice, where it is unlikely all edge insertions will be invalidating.

The data, presented in Figures 5, 6 and 7, was generated on a 900Mhz Athlon based machine with 1GB of main memory. Note, we have used some (clearly marked) scaling factors to help bring out features of the data. The implementation itself was in C++ and took the form of an extension to the *Boost Graph Library*. The executables were compiled using gcc 3.2, with optimisation level "-O2" and timing was performed using the `gettimeofday` function. Our implementation of AHRSZ employs the $O(1)$ amortised (not $O(1)$ worse-case) time structure of Dietz and Sleator [7]. This seems reasonable as they themselves state it likely to be more efficient in practice.

### 4.1 Discussion

The clearest observation from Figures 5 and 6 is that PK and AHRSZ have similar behaviour, while MNR is quite different. This seems surprising as we expected the theoretical differences between PK and AHRSZ to be measured. One plausible explanation is that the uniform nature of our random graphs makes the work saved by AHRSZ (over PK) reasonably constant. Thus, it is outweighed by the gains from simplicity and efficiency offered by PK.

**Figure 5:** These graphs confirm our theoretical evaluation of MNR, whose observed behaviour strongly relates to that of $|AR_{xy}|$. Furthermore, we expected the average size of $AR_{xy}$ to increase linearly with $|V|$ as $|AR_{xy}| \leq |V|$. Likewise, the graphs for PK and AHRSZ correspond with those of $||\delta_{xy}||$. The curve for $||\delta_{xy}||$ is perhaps the most interesting here. With outdegree 1, it appears to level off and we suspect this would be true at outdegree 10, if larger values of $|V|$ were shown. We know the graphs become sparser when $|V|$ gets larger as, by maintaining constant outdegree, $|E|$ is increasing linearly (not quadratically) with $|V|$. This means, for a fixed sized affected region, $|\delta_{xy}|$ goes down as $|V|$ goes up. However, the average size of the affected region is also going up and, we believe, these two factors cancel each other out after a certain point.

**Figure 6:** From these graphs, we see that MNR is worst (best) overall for sparse (dense) graphs. Furthermore, the graphs for MNR are interesting as they level off where $|AR_{xy}|$ does not appear to. This is particularly evident from the log plot, where $|AR_{xy}|$ is always decreasing. This makes sense as the complexity of MNR is dependent on both $|AR_{xy}|$ and $||\delta_{xy}||$. So, for low outdegree MNR is dominated by $|AR_{xy}|$, but soon $||\delta_{xy}||$ becomes more significant at which point the behaviour of MNR follows this instead. This is demonstrated most clearly in the graph with high outdegree. Note, when its behaviour matches PK, MNR is always a constant factor faster as it performs one depth-first search and not two. Moving on to $|AR_{xy}|$, if we consider that the probability of a path existing between any two nodes must increase with outdegree, then the chance of inserting an invalidating edge must decrease accordingly. Furthermore, as each non-invalidating edge corresponds to a zero value of $|AR_{xy}|$ in our average, we can see why $|AR_{xy}|$ goes down with outdegree. Another interesting feature of the data is that we observe both a positive and negative gradient for $|\delta_{xy}|$. Again, this is highlighted in the log graph, although it can be observed in the other. Certainly, we expect $|\delta_{xy}|$ to increase with outdegree, as the average size of the subgraph reachable from $y$ (the head of an invalidating edge) must get larger. Again, this is because the probability of two nodes being connected by some path increases. However, $|\delta_{xy}|$ is also governed by the size of the affected region. Thus, as $|AR_{xy}|$ has a negative gradient we must eventually expect $\delta_{xy}$ to do so as well. Certainly, when $|AR_{xy}| \approx |\delta_{xy}|$, this must be the case. In fact, the data suggests the downturn happens some way before this. Note that, although $|\delta_{xy}|$ decreases, the increasing outdegree appears to counterbalance this, as we observe that $||\delta_{xy}||$ does not exhibit a negative gradient. In general, we would have liked to examine even higher outdegrees, but the time required for this has been a prohibitive factor.

**Figure 7:** These graphs compare the simple algorithm from Figure 3, with the offline topological sort implemented using depth-first search, to those we are studying. They show a significant advantage is to be gained from using the online algorithms when the batch size is small. Indeed, the data suggests that they compare favourably even for sizeable batch sizes. It is important to realise here that, as the online algorithms can only process one edge at a time, their graphs are flat since they obtain no advantage from seeing the edge insertions in batches.

## 5    Conclusion

We have presented a new algorithm for maintaining the topological order of a graph online, provided a complexity analysis, correctness proof and shown it performs better, for sparse graphs, than any previously known. Furthermore, we have provided the first empirical comparison of algorithms for this problem over a large number of randomly generated acyclic digraphs.

For the future, we would like to investigate performance over different classes of random graphs (e.g. using the *locality factor* from [10]). We are also aware that random graphs may not reflect real life structures and, thus, experimentation on physically occurring structures would be useful. Another area of interest is the related problem of dynamically identifying strongly connected components and we have shown elsewhere how MNR can be modified for this purpose [17]. We refer the reader to [16], where a more thorough examination of the work in this paper and a number of related issues can be found. Finally, Irit Katriel has since shown that algorithm PK is worse-case optimal, with respect to the number of nodes reordered over a series of edge insertions [13].

**Acknowledgements.** Special thanks must go to Irit Katriel for her excellent comments and observations on this work. We also thank Umberto Nanni and some anonymous referees for their helpful comments on earlier versions of this paper.

# References

1. B. Alpern, R. Hoover, B. K. Rosen, P. F. Sweeney, and F. K. Zadeck. Incremental evaluation of computational circuits. In *Proc. 1st Annual ACM-SIAM Symposium on Discrete Algorithms*, pages 32–42, 1990.
2. S. Baswana, R. Hariharan, and S. Sen. Improved algorithms for maintaining transitive closure and all-pairs shortest paths in digraphs under edge deletions. In *Proc. ACM Symposium on Theory of Computing*, 2002.
3. A. M. Berman. *Lower And Upper Bounds For Incremental Algorithms*. PhD thesis, New Brunswick, New Jersey, 1992.
4. T. H. Cormen, C. E. Leiserson, R. L. Rivest, and C. Stein. *Introduction to Algorithms*. MIT Press, 2001.
5. C. Demetrescu, D. Frigioni, A. Marchetti-Spaccamela, and U. Nanni. Maintaining shortest paths in digraphs with arbitrary arc weights: An experimental study. In *Proc. Workshop on Algorithm Engineering*, pages 218–229. LNCS, 2000.
6. C. Demetrescu and G. F. Italiano. Fully dynamic transitive closure: breaking through the $O(n^2)$ barrier. In *Proc. IEEE Symposium on Foundations of Computer Science*, pages 381–389, 2000.
7. P. Dietz and D. Sleator. Two algorithms for maintaining order in a list. In *Proc. ACM Symposium on Theory of Computing*, pages 365–372, 1987.
8. H. Djidjev, G. E. Pantziou, and C. D. Zaroliagis. Improved algorithms for dynamic shortest paths. *Algorithmica*, 28(4):367–389, 2000.
9. D. Frigioni, A. Marchetti-Spaccamela, and U. Nanni. Fully dynamic shortest paths and negative cycle detection on digraphs with arbitrary arc weights. In *Proc. European Symposium on Algorithms*, pages 320–331, 1998.
10. Y. Ioannidis, R. Ramakrishnan, and L. Winger. Transitive closure algorithms based on graph traversal. *ACM Transactions on Database Systems*, 18(3):512–576, 1993.
11. G. F. Italiano, D. Eppstein, and Z. Galil. Dynamic graph algorithms. In *Algorithms and Theory of Computation Handbook, CRC Press*. 1999.
12. R. M. Karp. The transitive closure of a random digraph. *Random Structures & Algorithms*, 1(1):73–94, 1990.
13. I. Katriel. On algorithms for online topological ordering and sorting. Research Report MPI-I-2004-1-003, Max-Planck-Institut für Informatik, 2004.
14. V. King and G. Sagert. A fully dynamic algorithm for maintaining the transitive closure. In *Proc. ACM Symposium on Theory of Computing*, pages 492–498, 1999.
15. A. Marchetti-Spaccamela, U. Nanni, and H. Rohnert. Maintaining a topological order under edge insertions. *Information Processing Letters*, 59(1):53–58, 1996.
16. D. J. Pearce. *Some directed graph algorithms and their application to pointer analysis (work in progress)*. PhD thesis, Imperial College, London, 2004.

17. D. J. Pearce, P. H. J. Kelly, and C. Hankin. Online cycle detection and difference propagation for pointer analysis. In *Proc. IEEE workshop on Source Code Analysis and Manipulation*, 2003.
18. G. Ramalingam. *Bounded incremental computation*, volume 1089 of *Lecture Notes in Computer Science*. Springer-Verlag, 1996.
19. G. Ramalingam and T. Reps. On competitive on-line algorithms for the dynamic priority-ordering problem. *Information Processing Letters*, 51(3):155–161, 1994.
20. T. Reps. Optimal-time incremental semantic analysis for syntax-directed editors. In *Proc. Symp. on Principles of Programming Languages*, pages 169–176, 1982.
21. J. Zhou and M. Müller. Depth-first discovery algorithm for incremental topological sorting of directed acyclic graphs. *Information Processing Letters*, 88(4):195–200, 2003.

# Approximating Interval Coloring and Max-Coloring in Chordal Graphs*

Sriram V. Pemmaraju, Sriram Penumatcha, and Rajiv Raman

The Department of Computer Science, The University of Iowa, Iowa City, IA
52240-1419, USA.
{sriram,spenumat,rraman}@cs.uiowa.edu

**Abstract.** We consider two coloring problems: interval coloring and max-coloring for chordal graphs. Given a graph $G = (V, E)$ and positive integral vertex weights $w : V \to \mathbf{N}$, the *interval coloring* problem seeks to find an assignment of a real interval $I(u)$ to each vertex $u \in V$ such that two constraints are satisfied: (i) for every vertex $u \in V$, $|I(u)| = w(u)$ and (ii) for every pair of adjacent vertices $u$ and $v$, $I(u) \cap I(v) = \emptyset$. The goal is to minimize the span $|\cup_{v \in V} I(v)|$. The *max-coloring problem* seeks to find a proper vertex coloring of $G$ whose color classes $C_1, C_2, \ldots, C_k$, minimize the sum of the weights of the heaviest vertices in the color classes, that is, $\sum_{i=1}^{k} max_{v \in C_i} w(v)$. Both problems arise in efficient memory allocation for programs. Both problems are NP-complete even for interval graphs, though both admit constant-factor approximation algorithms on interval graphs. In this paper we consider these problems for *chordal graphs*. There are no known constant-factor approximation algorithms for either interval coloring or for max-coloring on chordal graphs. However, we point out in this paper that there are several simple $O(\log(n))$-factor approximation algorithms for both problems. We experimentally evaluate and compare three simple heuristics: first-fit, best-fit, and a heuristic based on partitioning the graph into vertex sets of similar weight. Our experiments show that in general first-fit performs better than the other two heuristics and is typically very close to OPT, deviating from OPT by about 6% in the worst case for both problems. Best-fit provides some competition to first-fit, but the graph partitioning heuristic performs significantly worse than either. Our basic data comes from about 10000 runs of the each of the three heuristics for each of the two problems on randomly generated chordal graphs of various sizes and sparsity.

## 1 Introduction

*Interval coloring.* Given a graph $G = (V, E)$ and positive integral vertex weights $w : V \to \mathbf{N}$, the *interval coloring* problem seeks to find an assignment of an interval $I(u)$ to each vertex $u \in V$ such that two constraints are satisfied: (i) for every vertex $u \in V$, $|I(u)| = w(u)$ and (ii) for every pair of adjacent vertices

---

* The first and the third author have been partially supported by National Science Foundation Grant DMS-0213305.

C.C. Ribeiro and S.L. Martins (Eds.): WEA 2004, LNCS 3059, pp. 399–416, 2004.

$u$ and $v$, $I(u) \cap I(v) = \emptyset$. The goal is to minimize the span $| \cup_v I(v)|$. The interval coloring problem has a fairly long history dating back, at least to the 70's. For example, Stockmeyer showed in 1976 that the interval coloring problem is NP-complete even when restricted to interval graphs and vertex weights in $\{1, 2\}$ (see problem SR2 in Garey and Johnson [1]). The main application of the interval coloring problem is in the *compile-time memory allocation problem*. Fabri [2] made this connection in 1979. In order to reduce the total memory consumption of source-code objects (simple variables, arrays, structures), the compiler can make use of the fact that the memory regions of two objects are allowed to overlap provided that the objects do not "interfere" at run-time. This problem can be abstracted as the interval coloring problem, as follows. The source-code objects correspond to vertices of our graph, run-time interference between pairs of source code objects is represented by edges of the graph, the amount of memory needed for each source-code object is represented by the weight of the corresponding vertex, and the assignment of memory regions to source code objects is represented by the assignment of intervals to vertices of the graph. Minimizing the size of the union of intervals corresponds to minimizing the amount of memory allocation.

If we restrict our attention to straight-line programs, that is, programs without loops or conditional statements, then the compile-time memory allocation problem can be modeled as the interval coloring problem for interval graphs. Since the interval coloring problem is NP-complete for interval graphs research has focused approximation algorithms. The current best approximation factor is due to Buchsbaum et.al. [3], who give a $(2 + \varepsilon)$-algorithm for the problem.

*Max-coloring.* Like interval coloring, the *max-coloring problem* takes as input a vertex-weighted graph $G = (V, E)$ with weight function $w : V \to \mathbf{N}$. The problem requires that we find a proper vertex coloring of $G$ whose color classes $C_1, C_2, \ldots, C_k$, minimize the sum of the weights of the heaviest vertices in the color classes, that is, $\sum_{i=1}^{k} max_{v \in C_i} w(v)$. The max-coloring problem models the problem of minimizing the total buffer size needed for memory management in different applications. For example, [4] uses max-coloring to minimize buffer size in digital signal processing applications. In [5], max-coloring models the problem of minimizing buffer size needed by memory managers for wireless protocol stacks like GPRS or 3G. In general, programs that run with stringent memory or timing constraints use a dedicated memory manager that provides better performance than the general purpose memory management of the operating system. The most commonly used memory manager design for this purpose is the *segregated buffer pool*. This consists of a fixed set of buffers of various sizes with buffers of the same size linked together in a linked list. As each memory request arrives, it is satisfied by a buffer whose size is at least as large as the size of the memory request. The assignment of buffers to memory requests can be viewed as an assignment of colors to the requests – all requests that are assigned a buffer are colored identically. Requests that do not interfere with each other can be assigned the same color/buffer. Thus the problem of minimizing the total size of the buffer pool corresponds to the max-coloring problem.

[5] shows that the max-coloring problem is NP-complete for interval graphs and presents the first constant-factor approximation algorithm for the max-coloring problem for interval graphs. The paper makes a connection between max-coloring and on-line graph coloring and using a known result of Kierstead and Trotter [6] on on-line coloring interval graphs, they obtain a 2-approximation algorithm for interval graphs and a 3-approximation algorithm for circular arc graphs.

*Connections between interval coloring and max-coloring.* Given a coloring of a vertex weighted graph $G = (V, E)$ with color classes $C_1, C_2, \ldots, C_k$, we can construct an assignment of intervals to vertices as follows. For each $i$, $1 \leq i \leq k$, let $v_i \in C_i$ be the vertex with maximum weight in $C_i$. Let $H(1) = 0$, and for each $i$, $2 \leq i \leq k$, let $H(i) = \sum_{j=1}^{i-1} w(v_j)$. For each vertex $v \in C_i$, we set $I(v) = (H(i), H(i) + w(v))$. Clearly, no two vertices in distinct color classes have overlapping intervals and therefore this is a valid interval coloring of $G$. We say that this is the interval coloring *induced* by the coloring $C_1, C_2, \ldots, C_k$. The span of this interval coloring is $\sum_{i=1}^{k} w(v_i)$, which is the same as the weight of the coloring $C_1, C_2, \ldots, C_k$ viewed as a max-coloring. In other words, if there is a max-coloring of weight $W$ for a vertex weighted graph $G$, then there is an interval coloring of $G$ of the same weight.

However, in [5] we show an instance of a vertex weighted interval graph on $n$ vertices for which the weight of an optimal max-coloring is $\Omega(\log n)$ times the weight of the heaviest clique. This translates into an $\Omega(\log n)$ gap between the weight of an optimal max-coloring and the span of an optimal interval coloring because an optimal interval coloring of an interval graph has span that is within $O(1)$ of the weight of a heaviest clique [3].

In general, algorithms for max-coloring can be used for interval coloring with minor modifications to make the interval assignment more "compact." These connections motivate us to study interval coloring and max-coloring in the same framework.

*Chordal graphs.* For both the interval coloring and max-coloring problems, the assumption that the underlying graph is an interval graph is somewhat restrictive since most programs contain conditional statements and loops. In this paper we consider a natural generalization of interval graphs called chordal graphs. A graph is a *chordal graph* if it has no induced cycles of length 4 or more. Alternately, every cycle of length 4 or more in a chordal graph has a chord.

The approximability of interval coloring and max-coloring on chordal graphs is not very well understood yet. As we point out in this paper, there are several $O(\log(n))$-factor approximation algorithms for both problems on chordal graphs, however the existence of constant-factor approximation algorithms for these problems is open.

There are many alternate characterizations of chordal graphs. One that will be useful in this paper is the existence of a perfect elimination ordering of the vertices of any chordal graph. An ordering $v_n, v_{n-1}, \ldots, v_1$ of the vertex set of a graph is said to be a *perfect elimination ordering* if when vertices are deleted in this order, for each $i$, the neighbors of vertex $v_i$ in the remaining graph, $G[\{v_1, v_2, \ldots, v_i\}]$ form a clique. A graph is a chordal graph iff it has a perfect elimination ordering. Tarjan and Yannakakis [7] describe a simple linear-time algorithm called *maximum cardinality search* that can be used to determine if a given graph has a perfect elimination ordering and to construct such an ordering if it exists. Given a perfect elimination ordering of a graph $G$, the graph can be colored by considering vertices in reverse perfect elimination order and assigning to each vertex the minimum available color. It is easy to see that this greedy coloring algorithm uses exactly as many colors as the size of the largest clique in the graph and therefore produces an optimal vertex coloring.

Every interval graph is also a chordal graph (but not vice versa). To see this, take an interval representation of an interval graph and order the intervals in left-to-right order of their left endpoints. It is easy to verify that this gives a perfect elimination ordering of the interval graph.

*The rest of the paper.* In this paper, we consider three simple heuristics for the interval coloring and max-coloring problems and experimentally evaluate their performance. These heuristics are:

- **First fit.** Vertices are considered in decreasing order of weight and each vertex is assigned the first available color or interval.
- **Best fit.** Vertices are considered in reverse perfect elimination order and each vertex is assigned the color class or interval it "fits" in best.
- **Graph partitioning.** Vertices are partitioned into groups with similar weight and we use the greedy coloring algorithm to color each subgraph with optimal number of colors. The interval assignment induced by this coloring is returned as the solution to the interval coloring problem.

First fit and best-fit are fairly standard heuristics for many resource allocation problems and have been analyzed extensively for problems such as the bin packing problem. Using old results and a few new observations, we point out that the first fit heuristic and the graph partitioning heuristic provide an $O(\log n)$ approximation guarantee. The best-fit heuristic provides no such guarantee and it is not hard to construct an example of a vertex weighted interval graph for which the best-fit heuristic returns a solution to the max-coloring problem whose weight is $\Omega(\sqrt{n})$ times the weight of the optimal solution.

Our experiments show that in general first-fit performs better than the other two heuristics and is typically very close to OPT, deviating from OPT by about 6% in the worst case for both problems. Best-fit provides some competition to first-fit, but the graph partitioning heuristic performs significantly worse than either. Our basic data comes from about 10000 runs of the each of the three heuristics for each of the two problems on randomly generated chordal graphs of various sizes and sparsity.

Our experiments also reveal that best-fit performs better on chordal graphs that are "irregular", while the performance of first-fit deteriorates slightly. In all other cases, first-fit is the best algorithm. Here, "regularity" refers to the variance in the sizes of maximal cliques – greater this variance, more irregular the graph.

## 2   The Algorithms

In this section we describe three simple algorithms for the interval coloring and max-coloring problems.

### 2.1   Algorithm 1: First-Fit in Weight Order

For the interval coloring problem, we preprocess the vertices and "round up" their weights to the nearest power of 2. Then, for both problems we order the vertices of the graph in non-increasing order of weights. Let $v_1, v_2, \ldots, v_n$ be this ordering. We process vertices in this order and use a "first-fit heuristic" to assign intervals and colors to vertices to solve the interval coloring and max-coloring problem respectively.

The algorithm for interval coloring is as follows. To each vertex we assign a real interval with non-negative endpoints. To vertex $v_1$, we assign $(0, w(v_1))$. When we get to vertex $v_i$, $i > 1$, each vertex $v_j$, $1 \leq j \leq i-1$ has been assigned an interval $I(v_j)$. Let $U_i$ be the union of the intervals already assigned to neighbors of $v_i$. Then $(0, \infty) - U_i$ is a non-empty collection of disjoint intervals. Because the weights are powers of 2 and vertices are considered in non-increasing order of weights, every interval in $(0, \infty) - U_i$ has length at least $w(v_i)$. Of these, pick an interval $I = (a, b)$ with smallest right endpoint and assign the interval $(a, a + w(v_i))$ to $v_i$. This is $I(v_i)$.

For a solution to the max-coloring problem, we assume that the colors to be assigned to vertices are natural numbers, and assign to each vertex $v_i$ the smallest color not already assigned to a neighbor of $v_i$. We denote the two algorithms described above by FFI (short for first-fit by weight order for interval coloring) and FFM (short for first-fit by weight order for max-coloring) respectively.

We now observe that both algorithms provide an $O(\log(n))$-approximation guarantee. The following result is a generalization of the result from [8].

**Theorem 1.** *Let $C$ be a class of graphs and suppose there is a function $\alpha(n)$ such that the first-fit on-line graph coloring algorithm colors any $n$-vertex graph $G$ in $C$ with at most $\alpha(n) \cdot \chi(G)$ colors. Then, for any $n$-vertex graph $G$ in $C$ the* FFI *algorithm produces a solution with span at most $2\alpha(n) \cdot OPT_I(G)$, where $OPT_I(G)$ is the optimal span of any feasible assignment of intervals to vertices.*

The following is a generalization of the result from [5].

**Theorem 2.** *Let $C$ be a class of graphs and suppose there is a function $\alpha(n)$ such that the first-fit on-line graph coloring algorithm colors any $n$-vertex graph $G$ in $C$ with at most $\alpha(n) \cdot \chi(G)$ colors. Then, for any $n$-vertex graph $G$ in $C$ the* FFM *algorithm produces a solution with weight at most $\alpha(n) \cdot OPT_M(G)$, where $OPT_M(G)$ is the optimal weight of any proper of vertex coloring of $G$.*

Irani [9] has shown that the first-fit graph coloring algorithm uses at most $O(\log(n)) \cdot \chi(G)$ colors for any $n$-vertex chordal graph $G$. This fact together with the above theorems implies that FFI and FFM provide $O(\log(n))$-approximation guarantees.

An example that is tight for both algorithms is easy to construct. Let $T_0, T_1, T_2, \ldots$ be a sequence of trees where $T_0$ is a single vertex and $T_i$, $i > 0$, is constructed from $T_{i-1}$ as follows. Let $V(T_{i-1}) = \{u_1, u_2, \ldots, u_k\}$. To construct $T_i$, start with $T_{i-1}$ and add vertices $\{v_1, v_2, \ldots, v_k\}$ and edges $\{u_i, v_i\}$ for all $i = 1, 2, \ldots, k$. Thus the leaves of $T_i$ are $\{v_1, v_2, \ldots, v_k\}$ and every other vertex in $T_i$ has a neighbor $v_j$ for some $j$. Now consider a tree $T_n$ in this sequence. Clearly, $|V(T_n)| = 2^n$. Assign to each vertex in $T_n$ a unit weight. To construct an ordering on the vertices of $T_n$ first delete the leaves of $T_n$. This leaves the tree $T_{n-1}$. Recursively construct the ordering on vertices of $T_{n-1}$, and prepend to this the leaves of $T_n$ in some order. It is easy to see that first-fit coloring algorithm that considers the vertices of $T_n$ in this order uses $n$ colors. As a result, both FFI and FFM have cost $n$, whereas $OPT$ in both cases is 2. See Figure 1 for $T_0, T_1, T_2$, and $T_3$.

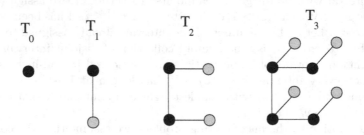

**Fig. 1.** The family of tight examples for FFI and FFM.

## 2.2    Algorithm 2: Best-Fit in Reverse Perfect Elimination Order

A second pair of algorithms that we experiment with are obtained by considering vertices in reverse perfect elimination order and using a "best-fit" heuristic to assign intervals or colors. Let $v_1, v_2, \ldots, v_n$ be the reverse of a perfect elimination ordering of the vertices of $G$. Recall that if vertices are considered in reverse perfect elimination order and colored, using the smallest color at each step, we get an optimal coloring of the given chordal graph. This essentially implies that the example of a tree with unit weights that forced FFI and FFM into worst case behavior will not be an obstacle for this pair of algorithms.

The algorithm for interval coloring is as follows. As before, to each vertex we assign a real interval with non-negative endpoints and to vertex $v_1$, we assign $(0, w(v_1))$. When we get to vertex $v_i$, $i > 1$, each vertex $v_j$, $1 \leq j \leq i - 1$ has been assigned an interval $I(v_j)$. Let $M = |\cup_{j=1}^{i-1} I(v_j)|$ and let $U_i$ be the union of the intervals $I(v_j)$, where $1 \le j \le i-1$ and $v_j$ is a neighbor of $v_i$. If $U_i = (0, M)$, then $v_i$ is assigned the interval $(M, M + w(v_i))$. Otherwise, if $U_i \ne (0, M)$, then $(0, M) - U_i$ is a non-empty collection of disjoint intervals. However, since the vertices were not processed in weight order, we are no longer guaranteed that there is any interval in $(0, M) - U_i$ with length at least $w(v_i)$. There are two cases.

**Case 1.** If there is an interval in $(0, M) - U_i$ of length at least $w(v_i)$, then pick an interval $I \in (0, M) - U_i$ of smallest length such that $|I| \ge w(v_i)$. Suppose $I = (a, b)$. Then assign the interval $(a, a + w(v_i))$ to $v_i$.

**Case 2.** Otherwise, if all intervals in $(0, M) - U_i$ have length less than $w(v_i)$, pick the largest interval $I = (a, b)$ in $(0, M) - U_i$ (breaking ties arbitrarily) and assign $(a, a + w(v_i))$ to $v_i$. Note that this assignment of an interval to $v_i$ causes the interval assignment to become infeasible. This is because there is some neighbor of $v_i$ that has been assigned an interval with left endpoint $b$ and $(a, a + w(v_i))$ intersects this interval. To restore feasibility, we increase the endpoints of all intervals "above" $b$ by $\Delta = (a + w(v_i)) - b$. In other words, for every vertex $v_j$, $1 \le j \le i$, if $I(v_j) = (c, d)$, where $c \ge b$ then $I(v_j)$ is reset to the interval $(c + \Delta, d + \Delta)$.

Consider the chordal graph shown in Figure 2. The numbers next to vertices are vertex weights and the letters are vertex labels. The ordering of vertices $A, B, C, D, E$ is a reverse perfect elimination ordering. By the time we get to processing vertex $E$, the assignment of intervals to vertices is as shown in the middle in Figure 2. When $E$ is processed, we look for "space" to fit it in and find the interval $(10, 15)$, which is not large enough for $E$. So we move the interval $I(D)$ up by 5 units to make space for $I(E)$ and obtain the assignment shown on the right.

A similar "best-fit" solution to the max-coloring problem is obtained as follows. Let $k$ be the size of a maximum clique in $G$. Start with a pallete of colors $C = \{1, 2, \ldots, k\}$ and an assignment of color 1 to vertex $v_1$. Let $AC(v_i) \subseteq C$ be the colors available for $v_i$. For each color $j$, let $W_j$ denote the maximum weight

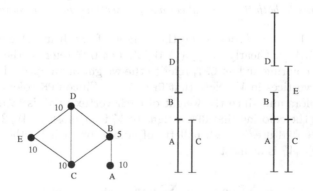

**Fig. 2.** The best-fit heuristic in action for interval coloring.

among all vertices colored $j$; for an empty color class $j$, $W_j = 0$. From the subset of $AC(v_i)$ of available colors whose weights are atleast as large as $w(v_i)$, pick the color class $j$ for $v_i$, whose weight is the smallest. If no such color exists, color vertex $v_i$ with a color $j \in AC(v_i)$ for which $W_j$ is maximum, with ties broken arbitrarily. This ensures that the color we assign to $v_i$ minimizes the increase in the weight of the coloring.

We will call these "best-fit" algorithms for interval coloring and max-coloring, BFI and BFM respectively. It is not hard to construct an example of a vertex weighted interval graph for which the BFM returns a solution whose weight is $\Omega(\sqrt{n})$ times OPT. This example does not appear in this paper due to lack of space.

## 2.3   Algorithm 3: Via Graph Partitioning

Another pair of algorithms for interval coloring and max-coloring can be obtained by partitioning the vertices of the given graph into groups with similar weight. Let $W$ be the maximum vertex weight. Fix an integer $1 \le k \le \log_2 W$ and partition the range $[1, W]$ into $(k + 1)$ subranges:

$$\left[1, \frac{W}{2^k}\right], \left(\frac{W}{2^k}, \frac{W}{2^{k-1}}\right], \ldots, \left(\frac{W}{2^2}, \frac{W}{2}\right]\left(\frac{W}{2}, W\right].$$

For $i$, $1 \le i \le k$, let $R_i = (W/2^i, W/2^{i-1}]$ and let $R_{k+1} = [1, W/2^k]$. Partition the vertex set $V$ into subsets $V_i$, $1 \le i \le (k+1)$ defined as $V_i = \{v \in V \mid w(v) \in R_i\}$. For each $i$, $1 \le i \le (k + 1)$ let $G_i$ be the induced subgraph $G[V_i]$. We ignore the weights and color each subgraph $G_i$ with the fewest number of colors, using a fresh pallette of colors for each subgraph $G_i$. For the max-coloring problem, we simply use this coloring as the solution. The solution to the interval coloring problem is simply the interval assignment induced by the coloring.

We will call these graph partitioning based algorithms for interval coloring and max-coloring, GPI and GPM respectively.

**Theorem 3.** *If we set $k = 2\log(n)$, then GPI and GPM produce $(4 \cdot \log(n) + o(1))$-approximations to both the interval coloring as well as the max-coloring problems.*

*Proof.* For $i$, $1 \le i \le k$, let $\alpha_i$ be the weight of the heaviest clique in $G[V_i]$. Let $\chi_i = \chi(G[V_i])$. Clearly, $\alpha_i \ge \chi_i \cdot W/2^i$. Let OPT refer to the weight of an optimal max-coloring and let $\text{OPT}_i$ refer to the weight of an optimal max-coloring restricted to vertices in $V_i$. Note that $\text{OPT}_i \ge \alpha_i$. Since GPM colors each $V_i$ with exactly $\chi_i$ colors and since the weight of each vertex in $V_i$ is at most $W/2^{i-1}$, the weight of the coloring that GPM assigns to $V_i$ is at most $\chi_i \cdot W/2^{i-1} \le 2 \cdot \alpha_i \le 2 \cdot \text{OPT}_i$. Since GPM uses a fresh pallette of colors for each $V_i$, the weight of the coloring of $\cup_{i=1}^{k} V_i$ is at most

$$2 \cdot \sum_{i=1}^{k} \text{OPT}_i \le 2 \cdot \sum_{i=1}^{k} \text{OPT} = 4\log(n) \cdot \text{OPT}.$$

Since $k = 2\log(n)$, $W/2^k = W/n^2$. Therefore, any coloring of $V_{k+1}$ adds a weight of atmost $W/n$ to the coloring of the rest of the graph. Since $W \leq \mathsf{OPT}$, GPM colors the entire graph with weight atmost $(4 \cdot \log(n) + 1/n)\mathsf{OPT}$.

The lower bound on $\alpha_i$ that was used in the above proof for max-coloring also applies to interval coloring and we get the same approximation factor for interval coloring.

## 3   Overview of the Experiments

### 3.1   How Chordal Graphs Are Generated

We have implemented an algorithm that takes in parameters $n$ (a positive integer) and $\alpha$ (a real number in $[0, 1]$) and generates a random chordal graph with $n$ vertices, whose sparsity is characterized by $\alpha$. The smaller the value of $\alpha$ the more sparse the graph. In addition, the algorithm can run in two modes; in mode 1 it generates somewhat "regular" chordal graphs and in mode 2 it generates somewhat "irregular" chordal graphs.

The algorithm generates chordal graphs with $n, (n-1), \ldots, 2, 1$ as a perfect elimination ordering. In the $i$th iteration of the algorithm vertex $i$ is connected to some subset of the vertices in $\{1, 2, \ldots, i-1\}$. Let $G_{i-1}$ be the graph containing vertices $1, 2, \ldots, (i-1)$, generated after iteration $(i-1)$. Let $\{C_1, C_2, \ldots, C_t\}$ be the set of maximal cliques in $G_{i-1}$. It is well known that any chordal graph on $n$ vertices has at most $n$ maximal cliques. So we explicitly maintain the list of maximal cliques in $G_{i-1}$. We pick a maximal clique $C_j$ and a random subset $S \subseteq C_j$ and connect $i$ to the vertices in $S$. This ensures that the neighbors of $i$ in $\{1, 2, \ldots, i-1\}$ form a clique, thereby ensuring that $n, (n-1), \ldots, 2, 1$ is a perfect elimination ordering.

We use the parameter $\alpha$ in order to pick the random subset $S$. For each $v \in C_j$, we independently add $v$ to set $S$ with probability $\alpha$. This makes the expected size of $S$ equal $\alpha \cdot |C_j|$. The algorithm also has a choice to make on how to pick $C_j$. One approach is to choose $C_j$ uniformly at random from the set $\{C_1, C_2, \ldots, C_t\}$. This is mode 1 and it leads to "regular" random chordal graphs, that is, random chordal graphs in which the sizes of maximal cliques show small variance. Another aproach is to choose a maximal clique with largest size from among $\{C_1, C_2, \ldots, C_t\}$. This is mode 2 and it leads to more "irregular" random chordal graphs, that is, random chordal graphs in which there are a small number of very large maximal cliques and many small maximal cliques. Graphs generated in the two modes seem to be structurally quite different. This is illustrated in Table 1, where we show information associated with 10 instances of graphs with $n = 250$ and $\alpha = 0.9$ generated in mode 1 and in mode 2. Each column corresponds to one of the 10 instances and comparing corresponding mode 1 and mode 2 rows easily reveals the the fairly dramatic difference in these graphs. For example, the mean clique size in mode 1 is about 8.5, while it is about 22 in mode 2. Even more dramatic is the large difference in the variance of the clique sizes and this justifies our earlier observation that mode 2 chordal graphs tend to have a few large cliques and many small cliques, relative to mode 1 chordal graphs.

**Table 1.** Properties of 20 instances of graphs with $n = 250$ and $\alpha = 0.9$. Ten of these were generated in Mode 1 and the other ten in mode 2.

| | MODE 1 | | | | | | | | | |
|---|---|---|---|---|---|---|---|---|---|---|
| No. of maximal cliques | 149 | 126 | 110 | 126 | 147 | 116 | 119 | 119 | 128 | 149 |
| Size of largest clique | 13 | 14 | 12 | 12 | 14 | 11 | 12 | 12 | 12 | 14 |
| Size of smallest clique | 4 | 3 | 5 | 3 | 5 | 4 | 4 | 4 | 3 | 5 |
| Mean clique size | 8.58 | 7.35 | 8.35 | 7.83 | 9.41 | 7.51 | 7.10 | 7.31 | 7.98 | 9.53 |
| Variance | 4.06 | 3.83 | 3.23 | 2.35 | 3.32 | 1.99 | 2.36 | 2.82 | 3.61 | 3.23 |

| | MODE 2 | | | | | | | | | |
|---|---|---|---|---|---|---|---|---|---|---|
| No. of maximal cliques | 220 | 216 | 216 | 218 | 218 | 219 | 213 | 216 | 219 | 219 |
| Size of largest clique | 29 | 33 | 33 | 31 | 14 | 31 | 30 | 36 | 30 | 30 |
| Size of smallest clique | 5 | 3 | 5 | 7 | 4 | 5 | 7 | 5 | 4 | 4 |
| Mean clique size | 20.13 | 22.34 | 22.37 | 24.00 | 21.15 | 21.83 | 25.17 | 23.70 | 22.80 | 20.89 |
| Variance | 28.43 | 30.02 | 30.69 | 29.55 | 33.98 | 31.73 | 48.72 | 32.66 | 25.81 | 29.68 |

## 3.2   How Weights Are Assigned

Once we have generated a chordal graph $G$ we assign weights to the vertices as follows. This process is paramaterized by $W$, the maximum possible weight of a vertex. Let $k$ be the chromatic number of $G$ and let $\{C_1, C_2, \ldots, C_k\}$ be a $k$-coloring of $G$. Since $G$ is a chordal graph, it contains a clique of size $k$. Let $Q = \{v_1, v_2, \ldots, v_k\}$ be a clique in $G$ with $v_i \in C_i$. For each $v_i$, pick $w(v_i)$ uniformly at random from the set of integers $\{1, 2, \ldots, W\}$. Thus the weight of $Q$ is $\sum_{i=1}^{k} w(v_i)$. For each vertex $v \in C_i - \{v_i\}$, pick $w(v)$ uniformly at random from $\{1, 2, \ldots, w(v_i)\}$. This ensures that $\{C_1, C_2, \ldots, C_k\}$ is a solution to max-coloring with weight $\sum_{i=1}^{k} w(v_i)$ and the interval assignment induced by this coloring is an interval coloring of span $\sum_{i=1}^{k} w(v_i)$. Since $\sum_{i=1}^{k} w(v_i)$ is also the weight of the clique $Q$, which is a lower bound on OPT in both cases, we have that OPT $= \sum_{i=1}^{k} w(v_i)$ in both cases. The advantage of this method of assigning weights is that it is simple and gives us the value of OPT for both problems. The disadvantage is that, in general OPT for both problems can be strictly larger than the weight of the heaviest clique and thus by generating only those instances for which OPT equals the weight of the heaviest clique, we might be missing a rich class of problem instances.

We also tested our algorithms on instances of chordal graphs for which the weights were assigned uniformly at random. For these algorithms, we use the maximum weighted clique as a lower bound for OPT.

**Table 2.** Results of our main experiments on mode 1 random chordal graphs, evaluating heuristics for the max-coloring problem.

| $|V|$ | $|E|$ | OPT | Best Fit | First Fit | Partition | $\frac{(BF-OPT)}{OPT}$ | $\frac{(FF-OPT)}{OPT}$ | $\frac{(GPM-OPT)}{OPT}$ |
|---|---|---|---|---|---|---|---|---|
| 40 | 64.4333 | 2292.16 | 2416.64 | 2309.09 | 3184.84 | 5.43108 | 0.738751 | 38.9454 |
| 80 | 139.178 | 2474.74 | 2604.09 | 2525.5 | 3569.16 | 5.22658 | 2.05094 | 44.2232 |
| 120 | 223.756 | 2605.77 | 2827.46 | 2660.5 | 3956.09 | 8.50763 | 2.10047 | 51.8205 |
| 160 | 304.822 | 2682.98 | 2941.02 | 2733.71 | 4095.01 | 9.61784 | 1.89093 | 52.6293 |
| 200 | 394.556 | 2712.4 | 2938.49 | 2761.17 | 4107.06 | 8.33538 | 1.79792 | 51.4178 |
| 240 | 451.978 | 2835.52 | 3161.7 | 2882.82 | 4394.1 | 11.5033 | 1.66812 | 54.9662 |
| 280 | 525.389 | 2863.43 | 3145.16 | 2928.19 | 4491.21 | 9.83862 | 2.26147 | 56.8471 |
| 320 | 626.744 | 2775.11 | 3046.61 | 2819.38 | 4335.47 | 9.78339 | 1.59513 | 56.2268 |
| 360 | 715.389 | 2913.18 | 3166.28 | 2992.4 | 4667.89 | 8.68811 | 2.71944 | 60.2336 |
| 400 | 821.822 | 3150.07 | 3427.57 | 3197.59 | 5033.11 | 8.80934 | 1.50861 | 59.7779 |
| 440 | 894.922 | 2983.16 | 3292.34 | 3065.1 | 4864.44 | 10.3645 | 2.7469 | 63.0637 |
| 480 | 973.078 | 3060.02 | 3395.98 | 3117.88 | 4841.14 | 10.9789 | 1.89069 | 58.2062 |
| 520 | 1061.74 | 3053.21 | 3393.28 | 3120.32 | 4883.03 | 11.138 | 2.19805 | 59.9311 |

## 3.3 Main Observations

For our main experiment we generated instances of random chordal graphs with number of vertices $n = 10, 20, 30, \ldots, 550$. For each value of $n$, we used values of $\alpha = 0.1, 0.2, \ldots, 0.9$. For each of the $55 \times 9$ $(n, \alpha)$ pairs, we generated 10 random vertex weighted chordal graphs. We ran each of the three heuristics for the two problems and averaged the weight and span of the solutions over the 10 instances for each $(n, \alpha)$ pairs. Thus each heuristic was evaluated on 4950 instances, for each problem. The vertex weights are assigned as described above, with the maximum weight $W$ fixed at 1000. We first conducted this experiment for the max-coloring problem on mode 1 and mode 2 chordal graphs and and then repeated them for the interval coloring problem.

We then generated the same number of instances, but this time assigning to each vertex, a weight chosen uniformly from $[0, 1000]$. We repeated each of the three heuristics for the two problems on these randomly generated instances. For these instances, we used the maximum weight clique as a lower bound to OPT. The initial prototyping was done in the discrete mathematics system Combinatorica. However, the final version of the algorithms were written in C++, and run on on a desktop running linux release 9. The running time of the programs is approximately 106.5 seconds of user time (measured using the Linux `time` command), for all six algorithms on 4950 instances for mode-1 graphs, including time to generate random instances.

Our data is presented in the following tables.[1] First we have two tables for our experiments for the max-coloring problem. The first two tables (Table 2 and

---

[1] In each table, we only show representative values due to lack of space. The complete data is available at
http://www.cs.uiowa.edu/~rraman/chordalGraphExperiments.html.

**Table 3.** Results of our main experiments on mode 2 random chordal graphs, evaluating heuristics for the max-coloring problem.

| $\lvert V\rvert$ | $\lvert E\rvert$ | OPT | Best Fit | First Fit | Partition | $\frac{(BF-OPT)}{OPT}$ | $\frac{(FF-OPT)}{OPT}$ | $\frac{(GPM-OPT)}{OPT}$ |
|---|---|---|---|---|---|---|---|---|
| 40 | 109.689 | 3094.16 | 3095.61 | 3151.24 | 3848.27 | 0.0470421 | 1.84506 | 24.3721 |
| 80 | 302.489 | 3927.78 | 3938.29 | 4023.13 | 5035 | 0.26761 | 2.42772 | 28.1895 |
| 120 | 524.8 | 4459.87 | 4471.22 | 4614.27 | 5758.13 | 0.254616 | 3.46199 | 29.11 |
| 160 | 799.044 | 4891.27 | 4903.94 | 5058 | 6387.23 | 0.259192 | 3.4088 | 30.5844 |
| 200 | 1052.07 | 5290.47 | 5293.46 | 5457.12 | 6976.49 | 0.0564958 | 3.15011 | 31.8691 |
| 240 | 1363.43 | 5390.86 | 5398.32 | 5556.09 | 7109.43 | 0.138506 | 3.06507 | 31.8795 |
| 280 | 1655.84 | 5749.18 | 5763.03 | 5937.57 | 7567.82 | 0.241001 | 3.2768 | 31.6331 |
| 320 | 1953.24 | 5776.97 | 5779.74 | 5982 | 7655.43 | 0.0480837 | 3.54915 | 32.5165 |
| 360 | 2261.96 | 5899.47 | 5916.26 | 6165.27 | 7847.38 | 0.284583 | 4.50549 | 33.0184 |
| 400 | 2639.39 | 5987.34 | 5992 | 6191.81 | 7905.62 | 0.0777566 | 3.41498 | 32.0389 |
| 440 | 2956.79 | 6161.91 | 6167.82 | 6390.23 | 8192.71 | 0.0959298 | 3.70538 | 32.9573 |
| 480 | 3243.42 | 6236.88 | 6251.6 | 6471.51 | 8228.43 | 0.236051 | 3.76203 | 31.9319 |
| 520 | 3645.54 | 6296.03 | 6302.76 | 6558.19 | 8388.33 | 0.106769 | 4.16382 | 33.232 |

**Table 4.** This table shows aggregate performance over all 4950 runs of the three heuristics for max-coloring, separately for mode 1 and mode 2 graphs. The first three rows correspond to the experiments where the value of OPT is known. The next three rows correspond to the runs where the weights were assigned randomly. In this case, the algorithms are compared against the weight of the maximum weight clique (LB). The row "Equals OPT(LB)" lists the number of times each heuristic produces a coloring with weight equal to OPT(LB), the row "Equals $\chi$" lists the number of times each heuristic produces a coloring using minimum number of colors, and the row "% Deviation" lists the percentage deviation of the weight of the solution produced from OPT(LB), averaged over the 4950 runs.

| | MODE 1 | | | MODE 2 | | |
|---|---|---|---|---|---|---|
| | BFM | FFM | GPM | BFM | FFM | GPM |
| Equals OPT | 1216 | 2965 | 0 | 4791 | 1515 | 2 |
| Equals $\chi$ | 4950 | 3393 | 1 | 4950 | 1675 | 6 |
| % Deviation | 10.70 | 1.93 | 58.66 | 0.20 | 4.47 | 39.43 |
| | Random Weights | | | | | |
| Equals LB | 53 | 71 | 0 | 38 | 57 | 0 |
| Equals $\chi$ | 4950 | 2666 | 3 | 4950 | 441 | 7 |
| % Deviation | 24.64 | 17.43 | 78.67 | 36.20 | 25.88 | 59.13 |

3) for mode 1 and mode 2 chordal graphs respectively. This is followed by table 4 that summarizes the performance of the three heuristics for the max-coloring problem for both mode 1 and mode 2 chordal graphs over all the runs. After this we present three tables (Table 5, 6, and 7) that contains corresponding information for the interval coloring problem. The data for the experiments on graphs with randomly assigned vertex weights is presented in the appendix. Based on all this data, we make 4 observations.

**Table 5.** Results of our main experiments on mode 1 random chordal graphs, evaluating heuristics for the interval coloring problem.

| $|V|$ | $|E|$ | OPT | Best Fit | First Fit | Partition | $\frac{(BFI-OPT)}{OPT}$ | $\frac{(FFI-OPT)}{OPT}$ | $\frac{(GPI-OPT)}{OPT}$ |
|---|---|---|---|---|---|---|---|---|
| 40 | 64.4333 | 2292.16 | 2377.64 | 2336.58 | 2443.47 | 3.72963 | 1.93801 | 6.60126 |
| 80 | 139.178 | 2474.74 | 2612.68 | 2531.07 | 2709.23 | 5.57364 | 2.27588 | 9.47528 |
| 120 | 223.756 | 2605.77 | 2803.87 | 2680.47 | 2839.19 | 7.60237 | 2.86672 | 8.95791 |
| 160 | 304.822 | 2682.98 | 3032.76 | 2767.01 | 2957.28 | 13.0369 | 3.13209 | 10.2237 |
| 200 | 394.556 | 2712.4 | 3055.42 | 2808.32 | 2975.33 | 12.6464 | 3.53643 | 9.69375 |
| 240 | 451.978 | 2835.52 | 3215.13 | 2926.63 | 3106.92 | 13.3877 | 3.2132 | 9.57143 |
| 280 | 525.389 | 2863.43 | 3291.52 | 2955.3 | 3175.16 | 14.9502 | 3.20827 | 10.8863 |
| 320 | 626.744 | 2775.11 | 3235.78 | 2894.63 | 3074.89 | 16.5999 | 4.30693 | 10.8024 |
| 360 | 715.389 | 2913.18 | 3374.4 | 3026.26 | 3317.4 | 15.8323 | 3.8816 | 13.8756 |
| 400 | 821.822 | 3150.07 | 3603.43 | 3304.29 | 3573.03 | 14.3923 | 4.89584 | 13.4272 |
| 440 | 894.922 | 2983.16 | 3496.91 | 3112.83 | 3325.01 | 17.2219 | 4.347 | 11.4595 |
| 480 | 973.078 | 3060.02 | 3678.7 | 3182.17 | 3378.76 | 20.2181 | 3.99162 | 10.416 |
| 520 | 1061.74 | 3053.21 | 3846.96 | 3194.8 | 3443.17 | 25.997 | 4.63738 | 12.772 |

**Table 6.** Results of our main experiments on mode 2 random chordal graphs, evaluating heuristics for the interval coloring problem.

| $|V|$ | $|E|$ | OPT | Best Fit | First Fit | Partition | $\frac{(BFI-OPT)}{OPT}$ | $\frac{(FFI-OPT)}{OPT}$ | $\frac{(GPI-OPT)}{OPT}$ |
|---|---|---|---|---|---|---|---|---|
| 40 | 109.689 | 3094.16 | 3138.77 | 3216.93 | 3377.4 | 1.44179 | 3.96805 | 9.15418 |
| 80 | 302.489 | 3927.78 | 4074.87 | 4175.86 | 4441.82 | 3.74484 | 6.31598 | 13.0874 |
| 120 | 524.8 | 4459.87 | 4623.33 | 4811.37 | 5110.5 | 3.66528 | 7.8814 | 14.5886 |
| 160 | 799.044 | 4891.27 | 5002.19 | 5306.62 | 5711.9 | 2.26776 | 8.49178 | 16.7775 |
| 200 | 1052.07 | 5290.47 | 5473.74 | 5793.81 | 6204.62 | 3.4643 | 9.51418 | 17.2793 |
| 240 | 1363.43 | 5390.86 | 5600.83 | 5896.01 | 6317.99 | 3.89507 | 9.3706 | 17.1983 |
| 280 | 1655.84 | 5749.18 | 5988.46 | 6283.7 | 6728.99 | 4.16195 | 9.29737 | 17.0426 |
| 320 | 1953.24 | 5776.97 | 6017.8 | 6364.53 | 6816.09 | 4.16885 | 10.1709 | 17.9873 |
| 360 | 2261.96 | 5899.47 | 6249.91 | 6552.94 | 7039.04 | 5.94027 | 11.0769 | 19.3166 |
| 400 | 2639.39 | 5987.34 | 6255.47 | 6616.7 | 7155.96 | 4.47815 | 10.5114 | 19.518 |
| 440 | 2956.79 | 6161.91 | 6446.57 | 6807.86 | 7327.6 | 4.6196 | 10.4829 | 18.9177 |
| 480 | 3243.42 | 6236.88 | 6569.51 | 6833.61 | 7423.2 | 5.33333 | 9.56782 | 19.0211 |
| 520 | 3645.54 | 6296.03 | 6573.81 | 6980.31 | 7555.84 | 4.41195 | 10.8684 | 20.0096 |

1. First fit in decreasing order of weights is clearly a heuristic that returns solutions very close to OPT for both problems. Overall percentage deviations from OPT, each being over 4950 runs are 1.93, 4.47, 2.29 and 9.52 - the first two are for max-coloring on mode 1 and mode 2 chordal graphs respectively, and the next two are for interval coloring on mode 1 and mode 2 graphs. Even in the experiments on graphs with randomly assigned vertex weights, first-fit performs better than the other algorithms overall. The average deviations being less than 26% from the maximum weight clique for both problems (Tables 4, 7 and Tables 8 and 9 in the appendix). Note that in these cases,

**Table 7.** This table shows aggregate performance over all 4950 runs of the three heuristics for interval coloring, separately for mode 1 and mode 2 graphs. The first three rows correspond to the case where the value of OPT is known. The next three rows correspond to the case where the weights were assigned randomly. In the latter case, the performance of the algorithms are compared with the weight of the maximum weight clique (LB). The row "Equals OPT(LB)" lists the number of times each heuristic produces a coloring with weight equal to OPT(LB) and the row "% Deviation" lists the percentage deviation of the weight of the solution produced from OPT(LB), averaged over the 4950 runs.

|  | MODE 1 | | | MODE 2 | | |
|---|---|---|---|---|---|---|
|  | BFI | FFI | GPI | BFI | FFI | GPI |
| Equals OPT | 2878 | 3368 | 2044 | 2459 | 1353 | 282 |
| % Deviation | 9.51 | 2.29 | 7.89 | 6.70 | 9.52 | 17.45 |
| Random Weights | | | | | | |
| Equals LB | 1755 | 1568 | 939 | 467 | 211 | 110 |
| % Deviation | 26.87 | 10.99 | 16.91 | 25.21 | 24.52 | 30.01 |

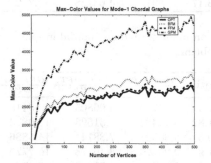 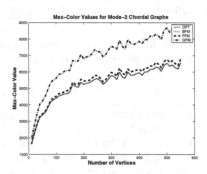

**Fig. 3.** Graphs showing values for max-coloring mode 1 and mode 2 chordal graphs. The x-axis corresponds to the number of vertices in the graph, and the y-axis corresponds to the max-color value. The solid line shows the value of OPT for the different sizes of the graph, and the dashed line corresponds to the value of the coloring produced by first-fit; the dotted line, the performance of best-fit; and the dashed-dotted line, the performance of the graph partitioning heuristic.

the percentage deviations are an exaggeration of the actual amount, since the maximum weight clique can be quite small compared to OPT.

2. The graph partitioning heuristic is not competitive at all, relative to first-fit or best-fit, in any of the cases, despite the $O(\log n)$-factor approximation guarantee it provides.

3. Between the first-fit heuristic and the best-fit in reverse perfect elimination order heuristic, first-fit seems to do better in an overall sense. However, they exhibit opposite trends in going from mode 1 to mode 2 graphs. Specifically, first-fit's performance worsens with the percentage deviation changing as $1.93 \rightarrow 4.47$ for max-coloring, while best-fit's performance improves with the percentage deviation going from $10.70 \rightarrow 0.20$. Table 4 and the graphs in Figure 3, show the performance of the three algorithms on mode 1 and

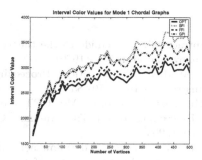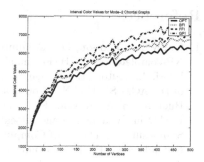

**Fig. 4.** Graph showing values for interval coloring mode 1 and mode 2 chordal graphs. The x-axis corresponds to the number of nodes in the graph, and the y-axis corresponds to the interval color value. The solid line corresponds to OPT; the dotted line, to the performance of best-fit; the dashed line to the performance of first-fit; and the dashed-dotted line, to that of the graph partitioning heuristic.

mode 2 graphs. We see the same trend for interval coloring as well. The performance of first-fit worsens as we go from mode 1 to mode 2 graphs, while best-fit's performance improves. Table 7 and the graphs in Figure 4 show the deviations from OPT for interval coloring. In order to verify this trend, we tested the algorithms on mode 1 and mode 2 graphs, with randomly assigned vertex weights. First-fit's performance continues to show this trend of deteriorating performance in going from mode 1 to mode 2 graphs for both max-coloring and interval coloring. However the performance of best-fit is quite different. It's performance on max-coloring worsens as we go from mode 1 to mode 2 graphs, while it improves slightly for interval coloring.

4. Best-fit heuristic seems to be at a disadvantage because it is constrained to use as many colors as the chromatic number. First-fit uses more colors than the chromatic number a fair number of times. Examining Table 4 we note that for max-coloring, first-fit uses more colors than OPT about 31.45% and 66.16% of the time for mode 1 and mode 2 graphs respectively.

## 4   Conclusion

Our goal was to evaluate the performance of three simple heuristics for the max-coloring problem and for the interval coloring problem. These heuristics were first-fit, best-fit, and a heuristic based on graph partitioning. First-fit outperformed the other algorithms in general, and our recommendation is that this be the default heuristic for both problems. Despite the logarithmic approximation guarantee it provides, the heuristic based on graph partitioning is not competitive in comparison to first-fit or best-fit. Best-fit seems to perform better on graphs that are more "irregular" and offers first-fit competition for such graphs. We have also experimented with other classes of chordal graphs such as trees and sets of disjoint cliques. Results from these experiments are available at http://www.cs.uiowa.edu/~rraman/chordalGraphExperiments.html

# References

1. Garey, M., Johnson, D.: Computers and Intractability: A Guide to the theory of NP-completeness. W.H. Freeman and Company, San Fransisco (1979)
2. Fabri, J.: Automatic storage optimization. ACM SIGPLAN Notices: Proceedings of the ACM SIGPLAN '79 on Compiler Construction **14** (1979) 83–91
3. Buchsbaum, A., Karloff, H., Kenyon, C., Reingold, N., Thorup, M.: OPT versus LOAD in dynamic storage allocation. In: Proceedings of the 35th Annual ACM Symposium on Theory of Computing (STOC). (2003)
4. Govindarajan, R., Rengarajan, S.: Buffer allocation in regular dataflow networks: An approach based on coloring circular-arc graphs. In: Proceedings of the 2nd International Conference on High Performance Computing. (1996)
5. Pemmaraju, S., Raman, R., Varadarajan, K.: Buffer minimization using max-coloring. In: Proceedings of 15th ACM-SIAM Symposium on Discrete Algorithms (SODA). (2004) 555–564
6. Kierstead, H., Trotter, W.: An extremal problem in recursive combinatorics. Congressus Numerantium **33** (1981) 143–153
7. Tarjan, R., Yannakakis, M.: Simple linear-time algorithms to test chordality of graphs, test acyclicity of hypergraphs, amd selectively reduce acyclic hypergraphs. SIAM Journal on Computing **13** (1984) 566–579
8. Chrobak, M., Ślusarek, M.: On some packing problems related to dynamic storage allocation. Informatique théorique et Applications/Theoretical Informatics and Applications **22** (1988) 487–499
9. Irani, S.: Coloring inductive graphs on-line. Algorithmica **11** (1994) 53–72

# Appendix

## Max-Coloring Mode-1 and Mode-2 Graphs with Random Weights

**Table 8.** Results of our experiments on mode 1 random chordal graphs, with random weights evaluating heuristics for the max-coloring problem. Here LB refers to the maximum weight clique.

| $|V|$ | $|E|$ | LB | Best Fit | First Fit | Partition | $\frac{(BF-OPT)}{OPT}$ | $\frac{(FF-OPT)}{OPT}$ | $\frac{(GPM-OPT)}{OPT}$ |
|---|---|---|---|---|---|---|---|---|
| 80 | 144.656 | 3057.64 | 3759.9 | 3499.66 | 4668.01 | 22.9672 | 14.4559 | 52.6669 |
| 120 | 217.022 | 3177.24 | 3946.64 | 3668.24 | 5015.84 | 24.216 | 15.4536 | 57.8678 |
| 160 | 310.744 | 3263.64 | 4123.96 | 3863.77 | 5306.64 | 26.3604 | 18.3881 | 62.5987 |
| 200 | 369.5 | 3436.68 | 4236.58 | 4049.02 | 5544.8 | 23.2754 | 17.8179 | 61.3419 |
| 240 | 465.822 | 3555.23 | 4466.49 | 4171.19 | 5732.43 | 25.6314 | 17.3253 | 61.2393 |
| 280 | 543.056 | 3670.51 | 4664.58 | 4319.58 | 5981.14 | 27.0825 | 17.6833 | 62.9513 |
| 320 | 616.833 | 3683.39 | 4599.39 | 4257.07 | 5971.09 | 24.8684 | 15.5747 | 62.1086 |
| 360 | 725.544 | 3675.27 | 4721.42 | 4376.39 | 6133.7 | 28.4648 | 19.0768 | 66.8913 |
| 400 | 799.322 | 3718.68 | 4700.19 | 4412.16 | 6158.34 | 26.3941 | 18.6485 | 65.6058 |
| 440 | 853.956 | 3766.96 | 4733.96 | 4428.67 | 6248.27 | 25.6706 | 17.5662 | 65.8705 |
| 480 | 953.733 | 3708.09 | 4750.59 | 4444.38 | 6195.4 | 28.1142 | 19.8563 | 67.078 |

**Table 9.** Results of our main experiments on mode 2 random chordal graphs, with random weights evaluating heuristics for the max-coloring problem.

| $|V|$ | $|E|$ | LB | Best Fit | First Fit | Partition | $\frac{(BFI-OPT)}{OPT}$ | $\frac{(FFI-OPT)}{OPT}$ | $\frac{(GPI-OPT)}{OPT}$ |
|---|---|---|---|---|---|---|---|---|
| 80 | 300.522 | 4545.92 | 5842.42 | 5302.06 | 6248.58 | 28.5201 | 16.6332 | 37.4546 |
| 120 | 534.711 | 5007.94 | 6674.97 | 6027.22 | 7103.28 | 33.2876 | 20.3532 | 41.8402 |
| 160 | 799.078 | 5477.86 | 7339.93 | 6594.4 | 7700.49 | 33.9928 | 20.3829 | 40.5749 |
| 200 | 1032.77 | 5613.7 | 7683.96 | 6792.93 | 8030.8 | 36.8786 | 21.0063 | 43.0572 |
| 240 | 1351.2 | 5877.72 | 8096.71 | 7077.89 | 8455.37 | 37.7525 | 20.4189 | 43.8545 |
| 280 | 1670.63 | 6349.47 | 8686.23 | 7656.33 | 9033.13 | 36.8026 | 20.5823 | 42.266 |
| 320 | 1977.26 | 6538.11 | 8888.31 | 7837.77 | 9316.49 | 35.9462 | 19.8782 | 42.4951 |
| 360 | 2259.06 | 6500.41 | 9195.9 | 7919.62 | 9315.26 | 41.4664 | 21.8326 | 43.3026 |
| 400 | 2579.76 | 6623.37 | 9270.36 | 8096.2 | 9643.78 | 39.9644 | 22.2369 | 45.6024 |
| 440 | 2991.97 | 6919.62 | 9724.36 | 8390.56 | 9996.6 | 40.533 | 21.2574 | 44.4674 |
| 480 | 3315.63 | 7062.68 | 9891.43 | 8462.93 | 10052.4 | 40.0522 | 19.8261 | 42.3314 |

## Interval Coloring Mode-1 and Mode-2 Graphs with Random Weights

**Table 10.** Results of our main experiments on mode 1 random chordal graphs, with random weights evaluating heuristics for the interval coloring problem.

| $|V|$ | $|E|$ | LB | Best Fit | First Fit | Partition | $\frac{(BFI-OPT)}{OPT}$ | $\frac{(FFI-OPT)}{OPT}$ | $\frac{(GPI-OPT)}{OPT}$ |
|---|---|---|---|---|---|---|---|---|
| 80 | 138.911 | 3025.34 | 3730.73 | 3361.86 | 3572.19 | 23.316 | 11.1231 | 18.0754 |
| 120 | 221.489 | 3246.92 | 4213.96 | 3573.64 | 3803.58 | 29.7831 | 10.0625 | 17.1441 |
| 160 | 304.589 | 3349.5 | 4345.77 | 3773.23 | 4031.22 | 29.7437 | 12.6506 | 20.353 |
| 200 | 388.922 | 3487.59 | 4702.46 | 3960.63 | 4250.49 | 34.834 | 13.5637 | 21.8747 |
| 240 | 465.622 | 3564.61 | 4774.09 | 3985.03 | 4317.4 | 33.9301 | 11.7943 | 21.1184 |
| 280 | 547.156 | 3621.37 | 4935.97 | 4079.52 | 4346.26 | 36.3012 | 12.6515 | 20.017 |
| 320 | 646.844 | 3632.78 | 5053.3 | 4158.73 | 4420.82 | 39.1029 | 14.4781 | 21.6926 |
| 360 | 708.389 | 3636.84 | 5215.73 | 4140.89 | 4413.28 | 43.4137 | 13.8594 | 21.3491 |
| 400 | 800.833 | 3790.56 | 5389 | 4357.7 | 4544.91 | 42.1691 | 14.962 | 19.9009 |
| 440 | 890.633 | 3751.67 | 5564.04 | 4290.47 | 4549.29 | 48.3086 | 14.3616 | 21.2605 |
| 480 | 972.989 | 3770.38 | 5594.34 | 4365.38 | 4627.62 | 48.3762 | 15.7809 | 22.7363 |
| 520 | 1034.57 | 3785.07 | 5645.89 | 4345.94 | 4671.27 | 49.1622 | 14.8182 | 23.4131 |

**Table 11.** Results of our main experiments on mode 2 random chordal graphs with random weights, evaluating heuristics for the interval coloring problem.

| $|V|$ | $|E|$ | LB | Best Fit | First Fit | Partition | $\frac{(BFI-OPT)}{OPT}$ | $\frac{(FFI-OPT)}{OPT}$ | $\frac{(GPI-OPT)}{OPT}$ |
|---|---|---|---|---|---|---|---|---|
| 80 | 305.778 | 4599.7 | 5216.71 | 5446.9 | 5729.69 | 13.4142 | 18.4186 | 24.5666 |
| 120 | 534.244 | 5080.24 | 5892.5 | 6148.76 | 6414.76 | 15.9885 | 21.0327 | 26.2686 |
| 160 | 791.522 | 5379.14 | 6460.83 | 6590.64 | 6915.19 | 20.1089 | 22.5222 | 28.5556 |
| 200 | 1076.37 | 5854.59 | 6830.21 | 7224.81 | 7488.83 | 16.6642 | 23.4042 | 27.9139 |
| 240 | 1338.13 | 5969.88 | 7236.98 | 7329.67 | 7760.87 | 21.2249 | 22.7775 | 30.0004 |
| 280 | 1685.17 | 6280.87 | 7541.18 | 7765.64 | 8109.21 | 20.0659 | 23.6397 | 29.1097 |
| 320 | 1911.62 | 6281.07 | 7692.9 | 7842.99 | 8220.39 | 22.4776 | 24.8671 | 30.8757 |
| 360 | 2308.61 | 6597.4 | 8010.04 | 8071.09 | 8564.29 | 21.4121 | 22.3374 | 29.8131 |
| 400 | 2584.28 | 6800 | 8210.2 | 8375.16 | 8740.73 | 20.7382 | 23.1641 | 28.5402 |
| 440 | 2958.31 | 6940.89 | 8431.81 | 8543.13 | 9072.14 | 21.4803 | 23.0841 | 30.7058 |
| 480 | 3251.54 | 7010.97 | 8681.73 | 8574.38 | 9276.71 | 23.8308 | 22.2995 | 32.3171 |
| 520 | 3625.52 | 7366.5 | 9047.21 | 9139.46 | 9543.09 | 22.8156 | 24.0678 | 29.5471 |

# A Statistical Approach for Algorithm Selection[*]

Joaquín Pérez[1], Rodolfo A. Pazos[1], Juan Frausto[2], Guillermo Rodríguez[3],
David Romero[4], and Laura Cruz[5]

[1] Centro Nacional de Investigación y Desarrollo Tecnológico (CENIDET), AP 5-164,
Cuernavaca, 62490, Mexico.
{jperez,pazos}@sd-cenidet.com.mx
[2] ITESM, Campus Cuernavaca, AP C-99 Cuernavaca, 62589, Mexico.
juan.frausto@itesm.mx
[3] Instituto de Investigaciones Eléctricas, Mexico.
gro@iie.org.mx
[4] Instituto de Matemáticas, UNAM, Mexico.
david@matcuer.unam.mx
[5] Instituto Tecnológico de Ciudad Madero, Mexico.
lcruzreyes@prodigy.net.mx

**Abstract.** This paper deals with heuristic algorithm characterization, which is applied to the solution of an NP-hard problem, in order to select the best algorithm for solving a given problem instance. The traditional approach for selecting algorithms compares their performance using an instance set, and concludes that one outperforms the other. Another common approach consists of developing mathematical models to relate performance to problem size. Recent approaches try to incorporate more characteristics. However, they do not identify the characteristics that affect performance in a critical way, and do not incorporate them explicitly in their performance model. In contrast, we propose a systematic procedure to create models that incorporate critical characteristics, aiming at the selection of the best algorithm for solving a given instance. To validate our approach we carried out experiments using an extensive test set. In particular, for the classical bin packing problem, we developed models that incorporate the interrelation among five critical characteristics and the performance of seven heuristic algorithms. As a result of applying our procedure, we obtained a 76% accuracy in the selection of the best algorithm.

## 1 Introduction

For the solution of NP-hard combinatorial optimization problems, non-deterministic algorithms have been proposed as a good alternative for very large instances [1]. On the other hand, deterministic algorithms are considered adequate for small instances of these problems [2]. As a result, many deterministic and non-deterministic algorithms have been devised for NP-hard optimization

---

[*] This research was supported in part by CONACYT and COSNET.

C.C. Ribeiro and S.L. Martins (Eds.): WEA 2004, LNCS 3059, pp. 417–431, 2004.
© Springer-Verlag Berlin Heidelberg 2004

problems. However, no adequate method is known nowadays for selecting the most appropriate algorithm to solve them.

The problem of choosing the best algorithm for a particular instance is far away from being easily solved due to many issues. Particularly, it is known that in real-life situations no algorithm outperforms the other in all circumstances [3]. But until now theoretical research has suggested that problem instances can be grouped in classes and there exists an algorithm for each class that solves the problems of that class most efficiently [4]. Consequently, few researches have tried to identify the algorithm dominance regions considering more than one problem characteristic. However, they do not identify systematically the characteristics that affect performance in a critical way and do not incorporate them explicitly in a performance model.

For several years we have been working on the problem of data-object distribution on the Internet [5], which can be seen as a generalization of the bin packing problem. We have designed solution algorithms and carried out a large number of experiments with them. As expected, no algorithm showed absolute superiority; hence our interest in developing an automatic method for algorithm selection. For this purpose, we propose a procedure for the systematic characterization of algorithm performance.

The proposed procedure in this paper consists of four main phases: modeling problem characteristics, grouping instance classes dominated by an algorithm, modeling the relationship between the characteristics of the grouped instances and the algorithm performance, and applying the relationship model to algorithm selection for a given instance.

This paper is organized as follows. An overview of the main works on algorithm selection is presented in Section 2. Then, Section 3 describes a general mechanism to characterize algorithm performance and select the algorithm with the best expected performance. An application problem and its solution algorithms are described in Section 4, in particular we use the bin packing (BP) problem and seven heuristic algorithms (deterministic and non-deterministic). Details of the application of our characterization mechanism to the solution of BP instances are described in Section 5.

## 2   Related Work

Recent approach for modeling algorithm performance tries to incorporate more than one problem characteristic in order to obtain a better problem representation. The objective of this is to increase the precision of the algorithm selection process. The works described below follow this approach.

Borghetti developed a method to correlate each instance characteristic to algorithm performance [6]. The problem dealt with was reasoning over a Bayesian Knowledge Base (BKB), which was solved with a genetic algorithm (GA) and a best first search algorithm (BFS). For the BKB problem, two kinds of critical characteristics were identified: topological and probabilistic. In this case the countable topological characteristics are number of nodes, number of arcs and number of random variables. An important shortcoming of this method is that it does not consider the combined effect of all the characteristics.

Minton proposed a method that allows specializing generic algorithms for particular applications [7]. The input consists of a description of the problem and a training instance set, which guides the search through the design space, constituted by heuristics that contend for their incorporation into the generic algorithm. The output is a program adjusted to the problem and the distribution of the instances.

Fink developed an algorithm selection technique for decision problems, which is based on the estimation of the algorithm gain, obtained from the statistical analysis of their previous performance [8]. Although the estimation can be enriched with new experiences, its efficiency depends on the user's ability to define groups of similar problem instances and to provide an appropriate metric of the problem size. The relationship among the problem characteristics given by the user and the algorithm performance is not defined formally.

The METAL group proposed a method to select the most appropriate classification algorithm for a set of similar instances [9]. They identify groups of old instances that exhibit similar characteristics to those of a new instance group. The algorithm performance of old instances is known and is used to predict the best algorithms for the new instance group. The similarity among instance groups is obtained considering three types of problem characteristics: general, statistics and derived from information theory. Since they do not propose a model for relating the problem characteristics to performance, the identification process of similar instances is repeated with each new group of data, so the processing time for algorithm selection can be high.

Rice introduced the poly-algorithms concept [10]. This paper refers to the use of a function that allows selecting, from an algorithms set, the best one for solving a given situation. After this work, other researchers have formulated different functions, for example those presented in [11,12]. In contrast with these works, in our solution approach we integrate three aspects: 1) self-adaptation of functions to incorporate new knowledge; 2) systematic method with statistical foundation to obtain the most appropriate functions; and 3) modeling of the interrelation of critical variables for algorithmic performance.

Table 1 presents the most important works that consider several problem characteristics. Column 2 indicates if problem characteristics are modeled. Column 3 is used to indicate if problem characteristics are incorporated explicitly into a performance model. Column 4 shows the granularity of the prediction, i.e., if the prediction can be applied for selecting the best algorithm for only one instance. Finally, column 5 indicates if the prediction model has been applied to algorithm selection.

Notice from Table 1, that no work includes all aspects required to characterize algorithm performance aiming at selecting the best algorithm for a given instance. The use of instance characterization to integrate groups of similar instances is an emerging and promising area for identifying dominance regions of algorithms. The works of Borghetti and the METAL group have made important advances in this area. However, the first one does not combine the identified characteristics, hindering the selection of algorithms; and the second is not applicable to the solution of only one instance. In contrast, our method is the only one that considers the four main aspects of algorithm characterization. Next section describes the method in detail.

**Table 1.** Related work on algorithm performance characterization

| Research | Problem characteristics modeling | Characteristics into performance model | Performance prediction for one instance | Selection |
|---|---|---|---|---|
| Fink | | | √ | √ |
| Borghetti | √ | | √ | |
| METAL | √ | | | √ |
| Our proposal | √ | √ | √ | √ |

## 3  Automation of Algorithm Selection

In this section a statistical method is presented for characterizing algorithm performance. This characterization is used to select the best algorithm for a specific instance. Section 3.1 describes the general software architecture, which is based on the procedure described in Section 3.2.

### 3.1  Architecture of the Characterization and Selection Process

The software architecture proposed for performance characterization and its application to algorithm selection is shown in Figure 1 and consists of five basic modules: 1) Statistical Sampling, 2) Instance Solution, 3) Characteristics Modeling, 4) Clustering Method, and 5) Classification Method.

Initially, the Statistical Sampling module generates a set of representative instances of the optimization problem. This set grows each time a new instance is solved.

The problem instances generated by the Statistical Sampling module are solved by the Instance Solution module, which has a configurable set of heuristic algorithms. For each instance, the performance statistics of each algorithm are obtained. The usual metrics for quantifying the final performance are the following: the percentage of deviation of the optimal solution value, the processing time and their corresponding standard deviations.

In the Characteristics Modeling module, the parameters values of each instance are transformed into indicators of the problem critical characteristics; i.e., those that impact the algorithms performance.

The Clustering module integrates groups constituted by the instances for which an algorithm had a performance similar or better than the others. Each group defines a region dominated by an algorithm. The similarity among the members of each group is determined through the instance characteristics and the algorithms performance.

The Classification module relates a new instance to one of the previously created groups. It is expected that the heuristic algorithm associated to the group obtains the best performance when solving the given instance. The new solved instances are incorporated to the characterization process for increasing the selection quality.

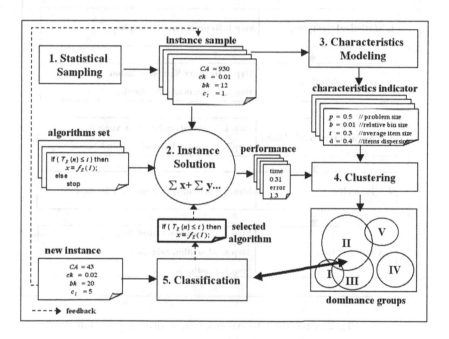

**Fig. 1.** Architecture of the characterization and selection process

## 3.2  Procedure to Systematize Algorithm Selection

We propose a procedure for systematizing the creation of mathematical models that incorporate problem characteristics aiming at selecting the best algorithm for a given instance. This procedure includes the steps needed to develop the architecture of the characterization and selection process, described before (see Figure 1). In Figure 2 the steps of the proposed procedure are associated to the architecture.

*Step 1. Representative Sampling.* Develop a sampling method to generate representative problem instances. This instances base will be used to determine the relationship between instance characteristics and algorithm performance.

*Step 2. Instance Sample Solution.* Provide a set of solution algorithms. Each instance of the sample must be solved with each available algorithm. The average performance of each algorithm for each instance must be calculated carrying out 30 experiments. Examples of performance measures are: the ratio of the best solution value with respect to a lower bound of the optimal solution value, and processing time.

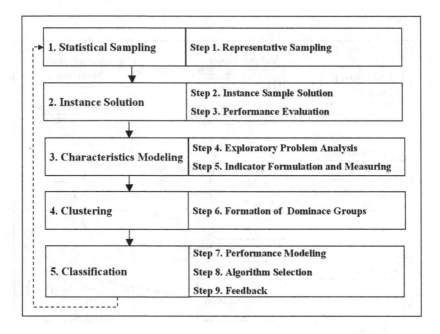

**Fig. 2.** Procedure to systematize algorithm selection

*Step 3. Performance Evaluation.* Develop a method to evaluate the performance of the solution algorithms, and determine the best algorithm for each sample instance. An alternative for considering all the different performance metrics is using their weighted average. Another is choosing the algorithm with the best quality and, in case of tie, choosing the fastest algorithm.

*Step 4. Exploratory Problem Analysis.* Establish hypothesis about the possible critical variables. These variables represent the instance characteristics that can be used as good indicators of algorithm performance on these instances.

*Step 5. Indicators Formulation and Measurement.* Develop indicator functions. These metrics are established to measure the values of the critical characteristics of the instances. The indicators are obtained using the instance parameter values in the indicator functions. Both, specific features of the instances and algorithm performance information, affect the algorithm selection strategy. In this step, a way to extract relevant features from the problem parameters must be found.

*Step 6. Creation of Dominance Groups.* Develop a method to create instance groups dominated, each one, by an algorithm or an algorithm set. The similarity among members of each group is determined through: indicators of instance characteristics and the algorithm with the best performance for each one. The output of this method is a group set, where each group is associated with an instance set and the algorithm with the best performance for the set.

*Step 7. Performance Modeling.* Develop a method to model the relationship between problem characteristics and algorithm performance. The relationship is learned from the groups created in Step 6.

*Step 8. Algorithm Selection.* Develop a method to use the performance model. For each new instance, its characteristic indicators must be calculated in order to determine which group it belongs to, using the relationship learned. The algorithm associated to this group is the expected best algorithm for the new instance.

*Step 9. Feedback.* The results of the new instance, solved with all algorithms, are used to feedback the procedure. If the prediction is right, the corresponding group is reinforced, otherwise a classifying adjustment is needed.

# 4   Application Problem

The bin packing problem is used for exemplifying our algorithm selection methodology. In this section a brief description of the one-dimensional bin packing problem and its solution algorithms is made.

## 4.1   Problem Description

The Bin Packing problem is an NP-hard combinatorial optimization problem, in which there is a given sequence of $n$ items $L = \{a_1, a_2, ..., a_n\}$ each one with a given size $0 < s(a_i) \leq c$, and an unlimited number of bins each of capacity $c$. The question is to determine the smallest number of bins $m$ into which the objects can be packed. In formal words, determine an $L$ minimal partition $B_1, B_2, ..., B_m$ such that in each bin $B_i$ the aggregate size of all the items in $Bi$ does not exceed $c$. This constraint is expressed in (1).

$$\sum_{a_i \in B_j} s(a_i) \leq c \qquad \forall j, 1 \leq j \leq m \tag{1}$$

In this work, we consider the discrete version of the one-dimensional bin packing problem, in which the bin capacity is an integer $c$, the number of items is $n$, and for simplicity, each item size is $s_i$, which is chosen from the set $\{1, 2, ..., c\}$.

## 4.2   Heuristic Solution Algorithms

An optimal solution can be found by considering all the ways to partition a set of $n$ items into $n$ or fewer subsets, unfortunately the number of possible partitions is larger than $(n/2)^{n/2}$ [13]. The heuristic algorithms presented in this section use deterministic and non-deterministic strategies for obtaining suboptimal solutions with less computational effort.

**Deterministic Algorithms.** These algorithms always follow the same path to arrive at the solution. For this reason, they obtain the same solution in different executions. The approximation deterministic algorithms for bin packing are very simple and run fast. A theoretical analysis of approximation algorithms is presented in [14,15,16]. In these surveys, the most important results for the one-dimensional bin packing problem and variants are discussed.

*First Fit Decreasing* (FFD). With this algorithm the items are first placed in a list sorted in non-increasing weight order. Then each item is picked orderly from the list and placed into the first bin that has enough unused capacity to hold it. If no partially filled bin has enough unused capacity, the item is placed in empty bin.

*Best Fit Decreasing* (BFD). The only difference with FFD is that the items are not placed in the first bin that can hold them, but in the best-filled bin that can hold them.

*Match to First Fit* (MFF). It is a variation of FFD. It asks the user to type the number of complementary bins. Each of these auxiliary bins is intended for holding items in a unique range of sizes. As the list is processed, each item is examined to check if it can be packed in a new bin, with items of a proper complementary bin; or packed in a partially-filled bin; or packed alone in a complementary bin. Finally, the items that are in the complementary bins are packed according to the basic algorithm.

*Match to Best Fit*  (MBF). It is a variation of BFD and similar to MFF, except for the basic algorithm used.

*Modified Best Fit Decreasing* (MBFD). The algorithm asks for a percentage value. This is the amount of bin capacity that can be left empty and qualify as a "good fit". All the items over 50% of the bin capacity are placed definitely in their own bin. With each partially filled bin, a special procedure to find a "good fit" item combination is followed. Finally, all remaining items are packed according to BFD.

**Non-Deterministic Algorithms.** These algorithms generally do not obtain the same solution in different executions. Approximation non-deterministic algorithms are considered general purpose algorithms.

*Ant Colony Optimization* (ACO). It is inspired on the ability of real ants to find the shortest path between their nest and a food source using a pheromone trail. For every ant build an items list partition starting with an empty bin. Each new bin is filled with "selected items" until no remaining item fits in it. A "selected item" is chosen stochastically using mainly a pheromone trail, which indicates the advantage of having a new item of size j with the item sizes already packed. The pheromone trail evaporates a little after each iteration and is reinforced by good solutions [17].

*Threshold Accepting* (TA). In this algorithm, to each $x \in X$ (where $X$ represents the set of all feasible solutions) a neighborhood $H(x) \subset X$ is associated. Thus, given a current feasible solution $x$, and a control parameter $T$ (called *temperature*), a neighboring feasible solution $y \in H(x)$ is generated; if $(z(y) - z(x)) < T$, then $y$ is accepted as the new current solution, otherwise the current solution remains unchanged. The value of $T$ is decreased each time *thermal equilibrium* is reached. This condition is verified when a set $S$ of feasible solution is formed. The value of $T$ is reduced, by repeatedly multiplying it by a *cooling factor* $\mu < 1$ until the system is *frozen* [18].

## 5 Implementation

This section shows the application of our procedure for algorithm characterization and selection. The procedure was applied to the one-dimensional bin packing problem (BP) and seven heuristic algorithms to solve it.

### 5.1 Statistical Sampling

In order to ensure that all problem characteristics were represented in the instances sample, stratified sampling and a sample size derived from survey sampling were used. The formation of strata is a technique that allows reducing the variability of the results, increasing the representativeness of the sample, and can help ensure consistency especially in handling clustered data [19].

Specifically, the following procedure was used: calculation of the sample size, creation of strata, calculation of the number of instances for each stratum, and random generation of the instances for each stratum. With this method 2,430 random instances were generated.

The task of solving 2,430 random instances requires a great amount of time: actually it took five days with four workstations. However, it is important to point out that this investment of time is only necessary once, at the beginning of the process of algorithms characterization, to create a sample of minimum size. We consider that this is a reasonable time for generating an initial sample whose size, validated statistically, increases the confidence level in the results. Besides, the initial quality of the prediction of the best algorithm for a particular instance can be increased through feedback.

### 5.2 Instance Solution

In order to learn the relationship between algorithm performance and problem characteristics, the random instances (which were used for training purpose) were solved. For testing the learned performance model, standard instances that are accepted by the research community were solved. For most of them, the optimal solution is known; otherwise the best-known solution is available. The experimental performances obtained for standard instances were used to validate the performances predicted by our model. Additionally, we confirmed the quality of our heuristic algorithm with the known solution.

2,430 random instances were generated using the method described in Section 5.1, and 1369 standard instances were considered. These instances were solved with the seven heuristic algorithms described in Section 4. The performance results obtained were: execution time, theoretical ratio and their corresponding standard deviation. Theoretical ratio is one of the usual performance metrics for bin packing and it is the ratio between the obtained solution and the theoretical optimum (it is a lower bound of the optimal value and equals the sum of all the item sizes divided by the bin capacity).

For each sample instance, all the algorithms were evaluated in order to determine the best algorithm. For a given instance, the algorithm with the smallest performance value was chosen, assigning the largest priority to the theoretical ratio.

## 5.3   Characteristics Modeling

In this step relevant features of the problem parameters were identified. Afterwards, expressions to measure the values of identified critical characteristics were derived.

In particular, four critical characteristics that affect algorithm performance were identified. The critical characteristics identified using the most common recommendation were instance size and item size dispersion. The critical characteristics identified using parametric analysis were capacity constraint and bin usage.

Once the critical characteristics were identified, expressions to measure their influence on algorithm performance were derived. Expressions (2) through (6) show five indicators derived from the analysis of the critical variables.

*Instance size.* The $p$ indicator in (2) expresses a relationship between instance size and the maximum size solved. The instance size is the number of items $n$, and the maximum size solved is $maxn$. The value of $maxn$ was set to 1000, which corresponds to the number of items of the largest instance solved in the specialized literature that we identified.

*Capacity constraint.* The $t$ indicator in (3) expresses a relationship between the average item size and the bin size. The size of item i is $s_i$ and the bin size is $c$. This metric quantifies the proportion of the bin c that is occupied by an item of average size.

*Item size dispersion.* Two indicators were derived for this variable. The $d$ indicator in (4) expresses the dispersion degree of the item size values. It is measured using the standard deviation of $t$. The $f$ indicator in (5) expresses the proportion of items whose sizes are factors of the bin capacity. In other words, an item is a factor when the bin capacity $c$ is multiple of its corresponding item size $s_i$. Instances with many factors are considered easy to solve.

*Bin usage.* The objective function has only one term, from which the $b$ indicator was derived. The $b$ indicator is shown in expression (6). This indicator expresses the proportion of the total size that can fit in a bin of capacity $c$. The inverse of this metric is used to calculate the theoretical optimum.

$$p = \frac{n}{maxn} \tag{2}$$

$$t = \frac{\sum_i (s_i/c)}{n} \qquad 1 \leq i \leq n \tag{3}$$

$$d = \sigma(t) \tag{4}$$

$$f = \frac{\sum_i factor(c, s_i)}{n} \qquad 1 \leq i \leq n \tag{5}$$

$$b = \frac{c}{\sum_i s_i} \qquad 1 \leq i \leq n \tag{6}$$

The factor analysis technique was used to confirm if derived indicators were critical too. Table 2 shows the characteristic indicators and the best algorithm for a small instance set, which were selected from a sample with 2,430 random instances.

## 5.4   Clustering

K-means was used as a clustering method to create instance groups dominated, each one, by an algorithm. The cluster analysis was carried out using the commercial software SPSS version 11.5 for Windows. The similarity among members of each group was determined through: characteristics indicators of the instances and the algorithm with the best performance for each one (see Table 2). Five groups were obtained; each group was associated with an instance set and an algorithm with the best performance for it. Two algorithms had poor performance and were outperformed by the other five algorithms. This dominance result applies only to the instance space explored in this work.

## 5.5   Classification

In this investigation discriminant analysis was used as a machine learning method to find out the relationship between the problem characteristics and algorithm performance.

The discriminant analysis extracts from data of group members, a group classification criterion named discriminant functions, which will be used later for classifying each new observation in the corresponding group. The percentage of new correctly classified observations is an indicator of the effectiveness of the discriminant functions. If these functions are effective on the training sample, it

**Table 2.** Example of random intances with their characteristics and the best algorithm

| Instance | Characteristics indicators | | | | | Best algorithm |
|---|---|---|---|---|---|---|
| | Problem size $p$ | Bin size $b$ | Item size $t$ | Factors $f$ | Item dispersion $d$ | |
| E1i10.txt | 0.078 | 0.427 | 0.029 | 0.000 | 0.003 | FFD |
| E50i10.txt | 0.556 | 0.003 | 0.679 | 0.048 | 0.199 | ACO |
| E147i10.txt | 0.900 | 0.002 | 0.530 | 0.000 | 0.033 | TA |
| E162i10.txt | 0.687 | 0.001 | 0.730 | 0.145 | 0.209 | BFD |
| E236i10.txt | 0.917 | 0.002 | 0.709 | 0.000 | 0.111 | TA |

is expected that with new observations whose corresponding group is unknown, they will classify well.

The analysis was made using the commercial software SAS version 4.10 for Windows. For obtaining the classification criterion, five indicators (see section 5.3) were used as independent variable, and the number of the best algorithm as dependent variable (or class variable). The discriminant classifier was trained with 2,430 bin packing instances generated with the procedure described in section 5.1, and validated with a resubstitution method, which uses the same training instances.

**Table 3.** Clasification results with 2,430 random instances

| Origin Group | Target Group | | | | | |
|---|---|---|---|---|---|---|
| | FFD | BFD | MBF | TA | ACO | Total |
| FFD | 600 | 190 | 374 | 59 | 243 | 1466 |
| | (40.9%) | (13.0%) | (25.5%) | (4.0%) | (16.6%) | (100%) |
| BFD | 1 | 19 | 0 | 0 | 1 | 21 |
| | (4.8%) | (90.5%) | (0.0%) | (0.0%) | (4.8%) | (100%) |
| MBF | 0 | 1 | 20 | 0 | 2 | 23 |
| | (0.0%) | (4.4%) | (87.0%) | (0.0%) | (8.7%) | (100%) |
| TA | 0 | 0 | 0 | 242 | 0 | 242 |
| | (0.0%) | (0.0%) | (0.0%) | (100%) | (0.0%) | (100%) |
| ACO | 81 | 73 | 103 | 2 | 419 | 678 |
| | (11.9%) | (10.8%) | (15.2%) | (0.3%) | (61.8%) | (100%) |
| Error Rate | 0.59 | 0.09 | 0.13 | 0.00 | 0.38 | 0.24 |

Table 3 presents the validation results of the obtained classifier, which are similar to those generated by SPSS. This table shows the relationship between two groups: number of instances than belong to the origin group and were classified into the target group, the corresponding percentage is included. In this case, each algorithm defines a group. On column 7, the total number of instances that belong to the origin group is shown. The final row shows the error rate for each target group, which is the proportion of misclassified instances. The average error was 24%.

## 5.6  Selection

To validate the effectiveness of the discriminant classifier mentioned in Section 5.5, we considered four types of standard bin packing instances with known solution (optimal or the best know). The Beasley's OR-Library contains two types of bin packing problems: u instances, t instances. The Operational Research Library contains problems of two kinds: N instances and hard instances. All of these instances are thoroughly described in [20].

Table 4 presents a fraction of the 1,369 instances collected. For each instance, the indicators and best algorithm are shown, and they were obtained as explained in Sections 5.2 and 5.3. These results were used for testing the classifier trained with random instances. The classifier predicted the right algorithm for 76% of the standard instances. Notice the consistency of this result with the error shown in Table 3.

**Table 4.** Example of standard intances with its characteristics and the best algorithm

| Instance | Characteristics indicators | | | | | Best algorithm |
|---|---|---|---|---|---|---|
| | Problem size $p$ | Bin size $b$ | Item size $t$ | Factors $f$ | Item dispersion $d$ | |
| Hard0.txt | 0.200 | 0.183 | 0.272 | 0.000 | 0.042 | ACO |
| N3c1w1_t.txt | 0.200 | 0.010 | 0.499 | 0.075 | 0.306 | FFD |
| T60_19.txt | 0.060 | 0.050 | 0.333 | 0.016 | 0.075 | MBF |
| U1000_19.txt | 1.000 | 0.003 | 0.399 | 0.054 | 0.155 | ACO |
| N1w2b2r0.txt | 0.050 | 0.102 | 0.195 | 0.020 | 0.057 | MBF |

# 6   Conclusions and Future Work

In this article, we propose a new approach to solve the selection algorithm problem in an innovative way. The main contribution is a systematic procedure to create mathematical models that relate algorithm performance to problem characteristics, aiming at selecting the best algorithm to solve a specific instance. With this approach it is possible to incorporate more than one characteristic into the models of algorithm performance, and get a better problem representation than other approaches.

For test purposes 2,430 random instances of the bin packing problem were generated. They were solved using seven different algorithms and were used for training the algorithm selection system. Afterwards, for validating the system, 1,369 standard instances were collected, which have been used by the research community. The experimental results showed an accuracy of 76% in the selection of the best algorithm for all standard instances. Since a direct comparison can not be made versus the methods mentioned in Section 2.2, this accuracy has to be compared with that of a random selection from the seven algorithms: 14.2%. For the instances of the remaining percentage, the selected algorithms generate a solution close to the optimal.

An additional contribution is the systematic identification of five characteristics that most influence algorithm performance. The detection of these critical characteristics for the bin packing problem was crucial for obtaining the results accuracy. We consider that the principles followed in this research can be applied for identifying critical characteristics of other NP-hard problems.

Currently, the proposed procedure is being tested for solving a design model of Distributed Data-objects (DD) on the Internet, which can be seen as a generalization of bin packing (BP). Our previous work shows that DD is much heavier than BP.

For future work we are planning to consolidate the software system based on our procedure and incorporate an adaptive module that includes new knowledge generated from the exploitation of the system with new instances. The objective is to keep the system in continuous self-training. In particular, we are interested in working with real-life instances of the design model of Distributed Data-objects (DD) on the Internet and incorporate new problems as they are being solved.

# References

1. Garey, M. R., Johnson, D. S.: Computers and Intractability, a Guide to the Theory of NP-completeness. W. H. Freeman and Company, New York (1979)
2. Papadimitriou, C., Steiglitz, K.: Combinatorial Optimization, Algorithms and Complexity. Prentice-Hall, New Jersey (1982)
3. Bertsekas: Linear Network Optimization, Algorithms and Codes. MIT Press, Cambridge, MA (1991)
4. Wolpert, D. H., Macready, W. G.: No Free Lunch Theorems for Optimizations. IEEE Transactions on Evolutionary Computation, Vol. 1 (1997) 67-82
5. Pérez, J., Pazos, R.A., Fraire, H., Cruz, L., Pecero, J.: Adaptive Allocation of Data-Objects in the Web Using Neural Networks. Lectures Notes in Computer Science, Vol. 2829. Springer-Verlag, Berlin Heidelberg New York (2003) 154-164
6. Borghetti, B. J.: Inference Algorithm Performance and Selection under Constrained Resources. MS Thesis. AFIT/GCS/ENG/96D-05 (1996)
7. Minton, S.: Automatically Configuring Constraint Satisfaction Programs: A Case Study. Journal of Constraints, Vol. 1, No. 1 (1996) 7-43
8. Fink, E.: How to Solve it Automatically, Selection among Problem-solving Methods. Proceedings of the Fourth International Conference on AI Planning Systems AIPS'98 (1998) 128-136

9. Soares, C., Brazdil, P.: Zoomed Ranking, Selection of Classification Algorithms Based on Relevant Performance Information. In: Zighed D.A., Komorowski J., and Zytkow J. (Eds.): Principles of Data Mining in Knowledge Discovery 4th European Conference (PKDD 2000), LNAI 1910. Springer Verlag, Berlin Heidelberg New York (2000) 126-135

10. Rice, J.R.,: On the Construction of Poly-algorithms for Automatic Numerical Analysis. In: Klerer, M., Reinfelds, J. (Eds.): Interactive Systems for Experimental Applied Mathematics. Academic Press (1968) 301-313

11. Li, J., Skjellum, A., Falgout, R.D.: A Poly-Algorithm for Parallel Dense Matrix Multiplication on Two-Dimensional Process Grid Topologies. Concurrency, Practice and Experience, Vol. 9, No. 5. (1997) 345-389

12. Brewer, E.A.: High-Level Optimization Via Automated Statistical Modeling. Proceedings of Principles and Practice of Parallel Programming (1995) 80-91

13. Basse S.: Computer Algortihms, Introduction to Design and Analysis. Editorial Addison-Wesley Publishing Compañy (1998)

14. Coffman, E.G. Jr., Garey, M.R., Johnson, D.S.: Approximation Algorithms for Bin-Packing, a Survey. In Approximation Algorithms for NP-hard Problems. PWS, Boston (1997) 46-93

15. Coffman, J.E.G., Galambos, G., Martello, S., Vigo, D.: Bin Packing Approximation Algorithms; Combinatorial Analysis. In: Du, D.-Z, Pardalos, P.M. (eds.): Handbook of Combinatorial Optimization. Kluwer Academic Publishers, Boston, MA (1998)

16. Lodi, A., Martello, S., Vigo, D.: Recent Advances on Two-dimensional Bin Packing Problems. Discrete Applied Mathematics, Vol. 123, No. 1-3. Elsevier Science B.V., Amsterdam (2002)

17. Ducatelle, F., Levine, J.: Ant Colony Optimisation for Bin Packing and Cutting Stock Problems. Proceedings of the UK Workshop on Computational Intelligence. Edinburgh (2001)

18. Pérez, J., Pazos, R.A., Vélez, L. Rodríguez, G.: Automatic Generation of Control Parameters for the Threshold Accepting Algorithm. Lectures Notes in Computer Science, Vol. 2313. Springer-Verlag, Berlin Heidelberg New York (2002) 119-127.

19. Micheals, R.J., Boult, T.E.: A Stratified Methodology for Classifier and Recognizer Evaluation. IEEE Workshop on Empirical Evaluation Methods in Computer Vision (2001)

20. Ross P., Schulenburg, S., Marin-Blázquez J.G., Hart E.: Hyper-heuristics, Learning to Combine Simple Heuristics in Bin-packing Problems. Proceedings of the Genetic and Evolutionary Computation Conference. Morgan Kaufmann (2002) 942-948

# An Improved Time-Sensitive Metaheuristic Framework for Combinatorial Optimization

Vinhthuy Phan[1] and Steven Skiena[2]

[1] University of Memphis, Memphis, TN 38152, USA.
vphan@memphis.edu
[2] SUNY Stony Brook, Stony Brook, NY 11794, USA.
skiena@cs.sunysb.edu

**Abstract.** We introduce a metaheuristic framework for combinatorial optimization. Our framework is similar to many existing frameworks (e.g. [27]) in that it is modular enough that important components can be independently developed to create optimizers for a wide range of problems. Ours is different in many aspects. Among them are its combinatorial emphasis and the use of simulated annealing and incremental greedy heuristics. We describe several annealing schedules and a hybrid strategy combining incremental greedy and simulated annealing heuristics. Our experiments show that (1) a particular annealing schedule is best on average and (2) the hybrid strategy on average outperforms each individual search strategy. Additionally, our framework guarantees the feasibility of returned solutions for combinatorial problems that permit infeasible solutions. We, further, discuss a generic method of optimizing efficiently bottle-neck problems under the local-search framework.

## 1 Introduction

Combinatorial optimization is important in many practical applications. Unfortunately, most combinatorial optimization problems are usually found to be NP-hard and thus impractical to solve optimally when their sizes get large. Further, provable approximation algorithms for them and especially their variants need to be designed and implemented carefully and specifically for each application. This approach is sometime not affordable and consequently metaheuristics such as simulated annealing, genetic algorithms, tabu search, etc. become more attractive due to the relative ease with which they can be adapted to problems with complicated and application-specific constraints. Systems and methodologies [8,9,27] based on local-search heuristics have been developed to provide flexible frameworks for creating optimizers.

In this paper, we report an improved version of *Discropt* (first proposed in [23]), a general-purpose metaheuristic optimizer. It is designed based on the local-search model, implemented using C++ in such a way that optimizers can be constructed with a minimal user-effort by putting built-in components together and/or modifying them appropriately. *Discropt* is different from other local-search frameworks [8,27] in a number of aspects in terms of built-in features,

choice and implementation of search strategies. First, it supports as built-ins
(1) three important *solution types*, namely permutation, subset, and set partition, which can be used to model many combinatorial problems, and (2) two
fundamentally different search approaches, simulated annealing and incremental greedy construction. Second, for problems which permit infeasible solutions
(e.g. vertex coloring, vertex cover, TSP on incomplete graphs), *Discropt* guarantees the feasibility of returned solutions in a non-trivial manner. Third, *Discropt*
is aware of running times given as inputs by requiring search heuristics adapt
themselves efficiently to different running times. Running-time awareness may
be essential many cases; consider the problem *vertex coloring*, in which the time
alloted to solve an instance can depend greatly on the intended application: in
*register allocation* for compiler optimization [2,3], solutions are expected within
a few seconds or less, whereas in *frequency channel assignment* [26], the economic significance of having a good solution may make it desirable to spend
more time on optimizing. This is similar to real-time optimizers [4,15,16], which
are formulated in the form of intelligent search strategies.

This paper is organized as follows. *Discropt*'s architecture is briefly discussed
and compared to similar frameworks such as HotFrame, EasyLocal++, and iOpt
[27] in Section 2. Recent improvements and an experimental comparison between
the current and previous versions, together an experimental constrast between
*Discropt* and iOpt [28] are discussed in Section 3. *Discropt*'s important features
are discussed next. They are: (1) time-sensitive search heuristics in Section 4,
(2) the evaluation of the cost of moving between solutions and minimization of
bottleneck functions in Section 5, and (3) systematic combination of cost and
feasibility in Section 6.

## 2  Overview of *Discropt*

*Discropt* is a framework based on the local search paradigm, in which the process
of minimizing an optimization problem is modeled as the search for the best
solution in a solution space. This process selects an initial solution arbitrarily,
and iterates the following procedure: from any solution, a neighboring solution is
generated and evaluated. If its objective cost is acceptable, it replaces the older
solution, and the search proceeds until a stopping criterion is met. This abstract
process can completely constructed by four components:

- Solution type. *Discropt* supports three primitive types *permutation, subset,*
  and *set partition*. Take the *traveling saleman* problem as an example. The
  solution type is most appropriately a permutation (of vertices). On the other
  hand, in the *vertex cover* problem the solution type should be represented as
  a subset (of vertices), and in *vertex coloring* problem it should be a partition
  (of vertices into colors).
- Neighborhood operator. This component generates randomly neighboring
  solutions of a given solution. Such generation is needed in order to traverse
  the solution space. Each solution type must have at least one operator defined
  specifically for it. In *Discropt* , the neighborhood operator for a permutation

$s$ generates a random permutation that differs from $s$ in exactly two indices; the operator for a subset $s$ (viewing as a bit-vector of 0's and 1's) generates a random subset that differs from $s$ in exactly one bit; and the operator for a partition $s$ generates a random partition that differs from $s$ in terms of part-membership in exactly one element.

- Search. The search component decides whether or not to replace a solution with its randomly generated neighboring solution. It also needs to decide when to stop searching based on the current progress and the amount of running time left. *Discropt* supports two main search methods and a hybrid strategy combing the two methods. The two main methods include variants of the simulated annealing heuristic and an incremental greedy construction.
- Objective function. This component evaluates the cost of each solution and the cost of moving from one solution to another.

Although *Discropt* is designed in such a way users can modify each component in ways they see best for their applications, it also provides built-in primitives for each of these components to minimize users' effort in creating an operational system for many combinatorial problems. Many existing local-search frameworks [27] share *Discropt* 's objective of providing a flexible and broadly-applicable platform by realizing the local-search model in an object oriented setting. *Discropt* interprets the polymorphic characters of the local-search paradigm by implementing dynamically-bound search algorithms (through C++ virtual functions) and statictally-bound solution types and neighborhood operators (through C++ templates). This design choice is similar to that of Easy-Local++ [11] and different from that of HotFrame [7]. It allows both efficiency by defining a solution type statically using templates and flexibility of dynamically choosing and mixing search methods by invoking appropriate virtual search functions. In addition to the general local-search architecture, *Discropt* is similar to Hotframe in having an explicit evaluation of the cost of *moving from one solution to another*. This function in many cases improves the efficiency of cost evaluation by a linear factor (see Section 5).

While these systems are all applicable to a wide range of problems, their effectiveness lies in how much effort each user puts in tailoring the components suitably to his or her problem. Thus, their differences in terms of built-in features may make one framework more appropriate for certain needs than the others. In this aspect, *Discropt* is different from EasyLocal and HotFrame in our emphasis on the combinatorial structure of intended problems; we support primitive combinatorial solution types (permutation, subset, and set partition) and specifically designed neighborhood operators for each type. Although Iopt [28] does support two solution types, *vector of variables* and *set of sequences*, we think that many combinatorial problems are more suitably represented as one of the three primitive types supported by *Discropt* . Another different aspect is that *Discropt* uses two fundamentally different search methods: simulated annealing and incremental greedy construction, together with a hybrid strategy combining these two heuristics. Early experimental data [23] showed that there were problems for which simulated annealing clearly outperformed incremental greedy construction, and vice versa. This suggests an inclusion of these two as primitive search methods and a development of a hybrid strategy to take advantage of them both. Further, *Discropt* fully realizes incremental greedy construction as a local search method and incorporates the evaluation of the cost of *moving from one solution to a neighboring solution*. This architectural aspect has a significant impact on the efficiency of the incremental greedy construction in the local search framework. Finally, *Discropt* is designed to be *senstive to running time inputs*. This means that both of these search methods must adjust their own parameters to do their best within a given amount of running time.

## 3   Recent Improvements and Experimental Results

Since the preliminary version [23], we have made the following improvements:

- *Improved Annealing Schedules* – We compared several cooling schedules for the simulated annealing heuristic to identify one that works well over all implemented problems. We found that our new annealing schedule, where the positive/negative acceptance rate is explicitly fitted to a time schedule, performed more robustly on the problems we studied.

- *When to Go Greedy* – We generalized the greedy heuristic to work efficiently in a time-sensitive local-search environment, by treating the *extensions* of each partially constructed solution as its *neighbors* in the local-search context. We developed an improved combined search strategy heuristic that provides a more sophisticated composition of the simulated annealing and greedy heuristics. The idea of combining different search strategies cannot be employed in systems based on a single search strategy.

- *Bottleneck Optimization through $L_p$ Norms* – We provide a general optimization scheme for bottleneck optimization problems (such as graph bandwidth optimization) that improves the efficiency of local search heuristics for this class of problems. We show that optimizing under an appropriate $L_p$ norm can provide better bottleneck solutions than optimizing under the actual bottleneck metric, due to faster objective function evaluation.

- *Guaranteed Feasibility through Local Improvement* – Infeasible solutions (e.g vertex cover and graph coloring) give rise to another issue in using local search. Even though it is essential that a feasible solution be reported at the end, restricting the search exclusively to feasible solutions results in inefficient optimization. We present a systematic method which combines user-defined objective cost and feasibility functions in such a way as to guarantee that only feasible solutions are returned for any *feasibilizable* problem.

  Feasibility can be enforced specifically for each problem by specifying appropriate constraints [9,22]. Our method, however, is more universal; it can be applied to any problem that fit the local-search framework, although fine tuning for each may be needed to achieve better performance.

To illustrate the ease to construct an operational system, we implemented optimizers for the following popular combinatorial problems:

- Permutation problems: Shortest Common Superstring, Traveling Salesman, and Minimum Bandwidth.
- Subset problems: Max Cut, Max Satisfiability and Min Vertex Cover.
- Set partition problems: Vertex Coloring and Clustering. These are implemented recently.

Table 1 presents the historical performance of our optimizer on permutation and subset problems (partition problems were not yet implemented in the early version) over different running time inputs. Shown in Table 1 are the costs of the returned solutions at each running time input, on a 1GHz machine with 768MB of RAM; lower costs are better since all problems are formulated to be minimized. They qualitatively show that *Discropt* now performs substantially better than the February 2002 version reported in [23], across almost all problems and time scales. The lone exception is on vertex cover, where direct comparison is invalid because the previous system potentially returned infeasible solutions; our current version provably guarantees the feasibility of returned solutions.

**Table 1.** Snapshot of DISCROPT's performance in three different versions. Running time vary from 5 to 120 seconds. Hightlighted lower costs denote better solutions.

| Problem | | Running times in seconds | | | | | | | | | |
|---|---|---|---|---|---|---|---|---|---|---|---|
| | | 5 | 10 | 15 | 20 | 25 | 30 | 45 | 60 | 90 | 120 |
| VertCover | 08/01 | 2260 | 2315 | 642 | 654 | 644 | 650 | 641 | 643 | 598 | 586 |
| | 02/02 | 1100 | **571** | **554** | **541** | **546** | **528** | **531** | **521** | **521** | **523** |
| | 09/03 | 580 | 579 | 577 | 574 | 574 | 573 | 574 | 574 | 573 | 572 |
| Bandwidth | 08/01 | 604 | 616 | 611 | 603 | 595 | 622 | 620 | 609 | 595 | 602 |
| | 02/02 | **633** | **620** | **614** | 615 | 613 | 609 | 607 | 608 | 603 | 602 |
| | 09/03 | 984 | 841 | 620 | **599** | **537** | **597** | **525** | **539** | **515** | **526** |
| MaxCut | 08/01 | 3972 | 3972 | 3972 | 3972 | 3969 | 3972 | 3972 | 3885 | 1373 | 1349 |
| | 02/02 | **1401** | 1385 | 1346 | 1462 | 1452 | 1441 | 1971 | 1689 | 1276 | 1240 |
| | 09/03 | 1625 | **1206** | **1150** | **1155** | **1130** | **1117** | **1110** | **1098** | **1088** | **1077** |
| SCS | 08/01 | 3688 | 1152 | 1158 | 1225 | 1120 | 1035 | 1013 | 994 | 902 | 982 |
| | 02/02 | 614 | 596 | 570 | 562 | 544 | 541 | 515 | 505 | 502 | 505 |
| | 09/03 | **527** | **500** | **500** | **500** | **500** | **500** | **500** | **500** | **500** | **500** |
| TSP | 08/01 | 1043 | 846 | 730 | 642 | 687 | 568 | 551 | 508 | 577 | 504 |
| | 02/02 | 933 | 633 | 472 | 411 | 406 | 397 | 378 | 381 | 375 | 374 |
| | 09/03 | **377** | **360** | **357** | **354** | **345** | **346** | **348** | **345** | **343** | **345** |
| MaxSat | 08/01 | 572 | 705 | 735 | 533 | 403 | 518 | 344 | 518 | 352 | 318 |
| | 02/02 | 636 | **359** | 465 | **503** | 398 | 374 | **262** | **206** | 255 | 236 |
| | 09/03 | **462** | 778 | **462** | 553 | **276** | **356** | 344 | 356 | **155** | **209** |

Table 2 shows the average performance of five heuristics, which are incremental greedy construction and variants of simulated annealing based on the constant-decay schedule (sa1), the variation of cost at each annealing trial (sa2), the fit of acceptance rate to a decreasing curve (sa3), and the combined heuristic. The results (each is a ratio of the smallest cost known to the cost found for each instance) are averaged across 2 instances of all of the permutation and subset problems mentioned above and the Vertex Coloring problem. Our measure of performance is the ratio of the smallest cost known for each instance to the cost of the returned solution for each search heuristic and each running time. This number is a value between 0 and 1 for each heuristic/problem instance pair. We conclude that among the annealing schedules, sa3 performs the best on average. This agrees with our hypothesis (Section 4.2) that direct manipulation of acceptance rate would be more effective in time-sensitive applications. More interestingly, on average the *combined* heuristic (which attempts to select the most appropriate strategy for a given problem) outperforms all individual heuristics of which it is composed.

**Table 2.** Average performance ($\frac{1}{14} \cdot \sum \frac{\Omega_i}{p(i,t)}$) over 2 instances on each of 7 problems. $\Omega_i$ a lower bound for instance $i$; $p(i,t)$ the performance by a heuristic at the given time $t$ on instance $i$. Because $\Omega_i \leq p(i,t)$, bigger fractions are better (maximum is 1).

| Heuristic | 5 | 10 | 15 | Running times in seconds 20 | 25 | 30 | 45 | 60 | 90 | 120 |
|---|---|---|---|---|---|---|---|---|---|---|
| greedy | 0.604 | 0.650 | 0.658 | 0.672 | 0.668 | 0.679 | 0.687 | 0.691 | 0.700 | 0.703 |
| sa1 | 0.652 | 0.692 | 0.688 | 0.697 | 0.720 | 0.729 | 0.737 | 0.760 | 0.780 | 0.785 |
| sa2 | 0.653 | 0.699 | 0.700 | 0.713 | 0.723 | 0.748 | 0.761 | 0.781 | 0.771 | 0.782 |
| sa3 | **0.701** | 0.729 | 0.756 | 0.762 | 0.762 | 0.766 | 0.795 | 0.804 | 0.821 | 0.826 |
| combined | 0.633 | **0.757** | **0.784** | **0.805** | **0.819** | **0.819** | **0.844** | **0.843** | **0.870** | **0.861** |

Tables 3 and 4 show a qualitative comparison between iOpt's [28] and *Discropt*'s performance on the Vertex Coloring problem. In Table 4, each tripple shows the number of colors of returned solutions for each heuristic Hill Climbing, Simulated Annealing (SA1), and Simulated Annealing with an annealing schedule in which the acceptance rate is fit to a curve (SA3). For example, the solutions produced by Hill Climbing on instance dsjc125.1.col have costs 9,8,8,8 at 2,4,8,16 second running times respectively. It should be noted that in iOpt the solution type is implemented in as a *vector of variables*, and in *Discropt* it is implemented as a *partition* (of vertices into different colors). Further, the results are produced on different machines, compilers, and possibly different objective functions. They, however, show a qualitative comparison and constrast between the two systems and another look of how *Discropt* behaves at varying running times. We observe that *Discropt*'s usage of memory is more frugal.

**Table 3.** Performance of iOpt on three Vertex Coloring instances on an 866Mhz PC. Memory taken during each run is maximally about 30MB of RAM.

| Problem | Hill Climbing | | Simul. Ann. | | Tabu Search | |
|---|---|---|---|---|---|---|
| | colors | time | colors | time | colors | time |
| dsjc125.1.col | 7 | 2.37 | 6 | 166.06 | 5 | 44.00 |
| dsjc125.5.col | 21 | 70.18 | 20 | 334.75 | 18 | 557.00 |
| dsjc125.9.col | 46 | 492.81 | 46 | 347.40 | 44 | 2904.75 |

**Table 4.** Performance of Discropt on three Vertex Coloring instances on an 2Ghz PC. Memory taken during each run is maximally about 1.3MB of RAM.

| Problem | number of colors for HC, SA1, SA3 | | | |
|---|---|---|---|---|
| | 2 seconds | 4 seconds | 8 seconds | 16 seconds |
| dsjc125.1.col | 9,7,7 | 8,7,7 | 8,6,7 | 8,6,7 |
| dsjc125.5.col | 24,20,21 | 24,20,20 | 23,20,20 | 22,20,20 |
| dsjc125.9.col | 52,47,45 | 52,46,45 | 52,46,45 | 52,45,44 |

## 4   Time-Sensitive Search Heuristics

Real-time search strategies have been studied in the field of Artificial Intelligence [15,16], and non-local-search combinatorial optimization [4]. Existing local-search time-sensitive heuristics in the literature have generally been qualitative in the sense that there is a set of parameters whose chosen values translate into a qualitative longer or shorter running time. Lam and Delosme [17,18,19] proposed an annealing schedule for the simulated annealing heuristic in which how long a search runs depends qualitatively on $\lambda$, the upper-bound of expected fitness of adjacent annealing trials. Hu, Kahng, and Tsao [13] suggested a variant of the threshold accepting heuristic based on the so-called old bachelor acceptance criteria, in which the trial length and hence running time is qualitatively controlled by users. A general theoretical model for evaluating time-sensitive search heuristics is discussed in [24].

*Discropt*'s search heuristics are based on the three main methods: (1) a general incremental greedy construction heuristic (2) a global simulated annealing heuristic, and (3) a combined hybrid-strategy that appropriately allocates time among different heuristics. The generic view of our metaheuristics is abstractly an iteration of the following process: start with an initial solution, generate a random neighboring solution, decide whether or not to move to that neighbor or stop according to the strategy defined by the metaheuristic.

*Discropt* employs a time-sensitive policy, suggesting that each heuristic should adjust its parameters so that the search performs as well as possible within a given running time. Greedy and simulated annealing implement this policy by varying the size of the selection pool (greedy) and the rate of converging to a final temperature (simulated annealing) in order to converge to local minima when the given running time expires. Converging to local minima in a non-trivial manner is arguably performing the best within a given running time. Further, by varying the parameters as described, better minima are achieved at longer running times. In other words, performance is expectedly increasing with running times.

## 4.1    Time-Sensitive Incremental Greedy Construction

The incremental greedy construction heuristic (see Algorithm 1) starts from an empty solution and at each step selects the best element among a pool of candidates according to some problem-specific criteria. The size of each pool of candidate is determined by how much time remains and how much time was spent on average per item in the last selection. *Discropt* treats incremental greedy construction as a local-search technique by supporting random generation of extensions of *partial* solutions. Thus, the process of incrementally extending partial solutions to larger ones is analogous to the process of moving from one solution to another. Since incremental greedy construction is viewed as a local-search heuristic, we can take advantage of strategies applied to local-search heuristics, including combining it with other local-search heuristics (see Section 4.3), and optimizing bottle-neck problems efficiently (see Section 5).

Our notion of incremental construction is similar to but more general than the Brelaz [1] graph coloring heuristic, and is related to the generalized greedy heuristic of Feo and Resende [6]. Greedy techniques have also been studied in intelligent search strategies [15].

---

**Algorithm 1** Time-sensitive, local-search, incremental greedy construction

---

Let $t$ be the given running time input.
Let $C$ be the pool of all possible elements.
$s = \emptyset$.
**while** $C \neq \emptyset$ **do**
    Use $t$ to determine $C_i \subset C$ be a random subset of $C$ with $i$ elements.
    Use $t$ to determine $P_i \subset P$, $P$ is the set of all possible positions of $s$.
    Based on $s$, select the best $e \in C_i$ and $p \in P_i$.
    **if** $s \cup e$ is feasible **then**
        Insert $e$ in $s$ at position $p$
        $C = C - e$.
    **end if**
    Estimate the remaining time and update $t$.
**end while**

---

## 4.2    Time-Sensitive Simulated Annealing

In simulated annealing, moving from a solution to a neighboring solution is determined with probability $\exp(-\frac{\Delta_{ij}}{T_k})$, where $\Delta_{ij}$ is the difference in cost between the two solutions and $T_k$ is the current temperature. A *final temperature*, $T_f$, is derived by statistical sampling in such a way that it is the temperature at which neighborhood movement is strictly descendant; at $T_f$, its behavior is identical to hill climbing. An annealing schedule is characterized by the manner in which temperature is reduced. This reduction must be carried out so that $T_f$ is the temperature at which time is about to expire, and each manner of reducing temperature periodically yields a different annealing schedule. We explore three classes of schedule in an attempt to identify which works best in a time-sensitive setting:

- *Traditional Annealing Schedules* – These simple schedules reduce the temperature by a constant fraction at regular intervals so that the final temperature $T_f$ is reached as time expires.

- *Variance-Based Schedules* – These schedules reduce temperature by taking into account the variance of cost at each annealing trial (i.e. the period during which $T_k$ is kept constant). The idea is that if the costs during each trial varies substantially (high variance), then $T_k$ is reduced slowly, and vice versa. This schedule is similar to that of Huang, Romeo, and Sangiovanni-Vincetelli [14], with an additional effort to make sure $T_f$ is reached at the end.

- *Forced-Trajectory Schedules* – This class of schedules appears to be novel. Schedules that are not aware of running times indirectly control the rate of accepting solutions by manipulating temperature. In a time-sensitive context, we suspect that this indirect mechanism may not be robust enough. These schedules forcefully fit the acceptance rate (i.e. the ratio of upward to downward movements) to a monotonically decreasing curve. The rate of accepting solutions is high initially, say 0.75, but decreases to 0 as time expires. Figure 1 shows examples of such curves with different rates.

Experimentally, we found that our forced-trajectory schedule outperformed the other two classes of schedule, averaging over instances of Vertex Coloring and all of the permutation and subset problems mentioned before; see experimental results in Section 3.

## 4.3   Combination of Time-Sensitive Heuristics

Greedy and local search strategies have been previously combined together for specific applications [20,21,25]. We found experimentally in [23] that simulated annealing outperformed incremental greedy construction on a number of problems, and vice versa, which seemed to be in accordance with the no-free-luch theorems [29,30]. Specifically, the greedy heuristic was found to be superior minimizing the shortest common superstring problem, whereas simulated annealing excelled in minimizing the bandwidth reduction problem. These results empirically agree with the Wolpert and Macready no-free-lunch theorems [29,30]. When

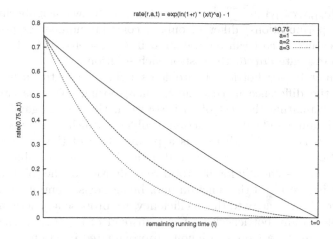

**Fig. 1.** Three different forced-trajectory annealing schedules.

no search strategy dominates the rest or when it is unknown if one strategy is should be selected, it makes sense to take advantage of them both by combining them strategically. Our hybrid strategy can be outlined as follows:

1. Predict longer-run performance based on short-run performance. Given short running times to each heuristic, we pick the seemingly dominant heuristic at the given running time. Which search heuristic is dominant is roughly estimated as follows:
   - Hill-climb dominates if its short-run result is better than those of both greedy and simulated annealing.
   - Otherwise, greedy dominates if its short-run performance is best.
   - Otherwise, simulated annealing dominates.
2. Combine greedy and simulated annealing by running simulated annealing on a local optimum obtained by greedy whenever no heuristic dominates the others.

As expected, we found empirically this hybrid metaheuristic outperformed each individual one, averaging over all studied problems. See Section 3.

## 5   Optimization of Bottleneck Problems

*Discropt* requires a function to evaluate the *cost of moving from one solution to a neighbor* be implemented in addition to an objective function. This implementation allows the evaluation of a solution's cost by evaluating the cost of moving to it from a neighbor with known cost. In many cases when the neighborhood structure is refined enough so that evaluating the *difference in cost* between solution is actually simpler than evaluating the cost of each solution, this indirect evaluation of cost improves efficiency by a linear factor. An example is the TSP problem, whose objective function is defined as:

$f(\pi) = d(c_{\pi(n)}, c_{\pi(1)}) + \sum_{i=1}^{n-1} d(\pi(c_i), \pi(c_{i+1}))$. If two neighboring solutions (which are permutations) differ in only a constant number of positions (e.g. 2), the difference in cost can be evaluated in $O(1)$ steps, instead of $O(n)$ steps required to evaluate *directly* the cost of each solution.

Unfortunately, for bottleneck problems (which have the form $\max\{\cdots\}$), evaluating the difference in cost among neighboring solutions can be as expensive as evaluating the cost of each solution; there is no gain! An example is the minimum bandwidth reduction problem, which is defined as $f(\pi) = \max_{(u,v)\in E} |\pi(u) - \pi(v)|$, where $\pi$ is a permutation of the vertices. Changing a permutation a little can result in a drastic change in the value of the maximum span, thus essentially forcing total re-evaluation of the solution cost. This inefficiency is a result of the function max being not-so-continuous over the solution space. As a remedy to this inefficiency, we propose a generic method that is applicable to any bottleneck problem: instead of optimizing the bottleneck problem, optimizing a more continuous approximated function. Since we have

$$\max\{x_1, \cdots, x_n\} = \lim_{p \to \infty} (\sum_{i=1}^{n} x^p)^{\frac{1}{p}}$$

instead of optimizing a $\max\{\cdots\}$, we optimize a $\sum\{\cdots\}$. It can be shown, similarly to the case of TSP, that evaluating the difference in cost of two neighboring solutions when the objective function is a $\sum\{\cdots\}$ is much more efficient than evaluating directly the cost of each solution. In case of minimum bandwidth reduction, the new function to be minimized is $(\sum_{(u,v)\in E}(\pi(u)-\pi(v))^p)^{\frac{1}{p}}$. Table 5 shows a comparison of minimizing the bottleneck function to minimizing p-$\sum$'s for various $p$. As shown, minimizing any p-$\sum$ yields much better results due to efficient cost evaluation.

Other *non-local-search* algorithms have benefited from optimizing an $L_p$ function instead of the originally stated $L_q$. These include Csirik, et. al. [5], who obtained several approximation results by optimizing an $L_2$ function for bin packing, an NP-hard $L_1$ problem [10]. Similarly, Gonzalez [12] obtained a 2-approximation to the NP-hard $L_2$ clustering problem by optimizing a reformulated $L_\infty$ version.

**Table 5.** Simulated Annealing on Bandwidth Reduction with different $L_p$'s.

| optimized metric | Actual bandwidths ($\max\{c_i\}$) at different running times | | | | | | | | | |
|---|---|---|---|---|---|---|---|---|---|---|
| | 5s | 10s | 15s | 20s | 25s | 30s | 45s | 60s | 90s | 120s |
| $\mathbf{L_\infty} = \max\{c_i\}$ | 860 | 806 | 817 | 743 | 747 | 735 | 704 | 675 | 651 | 623 |
| $L_2$ | 525 | 542 | 552 | 568 | 551 | 504 | 514 | 502 | 507 | 506 |
| $L_4$ | 349 | 322 | 336 | 339 | 328 | 316 | 313 | 320 | 308 | 302 |
| $\mathbf{L_6} = \sum c_i^6$ | 310 | **290** | **288** | **276** | **275** | 290 | 278 | **278** | 267 | 268 |
| $L_8$ | **302** | 317 | 293 | 283 | 318 | 295 | **276** | 282 | **267** | 313 |
| $L_{10}$ | 333 | 283 | 299 | 296 | 333 | 352 | 314 | 288 | 368 | **252** |

# 6  Separation of Cost and Feasibility

*Discropt* uses a systematic strategy to combine cost and feasibility, given that they have been defined separately as two different functions to be simultaneously optimized. Problems such as the graph coloring contain infeasible solutions (invalid colorings), which are unacceptable as returned solutions. One resolution to this issue would be an incorporation of a penalty mechanism in the definition of the objective funtion. For example, penalize an invalid coloring proportionally to the number of *violated* edges. Any such method must ensure that returned solutions are feasible. In *Discropt* , cost and feasibility can optionally be defined as two separate functions, and the objective function to be minimized is defined in terms of these two functions. This approach guarantees two things: (1) returned solutions are local optima with respect to the new objective function and (2) local optima are provably feasible. To accomplish this, we keep track of the current maximum rate of change of cost with respect to feasibility, and a function $\epsilon(t)$ that allows a qualitative emphasis be placed on cost initially and feasibility eventually, by maintaining the rate of $\epsilon(t) \to 0$.

A problem with cost function $c$ and feasibility function $f$ is called *feasibilizable* if any infeasible solution has a "more" feasible neighboring solution, i.e. one with a lower $f$-value. For example, graph coloring under the *swap neighborhood* is feasibilizable, because we can always assign an offending vertex a new color, thus guaranteeing the existence of a more feasible neighbor of an infeasible solution.

**Proposition 1.** *Let $P$ be a feasibilizable problem with cost function $c$ and feasibility function $f$. Define the linear objective function $g$*

$$g(A, c, f) = c(A) + (\delta(t) + \epsilon(t)) \cdot f(A)$$

*that evaluates the objective function of a solution $A$ by combining the given cost and feasibility functions, using the dynamically updated $\delta$ and $\epsilon$ which are functions of time and defined as*

$$\delta(t) = \max \frac{|c(A) - c(B)|}{|f(A) - f(B)|}$$

$$\lim_{t \to \infty} \epsilon(t) = 0$$

*for all solutions $A$ and $B$ which are neighbors encountered during the search. Then, any solution $U$ that is locally optimal with respect to $g$ is feasible.*

*Proof.* If $U$ is not feasible, there is a neighbor $V$ of $U$ such that $f(U) > f(V)$ and $g(U) = c(U) + (\delta(t) + \epsilon(t)) \cdot f(U) < c(V) + (\delta(V) + \epsilon(V)) \cdot f(V) = g(V)$, since $U$ is locally optimal with respect to $g$. This implies

$$0 < (\delta(t) + \epsilon(t)) \cdot (f(U) - f(V)) < c(V) - c(U)$$

which implies $0 < \delta(t) + \epsilon(t) < \frac{c(V)-c(U)}{f(U)-f(V)}$ as $t \to \infty$ (i.e. after a number of steps). This means $\delta(t) < \frac{c(V)-c(U)}{f(U)-f(V)}$, contradicting the definition of $\delta(t)$.

□

**Acknowledgements.** We would like to thank the referees for their helpful comments.

# References

1. D. Brelaz. New methods to color the vertices of a graph. *Communications of the ACM*, 22(4):251–256, 1979.
2. P. Briggs, K. Cooper, K. Kennedy, and L. Torczon. Coloring heuristics for register allocation. *ASCM Conference on Program Language Design and Implementation*, pages 275–284, 1989.
3. F. Chow and J. Hennessy. The priority-based coloring approach to register allocation. *ACM Transactions on Programming Languages and Systems*, 12(4):501–536, 1990.
4. Lon-Chan Chu and Benjamin W. Wah. Optimization in real time. In *IEEE Real-Time Systems Symposium*, pages 150–159, 1991.
5. Janos Csirik, David S. Johnson, Claire Kenyon, James B. Orlin, Peter W. Shor, and Richard R. Weber. On the sum-of-squares algorithm for bin packing. In *In Proceedings of the 32nd Annual ACM Symposium on the Theory of Computing*, pages 208–217, 2000.
6. T. Feo and M. Resende. Greedy randomized adaptive search procedures. *Journal of Global Optimization*, 6:109–133, 1995.
7. A. Fink and S. Voss. *in [27]*, chapter HotFrame: A Heuristic Optimization Framework. Kluwer, 2002.
8. M. Fontoura, C. Lucena, A. Andreatta, S.E. Carvalho, and C. Ribeiro. Using uml-f to enhance framework development: a case study in the local search heuristics domain. *J. Syst. Softw.*, 57(3):201–206, 2001.
9. P. Galinier and J. Hao. A general approach for constraint solving by local search. In *Proceedings of the Second International Workshop on Integration of AI and OR Techniques in Constraint Programming for Combinatorial Optimization Problems (CP-AI-OR'00), Paderborn, Germany*, March 2000.
10. M. Garey and D. Johnson. *Computers and Intractability, A Guide to the Theory of NP-Completeness*. W.H. Freedman and Company, 1979.
11. L. Di Gaspero and A. Schaerf. *in [27]*, chapter Writing Local Search Algorithms Using EasyLocal++. Kluwer, 2002.
12. T. Gonzalez. Clustering to minimize the maximum intercluster distance. *Theoretical Computer Science*, 38(2-3):293–306, 1985.
13. T. Hu, A. Kahng, and C. Tsao. Old bachelor acceptance: A new class of nonmonotone threshold accepting methods. *ORSA Journal on Computing*, 7(4):417–425, 1995.
14. M. Huang, F. Romeo, and Sangiovanni-Vincentelli. An efficient general cooling schedule for simulated annealing. In *ICCAD*, pages 381–384, 1986.
15. S. Koenig. Agent-centered search. *Artificial Intelligence Magazine*, 22(4):109–131, 2001.
16. R. E. Korf. Real-time heuristic search. *Artificial Intelligence*, 42(3):189–211, 1990.
17. J. Lam and J.-M. Delosme. An efficient simulated annealing schedule: derivation. Technical Report 8816, Yale University, 1988.
18. J. Lam and J.-M. Delosme. An efficient simulated annealing schedule: implementation and evaluation. Technical Report 8817, Yale University, 1988.

19. J. Lam and J.-M. Delosme.  Performance of a new annealing schedule.  In *1EEE/ACM Proc. of 25th. Design Automation Conference (DAC)*, pages 306–311, 1988.

20. P. Merz and B. Freisleben. A genetic local search approach to the quadratic assignment problem. In Thomas Bäck, editor, *Proceedings of the Seventh International Conference on Genetic Algorithms (ICGA97)*. Morgan Kaufmann, 1997.

21. P. Merz and B. Freisleben. Greedy and local search heuristics for the unconstrained binary quadratic programming problem. Technical Report 99-01, University of Siegen, Germany, 1999.

22. Laurent Michel and Pascal Van Hentenryck. A constraint-based architecture for local search. In *Proceedings of the 17th ACM SIGPLAN conference on Object-oriented programming, systems, languages, and applications*, pages 83–100. ACM Press, 2002.

23. V. Phan, S. Skiena, and P. Sumazin. A time-sensitive system for black-box optimization. In *4th Workshop on Algorithm Engineering and Experiments*, volume 2409 of *Lecture Notes in Computer Science*, pages 16–28, 2002.

24. V. Phan, S. Skiena, and P. Sumazin. A model for analyzing black box optimization. In *Workshop on Algorithms and Data Structures*, to be published in Lecture Notes in Computer Science, 2003.

25. Bart Selman and Henry A. Kautz.  An empirical study of greedy local search for satisfiability testing. In *Proceedings of the Eleventh National Conference on Artificial Intelligence(AAAI-93)*, Washington DC, 1993.

26. K. Smith and M. Palaniswami.  Static and dynamic channel assignment using neural networks. *IEEE Journal on Selected Areas in Communications*, 15(2):238–249, 1997.

27. S. Voss and D. Woodruff, editors. *Optimization Software Class Libraries*. Kluwer, 2002.

28. C. Voudouris and R. Dorne. *in [27]*, chapter Integrating Heuristic Search and One-Way Constraints in the iOpt Toolkit. Kluwer, 2002.

29. D. Wolpert and W. Macready. No free lunch theorems for search. Technical Report SFI-TR-95-02-010, Santa Fe, 1995.

30. D. Wolpert and W. Macready. No free lunch theorems for optimization. *IEEE Transactions on Evolutionary Computation*, 1(1):67–82, April 1997.

# A Huffman-Based Error Detecting Code

Paulo E.D. Pinto[1], Fábio Protti[2], and Jayme L. Szwarcfiter[3]*

[1] Instituto de Matemática e Estatistica, Universidade Estadual do Rio de Janeiro,
Rio de Janeiro, Brazil.
pauloedp@ime.uerj.br

[2] Instituto de Matemática and Núcleo de Computação Eletrônica, Universidade
Federal do Rio de Janeiro, Caixa Postal 2324, 20001-970, Rio de Janeiro, Brazil.
fabiop@nce.ufrj.br

[3] Instituto de Matemática, Núcleo de Computação Eletrônica and COPPE-Sistemas,
Universidade Federal do Rio de Janeiro, Caixa Postal 68511, 21945-970, Rio de
Janeiro, Brazil.
jayme@nce.ufrj.br

**Abstract.** Even codes are Huffman based prefix codes with the additional property of being able to detect the occurrence of an odd number of 1-bit errors in the message. They have been defined motivated by a problem posed by Hamming in 1980. Even codes have been studied for the case in which the symbols have uniform probabilities. In the present work, we consider the general situation of arbitrary probabilities. We describe an exact algorithm for constructing an optimal even code. The algorithm has complexity $O(n^3 \log n)$, where $n$ is the number of symbols. Further we describe an heuristics for constructing a nearly optimal even code, which requires $O(n \log n)$ time. The cost of an even code constructed by the heuristics is at most 50% higher than the cost of a Huffman code, for the same probabilities. That is, less than 50% higher than the cost of the corresponding optimal even code. However, computer experiments have shown that, for practical purposes, this value seems to be much less: at most 5%, for $n$ large enough. This corresponds to the overhead in the size of the encoded message, for having the ability to detect an odd number of 1-bit errors.

## 1 Introduction

Huffman codes [4] form one of the most traditional methods of coding. One of the important aspects of these codes is the possibility of handling encodings of variable sizes. A great number of extensions an variations of the classical Huffman codes have been described throught the time. For instance, Faller [1], Gallager [2], Knuth [6] and Milidiú, Laber and Pessoa [10] adressed adaptive methods for the construction of Huffman trees. Huffman trees with minimum height were described by Schwartz [12]. The consctruction of Huffman type trees with length constraints was considered by Turpin and Moffat [13], Larmore and Hirschberg

---

* Partially supported by Conselho Nacional de Desenvolvimento Científico e Tecnológico and Fundação de Amparo à Pesquisa do Estado do Rio de Janeiro.

C.C. Ribeiro and S.L. Martins (Eds.): WEA 2004, LNCS 3059, pp. 446–457, 2004.
© Springer-Verlag Berlin Heidelberg 2004

**Table 1.** Example of a Hamming-Huffman Code.

| Symbol | Encoding |
|--------|----------|
| a | 000 |
| b | 0110 |
| c | 1010 |
| d | 1100 |
| e | 1111 |

[7] and Milidiú and Laber [8,9]. On the other hand, Hamming formulated algo-rithms for the construction of error detecting codes [3]. Further, Hamming [3] posed the problem of describing an algorithm that would combine advantages of Huffman codes with the noise protection of Hamming codes. The idea is to define a prefix code in which the encoding would contain redundancies that would allow the detection of certain kinds of errors. This is equivalent to forbid some encodings which, when present in the reception, would signal an error. Such a code is a Hamming-Huffman code and its representing binary tree, a Hamming-Huffman tree. In a Huffman tree, all leaves correspond to encodings. In a Hamming-Huffman tree, there are encoding leaves and error leaves. Hit-ting an error leaf in the decoding process indicates the existence of an error. The problem posed by Hamming is to detect the occurrence of an error of one bit, as ilustrated in the following example given by Hamming [3], p.76. Table 1 shows the symbols and their corresponding encodings. Figure 1 depicts the cor-responding Hamming-Huffman tree. Error leaves are represented by black nodes. An error of one bit in the above encodings would lead to an error leaf.

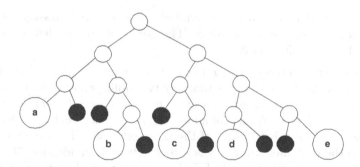

**Fig. 1.** A Hamming-Huffman tree

Motivated by the above problem, we have recently proposed [11] a special prefix code, called even code, which has the property of detecting the occur-rence of any odd number of 1-bit errors in the message. In [11], the study was restricted to codes corresponding to symbols having uniform probabilities. In the present paper, we consider the general situation of arbitrary probabilities.

First, we describe an exact algorithm for constructing an optimal even code, for a given set of symbols, each one with a given probability. The algorithm employs dynamic programming and its complexity is $O(n^3 \log n)$, where $n$ is the number of symbols. Next, we propose a simple heuristics for approximating an optimal code, based on Huffman's algorithm. The time required for computing an even code using the heuristics is $O(n \log n)$. We show that the cost of an even code constructed by the heuristics is at most 50% higher than the cost of a Huffman code for the same probabilities. That is, less than 50% higher than the corresponding optimal even code. However, for practical purposes, this value seems to be much less. In fact, we have performed several computer experiments, obtaining values always less than 5%, except for small $n$. This corresponds to the overhead in the size of the encoded message, for having the ability to detect an odd number of 1-bit errors.

The plan of the paper is as follows. In Section 2 we describe the Exact Algorithm for constructing an optimal even code. The heuristics is formulated in Section 3. In Section 4 we present comparisons between the code obtained by the heuristics and a corresponding Huffman code for the same probabilities. The comparisons are both analytical and experimental.

The following definitions are of interest.

Let $S = \{s_1, \dots, s_n\}$ be a set of elements, called *symbols*. Each $s_i \in S$ has an associated probability $f_i$. Throught the paper, we assume $f_i \leq f_{i+1}$. An *encoding* $e_i$ for a symbol $s_i \in S$ is a finite sequence of 0's and 1's, associated to $s_i$. Each 0 and 1 is a *bit* of $e_i$. The *parity* of $e_i$ is the parity of the quantity of 1's contained in $e_i$. A subsequence of $e_i$ starting from its first bit is a *prefix* of $e_i$. The set of encodings for all symbols of S is a *code* C for S. A code in wich every encoding does not coincide with a prefix of any other encoding is a *prefix code*.

A *message M* is a sequence of symbols. The *encoded message of M* is the corresponding sequence of encodings. The *parity* of an encoded message is the number of 1's contained in it.

A *binary tree* is a rooted tree $T$ in which every node $z$, other than the root, is labelled *left* or *right* in such a way that any two siblings have different labels. Say that $T$ is *trivial* when it consists of a single node. A *binary forest* is a set of binary trees. A *path* of $T$ is a sequence of nodes $z_1, \dots, z_t$, such that $z_q$ is the parent of $z_{q+1}$. The value $t - 1$ is the *size* of the path, whereas all $z_i$ are *descendants* of $z_1$. If $z_1$ is the root then $z_1, \dots, z_t$ is a *root path* and, in addition, if $z_t$ is a leaf, then $z_1, \dots, z_t$ is a *root-leaf path* of $T$. The *depth* of a node is the size of the root path to it. For a node $z$ of $T$, $T(z)$ denotes the *subtree of T rooted at z*, that is, the binary tree whose root is $z$ and containing all descendants of $z$ in $T$. The *left subtree* of $z$ is the subtree $T(z')$, where $z'$ is the left child of $z$. Similarly, define the *right subtree* of $z$. The left and right subtrees of the root of $T$ are denoted by $T_L$ and $T_R$, respectively. A *strictly binary tree* is one in which every node is a leaf or has two children. A *full binary tree* is a strictly binary tree in which all root-leaf paths have the same size. The edges of $T$ leading to left children are labelled 0, whereas those leading to right children are labelled 1. The *parity* of a node $z$ is the parity of the quantity of 1's among the edges forming the root path to $z$. A node is *even* or *odd*, according to its parity, respectively.

A *(binary tree) representation* of a code C is a binary tree $T$ such that there exists a one-to-one correspondence between encodings $e_i \in$ C and root-leaf paths $p_i$ of $T$ in such a way that $e_i$ is precisely the sequence of labels, 0 or 1, of the edges forming $p_i$. A code admits a binary tree representation if and only if it is a prefix code. Let $d_i$ be the depth of the leaf of $T$ associated to the symbol $s_i$. Define the *cost* as the sum $c(T) = \sum_{i=1}^{n} f_i.d_i$. Hence, the cost of a trivial tree is 0. An *optimal* code (tree) is one with the least cost. A *full representation tree* of C is a binary tree $T^*$ obtained from the representation tree $T$ of C, by adding a new leaf as the second child of every node having exactly one child. The original leaves of $T$ are the *encoding leaves*, whereas the newly introduced leaves, are the *error leaves*. Clearly, in case of Huffman trees, there are no error leaves.

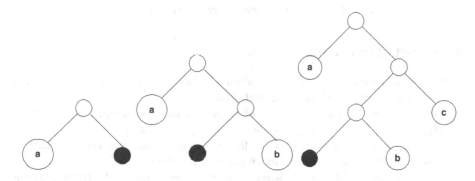

**Fig. 2.** Examples of even trees

An *even (odd) code* is a prefix code in which all encodings are even (odd). Similarly, an *even (odd) tree* is a tree representation of an even (odd) code. Examples of even trees for up to three symbols appear in Figure 2, while Figure 3 depicts an optimal even tree for 11 symbols of uniform probabilities.

It is simple to conclude that even codes detect the occurrence of an odd number of 1-bit errors in a message as follows. We know that all encodings are even, so the encoded message is also even. By introducing an odd number of errors, the encoded message becomes odd. Since the encodings are even, the latter implies that in the full tree representation of the code, an error leaf would be hit during the decodification process, or otherwise the process terminates at some odd node of the tree. It should be noted that odd codes do not have this property. For example, if we have a code C = {1, 01} and a message 01, if the first bit is changed, resulting 11, the message would be wrongly decoded without signaling error.

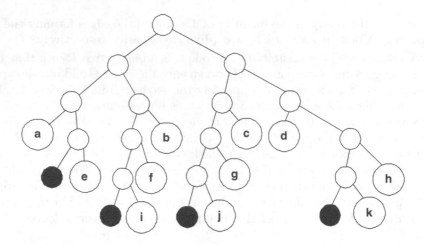

**Fig. 3.** Optimal even tree for 11 symbols

## 2   Exact Algorithm

In this section, we describe an exact algorithm for constructing an optimal even tree for symbols with arbitrary probabilities.

Let $S = \{s_1, \ldots, s_n\}$ be a set of symbols, each $s_i$ having probability $f_i$ satisfying $f_i \leq f_{i+1}$. For $m \leq n$, denote $S_m = \{s_1, \ldots, s_m\}$.
Our aim is to find an even code C for S, having minimum cost. In fact, we propose a solution for a slightly more general problem.

A *parity forest* $F$ for $S_m$ is a set of $p$ even trees and $q$ odd trees, and such that their leaves correspond to the symbols of $S_m$, for $0 \leq p, q \leq m$, with $p$ or $q \neq 0$. Define the *cost* of $F$ as the sum of the costs of its trees. Say that $F$ is $(m, p, q)$-optimal when its cost is the least among all forests for $S_m$ having $p$ even trees and $q$ odd trees. Denote by $c(m, p, q)$ the cost of an $(m, p, q)$-optimal forest. In terms of this notation, the solution of our problem is $c(n, 1, 0)$.

First, define the funtion

$$A_i = \begin{cases} \sum_{j=1}^{i} f_j, & \text{if } i > 0 \\ 0, & \text{otherwise} \end{cases}$$

The following theorem describes the computation of $c(m, p, q)$.

**Theorem 1.** *Let $m, p, q$ be integers such that $0 \leq m \leq n$, $m \geq p \geq 0$ and $m \geq q \geq 1$. Let $d_{max} = \lfloor \log \frac{m}{p+q} \rfloor + 1$. Then:*

(1) *if $m \leq p + q$ then $c(m, p, q) = A_{m-p}$*

(2) *if $m > p + q$ and $p = 0$ then $c(m, p, q) = c(m, q, q) + A_m$*

(3) *if $m > p + q$ and $p \neq 0$, then $c(m, p, q)$ is equal to*
$$\min\{c(m - 1, p - 1, q), \min_{1 \leq d \leq d_{max}} \{c(m - 1, (p + q)2^d - 1, (p + q).2^d) + d.A_m\}\}$$

**Proof:** By induction, we show that cases (1)-(3) correctly compute $c(m, p, q)$, for $0 \leq m \leq n$, $m \geq p \geq 0$ and $m \geq q \geq 1$. When $m = 0$, case (1) implies that $c(0, p, q) = 0$, which is correct since there are no symbols. For $m > 0$, let $F$ be an $(m, p, q)$-optimal forest for $S_m$. Consider the alternatives.

(1) $m \leq p + q$: In this case, the $p$ even trees of $F$ contain the $p$ symbols of highest probabilities, respectively. In addition, when $m > p$, the remaining $m - p$ symbols are assigned to the leaves of $m - p$ odd trees, respectively, each of these trees consisting of exactly one edge. Then $c(m, p, q) = A_{m-p}$.

(2) $m > p + q$ and $p = 0$: Then $F$ consists of $q$ odd trees, all of them empty and non trivial. We know that the $q$ left subtrees of the trees of $F$ are odd trees, while the $q$ right subtrees are even trees. Furthermore, the children of the roots of $F$ are roots of an $(m, q, q)$-optimal forest. Hence $c(m, 0, q) = c(m, q, q) + A_m$.

(3) $m > p + q$ and $p \neq 0$: Apply the following decomposition. Let $d$ be the minimum depth of a leaf of $F$. Clearly, the subset of nodes of depth $< d$ induces a full binary forest in $F$. Because $F$ is optimal, the leaf containing $s_m$ has depth $d$. If $d = 0$, $s_m$ is assigned to a trivial tree of $F$. In this situation, the forest formed by the remaining $p - 1$ even trees and $q$ odd trees contain the remaining $m - 1$ symbols, and it must be optimal. Consequently, $c(m, p, q) = c(m - 1, p - 1, q)$. Consider $d > 0$. Let $F'$ be the forest obtained from $F$ by removing all nodes in levels less than $d$. Clearly, $F'$ contains $(p + q).2^d$ trees, equally divided into even and odd trees. We know that $s_m$ has been assigned to a trivial tree $T'$ of $F'$. Clearly, the forest $F - T$ contains $(p+q).2^d - 1$ even trees and $(p+q).2^d$ odd trees. Since $F$ is $(m, p, q)$-optimal, $F' - T$ must be $(m - 1, (p + q).2^d - 1, (p + q).2^d)$-optimal. Regarding $F$, all nodes of $F'$ have been shifted $d$ levels. Therefore $c(m, p, q) = c(m - 1, (p + q).2^d - 1, (p + q).2^d) + d.A_m$.

Next, we determine the interval $[d_{min}, d_{max}]$ of possible variation of $d$. Clearly, $d_{min} = 0$, when the forest contains a trivial tree. Next, we find $d_{max}$. The maximum depth in a $(m, p, q)$-optimal forest $F$, such that no leaf of $F$ has depth less than $d_{max}$ corresponds to a forest in which the $p$ even trees are trivial and the $q$ odd ones are formed by one edge each, with possible empty trees. There are $(p + q).2^{d_{max}}$ nodes with depth $d_{max}$, half even and half odd. Consequently, $(p + q).2^{d_{max}-1} \leq m \leq 2.(p + q).2^{d_{max}-1}$. That is $d_{max} - 1 \leq \log \frac{m}{p+q} \leq d_{max}$, implying $d_{max} = \lfloor \log \frac{m}{p+q} \rfloor + 1$.

Considering that the situations $d = 0$ and $d \neq 0$ have been handled separately and that $f$ is $(m, p, q)$-optimal, we obtain that $c(m, p, q)$ is the minimum between $c(m - 1, p - 1, q)$ and $\min_{1 \leq d \leq d_{max}} \{c(m - 1, (p + q)2^d - 1, (p + q).2^d) + d.A_m\}$.

$\square$

Theorem 1 leads to a dynamic programming algorithm for determining $c(m, p, q)$, for all $0 \leq m \leq n$, $m \geq p \geq 0$ and $m \geq q \geq 1$.

Start by evaluating the function $A_i$ for $1 \leq i \leq n$. The first cost to be computed is $c(0, n, n)$, which is 0, since $c(0, p, q) = 0$, by (1). The parameter $m$ varies increasingly, $0 \leq m \leq n$. For each such $m$, vary $p$ and $q$ decreasingly, $m \leq p \leq 0$ and $m \leq q \leq 1$. For each such triple $m, p, q$, compute $c(m, p, q)$ applying (1), (2) or (3), according to their values. The computation stops when $c(n, 0, 1)$ is calculated, as it is equal to our target $c(n, 1, 0)$. Observe that $c(m, p, q) =$

$c(m, q, p)$, in general. There are $O(n^3)$ subproblems. The evaluation of each one is performed in constant time, if by equations (1) or (2), or in $O(\log n)$ time when the evaluation is by (3). Consequently, the time complexity is $O(n^3 \log n)$. The space requirements are $O(n^2)$, since for each $m$ it suffices to maintain the subproblems corresponding to $m$ and $m - 1$.

## 3   Heuristics

In this section we describe two heuristics to obtain even codes. Heuristics 1 is very simple and is based on a slight modification of the classical Huffman algorithm [4]. Heuristics 2 simply adds some possible improvements to the previous one. As we shall see, those improvements allow to yield even codes very close to the optimal ones.

### 3.1   Heuristics 1

Given $n$ symbols with probabilities $f_1, f_2, \ldots, f_n$, Heuristics 1 consists of two steps:

**Step 1.** *Run Huffman's algorithm in order to obtain a Huffman tree $T_H$ for the $n$ symbols.*

**Step 2.** *Convert $T_H$ into an even tree $T_{Heu1}$ in the following way: for each odd leaf $z$ corresponding to a symbol $s_i$, create two children $z_L$ and $z_R$ such that:*
*- the left child $z_L$ is an error leaf;*
*- the right child $z_R$ is the new encoding leaf corresponding to $s_i$. We call $z_R$ an* augmented leaf.

Observe that the overall running time of Heuristics 1 is $O(n \log n)$, since it is dominated by Step 1. Step 2 can be easily done in $O(n)$ time.

### 3.2   Heuristics 2

Now we present three possible ways to improve the heuristics previously described. As we shall see, these improvements do not increase the running time in practice, while producing a qualitative jump in performance with respect to the cost of the generated code.

**Improvement I.** During Step 1 (execution of Huffman's algorithm), add the following test:

*Among the candidate pairs of the partial trees to be merged at the beginning of a new iteration, give preference to a pair of trees $T_1$ and $T_2$ such that $T_1$ is trivial and $T_2$ is not.*

In other words, the idea is to avoid merging trivial trees as much as possible. The reason why this strategy is employed is explained in the sequel.

In $T_H$, there exist two sibling leaves for each merge operation of trivial trees occurring along the algorithm. Of course, one of the siblings is guaranteedly an odd leaf. When we force a trivial tree to be merged with a not trivial one, we minimize the number of pairs of sibling leaves in $T_H$, and thus the number of those "guaranteedly odd" leaves. In many cases, this strategy lowers the additional cost needed to produce the even tree in Step 2.

Let us denote by $T_{H_1}$ the Huffman tree obtained by Improvement I. It is worth remarking that this improvement does not affect the essence of Huffman's algorithm, since $T_{H_1}$ is a plausible tree.

Moreover, it is possible to implement Improvement I in constant time by keeping during the execution of Huffman's algorithm two heaps $H'$ and $H''$ where the nodes of $H'$ contain trivial trees and the nodes of $H''$ the remaining ones. At the beginning of the algoritm, $H'$ contains $n$ nodes and $H''$ is empty. When starting a new iteration, simply test whether the roots of $H'$ and $H''$ form a candidate pair of partial trees to be merged; if so, merge them.

**Improvement II.** Change $T_{H_1}$ by repeatedly applying the following operation in increasing depth order:

*If there exist two nodes $z', z''$ at the same depth of $T_H$ such that $z'$ is an odd leaf and $z''$ is an even internal node, exchange the positions of $z'$ and $z''$.*

Observe each single application of the above operation decreases the number of odd leaves in $T_H$ by one unit. Each time we find $k$ odd leaves and $\ell$ even internal nodes at some depth $i$, we perform $\min\{k, \ell\}$ changes and proceed to depth $i + 1$.

It is clear that the number of changes is bounded by the number of leaves of $T_H$. Since a single change can be done by modifying a constant number of pointers, the overall complexity of Improvement II is $O(n)$.

Denote by $T_{H_2}$ the Huffman tree obtained by Improvement II. Again, the essence of Huffman's algorithm is not affected, since $T_{H_2}$ is still plausible.

**Improvement III.** Apply Step 2 on $T_{H_2}$. Let $T$ be the even tree obtained. Then redistribute the symbols among the leaves of $T$ as follows:

*Whenever there exist two leaves $z_i, z_j$ (of any parities) in $T$ with depths $d_i \leq d_j$, representing symbols $s_i, s_j$ with probabilities $f_i \leq f_j$, respectively, then exchange the symbols assigned to $z_i$ and $z_j$.*

Observe that each single re-assignment performed above reduces the cost of the resulting even tree by $(d_j - d_i)(f_j - f_i)$.

The entire process can be implemented in the following way: after applying Step 2, let $z_i$ be the leaf of the tree representing the symbol $s_i$ with probability $f_i$, for $i = 1, \ldots, n$. Assume that the depth of $z_i$ is $d_i$. Run first a bucket sort on the values $d_i$, and then assign the leaf $z_i$ with depth $d_i$ to the symbol $s_{n-i+1}$ with probability $f_{n-i+1}$. (Recall that $f_1 \leq f_2 \leq \ldots f_n$.) The time required for this operation is therefore $O(n)$. Consequently, the overall time bound for Heuristics II is $O(\log n)$, with $O(n)$ space.

# 4   Comparisons

In this section, we first present an analytical upper bound for the cost of the even tree generated by Heuristics 2 with respect to the cost of the corresponding Huffman tree. Then we will exhibit some experimental results.

The terminology employed in this section is the following: given $n$ symbols with probabilities $f_1, f_2, \ldots, f_n$, $T_H$ is the Huffman tree for these symbols; $T_E$ is the corresponding optimal even tree for these symbols; $T_{Heu1}$ is the even tree obtained by applying Heuristics 1; and $T_{Heu2}$ is the even tree obtained by applying Heuristics 2. Observe that

$$c(T_H) \leq c(T_E) \leq c(T_{Heu2}) \leq c(T_{Heu1}).$$

### 4.1  An Analytical Bound

**Theorem 2.** $c(T_{Heu2}) \leq 1.5\, c(T_H)$.

**Proof:**
This bound is due to Improvements II and III. Let $n_{ol}(k), n_{oi}(k), n_{el}(k), n_{ei}(k)$ be the number of odd leaves, odd internal nodes, even leaves and even internal nodes at depth $k$ of $T_{H_2}$. Then either $n_{ol}(k) = 0$ or $n_{ei}(k) = 0$ (1). Moreover, it is clear that $n_{ol}(k) + n_{oi}(k) = n_{el}(k) + n_{ei}(k)$ (2).

We claim that $n_{ol}(k) \leq n_{el}(k)$. Otherwise, if $n_{ol}(k) > n_{el}(k)$, then $n_{ol}(k) > 0$, which implies $n_{ei}(k) = 0$ from (1). But in this case $n_{ol}(k) + n_{oi}(k) = n_{el}(k)$ from (2), that is, $n_{ol}(k) \leq n_{el}(k)$, a contradiction.

By summing up $n_{ol}(k)$ and $n_{el}(k)$ for all values of $k$, we conclude that the number of odd leaves is less than or equal to the number of even leaves in $T_{H_2}$ That is, the number of odd leaves is at most $\lfloor \frac{n}{2} \rfloor$ and i s less than or equal to the number of merge operations between two trivial trees in Huffman's algorithm. Next, Step 2 puts odd leaves one level deeper, in order to convert $T_{H_2}$ into an even tree.

Now, when applying Improvement III, the probabilities are redistributed in the tree, in such a way that two leaves $z_i, z_j$ with depths $d_i, d_j$ and probabilities $f_i, f_j$ satisfy the condition $f_i \geq f_j \Leftrightarrow d_i \leq d_j$. Consequently, there exists a one-to-one correspondence between the set of augmented leaves and a subset of the even leaves such that if $z_i$ is an augmented leaf and $z_j$ is its corresponding even leaf then $f_i \geq f_j$. Thus:

$$c(T_{Heu2}) \leq c(T_H) + \frac{\sum_{i=1}^{n} f_i}{2} \leq c(T_H) + c(T_H)/2 = 1.5\, c(T_H) \qquad \square$$

### 4.2  Experimental Results

The experimental results are summarized in Tables 2 to 4. The tables present the costs of the trees obtained by the algorithms, for several values of $n$. They were obtained from a program written in Pascal, running on a Pentium IV computer with 1.8 GHz and 256M RAM.

In Tables 2 and 3 we compare $c(T_H)$, $c(T_E)$, $c(T_{Heu2})$ and $c(T_{Heu1})$, for $n$ in the range 64 to 1024. In the first case (Table 2) we use uniform probabilities, and in the second one (Table 3), arbitrary probabilities, obtained from the standard

**Table 2.** Comparisons with uniform probabilities

| $n$ | $c(T_H)$ | $c(T_E)$ | $c(T_{Heu1})$ | $c(T_{Heu2})$ | $\frac{c(T_E)-c(T_H)}{c(T_H)}$ % | $\frac{c(T_{Heu1})-c(T_H)}{c(T_H)}$ % | $\frac{c(T_{Heu2})-c(T_H)}{c(T_H)}$ % |
|---|---|---|---|---|---|---|---|
| 64 | 6.00 | 6.50 | 6.50 | 6.50 | 8.33 | 8.33 | 8.33 |
| 128 | 7.00 | 7.50 | 7.50 | 7.50 | 7.14 | 7.14 | 7.14 |
| 192 | 7.67 | 8.00 | 8.17 | 8.00 | 4.35 | 6.52 | 4.35 |
| 256 | 8.00 | 8.50 | 8.50 | 8.50 | 6.25 | 6.25 | 6.25 |
| 320 | 8.40 | 8.80 | 8.90 | 8.80 | 4.76 | 5.95 | 4.76 |
| 384 | 8.67 | 9.00 | 9.17 | 9.00 | 3.85 | 5.77 | 3.85 |
| 448 | 8.86 | 9.29 | 9.36 | 9.29 | 4.84 | 5.65 | 4.84 |
| 512 | 9.00 | 9.50 | 9.50 | 9.50 | 5.56 | 5.56 | 5.56 |
| 576 | 9.22 | 9.67 | 9.72 | 9.67 | 4.82 | 5.42 | 4.82 |
| 640 | 9.40 | 9.80 | 9.90 | 9.80 | 4.26 | 5.32 | 4.26 |
| 704 | 9.55 | 9.91 | 10.05 | 9.91 | 3.81 | 5.24 | 3.81 |
| 768 | 9.67 | 10.00 | 10.17 | 10.00 | 3.45 | 5.17 | 3.45 |
| 832 | 9.77 | 10.15 | 10.27 | 10.15 | 3.94 | 5.12 | 3.94 |
| 896 | 9.86 | 10.29 | 10.36 | 10.29 | 4.35 | 5.07 | 4.35 |
| 960 | 9.93 | 10.40 | 10.43 | 10.40 | 4.70 | 5.03 | 4.70 |
| 1024 | 10.00 | 10.50 | 10.50 | 10.50 | 5.00 | 5.00 | 5.00 |

**Table 3.** Comparisons with arbitrary probabilities

| $n$ | $c(T_H)$ | $c(T_E)$ | $c(T_{Heu1})$ | $c(T_{Heu2})$ | $\frac{c(T_E)-c(T_H)}{c(T_H)}$ % | $\frac{c(T_{Heu1})-c(T_H)}{c(T_H)}$ % | $\frac{c(T_{Heu2})-c(T_H)}{c(T_H)}$ % |
|---|---|---|---|---|---|---|---|
| 64 | 5.82 | 5.97 | 6.38 | 6.03 | 2.65 | 9.69 | 3.66 |
| 128 | 6.76 | 6.88 | 7.28 | 6.92 | 1.86 | 7.75 | 2.43 |
| 192 | 7.33 | 7.52 | 7.84 | 7.59 | 2.65 | 6.96 | 3.54 |
| 256 | 7.75 | 7.87 | 8.25 | 7.90 | 1.61 | 6.47 | 1.95 |
| 320 | 8.08 | 8.20 | 8.57 | 8.30 | 1.51 | 6.13 | 2.71 |
| 384 | 8.34 | 8.54 | 8.85 | 8.60 | 2.48 | 6.11 | 3.12 |
| 448 | 8.52 | 8.68 | 9.04 | 8.73 | 1.88 | 6.00 | 2.40 |
| 512 | 8.72 | 8.84 | 9.22 | 8.87 | 1.34 | 5.76 | 1.71 |
| 576 | 8.92 | 9.03 | 9.39 | 9.09 | 1.24 | 5.30 | 1.98 |
| 640 | 9.10 | 9.22 | 9.60 | 9.31 | 1.35 | 5.55 | 2.34 |
| 704 | 9.20 | 9.33 | 9.69 | 9.41 | 1.43 | 5.42 | 2.37 |
| 768 | 9.33 | 9.53 | 9.83 | 9.58 | 2.05 | 5.27 | 2.67 |
| 832 | 9.44 | 9.63 | 9.93 | 9.67 | 1.98 | 5.22 | 2.41 |
| 896 | 9.57 | 9.73 | 10.08 | 9.77 | 1.70 | 5.31 | 2.14 |
| 960 | 9.64 | 9.78 | 10.14 | 9.82 | 1.47 | 5.22 | 1.87 |
| 1024 | 9.75 | 9.87 | 10.24 | 9.89 | 1.27 | 5.11 | 1.49 |

Pascal generation routine for random numbers in the range 1 to 10000 (we found no significant variations changing this range). All the probabilities were further normalized so that the total sum is 1. In Table 4 we compare the heuristics with Huffman's algorithm for $n$ in the range 1000 to 100000.

**Table 4.** Comparisons with arbitrary probabilities

| $n$ | $c(T_H)$ | $c(T_{Heu1})$ | $c(T_{Heu2})$ | $\frac{c(T_{Heu1})-c(T_H)}{c(T_H)}$ % | $\frac{c(T_{Heu2})-c(T_H)}{c(T_H)}$ % |
|---|---|---|---|---|---|
| 1000 | 9.71 | 10.22 | 9.87 | 5.23 | 1.61 |
| 2000 | 10.72 | 11.21 | 10.87 | 4.58 | 1.45 |
| 3000 | 11.31 | 11.81 | 11.57 | 4.46 | 2.29 |
| 4000 | 11.72 | 12.22 | 11.87 | 4.27 | 1.34 |
| 5000 | 12.05 | 12.55 | 12.26 | 4.13 | 1.73 |
| 10000 | 13.04 | 13.55 | 13.25 | 3.85 | 1.56 |
| 15000 | 13.61 | 14.11 | 13.80 | 3.65 | 1.37 |
| 20000 | 14.05 | 14.55 | 14.25 | 3.55 | 1.46 |
| 25000 | 14.36 | 14.86 | 14.61 | 3.50 | 1.77 |
| 30000 | 14.62 | 15.12 | 14.81 | 3.44 | 1.26 |
| 35000 | 14.85 | 15.35 | 15.01 | 3.38 | 1.11 |
| 40000 | 15.05 | 15.55 | 15.25 | 3.33 | 1.35 |
| 45000 | 15.21 | 15.71 | 15.45 | 3.28 | 1.54 |
| 50000 | 15.36 | 15.86 | 15.61 | 3.25 | 1.65 |
| 55000 | 15.49 | 15.99 | 15.71 | 3.23 | 1.43 |
| 60000 | 15.62 | 16.12 | 15.80 | 3.21 | 1.19 |
| 65000 | 15.74 | 16.24 | 15.89 | 3.18 | 0.95 |
| 70000 | 15.85 | 16.35 | 16.01 | 3.16 | 1.03 |
| 75000 | 15.95 | 16.45 | 16.14 | 3.14 | 1.16 |
| 80000 | 16.04 | 16.54 | 16.25 | 3.12 | 1.26 |
| 85000 | 16.13 | 16.63 | 16.35 | 3.09 | 1.36 |
| 90000 | 16.21 | 16.71 | 16.45 | 3.09 | 1.45 |
| 95000 | 16.29 | 16.79 | 16.54 | 3.06 | 1.53 |
| 100000 | 16.36 | 16.86 | 16.61 | 3.05 | 1.55 |

The main result observed in Table 2 is that, for uniform probabilities, Heuristics 2 equals the Exact Algorithm, while Heuristics 1 does not. The main explanation for this fact is that, when the Huffman tree is a complete binary tree, the improvements of Heuristics 2 apply very well. It can also be observed the small difference between all methods and Huffman's algorithm, and the decrease of the relative costs when $n$ increases. It can still be confirmed a theoretical result stated in [11]: the cost difference between the optimal even tree and the Huffman tree lays in the interval $[1/3, 1/2]$, being maximum $(1/2)$ when the number of symbols is $n = 2^k$ for some integer $k$, and minimum $(1/3)$ when $n = 3.2^k$.

Now, examine the results presented in Table 3, for arbitrary probabilities. First, compare data from Tables 2 and 3. We can see that all data in columns 2 to 5 in Table 3 are smaller than the corresponding ones in Table 2, fact that is more related to the situation of arbitrary probabilities than to the methods. The relative difference between $c(T_E)$ and $c(T_H)$ decreases considerably as $n$ increases. The same facts happened for Heuristics 2, suggesting that it is also well applied for this situation, although it does not equal the optimal solution.

However, for Heuristics 1, the behavior is quite different. Both the absolute value of the difference to $c(T_H)$ and the relative value increased. So, we have a great advantage for Heuristics 2, in this situation.

Table 4 illustrates the costs obtained for large values of $n$ and arbitrary probabilities. The costs compared are $c(T_H)$, $c(T_{Heu1})$ and $c(T_{Heu2})$. The main results obtained from Table 3 are confirmed, that is, Heuristics 2 is far better than Heuristics 1. Moreover, the relative differences of costs from the two heuristics to Huffman's algorithm again decrease. Those differences become negligible for large values of $n$.

Finally, from the three tables, we can confirm how loose was the upper theoretical bound presented in Subsection 4.1, since all the relative differences between the costs of the even trees obtained by Heuristics 2 and the Huffman trees were at most 5%, for $n$ large enough. It seems to be interesting to search for tighter bounds for this situation.

# References

1. N. Faller. *An adaptative Method for Data Compression* . Record of the 7th Asilomar Conference on Circuits, Systems and Computers, Naval Postgraduate School, Monterrey, Ca., pp. 593-597, 1973.
2. R. G. Gallager. *Variations on a Theme by Huffman* . IEEE Transactions on Information Theory, 24(1978), pp. 668-674.
3. R. W. Hamming. *Coding And Information Theory*. Prentice Hall, 1980.
4. D. A. Huffman. *A Method for the Construction of Minimum Redundancy Codes*. Proceedings of the IRE, 40:1098-1101, 1951.
5. D. E. Knuth. *The Art of Computer Programming, V.1: Fundamental Algorithms*, Addison Wesley, 2nd Edition, 1973.
6. D. E. Knuth. *Dynamic Huffman Coding*. Journal of Algorithms, 6(1985), pp. 163-180.
7. L. L. Larmore and D. S. Hirshberg. *A fast algorithm for optimal length-limited Huffman codes*. JACM, Vol. 37 No 3, pp. 464-473, Jul. 1990.
8. R. L. Milidiú and E. S. Laber. *The Warm-up Algorithm: A Lagrangean Construction of Length Restricted Huffman Codes*. Siam Journal on Computing, Vol. 30 No 5, pp. 1405-1426, 2000.
9. R. L. Milidiú and E. S. Laber. *Improved Bounds on the Ineficiency of Length Restricted Codes*. Algorithmica, Vol. 31 No 4, pp. 513-529, 2001.
10. R. L. Milidiú, E. S. Laber and A. A. Pessoa. *Improved Analysis of the FGK Algorithm*. Journal of Algorithms, Vol. 28, pp. 195-211, 1999.
11. P. E. D. Pinto, F. Protti and J. L. Szwarcfiter. *Parity codes*. Technical Report NCE 01/04, Universidade Federal do Rio de Janeiro. Rio de Janeiro, 2004. Submitted.
12. E. S. Schwartz. *An Optimum Encoding with Minimal Longest Code and Total Number of Digits*. Information and Control, 7(1964), pp. 37-44.
13. A. Turpin and A. Moffat. *Practical length-limited coding for large alphabeths*. Computer J., Vol. 38, No 5, pp. 339-347, 1995.

# Solving Diameter Constrained Minimum Spanning Tree Problems in Dense Graphs

Andréa C. dos Santos[1], Abílio Lucena[2], and Celso C. Ribeiro[3]

[1] Department of Computer Science, Catholic University of Rio de Janeiro, Rua Marquês de São Vicente 225, Rio de Janeiro, RJ 22453-900, Brazil.
cynthia@inf.puc-rio.br
[2] Federal University of Rio de Janeiro, Departamento de Administração, Av. Pasteur 250, Rio de Janeiro, RJ 22290-240, Brazil.
lucena@facc.ufrj.br
[3] Department of Computer Science, Universidade Federal Fluminense, Rua Passo da Pátria 156, Niterói, RJ 24210-240, Brazil.
celso@inf.puc-rio.br

**Abstract.** In this study, a lifting procedure is applied to some existing formulations of the Diameter Constrained Minimum Spanning Tree Problem. This problem typically models network design applications where all vertices must communicate with each other at minimum cost, while meeting or surpassing a given quality requirement. An alternative formulation is also proposed for instances of the problem where the diameter of feasible spanning trees can not exceed given odd numbers. This formulation dominated their counterparts in this study, in terms of the computation time required to obtain proven optimal solutions. First ever computational results are presented here for complete graph instances of the problem. Sparse graph instances as large as those found in the literature were solved to proven optimality for the case where diameters can not exceed given odd numbers. For these applications, the corresponding computation times are competitive with those found in the literature.

## 1 Introduction

Let $G = (V, E)$ be a finite undirected connected graph with a set $V$ of vertices and a set $E$ of edges. Assume that a cost $c_{ij}$ is associated with every edge $[i, j] \in E$, with $i < j$. Denote by $T = (V, E')$ a spanning tree of $G$, with $E' \subseteq E$. For every pair of distinct vertices $i, j \in V$, there exists a unique path $\mathcal{P}_{ij}$ in $T$ linking $i$ and $j$. Denote by $d_{ij}$ the number of edges in $\mathcal{P}_{ij}$ and by $d = \max\{d_{ij} : i, j \in V\}$ the *diameter* of $T$. Given a positive integer $2 \leq D \leq |V| - 1$, the Diameter Constrained Minimum Spanning Tree Problem (DCMST) is to find a minimum cost spanning tree $T$ with $d \leq D$.

DCMST has been shown to be $NP$-hard when $D \geq 4$ [6]. The problem typically models network design applications where all vertices must communicate with each other at minimum cost, while meeting or surpassing a given quality requirement [7]. Additional applications are found in data compression [3] and distributed mutual exclusion in parallel computing [4,11].

C.C. Ribeiro and S.L. Martins (Eds.): WEA 2004, LNCS 3059, pp. 458–467, 2004.
© Springer-Verlag Berlin Heidelberg 2004

DCMST formulations in the literature implicitly use a property of feasible diameter constrained spanning trees, pointed out by Handler [8]. Consider first the case where $D$ is even. Handler noted that a *central vertex* $i \in V$ must exist in a feasible tree $T$, such that no other vertex of $T$ is more than $D/2$ edges away from $i$. Conversely, if $D$ is odd, a *central edge* $e = [i, j] \in E$ must exist in $T$, such that no vertex of $T$ is more than $(D - 1)/2$ edges away from the closest extremity of $(i, j)$. Another feature shared by these formulations is that, in addition to the use of natural space variables (i.e. variables associated with the edges of $G$, for this application), the central vertex (resp. edge) property of $T$ is enforced through the use of an auxiliary network flow structure. In doing so, connectivity of $T$ is naturally enforced by these structures.

The formulation proposed in [1,2] for even $D$ relies on an artificial vertex to model central spanning tree vertices. For odd $D$, however, the corresponding formulation in [1,2] do not use either artificial vertices or edges. Similarly, formulations in [7], irrespective of $D$ being odd or even, do not rely on artificial vertices or edges. Another distinction between formulations in [1,2] and those in [7] is that the former contains multicommodity network flow structures, while the latter contains single commodity network flow ones. As a result, tighter linear programming relaxations are obtained in [7], albeit at a much larger computer memory requirement.

Achuthan et al. [1,2] do not present computational results for their DCMST formulation. Gouvea and Magnanti [7] used the Mixed Integer Programming (MIP) solver CPLEX 5.0 to test their formulation uniquely on fairly sparse graphs.

In this paper, we introduce an alternative form of enforcing the central edge property for the odd $D$ case of DCMST. The proposed model is based on the use of an artificial vertex. We also apply a lifting procedure to strengthen the formulations in [1,2]. Original formulations and lifted versions of them were tested, under the MIP solver CPLEX 9.0, on complete graph instances as well as on sparse graph ones. For the computational results obtained, lifted versions of the formulations invariably required significantly less computation time to prove optimality than their unlifted counterparts. That feature was further enhanced for the odd $D$ case, with the use of the artificial vertex model.

In Section 2, a summary of the main results for formulations in [1,2] is presented. In Section 3, the artificial vertex DCMST formulation for odd $D$ is described. Strengthened (i.e. lifted) versions of the formulations in [1,2] are presented in Section 4. Computational experiments for dense and sparse graph instances are reported in Section 5. In these experiments, denser instances than previously attempted in the literature were solved to proven optimality. Concluding remarks are made in Section 6.

## 2   Formulations

Formulations in this study make use of a directed graph $G' = (V, A)$. Graph $G'$ is obtained from the original undirected graph $G = (V, E)$, as follows. For every edge $e = [i, j] \in E$, with $i < j$, there exist two arcs $(i, j)$ and $(j, i) \in A$, with costs $c_{ij} = c_{ji}$. Let $L = D/2$ if $D$ is even and $L = (D - 1)/2$, otherwise.

The very first formulations for DCMST were proposed by Achuttan et al. [1, 2]. Distinct formulations are presented by the authors for even $D$ and odd $D$ cases of the problem. Consider first the case where $D$ is even and introduce an artificial vertex, denoted by $r$, into $G'$. Let $G'' = (V', A')$ be the resulting graph with $V' = V \cup \{r\}$ and $A' = A \cup \{(r, 1), \ldots, (r, |V|)\}$. Associate a binary variable $x_{ij}$ with every arc $(i, j) \in A'$ and a non-negative variable $u_i$ with every vertex $i \in V'$. Binary variables $x_{ij}$ are used to identify a spanning tree, while variable $u_i$ denotes the number of arcs in a path from $r$ to $i \in V$. For even $D$, DCMST is formulated as follows:

$$\min \sum_{(i,j) \in A} c_{ij} x_{ij} \tag{1}$$

$$\sum_{j \in V} x_{rj} = 1 \tag{2}$$

$$\sum_{(i,j) \in A'} x_{ij} = 1 \quad \forall j \in V \tag{3}$$

$$u_i - u_j + (L + 1)x_{ij} \leq L \quad \forall (i, j) \in A' \tag{4}$$

$$x_{ij} \in \{0, 1\} \quad \forall (i, j) \in A' \tag{5}$$

$$0 \leq u_i \leq L + 1 \quad \forall i \in V'. \tag{6}$$

Equation (2) ensures that the artificial vertex $r$ is connected to exactly one vertex in $V$, i.e. the central spanning tree vertex. Constraints (3) establish that exactly one arc must be incident to each vertex of $V$. Constraints (4) and (6) ensure that paths from the artificial vertex $r$ to each vertex $i \in V$ have at most $L+1$ arcs. Constraints (5) are the integrality requirements. Edges $[i, j] \in E$ such that $x_{ij} = 1$ or $x_{ji} = 1$ in a feasible solution to (2)-(6) define a spanning tree $T$ of $G$ with diameter less than or equal to $D$.

We now consider the odd $D$ case of DCMST. Let $z_{ij}$ be a binary variable associated with each edge $[i, j] \in E$, with $i < j$. Whenever $z_{ij} = 1$, edge $[i, j]$ is selected as the central spanning tree edge. Otherwise, $z_{ij} = 0$. For $D$ odd, DCMST is formulated as follows:

$$\min \sum_{(i,j) \in A} c_{ij} x_{ij} + \sum_{[i,j] \in E} c_{ij} z_{ij} \tag{7}$$

$$\sum_{[i,j] \in E} z_{ij} = 1 \tag{8}$$

$$\sum_{(i,j)\in A} x_{ij} + \sum_{[i,j]\in E} z_{ij} + \sum_{[j,i]\in E} z_{ji} = 1 \quad \forall j \in V \tag{9}$$

$$u_i - u_j + (L+1)x_{ij} \le L \quad \forall (i,j) \in A \tag{10}$$

$$x_{ij} \in \{0,1\} \quad \forall (i,j) \in A \tag{11}$$

$$z_{ij} \in \{0,1\} \quad \forall [i,j] \in E \tag{12}$$

$$0 \le u_i \le L \quad \forall i \in V. \tag{13}$$

Equation (8) ensures that there must be exactly one central edge. Constraints (9) establish that for any vertex $i \in V$ either there is an arc incident to it or else vertex $i$ must be one of the extremities of the central spanning tree edge. Constraints (10) and (13) ensure that spanning tree paths from the closest extremity of the central edge to every other vertex $i \in V$ have at most $L$ arcs. Constraints (11) and (12) are the integrality requirements. In a feasible solution to (8)–(13), the central edge together with those edges $[i,j] \in E$ such that $x_{ij} = 1$ or $x_{ji} = 1$ define a spanning tree $T$ of $G$ with diameter less than or equal to $D$.

## 3   An Alternative Formulation for the Odd $D$ Case

The formulation in [1,2] for $D$ odd selects one edge in $E$ to be central and si-multaneously builds an auxiliary network flow problem around that edge. Flow emanating from the central edge is then controlled to enforce the diameter con-straint. Figure 1 (a) illustrates a solution obtained for $D = 3$. Notice that edge $[p,q]$ plays the central edge role and that any spanning tree leaf is no more than $L = (D-1)/2 = 1$ edges away from edge $[p,q]$.

An alternative formulation which uses an artificial vertex $r$, as for the even $D$ case, is also possible here. Recall that, for $D$ even, the artificial vertex $r$ is connected to exactly one vertex of $V$, i.e. the central spanning tree vertex. Now, the artificial vertex $r$ will be connected to exactly two vertices. Namely those two vertices incident on the central edge. This situation is modeled by implicitly enforcing selection of a central edge $[p,q] \in E$ by explicitly forcing *artificial* edges $[p,r]$ and $[q,r]$ to appear in the solution. An illustration of this scheme appears in Figure 1 (b). A feasible spanning tree $T$ of $G$ is obtained by eliminating the two edges incident on $r$ and connecting their extremities through the central edge $[p,q]$.

The motivation behind our formulation for $D$ odd is to highlight a structure that has already been well studied from a polyhedral viewpoint. In doing so, we expect to strengthen the overall DCMST formulation through the use of facet defining inequalities for that structure.

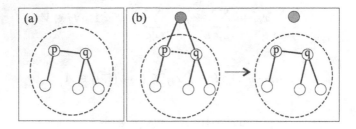

**Fig. 1.** Solutions to DCMST in the odd case with $D = 3$.

Consider the same notation and variables introduced in Section 2 for $D$ odd. Additional variables $x_{ri}$, for every $i \in V$, are introduced to represent the edges associated with artificial vertex $r$. An edge $[i, j] \in E$ is selected as the central spanning tree edge if and only if edges $[r, i]$ and $[r, j]$ are also selected. This condition is enforced through the nonlinear equation $z_{ij} = x_{ri} \cdot x_{rj}$ or convenient linearizations of it. A valid formulation for DCMST when $D$ is odd is given by:

$$\min \sum_{(i,j)\in A} c_{ij}x_{ij} + \sum_{[i,j]\in E} c_{ij}z_{ij} \tag{14}$$

$$\sum_{j\in V} x_{rj} = 2 \tag{15}$$

$$\sum_{(i,j)\in A'} x_{ij} = 1 \quad \forall j \in V \tag{16}$$

$$\sum_{[i,j]\in E} z_{ij} = 1 \tag{17}$$

$$z_{ij} \geq x_{ri} + x_{rj} - 1 \quad \forall [i, j] \in E \tag{18}$$

$$z_{ij} \leq x_{ri} \quad \forall [i, j] \in E \tag{19}$$

$$z_{ij} \leq x_{rj} \quad \forall [i, j] \in E \tag{20}$$

$$u_i - u_j + (L + 1)x_{ij} \leq L \quad \forall (i, j) \in A' \tag{21}$$

$$0 \leq u_i \leq L + 1 \quad \forall i \in V \tag{22}$$

$$x_{ij} \in \{0,1\} \quad \forall (i,j) \in A' \tag{23}$$

$$z_{ij} \in \{0,1\} \quad \forall [i,j] \in E. \tag{24}$$

Equation (15) establishes that the artificial central vertex $r$ is connected to exactly two vertices of $V$. Constraints (16) establish that there is exactly one arc incident to each vertex of $V$. Constraints (17) to (20) give a linearization of $z_{ij} = x_{ri} \cdot x_{rj}$ for every edge $[i,j] \in E$. Finally, constraints (21) and (22) ensure that the paths from the artificial vertex $r$ to each vertex $i \in V$ have at most $L+1$ arcs. Constraints (23) and (24) are the integrality requirements.

We now derive valid inequalities for the formulation (14)-(24). If constraint (15) is multiplied by variable $x_{ri}$, for $i \in V$,

$$\sum_{j \in V, j \neq i} x_{ri} \cdot x_{rj} + x_{ri} \cdot x_{ri} = 2 \cdot x_{ri} \tag{25}$$

results. Bearing in mind that all variables in (25) are binary 0-1 and consequently $x_{ri} = x_{ri} \cdot x_{ri}$ holds, it is valid to write

$$\sum_{j \in V, j \neq i} x_{ri} \cdot x_{rj} = x_{ri}. \tag{26}$$

However, since $z_{ij} = x_{ri} \cdot x_{rj}$ for every edge $[i,j] \in E$ and $x_{ri}$ and $x_{rj}$ cannot simultaneously be equal to 1 if an edge $[i,j]$ does not exist in $E$, valid constraints for (14)–(24) are

$$\sum_{[i,j] \in E} z_{ij} + \sum_{[j,i] \in E} z_{ji} = x_{ri} \quad \forall i \in V. \tag{27}$$

Constraints (27) are redundant for formulation (14)–(24) but are not necessarily so for its linear programming relaxation.

Additional valid inequalities for (14)–(24) can be found if one concentrates on inequalities (18)–(20) and the underlying Boolean quadric polytope [10].

## 4 Lifting

In this section, following the work of Desrochers and Laporte [5], we lift the Miller-Tucker-Zemlin [9] inequalities $u_i - u_j + (L+1)x_{ij} \leq L$, $\forall (i,j) \in A'$. In doing so, strengthened versions of DCMST formulations presented in previous sections are obtained. The idea of lifting consists in adding a valid nonnegative term $\alpha_{ji}x_{ji}$ to the above inequalities, transforming them into

$$u_i - u_j + (L+1)x_{ij} + \alpha_{ji}x_{ji} \leq L. \tag{28}$$

The larger is the value of $\alpha_{ji}$, the larger will be the reduction in the original solution space. If $x_{ji} = 0$, then $\alpha_{ji}$ may take any value. Suppose now $x_{ji} = 1$.

Then, $u_i = u_j + 1$ since the path from the central vertex in the even case (resp. from the closest extremity of the central edge in the odd case) to vertex $i \in V$ visits $j$ before visiting $i$. Moreover, $x_{ji} = 1$ implies $x_{ij} = 0$ due to constraints (3) or (9). By substitution in (28), we obtain $u_i - u_j + (L+1)x_{ij} + \alpha_{ji}x_{ji} \leq L \Rightarrow 1 + u_j - u_j + \alpha_{ji} \leq L \Rightarrow \alpha_{ji} \leq L - 1$. To maximize the value of $\alpha_{ji}$, we take $\alpha_{ji} = L - 1$. Then,

$$u_i - u_j + (L+1)x_{ij} + (L-1)x_{ji} \leq L \tag{29}$$

is a valid inequality for all $(i,j) \in A'$ (resp. for all $(i,j) \in A$) for $D > 2$ in the even case (resp. for $D > 3$ in the odd case).

We now derive improved generalized upper bounds for the variables $u_i$, for $i \in V$. In the even case, there is an artificial vertex $r$ such that $u_r = 0$. The central vertex connected to $r$ will necessarily be the first vertex to be visited in any path emanating from $r$. Then,

$$u_i \leq (L+1) - Lx_{ri} \quad \forall i \in V. \tag{30}$$

Moreover, $u_i \leq L$ for any vertex $i \in V$ which is not a leaf of the spanning tree. Then,

$$u_i \leq (L+1) - x_{ij} \quad \forall (i,j) \in A \tag{31}$$

holds.

We now consider the odd case. If an edge $[i,j] \in E$ is the central one, then $z_{ij} = 1$, $u_i = 0$, and $u_j = 0$. In consequence,

$$u_i \leq L - Lz_{ij} \quad \forall [i,j] \in E. \tag{32}$$

Analogously to the odd case, $u_i < L$ for any vertex $i \in V$ which is not a leaf of the spanning tree. Then,

$$u_i \leq L - x_{ij} \quad \forall (i,j) \in A. \tag{33}$$

Inequalities (30) and (31) define improved generalized upper bounds for the even $D$ case, while inequalities (32) and (33) correspond to new generalized upper bounds for the odd $S$ case.

We now derive improved generalized lower bounds for the variables $u_i$, for $i \in V$. In the even case, $u_i \geq 1 \geq x_{ri}$ for any vertex $i \in V$. If $i$ is not directly connected to the central vertex, then $x_{ri} = 0$ and $u_i \geq 2$. If these two conditions are taken into account simultaneously, we have the first improvement in the lower bounds:

$$u_i \geq x_{ri} + 2 \sum_{j \in V : j \neq r} x_{ji} \quad \forall i \in V. \tag{34}$$

The above condition is simpler in the odd case, where no central vertex exists:

$$u_i \geq \sum_{j \in V, j \neq i} x_{ji} \quad \forall i \in V. \tag{35}$$

# 5   Computational Results

Computational experiments were performed on a Pentium IV machine with a 2.0 GHz clock and 512 MB of RAM memory, using MIP solver CPLEX 9.0 under default parameters. In these experiments, the alternative formulation proposed for the odd $D$ case of DCMST was reinforced with valid inequalities (27).

Test instances were generated as follows. For a graph with a number $n = |V|$ of vertices, the uniform distribution was used to draw $n$ points with integer coordinates in a square of sides 100 on the Euclidean plane. Vertices were associated to points and edge costs were taken as the truncated Euclidean distance between corresponding pairs of points. Sparse graph instances with $m = |E|$ edges were generated as in [7]. The minimum cost spanning star is computed first and all of its $n - 1$ edges are selected. The remaining $m - n - 1$ edges are taken as the least cost edges not already contained in the minimum cost star. In all 19 odd $D$ instances and 18 even $D$ instances were generated. For each of the two cases, 12 complete graph instances (with up to 25 vertices) were generated. Test instance details are summarized in Tables 1 and 2.

Table 1 gives numerical results for odd $D$ instances. For each instance, the number of vertices, the number of edges, and the value of the diameter $D$ are given. These entries are followed by the results obtained with the original formulation in [1,2] (A), the original formulation with lifting (B), and the new artificial central vertex formulation with lifting (C). For each formulation, the CPU time required to prove optimality is given in seconds together with the number of

**Table 1.** Numerical results for the odd $D$ case.

| | | | (A) | | (B) | | (C) | |
|---|---|---|---|---|---|---|---|---|
| $\|V\|$ | $\|E\|$ | $D$ | time (s) | nodes | time (s) | nodes | time (s) | nodes |
| 10 | 45 | 5 | 0.14 | 85 | 0.16 | 62 | 0.13 | 28 |
| 10 | 45 | 7 | 0.14 | 190 | 0.17 | 88 | 0.17 | 49 |
| 10 | 45 | 9 | 0.05 | 40 | 0.10 | 73 | 0.08 | 10 |
| 15 | 105 | 5 | 40.95 | 73331 | 22.27 | 18640 | 31.22 | 21814 |
| 15 | 105 | 7 | 85.13 | 163355 | 25.02 | 23226 | 19.23 | 16118 |
| 15 | 105 | 9 | 76.14 | 146948 | 10.56 | 9597 | 7.80 | 6174 |
| 20 | 190 | 5 | 1686.17 | 1753713 | 272.00 | 132678 | 216.36 | 95813 |
| 20 | 190 | 7 | 31.16 | 36168 | 8.39 | 3532 | 5.05 | 1519 |
| 20 | 190 | 9 | 884.36 | 982551 | 151.5 | 85204 | 91.33 | 46559 |
| 25 | 300 | 5 | – | – | 73857.81 | 24309253 | 51551.80 | 17235497 |
| 25 | 300 | 7 | 24513.01 | 15083987 | 20160.13 | 6225967 | 16617.61 | 5272473 |
| 25 | 300 | 9 | 177024.22 | 66213391 | 9172.04 | 2438057 | 40183.44 | 12040315 |
| 20 | 50 | 5 | 39.78 | 93283 | 7.74 | 9386 | 5.58 | 3707 |
| 20 | 50 | 7 | 39.06 | 68843 | 3.16 | 4908 | 2.23 | 1915 |
| 20 | 50 | 9 | 70.12 | 128376 | 26.16 | 45561 | 40.28 | 57603 |
| 40 | 100 | 5 | 3596.34 | 3506756 | 780.74 | 688912 | 20.67 | 13993 |
| 40 | 100 | 7 | 12581.45 | 12272065 | 976.98 | 849861 | 207.64 | 171846 |
| 40 | 100 | 9 | 27735.56 | 23763469 | 4584.19 | 3530233 | 23359.05 | 15583400 |
| 60 | 150 | 5 | – | – | 215161.75 | 116775357 | 11644.75 | 5003876 |

**Table 2.** Numerical results for the even $D$ case.

| $|V|$ | $|E|$ | $D$ | (A) time (s) | (A) nodes | (B) time (s) | (B) nodes |
|---|---|---|---|---|---|---|
| 10 | 45 | 4 | 0.95 | 1849 | 0.77 | 1300 |
| 10 | 45 | 6 | 0.13 | 55 | 0.08 | 7 |
| 10 | 45 | 10 | 0.06 | 29 | 0.08 | 17 |
| 15 | 105 | 4 | 65.8 | 73024 | 24.17 | 26711 |
| 15 | 105 | 6 | 53.19 | 66834 | 32.53 | 41882 |
| 15 | 105 | 10 | 38.41 | 61822 | 8.95 | 11790 |
| 20 | 190 | 4 | 7462.1 | 4877014 | 1888.02 | 1091803 |
| 20 | 190 | 6 | 1630.58 | 1210813 | 593.91 | 412770 |
| 20 | 190 | 10 | 2729.48 | 2285819 | 172.81 | 144382 |
| 25 | 300 | 4 | – | – | 158836.45 | 64913343 |
| 25 | 300 | 6 | 43044.61 | 17498605 | 5194.90 | 2119161 |
| 25 | 300 | 10 | 1031.36 | 565737 | 459.88 | 205747 |
| 20 | 50 | 4 | 62.47 | 115144 | 0.64 | 615 |
| 20 | 50 | 6 | 221.31 | 446477 | 10.81 | 16396 |
| 20 | 50 | 10 | 619.52 | 1443014 | 74.15 | 173234 |
| 40 | 100 | 4 | 8957.38 | 8110166 | 54.14 | 51476 |
| 40 | 100 | 6 | 205940.95 | 167119305 | 909.95 | 1012212 |
| 40 | 100 | 10 | – | – | 146019.52 | 155646590 |

nodes visited in the branch-and-bound tree. Table 2 gives the same results for the even $D$ case, except for the central vertex formulation which does not apply in this case.

In spite of the considerable duality gaps associated with the formulations tested here, the lifted formulations we suggest are capable of solving, to proven optimality, sparse instances as large as those found in the literature in competitive CPU times.

No results appear in the literature for complete graph DCMST instances. A possible explanation for that is the large computer memory demands required by the other existing DCMST formulations [1,2]. The very first computational results for complete graph DCMST instances are thus introduced in this study.

From the computational results presented, it should also be noticed that our alternative odd $D$ case formulation dominates their counterparts in this study in terms of the CPU time required to prove optimality.

## 6   Conclusions

In this study, DCMST formulations proposed in [1,2] were strengthened through the use of a lifting procedure. In doing so, substantial duality gap reductions were attained for the computational experiments carried out. Additionally, we also propose an artificial central vertex strategy for modeling the odd $D$ case of the problem. For the computational tests carried out, the new formulation dominated its odd $D$ counterparts in in terms of total CPU time required to prove optimality. The same idea could also be extended to other existing formulations such as those presented in [7].

For sparse graphs instances, the strongest model proposed in this study was capable of solving, to proven optimality, instances as large as those previously solved in the literature [7]. It is worth mentioning here that the models suggested by Gouvea and Magnanti typically produce very small duality gaps. However, they are quite demanding in terms of computer memory requirements (particularly the model involving variables with four indices). In consequence, they do not appear adequate to directly tackling dense graph instances of the problem.

In spite of the considerable duality gaps observed in our computational experiments, our approach was capable of solving, to proven optimality, complete graph instances with up to 25 vertices. These are the first results ever presented for dense graph DCMST instances.

We conclude by pointing out that the alternative odd $D$ formulation introduced here can be further strengthened with valid inequalities associated with the Boolean quadric polytope.

# References

1. N.R. Achuthan, L. Caccetta, P.A. Caccetta, and J.F. Geelen. Algorithms for the minimum weight spanning tree with bounded diameter problem. In K.H. Phua, C.M. Wand, W.Y. Yeong, T.Y. Leong, H.T. Loh, K.C. Tan, and F.S. Chou, editors, *Optimisation Techniques and Applications*, volume 1, pages 297–304. World Scientific, 1992.
2. N.R. Achuthan, L.Caccetta, P.A. Caccetta, and J.F. Geelen. Computational methods for the diameter restricted minimum weight spanning tree problem. *Australasian Journal of Combinatorics*, 10:51–71, 1994.
3. A. Bookstein and S.T. Klein. Compression of correlated bitvectors. *Information Systems*, 16:110–118, 2001.
4. N. Deo and A. Abdalla. Computing a diameter-constrained minimum spanning tree in parallel. In G. Bongiovanni, G. Gambosi, and R. Petreschi, editors, *Algorithms and Complexity*, volume 1767, pages 17–31. 2000.
5. M. Desrochers and G. Laporte. Improvements and extensions to the Miller-Tucker-Zemlin subtour elimination constraints. *Operations Research Letters*, 10:27–36, 1991.
6. M.R. Garey and D.S. Johnson. *Computers and intractability: A guide to the theory of NP-Completeness*. W.H. Freeman, New York, 1979.
7. L. Gouveia and T.L. Magnanti. Modelling and solving the diameter-constrained minimum spanning tree problem. Technical report, DEIO-CIO, Faculdade de Ciências, 2000.
8. G.Y. Handler. Minimax location of a facility in an undirected graph. *Transportation Science*, 7:287–293, 1978.
9. C.E. Miller, A.W. Tucker, and R.A. Zemlin. Integer programming formulations and traveling salesman problems. *Journal of the ACM*, 7:326 – 329, 1960.
10. M. Padberg. The boolean quadric polytope: Some characteristics and facets. *Mathematical Programming*, 45:139–172, 1988.
11. K. Raymond. A tree-based algorithm for distributed mutual exclusion. *ACM Transactions on Computers*, 7:61–77, 1989.

# An Efficient Tabu Search Heuristic for the School Timetabling Problem

Haroldo G. Santos[1], Luiz S. Ochi[1], and Marcone J.F. Souza[2]

[1] Computing Institute, Fluminense Federal University, Niterói, Brazil.
{hsantos,satoru}@ic.uff.br
[2] Computing Department, Ouro Preto Federal University, Ouro Preto, Brazil.
marcone@iceb.ufop.br

**Abstract.** The School Timetabling Problem (STP) regards the weekly scheduling of encounters between teachers and classes. Since this scheduling must satisfy organizational, pedagogical and personal costs, this problem is recognized as a very difficult combinatorial optimization problem. This work presents a new Tabu Search (TS) heuristic for STP. Two different memory-based diversification strategies are presented. Computational experiments with real world instances, in comparison with a previously proposed TS found in literature, show that the proposed method produces better solutions for all instances, as well as observed increased speed in the production of good quality solutions.

## 1 Introduction

The School Timetabling Problem (STP) embraces the scheduling of sequential encounters between teachers and students so as to insure that requirements and constraints are satisfied. Typically, the manual solution of this problem extends for various days or weeks and normally produces unsatisfactory results due to the fact that lesson periods could be generated which are inconsistent with pedagogical needs or could even serve as impediments for certain teachers or students. STP is considered a NP-hard problem [5] for nearly all of its variants, justifying the usage of heuristic methods for its resolution. In this manner, various heuristic and metaheuristic approaches have been applied with success in the solution of this problem, such as: Tabu Search (TS) [12,4,10], Genetic Algorithms [13] and Simulated Annealing (SA) [2].

The application of TS to the STP is specially interesting, since this method is, as local search methods generally are, very well suited for the interactive building of timetables, a much recognized quality in timetable building systems. Furthermore, TS based methods often offer the best known solutions to many timetabling problems, when compared to other metaheuristics [3,11]. The diversification strategy is an important aspect in the design of a TS algorithm. Since the use of a tabu list is not enough to prevent the search process from becoming trapped in certain regions of the search space, other mechanisms have been proposed. In particular, for the STP, two main approaches have been used: adaptive relaxation [10,4] and random restart [12]. In adaptive relaxation the costs

C.C. Ribeiro and S.L. Martins (Eds.): WEA 2004, LNCS 3059, pp. 468–481, 2004.

involved in the objective function are dynamically changed to bias the search process to newly, unvisited, regions of the search space. In random restart a new solution is generated and no previous information is utilized.

This work employs a TS algorithm that uses an informed diversification strategy, which takes into account the history of the search process to bias the selection of diversification movements. Although it uses only standard TS components, it provides better results than more complex previous proposals [12].

The article is organized as follows: section 2 presents related works; section 3 introduces the problem to be treated; section 4 presents the proposed algorithm; section 5 describes the computational experiments and their results; and finally, section 6 formulates conclusions and future research proposals.

## 2    Related Works

Although the STP is a classical combinatorial optimization problem, no widely accepted model is used in the literature. The reason is that the characteristics of the problem are highly dependent on the educational system of the country and the type of institution involved. As such, although the basic search problem is the same, variations are introduced in different works [3,4,10,12]. Described afterwards, the problem considered in this paper derives from [12] and considers the timetabling problem encountered in typical Brazilian high schools. In [12], a GRASP-Tabu Search (GTS-II) metaheuristic was developed to tackle this problem. The GTS-II method incorporates a specialized improvement procedure named "Intraclasses-Interclasses", which uses a shortest-path graph algorithm. At first, the procedure is activated aiming to attain the feasibility of the constructed solution, after which, it then aims to improve the feasible solution. The movements made in the "Intraclasses-Interclasses" also remain with the tabu status for a given number of iterations. Diversification is implemented through the generation of new solutions, in the GRASP constructive phase. In [11] three different metaheuristics that incorporate the "Intraclasses-Interclasses" were proposed: Simulated Annealing, Microcanonical Optimization (MO) and Tabu Search. The TS proposal significantly outperformed both SA and MO.

## 3    The Problem Considered

The problem considered deals with the scheduling of encounters with teachers and classes over a weekly period. The schedule is made up of $d$ days of the week with $h$ daily periods, defining $p = d \times h$ distinct periods. There is a set $T$ with $t$ teachers that teach a set $S$ of $s$ subjects to a set $C$ of $c$ classes, which are disjoint sets of students with the same curriculum. The association of teachers to subjects in certain classes is previously fixed and the workload is informed in a matrix of requirements $R_{t \times c}$, where $r_{ij}$ indicates the number of lessons that teacher $i$ shall teach for class $j$. Classes are always available, and must have their time schedules, of size $p$, completely filled out, while teachers indicate a set of available periods. Also, teachers may request a number of double lessons per class. These lessons are lessons which must be allocated in two consecutive periods on the same day. This way a solution to the STP problem must satisfy the following constraints:

1. no class or teacher can be allocated for two lessons in the same period;
2. teachers can only be allocated respecting their availabilities;
3. each teacher must fulfill his/her weekly number of lessons;
4. for pedagogical reasons no class can have more than two lesson periods with the same teacher per day.

Also, there are the following desirable features that a timetable should present:

1. the time schedule for each teacher should include the least number possible of days;
2. double lessons requests must be satisfied whenever possible;
3. "gaps" in the time schedule of teachers should be avoided, that is: periods of no activity between two lesson periods.

## 3.1    Solution Representation

A timetable is represented as a matrix $Q_{t \times p}$, in a such way that each row represents the complete weekly timetable for a given teacher. As such, the value $q_{ik} \in \{0, 1, \cdots, c\}$, indicates the class for which the teacher $i$ is teaching during period $k$ ($q_{ik} \in \{1, \cdots, c\}$), or if the teacher is available for allocation ($q_{ik} = 0$). The advantage of this representation is that it eliminates the possibility for the occurrence of conflicts in the timetable for teachers. The occurrence of conflicts in classes happens when in a given period $k$ more than one teacher is allocated to that class. Allocations are only allowed in periods with teacher availability. A partial sample of a timetable with 5 teachers can be found in Figure 1, with value "X" indicating teacher unavailability.

| Teacher \ Period | 1 | 2 | 3 | 4 | 5 | $\cdots d \times h$ |
|:---:|:---:|:---:|:---:|:---:|:---:|:---:|
| 1 | 1 | 0 | 0 | 2 | 2 | $\cdots$ |
| 2 | 0 | X | X | 0 | 1 | $\cdots$ |
| 3 | X | X | 1 | 0 | 3 | $\cdots$ |
| 4 | 0 | 1 | 0 | 1 | 0 | $\cdots$ |
| 5 | 0 | 0 | 2 | 3 | X | $\cdots$ |

**Fig. 1.** Fragment of generated timetable

## 3.2   Objective Function

In order to treat STP as an optimization problem, it is necessary to define an objective function that determines the degree of infeasibility and satisfaction of requirements; that is, pretends to generate feasible solutions with a minimal number of unsatisfied requisites. Thus, a timetable $Q$ is evaluated with the following objective function, which should be minimized:

$$f(Q) = \omega \times f_1(Q) + \delta \times f_2(Q) + \rho \times f_3(Q) \qquad (1)$$

where $f_1$ counts, for each period $k$, the number of times that more than one teacher teaches the same class in period $k$ and the number of times that a class has no activity in $k$. The $f_2$ portion measures the number of allocations that disregard the daily limits of lessons of teachers in classes (constraint 4). As such, a timetable can only be considered feasible if $f_1(Q) = f_2(Q) = 0$. The importance of the costs involved defines a hierarchy so that: $\omega > \delta \gg \rho$. The $f_3$ component in the objective function measures the satisfaction of personal requests from teachers, namely: double lessons, non-existence of "gaps" and timetable compactness, as follows:

$$f_3(Q) = \sum_{i=1}^{t} \alpha_i \times b_i + \beta_i \times v_i + \gamma_i \times c_i \qquad (2)$$

where $\alpha_i$, $\beta_i$, and $\gamma_i$ are weights that reflect, respectively, the relative importance of the number of "gaps" $b_i$ , the number of week days $v_i$ each teacher is involved in any teaching activity during the same shift, and the non-negative difference $c_i$ between the minimum required number of double lessons and the effective number of double lessons in the current agenda of teacher $i$.

## 4   Tabu Search for the School Timetabling Problem

Tabu Search (TS) is an iterative method for solving combinatorial optimization problems. It explicitly makes use of memory structures to guide a hill-descending heuristic to continue exploration without being confounded by the absence of improvement movements. This technique was independently proposed by Glover [6] and Hansen [8]. For a detailed description of TS, the reader is referred to [7]. This section presents a brief explanation of TS principles. They are followed by specifications of the customized TS implementation proposed in this paper.

Starting from an initial solution $x$, the method systematically explores the neighborhood $\mathcal{N}(x)$ and selects the best admissible movement $m$, such that the application of $m$ in the current solution $x$ (denoted by $x \oplus m$) produces the new current solution $x' \in \mathcal{N}(x)$. When no improvement movements are found, movements that deteriorate the cost function are also permitted. Thus, to try to avoid cycling, a mechanism called short-term memory is employed. The objective of short-term memory is try to forbid movements toward already visited solutions, which is usually achieved by the prohibition of the last reversal movements. These movements are stored in a tabu list and remains forbidden (with tabu status), for a given number of iterations, called *tabu tenure*. Since this strategy can be too restrictive, and in order to not disregard high quality solutions, movements with tabu status can be accepted if the cost of the new solution produced satisfy an *aspiration criterion*. Also, intensification and diversification procedures can be used. These procedures, respectively, aim to deeply investigate promising regions of the search space and to ensure that no region of the search space remains neglected. Following is a description of the constructive algorithm and the customized TS implementation proposed in this paper.

## 4.1   Constructive Algorithm

The constructive algorithm basically consists of a greedy randomized constructive procedure [9]. While in other works the option for a randomized construction is to allow diversification, through multiple initializations, in this case the only purpose is to have control of the randomization degree of initial solution. To build a solution, step-by-step, the principle of allocating first the *most urgent* lessons in the *most appropriate* periods is used. In this case, the urgency degree $\theta_{ij}$ of allocating a lesson from teacher $i$ for class $j$ is computed considering the available periods $V_i$ from teacher $i$, the available periods $W_j$ from class $j$ and the number of unscheduled lessons $\zeta_{ij}$ of teacher $i$ for class $j$, as follows: $\theta_{ij} = \frac{\zeta_{ij}}{|V_i \cap W_j| + 1}$. Also, let $L$ be the set of required lessons, such that $r_{ij} > 0, \forall l_{ij} \in L$. The algorithm then builds a restricted candidate list ($RCL$) with the most urgent lessons, in a such a way that $RCL = \{l_{ij} \in L \mid \theta_{ij} \geq \overline{\theta} - (\overline{\theta} - \underline{\theta}) \times \alpha\}$, where $\overline{\theta} = \max\{\theta_{ij} \mid i \in T, j \in C\}$ and $\underline{\theta} = \min\{\theta_{ij} \mid i \in T, j \in C\}$. At each iteration, a lesson $l_{ij}$ for allocation is randomly selected from $RCL$. The lecture is allocate in a free period of its corresponding teacher, attempting to maintain the timetable free of conflicts in classes. Allocations are made first in periods with the least teacher availability. The $\alpha$ parameter allows tuning the randomization degree of the algorithm, varying from the pure greedy algorithm ($\alpha = 0$) to a completely random ($\alpha = 1$) selection of the teacher and class to allocation. At each step, the number of unscheduled lessons and the urgency degrees are recomputed. The process continues till no more unscheduled lessons are found (i.e., $\zeta_{ij} = 0, \forall i \in T, j \in C$).

## 4.2   Tabu Search Components

The TS procedure (Figure 2) starts from the initial timetable $Q$ provided by the constructive algorithm and, at each iteration, fully explores the neighborhood $\mathcal{N}(Q)$ to select the next movement $m$. The movement definition used here is the same as in [10], and involves the swap of two values in the timetable of a teacher $i \in T$, which can be defined as the triple $\langle i, p_1, p_2 \rangle$, such that $q_{ip_1} \neq q_{ip_2}$, $p_1 < p_2$ and $p_1, p_2 \in \{1, \cdots, p\}$. Clearly, any timetable can be reached through a sequence of these movements that is at most, the number of lessons in the requirement matrix. Once a movement $m$ is selected, its reversal movement will be kept in the tabu list during the next $tabuTenure(m)$ iterations, which is randomly selected

```
procedure TSDS(Q, divActivation, iterationsDiv, minTT, maxTT)
 1: Q* = Q; TabuList = ∅;
 2: noImprovementIterations = 0; iteration = 0;
 3: initializeMovementFrequencies();
 4: repeat
 5:    Δ = ∞; iteration + +;
 6:    for all movement m such that (Q ⊕ m) ∈ N(Q)  do
 7:       penalty = 0;
 8:       if noImprovementIterations mod divActivation < iterationsDiv
          and iteration ≥ divActivation  then
 9:          penalty = computePenalty(m);
10:       end if
11:       if (f(Q) − f(Q ⊕ m) + penalty < Δ and (m ∉ TabuList)) or
          (f(Q ⊕ m) < f(Q*))  then
12:          bestMov = m;
13:          Δ = f(Q) − f(Q ⊕ m);
14:       end if
15:    end for
16:    Q = Q ⊕ bestMov;
17:    tabuTenure(bestMov) = random(minTT, maxTT);
18:    updateTabuList(bestMov, iteration);
19:    computeMovementFrequency(bestMov);
20:    if (f(Q) < f(Q*)) then
21:       Q* = Q; noImprovementIterations = 0;
22:       initializeMovementFrequencies();
23:    else
24:       noImprovementIterations++;
25:    end if
26: until termination criterion reached;
end TSDS;
```

**Fig. 2.** Pseudo-code for TSDS algorithm

(line 17) within the interval $[minTT, maxTT]$. Insertions and removals in tabu list can be made at every iteration (line 18). The aspiration criterion defined is that the movement will loose its tabu status if its application produces the best solution found so far (line 11).

Since short-term memory is not enough to prevent the search process from becoming entrenched in certain regions of the search space, some diversification strategy is needed. In the proposed method, long-term memory is used to guide the diversification procedure as follows: frequency of movements involving each teacher and class are computed. While the diversification procedure is active, the selection of movements emphasizes the execution of few explored movements, through the incorporation of penalties (line 9) in the evaluation of movements. Each time a movement is done, movement frequencies will be updated (line 19). These frequencies are zeroed each time that the best solution found so far is updated (line 22). Following is an explanation of how the penalties in function *computePenalty* (line 9) are computed. Considering that the counts of movements made with each teacher and class are stored in a matrix $Z_{t \times c}$, the penalty for a given movement takes into account the transition ratio $\epsilon_{ij}$ of teacher $i$ and class $j$, which can be computed as follows:

$$\epsilon_{ij} = \frac{z_{ij}}{\bar{z}} \tag{3}$$

where $\bar{z} = \max\{z_{ij} \mid i \in T, \ j \in C\}$. Since a movement can involve two lesson periods, or a lesson period and a free period, the penalty $\psi_{ia_1a_2}$ associated with a movement in the timetable of teacher $i$, in periods $p_1$ and $p_2$ with allocations $a_1 = q_{ip_1}$ and $a_2 = q_{ip_2}$, respectively, considering the cost of the best solution found so far $Q^*$ is:

$$\psi_{ia_1a_2} = \begin{cases} \epsilon_{ia_1} \times f(Q^*) & \text{if } a_1 \neq 0 \text{ and } a_2 = 0 \\ \epsilon_{ia_2} \times f(Q^*) & \text{if } a_1 = 0 \text{ and } a_2 \neq 0 \\ (\epsilon_{ia_1} + \epsilon_{ia_2})/2 \times f(Q^*) & \text{if } a_1 \neq 0 \text{ and } a_2 \neq 0 \end{cases}$$

Another penalty function proposed in this paper also considers the teacher workload to promote diversification. In this case, the objective is to favor movements involving teachers whose timetable changes would probably produce bigger modifications in the solution structure. The value of the penalty function $\tau_{ia_1a_2}$ for allocations $a_1$ and $a_2$ of teacher $i$ is:

$$\tau_{ia_1a_2} = \frac{\psi_{ia_1a_2}}{\sum_{j=1}^{c} r_{ij} / \max_{l=1}^{t} \left( \sum_{j=1}^{c} r_{lj} \right)} \tag{4}$$

The diversification strategy is applied whenever signals that regional entrenchment may be in action are detected. In this case, the number of non-improvement iterations is evaluated before starting the diversification strategy (line 8). The number of non-improvement iterations necessary to start the diversification process (*divActivation*) and the number of iterations that the process will remain active (*iterationsDiv*) are input parameters. Movements performed in this phase can be viewed as *influential movements* [7], since these movements try to modify the solution structure in a influential (non-random) manner. The function *computePenalty* (line 9) can use one of the penalty functions previously presented. In the following sections, the implementation that considers the penalty function which only takes into account the frequency ratio of transitions will be referred as TSDS, while the implementation that uses the penalty function which also takes into account the workload of teachers will be referred as TSDSTL. For comparison purposes, an implementation without the diversification strategy (TS), also will be considered in the next section.

## 5   Computational Experiments and Discussion

Experiments were done in the set of instances originated from [12], and the data referred to Brazilian high schools, with 25 lesson periods per week for each class, in different shifts. In Table 1 some of the characteristics of the instances

**Table 1.** Characteristics of problem instances.

| Instance | Teachers | Classes | Total Lessons | Double Lessons | Sparseness Ratio ($sr$) |
|:---:|:---:|:---:|:---:|:---:|:---:|
| 1 | 8 | 3 | 75 | 21 | 0.43 |
| 2 | 14 | 6 | 150 | 29 | 0.50 |
| 3 | 16 | 8 | 200 | 4 | 0.30 |
| 4 | 23 | 12 | 300 | 66 | 0.18 |
| 5 | 31 | 13 | 325 | 71 | 0.58 |
| 6 | 30 | 14 | 350 | 63 | 0.52 |
| 7 | 33 | 20 | 500 | 84 | 0.39 |

can be verified, such as dimension and sparseness ratio ($sr$), which can be computed considering the total number of lessons ($\#lessons$) and the total number of unavailable periods ($\#u$): $sr = \frac{t \times p - (\#lessons + \#u)}{t \times p}$. Lower sparseness values indicate more restrictive problems and likewise, more difficult resolution.

The algorithms were coded in C++. The implementation of GTS-II was the same presented in [12], and was implemented in C. The compiler used was GCC 3.2.3 using flag -O2. The experiments were performed in a micro-computer with an AMD Athlon XP 1533 MHz processor, 512 megabytes of RAM running the Linux operating system.

The weights in the objective function were defined as in [12]: $\omega = 100$, $\delta = 30$, $\rho = 1$, $\alpha_i = 3$, $\beta_i = 3$ and $\gamma_i = 1$, $\forall i = 1, \cdots, t$.

In the first set of experiments, the objective was to verify the average solution cost produced by each algorithm, within some time limits. The results (Table 2) consider the average best solution found in 20 independent executions, with the following time limits to instances $1, \cdots, 7$, respectively: $\{90, 280, 380, 870, 1930, 1650, 2650\}$. The parameters for GTS-II and the time limits are the same proposed in [12]. The parameters for TSDS and its variations are: $\alpha = 0.1$ (constructive algorithm), $minTT = 20$, $maxTT = 25$, $divActivation = 500$ and $iterationsDiv = 10$. Best results are shown in bold.

**Table 2.** Average results with fixed time limits.

| Instance | GTS-II | TSDSTL | TSDS | TS |
|:---:|:---:|:---:|:---:|:---:|
| 1 | 204.80 | 203.42 | **203.37** | 207.05 |
| 2 | 350.10 | **344.84** | 345.36 | 349.26 |
| 3 | 455.70 | 439.94 | **439.05** | 455.58 |
| 4 | 686.30 | **669.69** | 672.15 | 670.92 |
| 5 | 796.30 | 782.74 | **780.74** | 782.84 |
| 6 | 799.10 | 783.38 | **781.77** | 787.85 |
| 7 | 1,076.20 | 1,060.84 | **1,059.05** | 1,071.21 |

**Table 3.** Average costs of objective function components obtained by the constructive algorithm and at the end of the tabu search heuristic TSDS.

| Constructive Algorithm | | | | | TSDS | | | | |
|---|---|---|---|---|---|---|---|---|---|
| $f_1(Q^*)$ | $f_2(Q^*)$ | #d (%d) | #g (%g) | cr | $f_1(Q^*)$ | $f_2(Q^*)$ | #d (%d) | #g(%g) | cr |
| 0.0 | 0.5 | 15.1 (71.5) | 17.2 (22.9) | 1.6 | 0.0 | 0.0 | 2.0 (9.5) | 4.0 (5.3) | 1.2 |
| 0.0 | 0.0 | 24.3 (83.8) | 24.8 (16.5) | 1.3 | 0.0 | 0.0 | 8.2 (28.3) | 1.2 (0.8) | 1.0 |
| 0.3 | 2.5 | 2.0 (50.0) | 31.2 (15.6) | 1.4 | 0.0 | 0.0 | 0.4 (8.8) | 5.0 (2.5) | 1.1 |
| 4.3 | 0.9 | 35.5 (53.8) | 21.0 (7.0) | 1.2 | 0.0 | 0.0 | 19.9 (30.2) | 3.7 (1.2) | 1.0 |
| 0.0 | 0.2 | 54.1 (76.1) | 46.4 (14.3) | 1.5 | 0.0 | 0.0 | 13.6 (19.2) | 4.1 (1.2) | 1.1 |
| 0.2 | 0.0 | 53.7 (85.2) | 53.4 (15.3) | 1.4 | 0.0 | 0.0 | 13.5 (21.4) | 9.9 (2.8) | 1.0 |
| 0.5 | 0.2 | 69.6 (82.9) | 74.1 (14.8) | 1.3 | 0.0 | 0.0 | 24.4 (29.0) | 10.7 (2,1) | 1.0 |

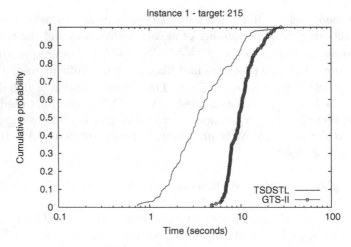

**Fig. 3.** Empirical probability distribution of finding target value in function of time for instance 1

**Table 4.** Time (in seconds) for 25%, 50% and 75% of runs achieve the target solution values.

| | GTS-II | | | TSDS | | |
|---|---|---|---|---|---|---|
| Instance | 25% | 50% | 75% | 25% | 50% | 75% |
| 1 | 7.64 | 9.57 | 12.15 | 2.13 | 3.36 | 6.39 |
| 2 | 21.39 | 26.57 | 34.68 | 9.03 | 13.48 | 19.71 |
| 3 | 28.57 | 46.84 | 85.41 | 16.29 | 27.66 | 46.47 |
| 4 | 49.22 | 92.57 | 146.50 | 2.65 | 3.40 | 5.45 |
| 5 | 47.79 | 62.85 | 102.20 | 27.63 | 37.85 | 54.51 |
| 6 | 35.81 | 48.00 | 72.12 | 25.20 | 33.97 | 44.38 |
| 7 | 92.41 | 150.72 | 287.48 | 89.57 | 118.82 | 155.72 |

As can be seen in Table 2, although only minor differences can be observed among the two implementations that use different penalty functions in the diversification strategy, results show that versions using the informed diversification

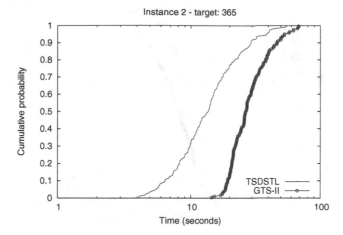

**Fig. 4.** Empirical probability distribution of finding target value in function of time for instance 2

strategy perform significantly better than GTS-II and TS. In order to evaluate the quality of the solutions obtained by the proposed method, and to verify how significant is the improvement of TSDS over the solution received from the constructive algorithm, Table 3 presents the average costs involved in each objective function component, for the solution provided by the constructive algorithm and for the improved solution from TSDS. Columns #d (%d), #g (%g) and cr are related to the $f_3$ component of the objective function, in the following way: #d (%d) indicates the unsatisfied double lessons (and the percentage of unsatisfied double lessons, considering the number of double lesson requests), #g (%g) indicates the number of "gaps" in the timetable of teachers (and the percentage considering the total number of lessons) and cr measures the compactness ratio of timetable of teachers. To compute cr, the summation of the actual number of days $ad$ that each teacher must attend to some lesson in the school and the lower bound for this value $\underline{ad}$ are used. The $\underline{ad}$ value considers the minimum number of days $md_i = \lceil \frac{\sum_{j=1}^{c} r_{ij}}{h} \rceil$ that each teacher $i$ must attend some lecture in the school, such that $\underline{ad} = \sum_{i=1}^{t} md_i$. This way, $cr = ad/\underline{ad}$. Values close to one indicate that the timetable is as compact as it can be. As can be seen in Table 3, the solution provided by the constructive algorithm usually contains some type of infeasibility. These problems were always solved by the TSDS algorithm, in a way that no infeasible timetable was produced. Regarding the preferences of teachers, the timetable compactness, which has the highest weight in the $f_3$ component of the objective function, it can be seen that in most cases the optimal value was reached ($cr = 1$). Also, small percentage values of "gaps" and unsatisfied double lessons were obtained.

In another set of experiments, the objective was to verify the empirical probability distribution of reaching a given sub-optimal target value (i.e. find a solution with cost at least as good as the target value) in function of time in different

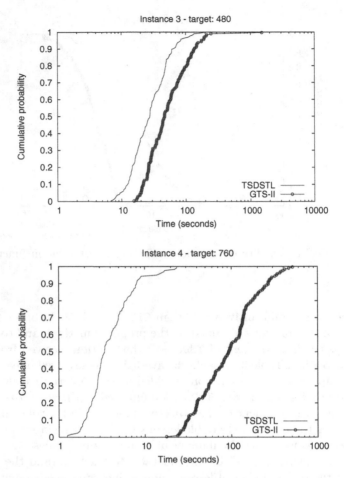

**Fig. 5.** Empirical probability distribution of finding target values in function of time for instances 3 and 4

instances. The sub-optimal values were chosen in a way that the slowest algorithm could terminate in a reasonable amount of time. In these experiments, TSDS and GTS-II were evaluated and the execution times of 150 independent runs for each instance were computed. The experiment design follows the proposal of [1]. The results of each algorithm were plotted associating with the $i$-th smallest running time $t_i$ a probability $p_i = (i - \frac{1}{2})/150$, which generates points $z_i = (t_i, p_i)$, for $i = 1, \cdots, 150$. As can be be seen in Figures 3 to 6 the TSDS heuristic achieves high probability values ($\geq 50\%$) of reaching the target values in significantly smaller times than GTS-II. This difference is enhanced mainly in instance 4, which presents a very low sparseness ratio. This result may be related to the fact that the "Intraclasses-Interclasses" procedure of GTS-II works with movements that use free periods, which are hard to find in this instance. Another analysis, taking into account all test instances, shows that at the time

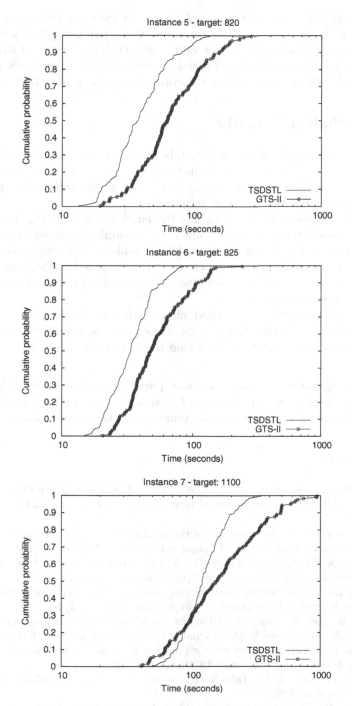

**Fig. 6.** Empirical probability distribution of finding target values in function of time for instances 5, 6 and 7

when 95% of TSDS runs have achieved the target value, in average, only 64% of GTS-II runs have achieved the target value. Considering the time when 50% of TSDS runs have achieved the target value, only 11%, in average, of GTS-II runs have achieved the target value. Table 4 presents the execution times needed by GTS-II and TSDS to achieve different probabilities of reaching the target values.

## 6   Concluding Remarks

This paper presented a new tabu search heuristic to solve the school timetabling problem. Experiments in real world instances showed that the proposed method outperforms significantly a previously developed hybrid tabu search algorithm, and it has the advantage of a simpler design.

Contributions of this paper include the empirical verification that although informed diversification strategies are not commonly employed in tabu search implementations for the school timetabling problem, its incorporation can significantly improve the method robustness. The proposed method not only produced better solutions for all test instances but also performed faster than a hybrid tabu search approach.

Although the proposed method offers quite an improvement, future researches may combine the "Intraclasses-Interclasses" procedure with an informed diversification strategy, which could lend to even better results .

**Acknowledgements.** This work was partially supported by CAPES and CNPq. The authors would like thank Olinto C. B. Araújo, from DENSIS-FEE-UNICAMP, Brazil for their valuable comments on the preparation of this paper.

## References

1. Aiex, R. M., Resende, M. G. C., Ribeiro, C. C.: Probability distribuition of solution time in GRASP: an experimental investigation, Journal of Heuristics, **8** (2002), 343–373
2. Abramson, D.: Constructing school timetables using simulated annealing: sequential and parallel algorithms. Management Science. **37** (1991) 98–113.
3. Colorni, A., Dorigo, M., Maniezzo, V.: Metaheuristics for High-School Timetabling. Computational Optimization and Applications. **9** (1998) 277–298.
4. Costa, D.: A Tabu Search algorithm for computing an operational timetable. European Journal of Operational Research Society. **76** (1994) 98–110.
5. Even, S., Itai, A., Shamir, A.: On the complexity of timetabling and multicommodity flow problems. SIAM Journal of Computation. **5** (1976) 691–703.
6. Glover, F.: Future paths for integer programming and artificial intelligence. Computers & Operations Research. **13** (1986) 533–549.
7. Glover, F., Laguna, M.: Tabu Search. Kluwer Academic Publishers, Boston Dordrecht London (1997)
8. Hansen, P.: The steepest ascent mildest descent heuristic for combinatorial programming. Congress on Numerical Methods in Combinatorial Optimization. Capri (1986)

9. Resende, M.G.C., Ribeiro. C.C.: Greedy randomized adaptive search procedures. Handbook of Metaheuristics. Kluwer. (2003) 219–249
10. Schaerf, A.: Tabu search techniques for large high-school timetabling problems. Report CS-R9611. Centrum voor Wiskunde en Informatica, Amsterdam (1996)
11. Souza, M.J.F.: Programação de Horários em Escolas: Uma Aproximação por Meta-heurísticas, D.Sc. Thesis (in Portuguese), Universidade Federal do Rio de Janeiro - Rio de Janeiro (2000)
12. Souza, M.J.F., Ochi, L.S., Maculan, N.: A GRASP-Tabu search algorithm for solving school timetabling problems. In: Resende, M.G.C., Souza, J.P. (eds.): Metaheuristics: Computer Decision-Making. Kluwer Academic Publishers, Boston (2003) 659–672
13. Wilke, P, Gröbner, M., Oster, N.: A hybrid genetic algorithm for school timetabling. In: AI 2002: McKay B. and Slaney J. (eds.): Advances in Artificial Intelligence. Springer Lecture Notes in Computer Science, Vol. 2557. Springer-Verlag, New York (2002) 455–464

# Experimental Studies of Symbolic Shortest-Path Algorithms

Daniel Sawitzki*

University of Dortmund, Computer Science 2, D-44221 Dortmund, Germany.
`daniel.sawitzki@cs.uni-dortmund.de`

**Abstract.** Graphs can be represented symbolically by the Ordered Binary Decision Diagram (OBDD) of their characteristic function. To solve problems in such implicitly given graphs, specialized symbolic algorithms are needed which are restricted to the use of functional operations offered by the OBDD data structure. In this paper, two symbolic algorithms for the single-source shortest-path problem with nonnegative positive integral edge weights are presented which represent symbolic versions of Dijkstra's algorithm and the Bellman-Ford algorithm. They execute $\mathcal{O}\big(N \cdot \log(NB)\big)$ resp. $\mathcal{O}\big(NM \cdot \log(NB)\big)$ OBDD-operations to obtain the shortest paths in a graph with $N$ nodes, $M$ edges, and maximum edge weight $B$. Despite the larger worst-case bound, the symbolic Bellman-Ford-approach is expected to behave much better on structured graphs because it is able to handle updates of node distances effectively in parallel. Hence, both algorithms have been studied in experiments on random, grid, and threshold graphs with different weight functions. These studies support the assumption that the Dijkstra-approach behaves efficient w. r. t. space usage, while the Bellman-Ford-approach is dominant w. r. t. runtime.

## 1 Introduction

Algorithms on graphs $G$ with node set $V$ and edge set $E \subseteq V^2$ typically work on adjacency lists of size $\Theta(|V| + |E|)$ or on adjacency matrices of size $\Theta(|V|^2)$. These representations are called *explicit*. However, there are application areas in which problems on graphs of such large size have to be solved that an explicit representation on today's computers is not possible. In the area of logic synthesis and verification, state-transition graphs with for example $10^{27}$ nodes and $10^{36}$ edges occur. Other applications produce graphs which are representable in explicit form, but for which even runtimes of efficient polynomial algorithms are not practicable anymore. Modeling of the WWW, street, or social networks are examples of this problem scenario.

However, we expect the large graphs occurring in application areas to contain regularities. If we consider graphs as Boolean functions, we can represent them

---

* Supported by the Deutsche Forschungsgemeinschaft (DFG) as part of the Research Cluster "Algorithms on Large and Complex Networks" (1126).

by *Ordered Binary Decision Diagrams* (*OBDDs*) [4,5,19]. This data structure is well established in verification and synthesis of sequential circuits [9,11,19] due to its good compression of regular structures. In order to represent a graph $G = (V, E)$ by an OBDD, its edge set $E$ is considered as a *characteristic Boolean function* $\chi_E$, which maps binary encodings of $E$'s elements to 1 and all others to 0. This representation is called *implicit* or *symbolic*, and is not essentially larger than explicit ones. Nevertheless, we hope that advantageous properties of $G$ lead to small, that is sublinear OBDD-sizes.

Having such an OBDD-representation of a graph, we are interested in solving problems on it without extracting too much explicit information from it. Algorithms that are mainly restricted to the use of functional operations are called *implicit* or *symbolic algorithms* [19]. They are considered as heuristics to save time and/or space when large structured input graphs do not fit into the internal memory anymore. Then, we hope that each OBDD-operation processes many edges in parallel. The runtime of such methods depends on the number of executed operations as well as on the efficiency of each single one. The latter in turn depends on the size of the operand OBDDs.

In general, we want heuristics to perform well on special input subsets, while their worst-case runtime is typically worse than for optimal algorithms. Symbolic algorithms often have proved to behave better than explicit methods on interesting graphs and are well established in the area of logic design and verification. Most papers on OBDD-based algorithms prove their usability just by experiments on benchmark inputs from a special application area [10,11,13]. In less application-oriented works considering more general graph problems, mostly the number of OBDD-operations is bounded as a hint on the real over-all runtime [3,8,7,14]. Only a few of them contain analyses of the over-all runtime and space usage for special cases like grids [15,16,20].

Until now, symbolic shortest-path algorithms only existed for graph representations by algebraic decision diagrams (ADDs) [1], which are difficult to analyze and are useful only for a small number of different weight values. A new OBDD-based approach to the all-pairs shortest-paths problem [18] aims at polylogarithmic over-all runtime on graphs with very special properties. In contrast to these results, the motivation of this paper's research was to transform popular methods for the single-source shortest-path problem into symbolic algorithms, and to compare their performance in experiments. This has been done for Dijkstra's algorithm as well as for the Bellman-Ford algorithm.

This paper is organized as follows: Section 2 introduces the principles of symbolic graph representation by OBDDs. Section 3 presents the symbolic shortest-path algorithms studied in this paper. Section 4 discusses experimental results of the algorithms on random, grid, and threshold graphs. Finally, Sect. 5 gives conclusions on the work.

## 2   Symbolic Graph Representation

We denote the class of Boolean functions $f\colon \{0,1\}^n \to \{0,1\}$ by $B_n$. The $i$th character of a binary number $x \in \{0,1\}^n$ is denoted by $x_i$ and $|x| := \sum_{i=0}^{n-1} x_i 2^i$ identifies its value.

Consider a directed graph $G = (V, E)$ with node set $V = \{v_0, \ldots, v_{2^n-1}\}$ and edge set $E \subseteq V^2$. $G$ can be represented by a *characteristic* Boolean function $\chi_E \in B_{2n}$ which maps pairs $(x,y) \in \{0,1\}^{2n}$ of binary node numbers of length $n$ to 1 iff $(v_{|x|}, v_{|y|}) \in E$. We can capture more complex graph properties by adding further arguments to characteristic functions. An additional weight function $c\colon E \to \{0, \ldots, 2^m - 1\}$ is modeled by $\chi_C \in B_{2n+m}$ which maps triples $(x,y,d)$ to 1 iff $(v_{|x|}, v_{|y|}) \in E$ and $c(v_{|x|}, v_{|y|}) = |d|$.

A Boolean function $f \in B_n$ defined on variables $x_0, \ldots, x_{n-1}$ can be represented by an OBDD $\mathcal{G}_f$ [4,5,19]. An OBDD is a directed acyclic graph consisting of *internal nodes* and *sink nodes*. Each internal node is labeled with a Boolean variable $x_i$, while each sink node is labeled with a Boolean constant. Each internal node is left by two edges one labeled by 0 and the other by 1. A *function pointer* $p$ marks a special node that represents $f$. Moreover, a permutation $\pi \in \Sigma_n$ called *variable order* must be respected by the internal nodes' labels on every path in the OBDD.

For a given variable assignment $a \in \{0,1\}^n$, we compute the corresponding function value $f(a)$ by traversing $\mathcal{G}_f$ from $p$ to a sink labeled with $f(a)$ while leaving a node $x_i$ via its $a_i$-edge. The size $\mathrm{size}(\mathcal{G}_f)$ of $\mathcal{G}_f$ is measured by the number of its nodes. An OBDD is called *complete*, if every path from $p$ to a sink has length $n$. This has not to be the case in general, because OBDDs may skip a variable test. We adopt the usual assumption that all OBDDs occurring in symbolic algorithms are minimal, since all OBDD-operations exclusively produce minimized diagrams, which are known to be canonical. There is an upper bound of $(2 + o(1))2^n/n$ for the OBDD-size of every $f \in B_n$; hence, a graph's edge set $E \subseteq V^2$ has an OBDD of worst-case size $\mathcal{O}(V^2/\log |V|)$.

**OBDD-operations.** The satisfiability of $f$ can be decided in time $\mathcal{O}(1)$. The negation $\overline{f}$ as well as the replacement of a function variable $x_i$ by a constant $c$ (i.e., $f_{|x_i=c}$) is computable in time $\mathcal{O}(\mathrm{size}(\mathcal{G}_f))$. Whether two functions $f$ and $g$ are equivalent (i.e., $f = g$) can be decided in time $\mathcal{O}(\mathrm{size}(\mathcal{G}_f) + \mathrm{size}(\mathcal{G}_g))$. These operations are called *cheap*. Further essential operations are the *binary synthesis* $f \otimes g$ for $f, g \in B_n$, $\otimes \in B_2$ (e.g., "$\wedge$" and "$\vee$") and the *quantification* $(\mathcal{Q}x_i)f$ for a quantifier $\mathcal{Q} \in \{\exists, \forall\}$. In general, the result $\mathcal{G}_{f \otimes g}$ has size $\mathcal{O}(\mathrm{size}(\mathcal{G}_f) \cdot \mathrm{size}(\mathcal{G}_g))$, which is also the general runtime of this operation. The computation of $\mathcal{G}_{(\mathcal{Q}x_i)f}$ can be realized by two cheap operations and one binary synthesis in time and space $\mathcal{O}(\mathrm{size}^2(\mathcal{G}_f))$.

**Notation.** The characteristic functions used for symbolic representation are typically defined on several subsets of Boolean variables, each representing a different argument. For example, a weighted graph's function $\chi_C$ is defined on two binary node numbers $x = x_{n-1} \ldots x_0$ and $y = y_{n-1} \ldots y_0$ and a binary weight value $d = d_{m-1} \ldots d_0$. We assume w. l. o. g. that all arguments consist of the same number of $n$ variables. Moreover, both a function $\chi_S \in B_{kn}$ defined on $k$ arguments $x^{(1)}, \ldots, x^{(k)} \in \{0,1\}^n$ as well as its OBDD-representation $\mathcal{G}_{\chi_S}$ will be denoted by $S(x^{(1)}, \ldots, x^{(k)})$ in this paper. We use an *interleaved* variable order $\pi = (x_0^{(1)}, x_0^{(2)}, \ldots, x_0^{(k)}, x_1^{(1)}, \ldots, x_{n-1}^{(k)})$, which enables to *swap* [19] arguments in time $\mathcal{O}(\text{size}(\mathcal{G}_{\chi_S}))$ (e. g., $F(x,y) := G(y,x)$).

The symbolic algorithms will be described in terms of functional assignments like "$F(x) := G(x) \wedge H(x)$." The quantification $(\exists y_{n-1} \ldots \exists y_0) \, F(x,y)$ over the $n$ bits of an argument $y$ will be denoted by $(\exists y) \, F(x,y)$. Although this seems to be one OBDD-operation, this corresponds to $\mathcal{O}(n)$ quantification operations. Identifiers with braced superscripts mark additional arguments of characteristic functions occurring only temporarily in quantified formulas (e. g., $d^{(1)}$). Furthermore, the functional assignments will contain tool functions for comparisons of weighted sums like $F(x,y,z) := (|x| + |y| = |z|)$. These can be composed from *multivariate threshold functions*.

**Definition 1.** *Let $f \in B_{kn}$ be defined on variables $x^{(1)}, \ldots, x^{(k)} \in \{0,1\}^n$. Moreover, let $\mathcal{W}, T \in \mathbb{Z}$, and $w_1, \ldots, w_k \in \{-\mathcal{W}, \ldots, \mathcal{W}\}$. $f$ is called $k$-variate threshold function iff it is*

$$f(x^{(1)}, \ldots, x^{(k)}) = \left( \sum_{i=1}^k w_i \cdot |x^{(i)}| \geq T \right).$$

$\mathcal{W}$ *is called the* maximum absolute weight *of $f$.*

Besides the greater or equal comparison, the relations $>$, $\leq$, $<$, and $=$ can be realized by binary syntheses of multivariate threshold functions, too. For a constant number $k$ of arguments and a constant maximum absolute weight $\mathcal{W}$, such a comparison function $f \in B_{kn}$ has a compact OBDD of size $\mathcal{O}(n)$ [20].

## 3   Symbolic Shortest-Path Algorithms

In this section, symbolic versions of two popular shortest-path algorithms are presented: Dijkstra's algorithm [6] and the Bellman-Ford algorithm [2]. We assume that the reader is familiar with these two methods, and describe their symbolic versions in separate sections. Both solve the single-source shortest-path problem in symbolically represented directed graphs $G = (V, E, s, c)$ with node set $V = \{v_0, \ldots, v_{N-1}\}$, edge set $E \subseteq V^2$ of cardinality $M$, source node $s \in V$, edge weight function $c: E \to \mathbb{N}_0$, and $B := \max\{c(e) \mid e \in E\}$. The maximum path length from $s$ to any node $v$ is $B(N-1) =: L$. Let $n := \lceil \log(L+1) \rceil = \Theta(\log N + \log B)$ be the number of bits necessary to encode one node number or distance value. The algorithms receive the input graph in form of two OBDDs for the characteristic functions $C(x,y,d)$ and $s(x)$ with

$$C(x,y,d) = 1 \Leftrightarrow \left[ (v_{|x|}, v_{|y|}) \in E \right] \wedge \left[ c(v_{|x|}, v_{|y|}) = |d| \right],$$
$$s(x) = 1 \Leftrightarrow v_{|x|} = s.$$

The output is the distance function dist: $V \to \mathbb{N}_0 \cup \{\infty\}$ which maps a node $v \in V$ to the length of a shortest path from $s$ to $v$, given as an OBDD $DIST(x,d)$ with

$$DIST(x,d) = 1 \Leftrightarrow \text{dist}(v_{|x|}) = |d| \ .$$

Both algorithms maintain a temporary distance function $\Delta: E \to \mathbb{N}_0 \cup \{\infty\}$ represented by an OBDD $D(x,d)$, which is updated until it equals dist.

## 3.1   The Dijkstra-Approach

Dijkstra's algorithm [6] stores a node set $A \subseteq V$ of nodes for which the shortest-path length is already known. At the beginning, it is $A = \{s\}$, $\Delta(s) = 0$, and $\Delta(v) = \infty$ for all nodes $v \neq s$. In each iteration, we add one node to $A$. Let $u$ be the last node added to $A$. For each edge $(u,v)$ it is checked whether $\Delta(u) + c(u,v) < \Delta(v)$. If this is the case, we update $\Delta(v)$ to $\Delta(u) + c(u,v)$. After this relaxation step, we add a node $v^{\min} \in V \setminus A$ to $A$ whose value $\Delta(v^{\min})$ is minimal. If $\{v \in V \setminus A \mid \Delta(v) \neq \infty\} = \emptyset$, the actual distances $\Delta$ correspond to dist and we terminate. If the nodes are stored in a priority heap with access time $\mathcal{O}(\log N)$, this explicit algorithm needs time $\mathcal{O}((N + M) \cdot \log N)$.

This approach is now transformed into a symbolic algorithm that works with corresponding OBDDs $A(x)$ and $D(x,d)$ for the characteristic functions of $A$ and $\Delta$. At the beginning, they are initialized to the source node:

$$A(x) := s(x) \ ,$$
$$D(x,d) := s(x) \wedge (|d| = 0) \ .$$

$x^{\min}$ and $d^{\min}$ are bit strings representing the node $v^{\min} = v_{|x^{\min}|}$ lastly added to $A$ with $\Delta(v^{\min}) = |d^{\min}|$. Initially, $x^{\min}$ represents $s$ and it is $v_{|x^{\min}|} = s$ and $|d^{\min}| = 0$.

Now all edges leaving $v^{\min}$ have to be relaxed. We introduce three helping functions which will be used to update $D(x,d)$: Function $RELAX(x,d)$ represents pairs $(x,d)$ such that $(v^{\min}, v_{|x|}) \in E$ and $\Delta(v^{\min}) + c(v^{\min}, v_{|x|}) = |d|$. $D_1(x,d)$ and $D_2(x,d)$ represent the two possibilities for nodes $v_{|x|} \notin A$: 1. It is $D_1(x,d) = 1$ iff distance $|d|$ is the relaxed distance of $v_{|x|}$ not being larger than the actual distance $\Delta(v_{|x|})$. 2. It is $D_2(x,d) = 1$ iff distance $|d|$ is the actual distance $\Delta(v_{|x|})$ not being larger than the relaxed distance of $v_{|x|}$. Case 1 represents the update of $\Delta(v_{|x|})$, while Case 2 represents its retention. Finally, the new $D(x,d)$ equals the actual $D(x,d)$ for nodes $v_{|x|} \in A$, while for nodes $v_{|x|} \notin A$ Case 1 ($D_1(x,d)$) or Case 2 ($D_2(x,d)$) applies. This leads to the following symbolic formulation:

$$RELAX(x,d) := (\exists d^{(1)}) \left[ C(x^{\min}, x, d^{(1)}) \wedge (|d| = |d^{\min}| + |d^{(1)}|) \right] \ ,$$
$$D_1(x,d) := RELAX(x,d) \wedge \overline{(\exists d^{(1)}) \left[ D(x, d^{(1)}) \wedge (|d^{(1)}| < |d|) \right]} \ ,$$
$$D_2(x,d) := D(x,d) \wedge \overline{(\exists d^{(1)}) \left[ RELAX(x, d^{(1)}) \wedge (|d^{(1)}| < |d|) \right]} \ ,$$
$$D(x,d) := \left[ A(x) \wedge D(x,d) \right] \vee \left[ \overline{A(x)} \wedge \left[ D_1(x,d) \vee D_2(x,d) \right] \right] \ .$$

At next, we select the new minimal node $v^{\min}$. If there are several nodes with minimal value $\Delta(v)$, we select the node $v_{|x|}$ with the smallest node number $|x|$. Hence, we need a comparison function for two node–distance-pairs denoted by $LESS(x, d, y, d')$.

$$LESS(x, d, y, d') := (|d| < |d'|) \vee [(d = d') \wedge (|x| < |y|)]$$

The facts on multivariate threshold functions in Sect. 2 imply that comparisons like $LESS(x, d, y, d')$ and $(|d| = |d^{\min}| + |d^{(1)}|)$ have OBDD-size $\mathcal{O}(n)$. Now we define the selection function $SEL(x, d)$.

$$SEL(x, d) := \overline{A(x)} \wedge D(x, d)$$

$$\wedge (\exists x^{(1)}, d^{(1)}) \left[ \overline{A(x^{(1)})} \wedge D(x^{(1)}, d^{(1)}) \wedge LESS(x^{(1)}, d^{(1)}, x, d) \right]$$

The interpretation of this functional assignment is that the node–distance-pair $(v_{|x|}, |d|)$ is selected iff $v_{|x|} \notin A$, $\Delta(v_{|x|}) = |d|$, and there is no other node-distance pair $(v_{|x^{(1)}|}, |d^{(1)}|)$ with these properties and $|d^{(1)}| < |d|$ or $(d = d') \wedge (|x| < |y|)$. If $SEL(x, d) \equiv 0$, all nodes reachable from $s$ have been added to $A$ and we may terminate with output $DIST(x, d) = D(x, d)$. Otherwise, $SEL(x, d)$ contains exactly one satisfying assignment for $x$ and $d$. This can be extracted in linear time w. r. t. size($SEL$) [19]. Finally, we just need to add $x^{\min}$ to $A(x)$.

$$(x^{\min}, d^{\min}) := SEL^{-1}(1) \ ,$$
$$A(x) := A(x) \vee (x = x^{\min})$$

Afterwards, we jump to the relaxation step. The correctness of this symbolic procedure follows from the correctness of Dijkstra's algorithm, while we now consider the number of executed OBDD-operations.

**Theorem 1.** *The symbolic Dijkstra-approach computes the output OBDD $DIST(x, d)$ by $\mathcal{O}(N \cdot \log(NB))$ OBDD-operations.*

*Proof.* All nodes reachable from $s$ are added to $A(x)$. That is, at most $N$ relaxation and selection iterations are executed. In each iteration, the algorithm performs a constant number of cheap operations, argument swaps, binary syntheses, and quantifications over node or distance arguments. Each of the latter corresponds to $\mathcal{O}(n) = \mathcal{O}(\log(NB))$ quantifications over single Boolean variables. Altogether, $\mathcal{O}(N \cdot \log(NB))$ OBDD-operations are executed. $\square$

We have also studied a parallelized symbolic version of Dijkstra's algorithm, which selects not only one distance-minimal node to be handled in each iteration, but a maximal set of independent nodes not interfering by adjacency. Experiments showed that the parallelization could compensate the overhead caused by the more complex symbolic formulation only for graphs of very special structure, why this approach is not discussed in this work.

## 3.2   The Bellman-Ford-Approach

In contrast to Dijkstra's algorithm, the Bellman-Ford algorithm [2] does not select special edges to relax, but performs $N$ iterations over all edges $(u, v) \in E$ to check the condition $\Delta(u) + c(u, v) < \Delta(v)$ and to update $\Delta(v)$ eventually. Therefore, its explicit runtime is $\mathcal{O}(NM)$. In contrast to Dijkstra's algorithm, Bellman-Ford is able to handle graphs with negative edge weights if they do not contain negative cycles. Furthermore, it is easy to parallelize, which motivated the development of a symbolic version: Few OBDD-operations hopefully perform many edge relaxations at once.

Again, the actual distance function $D(x, d)$ is only known for the source $s$ at the beginning:

$$D(x, d) := s(x) \wedge (|d| = 0) \ .$$

We again need a function $RELAX(x, y, d)$ representing the candidates for edge relaxation. Let $RELAX(x, y, d) = 1$ iff $\Delta(v_{|x|}) + c(v_{|x|}, v_{|y|}) = |d|$ and $|d|$ is not larger than the actual $\Delta(v_{|y|})$.

$$RELAX(x, y, d) := (\exists d^{(1)}, d^{(2)}) \left[ D(x, d^{(1)}) \wedge C(x, y, d^{(2)}) \wedge (|d| = |d^{(1)}| + |d^{(2)}|) \right]$$
$$\wedge \overline{(\exists d^{(1)}) \left[ D(y, d^{(1)}) \wedge (|d^{(1)}| \le |d|) \right]}$$

If $RELAX(x, y, d) \equiv 0$, no relaxations are applicable and $D(x, d) = DIST(x, d)$ represents the correct output—we may terminate. Otherwise, we use the comparison function $LESS(x, d, y, d')$ to choose the subset that is minimal w. r. t. distance $|d|$ and, secondly, the node number $|x|$:

$$SEL(x, y, d) := RELAX(x, y, d)$$
$$\wedge \overline{(\exists x^{(1)}, d^{(1)}) \left[ RELAX(x^{(1)}, y, d^{(1)}) \wedge LESS(x^{(1)}, d^{(1)}, x, d) \right]} \ .$$

Finally, we compute the symbolic set $U(x, d)$ of node-distance pairs that have to be updated in $D(x, d)$ because they were part of a selected relaxation:

$$U(x, d) := (\exists x^{(1)}) \, SEL(x^{(1)}, x, d) \ ,$$
$$D(x, d) := U(x, d) \vee \left[ \overline{U(x, d)} \wedge D(x, d) \right] \ .$$

In this way, the new distances of $U(x, y)$ are taken over into $D(x, d)$, while the other nodes keep their distance value. The new iteration starts with computing $RELAX(x, y, d)$. Again, the correctness follows from the correctness of the explicit Bellman-Ford algorithm.

**Theorem 2.** *The symbolic Bellman-Ford-approach computes the output OBDD* $DIST(x, d)$ *by* $\mathcal{O}(NM \cdot \log(NB))$ *OBDD-operations.*

*Proof.* Every implementation of the Bellman-Ford-algorithm performs at most $\mathcal{O}(NM)$ edge relaxations. In each iteration, the symbolic method relaxes at least one edge, and executes a constant number of cheap operations, argument swaps, binary syntheses, and quantifications over node and distance arguments. Each of the latter corresponds to $\mathcal{O}(n) = \mathcal{O}(\log(NB))$ quantifications over single Boolean variables. Altogether, $\mathcal{O}(NM \cdot \log(NB))$ OBDD-operations are executed. □

## 3.3   Computing the Predecessor Nodes

Besides dist, explicit shortest-path algorithms return for each node $v \in V$ a predecessor node $\operatorname{pred}(v) =: u$, such that there is a shortest path from $s$ to $v$ which uses the edge $(u, v)$. Analogue, the symbolic approaches can be modified such that they also compute the predecessor nodes on shortest paths.

The following method computes these just from the final $DIST(x, d)$ and is independent of the considered symbolic algorithm. It uses the helping function $P(x, y, d^{(1)}, d^{(2)}, d^{(3)})$ which is satisfied iff $\operatorname{dist}(v_{|x|}) = |d^{(1)}|$, $c(v_{|x|}, v_{|y|}) = |d^{(2)}|$, $\operatorname{dist}(v_{|y|}) = |d^{(3)}|$, and $(|d^{(1)}| + |d^{(2)}| = |d^{(3)}|)$. By existential quantification over the distances $d^{(1)}$, $d^{(2)}$, and $d^{(3)}$, we obtain the function $PREDS(x, y)$, which represents exactly all edges being part of some shortest path (i. e., for which $v_{|x|}$ is a predecessor of $v_{|y|}$).

$$
P(x, y, d^{(1)}, d^{(2)}, d^{(3)}) := DIST(x, d^{(1)}) \wedge C(x, y, d^{(2)})
$$
$$
\wedge \, DIST(y, d^{(3)}) \wedge (|d^{(1)}| + |d^{(2)}| = |d^{(3)}|) \ ,
$$
$$
PREDS(x, y) := (\exists d^{(1)}, d^{(2)}, d^{(3)}) \, P(x, y, d^{(1)}, d^{(2)}, d^{(3)})
$$

If we are only interested in an arbitrary predecessor of a concrete node $v_{|y^*|}$, we may omit the computation of $PREDS(x, y)$ by replacing argument $y$ of $P$ by $y^*$ and extracting an arbitrary satisfying variable assignment $x^*$ of $P$. Therefore, computing $DIST(x, d)$ is the essential part of symbolic shortest-path algorithms, which has been analyzed by means of the experiments documented in Sect. 4.

*Remark 1.* The worst-case behavior of a particular OBDD-operation executed by a symbolic algorithm can be obtained from the general bound $(2 + o(1))2^n/n$ for the OBDD-size of any function $f \in B_n$ together with the worst-case bounds for runtime and space in Sect. 2.

Analogue to Theorem 2 in [18], it can be shown that constant width bounds of input OBDD $C(x, y, d)$ and output OBDD $DIST(x, d)$ imply a polylogarithmic upper bound on time and space for each operation. However, we did not want to restrict ourselves to such special cases and applied the Dijkstra-approach as well as the Bellman-Ford-approach in experiments to obtain more general results.

# 4   Experimental Results

Although the symbolic Bellman-Ford-approach has a higher worst-case bound for the number of OBDD-operations than the Dijkstra-approach, we hope that each of its iterations relaxes many edges in parallel leading to a sublinear operation number. On the other hand, representing symbolic sets like $RELAX(x, y, d)$ may involve many little structured information causing larger OBDDs than Dijkstra. That is, we expect the Bellman-Ford method to need more space, while hoping that the smaller operation count results in less over-all runtime.

In order to check these hypotheses, the symbolic shortest-path algorithms have been applied in experiments on random, grid, and threshold graphs. Because the OBDD-size size$(\mathcal{G})$ of a symbolic algorithm's input graph $G$ is a natural lower bound for its resource usage, we investigate experimental behaviors also w. r. t. these input sizes. This allows to measure performances independently of how well an input $G$ is suited for OBDD-representation.

**Experiment setting.** Both symbolic algorithms have been implemented[1] in C++ using the OBDD package CUDD 2.3.1 by Fabio Somenzi[2]. An interleaved variable order with increasing bit significance has been used for the Boolean variables of each function argument. The experiments took place on a PC with Pentium 4 2GHz processor and 512 MB of main memory. The runtime has been measured by seconds of process time, while the space usage is given as the maximum number of OBDD-nodes present at any time during an algorithm execution. The latter is of same magnitude as the over-all space usage and independent of the used computer system.

## 4.1   Random Graphs

Random graphs possess no regular structure and, therefore, are pathological cases for symbolic representations—they have expected OBDD-size $\Theta(N^2/\log N)$. Just for dense graphs some compression is achieved because, intuitively spoken, the OBDD stores the smaller number of missing edges instead of all existing ones. However, we cannot hope symbolic methods to beat explicit algorithms on random graphs. But even in such worst cases their runtime and space usage may be only linear w. r. t. the (correspondingly large) input OBDD-sizes, which is the best we can expect from symbolic methods in general.

Both presented symbolic shortest-path algorithms have been tested on random graphs with 100, 200, 300, and 400 nodes and edge probabilities from 0.05 to 1 in steps of 0.05 influencing the observed edge density. Node 0 served as source. Moreover, three edge weight functions of different regularity have been considered. The documented experimental results are the averages of results of 10 independent experiments for each parameter setting. The particular results merely deviated from their averages.

**Constant edge weights.** At first, the constant edge weight function $c(e) = 1$ has been considered. This structural assumption causes a slightly sublinear growth of the symbolic representation's OBDD-size w. r. t. the edge probability (see Fig. 1(a)). Figures 1(b) to 1(e) show the observed runtimes and space usage of both algorithms w. r. t. the edge probabilities, where "ParBF" identifies the symbolic (parallelized) Bellman-Ford-approach.

As expected, the Dijkstra method uses less space, while Bellman-Ford has lower runtimes. Figures 1(f) and 2(a) integrate all runtimes resp. space usage into one plot w. r. t. the input graphs' OBDD-sizes, which constitutes the Dijkstra–space resp. Bellman-Ford–time connection: The space usage of the Dijkstra-approach grows linearly with the input graph's OBDD-size with same offset and

---

[1] Implementation and experiments available at http://thefigaro.sourceforge.net/.
[2] CUDD is available at http://vlsi.colorado.edu/.

gradient for all considered numbers of nodes $N$, while this is not the case for the Bellman-Ford-approach. For the runtime, the situation is vice versa: Only the Bellman-Ford approach shows unique linear runtime-growth.

**Difference edge weights.** In order to proceed to a less simple weight function than constant weights, *difference edge weights* have been considered:

$$c(v_a, v_b) := |a - b| \bmod 200 \ .$$

The modulo-operation was used to bound the gap between maximum weights for the different numbers of nodes $N = 100$ to $400$.

This weight function can be composed of multivariate threshold functions, and has OBDD-size $\mathcal{O}(\log N)$. Accordingly, the OBDD-sizes of random graphs with difference weights are not essentially larger than for constant weights (see Fig. 2(b)). Again, Dijkstra dominates w.r.t. space usage and Bellman-Ford dominates w.r.t. runtime, while the general resource usage is higher than for constant weights (see Tables 2(c) and 2(d)). The dependence of time and space on the input OBDD-sizes is given by Figs. 2(f) and 3(a): While Dijkstra's space still grows linearly with the same offset and gradient for all node numbers, the Bellman-Ford's runtime behavior now differs for different $N$.

**Random edge weights.** Finally, random graphs with random edge weights between 1 and 200 have been considered in experiments. Figure 2(e) shows that their OBDD-sizes grow linearly with the edge density, because the random weights prohibit the space savings observed for the two other weight functions. The runtimes w.r.t. edge probabilities and numbers of nodes $N$ are given by Tables 2(c) and 2(d), while the dependence of time and space on the input OBDD-sizes is given by Figs. 2(f) and 3(a). The general resource usage further increased in comparison to difference weights. The missing structure of the inputs leads to nearly the same runtime for Dijkstra and Bellman-Ford—the latter is not able to compensate the larger space requirements by less operations anymore. In contrast, the advantage of the Dijkstra-approach still remains: Its space usage grows linearly with the same offset and gradient for all considered edge probabilities $p$ and numbers of nodes $N$.

### 4.2   Grid and Threshold Graphs

In contrast to random graphs, the grid and threshold graphs considered in this section are examples of structured inputs with logarithmic OBDD-size $\mathcal{O}(\log N)$ [16,20], whose OBDDs can be constructed efficiently. Hence, we hope a useful symbolic algorithm to use only polylogarithmic resources in these cases.

**Grid graphs.** Both algorithms have been applied to $2^{n/2} \times 2^{n/2}$-grid graphs, which are quadratic node matrices of $2^n$ nodes $(i, j)$, $i, j \in \{0, \ldots, 2^{n/2} - 1\}$, with vertical edges $((i, j), (i + 1, j))$ and horizontal edges $((i, j), (i, j + 1))$. Grids of

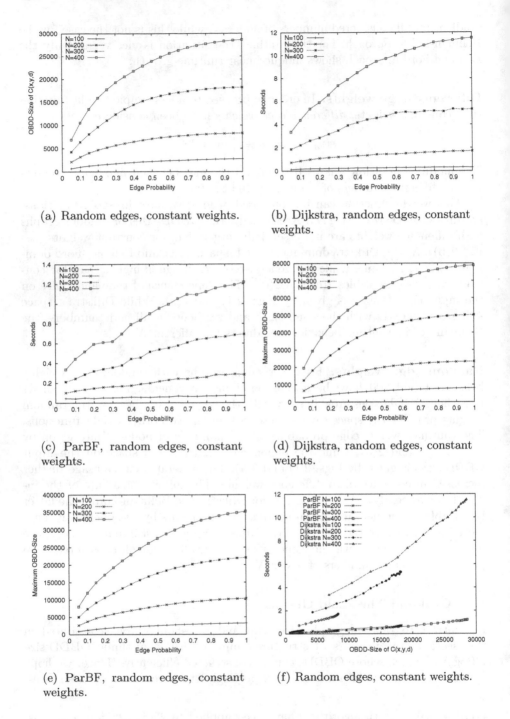

(a) Random edges, constant weights.

(b) Dijkstra, random edges, constant weights.

(c) ParBF, random edges, constant weights.

(d) Dijkstra, random edges, constant weights.

(e) ParBF, random edges, constant weights.

(f) Random edges, constant weights.

**Fig. 1.** Experimental results on random graphs.

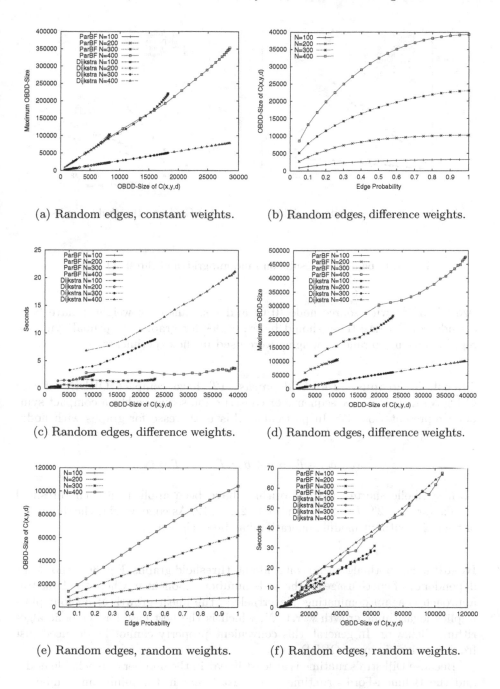

(a) Random edges, constant weights.

(b) Random edges, difference weights.

(c) Random edges, difference weights.

(d) Random edges, difference weights.

(e) Random edges, random weights.

(f) Random edges, random weights.

**Fig. 2.** Experimental results on random graphs.

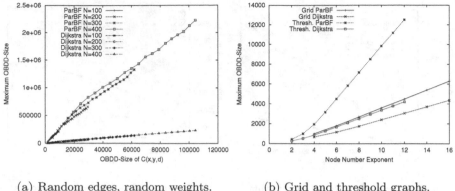

(a) Random edges, random weights.     (b) Grid and threshold graphs.

**Fig. 3.** Experimental results on random, grid, and threshold graphs.

size $2^2$ to $2^{16}$ with source node $(0,0)$ and constant edge weight 1 have been considered. Because these should be examples for graphs of optimal symbolic representation, no random weights were used in the experiments.

**Threshold graphs.** *Threshold graphs* [12] have compact OBDDs of size $\mathcal{O}(\log N)$ if their degree sequence or construction sequence has a compact symbolic representation [17]. In particular, this is the case for graphs with nodes $v_0, \ldots, v_{N-1}$ and edges

$$(v_a, v_b) \in E \Leftrightarrow a + b \geq T \qquad , T \in \mathbb{N} \ .$$

Both symbolic shortest-path algorithms have been applied on such threshold graphs for $N = 2^n$, $T = 2^{n-1}$, and $n := 2, \ldots, 12$. As edge weight, the difference of Sect. 4.1 without modulo-operation has been chosen.

**Results.** For both algorithms on grid and threshold graphs, Fig. 3(b) shows the dependency of space usage on the node number exponent $n$, where the Dijkstra-approach is again dominating. Nevertheless, the linear growth of *all* four plots implies logarithmic growth w.r.t. $N$, which is the best case for symbolic algorithms' behavior. In general, this convenient property cannot be deduced just from logarithmic input OBDD-size.

Because Dijkstra's runtime is at least linear in the number of reachable nodes and the Bellman-Ford's runtime is at least linear in the minimum number of edges on any $s$–$v$-path, no polylogarithmic runtime can be obtained on grid graphs. Despite this theoretical fact, Bellman-Ford performed very efficiently on grids, while Dijkstra's runtime got very inefficient for the exponentially growing grid sizes (see Table 3(a)). Moreover, in experiments on grids with a number

**Table 1.** Experimental runtime results on random graphs.

| $p/N$ | 100 | 200 | 300 | 400 |
|---|---|---|---|---|
| 0.1 | 0.332 | 1.49 | 3.828 | 7.715 |
| 0.2 | 0.34 | 1.611 | 4.423 | 10.405 |
| 0.3 | 0.361 | 1.891 | 5.362 | 12.934 |
| 0.4 | 0.371 | 2.07 | 6.232 | 15.192 |
| 0.5 | 0.394 | 2.121 | 6.869 | 16.843 |
| 0.6 | 0.401 | 2.256 | 7.508 | 18.468 |
| 0.7 | 0.393 | 2.342 | 8.191 | 19.621 |
| 0.8 | 0.406 | 2.409 | 8.529 | 20.347 |
| 0.9 | 0.415 | 2.479 | 8.724 | 20.794 |
| 1 | 0.415 | 2.497 | 8.912 | 21.156 |

(a) Dijkstra, random edges, difference weights.

| $p/N$ | 100 | 200 | 300 | 400 |
|---|---|---|---|---|
| 0.1 | 0.148 | 0.693 | 1.691 | 3.032 |
| 0.2 | 0.148 | 0.535 | 1.433 | 2.961 |
| 0.3 | 0.165 | 0.504 | 1.313 | 2.601 |
| 0.4 | 0.154 | 0.538 | 1.441 | 2.593 |
| 0.5 | 0.155 | 0.504 | 1.472 | 2.624 |
| 0.6 | 0.13 | 0.483 | 1.33 | 3.267 |
| 0.7 | 0.135 | 0.509 | 1.484 | 3.541 |
| 0.8 | 0.133 | 0.506 | 1.541 | 3.245 |
| 0.9 | 0.135 | 0.553 | 1.525 | 3.773 |
| 1 | 0.134 | 0.508 | 1.75 | 3.693 |

(b) ParBF, random edges, difference weights.

| $p/N$ | 100 | 200 | 300 | 400 |
|---|---|---|---|---|
| 0.1 | 0.654 | 3.048 | 7.669 | 14.76 |
| 0.2 | 0.783 | 3.932 | 9.849 | 20.64 |
| 0.3 | 0.959 | 4.535 | 12.281 | 26.619 |
| 0.4 | 1.054 | 5.268 | 14.664 | 32.572 |
| 0.5 | 1.184 | 6.169 | 16.796 | 37.954 |
| 0.6 | 1.244 | 6.919 | 19.521 | 42.984 |
| 0.7 | 1.433 | 7.567 | 21.875 | 49.621 |
| 0.8 | 1.457 | 7.977 | 24.446 | 55.312 |
| 0.9 | 1.615 | 8.454 | 26.745 | 61.136 |
| 1 | 1.674 | 9.415 | 28.467 | 66.74 |

(c) Dijkstra, random edges, random weights.

| $p/N$ | 100 | 200 | 300 | 400 |
|---|---|---|---|---|
| 0.1 | 0.429 | 1.949 | 6.554 | 12.928 |
| 0.2 | 0.618 | 3.882 | 11 | 20.59 |
| 0.3 | 1.076 | 5.622 | 14.14 | 23.841 |
| 0.4 | 1.204 | 6.662 | 14.553 | 28.471 |
| 0.5 | 1.354 | 8.192 | 16.832 | 35.826 |
| 0.6 | 1.603 | 9.093 | 20.15 | 43.281 |
| 0.7 | 1.928 | 10.561 | 22.796 | 46.972 |
| 0.8 | 1.976 | 10.457 | 23.986 | 55.561 |
| 0.9 | 2.237 | 11.819 | 24.637 | 56.82 |
| 1 | 2.541 | 12.502 | 30.942 | 67.631 |

(d) ParBF, random edges, random weights.

of $20 \log N$ randomly added edges, both algorithms' runtime did not change essentially in comparison to unmodified grids.

Table 3(b) shows the observed runtimes on the considered threshold graphs. Due to the very small runtimes of the Bellman-Ford-approach, we cannot deduce any assumptions about its behavior besides that its again performing much more efficient than the Dijkstra-approach. Both on grids and threshold graphs with $n \geq 8$, it was even able to beat an explicit shortest-path algorithm implemented in LEDA[3] version 4.3.

---

[3] Available at http://www.algorithmic-solutions.com/.

**Table 2.** Experimental runtime results on grid and threshold graphs.

| $n$ | ParBF | Dijkstra |
|---|---|---|
| 4 | 0.00 | 0.01 |
| 6 | 0.01 | 0.03 |
| 8 | 0.04 | 0.20 |
| 10 | 0.12 | 0.97 |
| 12 | 0.28 | 5.29 |
| 14 | 0.71 | 26.40 |
| 16 | 1.76 | 127.02 |

(a)     Grid graphs.

| $n$ | ParBF | Dijkstra |
|---|---|---|
| 2 | 0.00 | 0.00 |
| 3 | 0.00 | 0.00 |
| 4 | 0.00 | 0.01 |
| 5 | 0.00 | 0.02 |
| 6 | 0.01 | 0.07 |
| 7 | 0.01 | 0.18 |
| 8 | 0.02 | 0.42 |
| 9 | 0.02 | 1.01 |
| 10 | 0.02 | 2.38 |
| 11 | 0.02 | 5.47 |
| 12 | 0.03 | 12.25 |

(b)   Threshold graphs.

# 5   Conclusions

Two symbolic algorithms for the single-source shortest-path problem on OBDD-represented graphs with nonnegative integral edge weights have been presented which execute $\mathcal{O}(N \cdot \log(NB))$ resp. $\mathcal{O}(NM \cdot \log(NB))$ OBDD-operations. Although Bellman-Ford's worst-case bound is the larger one, this symbolically parallelized approach was expected to have better runtime but higher space usage than the Dijkstra-approach. This was confirmed by experiments on random graphs with constant and difference weights as well as on grid and threshold graphs. Dijkstra's space usage was always of linear magnitude w. r. t. the size of its input OBDDs with a relative error of less than 0.06. For the Bellman-Ford-approach, this property was only observed on grid and threshold graphs as well as for the runtime on random graphs with constant edge weights.

Altogether, experiments both on pathological instances (random graphs) and structured graphs well-suited for symbolic representation (grid and threshold graphs) show that for each of the resources time resp. space at least one algorithm performs well or even asymptotically optimal w. r. t. the input OBDD-size. Hence, both shortest-path algorithms can be considered as useful symbolic methods with individual strengths.

**Acknowledgment.** Thanks to Ingo Wegener for proofreading and helpful discussions.

# References

[1] R.I. Bahar, E.A. Frohm, C.M. Gaona, G.D. Hachtel, E. Macii, A. Pardo, and F. Somenzi. Algebraic decision diagrams and their applications. In *ICCAD'93*, pages 188–191. IEEE Press, 1993.

[2] R. Bellman. On a routing problem. *Quarterly of Applied Mathematics*, 16:87–90, 1958.

[3] R. Bloem, H.N. Gabow, and F. Somenzi. An algorithm for strongly connected component analysis in $n \log n$ symbolic steps. In *FMCAS'00*, volume 1954 of *LNCS*, pages 37–54. Springer, 2000.

[4] R.E. Bryant. Symbolic manipulation of Boolean functions using a graphical representation. In *DAC'85*, pages 688–694. ACM Press, 1985.

[5] R.E. Bryant. Graph-based algorithms for Boolean function manipulation. *IEEE Transactions on Computers*, 35:677–691, 1986.

[6] E.W. Dijkstra. A note on two problems in connexion with graphs. *Numerische Mathematik*, 1:269–271, 1959.

[7] R. Gentilini, C. Piazza, and A. Policriti. Computing strongly connected components in a linear number of symbolic steps. In *SODA'03*, pages 573–582. ACM Press, 2003.

[8] R. Gentilini and A. Policriti. Biconnectivity on symbolically represented graphs: A linear solution. In *ISAAC'03*, volume 2906 of *LNCS*, pages 554–564. Springer, 2003.

[9] G.D. Hachtel and F. Somenzi. *Logic Synthesis and Verification Algorithms*. Kluwer Academic Publishers, Boston, 1996.

[10] G.D. Hachtel and F. Somenzi. A symbolic algorithm for maximum flow in 0–1 networks. *Formal Methods in System Design*, 10:207–219, 1997.

[11] R. Hojati, H. Touati, R.P. Kurshan, and R.K. Brayton. Efficient $\omega$-regular language containment. In *CAV'93*, volume 663 of *LNCS*, pages 396–409. Springer, 1993.

[12] N.V.R. Mahadev and U.N. Peled. *Threshold Graphs and Related Topics*. Elsevier Science, Amsterdam, 1995.

[13] I. Moon, J.H. Kukula, K. Ravi, and F. Somenzi. To split or to conjoin: The question in image computation. In *DAC'00*, pages 23–28. ACM Press, 2000.

[14] K. Ravi, R. Bloem, and F. Somenzi. A comparative study of symbolic algorithms for the computation of fair cycles. In *FMCAD'00*, volume 1954 of *LNCS*, pages 143–160. Springer, 2000.

[15] D. Sawitzki. Implicit flow maximization by iterative squaring. In *SOFSEM'04*, volume 2932 of *LNCS*, pages 301–313. Springer, 2004.

[16] D. Sawitzki. Implicit flow maximization on grid networks. Technical report, Universität Dortmund, 2004.

[17] D. Sawitzki. On graphs with characteristic bounded-width functions. Technical report, Universität Dortmund, 2004.

[18] D. Sawitzki. A symbolic approach to the all-pairs shortest-paths problem. Submitted, 2004.

[19] I. Wegener. *Branching Programs and Binary Decision Diagrams*. SIAM, Philadelphia, 2000.

[20] P. Woelfel. Symbolic topological sorting with OBDDs. In *MFCS'03*, volume 2747 of *LNCS*, pages 671–680. Springer, 2003.

# Experimental Comparison of Greedy Randomized Adaptive Search Procedures for the Maximum Diversity Problem

Geiza C. Silva, Luiz S. Ochi, and Simone L. Martins

Universidade Federal Fluminense, Departamento de Ciência da Computação, Rua Passo da Pátria 156, Bloco E, Niterói, RJ 24210-240, Brazil.
{gsilva,satoru,simone}@ic.uff.br

**Abstract.** The maximum diversity problem (MDP) consists of identifying optimally diverse subsets of elements from some larger collection. The selection of elements is based on the diversity of their characteristics, calculated by a function applied on their attributes. This problem belongs to the class of NP-hard problems. This paper presents new GRASP heuristics for this problem, using different construction and local search procedures. Computational experiments and performance comparisons between GRASP heuristics from literature and the proposed heuristics are provided and the results are analyzed. The tests show that the new GRASP heuristics are quite robust and find good solutions to this problem.

## 1 Introduction

The maximum diversity problem (MDP) [5,6,7] consists of identifying optimally diverse subsets of elements from some larger collection. The selection of elements is based on the diversity of their characteristics, calculated by a function applied on their attributes. The goal is to find the subset that presents the maximum possible diversity. There are many applications [10] that can be solved using the resolution of this problem, such as medical treatment, selecting jury panel, scheduling final exams, and VLSI design. This problem belongs to the class of NP-hard problems [6].

Glover et al. [6] presented mixed integer zero-one formulation for this problem, that can be solved for small instances by exact methods. Bhadury et al. [3] developed an exact algorithm using a network flow approach for the diversity problem of working groups for a graduate course.

Some heuristics are available to obtain approximate solutions. Weitz and Lakshminarayanan [12] developed five heuristics to find groups of students with the most possible diverse characteristics, such as nationality, age and graduation level. They tested the heuristics using instances based on real data and implemented an exact algorithm for solving them and the heuristic LCW (Lofti-Cerveny-Weitz method) was considered the best for solving these instances.

Constructive and destructive heuristics were presented by Glover et al. [7], who created instances with different size of population (maximum value was 30)

C.C. Ribeiro and S.L. Martins (Eds.): WEA 2004, LNCS 3059, pp. 498–512, 2004.
© Springer-Verlag Berlin Heidelberg 2004

and showed that the proposed heuristics obtained results close (2 %) to the ones obtained by the exact algorithm, but much faster.

Kochenberger and Glover [10] showed results obtained using a tabu search and Katayama and Naribisa [9] developed a memetic algorithm. Both report that computational experiments were carried out, but they did not compare the performance of their algorithms with exact or other heuristics procedures.

Ghosh [5] proposed a GRASP (Greedy Randomized Adaptive Search Procedure) that obtained good results for small instances of the problem. Andrade et al. [2] developed a new GRASP and showed results for instances randomly created with a maximum population of 250 individuals. This algorithm was able to find some solutions better than the ones found by the Ghosh algorithm.

GRASP [4] is an iterative process, where each iteration consists of two phases: construction and local search. In the construction phase a feasible solution is built, and its neighborhood is explored by a local search. The result is the best solution found over all iterations. In Section 2 we describe three construction procedures developed using the concept of reactive GRASP introduced by Prais and Ribeiro [11], and two local search strategies. In Section 3 we show computational results for different versions of GRASP heuristics created by the combination of a constructive algorithm and a local search strategy described in Section 2. Concluding remarks are presented in Section 4.

## 2   GRASP Heuristics

The construction phase of GRASP is an iterative process where, at each iteration, the elements $c \in C$ that do not belong to the solution are evaluated by a greedy function $g : C \to \Re_+$, that estimates the gain of including it in the partial solution. They are ordered by their estimated value in a list called restricted candidate list (RCL) and one of them is randomly chosen and included in the solution. The size of the RCL is limited by a parameter $\alpha$. For a maximization problem, only the elements whose $g$ values are in the range $[(1 - \alpha)g_{max}, g_{max}]$ are placed in RCL. This process stops when a feasible solution is obtained.

Prais and Ribeiro [11] proposed a new procedure called Reactive GRASP, for which the parameter $\alpha$ used in the construction phase is self adjusted for each iteration. For the first construction iteration, an $\alpha$ value is randomly selected from a discrete set $A = \{\alpha_1, \ldots, \alpha_m\}$. Each element $\alpha_i$ has a probability $p_i$ associated and, initially, a uniform distribution is applied, thus we have $p_i = 1/m, i = 1, \ldots, m$. Periodically the probability distribution $p_i, i = 1, \ldots, m$ is updated using information collected during the former iterations. The aim is to associate higher probabilities to values of $\alpha$ that lead to better solutions and lower ones to values of $\alpha$ that guide to worse solutions.

The solutions generated by the construction phase are not guaranteed to be locally optimal. Usually a local search is performed to attempt to improve each constructed solution. It works by successively replacing the current solution by a better one from its neighborhood, until no more better solutions are found. Normally, this phase demands great computational effort and execution time, so the construction phase plays an important role to diminish this effort by supplying good starting solutions for the local search. We implemented a technique widely used to accomplish this task, that leads to a more greedy construction. For each GRASP iteration, the construction algorithm is executed $X$ times generating $X$ solutions and only the best solution is selected to be used as the initial solution for the local search phase.

In the next subsections, we describe the construction and local search algorithms developed for the GRASP heuristics, using the concepts discussed in this section.

## 2.1   Construction Phase

Let $E = \{e_i : i \in N\}, N = \{1, 2, \ldots, n\}$ be a population of $n$ elements and $e_{il}$, $l \in L = \{1, 2, ..., l\}$ the $l$ values of the attributes of each element. In this paper, we measure the diversity between any two elements $i$ and $j$ by the Euclidean distance calculated as $d_{ij} = \sqrt{\sum_{k=1}^{l}(e_{ik} - e_{jk})^2}$. Let $M$ be a subset of $N$ and the overall diversity be $z(M) = \sum_{i<j:i,j\in M} d_{ij}$. The MDP problem consists of maximizing the cost function $z(M)$, subject to $|M| = m$.

We describe three construction algorithms developed to be used in GRASP heuristics where all of them use the techniques described before: Reactive GRASP and filtering of constructed solutions.

**K larger distances heuristic (KLD).** This algorithm constructs an initial solution by randomly selecting an element from a RCL of size K at each construction iteration. The RCL is created by selecting for each element $i \in N$, the $K$ elements $j \in N\setminus\{i\}$, that exhibit larger values of $d_{ij}$ and sum these $K$ values of $d_{ij}$, obtaining $s_i$. Then, we create a list of all elements $i$ sorted in descending order by their $s_i$ values and select the $K$ first elements to compose the RCL list.

The procedure developed to implement the reactive GRASP starts considering $m\_it$ to be the total number of GRASP iterations. In the first block of iterations $B_1 = 0.4m\_it$, we evaluate four different values for $K \in \{K_1, K_2, K_3, K_4\}$ and the evaluation is done by dividing the block into four equal intervals $c_i, i = 1, \ldots, 4$. We use the value $K_i$ for all iterations belonging to interval $c_j, i = j$. The values of $K_i$ are shown in Tab. 1, where $\mu = (n - m)/2$. After the execution of the last iteration of block $B_1$, we evaluate the quality of the solutions obtained for each $K_i$. We calculate the mean diversity value $zm_i = \sum_{1 \leq q \leq 0.1m\_it} z(sol_{iq})$

**Table 1.** $K$ values for block $B_1$

| $i$ | $c_i$ | $K$ |
|---|---|---|
| 1 | $[1, \ldots, 0.1m\_it]$ | $m + \mu - 0.2\mu$ |
| 2 | $(0.1m\_it, \ldots, 0.2m\_it]$ | $m + \mu - 0.1\mu$ |
| 3 | $(0.2m\_it, \ldots, 0.3m\_it]$ | $m + \mu + 0.1\mu$ |
| 4 | $(0.3m\_it, \ldots, 0.4m\_it]$ | $m + \mu + 0.2\mu$ |

**Table 2.** $K$ values for block $B_2$

| $i$ | $y_i$ | $K$ |
|---|---|---|
| 1 | $[0.4m\_it, \ldots, 0.64m\_it]$ | $lk_1$ |
| 2 | $(0.64m\_it, \ldots, 0.82m\_it]$ | $lk_2$ |
| 3 | $(0.82m\_it, \ldots, 0.94m\_it]$ | $lk_3$ |
| 4 | $(0.94m\_it, \ldots, m\_it]$ | $lk_4$ |

for the solutions $sol_{iq}, i = 1, \ldots 4; q = 1, \ldots, 0.1m\_it$ obtained using each $K_i$. The values $K_i$ are stored in a list $LK$ ordered by their $zm_i$ values.

Then for the next block of iterations $B_2 = 0.6m\_it$, we divide it into four intervals $y_i$, each one with different number of iterations, and use the $K_i$ values as shown in Tab. 2. In this way, the values $K_i$ that provide better solutions are used in a larger number of iterations.

At each GRASP iteration, we apply the filter technique for this heuristic by constructing 400 solutions and only the best solution is sent to the local search procedure.

The pseudo-code, including the description of the procedure for the construction phase using K larger Distances heuristic, is given in Fig. 1.

```
procedure constr_KLD(it_GRASP, m_it, numsol, n, m)
 1.    best_cost_sol ← 0;
 2.    K ← det_K(it_GRASP, m_it, LK, i);
 3.    num_sol [i] ← num_sol [i] + 1;
 4.    RCL ← Build_RCL(K);
 5.    for j = 1, . . . , max_sol_filter do
 6.       sol ← {};
 7.       for k = 1, . . . , m do
 8.          Randomly select an individual e* from RCL;
 9.          sol ← sol ∪ {e*};
10.          RCL ← RCL − {e*};
11.       end for;
12.       if (z(sol) > best_cost_sol) then do
13.          sol_constr ← sol;
14.          best_cost_sol ← z(sol);
15.       end if
16.    end for;
17.    sol_eval [i, num_sol[i]] ← z(sol_constr);
18.    if (it_GRASP == 0.4m_it) then do
19.       LK ← Build_LK(sol_eval);
20.    end if;
21.    return sol_constr.
```

**Fig. 1.** Construction procedure used to implement the KLD heuristic

In line 1, we initialize the cost of the best solution found in the execution of $max\_sol\_filter$ iterations. The value $K$ to be used to build the Restricted Candidate List (RCL) is calculated by the procedure $det\_K$ in line 2. This procedure defines the value for $K$ implementing the reactive GRASP described before. In line 3, the number of solutions found for a specific $K$ is updated and, in line 4, the RCL is built. From line 5 to line 16, the construction procedure is executed $max\_sol\_filter$ times and only the best solution is returned to be used as an initial solution by the local search procedure. From line 7 to line 11, a solution is constructed by the random selection of an element from RCL. In lines 12 to 15, we update the best solution found by the construction procedure and the cost of the solution found using the selected $K$ is stored in line 17. When the first block $B_1$ of iterations ends, the values $K_i$ are evaluated and put in the list $LK$ sorted in descending order, in line 19.

**K larger distances heuristic-v2 (KLD-v2).** This algorithm is similar to the previously described algorithm, the difference between them is the way that the Restricted Candidate List is built. In the former algorithm, the RCL is computed before the execution of the construction iterations and, for each iteration, the only modification made in the RCL is the removal of the element that is inserted in the solution.

In this algorithm, the RCL is built using an adaptive procedure, where the process to select the first element of the constructed solution is the same as of the KLD heuristic, which means that an element is randomly selected from the RCL built as described in line 4 of Fig. 1.

Let $M_c$ be a partial solution with $c, 1 \leq c < m$ elements and $i \in N \backslash M_c$ a candidate to be inserted in the next partial solution $M_{c+1}$. For each $i$, we select the $(K - c - 1)$ elements $j \in N \backslash (M_c \bigcup \{i\})$,that present larger values of $d_{ij}$ and calculate the sum of the $(K - c - 1)$ values of $d_{ij}$ obtaining $s_i$. To select the next element to be inserted, an initial candidate list is created based on the greedy function $gf(i)$ shown in (1), where the first term corresponds to the sum of distances from the candidate $i$ to the elements $j \in M_c$, and the second term stands for the sum of distances from element $i$ to the $(K - c - 1)$ elements that are not in the solution $M_c$ and present larger distances to $i$. The initial candidate list is formed by the elements $i$, sorted in descending order with respect to $gf(i)$, and the first $K$ elements are selected from this list to build the RCL.

$$gf(i) = \sum_{j \in M_c} d_{ij} + s_i \qquad (1)$$

The Reactive GRASP and the construction filter are implemented in the same way as in KLD. Once this construction algorithm demands much more execution time than KLD algorithm, only 2 solutions, instead of 400, are generated to be filtered.

**Most distant insertion heuristic (MDI).** Let $M_c$ be a partial solution with $c, (1 \leq c < m)$ elements, the partial solution $M_1$ is obtained by randomly selecting an element from all elements $i \in N$.

The second element $m_2$ is the element $j$, which presents the larger distance $d_{ij}, i \in M_1, j \in N \backslash M_1$. To obtain $M_c(c \geq 3)$ from $M_{c-1}$, the element to be inserted in the solution is randomly selected from a RCL. The RCL is built based on the function $dsum(j)$ showed in (2), where the first term of this function corresponds to the sum of distances between all elements $i \in M_{c-1}$. The second term is the sum between all elements $i \in M_{c-1}$ to a candidate $j$ that is not in the partial solution $M_{c-1}$.

$$dsum(j) = \sum_{1 \leq y \leq c-2} \sum_{y+1 \leq w \leq c-1} d_{yw} + \sum_{1 \leq v \leq c-1} d_{vj} \qquad (2)$$

An initial candidate list (ICL) is created containing the elements $j \in N \backslash M_{c-1}$, sorted in descending order by their $dsum(j)$ values. The first $\alpha \times n$ elements of ICL are selected to form the RCL.

For this algorithm, the reactive GRASP is implemented in the same way done for the K larger distances heuristic. The first block $B_1 = 0.4m\_it$ is divided into four intervals of the same size and four values for $\alpha \in \{\alpha_1, \alpha_2, \alpha_3, \alpha_4\}$ are evaluated. Table 3 shows the values of $\alpha$ used for each interval. The values $\alpha_i, i = 1, \ldots, 4$ are evaluated by calculating the mean diversity value $zm_i = \sum_{1 \leq q \leq 0.1m\_it} z(sol_{iq})$ for the solutions $sol_{iq}, i = 1, \ldots 4; q = 1, \ldots, 0.1m\_it$ obtained using each $\alpha_i$. The values $\alpha_i$ are stored in a list $L\alpha$ ordered by their $zm_i$ values.

**Table 3.** $\alpha$ values for block $B_1$

| $i$ | $c_i$ | $\alpha$ |
|---|---|---|
| 1 | $[1, \ldots, 0.1m\_it]$ | 0.03 |
| 2 | $(0.1m\_it, \ldots, 0.2m\_it]$ | 0.05 |
| 3 | $(0.2m\_it, \ldots, 0.3m\_it]$ | 0.07 |
| 4 | $(0.3m\_it, \ldots, 0.4m\_it]$ | 0.1 |

The next block of iterations $B_2 = 0.6m\_it$ is also divided into four intervals $y_i$, each one with distinct number of iterations and, for each one, a value of $\alpha$ is associated, as shown in Tab. 4.

We have also implemented the same procedure described above for filtering the constructed solutions. In this case, the number of solutions generated is $n$, so it depends on the population size of each instance.

Figure 2 shows the construction phase procedure using the MDI heuristic. In line 1, we initialize the value of the best solution found. The value $\alpha$ to be used to build the RCL is calculated by the procedure $det\_\alpha$ in line 2. This procedure selects $\alpha$ based on the reactive GRASP discussed before. In line 3, the

**Table 4.** $\alpha$ values for block $B_2$

| $i$ | $y_i$ | $\alpha$ |
|---|---|---|
| 1 | $[0.4m\_it, \ldots, 0.64m\_it]$ | $l\alpha_1$ |
| 2 | $(0.64m\_it, \ldots, 0.82m\_it]$ | $l\alpha_2$ |
| 3 | $(0.82m\_it, \ldots, 0.94m\_it]$ | $l\alpha_3$ |
| 4 | $(0.94m\_it, \ldots, m\_it]$ | $l\alpha_4$ |

number of solutions found for a specific $\alpha$ is updated and in line 4, the set that contains the candidates to be inserted in the solution is initialized to contain all elements belonging to $N$. From line 5 to line 24, the construction procedure is executed $max\_sol\_filter$ times and only the best solution is returned to be used as an initial solution by the local search procedure. In line 7, the first

---

```
procedure constr_MDI(it_GRASP, m_it, numsol, n, m)
 1.    best_cost_sol ← 0;
 2.    α ← det_α(it_GRASP, m_it, Lα, i);
 3.    num_sol[i] ← num_sol[i] + 1;
 4.    N_RCL ← N;
 5.    for j = 1,..., max_sol_filter do
 6.      sol ← {};
 7.      Randomly select an individual m₁ from N; sol ← sol ∪ {m1};
 8.      for all j ∈ N\M₁ do
 9.        Compute d_{m1j}
10.        m₂ ← l, |d_{m1l} = max(d_{m1j}), j ∈ N\M₁;
11.        sol ← sol ∪ {m₂};
12.      end for all;
13.      N_RCL ← N − M₂;
14.      for k = 3,..., m do
15.        RCL ← Build_RCL_α(N_RCL, α);
16.        Randomly select an individual e* from RCL;
17.        sol ← sol ∪ {e*};
18.        N_RCL ← N_RCL − {e*};
19.      end for;
20.      if (z(sol) > best_cost_sol) then do
21.        sol_constr ← sol;
22.        best_cost_sol ← z(sol);
23.      end if
24.    end for;
25.    sol_eval[i, num_sol[i]] ← z(sol_constr);
26.    if (it_GRASP == 0.4m_it) then do
27.      Lα ← Build_Lα(sol_eval);
28.    end if;
29.    return sol_constr;
```

**Fig. 2.** Construction procedure used to implement the MDI heuristic

element is selected and from line 8 to line 12, we determine the second element of the solution. From line 14 to line 19, the insertion of the other elements is performed. For each iteration, in line 15, a RCL is built and, in line 16, an element is randomly selected from it. In line 18, we update the candidates to be inserted in the next iteration. In lines 20 to 23, we update the best solution found by the construction procedure. The cost of the best solution found using the selected $\alpha$ is stored in line 25. When the first block $B_1$ of iterations finishes, the values $\alpha_i$ are evaluated and put in the list $L\alpha$ in line 27.

## 2.2   Local Search Phase

After a solution is constructed, a local search phase should be executed to attempt to improve the initial solution. In this paper, we use two different local search algorithms. The first one was developed by Ghosh [5] and the second one by us using the Variable Neighborhood Search (VNS) [8] heuristic.

**Ghosh Algorithm (GhA).** The neighborhood of a solution defined by Ghosh [5] is the set of all solutions obtained by replacing an element in the solution by other that does not belong to the set associated with the solution. The incumbent solution $M$ is initialized with the solution obtained by the construction phase. For each $i \in M$ and $j \in N\backslash M$, the improvement due to exchanging $i$ by $j$, $\Delta z(i,j) = \sum_{u \in M\backslash\{i\}}(d_{ju} - d_{iu})$ is computed. If for all $i$ and $j$, $\Delta z(i,j) < 0$, the local search is terminated, as no exchange will lead to a better solution. Otherwise, the elements of the pair $(i,j)$ that provides the maximum $\Delta z(i,j)$ are interchanged creating a new incumbent solution $M$ and the local search is performed again.

**SOM Algorithm (SOMA).** We have also implemented a local search using a VNS heuristic. In this case, we use the GhA algorithm until there is no more improvement in the solution. After that, we execute a local search based on a new neighborhood, which is defined as the set of all solutions obtained by replacing two elements in the solution by another two that are not in the solution. The incumbent solution $M$ is initialized with the solution obtained by the first phase of the local search. For each $(i,j) \in M$ and $(v,w) \in N\backslash M$, the improvement due to exchanging $(i,j)$ by $(v,w)$, $\Delta z((i,j),(v,w)) = \sum_{u \in M\backslash\{i,j\}}(d_{vu} + d_{wu} - d_{iu} - d_{ju})$ is computed. If for all pairs $(i,j)$ and $(v,w)$, $\Delta z((i,j),(v,w)) < 0$, as no exchange will improve the solution, the local search is terminated. Otherwise, the pairs $(i,j)$ and $(v,w)$ that provides the maximum $\Delta z((i,j),(v,w))$ are interchanged, a new incumbent solution $M$ is created and the local search is performed again.

We developed several GRASP heuristics combining the construction procedures with the local search strategies described above and the computational experiments implemented to evaluate the performance of these heuristics are presented in next section.

## 3    Computational Results

We tested nine GRASP procedures that are shown in Tab. 5.

The first GRASP procedure G1 is an implementation of the GRASP heuristic developed by Ghosh and the second one is a procedure that implements Ghosh construction heuristic but uses the new local search SOMA. G9 is the GRASP heuristic implemented by Andrade et al. [2]. Except for G9, which code was kindly provided to us by the authors, all other algorithms were implemented by us.

The algorithms were implemented in C++, compiled with g++ compiler version 3.2.2 and were tested on a PC AMD Athlon 1.3GHz with 256 Mbytes of RAM. Twenty instances for the problem were created with populations of sizes $n = 100$, $n = 200$, $n = 300$, $n = 400$ and $n = 500$, and subsets of sizes $m = 10\%n$, $m = 20\%n$, m=30%$n$ and $m = 40\%n$. The diversities in the set $\{d_{ij}; i < j; i, j \in N\}$ for each set of instances that have the same population size were randomly selected from a uniform distribution over $[0 \ldots 9]$.

In Tab. 6, we show the results of computing 500 iterations for each GRASP heuristic. The first and second columns identify two parameters of each instance: the size of the population and the number $m$ of elements to be selected. Each procedure was executed three times and for each one we show the average value of the solution cost and the best value found.

We can see that the proposed GRASP heuristics found better solutions than GRASP algorithms found in literature [2,5]. Algorithm G7, which implements the KLD-v2 for construction phase and GhA for local search, was the one that found better solutions for larger number of instances.

Table 7 reports the CPU times observed for the execution of the same instances. The first and second columns identify the two parameters of each instance. For each GRASP heuristic, the average time for three executions and the time obtained when the best solution was found are reported. Among the proposed heuristics, algorithm G5 is the most efficient related to execution time. Heuristic G7, for which we have the best quality solutions, demands more time

**Table 5.** GRASP procedures

| GRASP procedure | Construction heuristic | Local search heuristic |
|:---:|:---|:---|
| G1 | Ghosh | GhA |
| G2 | Ghosh | SOMA |
| G3 | MDI | GhA |
| G4 | MDI | SOMA |
| G5 | KLD | GhA |
| G6 | KLD | SOMA |
| G7 | KLD-v2 | GhA |
| G8 | KLD-v2 | SOMA |
| G9 | Andrade | Andrade |

**Table 6.** Solutions for GRASP heuristics

| n | m | G1 average | G1 best | G2 average | G2 best | G3 average | G3 best | G4 average | G4 best | G5 average | G5 best | G6 average | G6 best | G7 average | G7 best | G8 average | G8 best | G9 average | G9 best |
|---|---|---|---|---|---|---|---|---|---|---|---|---|---|---|---|---|---|---|---|
| 100 | 10 | 318,0 | 318,0 | 318,0 | 318,0 | 333,0 | 333,0 | 333,0 | 333,0 | 333,0 | 333,0 | 333,0 | 333,0 | 333,0 | 333,0 | 333,0 | 333,0 | 333,0 | 333,0 |
| 100 | 20 | 1178,0 | 1178,0 | 1178,0 | 1178,0 | 1195,0 | 1195,0 | 1195,0 | 1195,0 | 1195,0 | 1195,0 | 1195,0 | 1195,0 | 1195,0 | 1195,0 | 1195,0 | 1195,0 | 1191,7 | 1195,0 |
| 100 | 30 | 2457,0 | 2457,0 | 2457,0 | 2457,0 | 2457,0 | 2457,0 | 2457,0 | 2457,0 | 2457,0 | 2457,0 | 2457,0 | 2457,0 | 2457,0 | 2457,0 | 2457,0 | 2457,0 | 2454,7 | 2457,0 |
| 100 | 40 | 4142,0 | 4142,0 | 4142,0 | 4142,0 | 4142,0 | 4142,0 | 4142,0 | 4142,0 | 4142,0 | 4142,0 | 4142,0 | 4142,0 | 4142,0 | 4142,0 | 4142,0 | 4142,0 | 4140,3 | 4142,0 |
| 200 | 20 | 1233,0 | 1233,0 | 1233,0 | 1233,0 | 1245,7 | 1247,0 | 1247,0 | 1247,0 | 1247,0 | 1247,0 | 1247,0 | 1247,0 | 1247,0 | 1247,0 | 1247,0 | 1247,0 | 1243,0 | 1245,0 |
| 200 | 40 | 4443,0 | 4443,0 | 4442,3 | 4443,0 | 4446,7 | 4448,0 | 4446,0 | 4448,0 | 4446,3 | 4448,0 | 4449,3 | 4450,0 | 4447,0 | 4448,0 | 4448,0 | 4448,0 | 4443,3 | 4445,0 |
| 200 | 60 | 9437,0 | 9437,0 | 9437,0 | 9437,0 | 9434,3 | 9437,0 | 9437,0 | 9437,0 | 9437,0 | 9437,0 | 9437,0 | 9437,0 | 9437,0 | 9437,0 | 9437,0 | 9437,0 | 9424,0 | 9425,0 |
| 200 | 80 | 16225,0 | 16225,0 | 16225,0 | 16225,0 | 16224,7 | 16225,0 | 16225,0 | 16225,0 | 16182,3 | 16207,0 | 16170,0 | 16171,0 | 16225,0 | 16225,0 | 16225,0 | 16225,0 | 16200,0 | 16211,0 |
| 300 | 30 | 2666,0 | 2666,0 | 2666,0 | 2666,0 | 2684,0 | 2694,0 | 2679,0 | 2686,0 | 2686,0 | 2691,0 | 2683,0 | 2684,0 | 2688,5 | 2691,0 | 2692,0 | 2694,0 | 2681,3 | 2691,0 |
| 300 | 60 | 9652,0 | 9677,0 | 9644,3 | 9654,0 | 9678,0 | 9681,0 | 9678,3 | 9681,0 | 9667,7 | 9689,0 | 9658,0 | 9688,0 | 9682,5 | 9684,0 | 9676,0 | 9679,0 | 9658,0 | 9676,0 |
| 300 | 90 | 20725,0 | 20725,0 | 20725,0 | 20725,0 | 20721,3 | 20728,0 | 20722,3 | 20733,0 | 20640,0 | 20640,0 | 20640,0 | 20640,0 | 20726,0 | 20727,0 | 20727,0 | 20734,0 | 20673,7 | 20685,0 |
| 300 | 120 | 35880,0 | 35881,0 | 35880,3 | 35881,0 | 35880,0 | 35881,3 | 35877,0 | 35881,0 | 35871,0 | 35871,0 | 35871,0 | 35871,0 | 35878,0 | 35881,0 | 35881,0 | 35881,0 | 35853,5 | 35855,0 |
| 400 | 40 | 4615,0 | 4626,0 | 4611,0 | 4626,0 | 4648,0 | 4648,0 | 4654,7 | 4658,0 | 4635,3 | 4653,0 | 4634,7 | 4654,0 | 4647,0 | 4649,0 | 4654,0 | 4655,0 | 4629,7 | 4635,0 |
| 400 | 80 | 16900,5 | 16903,0 | 16905,3 | 16918,0 | 16946,7 | 16956,0 | 16944,7 | 16948,0 | 16916,3 | 16925,0 | 16902,0 | 16902,0 | 16930,0 | 16937,0 | 16938,0 | 16940,0 | 16873,3 | 16895,0 |
| 400 | 120 | 36298,7 | 36304,0 | 36301,0 | 36301,0 | 36301,0 | 36315,0 | 36301,7 | 36306,0 | 36175,0 | 36175,0 | 36175,0 | 36175,0 | 36272,5 | 36283,0 | 36290,0 | 36301,0 | 36187,7 | 36191,0 |
| 400 | 160 | 62442,3 | 62445,0 | 62432,0 | 62433,0 | 62450,0 | 62470,0 | 62439,3 | 62457,0 | 62313,0 | 62313,0 | 62313,0 | 62313,0 | 62478,0 | 62483,0 | 62451,5 | 62454,0 | 62345,0 | 62359,0 |
| 500 | 50 | 7079,3 | 7082,0 | 7079,0 | 7079,0 | 7110,7 | 7131,0 | 7105,3 | 7130,0 | 7124,0 | 7130,0 | 7116,0 | 7127,0 | 7112,5 | 7115,0 | 7107,0 | 7116,0 | 7082,7 | 7085,0 |
| 500 | 100 | 26219,0 | 26220,0 | 26219,0 | 26219,0 | 26213,0 | 26224,0 | 26220,0 | 26222,0 | 26197,0 | 26201,0 | 26221,0 | 26254,0 | 26229,0 | 26237,0 | 26229,5 | 26236,0 | 26129,0 | 26144,0 |
| 500 | 150 | 56571,0 | 56572,0 | 56571,5 | 56572,0 | 56549,0 | 56563,0 | 56554,7 | 56571,0 | 57055,7 | 58605,0 | 56571,7 | 56572,0 | 56572,0 | 56572,0 | 56571,0 | 56572,0 | 56360,0 | 56365,0 |
| 500 | 200 | 97255,0 | 97274,0 | 97322,3 | 97326,0 | 97316,3 | 97327,0 | 97325,5 | 97327,0 | 97213,0 | 97213,0 | 97333,0 | 97344,0 | 97316,5 | 97319,0 | 97310,0 | 97320,0 | 97166,7 | 97200,0 |

**Table 7.** CPU time for GRASP heuristics (seconds)

| n | m | G1 | | G2 | | G3 | | G4 | | G5 | | G6 | | G7 | | G8 | | G9 | |
|---|---|---|---|---|---|---|---|---|---|---|---|---|---|---|---|---|---|---|---|
| | | average | best | average | best | average | best | average | best | average | best | average | best | average | best | average | best | average | best |
| 100 | 10 | 18,0 | 16,9 | 19,3 | 18,5 | 22,3 | 21,4 | 23,1 | 22,3 | 10,4 | 9,6 | 11,8 | 11,0 | 28,0 | 27,1 | 30,2 | 29,3 | 0,4 | 0,4 |
| 100 | 20 | 47,4 | 45,9 | 64,7 | 62,9 | 53,3 | 52,3 | 60,5 | 59,7 | 28,2 | 27,4 | 37,0 | 36,0 | 64,8 | 63,7 | 79,5 | 78,4 | 2,8 | 2,7 |
| 100 | 30 | 88,3 | 87,2 | 123,4 | 123,2 | 91,0 | 90,3 | 113,6 | 112,3 | 57,4 | 56,9 | 79,1 | 77,9 | 111,7 | 111,0 | 148,3 | 147,8 | 7,5 | 7,4 |
| 100 | 40 | 130,94 | 132,6 | 198,7 | 196,9 | 138,8 | 137,6 | 187,1 | 187,0 | 89,2 | 88,3 | 126,2 | 125,6 | 165,6 | 164,9 | 234,4 | 234,0 | 14,6 | 14,8 |
| 200 | 20 | 147,5 | 145,3 | 203,4 | 201,9 | 224,3 | 224,2 | 247,6 | 247,9 | 71,7 | 71,0 | 101,0 | 99,6 | 263,6 | 262,8 | 301,1 | 300,3 | 6,3 | 6,2 |
| 200 | 40 | 561,9 | 561,6 | 825,3 | 827,5 | 635,6 | 633,9 | 835,6 | 833,1 | 318,3 | 317,4 | 510,3 | 507,3 | 784,7 | 781,1 | 986,3 | 982,4 | 39,3 | 39,2 |
| 200 | 60 | 1075,3 | 1067,3 | 1665,9 | 1675,4 | 1224,0 | 1215,7 | 1715,6 | 1723,6 | 626,6 | 628,6 | 1006,4 | 998,1 | 1958,8 | 1961,0 | 2807,0 | 2792,1 | 113,5 | 114,3 |
| 200 | 80 | 1647,0 | 1638,6 | 2629,7 | 2630,0 | 1850,0 | 1845,8 | 2673,7 | 2678,7 | 804,3 | 802,6 | 1329,0 | 1327,3 | 3051,8 | 3030,9 | 4502,5 | 4487,6 | 225,2 | 225,8 |
| 300 | 30 | 712,7 | 712,7 | 908,1 | 904,8 | 979,5 | 978,3 | 1189,9 | 1190,6 | 412,7 | 413,2 | 644,8 | 645,5 | 1159,4 | 1152,6 | 1405,3 | 1402,3 | 32,3 | 32,3 |
| 300 | 60 | 2477,2 | 2450,2 | 3557,2 | 3523,6 | 2954,4 | 2949,9 | 3998,7 | 4002,4 | 1690,5 | 1687,5 | 2800,1 | 2807,3 | 3450,4 | 3613,17 | 4774,9 | 4919,3 | 215,3 | 216,7 |
| 300 | 90 | 5603,2 | 5606,2 | 8151,1 | 8107,9 | 5966,6 | 5906,6 | 8582,0 | 8593,9 | 3260,6 | 3276,4 | 5164,7 | 5145,8 | 6829,5 | 6841,3 | 10514,6 | 10420,9 | 605,3 | 605,7 |
| 300 | 120 | 9740,2 | 9675,0 | 15672,3 | 15587,3 | 10329,5 | 10278,8 | 15842,3 | 15835,8 | 4453,0 | 4452,8 | 8216,8 | 8215,7 | 11458,5 | 11423,4 | 17517,72 | 17562,01 | 1184,6 | 1195,2 |
| 400 | 40 | 2469,3 | 2457,8 | 3323,3 | 3341,2 | 2790,0 | 2781,5 | 3371,0 | 3370,2 | 1208,7 | 1208,3 | 2044,7 | 2044,1 | 3202,7 | 3195,9 | 4066,2 | 4077,4 | 106,1 | 112,5 |
| 400 | 80 | 9651,9 | 9724,3 | 14833,6 | 14849,0 | 10048,6 | 10161,6 | 13772,0 | 13751,5 | 5266,5 | 5258,0 | 8077,1 | 8129,3 | 10464,3 | 10451,3 | 15275,1 | 15283,4 | 682,5 | 681,1 |
| 400 | 120 | 20210,3 | 20264,5 | 32205,1 | 32286,0 | 19201,3 | 19220,9 | 28881,7 | 28940,3 | 10097,5 | 10097,9 | 16620,7 | 16553,0 | 22038,36 | 22184,8 | 33594,6 | 33481,1 | 1346,0 | 1891,8 |
| 400 | 160 | 32217,8 | 32273,2 | 51497,7 | 51819,2 | 30010,7 | 30096,4 | 45161,1 | 45471,9 | 15750,3 | 15684,3 | 26943,4 | 27012,2 | 32843,6 | 32550,5 | 52042,6 | 52056,2 | 5064,3 | 5109,6 |
| 500 | 50 | 5122,4 | 5176,2 | 7304,3 | 7253,1 | 5625,8 | 5661,6 | 7452,6 | 7340,8 | 2934,9 | 2935,6 | 4555,8 | 4540,8 | 6988,6 | 7005,0 | 9171,3 | 9187,4 | 254,5 | 254,5 |
| 500 | 100 | 26128,9 | 26048,1 | 31542,5 | 31546,0 | 22781,3 | 22822,1 | 32888,1 | 32855,2 | 11921,7 | 11916,2 | 19200,8 | 19239,3 | 26513,3 | 26473,1 | 39343,5 | 39416,2 | 1611,3 | 1608,2 |
| 500 | 150 | 54392,6 | 54342,9 | 76068,63 | 75762,54 | 49617,7 | 49686,0 | 76443,5 | 76249,0 | 29051,4 | 27546,0 | 65082,3 | 65173,3 | 55898,1 | 55594,0 | 88158,7 | 88217,9 | 7922,9 | 8087,6 |
| 500 | 200 | 80252,09 | 80245,14 | 115137,3 | 115200,5 | 75998,8 | 75810,0 | 117932,6 | 117675,1 | 39043,3 | 38944,42 | 86247,0 | 86513,8 | 84678,4 | 84840,8 | 137503,3 | 138031,3 | 28842,6 | 29133,5 |

than G5 but is not the worst one, showing that this algorithm works very well for this problem.

We performed a deeper analysis for the results obtained for the GRASP heuristics G1, G5, G6, G7 and G8, which present better solutions and/or shorter execution times. We selected two instances: the first one has parameters $n = 200$ and $m = 40$, and the second one, $n = 300$ and $m = 90$. We executed each GRASP heuristic until a solution was found with a greater or equal cost compared to a target value. Two target values were used for each instance: the worst value obtained by these heuristics and an average of the values generated by them. Empirical probability distributions for the time to achieve a target value are plotted in Fig(s). 3 and 4. To plot the empirical distribution for each variant, we executed each GRASP heuristic 100 times using 100 different random seeds. In each execution, we measured the time to achieve a solution whose cost was

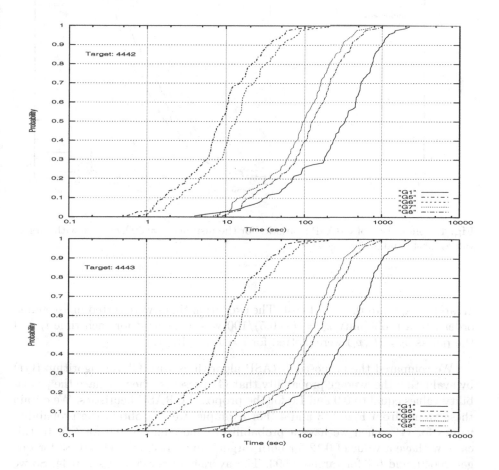

**Fig. 3.** Comparison of GRASP heuristics for the instance $n = 200, m = 40$ with targets values 4442 and 4443

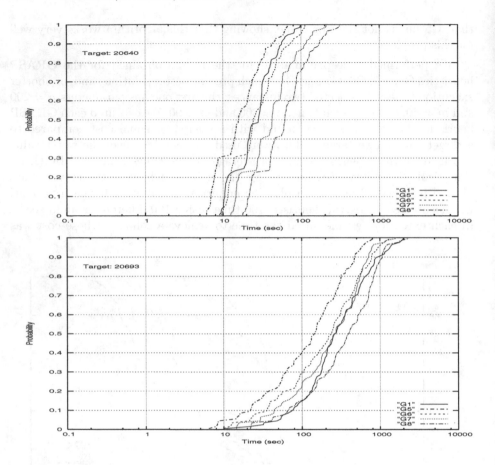

**Fig. 4.** Comparison of GRASP heuristics for the instance $n = 300, m = 90$ with targets values 20640 and 20693

greater or equal to the target cost. The execution times were sorted in ascending order and a probability $p_i = (i - 0.5)/100$ was associated for each time $t_i$ and the points $z_i = (t_i, p_i)$ were plotted for $i = 1, \ldots, 100$ [1].

We compared the proposed GRASP algorithms with Ghosh algorithm (G1) by evaluating the average probability that G1 presents when we have the probability values equal to 0.9 and 1.0 for the proposed GRASP heuristics. We obtain these values from Fig(s). 3 and 4. For example, we can obtain the probability values for G1, when we have a probability value equal to 0.9 for G5. In this case, we have a value of 0.12 for both target values 4442 and 4443, 0.83 for target 20640, and 0.7 for target 20693. The average of these values is 0.44. So we have evaluated these average values for G5, G6, G7 and G8 and the results are presented in Tab. 8.

**Table 8.** Comparison of convergence of solutions

| probability | G1-G5 | G1-G6 | G1-G7 | G1-G8 |
|---|---|---|---|---|
| 0.9 | 0.44 | 0.5 | 0.7 | 0.75 |
| 1.0 | 0.6 | 0.67 | 0.91 | 0.95 |

We can see that although the algorithm G1 presents a good convergence to the target values, the proposed algorithms G5, G6, G7 and G8 were able to improve this convergence.

## 4   Concluding Remarks

This paper presented some versions of GRASP heuristic to solve the maximum diversity problem (MDP). The main goal of this work was to analyse the influence of the construction and local search heuristics on the performance of GRASP techniques.

Experimental results show that the versions that use KLD or KLD-v2 construction algorithms and Gha or SOMA local search algorithms (G5, G6, G7 and G8) significantly improve the average performance of the best GRASP approaches proposed in the literature (G1 and G9).

Our experiments also show that if the execution time is restricted (limited to smaller value), version G5 is a good choice since it obtains reasonable results faster (see Fig(s). 3 and 4). On the other hand, if the execution time is not an issue, versions G7 and G8 tend to produce the best solutions (see Tabs. 6 and 7).

**Acknowledgments.** The authors acknowledge *LabPar* of PUC-RIO (Rio de Janeiro, Brazil) for making available their computational facilities on which some computational experiments were performed. We thank Paulo Andrade for allowing us to use the code developed by him for algorithm G9. We also acknowledge the Coordination of Improvement of Personnel of Superior Level (CAPES) support for providing a master scholarship to Geiza C. Silva.

## References

1. Aiex, R. M., Resende, M. G. C., Ribeiro, C. C.: Probability distribuition of solution time in GRASP: an experimental investigation, Journal of Heuristics, **8** (2002), 343–373
2. Andrade, P. M. F., Plastino, A., Ochi, L. S., Martins, S. L.: GRASP for the Maximum Diversity Problem, Proceedings of MIC 2003, (2003)
3. Bhadury J., Joy Mighty E., Damar, H.: Maximing workforce diversity in project teams: a network flow approach, Omega, **28** (2000), 143–153
4. Feo T. A., Resende, M. G. C.: Greedy randomized adaptive search procedures, Journal of Global Optimization **6** (1995), 109–133

5. Ghosh, J. B.: Computational aspects of maximum diversity problem, Operations Research Letters **19** (1996), 175–181
6. Glover, F., Hersh, G., McMillan C.: Selecting subsets of maximum diversity, MS/IS Report No. 77-9, University of Colorado at Boulder, (1977)
7. Glover, F., Kuo, C-C., Dhir,K. S.: Integer programming and heuristic approaches to the minimum diversity problem, Journal of Business and Management **4(1)**, (1996), 93–111
8. Hansen, P., Mladenović, N.: An introduction to variable neighborhood search, Metaheuristics, Advances and Trends in Local Search, Paradigms for Optimization, S. Voss et al. editors, (1999) 433–458
9. Katayama, K., Naribisa, H.: An evolutionary approach for the maximum diversity problem, Working Paper, Department of Information and Computer Engineering, Okayama University of Science, (2003)
10. Kochenberger, G., Glover, F.: Diversity data mining, Working Paper, The University of Mississipi, (1999)
11. Prais, M., Ribeiro, C. C.: Reactive GRASP: an aplication to a matrix decomposition problem in TDMA traffic assignment, INFORMS Journal on Computing **12** (2000), 164–176
12. Weitz, R., Lakshminarayanan, S.: An empirical comparison of heuristic methods for creating maximally diverse group, Journal of the operational Research Society **49** (1998), 635–646

# Using Compact Tries for Cache-Efficient Sorting of Integers

Ranjan Sinha

School of Computer Science and Information Technology, RMIT University,
Melbourne 3001, Australia.
rsinha@cs.rmit.edu.au

**Abstract.** The increasing latency between memory and processor speeds has made it imperative for algorithms to reduce expensive accesses to main memory. In earlier work, we presented cache-conscious algorithms for sorting strings, that have been shown to be almost two times faster than the previous algorithms, mainly due to better usage of the cache. In this paper, we propose two new algorithms, Burstsort and MEBurstsort, for sorting large sets of integer keys. Our algorithms use a novel approach for sorting integers, by dynamically constructing a compact trie which is used to allocate the keys to containers. These keys are then sorted within the cache. The new algorithms are simple, fast and efficient. We compare them against the best existing algorithms using several collections and data sizes. Our results show that MEBurstsort is up to 3.5 times faster than memory-tuned quicksort for 64-bit keys and up to 2.5 times faster for 32-bit keys. For 32-bit keys, on 10 of the 11 collections used, MEBurstsort was the fastest, whereas for 64-bit keys, it was the fastest for all collections.

## 1 Introduction

Sorting is one of the fundamental problems of computer science. Many applications are dependent on sorting, mainly for reasons of efficiency. It is also of great theoretical importance: several advances in data structures and algorithmic analysis have come from the study of sorting algorithms [6].

In recent years, the speed of CPU has increased by about 60% per year, but the speed of access to main-memory has decreased by only 7% per year [3]. Thus, there is an increasing latency gap between the processor speeds and access to main-memory and it appears that this trend is likely to continue. To reduce this problem, hardware developers have introduced hierarchies of memories – caches – between the processor and main-memory. The closer caches are to the processor, the smaller, faster and more expensive they get. Caches utilise the locality, temporal and spatial, inherent in most programs. As programs do not access all code or data uniformly, having those frequently accessed items closer to the processor is an advantage.

The prevalent approach to developing algorithms assumes the RAM model [1, 3,8], where all accesses to memory are given a unit cost. The focus is mainly on

C.C. Ribeiro and S.L. Martins (Eds.): WEA 2004, LNCS 3059, pp. 513–528, 2004.

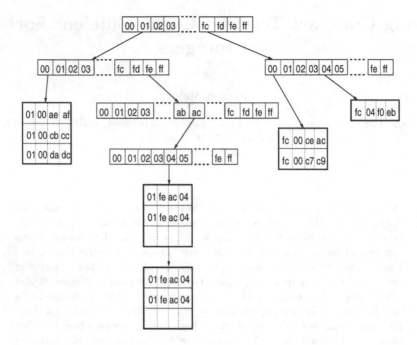

**Fig. 1.** Burstsort, with five trie nodes, five containers and ten keys. The keys are 0100aeaf$_h$, 0100cbcc$_h$, 0100dadc$_h$, 01feac04$_h$, 01feac04$_h$, 01feac04$_h$, 01feac04$_h$, fc00ceac$_h$, fc00c7c9$_h$ and fc04f0eb$_h$.

reducing the number of instructions used. As a result, many algorithms are tuned towards lowering the instruction count. They may not be particularly efficient for memory hierarchies, however.

Recently, there has been much work done on cache-conscious sorting [5,9,10, 12,14,15]. In 1996, LaMarca and Ladner [7] analyzed algorithms such as merge-sort, heapsort, quicksort and LSB radixsort in terms of cache misses and instructions and reported that it was practical to make them more efficient by improving the locality even at the cost of using more instructions. Their memory tuned algorithms are often used as a reference for comparing sorting algorithms. Since then, several cache-tuned implementations for well known algorithms have been developed, such as, Tiled-mergesort, Multi-mergesort, Memory-tuned quicksort, PLSB, EPLSB, EBT and CC-Radix. The most efficient of these algorithms are used in our experiments and considered in further detail later.

Tries are a much used data structure for algorithms on string keys and have been widely used for searching applications. But recently, Burstsort [12, 13], based on Burst-Tries [2] has shown excellent performance for sorting string keys, primarily by using the CPU-cache more effectively. A burst trie is a col-lection of small data structures, called containers, that are accessed by a normal

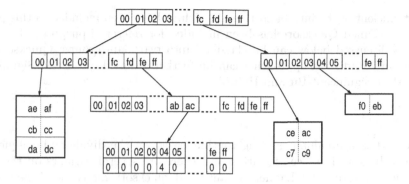

**Fig. 2.** MEBurstsort, with five trie nodes, three containers and ten keys. The keys are 0100aeaf$_h$, 0100cbcc$_h$, 0100dadc$_h$, 01feac04$_h$, 01feac04$_h$, 01feac04$_h$, fc00ceac$_h$, fc00c7c9$_h$ and fc04f0eb$_h$.

trie. The keys are stored in the containers and the first few bytes of the keys are used to construct the access trie.

In this paper, we propose two new algorithms, Burstsort and MEBurstsort, both based on a similar approach of using compact tries, but for integer keys. We believe this is the first case of using tries for the purpose of sorting integers. They reduce the number of times that keys need to be accessed from main-memory. This approach yields better dividends when key size is increased from 32 to 64 bits. We use several collections and data sizes to measure the performance of the sorting algorithms. Both artificially generated and real-world web collections are used in the experiments. Experiments include, measuring the running time of the algorithms and performing cache simulations to measure the number of instructions and cache misses.

Our results show that both Burstsort and MEBurstsort make excellent use of the cache while using low number of instructions. Indeed, they incur only 20% cache misses compared to MTQsort while the number of instructions is similar to the efficient radixsorts. For 32-bit keys, MEBurstsort is the fastest for 10 of 11 collections used and shows much the best performance for skewed data distributions. For 64-bit keys, Burstsort is comparable in performance to Sequential Counting Split radix sort (we believe, it is the fastest algorithm for 64-bit keys), whereas, MEBurstsort is the fastest for all collections. These results show that our approach of using a compact trie is practical, effective and adapts well to varied distributions and key sizes.

## 2   Background

Much of the effort on developing cache-efficient sorting algorithms has focused on restructuring existing techniques to take advantage of internal memory hierarchies. As observed from preliminary experiments on 32-bit keys, only the most efficient of the integer sorting algorithms have been included in this paper. Memory-Tuned Quicksort has been included for reference purposes. The algorithms discussed below can be classified into two main groups: Quicksort and Radixsort. Radixsort approaches can be further grouped into Least-Significant-Bit, Most-Significant-Bit and Hybrid.

## Quicksort

Quicksort is a dominant sorting routine based on the divide-and-conquer approach. It is an in-place algorithm and there are several variants on the basic scheme. LaMarca and Ladner [7] analyzed quicksort for cache-efficiency and suggested using insertion sort on small partitions when they are resident in the cache, instead of a final single pass over the entire data. This memory-tuned implementation of quicksort has become a standard reference for comparing cache-conscious sorting algorithms. For our experiments, we have used the implementation of LaMarca and Ladner [7] and labelled it as M T Q sort. We have also used an implementation by Xiao et al. [15] and labelled it as Q sort.

## Radix Sort

**Least Significant Bit (LSB).** Sedgewick [11] reports that the LSB approach is widely used due to its suitability for machine-language implementation as it is based on very simple data structures.

In LSB, the keys are sorted digit-by-digit starting from the least significant digits. The digits are usually sorted using counting sort, which has three phases in each pass: a count phase, a prefix sum phase, and a permute phase. Two arrays, source and destination (both the size of the dataset), and an auxiliary count array (dependent on the size of the alphabet) are used for this purpose. The count phase involves counting the number of keys that have identical digits at the position under consideration. The prefix sum phase is used to count the number of keys falling in a class and calculating the starting position for each class in the destination array. The permute phase involves moving the keys from the source array to the destination array using the count array as an index to the new location. Each pass requires that the source array is traversed twice, once each during the count phase and permute phase. The destination array is traversed once during the permute phase.

A simple modification to improve the locality of each pass is by pre-sorting small segments to group together keys with equal values. This approach was implemented by Rahman and Raman [10] and named Pre-sorting LSB radix sort. Each pass involves two sorts: pre-sort and global sort. Counting sort is used for both sorts. Though, it does improve the locality by pre-sorting, but the basic problem of it being a multi-pass algorithm would mean that the number of passes increases with the size of the key. For our experiments, we have used the implementation of Rahman and Raman [10] and labelled it as PLSB.

The permute phase is not cache efficient due to the random accesses to the destination array. To reduce these misses, Rahman and Raman [10] used a buffer

**Table 1.** Architectural parameters of the machine used for the experiments.

| Workstation | Pentium |
|---|---|
| Processor type | Pentium III Xeon |
| Clock rate (MHz) | 700 |
| L1 cache (KBytes) | 16 |
| L1 block size (Bytes) | 32 |
| L1 associativity | 4 |
| L1 miss latency (cycles) | 6 |
| L2 cache (KBytes) | 1,024 |
| L2 block size (Bytes) | 32 |
| L2 associativity | 8 |
| L2 miss latency (cycles) | 109 |
| TLB entries | 32 |
| TLB miss latency (cycles) | 5 |
| Memory Size (MBytes) | 2,048 |

to store keys in the same class and copy it in a block to the destination array. For our experiments, we have used the implementation of Rahman and Raman [10] and labelled it as EBT.

Extended-radix PLSB was developed to exploit the increase in locality offered by pre-sorting. This helps to reduce the number of passes without incurring many cache and TLB misses. We have used the implementation of Rahman and Raman [10] and labelled it as EPLSB.

For all the above variants of LSB radix sorts, we have used a radix size of 11 as it was found to be the most efficient for 32-bit keys and has also been used in previous such studies [10,5]. For some algorithms, a radix of 13 for 64-bit keys was found to be up to 10% faster due to the reduced number of passes.

**Most Significant Bit.** This is a type of distribution sorting where the classes are formed based on the value of the most significant bit. Depending upon the number of keys in each class, it proceeds recursively or uses a simple algorithm, such as insertion sort, to sort very small classes [6]. We have used the implementation of Rahman and Raman [9] and labelled it as MSBRadix. It is a multi-pass algorithm and unstable for equal keys. Based on the number of keys in a class, the radix is varied from a maximum of 16 to a minimum of 2 [9].

**Hybrid.** A hybrid approach uses both LSB and MSB methods. In Cache-Conscious radix sort (CCRadix), the data sets that do not fit within the cache (or the memory mapped by the TLB) is sorted on the most-significant-digit to dynamically divide the data set into smaller partitions that fit within the cache. Data sets of sizes less than the cache are sorted using LSB radix sort. Based upon data skew and size of the digit, there could be several reverse sorting calls. As noted by Jimenez-Gonzalez et al. [4], CCRadix does not scale to 64-bit keys.

**Table 2.** Random31 collection, 32-bit keys, sorting time for each method (milliseconds).

| | Qsort | MTQsort | PLSB | EPLSB | EBT | MSBRadix | CCRadix | Burstsort | MEBurstsort |
|---|---|---|---|---|---|---|---|---|---|
| SET 6 | 21,149 | 17,829 | 14,824 | 13,566 | 10,710 | 12,360 | 10,054 | 12,437 | 13,678 |
| SET 7 | 43,972 | 37,566 | 29,344 | 26,810 | 21,272 | 26,020 | 19,873 | 27,716 | 27,199 |

**Table 3.** Random20 collection, 32-bit keys, sorting time for each method (milliseconds).

| | Data set | | | | | | |
|---|---|---|---|---|---|---|---|
| | SET 1 | SET 2 | SET 3 | SET 4 | SET 5 | SET 6 | SET 7 |
| Qsort | 471 | 982 | 2,070 | 4,382 | 9,130 | 18,985 | 39,477 |
| MTQsort | 374 | 833 | 1,725 | 3,666 | 7,738 | 15,776 | 32,678 |
| PLSB | 456 | 916 | 1,832 | 3,668 | 7,333 | 14,666 | 29,157 |
| EPLSB | 423 | 839 | 1,675 | 3,341 | 6,674 | 13,319 | 26,275 |
| EBT | 317 | 634 | 1,268 | 2,541 | 5,099 | 10,193 | 20,137 |
| MSBRadix | 296 | 645 | 1,485 | 3,405 | 6,943 | 14,275 | 28,283 |
| CCRadix | 411 | 861 | 1,808 | 3,665 | 7,359 | 14,830 | 29,926 |
| Burstsort | 327 | 677 | 1,435 | 2,493 | 5,095 | 9,361 | 19,345 |
| MEBurstsort | 336 | 707 | 1,288 | 2,313 | 4,546 | 8,729 | 17,442 |

The parameters which showed the best performance for the uniform random distribution collection (Random31) was chosen for our experiments. We have used the implementation of Jimenez-Gonzalez et al. [5] and labelled it as CCRadix.

SCSRadix is used to sort 64-bit integer keys. It dynamically detects if a subset of the data has skew and skips the sorting of the subset. Based on the number of keys ($n$) and the number of bits that remain to be sorted ($b$), it chooses between insertion sort, CCRadix, LSB and Counting Split. Counting split is used for partitioning the dataset into smaller sub-buckets of similar size, whereupon depending upon $n$ and $b$, the other three algorithms are used. We believe, this is the fastest sorting routine for 64-bit keys. We have used the implementation of Jimenez-Gonzalez et al. [4] and labelled it as SCSRadix.

## 3    Sorting Integers with Compact Tries

Traditionally, trie data structures have been used for managing variable-length string keys and found applications in dictionary management, text compression and pattern matching [2]. In our earlier work, we investigated the practicality of using burst tries [2], a compact and efficient variant of tries, for sorting strings. In this section, we describe a similar approach for the purpose of sorting integer keys in a cache-efficient manner.

**Table 4.** Binomial collection, 32-bit keys, sorting time for each method (milliseconds).

| | Data set | | | | | | |
|---|---|---|---|---|---|---|---|
| | SET 1 | SET 2 | SET 3 | SET 4 | SET 5 | SET 6 | SET 7 |
| Qsort | 270 | 587 | 1,295 | 2,772 | 5,926 | 12,713 | 26,723 |
| MTQsort | 191 | 431 | 964 | 2,129 | 4,647 | 10,316 | 21,405 |
| PLSB | 452 | 908 | 1,815 | 3,626 | 7,259 | 14,489 | 28,805 |
| EPLSB | 415 | 822 | 1,644 | 3,307 | 6,554 | 13,632 | 25,784 |
| EBT | 298 | 596 | 1,193 | 2,389 | 4,772 | 9,537 | 18,821 |
| MSBRadix | 276 | 542 | 1,075 | 2,139 | 4,300 | 8,523 | 17,037 |
| CCRadix | 660 | 1,327 | 2,656 | 5,309 | 10,635 | 21,285 | 41,821 |
| Burstsort | 284 | 529 | 1,009 | 1,954 | 3,846 | 7,656 | 15,176 |
| MEBurstsort | 179 | 322 | 598 | 1,140 | 2,223 | 4,385 | 8,704 |

**Table 5.** Pascal collection, 64-bit keys, sorting time for each method (milliseconds).

| | Data set | | | | | |
|---|---|---|---|---|---|---|
| | SET 1 | SET 2 | SET 3 | SET 4 | SET 5 | SET 6 |
| Qsort | 838 | 1,842 | 4,018 | 8,570 | 18,201 | 39,197 |
| MTQsort | 662 | 1,487 | 3,292 | 7,061 | 14,933 | 31,761 |
| PLSB | 1,308 | 2,613 | 5,249 | 10,455 | 20,941 | 41,851 |
| EPLSB | 1,219 | 2,436 | 4,849 | 9,709 | 19,329 | 38,602 |
| EBT | 1,156 | 2,314 | 4,623 | 9,249 | 18,487 | 36,987 |
| MSBRadix | 1,019 | 2,012 | 3,994 | 7,963 | 15,910 | 31,813 |
| SCSRadix | 342 | 686 | 1,381 | 2,776 | 5,559 | 10,974 |
| Burstsort | 478 | 879 | 1,718 | 3,380 | 6,726 | 13,395 |
| MEBurstsort | 355 | 646 | 1,251 | 2,448 | 4,814 | 9,527 |

## Burstsort

The main principle behind Burstsort is to minimize the number of times that the keys need to be accessed from main-memory. This is achieved by dynamically constructing a compact trie that rapidly places the keys into containers. It divides the dataset based on both the data distribution and size of cache. Burstsort is a variant of most-significant-bit radixsort and needs to read the distinguishing bits in each key at most once. Our earlier work on string keys [12,13] have shown that such an approach has excellent performance.

There are two phases in the construction of a burst trie: insertion and traversal. The insertion phase inserts the keys into the trie. It is a single-pass traversal through the source array. The trie can grow in three ways: creation of a new trie node, increasing the size of the existing containers and the creation of new containers. When a container becomes too large, it is burst, resulting in the creation of a new trie node and new child containers. Once all the keys have been inserted, an in-order traversal of the trie is performed. The containers having more than one key are sorted on those bits that have not yet been read. We

**Table 6.** Random63 collection, 64-bit keys, sorting time for each method (milliseconds).

| | Data set | | | | | |
|---|---|---|---|---|---|---|
| | SET 1 | SET 2 | SET 3 | SET 4 | SET 5 | SET 6 |
| Qsort | 1,106 | 2,496 | 5,162 | 11,630 | 23,390 | 51,780 |
| MTQsort | 926 | 1,958 | 4,499 | 9,823 | 20,660 | 44,932 |
| PLSB | 1,369 | 2,749 | 5,487 | 11,007 | 22,564 | 44,005 |
| EPLSB | 1,268 | 2,529 | 5,039 | 10,104 | 20,184 | 40,463 |
| EBT | 1,180 | 2,363 | 4,729 | 9,499 | 18,946 | 38,693 |
| MSBRadix | 1,028 | 2,574 | 4,638 | 8,694 | 16,957 | 34,313 |
| SCSRadix | 549 | 1,189 | 2,694 | 5,600 | 11,286 | 22,708 |
| Burstsort | 556 | 1,046 | 2,060 | 4,365 | 10,377 | 22,003 |
| MEBurstsort | 538 | 1,037 | 2,050 | 4,288 | 9,569 | 18,649 |

have used MSBRadix for sorting containers due to its lower instruction count as compared to MTQsort. MSBRadix operates in-place thus making full-use of the L2 cache. The usage of other algorithms for sorting containers needs to be explored further; it would depend upon the depth of the memory hierarchy and their inherent latencies.

The data structures used for the trie nodes and the containers are arrays. The trie node structure is composed of four elements: pointer to a trie or container, counter of keys in container, level counter for growing container size, and a tail-pointer for the lowest level in the trie hierarchy. The tail-pointer is used to insert the keys at the end of the containers in order to maintain stability, though the current version is unstable due to using an unstable MSB Radix for sorting the containers.

For our experiments, the size of the container is restricted by the size of the L2 cache and is determined by the ratio of cache size to size of the key. Instead of allocating the space for the containers all at once, the container grows (using the realloc function call) from 16 to 262,144 for 32-bit keys and 16 to 131,072 for 64-bit keys. They are grown by a factor of 4, for example, 16, 64, 256, 1,024, 4,096, 16,384, 65,536 and 262,144 for 32-bit keys. Containers used in the lowest level are a linked list of arrays of size 128; the keys in these containers are not sorted as they are identical. A radix size of 8 bits has been used for the trie nodes.

An example of Burstsort for 32-bit keys is shown in Figure 1, the node is composed of 256 characters. The 32-bit integer keys are represented by hexadecimal numbers and stored in their entirety in the containers. The keys are $0100aeaf_h$, $0100cbcc_h$, $0100dadc_h$, $01feac04_h$, $01feac04_h$, $01feac04_h$, $01feac04_h$, $fc00ceac_h$, $fc00c7c9_h$ and $fc04f0eb_h$. The threshold value is assumed to be three, implying that the container bursts when there are three keys. The lowest level is a linked list of arrays of size three.

**Table 7.** Web collection, 32 and 64 bit keys, SET 6, sorting time for each method (milliseconds).

| | Qsort | MTQsort | PLSB | EPLSB | EBT | MSBRadix | CC/SCS Radix | Burstsort | MEBurstsort |
|---|---|---|---|---|---|---|---|---|---|
| 32-bit | 15,038 | 11,495 | 15,038 | 12,971 | 9,793 | 9,063 | 20,759 | 10,205 | 5,192 |
| 64-bit | 43,075 | 33,205 | 41,307 | 37,791 | 35,935 | 32,568 | 11,966 | 16,911 | 10,286 |

**MEBurstsort (Memory Efficient Burstsort)**

In Burstsort, the keys are stored in full. This may lead to redundancy as some of the information pertaining to each key is implicitly stored in the trie. Thus, MEBurstsort was developed for the purpose of eliminating this redundancy. Only that portion of the key is stored in the containers which has not already been read and thus cannot be gathered from a traversal of the trie. Once all the keys have been inserted, the trie is traversed depth-first, the keys in each container are reproduced in full and written back to the source array, where it is then sorted using the container sorting algorithm. A container at level $x$ has $sizeof(key) - (x + 1)$ bytes of the key, it is assumed that the root node is at 0th level. The linked list of arrays present at the lowest level in Burstsort have been eliminated, only the counters are required to keep track of the keys. As a result, it is not a stable algorithm, but stability is of significance only when sorting pointers to records.

This compressed storage, saves space in the containers, and it uses much less memory than Burstsort. Smaller containers makes it more cache-friendly. But as the keys are treated as variable-length bytes, bursting requires copying a key byte-by-byte, thus requiring more instructions. MEBurstsort has been observed to perform better than Burstsort for all collections.

An example of MEBurstsort is shown in Figure 2. The integer keys are represented by hexadecimal numbers. They are $0100aeaf_h$, $0100cbcc_h$, $0100dadc_h$, $01feac04_h$, $01feac04_h$, $01feac04_h$, $01feac04_h$, $fc00ceac_h$, $fc00c7c9_h$ and $fc04f0eb_h$. The threshold value is assumed to be three. As can be seen from the figure, the bytes of each key that have already been read are not stored in the containers.

## 4   Experiments

Several collections with a wide range of characteristics and sizes have been used in our experiments. Many of these collections have been used in previous such studies [15]. The collections are briefly described below.

**Random63.** uniformly distributed integers in the range 0 to $2^{63} - 1$ and generated using the random number generator random () from the C library.

**Random31.** uniformly distributed integers in the range 0 to $2^{31} - 1$ and generated using the random number generator random () from the C library.

**Random20.** uniformly distributed integers in the range 0 to $2^{20} - 1$ and generated using the random number generator random () from the C library.

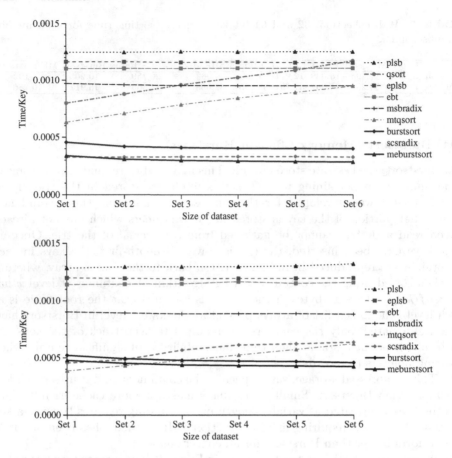

**Fig. 3.** Time/Key, 64-bit keys. Upper: Pascal, Lower: Sorted.

**Equilikely.** composed of integers in a specified range.
**Bernoulli.** a discrete probability distribution composed of integers 0 or 1
**Geometric.** a discrete probability distribution composed of integers 0, 1, 2, ...
**Pascal.** a discrete probability distribution composed of integers 0, 1, 2, ...
**Binomial.** a discrete probability distribution composed of integers 0, 1, 2, ...,
    N
**Poisson.** a discrete probability distribution composed of integers 0, 1, 2, ...
**Zero.** composed entirely of 0s.
**Sorted.** distinct integers sorted in ascending order
**Web.** integers in order of occurrence and drawn from a large web collection

For 32-bit keys, there are seven sets, designated as SET 1, SET 2, SET 3, SET 4, SET 5, SET 6 and SET 7. They represent data sizes from 1x1024x1024 to 64x1024x1024 keys. For 64-bit keys, there are six sets of sizes ranging from 1x1024x1024 to 32x1024x1024 keys. The set sizes are grown in multiples of two.

The goal of the experiments is to compare the performance of our algorithms with some of the best known algorithms in terms of running time, number of instructions and L2 cache misses. The implementation of the algorithms used in our experiments were assembled from the best sources we could identify and are confident that these are of high quality. All the algorithms were written in C.

The task is to sort an array of integers, the array is returned as output. The CPU time is measured by using the unix function gettimeofday (). The cost of generating the data collections in terms of time, number of instructions and L2 cache misses are not included in the results reported here. The configurations of the machine used for the experiments are presented in Table 1; `calibrator` [1] was used to measure some of the machine configurations. An open-source cache simulator, `valgrind` [2], has been used for simulating the cache, the configurations of our machine were used. The experiments were performed on a Linux operating system using the GNU `gcc` compiler with the highest compiler optimization O3. The experiments were performed under light load, that is, no other significant processes were running.

## 5   Results

All the graphs showing times, instructions and cache misses have been normalized by dividing by the number of keys. The timings in the tables are in milliseconds and are shown unnormalized.

In agreement with Jimenez-Gonzalez et al. [5], we found CCRadix to be the fastest sorting algorithm for 32-bit keys on the Random31 collection shown in Table 2. However, the performance of CCRadix is seen to deteriorate with the increase in the number of duplicates as shown in Table 3 for Random20 and even more so for small key values in skewed distributions such as the binomial collection in Table 4.

Table 3 shows Burstsort and MEBurstsort to be 1.68 and 1.87 times faster than MTQsort for the Random20 collection. The timings for the binomial collection (composed of only 11 distinct integers of small values) in Table 4, shows MEBurstsort to be 2.45 times faster than MTQsort. In the binomial collection, the first three bytes are identical for all keys. After the insertion of the threshold number of keys (a small fraction of the entire collection) into one container, the container is burst. For 32-bit keys, this bursting occurs three times in a loop until the threshold number of keys end up in the lowest level. The rest of the keys traverse the same path, and since the nodes along that path have already been created, much of the information pertaining to these keys can be read in just one access. Only the counters in the lowest level need to be incremented, so MEBurstsort effectively becomes an in-place algorithm. Both Burstsort and MEBurstsort are particularly efficient for this kind of skewed data. Since the main reason for the efficiency of our algorithms is the reduced number of times that the keys need to be accessed, we expected it to show even better relative

---

[1] http://homepages.cwi.nl/~manegold/Calibrator
[2] http://developer.kde.org/~sewardj

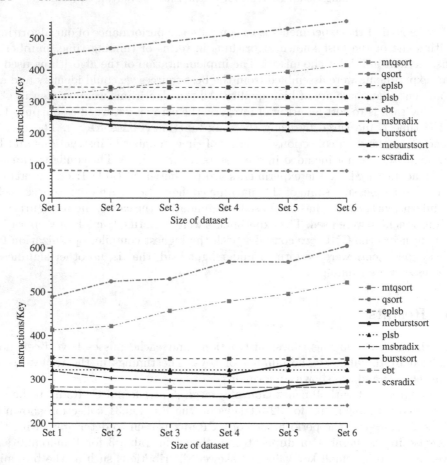

**Fig. 4.** Instructions/Key, 64-bit keys, 1 Mb cache, 8-way associativity, 32 bytes block size. Upper: Pascal, Lower: Random31.

performance for 64-bit keys, compared to the multi-pass LSB variants. Much of the focus below is on sorting 64-bit integer keys.

Table 5 shows the times for the Pascal collection on 64-bit keys, the results are stunning. Burstsort and MEBurstsort are 2.37 and 3.33 times faster than MTQsort and the LSB sorting routines such as EPLSB and EBT respectively. As Figure 3 shows, MEBurstsort and Burstsort is much the fastest and is competitive with SCSRadix (which is the fastest 64-bit algorithm to our knowledge). Even for the Random63 collection as shown in Table 6, both Burstsort and MEBurstsort shows the best timings though not as dramatic as for Pascal. Similar results have been reported for the real-world web collection. Interestingly, as shown in Figure 3, the time/key for the sorted collection improves with the increase in the datasize. The performance of our algorithms for most of the

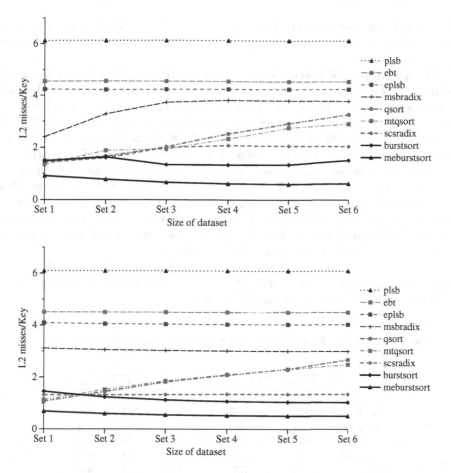

**Fig. 5.** Cache misses/Key, 64-bit keys, 1 Mb cache, 8-way associativity, 32 bytes block size. Upper: Random20, Lower: Pascal.

other collections such as Bernoulli, Geometric, Binomial, Poisson and Zero are similar to that of Pascal.

The normalized instructions per key are shown in Figure 4. SCSRadix uses the least number of instructions for the Pascal collection as it detects the skew. MTQsort due to its high complexity cost has the highest number of instructions. For the Pascal collection, MEBurstsort and Burstsort is second and third respectively. As discussed earlier, for skewed collections, MEBurstsort requires lesser number of instructions than Burstsort whereas Burstsort is more efficient for uniform distributions as seen from the lower graph in Figure 4.

Figure 5 shows the number of cache misses per key for 64-bit keys incurred by each algorithm for the collections Random20 and Pascal. Both Burstsort and MEBurstsort have the least number of cache misses and shows the effectiveness of our approach. MEBurstsort has less than one cache miss per key while

Burstsort incurs one to two cache misses per key. This is significant and the relative performance of our algorithms will continue to improve with more modern processors.

We have also investigated the use of memory by our algorithms. The amount of memory used for the largest set size of all collections and for 64-bit keys are shown in Table 8. Based on similar memory usage, the algorithms have been classified into four groups : in-place such as MTQsort and MSBRadix; LSB radix-sorts such as EBT, LSB, PLSB and EPLSB; CCRadix and SCSRadix; Burstsort and MEBurstsort. The memory usages for collections such as Geometric, Pascal, Binomial, Poisson and Zero are similar to Bernoulli. Burstsort uses as much memory as the LSBs for all collections except for the uniform distributions where it requires about 1.2 times more memory than the LSBs. Keep in mind that the buckets are grown by multiples of 4, smaller values will result in lesser memory usage. MEBurstsort uses up to 1.5 times less memory than Burstsort and for skewed data, it is effectively an in-place algorithm.

These results show that Burstsort and MEBurstsort are two of the best algorithms and have shown excellent performance on collections with varied characteristics for both 32-bit and 64-bit keys.

**Table 8.** Relative Memory usages (in megabytes) for 64-bit keys with SET 6.

|  | Burstsort | MEBurstsort | MTQsort | EBT | SCSRadix |
|---|---|---|---|---|---|
| Random63 | 947 | 665 | 512 | 768 | 800 |
| Random31 | 947 | 665 | 512 | 768 | 800 |
| Random20 | 849 | 565 | 512 | 768 | 800 |
| Equilikely | 853 | 566 | 512 | 768 | 800 |
| Bernoulli | 772 | 512 | 512 | 768 | 800 |
| Sorted | 759 | 578 | 512 | 768 | 800 |

# 6    Conclusions

In this paper, we have proposed two new algorithms, Burstsort and MEBurstsort, for sorting large collections of integer keys. They are based on a novel approach of using a compact trie for storing the keys. For the evaluation of these algorithms, we have compared them against some of the best known algorithms using several collections, both artificially generated and from the real-world.

Our experiments have shown Burstsort and MEBurstsort to be two of the fastest sorting algorithms for 64-bit keys, because they have a significantly lower number of cache misses, as well as a low instruction count. Even for 32-bit keys, they are the fastest for all collections except for Random31. They have a similar theoretical cost as the most-significant-digit radix sorts. They both adapt well to varying distributions such as skew, no skew, sorted, small and large values of keys. The experiments also confirm our expectations that the larger the size of keys, the better the relative performance of our algorithms. Thus, while Burstsort and MEBurstsort are up to 1.7 and 2.5 times faster than MTQsort for 32-bit keys, they are up to 2.37 and 3.5 times faster than MTQsort for 64-bit keys, respectively.

There is much scope to further improve these algorithms. They are expected to show better relative performance when applied to sorting pointers to keys. In preliminary work on sorting pointers to integer keys, Burstsort was found to be up to 3.72 times faster than MTQsort. In parallel work on strings, we have observed that randomization techniques are useful in lowering both instructions and cache misses even further; these techniques should be readily applicable for sorting integers. The effect of TLB on our algorithms needs to be investigated. Preliminary work on using different radix sizes in the trie nodes have shown promising results, such as, a radix of 11 shows better performance than a radix of 8 for 32-bit keys. Even without these improvements, Burstsort and MEBurstsort represent a novel and important advance for efficient sorting of integer keys.

**Acknowledgment.** I thank my supervisor Prof. Justin Zobel and colleague Vaughan Shanks for their assistance.

# References

1. A. Aho, J. E. Hopcroft, and J. D. Ullman. *The Design and Analysis of Computer Algorithms.* Addison-Wesley, Reading, Massachusetts, 1974.
2. S. Heinz, J. Zobel, and H. E. Williams. Burst tries: A fast, efficient data structure for string keys. *ACM Transactions on Information Systems*, 20(2):192–223, 2002.
3. J. L. Hennessy and D. A. Patterson. *Computer Architecture: A Quantitative Approach.* Morgan Kaufmann, San Mateo, CA, 2002.
4. D. Jimenez-Gonzalez, J. Navarro, and J. L. Larriba-Pey. The effect of local sort on parallel sorting algorithms. In *Proceedings of Parallel and Distributed Processing Workshop*, Las Palmas de Gran Canaria, Spain, January 2002.
5. D. Jimenez-Gonzalez, J. Navarro, and J. L. Larriba-Pey. Cc-radix: a cache conscious sorting based on radix sort. In *Proceedings of the 11th Euromicro Workshop on Parallel, Distributed and Network-based processing*, 2003.
6. D. E. Knuth. *The Art of Computer Programming, Volume 3: Sorting and Searching, Second Edition.* Addison-Wesley, Massachusetts, 1997.
7. A. LaMarca and R. E. Ladner. The influence of caches on the performance of sorting. *Jour. of Algorithms*, 31(1):66–104, 1999.
8. U. Meyer, P. Sanders, and J. F. Sibeyn, editors. *Algorithms for Memory Hierarchies, Advanced Lectures [Dagstuhl Research Seminar, March 10-14, 2002]*, volume 2625 of *Lecture Notes in Computer Science*. Springer, 2003.
9. N. Rahman and R. Raman. Analysing cache effects in distribution sorting. *ACM Jour. of Experimental Algorithmics*, 5:14, 2000.
10. N. Rahman and R. Raman. Adapting radix sort to the memory hierarchy. *ACM Jour. of Experimental Algorithmics*, 6(7), 2001.
11. R. Sedgewick. *Algorithms in C, third edition.* Addison-Wesley Longman, Reading, Massachusetts, 1998.

12. R. Sinha and J. Zobel. Cache-conscious sorting of large sets of strings with dynamic tries. In R. Ladner, editor, *5th ALENEX Workshop on Algorithm Engineering and Experiments*, pages 93–105, Baltimore, Maryland, January 2003.
13. R. Sinha and J. Zobel. Efficient trie-based sorting of large sets of strings. In M. Oudshoorn, editor, *Proceedings of the Australasian Computer Science Conference*, pages 11–18, Adelaide, Australia, February 2003.
14. R. Wickremesinghe, L. Arge, J. Chase, and J. S. Vitter. Efficient sorting using registers and caches. *ACM Jour. of Experimental Algorithmics*, 7(9), 2002.
15. L. Xiao, X. Zhang, and S. A. Kubricht. Improving memory performance of sorting algorithms. *ACM Jour. of Experimental Algorithmics*, 5:3, 2000.

# Using Random Sampling to Build Approximate Tries for Efficient String Sorting

Ranjan Sinha and Justin Zobel

School of Computer Science and Information Technology, RMIT University,
Melbourne 3001, Australia.
{rsinha,jz}@cs.rmit.edu.au

**Abstract.** Algorithms for sorting large datasets can be made more efficient with careful use of memory hierarchies and reduction in the number
of costly memory accesses. In earlier work, we introduced burstsort, a new
string sorting algorithm that on large sets of strings is almost twice as
fast as previous algorithms, primarily because it is more cache-efficient.
The approach in burstsort is to dynamically build a small trie that is
used to rapidly allocate each string to a bucket. In this paper, we introduce new variants of our algorithm: SR-burstsort, DR-burstsort, and
DRL-burstsort. These algorithms use a random sample of the strings to
construct an approximation to the trie prior to sorting. Our experimental results with sets of over 30 million strings show that the new variants reduce cache misses further than did the original burstsort, by up
to 37%, while simultaneously reducing instruction counts by up to 24%.
In pathological cases, even further savings can be obtained.

## 1 Introduction

In-memory sorting is a basic problem in computer science. However, sorting
algorithms face new challenges due to changes in computer architecture. Processor speeds have been increasing at 60% per year, while speed of access to main
memory has been increasing at only 7% per year, a growing processor-memory
performance gap that appears likely to continue. An architectural solution has
been to introduce one or more levels of fast memory, or cache, between the
main memory and the processor. Small volumes of data can be sorted entirely
within cache—typically a few megabytes of memory in current machines—but,
for larger volumes, each random access involves a delay of up to hundreds of
clock cycles.

Much of the research on algorithms has focused on complexity and efficiency
assuming a non-hierarchical RAM model, but these assumptions are not realistic
on modern computer architectures, where the levels of memory have different
latencies. While algorithms can be made more efficient by reducing the number
of instructions, current research [8,15,17] shows that an algorithm can afford to
increase the number of instructions if doing so improves the locality of memory accesses and thus reduces the number of cache misses. In particular, recent

C.C. Ribeiro and S.L. Martins (Eds.): WEA 2004, LNCS 3059, pp. 529–544, 2004.

work [8,13,17] has successfully adapted algorithms for sorting integers to memory hierarchies.

According to Arge et al. [2] "string sorting is the most general formulation of sorting because it comprises integer sorting (i.e., strings of length one), multikey sorting (i.e., equal-length strings) and variable-length key sorting (i.e., arbitrarily long strings)". String sets are typically represented by an array of pointers to locations where the variable-length strings are stored. Each string reference incurs at least two cache misses, one for the pointer and one or more for the string itself depending on its length and how much of it needs to be read.

In our previous work [15,16], we introduced burstsort, a new cache-efficient string sorting algorithm. It is based on the burst trie data structure [7], where a set of strings is organised as a collection of buckets indexed by a small access trie. In burstsort, the trie is built dynamically as the strings are processed. During the first phase, at most the distinguishing prefix—but usually much less—is read from each string to construct the access trie and place the string in a bucket, which is a simple array of pointers. The strings in each bucket are then sorted using an algorithm that is efficient both in terms of the space and the number of instructions for small sets of strings. There have been several recent advances made in the area of string sorting, but our experiments [15,16] showed burstsort to be much more efficient than previous methods for large string sets. (In this paper, for reference we compare against three of the best previous string sorting algorithms: MBM radixsort [9], multikey quicksort [3], and adaptive radixsort [1, 11].) However, burstsort is not perfect. A key shortcoming is that individual strings must be re-accessed as the trie grows, to redistribute them into sub-buckets. If the trie could be constructed ahead of time, this cost could be largely avoided, but the shape and size of the trie strongly depends on the characteristics of the data to be sorted.

Here, we propose new variants of burstsort: SR-burstsort, DR-burstsort, and DRL-burstsort. These use random sampling of the string set to construct an approximation to the trie that is built by the original burstsort. Prefixes that are repeated in the random sample are likely to be common in the data; thus it intuitively makes sense to have these prefixes as paths in the trie. As an efficiency heuristic, rather than thoroughly process the sample we simply process them in order, using each string to add one more node to the trie. In SR-burstsort, the trie is then fixed. In DR-burstsort, the trie can if necessary continue to grow as in burstsort, necessitating additional tests but avoiding inefficiency in pathological cases. In DRL-burstsort, total cache size is used to limit initial trie size.

We have used several small and large sets of strings, as described in our earlier work [15,16], for our experiments. SR-burstsort is in some cases slightly more efficient than burstsort, but in other cases is much slower. DR-burstsort and DRL-burstsort are more efficient than burstsort in almost all cases, though with larger collections the amount of improvement decreases. In addition, we have used a cache simulator to examine individual aspects of the performance, and have found that in the best cases both the number of cache misses and the number of instructions falls dramatically compared to burstsort. These new algorithms are the fastest known way to sort a large set of strings.

## 2   Background

In our earlier work [15,16] we examined previous algorithms for sorting strings. The most efficient of these were adaptive radixsort, multikey quicksort, and MBM radixsort. Adaptive radixsort was introduced by Andersson and Nilsson in 1996 [1,11]; it is an adaptation of the distributive partitioning developed by Dobosiewicz to standard most-significant-digit-first radixsort. The alphabet size is chosen based on the number of elements to be sorted, switching between 8 bits and 16 bits. In our experiments, we used the implementation of Nilsson [11].

Multikey quicksort was introduced by Sedgewick and Bentley in 1997 [3]. It is a hybrid of ternary quicksort and MSD radixsort. It proceeds character-wise and partitions the strings into buckets based upon the value of the character at the position under consideration. The partitioning stage proceeds by selecting a random pivot and comparing the first character of the strings with the first character of the pivot. As in ternary quicksort, the strings are then partitioned into three sets—less than, equal to, and greater than—which are then sorted recursively. In our experiments, we used an implementation by Sedgewick [3].

MBM radixsort (our nomenclature) is one of several high-performance MSD radixsort variants tuned for strings that were introduced by McIlroy, Bostic, and McIlroy [9] in the early 1990s. We used programC, which we found experimentally to be most efficient of these variants; we found it to be the fastest array-based, in-place sorting algorithm for strings.

*Burstsort.* Any data structure that maintains the data in order can be used as the basis of a sorting method. Burstsort is based on this principle. A trie structure is used to place the strings in buckets by reading at most the distinguishing prefix; this structure is built incrementally as the strings are processed. There are two phases; first is insertion of the strings into the burst trie structure, second is an in-order traversal, during which the buckets are sorted.

The trie is built by bursting a bucket once it becomes too large; a new node is created and the strings in the bucket are inserted into the node, creating new child buckets. A fixed threshold—the maximum number of strings that can be held in a bucket—is used to determine whether to burst. Strings that are completely consumed are managed in a special "end of string" structure.

During the second, traversal phase, if the number of strings in the bucket is more than one, then a sorting algorithm that takes the depth of the character of the strings into account is used to sort the strings in the bucket. We have used multikey quicksort [3] in our experiments.

The set of strings is recursively partitioned on their lead characters, then when a partition is sufficiently small it is sorted by a simple in-place method. However, there is a key difference between radixsorts and burstsort. In the first, trie-construction phase the standard radixsorts proceed character-wise, processing the first character of each string, then re-accessing each string to process the

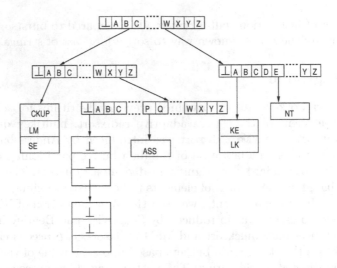

**Fig. 1.** A burst trie of four nodes and five buckets.

next character, and so on. Each trie node is handled once only, but strings are handled many times. In contrast, burstsort proceeds string-wise, accessing each string once only to allocate it to a bucket. Each node is handled many times, but the trie is much smaller than the data set, and thus the nodes can remain resident in cache.

Figure 1 shows an example of a burst trie containing eleven records whose keys are "backup", "balm", "base", "by", "by", "by", "by", "bypass", "wake", "walk", and "went" respectively. In this example, the alphabet is the set of letters from A to Z, and in addition an empty string symbol ⊥ is shown; the bucket structure used is an array. The access trie has four trie nodes and five buckets in all. The leftmost bucket has three strings, "backup", "balm" and "base", the second bucket has four identical strings "by", the fourth bucket has two strings "wake" and "walk", the rightmost bucket has only one string "went".

Experimental results comparing burstsort to previous algorithms are shown later. As can be seen, for sets of strings that are significantly larger than the available cache, burstsort is up to twice as fast. The gain is largely due to dramatically reduced numbers of cache misses compared to previous techniques.

*Randomised algorithms.* A randomised algorithm is one that makes random choices during its execution. According to Motwani and Raghavan [10], "two benefits of randomised algorithms have made them popular: simplicity and efficiency. For many applications, a randomised algorithm is the simplest available, or the fastest, or both."

One application of randomisation for sorting is to rearrange the input in order to remove any existing patterns, to ensure that the expected running time matches the average running time [4]. The best-known example of this is in

**Table 1.** Statistics of the data collections used in the experiments.

|  | Data set | | | | | |
|---|---|---|---|---|---|---|
|  | Set 1 | Set 2 | Set 3 | Set 4 | Set 5 | Set 6 |
| *Duplicates* | | | | | | |
| Size $Mb$ | 1.013 | 3.136 | 7.954 | 27.951 | 93.087 | 304.279 |
| Distinct Words ($\times 10^5$) | 0.599 | 1.549 | 3.281 | 9.315 | 25.456 | 70.246 |
| Word Occurrences ($\times 10^5$) | 1 | 3.162 | 10 | 31.623 | 100 | 316.230 |
| *No duplicates* | | | | | | |
| Size $Mb$ | 1.1 | 3.212 | 10.796 | 35.640 | 117.068 | 381.967 |
| Distinct Words ($\times 10^5$) | 1 | 3.162 | 10 | 31.623 | 100 | 316.230 |
| Word Occurrences ($\times 10^5$) | 1 | 3.162 | 10 | 31.623 | 100 | 316.230 |
| *Genome* | | | | | | |
| Size $Mb$ | 0.953 | 3.016 | 9.537 | 30.158 | 95.367 | 301.580 |
| Distinct Words ($\times 10^5$) | 0.751 | 1.593 | 2.363 | 2.600 | 2.620 | 2.620 |
| Word Occurrences ($\times 10^5$) | 1 | 3.162 | 10 | 31.623 | 100 | 316.230 |
| *Random* | | | | | | |
| Size $Mb$ | 1.004 | 3.167 | 10.015 | 31.664 | 100.121 | 316.606 |
| Distinct Words ($\times 10^5$) | 0.891 | 2.762 | 8.575 | 26.833 | 83.859 | 260.140 |
| Word Occurrences ($\times 10^5$) | 1 | 3.162 | 10 | 31.623 | 100 | 316.230 |
| *URL* | | | | | | |
| Size $Mb$ | 3.030 | 9.607 | 30.386 | 96.156 | 304.118 | — |
| Distinct Words ($\times 10^5$) | 0.361 | 0.923 | 2.355 | 5.769 | 12.898 | — |
| Word Occurrences ($\times 10^5$) | 1 | 3.162 | 10 | 31.623 | 100 | — |

quicksort, where randomisation of the input lessens the chance of quadratic running time. Input randomisation can also be used in cases such as binary search trees to eliminate the worst case when the input sequence is sorted.

Another application of randomisation is to process a small sample from a larger collection. In simple random sampling, each individual key in a collection has an equal chance of being selected. According to Olkem and Roten [12],

> Random sampling is used on those occasions when processing the entire dataset is unnecessary and too expensive ... The savings generated by sampling may arise either from reductions in the cost of retrieving the data ... or from subsequent postprocessing of the sample. Sampling is useful for applications which are attempting to estimate some aggregate property of a set of records.

# 3    Burstsort with Random Sampling

In earlier work [15], we showed that burstsort is efficient in sorting strings because of the low rate of cache miss compared to other string sorting methods. Cache misses occur when the string is fetched for the first time, during a burst, and

**Table 2.** Duplicates, sorting time for each method (milliseconds).

| Threshold | | Data set | | | | | |
|---|---|---|---|---|---|---|---|
| | | Set 1 | Set 2 | Set 3 | Set 4 | Set 5 | Set 6 |
| | Multikey quicksort | 62 | 272 | 920 | 3,830 | 14,950 | 56,070 |
| | MBM radixsort | 58 | 238 | 822 | 3,650 | 15,460 | 61,560 |
| | Adaptive radixsort | 74 | 288 | 900 | 3,360 | 12,410 | 51,870 |
| | SR-burstsort | 60 | 200 | 560 | 2,010 | 7,620 | 31,040 |
| 8192 | Burstsort | 58 | 218 | 630 | 2,220 | 7,950 | 29,910 |
| | DR-burstsort | 60 | 200 | 560 | 2,030 | 7,390 | 28,530 |
| | DRL-burstsort | 60 | 200 | 560 | 2,030 | 7,510 | 29,030 |
| 16384 | Burstsort | 60 | 210 | 630 | 2,270 | 7,970 | 28,490 |
| | DR-burstsort | 60 | 200 | 550 | 2,020 | 7,280 | 27,310 |
| 32768 | Burstsort | 60 | 210 | 630 | 2,380 | 8,250 | 28,530 |
| | DR-burstsort | 60 | 200 | 560 | 2,010 | 7,160 | 27,400 |
| 65536 | Burstsort | 60 | 210 | 640 | 2,480 | 8,590 | 29,620 |
| | DR-burstsort | 60 | 200 | 560 | 2,010 | 7,150 | 26,640 |
| 131072 | Burstsort | 60 | 220 | 660 | 2,550 | 9,190 | 31,260 |
| | DR-burstsort | 60 | 200 | 560 | 2,010 | 7,140 | 27,420 |

during the traversal phase when the bucket is sorted. Our results indicated that the threshold size should be selected such that the average number of cache misses per key during the traversal phase is close to 1.

Most cache misses occur while the strings are being inserted into the trie. One way in which cache misses could be reduced during the insertion phase is if the trie could be built beforehand, avoiding bursts and allowing strings to be placed in the trie with just one access, giving—if everything has gone well—a maximum of two accesses to a string overall, once during insertion and once during traversal. This is an upper bound, as some strings need not be referenced in the traversal phase and, as the insertion is a sequential scan, more than one string may fit into a cache line.

We propose building the trie beforehand using a random sample of the strings, which can be used to construct an approximation to the trie. The goal of the sampling is to get as close as possible to the shape of the tree constructed by burstsort, so the strings evenly distribute in the buckets, which can then be efficiently sorted in the cache. However, the cost of processing the sample should not be too great, or it can outweigh the gains. As a heuristic, we make just one pass through the sample, and use each string to suggest one additional trie node.

### Sampling process.

1. Create an empty trie root node $r$, where a trie node is an array of pointers (to either trie nodes or buckets).
2. Choose a sample size $R$, and create a stack of $R$ empty trie nodes.
3. A random sample of $R$ strings is drawn from the input data.

**Table 3.** Genome, sorting time for each method (milliseconds).

| Threshold | | Data set | | | | | |
|---|---|---|---|---|---|---|---|
| | | Set 1 | Set 2 | Set 3 | Set 4 | Set 5 | Set 6 |
| | Multikey quicksort | 72 | 324 | 1,250 | 4,610 | 16,670 | 62,680 |
| | MBM radixsort | 72 | 368 | 1,570 | 6,200 | 23,700 | 90,700 |
| | Adaptive radixsort | 92 | 404 | 1,500 | 4,980 | 17,800 | 66,100 |
| | SR-burstsort | 70 | 240 | 780 | 2,530 | 10,320 | 44,810 |
| 8192 | Burstsort | 70 | 258 | 870 | 2,830 | 8,990 | 31,540 |
| | DR-burstsort | 70 | 240 | 770 | 2,470 | 7,960 | 30,870 |
| | DRL-burstsort | 70 | 240 | 770 | 2,460 | 8,410 | 30,680 |
| 16384 | Burstsort | 70 | 290 | 910 | 2,760 | 8,720 | 30,280 |
| | DR-burstsort | 70 | 240 | 780 | 2,390 | 7,520 | 27,850 |
| 32768 | Burstsort | 80 | 280 | 940 | 3,000 | 9,520 | 31,140 |
| | DR-burstsort | 60 | 240 | 770 | 2,390 | 7,560 | 28,780 |
| 65536 | Burstsort | 70 | 310 | 1,010 | 3,130 | 9,820 | 32,860 |
| | DR-burstsort | 70 | 240 | 770 | 2,400 | 7,520 | 28,710 |
| 131072 | Burstsort | 80 | 300 | 1,070 | 3,400 | 10,940 | 36,630 |
| | DR-burstsort | 70 | 230 | 770 | 2,400 | 7,570 | 28,740 |

4. For each string $c_1 \ldots c_n$ in the sample,

    a) Use the string to traverse the trie until the current character corresponds to a null pointer. That is, set $p \leftarrow r$, and $i \leftarrow 1$, and, until $p[c_i]$ is null, continue by setting $p \leftarrow p[c_i]$ and incrementing $i$. For example, on insertion of "michael", if "mic" was already a path in the trie, a node is added for "h".

    b) If the string is not exhausted, that is, $i \leq n$, take a new node $t$ from the stack and set $p[c_i] \leftarrow t$.

The sampled strings are not stored in the buckets; to maintain stability, they are inserted when encountered during the main sorting process. The minimum number of trie nodes created is 1 if all the strings in the collection are identical and of length 1. The maximum number of trie nodes created is equal to the size of the sample and is more likely in collections such as the random collection.

The intuition behind this approach is that, if a prefix is common in the data then there will be several strings in the sample with that prefix. The sampling algorithm will then construct a branch of trie nodes corresponding to that prefix.

For example, in an English dictionary (from the utility ispell) of 127,001 strings, seven begin with "throu", 75 with "thro", 178 with "thr", 959 with "th", and 6713 with "t". Suppose we sample 127 times with replacement, corresponding to an expected bucket size of 1000. Then the probability of sampling "throu" is only 0.01, of "thro" is 0.07, of "thr" is 0.16, of "th" is 0.62, and of "t" is 0.999. With a bucket size of 1000, a burst trie would allocate a node corresponding to the path "t" and would come close to allocating a node for "th". Under sampling, it is almost certain that a node will be allocated for "t"—there is an even

**Table 4.** URLs, sorting time for each method (milliseconds). The fastest times in the burstsort family are shown in bold.

| Threshold | | Data set | | | | |
|---|---|---|---|---|---|---|
| | | Set 1 | Set 2 | Set 3 | Set 4 | Set 5 |
| | SR-burstsort | 100 | 360 | 1,310 | 5,350 | 19,420 |
| 8192 | Burstsort | 110 | 390 | 1,530 | 5,080 | 17,860 |
| | DR-burstsort | 110 | 370 | 1,450 | 4,860 | 17,130 |
| | DRL-burstsort | 100 | 370 | 1,450 | 4,850 | 17,610 |
| 16384 | Burstsort | 110 | 390 | 1,630 | 5,280 | 18,800 |
| | DR-burstsort | 110 | 380 | 1,530 | 4,890 | 17,350 |
| 32768 | Burstsort | 130 | 420 | 1,510 | 6,710 | 21,560 |
| | DR-burstsort | 110 | 370 | 1,380 | 5,890 | 18,670 |
| 65536 | Burstsort | 170 | 440 | 1,540 | 6,290 | 24,010 |
| | DR-burstsort | 110 | 370 | 1,380 | 5,410 | 19,360 |
| 131072 | Burstsort | 140 | 480 | 1,550 | 6,310 | 27,120 |
| | DR-burstsort | 110 | 370 | 1,340 | 5,330 | 19,830 |

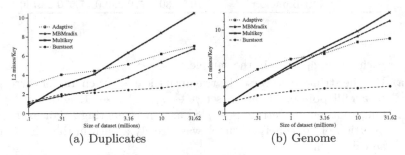

(a) Duplicates          (b) Genome

**Fig. 2.** L2 cache misses for the most efficient sorting algorithms, burstsort has a threshold of 8192.

chance that it would be one of the first 13 nodes allocated—and likely that a node would be allocated for "th". Nodes for the deeper paths are unlikely.

***SR-burstsort.*** In burstsort, the number of trie nodes created is roughly linear in the size of the set to be sorted. It is therefore attractive that the number of nodes allocated through sampling be a fixed percentage of the number of elements in the set; by the informal statistical argument above, the trie created in the initial phase should approximate the trie created by applying standard burstsort to the same data. In static randomised burstsort, or SR-burstsort, the trie structure created by sampling is then static. The structure grows only through addition of strings to buckets. The use of random sampling means that common prefixes will in the great majority of runs be represented in the trie and strings will distribute well amongst the buckets.

**Table 5.** Artificial sets, sorting time for each method (milliseconds).

| Threshold | | Collection | | |
|---|---|---|---|---|
| | | Artificial A | Artificial B | Artificial C |
| | SR-burstsort | 2,650 | 9,220 | 1,600 |
| 8192 | Burstsort | 2,740 | 10,130 | 1,430 |
| | DR-burstsort | 2,340 | 9,080 | 1,300 |
| 16384 | Burstsort | 2,510 | 10,110 | 1,460 |
| | DR-burstsort | 2,320 | 8,890 | 1,340 |
| 32768 | Burstsort | 2,910 | 10,540 | 1,880 |
| | DR-burstsort | 2,320 | 8,110 | 1,430 |
| 65536 | Burstsort | 3,760 | 11,210 | 2,610 |
| | DR-burstsort | 2,340 | 8,010 | 1,540 |
| 131072 | Burstsort | 5,190 | 11,820 | 3,810 |
| | DR-burstsort | 2,320 | 7,890 | 1,670 |
| 262144 | Burstsort | 7,900 | 13,200 | 5,660 |
| | DR-burstsort | 2,290 | 7,930 | 1,570 |

**Table 6.** Random, sorting time for each method (milliseconds).

| Threshold | | Data set | | | | | |
|---|---|---|---|---|---|---|---|
| | | Set 1 | Set 2 | Set 3 | Set 4 | Set 5 | Set 6 |
| | SR-burstsort | 50 | 170 | 570 | 1,930 | 7,060 | 29,410 |
| 8192 | Burstsort | 50 | 180 | 650 | 2,100 | 6,450 | 23,040 |
| | DR-burstsort | 50 | 180 | 580 | 2,050 | 6,910 | 30,790 |
| | DRL-burstsort | 50 | 180 | 570 | 2,050 | 6,470 | 23,340 |

For a set of $N$ strings, we need to choose a sample size. We use a relative trie size parameter $S$. For our experiments we used $S = 8192$, because this value was an effective bucket-size threshold in our earlier work. Then the sample size, and the maximum number of trie nodes that can be created, is $R = N/S$.

SR-burstsort proceeds as follows: use the sampling procedure above to build an access trie; insert the strings in turn into buckets; then traverse the trie and buckets to give the sorted result. No bursts occur. Buckets are a linked list of arrays of a fixed size (an implementation decision derived from preliminary experiments). The last element in each array is a pointer to the next array. In our experiments we have used an array size of 32.

SR-burstsort has several advantages compared to the original algorithm. The code is simpler, with no thresholds or bursting, thus requiring far fewer instructions during the insertion phase. Insertion also requires fewer string accesses. The nodes are allocated as a block, simplifying dynamic memory management.

However, bucket size is not capped, and some buckets may not fit entirely within the cache. The bucket sorting routine is selected mainly for its instruc-

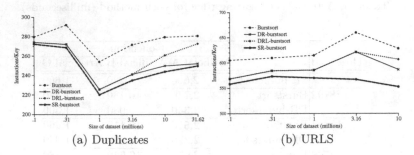

**Fig. 3.** Instructions per element on each data set, for each variant of burstsort for a threshold of 32768.

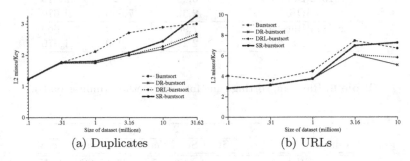

**Fig. 4.** L2 cache misses per element on each data set, for each variant of burstsort for a threshold of 32768.

tion and space efficiency for small sets of strings and not for cache efficiency. Moreover, small changes in the trie shape can lead to large variations in bucket size: omitting a single crucial trie node due to sampling error may mean that a very large bucket is created.

***DR-burstsort.*** An obvious next step is to eliminate the cases in SR-burstsort when the buckets become larger than cache and bucket sorting is not entirely cache-resident. This suggests dynamic randomised burstsort, or DR-burstsort. In this approach, an initial trie is created through sampling as before, but as in the original burstsort a limit is imposed on bucket size and buckets are burst if this limit is exceeded. DR-burstsort avoids the bad cases that arise in SR-burstsort due to sampling errors. The number of bursts should be small, but, compared to SR-burstsort, additional statistics must be maintained.

Thus DR-burstsort is as follows: using a relative trie size $S$, select a sample of $R = N/S$ strings and create an initial trie; insert the strings into the trie as for burstsort; then traverse as for burstsort or SR-burstsort. Buckets are represented as arrays of 16, 128, 1024, or 8192 pointers, growing from one size to the next as the number of strings to be stored increases, as we have described elsewhere for burstsort [16].

***DRL-burstsort.*** For the largest sets of strings, the trie is much too large to be cache resident. That is, there is a trade-off between whether the largest bucket can fit in cache and whether the trie can fit in cache. One approach is to stop bursts at some point, especially as bursts late in the process are not as helpful. We have not explored this approach, as it would be unsuccessful with sorted data.

Another approach is to limit the size of the initial trie to fit in cache, to avoid the disadvantages of extraneous nodes being created. This variant, DR-burstsort with limit or DRL-burstsort, is tested below. The limit used in our experiments depends on the size of the cache and the size of the trie nodes. In our experiments, we chose $R$ so that $R$ times node size is equal to the cache size.

## 4   Experiments

For realistic experiments with large sets of strings, we are limited to sources for which we have sufficient volumes of data. We have drawn on web data and genomic data. For the latter, we have parsed nucleotide strings into overlapping 9-grams. For the former, derived from the TREC project [5,6], we extracted both words—alphabetic strings delimited by non-alphabetic characters—and URLs. For the words, we considered sets with and without duplicates, in both cases in order of occurrence in the original data.

For the word data and genomic data, we created six subsets, of approximately $10^5$, $3.1623 \times 10^5$, $10^6$, $3.1623 \times 10^6$, $10^7$, and $3.1623 \times 10^7$ strings each. We call these SET 1, SET 2, SET 3, SET 4, SET 5, and SET 6 respectively. For the URL data, we created SET 1 to SET 5. In each case, only SET 1 fits in cache. In detail, the data sets are as follows.

**Duplicates.** Words in order of occurrence, including duplicates. The statistical characteristics are those of natural language text; a small number of words are frequent, while many occur once only.

**No duplicates.** Unique strings based on word pairs in order of first occurrence in the TREC web data.

**Genome.** Strings extracted from a collection of genomic strings, each typically thousands of nucleotides long. The strings are parsed into shorter strings of length 9. The alphabet is comprised of four characters, "a", "t", "g", and "c". There is a large number of duplicates and the data shows little locality.

**Random.** An artificially generated collection of strings whose characters are uniformly distributed over the entire ASCII range. The length of each string is random in the range 1–20.

**URL.** Complete URLs, in order of occurrence and with duplicates, from the TREC web data. Average length is high compared to the other sets of strings.

**Artificial A.** A collection of identical strings on an alphabet of one character. Each string is one hundred characters long and the size of the collection is one million.

**Artificial B.** A collection of strings with an alphabet of nine characters. The length of strings are varied randomly from one to hundred and the size of the collection is ten million.

**Artificial C.** A collection of strings whose length ranges from one to hundred. The alphabet size is one and the strings are ordered in increasing length arranged cyclically. The size of the collection is one million.

The cost of bursting increases with the size of the container as more strings need to be fetched from memory, leading to increases in the number of cache misses and of instructions. Each correct prediction of a trie node removes the need to burst a container. Another situation where bursting could be expensive is use of inefficient data structures such as binary search trees or linked lists as containers. Traversing a linked list could result in two memory accesses for each container element, one access to the string and one access to the list node. To show how sampling can be beneficial as bursting becomes more expensive, we have measured the running time, instruction count and cache misses as the size of the container is increased from 1024 to 131,072, or, for the artificial collections, up to 262,144.

The aim of the experiments is to compare the performance of our algorithms, in terms of the running time, number of instructions, and number of L2 cache misses. The time measured is to sort an array of pointers to strings; the array is returned as the output. We therefore report the CPU times, not elapsed times, and exclude the time taken to parse the collections into strings.

The experiments were run on a Pentium III Xeon 700 MHz computer with 2 Gb of internal memory, 1 Mb L2 cache with block size of 32 bytes, 8-way associativity and a memory latency of about 100 cycles. We have used the highest compiler optimization O3 in all our experiments. The total number of milliseconds of CPU time has been measured; the time taken for I/O or to parse the collection are not included as these are in common for all algorithms. For the cache simulations, we have used `valgrind` [14].

## 5   Results

We present results in three forms: time to sort each data set, instruction counts, and L2 cache misses. Times for sorting are shown in Tables 2 to 6. Instruction counts are shown in Figures 3 and 5. L2 cache misses are shown in Figures 3, 3 and 5; the trends for the other data sets are similar.

On duplicates, the sorting times for the burstsort methods are, for all cases but SET 1, faster than for the previous methods. These results are as observed in our previous work. The performance gap steadily grows with data set size, and the indications from all the results—instructions, cache misses, and timings—are that the improvements yielded by burstsort will continue to increase with both

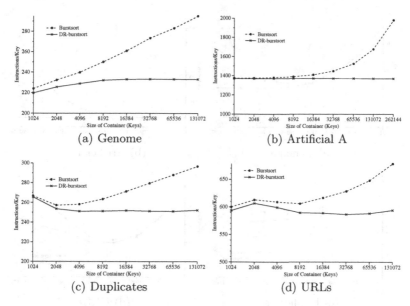

(a) Genome

(b) Artificial A

(c) Duplicates

(d) URLs

**Fig. 5.** Instructions per element for the largest data set, for each variant of burstsort.

changes in computer architecture and growing data volumes. Figure 3 shows the L2 cache misses in comparison to the best algorithms found in our earlier work.

Figures 3 and 3 show the number of instructions and L2 cache misses for a container size of 32768. Several overall trends can be observed. The number of instructions per string does not vary dramatically for any of the methods, though it does have perturbations due to characteristics of the individual data sets. SR-burstsort consistently uses fewer instructions than the other methods, while the original burstsort requires the most. Amongst the burstsorts, SR-burstsort is consistently the slowest for the larger sets due to more L2 cache misses than burstsort, despite requiring fewer instructions.

For most collections, either DR-burstsort or DRL-burstsort is the fastest sorting technique, and they usually yield similar results. Compared to burstsort, DR-burstsort uses up to 24% fewer instructions and incurs up to 37% fewer cache misses. However, there are exceptions, in particular DRL-burstsort has done much better than DR-burstsort on the random data; on this data, burstsort is by a small margin the fastest method tested. The heuristic in DRL-burstsort of limiting the initial trie to the cache size has led to clear gains in this case, in which the sampling process is error-prone.

Some of the data sets have individual characteristics that affect the trends. In particular, with the fixed length of the strings in the genome data, increasing the number of strings does not increase the number of distinct strings, thus the relative costs of sorting under the different methods changes with increasing data set size. In contrast, with duplicates the number of distinct strings continues to steadily grow.

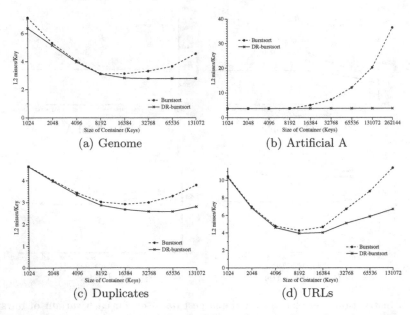

**Fig. 6.** L2 cache misses per element for the largest data set, for each variant of burstsort.

The sorting times shown in Tables 2 to 6 shows that as the size of the container increases, burstsort becomes more expensive. On the other hand, the cost of DR-burstsort does not vary much with increasing container size. Table 5 shows DR-burstsort can be as much as 3.5 times faster than burstsort. As shown in Figure 5, the number of instructions incurred by DR-burstsort can be up to 30% less than burstsort. Also, interestingly the number of instructions do not appear to vary much as the size of the container increases. Figure 5 similarly shows that the number of misses incurred by DR-burstsort can be up to 90% less than burstsort.

All of the new methods require fewer instructions than the original burstsort. More importantly, in most cases DR-burstsort and DRL-burstsort require fewer cache misses. This trend means that, as the hardware performance gap grows, the relative performance of our new methods will continue to improve.

## 6    Conclusions

We have proposed new algorithms—SR-burstsort, DR-burstsort, and DRL-burstsort—for fast sorting of strings in large data collections. They are a variant of our burstsort algorithm and are based on construction of a small trie that rapidly allocates strings to buckets. In the original burstsort, the trie was constructed dynamically; the new algorithms are based on taking a random sample of the strings and using them to construct an initial trie structure before any strings are inserted.

SR-burstsort, where the trie is static, reduces the need for dynamic memory management and simplifies the insertion process, leading to code with a lower instruction count than the other alternatives. Despite promising performance in preliminary experiments and the low instruction count, however, it is generally slower than burstsort, as there can easily be bad cases where a random sample does not correctly predict the trie structure, which leads to some buckets being larger than expected.

DR-burstsort and DRL-burstsort improve on the worst case of SR-burstsort by allowing the trie to be modified dynamically, at the cost of additional checks during insertion. They are faster than burstsort in all experiments with real data, due to elimination of the need for most of the bursts. The use of a limit in DRL-burstsort avoids poor cases that could arise in data with a flat distribution.

Our experimental results show that the new variants reduce cache misses even further than does the original burstsort, by up to 37%, while simultaneously reducing instruction counts by up to 24%. As the cost of bursting grows, the new variants reduce cache misses by up to 90%, while simultaneously reducing instruction counts by up to 30% and the time to sort is reduced by up to 72% as compared to burstsort.

There is scope to further improve these algorithms. Pre-analysis of collections to see whether the alphabet is restricted showed an improvement of 16% for genomic collections. Pre-analysis would be of value for customised sorting applications. Another variation is to choose the sample size based on analysis of collection characteristics. A further variation is to recursively apply SR-burstsort to large buckets. We are testing these options in current work.

Even without these improvements, however burstsort and its variants are a significant advance, dramatically reducing the costs of sorting a large set of strings. Cache misses and running time are as low as half that required by any previous method. With the current trends in computer architecture, the relative performance of our methods will continue to improve.

# References

1. A. Andersson and S. Nilsson. Implementing radixsort. *ACM Jour. of Experimental Algorithmics*, 3(7), 1998.
2. Lars Arge, Paolo Ferragina, Roberto Grossi, and Jeffrey Scott Vitter. On sorting strings in external memory. In *Proceedings of the 29th Annual ACM Symposium on Theory of Computing*, pages 540–548, El Paso, 1997. ACM Press.
3. J. Bentley and R. Sedgewick. Fast algorithms for sorting and searching strings. In *Proc. Annual ACM-SIAM Symp. on Discrete Algorithms*, pages 360–369, New Orleans, Louisiana, 1997. ACM/SIAM.
4. Rajiv Gupta, Scott A. Smolka, and Shaji Bhaskar. On randomization in sequential and distributed algorithms. *ACM Computing Surveys*, 26(1):7–86, 1994.
5. D. Harman. Overview of the second text retrieval conference (TREC-2). *Information Processing & Management*, 31(3):271–289, 1995.
6. D. Hawking, N. Craswell, P. Thistlewaite, and D. Harman. Results and challenges in web search evaluation. In *Proc. World-Wide Web Conference*, 1999.

7. S. Heinz, J. Zobel, and H. E. Williams. Burst tries: A fast, efficient data structure for string keys. *ACM Transactions on Information Systems*, 20(2):192–223, 2002.
8. A. LaMarca and R. E. Ladner. The influence of caches on the performance of sorting. In *Proc. Annual ACM-SIAM Symp. on Discrete Algorithms*, pages 370–379. ACM Press, 1997.
9. P. M. McIlroy, K. Bostic, and M. D. McIlroy. Engineering radix sort. *Computing Systems*, 6(1):5–27, 1993.
10. R. Motwani and P. Raghavan. *Randomized Algorithms*. Cambridge University Press, 1995.
11. S. Nilsson. *Radix Sorting & Searching*. PhD thesis, Department of Computer Science, Lund, Sweden, 1996.
12. F. Olken and D. Rotem. Random sampling from databases - a survey. *Statistics and Computing*, 5(1):25–42, March 1995.
13. N. Rahman and R. Raman. Adapting radix sort to the memory hierarchy. *ACM Jour. of Experimental Algorithmics*, 6(7), 2001.
14. J. Seward. Valgrind—memory and cache profiler, 2001. http://developer.kde.org/~sewardj/docs-1.9.5/cg_techdocs.html.
15. R. Sinha and J. Zobel. Cache-conscious sorting of large sets of strings with dynamic tries. In R. Ladner, editor, *5th ALENEX Workshop on Algorithm Engineering and Experiments*, pages 93–105, Baltimore, Maryland, January 2003.
16. R. Sinha and J. Zobel. Efficient trie-based sorting of large sets of strings. In M. Oudshoorn, editor, *Proceedings of the Australasian Computer Science Conference*, pages 11–18, Adelaide, Australia, February 2003.
17. L. Xiao, X. Zhang, and S. A. Kubricht. Improving memory performance of sorting algorithms. *ACM Jour. of Experimental Algorithmics*, 5:3, 2000.

# The Datapath Merging Problem in Reconfigurable Systems: Lower Bounds and Heuristic Evaluation

Cid C. de Souza[1], André M. Lima[1], Nahri Moreano[2], and Guido Araujo[1]

[1] Institute of Computing, State University of Campinas,
C.P. 6176, 13084-970 Campinas, Brazil.
{cid,andre.lima,guido}@ic.unicamp.br

[2] Department of Computing and Statistics, Federal University of Mato Grosso do Sul,
79070-900 Campo Grande, Brazil.
moreano@dct.ufms.br

**Abstract.** In this paper we investigate the datapath merging problem (DPM) in reconfigurable systems. DPM is in $\mathcal{NP}$-hard and it is described here in terms of a graph optimization problem. We present an Integer Programming (IP) formulation of DPM and introduce some valid inequalities for the convex hull of integer solutions. These inequalities form the basis of a branch-and-cut algorithm that we implemented. This algorithm was used to compute lower bounds for a set of DPM instances, allowing us to assess the performance of the heuristic proposed by Moreano et al. [1] which is among the best ones available for the problem. Our computational experiments confirmed the efficiency of Moreano's heuristic. Moreover, the branch-and-cut algorithm also was proved to be a valuable tool to solve small-sized DPM instances to optimality.

## 1 Introduction

It is well known that embedded systems must meet strict constraints of high-throughput, low power consumption and low cost, specially when designed for signal processing and multimedia applications [2]. These requirements lead to the design of application specific components, ranging from specialized functional units and coprocessors to entire application specific processors. Such components are designed to exploit the peculiarities of the application domain in order to achieve the necessary performance and to meet the design constraints.

With the advent of reconfigurable systems, the availability of large/cheap arrays of programmable logic has created a new set of architectural alternatives for the design of complex digital systems [3,4]. Reconfigurable logic brings together the flexibility of software and the performance of hardware [5,6]. As a result, it became possible to design application specific components, like specialized datapaths, that can be reconfigured to perform a different computation, according to the specific part of the application that is running. At run-time, as each portion of the application starts to execute, the system reconfigures the datapath

C.C. Ribeiro and S.L. Martins (Eds.): WEA 2004, LNCS 3059, pp. 545–558, 2004.
© Springer-Verlag Berlin Heidelberg 2004

so as to perform the corresponding computation. Recent work in reconfigurable computing research has shown that a significant performance speedup can be achieved through architectures that map the most time-consuming application kernel modules or inner-loops to a reconfigurable datapath [7,8,9].

The reconfigurable datapath should have as few and simple hardware blocks (functional units and registers) and interconnections (multiplexors and wires) as possible, in order to reduce its cost, area, and power consumption. Thus hardware blocks and interconnections should be reused across the application as much as possible. Resource sharing has also crucial impact in reducing the system reconfiguration overhead, both in time and space.

To design such a reconfigurable datapath, one must represent each selected piece of the application as a control/data-flow graph (CDFG) and merge them together, synthesizing a single reconfigurable datapath. The control/data-flow graph merging process enables the reuse of hardware blocks and interconnections by identifying similarities among the CDFGs, and produces a single datapath that can be dynamically reconfigured to work for each CDFG. Ideally, the resulting datapath should have the minimum area cost. Ultimately, this corresponds to minimize the amount of hardware blocks and interconnections in the reconfigurable datapath. The datapath merging problem (DPM) seeks such an optimal merging and is known to be in $\mathcal{NP}$-hard [10].

To minimize the area cost one has to minimize the total area required by both hardware blocks and interconnections in the reconfigurable datapath. However, since the area occupied by hardware blocks is typically much larger than that occupied by the interconnections, the engineers are only interested in solutions that use as few hardware blocks as possible. Clearly, the minimum quantity of blocks required for each type of hardware block is given by the maximum number of such block that is needed among all CDFGs passed at the input. The minimum amount of hardware blocks in the reconfigurable datapath can be computed as the sum of these individual minima. As a consequence, DPM reduces to the problem of finding the minimum number of interconnections necessary to implement the reconfigurable datapath.

Fig. 1 illustrates the concept of control/data-flow graph merging and the problem we are tackling. For simplicity, the multiplexors, who select the inputs for certain functional blocks, are not represented. The graphs $G'$ and $G$ represent two mappings of the CDFGs $G_1$ and $G_2$. In both these mappings, vertices $a_1$ and $a_5$ from $G_1$ are mapped onto vertices $b_1$ and $b_3$ from $G_2$, respectively, while vertex $a_4$ of $G_1$ has no counterpart in $G_2$. The difference between the two mappings is that, in $G'$ vertex $b_2$ of $G_2$ is mapped onto vertex $a_2$ of $G_1$, while it is mapped onto $a_3$ in $G$. The mappings $G'$ and $G$ are both feasible since they only match hardware blocks that are logically equivalent. Though their reconfigurable datapaths have the same amount of hardware blocks, in $G'$ no arcs are overlapped while in $G$ the arcs $(a_3, a_5)$ and $(b_2, b_3)$ coincide (see the highlighted arc in Fig. 1). In practical terms, this means that one less multiplexor is needed and, therefore, $G$ is a better solution for DPM than $G'$.

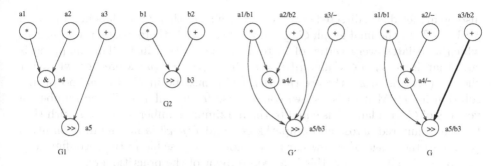

**Fig. 1.** Example of a DPM instance.

In this paper we present an Integer Programming (IP) formulation for DPM and introduce some valid inequalities for the convex hull of integer solutions. These inequalities form the basis of a branch-and-cut (**B&C**) algorithm that we implemented. The contributions of our work are twofold. First the **B&C** algorithm was able to compute lower bounds for a set of DPM instances, allowing us to assess the performance of the heuristic proposed by Moreano et al. [1], one of the best suboptimal algorithms available for DPM. Secondly, the **B&C** also proved to be a valuable tool to solve small-sized DPM instances to optimality.

The paper is organized as follows. The next section gives a formal description of DPM in terms of Graph Theory. Section 3 briefly discusses Moreano's heuristic. Section 4 presents an IP formulation for DPM, together with some classes of valid inequalities that can be used to tighten the original model. In Sect. 5 we report our computational experiments with the **B&C** algorithm and analyze the performance of Moreano's heuristic. Finally, in Sect. 6 we draw some conclusions and point out to future investigations.

## 2   A Graph Model for DPM

In this section we formulate DPM as a graph optimization problem. The input is assumed to be composed of $n$ datapaths corresponding to application loops of a computer program. The goal is to find a merging of those datapaths into a reconfigurable one that is able to work as each individual loop datapath alone and has as least hardware blocks (functional units and registers) and interconnections as possible. That is, the reconfigurable datapath must be capable of performing the computation of each loop, multiplexed in time.

The $i$-th datapath is modeled as a directed graph $G_i = (V_i, E_i)$, where the vertices in $V_i$ represent the hardware blocks in the datapath, and the arcs in $E_i$ are associated to the interconnections between the hardware blocks. The types of hardware blocks (e.g. adders, multipliers, registers, etc) are modeled through a labeling function $\pi_i : V_i \to \mathbb{T}$, where $\mathbb{T}$ is the set of labels representing hardware block types. For each vertex $u \in V_i$, $\pi_i(u)$ is the type of the hardware block associated to $u$. A reconfigurable datapath representing a solution of DPM can also be modeled as a directed graph $G = (V, E)$ together with a labeling function $\pi : V \to \mathbb{T}$. In the final graph $G$, given $i \in \{1, \ldots, n\}$, there exists a mapping $\mu_i$ which associates every vertex of $V_i$ to a distinct vertex in $V$. This mapping is such that, if $v \in V_i$, $u \in V$ and $\mu_i(v) = u$, then $\pi_i(v) = \pi(u)$. Moreover, whenever the arc $(v, v')$ is in $E_i$, the arc $(\mu_i(v), \mu_i(v'))$ must be in $E$. If $G$ is an optimal solution for DPM it satisfies two conditions: *(a)* for all $T \in \mathbb{T}$, the number of vertices of $G$ with label $T$ is equal to the maximum number of vertices with that label encountered across all datapaths $G_i$; and *(b)* $|E|$ is minimum. Condition *(a)* forces the usage of as few hardware blocks as possible in the reconfigurable datapath. As cited before, this is a requirement of the practitioners.

# 3   Moreano's Heuristic for DPM

Since DPM is $\mathcal{NP}$-hard, it is natural to devise suboptimal algorithms that can solve it fast, preferably in polynomial time. In Moreano et al. [1], the authors proposed a heuristic for DPM and give comparative results showing that it out-performs other heuristics presented in the literature. Moreano's heuristic (MH) is briefly described in this section. In Sect. 5, rather than assess the efficiency of MH using upper bounds generated with other methods, we compare its solutions with strong lower bounds computed via the IP model discussed in Sect. 4.

For an integer $k > 1$, define $k$-DPM as the DPM problem whose input is made of $k$ loop datapaths. Thus, the original DPM problem would be denoted by $n$-DPM but the former notation is kept for simplicity. MH is based on an algorithm for 2-DPM, here denoted by 2DPMalg, that is presented below.

Let $G_1 = (V_1, E_1)$ and $G_2 = (V_1, E_1)$ be the input graphs and $\pi_1$ and $\pi_2$ their respective labeling functions. A pair of arcs $\{(u, v), (w, z)\}$ in $E_1 \times E_2$ is said to form a *feasible mapping* if $\pi_1(u) = \pi_2(w)$ and $\pi_1(v) = \pi_2(z)$. The first step of 2DPMalg constructs the *compatibility graph* $H = (W, F)$ of $G_1$ and $G_2$. The graph $H$ is undirected. The vertices in $W$ are in one-to-one correspondence with the pairs of arcs in $E_1 \times E_2$ which form feasible mappings. Given two vertices $a$ and $b$ in $W$ represented by the corresponding feasible mappings, say $a = \{(u, v), (w, z)\}$ and $b = \{(u', v'), (w', z')\}$, the edge $(a, b)$ is in $F$ except if one of the following conditions hold: *(i)* $u = u'$ and $w \neq w'$ or *(ii)* $v = v'$ and $z \neq z'$ or *(iii)* $u \neq u'$ and $w = w'$ or *(iv)* $v \neq v'$ and $z = z'$. If the edge $(a, b)$ is in $F$, the feasible mappings that they represent are *compatible*, explaining why $H$ is called the compatibility graph. Now, as explained in [1], an optimal solution for 2-DPM can be computed by solving the maximum clique problem on $H$. The solution of DPM is easily derived from an optimal clique of $H$ since the feasible mappings associated to the vertices of this graph provide the proper matchings of the vertices of $G_1$ and $G_2$. However, it is well-known that the clique problem is $\mathcal{NP}$-hard. Thus, the approach used in MH is to apply a good heuristic available for cliques to solve 2-DPM. Later in Sect. 5, we discuss how this is done in practice.

Before we continue, let us give an example of the ideas discussed in the preceding paragraph. To this end, consider the graphs $G_1$ and $G_2$ in Fig. 2 representing an instance of 2-DPM. According to the notation used in this figure,

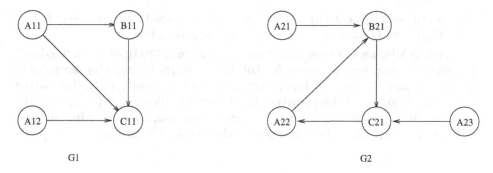

**Fig. 2.** Example of a 2-DPM instance.

each vertex $u$ in a graph $G_i$ is identified with a label $T_{ij}$, which denotes that $u$ is the $j$-th vertex of $G_i$ and $\pi_i(u) = T$. For instance, $A_{12}$ is the second vertex of $G_1$ which have type $A$. This notation is used to other figures representing DPM instances and solutions throughout. Figure 3 depicts the compatibility graph $H$ of $G_1$ and $G_2$. Consider, for example, the feasible mappings $(A_{11}, B_{11}), (A_{21}, B_{21})$ (vertex $w_1$ in $H$) and $(B_{11}, C_{11}), (B_{21}, C_{21})$ (vertex $w_5$ in $H$). For those mappings, no vertex from $G_1$ maps onto two distinct vertices in $G_2$ and vice-versa. As a result, these two mapping are compatible, and an edge $(w_1, w_5)$ is required in $H$. On the other hand, no edge exists in $H$ between vertices $w_2$ and $w_3$. The reason is that the mappings represented by these vertices are incompatible, since otherwise vertex $A_{11}$ in $G_1$ would map onto both $A_{22}$ and $A_{23}$ in $G_2$.

A maximum clique of the compatibility graph $H$ in Fig. 2 is given by vertices $w_1$, $w_4$ and $w_5$. An optimal solution $G$ for 2-DPM can be easily built from this clique. The resulting graph $G$ is shown in Fig. 3 and is obtained as follows. First, we consider the vertices of the clique. For instance, for vertex $w_1$ represents the feasible mapping $\{(A_{11}, B_{11}), (A_{21}, B_{21})\}$, we add to $G$ two vertices $u_1$ and $u_2$ corresponding respectively to the mapped vertices $\{A_{11}, A_{21}\}$ and $\{B_{11}, B_{21}\}$. Moreover, we also include in $G$ the arc $(u_1, u_2)$ to represent the feasible mapping associated to $w_1$. Analogous operations are now executed for vertices $w_4$ and $w_5$. The former vertex is responsible for the addition of vertices $u_4$ and $u_5$ and of arc $(u_4, u_5)$ in $G$ while the latter gives rise to the addition of arc $(u_2, u_4)$. Finally, we add to $G$ the vertex $u_3$ corresponding to the non-mapped vertex $A_{22}$ from

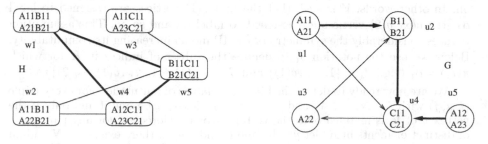

**Fig. 3.** Compatibility graph and an optimal solution for the 2-DPM instance of Fig. 2.

$G_2$, and the arcs $(u_1, u_4)$, $(u_4, u_3)$ and $(u_3, u_2)$ corresponding respectively to arcs $(A_{11}, C_{11})$ from $G_1$ and arcs $(C_{21}, A_{22})$ and $(A_{22}, B_{21})$ from $G_2$.

Back to MH, we now show how it uses algorithm 2DPMalg as a building-block for getting suboptimal solutions for DPM. MH starts by applying 2DPMalg to graphs $G_1$ and $G_2$ with labeling functions $\pi_1$ and $\pi_2$, respectively. The output is a graph $G$ and a labeling function $\pi$. At each iteration $i$, $i \in \{3, \ldots, n\}$, MH applies 2DPMalg to graphs $G$ and $G_i$ and their functions $\pi$ and $\pi_i$. After all these pairwise matchings have been completed, the graph $G$ is returned.

# 4   Integer Linear Programming Exact Solution

A natural question that arises when one solves a hard problem heuristically is how far the solutions are from the true optimum. For 2-DPM, algorithm 2DPMalg from Sect. 3 can be turned into an exact method, provided that an exact algorithm is used to find maximum cliques. However, this approach only works when merging two datapaths. A naive extension of the method to encompass the general case requires the solution of hard combinatorial problems on large-sized instances which cannot be handled in practice. As an alternative, in this section we derive an IP model for DPM. The aim is to compute that model to optimality via IP techniques whenever the computational resources available permit. When this is not the case, we would like at least to generate good lower bounds that allow us to assess the quality of the solutions produced by MH.

Let us denote by $\alpha_i$ the $i$-th type of hardware block and assume that $\mathbb{T}$ has $m$ elements, i.e., $\mathbb{T} = \{\alpha_1, \ldots, \alpha_m\}$. Moreover, for every $i \in \{1, \ldots, n\}$ and every $t \in \{1, \ldots, m\}$, let us define $b_{it}$ as the number of vertices in $V_i$ associated with a hardware block of type $\alpha_t$ and let $q(t) = \max\{b_{it} : 1 \leq i \leq n\}$. Then, the solutions of DPM are graphs with $k$ vertices, where $k = \sum_{t=1}^{m} q(t)$. In the remainder of the text, we denote by $K$ and $N$ the sets $\{1, \ldots, k\}$ and $\{1, \ldots, n\}$, respectively. Besides, we assume that for every hardware block of type $\alpha_t$ in $\mathbb{T}$ there exists $i \in N$ and $u \in V_i$ such that $\pi_i(u) = \alpha_t$.

When $V$ is given by $\{v_1, v_2, \ldots, v_k\}$, we can assume without loss of generality that $\pi(v_1) = \ldots = \pi(v_{q(1)}) = \alpha_1$, $\pi(v_{q(1)+1}) = \ldots = \pi(v_{q(1)+q(2)}) = \alpha_2$ and so on. In other words, $V$ is such that the first $q(1)$ vertices are assigned to label $\alpha_1$, the next $q(2)$ vertices are assigned to label $\alpha_2$ and so on. This assumption reduces considerably the symmetry of the IP model increasing its computability. Below we use the notation $J_t$ to denote the subset of indices in $K$ for which $\pi(v_i) = \alpha_t$ (e.g., $J_1 = \{1, \ldots, q(1)\}$ and $J_2 = \{q(1)+1, \ldots, q(1)+q(2)\}$).

We are now ready to define the binary variables of our model. For every triple $(i, u, j)$ with $i \in N$, $u \in V_i$ and $j \in J(\pi_i(u))$, let $x_{uij}$ be one if and only if the vertex $u$ of $V_i$ is mapped onto the vertex $v_j$ of $V$. Moreover, for any pair $(j, j')$ of distinct elements in $K$, let $y_{jj'}$ be one if and only if there exists $i \in N$ and an arc in $E_i$ such that one of its end-vertices is mapped onto vertex $v_j$ of $V$ while the other end-vertex is mapped onto $v_{j'}$. The IP model is then the following.

$$\min \qquad z = \sum_{\forall j} \sum_{\forall j' \neq j} y_{jj'} \tag{1}$$

$$x_{uij} + x_{u'ij'} - y_{jj'} \leq 1 \qquad \forall i \in N, \forall (u, u') \in E_i, \forall j \in J(\pi_i(u)),$$
$$\forall j' \in J(\pi_i(u')), j \neq j' \tag{2}$$

$$\sum_{u \in V_i \mid j \in J(\pi_i(u))} x_{uij} \leq 1 \qquad \forall i \in N, \forall j \in K \tag{3}$$

$$\sum_{j \in J(\pi_i(u))} x_{uij} = 1 \qquad \forall i \in N, \forall u \in V_i \tag{4}$$

$$y_{jj'} \in \{0, 1\} \qquad \forall j, j' \in K, j \neq j' \tag{5}$$

$$x_{uij} \in \{0, 1\} \qquad \forall i \in N, \forall u \in V_i, \forall j \in J(\pi_i(u)) \tag{6}$$

Equation (1) expresses the fact that an optimal solution to DPM is a graph with as few arcs as possible. Constraints (2) force the existence of arcs in the output graph. Constraints (3) avoid multiple vertices in one input graph to be mapped to a single vertex of the output graph. Finally, (4) guarantees that any vertex in any input graph is mapped to exactly one vertex of $V$.

Notice that (5) can be replaced by inequalities of the form $0 \leq y_{jj'} \leq 1$ for all $j \neq j'$ with $(j, j') \in K \times K$. This is so because the objective function together with (2) force the $y$ variables to assume values in the limits of the interval $[0, 1]$ and, therefore, to be integer-valued. This remark is important for computational purposes. The most successful algorithms implemented in commercial solvers for IP are based on branch-and-bound (**B&B**) algorithms. The size of the solution space increases exponentially with the number of integer variables in the model. Thus, relaxing the integrality constraints on the $y$ variables in our model, we reduce the search space and increase the chances of success of the algorithm.

The solution of hard combinatorial problems through IP algorithms relies largely on the quality of the dual bounds produced by the linear relaxation of the model at hand. To improve the dual bounds, the relaxation can be amended with additional constraints that are valid for integer solutions of the relaxation but not for all the continuous ones. This addition of valid inequalities tightens the relaxation for its feasibility set strictly decreases. The new constraints, typically chosen from a particular class of valid inequalities, can be either included *a priori* in the model, which is then solved by a standard **B&B** algorithm, or generated on the fly during the enumeration procedure whenever they are violated by the solution of the current relaxation. The latter method gives rise to **B&C** algorithms for IP. Quite often the use of **B&C** is justified by the number of potential inequalities that can be added to the model which, even for limited classes of valid inequalities, is exponentially large. On the other hand, when inequalities are generated on the fly, algorithms that search for violated inequalities are needed. These algorithms solve the so-called *separation problem* for classes of valid inequalities and are named *separation routines*. For a thorough presentation of the Theory of Valid Inequalities and IP in general, we refer to the book by Nemhauser and Wolsey [11]. In the sequel we present two classes of valid inequalities that we use to tighten the formulation given in (1)-(6).

### 4.1   The Complete Bipartite Subgraph (CBS) Inequalities

The idea is to strengthen (2) using (3). This is done through special subgraphs of
the input graphs. Given a directed graph $D$, we call a subgraph $H = (W_1, W_2, F)$
a CBS of $D$ if, for every pair of vertices $\{w_1, w_2\}$ in $W_1 \times W_2$, $(w_1, w_2)$ is in $F$.
Now, consider an input graph $G_i$, $i \in N$, of a DPM instance and two distinct
labels $\alpha_{t_1}, \alpha_{t_2} \in \mathbb{T}$. Let $H_i^{t_1,t_2} = (V_i^{t_1}, V_i^{t_2}, E_i^{t_1,t_2})$ be a CBS of $G_i$ such that all
vertices in $V_i^{t_1}$ ($V_i^{t_2}$) have label $\alpha_{t_1}$ ($\alpha_{t_2}$). Suppose that $H_i^{t_1,t_2}$ is maximal with
respect to vertex inclusion. Assume that $v_j$ and $v_{j'}$ are two vertices in $V$, the
vertex set of the resulting graph $G$, with labels $\alpha_{t_1}$ and $\alpha_{t_2}$, respectively. The
CBS inequality associated to $H_i^{t_1,t_2}$, $v_j$ and $v_{j'}$ is

$$\sum_{u \in V_i^{t_1}} x_{uij} + \sum_{v \in V_i^{t_2}} x_{vij'} - y_{jj'} \leq 1. \tag{7}$$

**Theorem 1.** *(7) is valid for all integer solutions of the system (2)-(6).*

*Proof.* Due to (3), the first summation in the left-hand side (LHS) of (7) cannot
exceed one. A similar result holds for the second summation. Thus, if an integer
solution exists violating (7), both summations in the LHS have to be one. But
then, there would be a pair of vertices $\{u, u'\}$ in $V_i^{t_1} \times V_i^{t_2}$ such that $u$ is mapped
onto vertex $v_j$ and $u'$ onto vertex $v_{j'}$. However, as $H_i^{t_1,t_2}$ is a CBS of $G_i$, $(v_j, v_{j'})$
must be an arc of $E$, the arc set of the output graph $G$, i.e., $y_{jj'}$ is one.     □

Clearly, if $(u, u')$ is not a maximal CBS of $G_i$, then (2) is dominated by some
inequality in (7) and, therefore, superfluous. Our belief is that the number of
CBS inequalities is exponentially large which, in principle, would not recommend
to add them all to the initial IP model. However, the DPM instances we tested
reveal that, in practical situations, this amount is actually not too large and the
CBS inequalities can all be generated via a backtracking algorithm. This allows
us to test **B&B** algorithms on models with all these inequalities present.

### 4.2   The Partition (PART) Inequalities

The next inequalities generalize (2). Consider the $i$-th input graph $G_i = (V_i, E_i)$,
$i \in \{1, \dots, n\}$. Let $u$ and $u'$ be two vertices in $V_i$ with labels $\alpha$ and $\alpha'$, respectively, with $\alpha \neq \alpha'$ and $(u, u') \in E_i$. Again assume that $G = (V, E)$ is the output
graph and that $v_j$ is a vertex of $V$ with label $\alpha$. Finally, suppose that $A$ and
$B$ form a partition of the set $J(\pi_i(u'))$ (see definition in Sect. 4). The **PART**
inequality corresponding to $u$, $u'$, $v_j$, $A$ and $B$ is

$$x_{uij} - \sum_{j' \in A} y_{jj'} - \sum_{j' \in B} x_{u'ij'} \leq 0 \tag{8}$$

**Theorem 2.** *(8) is valid for all integer solutions of the system (2)-(6).*

*Proof.* If $u'$ is mapped onto a vertex $v_{j'}$ of the resulting graph $G$ and $j'$ is in $B$, (8) reduces to $x_{uij} - \sum_{j' \in A} y_{jj'} \leq 1$ which is obviously true since $x_{uij} \leq 1$ and $y_{jj'} \geq 0$ for all $j' \in A$. On the other hand, if $u'$ is mapped onto a vertex $v_{j'}$ of $G$ with $j'$ in $A$, the last summation in (8) is null and the inequality becomes $x_{uij} - \sum_{j' \in A} y_{jj'} \leq 0$. If vertex $u$ is not mapped onto vertex $v_j$, the latter inequality is trivially satisfied. If not, then necessarily there must be an arc in $G$ joining vertex $v_j$ to some vertex $v_{j'}$ of $G$ with $j'$ in $A$. This implies that the second summation in (8) is at least one and, therefore, the inequality holds.    □

Notice that using (4) we can rewrite (8) as $x_{uij} - \sum_{j' \in A} y_{jj'} + \sum_{j' \in A} x_{u'ij'} \leq 1$ which, for $A = \{j'\}$, is nothing but (2). Moreover, since the size of $J(\pi_i(u'))$, in the worst case, is linear in the total number of vertices of all input graphs, there can be exponentially many PART inequalities. However, the separation problem for these inequalities can be solved in polynomial time. This is the ideal situation for, according to the celebrated Grötschel-Lovász-Schrijver theorem [12], the dual bound of the relaxation of (2)-(4) and all inequalities in (8) is computable in polynomial time using the latter inequalities as cutting-planes. A pseudo-code for the separation routine of (8) is shown in Fig. 4 and is now explained.

Given an input graph $G_i$, $i \in N$, consider two vertices $u$ and $u'$ such that $(u, u')$ is in $G_i$ and a vertex $v_j$ of $G$ whose label is identical to that of $u$. Now, let $(x^\star, y^\star)$ be an optimal solution of a linear relaxation during the **B&C** algorithm. The goal is to find the partition of the set $J(\pi_i(u'))$ that maximizes the LHS of (8). It can be easily verified that, with respect to the point $(x^\star, y^\star)$ and the input parameters $i$, $u$, $u'$ and $j$, the choice made in line 4 ensures that the LHS of (8) is maximized. Thus, if the value of LHS computed for the partition returned in line 7 is non positive, no constraint of the form (8) is violated, otherwise, $(A, B)$ is the partition that produces the most violated PART inequality for $(x^\star, y^\star)$. This routine is executed for all possible sets of input parameters. The number of such sets can be easily shown to be polynomial in the size of the input. Moreover, since the complexity of the routine is $O(J(\pi_i(u')))$ which, in turn, is $O(\sum_{i \in N} |V_i|)$, the identification of all violated PART inequalities can be done in polynomial time.

```
procedure part-separation-routine(x*, y*, i, u, u', j);
1     A ← ∅;
2     B ← ∅;
3     for all j' ∈ J(πᵢ(u')) do{
4         if (y*ⱼⱼ' ≤ x*u'ij') then A ← A ∪ {j'};
5         else B ← B ∪ {j'};
6     }
7     return (A, B);
end part-separation-routine.
```

**Fig. 4.** Separation routine for PART inequalities.

## 5   Computational Experiments

We now report on our computational tests with a set of benchmark instances generated from real applications from the MediaBench suite [13]. All programs were implemented in C++ and executed on a DEC machine equipped with an ALPHA processor of 675 MHz, 4 GB of RAM and running under a native Unix operating system. The linear programming solver used was CPLEX 7.0 and separation routines were coded as *callback functions* from the solver's callable library.

The program implementing heuristic MH resorts to the algorithm of Battiti and Protasi [14] to find solutions to the clique problem. The author's code, that was used in our implementation, can be downloaded from [15] and allows the setting of some parameters. Among them, the most relevant to us is the maximum computation time. Our tests reveal that running the code with this parameter set to one second produce the same results as if we had fixed it to 10 or 15 seconds. Unless otherwise specified, all results exhibited here were obtained for a maximum of computation time of one second. This means that, the MH heuristic as a whole had just a couple of seconds to seek a good solution.

The **B&B** and **B&C** codes that compute the IP models also had their computation times limited. In this case, the upper bound was set to 3600 seconds. **B&B** refers to the basic algorithm implemented in CPLEX having the system (1)-(6) as input. The results of **B&B** are identified by the "P" extension in the instance names. The **B&C** algorithms are based on a naive implementation. The only inequalities we generated on the fly are the PART inequalities. The separation routine from Fig. 4 is ran until it is unable to encounter an inequality that is violated by the solution of the current relaxation. So, new PART constraints are generated exhaustively, i.e., no attempt is done to prevent the well-known stalling effects observed in cutting plane algorithms. A simple rounding heuristic is also used to look for good primal bounds. The heuristic is executed at every node of the enumeration tree. The two versions of **B&C** differ only in the input model which may or may not include the set of CBS inequalities. As mentioned earlier, when used, the CBS inequalities are all generated a priori by a simple backtracking algorithm. The first (second) version **B&C** algorithm uses the system (1)-(6) (amended with CBS inequalities) as input and its results are identified by the "HC" ("HCS") extension in the instance names. It should be noticed that, both in **B&B** and in **B&C** algorithms, the generation of standard valid inequalities provided by the solver is allowed. If fact, Gomory cuts were added by CPLEX in all cases but had almost no impact on the dual bounds.

Table 1 summarizes the characteristics of the instances in our data set. Columns "$k$" and "$\ell$" refer respectively to the number of vertices and labels of the output graph $G$. Columns "$G_i$", $i \in \{1, \dots, 4\}$ display the features of each input graph of the instance. For each input graph, the columns "$k_i$", "$e_i$" and "$\ell_i$" denote the number of vertices, arcs and different labels respectively.

Table 2 exhibits the results we obtained. The first column contains the instance name followed by the extension specifying the algorithm to which the data in the row correspond. The second column reports the CPU time in seconds. We do not report on the specific time spent on generating cuts since it is negli-

**Table 1.** Characteristics of the instances.

| Instance name | $k$ | $\ell$ | $G_1$ | | | $G_2$ | | | $G_3$ | | | $G_4$ | | | $\sum k_i$ | $\sum e_i$ |
|---|---|---|---|---|---|---|---|---|---|---|---|---|---|---|---|---|
| | | | $k_1$ | $e_1$ | $\ell_1$ | $k_2$ | $e_2$ | $\ell_2$ | $k_3$ | $e_3$ | $\ell_3$ | $k_4$ | $e_4$ | $\ell_4$ | | |
| adpcm | 108 | 20 | 97 | 138 | 20 | 78 | 103 | 18 | – | – | – | – | – | – | 175 | 241 |
| epic_decode | 24 | 6 | 16 | 17 | 3 | 16 | 15 | 4 | 15 | 13 | 4 | 12 | 11 | 5 | 59 | 56 |
| epic_encode | 39 | 8 | 36 | 43 | 7 | 16 | 17 | 3 | 13 | 13 | 5 | 11 | 10 | 4 | 76 | 83 |
| g721 | 57 | 6 | 57 | 66 | 6 | 19 | 18 | 4 | – | – | – | – | – | – | 76 | 84 |
| gsm_decode | 92 | 8 | 91 | 102 | 8 | 65 | 69 | 8 | 20 | 20 | 5 | 19 | 18 | 4 | 195 | 209 |
| gsm_encode | 48 | 8 | 46 | 57 | 7 | 41 | 48 | 7 | 20 | 21 | 5 | 19 | 17 | 6 | 126 | 143 |
| jpeg_decode | 104 | 5 | 101 | 111 | 5 | 61 | 68 | 5 | – | – | – | – | – | – | 162 | 179 |
| jpeg_encode | 47 | 7 | 46 | 55 | 7 | 31 | 32 | 5 | 28 | 32 | 6 | – | – | – | 105 | 119 |
| mpeg2_decode | 34 | 6 | 32 | 31 | 4 | 24 | 29 | 6 | 15 | 15 | 4 | 11 | 10 | 4 | 82 | 85 |
| mpeg2_encode | 32 | 7 | 31 | 39 | 6 | 30 | 37 | 6 | 20 | 22 | 5 | 18 | 16 | 5 | 99 | 114 |
| pegwit | 41 | 7 | 40 | 46 | 7 | 27 | 28 | 6 | 27 | 30 | 5 | – | – | – | 94 | 104 |

gible compared to that of the enumeration procedure. Third and fourth columns contain respectively the dual and primal bounds when the algorithm stopped. Column "MH" displays the value of the solution obtained by Moreano's heuristic. To calculate these solutions, MH spent no more than 15 seconds in each problem and mpeg2_encode was the only instance in which the clique procedure was allowed to run for more than 10 seconds. Column "gap" gives the percentage gap between the value in column "MH" and that of column "DB" rounded up. Finally, the last two columns show respectively the total numbers of nodes explored in the enumeration tree and of PART inequalities added to the model. For each instance, the largest dual bound and the smallest gap is indicated in bold. Ties are broken by the smallest CPU time.

By inspecting Tab. 2, one can see that MH produces very good solutions. It solved 4 out of the 11 instances optimally. We took into account here that further testing with the IP codes proved that 60 is indeed the optimal value of instance pegwit. When a gap existed between MH's solution and the best dual bound, it always remained below 10%. Additional runs with larger instances showed that the gaps tend to increase, though they never exceeded 30%. However, this is more likely due to the steep decrease in performance of the IP codes than to MH. Instance epic_decode was the only case where an optimal solution was found that did not coincided with that generated by MH. Nevertheless, the gap observed in this problem can be considered quite small: 3.39%.

Comparing the gaps computed by the alternative IP codes, we see that the two B&C codes outperform the pure B&B code. Only in two instances the pure B&B code beat both code generation codes. The strength of inequalities PART and CBS can be assessed by checking the number of nodes explored during the enumeration. This number is drastically reduced when cuts are added as it can be observed, for instance, for problems adpcm and epic_decode where the use of cuts allowed the computation of the optimum at the root node while B&B explored thousands or hundreds of nodes. Instance g721 was also solved

**Table 2.** Computational results.

| Instance.extension | time | DB | PB | MH | gap | #nodes | #cuts |
|---|---|---|---|---|---|---|---|
| adpcm.P | 3604 | 165.50 | 170 | 168 | 1.20 | 9027 | 0 |
| adpcm.HC | 37 | **168.00** | 168 | 168 | **0.00** | 1 | 813 |
| adpcm.HCS | 78 | 168.00 | 168 | 168 | 0.00 | 1 | 848 |
| epic_decode.P | 5 | 33.00 | 33 | 33 | 0.00 | 143 | 0 |
| epic_decode.HC | 0 | **33.00** | 33 | 33 | **0.00** | 1 | 90 |
| epic_decode.HCS | 0 | **33.00** | 33 | 33 | **0.00** | 1 | 74 |
| epic_encode.P | 2221 | **59.00** | 59 | 61 | **3.39** | 299908 | 0 |
| epic_encode.HC | 3603 | 57.00 | 60 | 61 | 7.02 | 1055 | 4108 |
| epic_encode.HCS | 3082 | 59.00 | 59 | 61 | 3.39 | 1197 | 4016 |
| g721_.P | 460 | 70.00 | 70 | 70 | 0.00 | 50702 | 0 |
| g721_.HC | 880 | 70.00 | 70 | 70 | 0.00 | 371 | 1673 |
| g721_.HCS | 112 | **70.00** | 70 | 70 | **0.00** | 116 | 1259 |
| gsm_decode.P | 3616 | 105.75 | – | 120 | 13.21 | 1601 | 0 |
| gsm_decode.HC | 3614 | **112.09** | – | 120 | **6.19** | 0 | 3655 |
| gsm_decode.HCS | 3607 | 110.25 | – | 120 | 8.11 | 0 | 2002 |
| gsm_encode.P | 3604 | 67.12 | 80 | 72 | 5.88 | 23346 | 0 |
| gsm_encode.HC | 3606 | **68.24** | 73 | 72 | **4.35** | 97 | 6126 |
| gsm_encode.HCS | 3604 | 68.16 | 79 | 72 | 4.35 | 23 | 6094 |
| jpeg_decode.P | 3606 | 117.70 | 163 | 137 | 16.10 | 10651 | 0 |
| jpeg_decode.HC | 3620 | **126.20** | – | 137 | **7.87** | 1 | 4569 |
| jpeg_decode.HCS | 3623 | 125.10 | – | 137 | 8.73 | 1 | 3669 |
| jpeg_encode.P | 3608 | 61.00 | 83 | 71 | 16.39 | 63302 | 0 |
| jpeg_encode.HC | 3604 | 62.92 | – | 71 | 12.70 | 1 | 5419 |
| jpeg_encode.HCS | 3604 | **64.10** | – | 71 | **9.23** | 1 | 5620 |
| mpeg2_decode.P | 3613 | 47.03 | 51 | 51 | 6.25 | 220173 | 0 |
| mpeg2_decode.HC | 3603 | 46.09 | 51 | 51 | 8.51 | 310 | 4657 |
| mpeg2_decode.HCS | 3605 | **47.06** | 52 | 51 | **6.25** | 455 | 4555 |
| mpeg2_encode.P | 3608 | 46.50 | 60 | 53 | 12.77 | 86029 | 0 |
| mpeg2_encode.HC | 3603 | 47.59 | 55 | 53 | 10.42 | 68 | 7420 |
| mpeg2_encode.HCS | 3603 | **48.67** | 55 | 53 | **8.16** | 115 | 7778 |
| pegwit.P | 3607 | **58.22** | 60 | 60 | **1.69** | 81959 | 0 |
| pegwit.HC | 3604 | 55.69 | 62 | 60 | 7.14 | 289 | 6920 |
| pegwit.HCS | 3604 | 57.77 | 61 | 60 | 3.45 | 202 | 5843 |

to optimality by the **B&C** codes with much fewer nodes than **B&B**, however, when the CBS inequalities were not added a priori, this gain did not translate into an equivalent reduction in computation time. In the remaining cases, where optimality could not be proved, again we observed that **B&C** codes computed better dual bounds whereas the number of nodes visited were orders of magnitude smaller than that of **B&B**.

# 6   Conclusions and Future Research

In this paper we presented an IP formulation for DPM and introduced valid inequalities to tighten this model. Based on this study, we implemented **B&C** and **B&B** algorithms to assess the performance of Moreano's heuristic (MH) for DPM, which is reported as being one of the best available for the problem. Our computational results showed that MH is indeed very effective since it obtains high-quality solutions in a matter of just a few seconds of computation.

The cut generation codes also proved to be a valuable tool to solve some instances to optimality. However, better and less naive implementations are possible that may turn them more attractive. These improvements are likely to be achieved, at least in part, by adding tuning mechanisms that allow for a better trade off between cut generation and branching. For instance, in problems gsm_decode, jpeg_decode and jpeg_encode (see Tab. 2), the **B&C** codes seemed to get stuck in cut generation since they spent the whole computation time and were still at the root node. Other evidences of the need of such tuning mechanisms are given by instances pegwit and g721 were the pure **B&B** algorithm was faster than at least one of the **B&C** codes.

Of course, a possible direction of research would be to perform further polyhedral investigations since they could give rise to new strong valid inequalities for the IP model possibly resulting into better **B&C** codes. Another interesting investigation would be to find what actually makes a DPM instance into a hard one. To this end, we tried to evaluate which of the parameters displayed in Tab. 1 seemed to affect most the computation time of the IP codes. However, our studies were inconclusive. Probably, the structures of the input graphs play a more important role than the statistics that we considered here.

**Acknowledgments.** This work was supported by the Brazilian agencies FAPESP (grants 02/03584-9, 1997/10982-0 and 00/15083-9), CAPES (grants Bex04444/02-2 and 0073/01-6) and CNPq (grants 302588/02-7, 664107/97-4, 552117/02-1, 301731/03-9 and 170710/99-8).

# References

1. Moreano, N., Araujo, G., Huang, Z., Malik, S.: Datapath merging and interconnection sharing for reconfigurable architectures. In: Proceedings of the 15th International Symposium on System Synthesis. (2002) 38–43
2. Wolf, W.: Computers as Components – Principles of Embedded Computing System Design. Morgan Kaufmann Publishers (2001)
3. DeHon, A., Wawrzynek, J.: Reconfigurable computing: What, why, and implications for design automation. In: Proceedings of the Design Automation Conference (DAC). (1999) 610–615
4. Schaumont, P., Verbauwhede, I., Keutzer, K., Sarrafzadeh, M.: A quick safari through the reconfiguration jungle. In: Proceedings of the Design Automation Conference (DAC). (2001) 172–177

5. Compton, K., Hauck, S.: Reconfigurable computing: A survey of systems and software. ACM Computing Surveys **34** (2002) 171–210
6. Bondalapati, K., Prasanna, V.: Reconfigurable computing systems. Proceedings of the IEEE (2002)
7. Callahan, T., Hauser, J., Wawrzynek, J.: The Garp architecture and C compiler. IEEE Computer (2000) 62–69
8. Singh, H., Lee, M., Lu, G., Kurdahi, F., Bagherzadeh, N., Filho, E.: MorphoSys: An integrated reconfigurable system for data-parallel and computation-intensive applications. IEEE Transactions on Computers **49** (2000) 465–481
9. Schmit, H., et al.: PipeRench: A virtualized programmable datapath in 0.18 micron technology. In: Proceedings of the IEEE Custom Integrated Circuits Conference (CICC). (2002) 63–66
10. Moreano, N., Araujo, G., de Souza, C.C.: CDFG merging for reconfigurable architectures. Technical Report IC-03-18, Institute of Computing, University of Campinas SP, Brazil (2003)
11. Nemhauser, G.L., Wolsey, L.: Integer and Combinatorial Optimization. Wiley & Sons (1988)
12. Grötschel, M., Lovász, L., Schrijver, A.: The ellipsoid method and its consequences in combinatorial optimization. Combinatorica **1** (1981) 169–197
13. *MediaBench benchmark* http://cares.icsl.ucla.edu/MediaBench/.
14. Battiti, R., Protasi, M.: Reactive local search for the maximum clique problem. Algorithmica **29** (2001) 610–637
15. Clique code. http://rtm.science.unitn.it/intertools/clique/.

# An Analytical Model for Energy Minimization

Claude Tadonki and Jose Rolim

Centre Universitaire d'Informatique, University of Geneva, Department of Theoretical
Computer Science, Avenue du Géneral Dufour 24, 1211 Geneva 4, Switzerland.
{claude.tadonki,jose.rolim}@cui.unige.ch

**Abstract.** Energy has emerged as a critical constraint in mobile computing because the power availability in most of these systems is limited by the battery power of the device. In this paper, we focus on the memory energy dissipation. This is motivated by the fact that, for data intensive applications, a significant amount of energy is dissipated in the memory. Advanced memory architectures like the Mobile SDRAM and the RDRAM support multiple power states of memory banks, which can be exploited to reduce energy dissipation in the system. Therefore, it is important to design efficient controller policies that transition among power states. Since the addressed memory chip must be in the active state in order to perform a read/write operation, the key point is the tradeoff between the energy reduction due to the use of low power modes and the energy overheads of the resulting activations. The lack of rigorous models for energy analysis is the main motivation of this work. Assuming regular transitions, we derive a formal model that captures the relation between the energy complexity and the memory activities. Given a predetermined number of activations, we approximate the optimal repartition among available power modes. We evaluate our model on the RDRAM and analyze the behavior of each parameter together with the energy that can be saved or lost.

## 1 Introduction

Due to the growing popularity of embedded systems [13,14,15], energy has emerged as a new optimization metric for system design. As the power availability in most of these systems is limited by the battery power of the device, it is critical to reduce energy dissipation in these systems to maximize their operation cycle. Power limitation is also motivated by heat or noise limitations, depending on the target application.

The topic of energy reduction has been intensively studied in the literature and is being investigated at all levels of system abstraction, from the physical layout to software design. There have been several contributions on energy saving focused on scheduling/processors [6,7,8], data organizations [9,1], compilation [17,18,19,24], and the algorithmic level [21,22,24]. The research at the architecture level has led to new and advanced low energy architectures, like the Mobile SDRAM and the RDRAM, that support several low power features such as multiple power states of memory banks with dynamic transitions [11,

C.C. Ribeiro and S.L. Martins (Eds.): WEA 2004, LNCS 3059, pp. 559–569, 2004.

12], row/column specific activation, partial array refresh, and dynamic voltage/frequency scaling [20].

It is well known that the most important part of energy dissipation comes from memory activities [1,2], sometimes more that 90% [12]. Consequently, the topic of memory energy reduction is now into the spotlight. For the purpose of reducing the energy dissipation, contributions on cache memory optimization can be considered because of the resulting reduction of memory accesses [3,4, 5,22]. In order to benefit from the availability of different memory operating modes, effective memory controller policies should suit the tradeoff between the energy reduction obtained from the use of low power modes and the energy overhead of the consequent activations (*exit latency and synchronization time*) [25]. A combinatorial scheduling technique is proposed by Tadonki et al [23]. A *threshold* approach is considered by Fan et al. [25] in order to detect the appropriate instant for transitions into low power modes. A hardware-assisted approach for the detection and estimatation of idleness in order to perform power mode transitions is studied by Delaluz et al [12].

The goal of this paper is to design and evaluate a formal model for the energy minimization problem. This is important as a first step toward a design of an efficient power management policy. Our model clearly shows the relative impact of the storage cost and the activation overheads. The optimization problem derived from our model is a quadratic programming problem, that is well solved by standard routines. We consider only the transitions from low power modes to the active mode, thus in the paper, we say *activation* instead of *transition*. Given a predetermined amount of activations to be performed, our model gives the optimal assignment among power modes and the corresponding fraction of time that should be spent in each mode. It is clear that there is a correlation between the number of activations and the time we are allowed to spent in each mode. It is important to assume that the time we spend in a low power mode after a transition is bounded. Otherwise, we should transition to the lowest power mode and stay in that mode until the end of the computation. This is unrealistic in general because memory accesses will occur, very often at an unpredictable time. To capture this aspect, we consider a time slot for each power mode. Each transition to a given power mode implies that we will spent a period of time that is in a fix range (parameterizable). Once the parameters have been fixed, the resulting optimal energy becomes a function of the number of activations, which should be in a certain range in order to impact an energy reduction.

The rest of the paper is organized as follows. Section II presents our model for energy evaluation. In Section III, we formulate the optimization problem behind the energy minimization. An evaluation with the RDRAM is presented in section IV. We conclude in Section V.

## 2   A Model of Energy Evaluation

We assume that the energy spent for running an algorithm depends on three major types of operation:

- the operations performed by the processor (arithmetic and logical operations, comparisons, etc.);
- the operations performed on the memory (read/write operations, storage, and state transition);
- the data transfers at all levels of hardware system.

In this paper, we will focus only on the energy consumed by memory operations. We consider the memory energy model defined in [21], which we restate here. The memory energy $E(n)$ for problem size $n$ is defined as the sum of the memory access energy, the data storage energy, and state transition overheads. This yields the formula

$$E(n) = K_a \times C(n) + K_s \times S(n) \times A(n) + K_p \times P(n), \qquad (1)$$

where

- $K_a$ is the access energy cost per unit of data, and $C(n)$ represents the total number of memory accesses
- $K_s$ is the storage energy cost per unit of data per unit time, $S(n)$ is the space complexity, and $A(n)$ is the total time for which the memory is active
- $K_p$ is the energy overheads for each power transition, and $P(n)$ represents the total number of state transition.

As we can see, the model consider two memory state (*active* and *inactive*), and a single memory bank. Moreover, the storage cost in intermediate modes is neglected, otherwise we should have considered $T(n)$ (the total computation time) instead of $A(n)$ (the total active time). In our paper, we consider the general case with any given number of memory states, and several memory banks with an independent power control.

The main memory $\mathcal{M}$ is composed of $p$ banks, and each bank has $q$ possible inactive states. We denote the whole set of states by $\mathcal{S} = \{0, 1, 2, \cdots, q\}$, where 0 stands for the *active* state. For state transition, we consider only the activations (transition from a low power mode to the active node). This is justified by the fact that transitions to low power modes impact a negligible energy dissipation. The activation energy overheads is given by the vector $W = (w_0, w_1, \cdots, w_q)$, $w_0 = 0$. During the execution of an algorithm, a given bank $i$ spends a fraction $\alpha_{ij}$ of the whole time in state $j$, thus we have

$$\sum_{j=0}^{q} \alpha_{ij} = 1. \qquad (2)$$

About the storage cost, let $Q = (q_j), j = 0, \cdots, q$ denotes the vector of storage cost, means $q_j$ is the storage cost per unit data and per unit time when the memory is in power state $j$.

Concerning the activation complexity, note that since activations occur in a sequential processing, and the transition cost does not depend on the memory bank, we only need to consider the number of activations from each state $j$, we denote $x_j$. We then define the activation vector $x = (x_0, x_1, \cdots, x_q)$.

If we assume that memory banks are of same volume $\alpha$, we obtain the following memory energy formula for problem size $n$

$$E(n) = K_a \times C(n) + T(n) \times (\sum_{i=1}^{p}\sum_{j=0}^{q} \alpha_{ij}q_j) \times \alpha + \sum_{j=0}^{q} x_j w_j. \tag{3}$$

We define the vector $y = (y_0, y_1, \cdots, y_q)$ by

$$y_j = \sum_{i=1}^{p} \alpha_{ij}. \tag{4}$$

For a given state $j$, $y_j$ is the accumulation of the fractions of time each memory bank has spent in mode $j$. In case of a single memory bank, it is the fraction of the total execution time spent in the considered mode. The reader can easily see that

$$\sum_{j=1}^{q} y_j = p. \tag{5}$$

We shall consider the following straightforward equality

$$\sum_{i=1}^{p}\sum_{j=0}^{q} \alpha_{ij}q_j = \sum_{j=0}^{q}(\sum_{i=1}^{p} \alpha_{ij})q_j = yQ^T.$$

We define the vector $H = (H_0, H_1, \cdots, H_q)$ as the vector of activation delays, $H_j$ is the time overhead induced by an activation from state $j$ ($H_0 = 0$). The total time $T(n)$ is composed of

 − the cpu time $\tau(n)$
 − the memory accesses time $\delta C(n)$ ($\delta$ is the single memory access delay)
 − the activations overhead $Hx^T$

We can write

$$E(n) = K_a \times C + \alpha \times (\tau + \delta C + Hx^T) \times yQ^T + xW^T. \tag{6}$$

We make the following considerations

 − the power management energy overhead $xW^T$ is negligible [21].
 − the the additive part $K_a \times C(n)$ can be dropped since it doesn't depend on the power state management.

Thus, the objective to be minimized is (proportional to) the following

$$E(x, y) = [Hx^T + (\tau + \delta C)]yQ^T. \tag{7}$$

# 3  Optimization

## 3.1  Problem Formulation

Our goal is to study the energy reduction through the minimization of the objective (7). In order to be consistent and also avoid useless (or trivial) solutions, a number of constraints should be considered:

**Domain specification.** The variables $x$ and $y$ belong to $\mathcal{N}$ and $\mathcal{R}$ respectively, i.e.

$$x \in \mathbb{N}^q, \tag{8}$$
$$y \in \mathbb{R}^q. \tag{9}$$

**Time consistency.** As previously explained, we have

$$y \geq 0, \tag{10}$$
$$y_1 + y_2 + \cdots + y_q = p, \tag{11}$$

Another constraint that should be considered here is related to the fraction of time spent in the active mode ($y_0$). Indeed, the time spent in the active mode is greater than the total memory access time, which can be estimated from the number of memory accesses $C$, and the time of a single access $\delta$. Since, we consider fraction of time, we have

$$y_0 \geq \frac{\delta C}{R}, \tag{12}$$

where $\delta C$ is the total memory access time, and $R$ the total running time (without the power management overhead) which can be estimated from the time complexity of the program or from a profiling.

**Activations bounds.** It is reasonable to assume that each time a memory bank is activated, it will earlier or later be accessed. Thus, we have

$$\sum_{i=0}^{q} x_i \leq C. \tag{13}$$

However, except the ideal case of a highly regular and predictable memory access, several activations should be performed for a better use of power modes availability. This is well captured by a lower bound the number of activations. Thus, we have a lower bound and an upper bound in the number of activation. In our model we consider a fix amount of activations instead of a range. This gives,

$$x_1 + x_2 + \cdots + x_q = \rho C, \tag{14}$$

where $\rho$ is a scaling factor such that $0 \leq \rho \leq 1$.

**Compatibility between time and activation.** Recall that a memory bank is activated if and only if it will be accessed. Moreover, when a memory bank is put in a given low power mode, a minimum (resp. maximum) period of time is spent in that mode before transitioning to the active mode. This can be the fraction of time taken by the smallest job (or instruction depending on the granularity). We consider the set of time intervals $[\varphi_i, \eta_i]$ low power modes. Then, we have

$$\varphi_j x_j \leq y_j \leq \eta_j x_j \text{ for } j = 1, \cdots, q. \tag{15}$$

In addition, since any of every activation implies a minimum period of time, we denote $\gamma$, in the active mode, we also have

$$y_0 \geq \gamma(\sum_{j=0}^{q} x_j). \tag{16}$$

Using relation (14), relation (16) becomes

$$y_0 \geq \gamma \rho C. \tag{17}$$

We shall consider $\mu$ define by

$$\mu = \max\{\frac{\delta}{R}, \gamma\rho\}. \tag{18}$$

The inequalities (12) and (17) can be combined to

$$y_0 \geq \mu C. \tag{19}$$

We now analyze the model.

### 3.2  Model Analysis

We first note that transitioning from the active state to state $j$ for a period of time $\Delta t$ is advantageous (based of storage cost) if and only if we have

$$q_j(\Delta t + h_j) \leq q_0 \Delta t, \tag{20}$$

which gives the following threshold relation

$$\Delta t \geq (\frac{q_j}{q_0 - q_j})h_j. \tag{21}$$

The time threshold vector $D$ defined by

$$D_j = (\frac{q_j}{q_0 - q_j})h_j, j = 1, 2, \cdots, q \tag{22}$$

provides the minimum period of time that should be spent in each low power modes, and is also a good indicator to appreciate their relative impact. We propose to select the time intervals (15) for low power modes as follows

$$\varphi_j = \lambda_1 \frac{D_j}{R} \quad \varphi_j = \lambda_2 \frac{D_j}{R}, \tag{23}$$

where $1 \le \lambda_1 \le \lambda_2$.

Lastly, the active time threshold as defined in (17) should be greater than the memory accesses time. Then we should have

$$\gamma \ge \frac{\delta}{\rho R}. \tag{24}$$

We now solve the optimization problem provides by our model as described above.

### 3.3   Solving the Optimization Problem

According to our model, the optimization problem behind the energy reduction is the following

$$
\begin{aligned}
&\texttt{min } xH^T Q y^T + R Q y^T \\
&\texttt{subject to} \\
&\quad 1. \;\; x \in \mathbb{N}^q, \\
&\quad 2. \;\; y \in \mathbb{R}^q, \\
&\quad 3. \;\; y_1 + y_2 + \cdots + y_q = p, \\
&\quad 4. \;\; y_0 \ge \mu C, \\
&\quad 5. \;\; x_1 + x_2 + \cdots + x_q = \rho C, \\
&\quad 6. \;\; y \le \varphi x. \\
&\quad 7. \;\; y \ge \eta x.
\end{aligned}
$$

**Fig. 1.** Energy minimization problem

There are mainly two ways for solving the optimization problem formulated in figure 3.3. The first approach is to consider the problem as a mixed integer programming problem (MIP). For a given value of $x$, the resulting model becomes a linear programming (LP) problem. Thus, appropriate techniques like the standard LP based Branch and Bound can be considered. However, we think that this is an unnecessarily challenging computation. Indeed, a single transition does not have a significant impact on the overall energy dissipation as quantified by our model. Thus, we may consider a pragmatic approach where the variable $x$ is first assumed to be continuous, and next rounded down in order to obtain the required solution. This second approach yields a simple quadratic programming model that is easily solved by standard routines.

## 4   Experiments

We evaluate our model with the values provided in [25] for the RDRAM. Table 2 summarizes the corresponding values (vector D is calculated using the formula (22)).

```
Q = (300 180 30 3)
H = (0 16 60 6000)
p = 8
q = 4
δ = 60
D = (9.00 6.67 60.61)
```

**Fig. 2.** DRAM parameters

Our optimization is performed using MATLAB with the following code

```
function [X,Y,E] = Energy_Opt(H,Q,R,d,C,p,q,r,g,l1,l2)
% Matlab code to solve the energy minimization problem
% The quadratic objective is considered as follows
% 0.5 * X' * HH * X + ff' * X

% We form our objective coefficients
HH = [zeros(q, q), H' * Q; Q' * H , zeros(q, q)];
ff = R * [zeros(q, 1); Q'];

% Bound on the main variable Z = [X,Y]
LB = [zeros(q, 1) ; zeros(q, 1)];
UB = [inf * ones(q, 1) ; p * ones(q, 1)];
% Ajust the lower bound on Y1 (Y0 in the text)
LB(q+1) = max(d * C / t, g * r * C);

% Matrix of the equality constraints
Aeq = [ones(1,q), zeros(1,q); zeros(1,q) , ones(1,q)];
beq = [r * C; p];

% Matrix of the inequality constraints
a1 = [l1 * diag(D), - eye(q)]; b1 = zeros(q, 1);
a2 = [-l2 * diag(D),  eye(q)]; b2 = zeros(q, 1);
% Y0 is not bounded by X
a1(1,:)=[]; b1(1)  =[]; a2(1,:)=[]; b2(1)  =[];
% Forming the matrix
A  = [a1; a2];  b  = [b1; b2];

% OPTIMIZATION unsing the solver quadprog of MATLAB
[Z, E, EXITF,OUTPUT]  = quadprog(HH,ff,A,b,Aeq,beq,LB,UB);

% RETRIVING X AND Z from Z
X  = Z(1:q);
Y  = Z(q + 1: 2 * q);
```

We consider a problem (abstracted) where 75% of the total time is spent in memory accesses. We used R = 80000 and C = 1000. Note that our objective function is proportional to the time vector $y$ and the vector of storage coefficient $Q$. Thus, the measuring unit can be scaled as desired without changing the optimal argument. Figure 3 displays a selection of optimal activation repartition and the percentage of energy that is saved or lost. Figure 4 shows how the energy varies in relation with the number of activations.

| $\rho$ | $N_{act}$ | X | Y | $E_{opt}$ | Reduction |
|---|---|---|---|---|---|
| 0 | 0 | (0, 0, 0, 0) | (8, 0, 0, 0) | 1.92 | 0% |
| 0.01 | 10 | (0, 10, 0, 0) | (7.58, 0, 0.42, 0) | 1.84 | 4% |
| 0.02 | 20 | (0, 0, 7, 13) | (3, 0, 0.31, 4.69) | 1.43 | 25% |
| 0.05 | 50 | (0, 0, 41, 9) | (3, 0, 1.71, 3.29) | 1.30 | 32% |
| 0.10 | 100 | (0, 0, 97, 3) | (3, 0, 4.04, 0.96) | 1.04 | 46% |
| 0.11 | 110 | (0, 8, 100, 2) | (3, 0.1, 4.15, 0.74) | 1.024 | 47% |
| 0.125 | 125 | (0, 25, 100,0) | (3, 0.85, 4.15, 0) | 1.014 | 48% |
| 0.20 | 200 | (0, 100, 100, 0) | (3, 1.3, 3.7, 0)) | 1.08 | 44% |
| 0.21 | 210 | (0, 100, 100, 10) | (3, 1.3, 0.83, 2.87) | 1.72 | -11% |
| 0.22 | 220 | (0, 100, 100, 20) | (3, 1.3, 0.83, 2.87) | 2.41 | -26% |
| 0.25 | 250 | (0, 100, 100, 25) | (3, 1.3, 0.83, 2.87) | 2.76 | -44% |

**Fig. 3.** Experiments with our model on a RDRAM

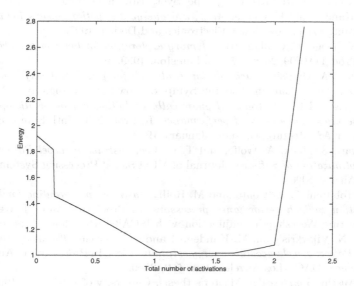

**Fig. 4.** Energy vs the number of activations

As we can see from Table 3, the best number of activation is 125 (12.5% of the number of memory accesses), with an energy reduction of 48% (taken the always active case as baseline). We also see that there is a critical value for the number of activations (200 in this case) under which we begin loosing energy. In addition, the optimal distribution of activations among low power modes depends on the total number of activations and the time we are allowed to stay in each mode.

## 5   Conclusion

We have formulated the problem of energy optimization in the context of several low power modes. We have shown that, in order to make a rewarding transition to a given low power mode, there is a minimum period of time that should be spent in that mode. From our experiments with a RDRAM, it follows that a reduction of 48% can be obtained by performing regular transitions. The optimal number of activations is determined experimentally. We think that our model can be used for a first evaluation of potential energy reduction before moving forward to any power management policy.

## References

1. F. Catthoor, S.Wuytack, E.D. Greef, F. Balasa, L. Nachtergaele, and A. Vande-cappelle, *Custom memory management methodology - exploration of memory organization for embedded multimedia system design*, Kluwer Academic Pub., June 1998.
2. A. R. Lebeck, X. Fan, H. Zeng, and C. S. Ellis, *Power aware page allocation*, Int. Conf. Arch. Support Prog. Lang. Ope. Syst., November 2000.
3. M. B. Kamble and K. Ghose, *Analytical energy dissipation models for low power caches*, Int. Symp. Low Power Electronics and Design, 1997.
4. W-T. Shiue and C. Chakrabarti, *Memory exploration for low power embedded systems*, Proc. DAC'99, New Orleans, Louisina, 1999.
5. C. Su and A. Despain, *Cache design trade-offs for power and performance optimization: a case study*, In Proc. Int. Symp. on Low Power Design, pp. 63-68, 1995.
6. D. Brooks and M. Martonosi, *Dynamically exploiting narrow width operands to improve processor power and performance*, In Proc. Fifth Intl. Symp. High-Perf. Computer Architecture, Orlando, January 1999.
7. V. Tiwari, S. Malik, A. Wolfe, and T. C. Lee, *Instruction Level Power Analysis and Optimization of Software*, Journal of VLSI Signal Processing Systems, Vol 13, No 2, August 1996.
8. M. C. Toburen, T. M. Conte, and M. Reilly, *Instruction scheduling for low power dissipation in high performance processors*, In Proc. the Power Driven Micro-Architecture Workshop in conjunction with ISCA'98, Barcelona, June 1998.
9. W. Ye, N. Vijaykrishnan, M. Kandemir, and M. J. Irwin, *The design and use of SimplePower: a cycle-accurate energy estimation tool*, In Proc. Design. Automation Conference (DAC), Los Angeles, June 5-9, 2000.
10. Todd Austin, *Simplescalar*, Master's thesis, University of Wisconsin, 1998.
11. 128/144-MBit Direct RDRAM Data Sheet, Rambus Inc., May 1999.

12. V. Delaluz and M. Kandemir and N. Vijaykrishnan and A. Sivasubramaniam and M. Irwin. Memory energy management using software and hardware directed power mode control. Tech. Report CSE-00-004, The Pennsylvania State University, April 2000.
13. W. Wolf. Software-Hardware Codesign of Embedded Systems. In , *Proceedings of the IEEE* , volume 82 , 1998.
14. R. Ernst. Codesign of Embedded Systems: Status and Trends . In , *IEEE Design and Test of Computers* , volume 15 , 1998.
15. Manfred Schlett. Trends in Embedded Microprocessors Design. In , *IEEE Computer*, 1998.
16. "Mobile SDRAM Power Saving Features," Technical Note TN-48-10, MICRON, http://www.micron.com
17. W. Tang, A. V. Veidenbaum, and R. Gupta. Architectural Adaptation for Power and Performance. In , *International Conference on ASIC*, 2001 .
18. L. Bebini and G. De Micheli. Sytem-Level Optimization: Techniques and Tools. In , *ACM Transaction on Design Automation of Electronic Systems*, 2000.
19. T. Okuma, T. Ishihara, H. Yasuura . Software Energy Reduction Techniques for Variable-Voltage Processors. In , *IEEE Design and Test of Computers* , 2001 .
20. J. Pouwelse, K. Langendoen, and H. Sips, "Dynamic Voltage Scaling on a Low-Power Microprocessor," UbiCom-Tech. Report, 2000.
21. M. Singh and V. K. Prasanna . Algorithmic Techniques for Memory Energy Reduction. In , *Worshop on Experimental Algorithms*, Ascona, Switzerland, May 26-28, 2003.
22. S. Sen and S. Chatterjee . Towards a Theory of Cache-Efficient Algorithms . In *SODA*, 2000 .
23. C. Tadonki, J. Rolim, M. Singh, and V. Prasanna. *Combinatorial Techniques for Memory Power State Scheduling in Energy Constrained Systems*, Workshop on Approximation and Online Algorithms (WAOA), WAOA2003, Budapest, Hungary, September 2003 .
24. D.F. Bacon, S.L. Graham, and O.J. sharp . Compiler Transformations for High-Performance Computing . *Hermes*, 1994 .
25. X. Fan, C. S. Ellis, and A. R. Lebeck. Memory Controller Policies for DRAM Power Management. *ISLPED'01*, August 6-7, Huntington Beach, California, 2001.

# A Heuristic for Minimum-Width Graph Layering with Consideration of Dummy Nodes

Alexandre Tarassov[1], Nikola S. Nikolov[1], and Jürgen Branke[2]

[1] CSIS Department, University of Limerick, Limerick, Ireland.
{alexandre.tarassov,nikola.nikolov}@ul.ie
[2] Institute AIFB, University of Karlsruhe, 76128 Karlsruhe, Germany.
branke@aifb.uni-karlsruhe.de

**Abstract.** We propose a new graph layering heuristic which can be used for hierarchical graph drawing with the minimum width. Our heuristic takes into account the space occupied by both the nodes and the edges of a directed acyclic graph and constructs layerings which are narrower that layerings constructed by the known layering algorithms. It can be used as a part of the Sugiyama method for hierarchical graph drawing. We present an extensive parameter study which we performed for designing our heuristic as well as for comparing it to other layering algorithms.

## 1 Introduction

The rapid development of Software Engineering in the last few decades has made *Graph Drawing* an important area of research. The Graph Drawing techniques find application in visualizing various diagrams, such as call graphs, precedence graphs, data-flow diagrams, ER diagrams, etc. In many of those applications it is required to draw a set of objects in a hierarchical relationship. Such sets are modeled by directed acyclic graphs (DAGs), i.e. directed graphs without directed cycles, and usually drawn by placing the graph nodes on parallel horizontal, concentric or radial levels with all edges pointing in the same direction.

There have been recognized a few different methods for hierarchical graph drawing. The more recent two are the evolutionary algorithm of Utech et al. [13] and the magnetic field model introduced by Sugiyama and Misue [11]. While they are an area of fruitful future research, an earlier method, widely known as the Sugiyama (or STT) method, has received most of the research attention and has become a standard method for hierarchical graph drawing. The STT method is a three phase algorithmic framework, originally proposed by Sugiyama, Tagawa, and Toda [12], and also based on work by Warfield [14] and Carpano [3]. At its first phase the nodes of a DAG are placed on horizontal levels; at the second phase the nodes are ordered within each level; and at the final third phase the $x-$ and $y-$ coordinates of all nodes and the eventual edge bends are assigned. The STT method can be employed for drawing any directed graph by reversing the direction of some edges in advance to ensure that there are no directed cycles in the graph and restoring the original direction at the end [6].

C.C. Ribeiro and S.L. Martins (Eds.): WEA 2004, LNCS 3059, pp. 570–583, 2004.

In order to assign DAG nodes to horizontal levels at the first phase of the STT method it is necessary to partition the node set into subsets such that nodes connected by a directed path belong to different subsets. In addition, it must be possible to assign integer ranks to the subsets such that for each edge the rank of the subset that contains the target of the edge is less than the rank of the subset that contains its source. Such an ordered partition of the node set of a DAG is known as a *layering* and the corresponding subsets are called *layers*. A DAG with a layering is called a *layered DAG*. Figure 1 gives an example of two alternative layerings of the same DAG. Algorithms which partition the node set of a DAG into layers are known as *layering algorithms*.

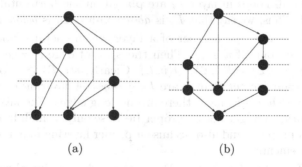

(a)                                 (b)

**Fig. 1.** Two alternative layerings of the same DAG. Each layer occupies a horizontal level marked by a dashed line. All edges point downwards.

In this paper we propose a new polynomial-time layering algorithm which approximately solves the problem of hierarchical graph drawing with the minimum width. It finds application in the cases when it is necessary to draw a DAG in a narrow drawing area and it is the first successful polynomial-time algorithm that solves this particular problem. We also present the extensive parameter study we performed to design our algorithm. In the next section we formally introduce the terminology related to DAG layering. Then, in Section 3 we present the minimum-width DAG layering problem and the initial rough version of our layering heuristic. In Section 4 we specify further our heuristic trough extensive parameter study and compare it to other well-known layering algorithms. We draw conclusions from this work in Section 5.

## 2    Mathematical Preliminaries

A *directed graph* $G = (V, E)$ is an ordered pair of a set of nodes $V$ and a set of edges $E$. Each edge $e$ is associated with an ordered pair of nodes $(u, v)$; $u$ is the *source* of $e$ and $v$ is the *target* of $e$. We denote this by $e = (u, v)$. We consider only directed graphs where different edges are associated with different node pairs.

The *in-degree* $d^-(v)$ of node $v$ is the number of edges with a target $v$ , and the *out-degree* $d^+(v)$ of $v$ is the number of edges with a source $v$. We denote the set of all immediate predecessors of node $v$ by $N_G^-(v)$ , and the set of all immediate successors of node $v$ by $N_G^+(v)$. That is, $N_G^-(v) = \{u : (u, v) \in E\}$ and $N_G^+(v) = \{u : (v, u) \in E\}$. The $k$-tuple of edges $p = ((u_1, u_2), (u_2, u_3), \ldots, (u_k, u_{k+1}))$ is called a *directed path* from node $u_1$ to node $u_{k+1}$ with length $k \geq 1$. If $u_1 = u_{k+1}$ then $p$ is a *directed cycle*. In the rest of this work we consider only directed acyclic graphs (DAGs), i.e. directed graphs without directed cycles.

Let $G$ be a DAG and let $\mathcal{L} = \{L_1, \ldots, L_h\}$ be a partition of the node set of $G$ into $h \geq 1$ subsets such that if $(u, v) \in E$ with $u \in L_j$ and $v \in L_i$ then $i < j$. $\mathcal{L}$ is called a *layering* of $G$ and the sets $L_1$, ..., $L_h$ are called *layers*. A DAG with a layering is called a *layered* DAG. We assume that in a visual representation of a layered DAG all nodes in layer $L_i$ are placed on the horizontal level with an $y$-coordinate $i$. Thus, we say that $L_j$ is *above* $L_i$ and $L_i$ is *below* $L_j$ if $i < j$.

Let $l(u, \mathcal{L})$ denotes the number of a layer which contains node $u \in V$, i.e. $l(u, \mathcal{L}) = i$ if and only if $u \in L_i$. Then the *span* of edge $e = (u, v)$ in layering $\mathcal{L}$ is defined as $s(e, \mathcal{L}) = l(u, \mathcal{L}) - l(v, \mathcal{L})$. Clearly, $s(e, \mathcal{L}) \geq 1$ for each $e \in E$; edges with a span greater than 1 are *long edges*. A layering of $G$ is *proper* if $s(e, \mathcal{L}) = 1$ for each $e \in E$, i.e. if there are no long edges. The layering found by a layering algorithm might not be proper because only a small fraction of DAGs can be layered properly and also because a proper layering may not satisfy other layering requirements.

In the STT method for drawing DAGs the node ordering algorithms applied after the layering phase assume that their input is a DAG with a proper layering. Thus, if the layering found at the layering phase is not proper then it must be transformed into a proper one. Normally, this is done by introducing so-called *dummy nodes* which subdivide long edges (see Figure 2).

It is desirable that the number of dummy nodes is as small as possible because a large number of dummy nodes significantly slows down the node ordering phase of the STT method. There are also aesthetic reasons for keeping the dummy node count small. A layered DAG with a small dummy node count would also have a small number of undesirable long edges and edge bends.

A layering algorithm may also be expected to produce a layering with specified either width and height, or aspect ratio. The *height* of a layering is the number of layers. Normally the nodes of DAGs from real-life applications have text labels and sometimes prespecified shape. We define the *width* of a node to be the width of the rectangle that encloses the node. If the node has no text label and no information about its shape or size is available we assume that its width is one unit. The *width of a layer* is usually defined as the sum of the widths of all nodes in that layer (including the dummy nodes) and the *width of a layering* is the maximum width of a layer. Usually the width and the height of a layering are used to approximate the dimensions of the final drawing.

The *edge density* between horizontal levels $i$ and $j$ with $i < j$ is defined as the number of edges $(u, v)$ with $u \in L_j \cup L_{j+1} \cup \ldots \cup L_h$ and $v \in L_0 \cup L_1 \cup \ldots \cup L_i$. The edge density of a layered DAG is the maximum edge density between adjacent layers (horizontal levels). Naturally, drawings with low edge density are clear and easier to comprehend.

## 3   Minimum-Width DAG Layering

Clearly, it is trivial to find a layering of a DAG with the minimum width if the width of a layer is considered equal to the sum of the widths of the original DAG nodes in that layer. In this case any layering with a single node per layer has the minimum width. However, such a definition of width does not approximate the width of the final drawing because the space occupied by long edges is not insignificant (see Figure 2). The contribution of the long edges to the layering width can be taken into account by assigning positive width to the dummy nodes and taking them into when computing the layering width. It is sensible to assume that the dummy nodes occupy smaller space than the original DAG nodes especially in DAGs which come from practical applications and may have large node labels.

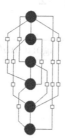

**Fig. 2.** A hierarchical drawing of a DAG. The black circles are the original DAG nodes and the smaller white squares are the dummy nodes along long edges. All edges point downwards.

It is NP-hard to find a layering with the minimum width when the contribution of the dummy nodes is taken into account [2]. The first attempt to solve this problem by a heuristic algorithm belongs to Branke et al. [1]. They proposed a polynomial-time heuristic which did not meet their expectations about quality when tested with relatively small graphs. To the best of our knowledge the only method that can be used for minimum-width DAG layering is the branch-and-cut algorithm of Healy and Nikolov which takes as an input an upper bound on the width and produces a layering subject to it (if feasible) [8]. Although exact, the algorithm of Healy and Nikolov is very complex to implement and its running time is exponential in the worst case.

In this work we design a simple polynomial-time algorithm which finds narrow layerings. We call it MinWidth. Similar to the algorithm of Branke et al. it is a heuristic and it does not guarantee the minimum width. Nevertheless it produces layerings which are narrower than the layerings produced by any of the known polynomial-time layering algorithms. In the remainder of this section we introduce the initial rough version of MinWidth which we tune and extensively test in Section 4.

## 3.1     The Longest-Path Algorithm

We base MinWidth on the longest-path algorithm displayed in Algorithm 1. The longest-path algorithm constructs layerings with the minimum height equal to the number of nodes in the longest directed path. It builds a layering layer by layer starting from the bottom layer labeled as layer 1. This is done with the help of two node sets $U$ and $Z$ which are empty at start. The value of the variable *current_layer* is the label of the layer currently being built. As soon as a node gets assigned to a layer it is also added to the set $U$. Thus, $U$ is the set of all nodes already assigned to a layer. $Z$ is the set of all nodes assigned to a layer below the current layer. A new node $v$ to be assigned to the current layer is picked among the nodes which have not been already assigned to a layer, i.e. $v \in V \setminus U$, and which have all their immediate successors assigned to the layers below the current one, i.e. $N_G^+(v) \subseteq Z$.

---

**Algorithm 1** The Longest-Path Algorithm($G$)

**Requires:** DAG $G = (V, E)$

$U \leftarrow \phi$
$Z \leftarrow \phi$
*currentLayer* $\leftarrow 1$
**while** $U \neq V$ **do**
　Select node $v \in V \setminus U$ with $N_G^+(v) \subseteq Z$
　**if** $v$ has been selected **then**
　　Assign $v$ to the layer with a number *currentLayer*
　　$U \leftarrow U \cup \{v\}$
　**end if**
　**if** no node has been selected **then**
　　*currentLayer* $\leftarrow$ *currentLayer* $+ 1$
　　$Z \leftarrow Z \cup U$
　**end if**
**end while**

---

## 3.2     A Rough Version of MinWidth

In the following, we will assume that all dummy nodes have the same width, $w_d$, although our considerations can be easily generalized to variable dummy node widths. We will also assume that $w(v)$ is the width of node $v$. We start with an initial rough version of MinWidth, displayed in Algorithm 2, which

---

**Algorithm 2** MinWidth(*G*)

---

**Requires:** DAG $G = (V, E)$

---

$U \leftarrow \phi;\ Z \leftarrow \phi$
*currentLayer* $\leftarrow 1$; *widthCurrent* $\leftarrow 0$; *widthUp* $\leftarrow 0$
**while** $U \neq V$ **do**
    Select node $v \in V \setminus U$ with $N_G^+(v) \subseteq Z$ and ConditionSelect
    **if** $v$ has been selected **then**
        Assign $v$ to the layer with a number *currentLayer*
        $U \leftarrow U \cup \{v\}$
        *widthCurrent* $\leftarrow$ *widthCurrent* $- w_d * d^+(v) + w(v)$
        Update *widthUp*
    **end if**
    **if** no node has been selected OR ConditionGoUp **then**
        *currentLayer* $\leftarrow$ *currentLayer* $+ 1$
        $Z \leftarrow Z \cup U$
        *widthCurrent* $\leftarrow$ *widthUp*
        Update *widthUp*
    **end if**
**end while**

---

contains a number of unspecified parameters. We specify them later in Section 4 by extensive parameter study.

We employ two variables widthCurrent and widthUp which are used to store the width of the current layer and the width of the layers above it respectively. The width of the current layer, widthCurrent, is calculated as the sum of the widths of the nodes already placed in that layer plus the sum of the widths of the potential dummy nodes along edges with a source in $V \setminus U$ and a target in $Z$ (one dummy node per edge). The variable widthUp provides an estimation of the width of *any* layer above the current one. It is the sum of the widths of the potential dummy nodes along edges with a source in $V \setminus U$ and a target in $U$ (one dummy node per edge).

When we select a node to be placed in a layer we employ an additional condition ConditionSelect. Our intention is to specify ConditionSelect so that the choice of node $v$ (among alternative candidates) will lead to as narrow a layering as possible. We propose to explore the following three alternatives as ConditionSelect:

- $A_1$: $v$ is the candidate with the maximum outdegree $d^+(v)$;
- $A_2$: $v$ is the candidate with the maximum $d^+(v) - d^-(v)$;
- $A_3$: $v$ or any immediate predecessor of $v$ has the maximum $d^+(v) - d^-(v)$ among all candidates and their immediate predecessors.

In $A_1$ we select the candidate with the maximum indegree because that choice will lead to the maximum possible improvement of widthCurrent. $A_2$ and $A_3$ are less greedy alternatives which do not make the best choice in terms of widthCurrent but look also at the effect to the upper layers. By choosing the candidate with the maximum $d^+(v) - d^-(v)$ $A_2$ makes the choice that will bring the best improvement to widthUp. The idea behind $A_3$ is to allow nodes which can bring big improvement to the width of some upper layer to do it without being blocked by their successors with low $d^+(v) - d^-(v)$. Thus, $A_3$ represents an alternative that tries to choose a node by looking ahead at the impact of that choice to the layering width.

In order to control the width of the layering we introduce a second modification to the longest-path algorithm. That is, we introduce an additional condition for moving up to a new layer, ConditionGoUp. The idea is to move to a new layer if the width of the current layer or of the layer above it becomes too large. In order to be able to check this we introduce the parameter $UBW$ against which we would like to compare the width of the current layer. Since widthUp represents only an approximation of the width of the layers above the current layer we propose to compare its width to $c \times UBW$ where $c \geq 1$, i.e. $c$ gives freedom to widthUp to be larger than widthCurrent because widthUp is just an estimation of the width of the upper layers. We do not consider $UBW$ and $c$ as input parameters, we would like to have their values (or narrow value ranges) hard-coded in MinWidth instead. We set up ConditionGoUp to be satisfied if either:

- widthCurrent $\geq UBW$ and $d^+(v) < 1$, or
- widthUp $\geq c \times UBW$.

We require $d^+(v) < 1$ for widthCurrent $\geq UBW$ to be taken into account because the initial value of widthCurrent is determined by the dummy nodes in the current layer and it gets smaller (or at least it does not change) when a regular node with a positive outdegree gets placed in the current layer. In that case the dummy nodes along edges with a source $v$ are removed from the current layer and get replaced by $v$. If $d^+(v) \geq 1$ then the condition widthCurrent $\geq UBW$ on its own is not a reason for moving to the upper layer because there is still a chance to add nodes to the current layer which will reduce widthCurrent. If $d^+(v) < 1$ then the assignment of $v$ to the current layer increases widthCurrent because it does not replace any dummy nodes. This is an indication that no further improvement of widthCurrent can be done.

In relation to the three alternatives, $A_1$, $A_2$, and $A_3$, we consider two alternative modes of updating the value of widthUp:

- Set widthUp at 0 when move to the upper layer; add $w_d \times d^-(v)$ to widthUp each time a node $v$ is assigned to the current layer;
- Do not change widthUp when move to the upper layer; add $w_d \times (d^-(v) - d^+(v))$ to widthUp each time a node $v$ is assigned to the current layer .

The first of the two modes builds up widthUp starting from zero 0 and taking into account only dummy nodes along edges between $V \setminus U$ and the current layer. We employ this update mode with $A_1$. The second mode approximates the width of the upper layers more precisely by keeping track of as many dummy nodes as possible. We employ it with $A_2$ and $A_3$ where the width of the upper layers plays more important role. We consider the three alternatives $A_1$, $A_2$, and $A_3$ with the corresponding widthUp update modes as parallel branches in the rough version of MinWidth and we choose one of them as a result of our experimental work.

In order to specify ConditionGoUp we need to set UBW and $c$. To specify ConditionSelect we need to select one of $A_1$, $A_2$ and $A_3$. We propose to run MinWidth for 5911 test DAGs and various sets of values of UBW and $c$ as well as for each of the alternatives $A_1$, $A_2$ or $A_3$ with the corresponding widthUp update mode. We expect that the extensive experiments will suggest the most appropriate values or ranges of values for UBW and $c$ as well as the winner among the alternatives $A_1$, $A_2$ or $A_3$.

## 4  Parameter Study

In our experimental work we used 5911 DAGs from the well-known Rome graph dataset [5]. The Rome graphs come from practical applications. They are graphs with node count between 10 and 100 nodes and typically each of them has twice as many edges as nodes. We run MinWidth with each of the three alternatives $A_1$, $A_2$, and $A_3$, for each of the 5911 DAGs and for each pair (UBW, $c$) with UBW = 1..50, and $c$ = 1..10. In total, we had about 9 million tasks. We executed the tasks in a computational grid environment with two computational nodes. One of the computational nodes was a PC with a Pentium III/800 MHz processor, and the other was a PC with a Pentium 4/2.4 GHz processor.

### 4.1  $A_1$, $A_2$, and $A_3$ Compared

For each of the 5911 input DAGs and each alternative - $A_1$, $A_2$, and $A_3$ - we chose the layering with the smallest width (taking into account the dummy nodes) and stored the pair of parameters (UBW, $c$) for which it was achieved. As we stated above, we explored any combination of UBW with = 1..50, and $c$ = 1..10.

Figures 3-8 compare various properties of the stored layerings. The $x$-axis in all pictures represents the number of original nodes in a graph. Since the Rome graphs have no node labels we assume that the width of all original and all dummy nodes is 1 unit if not specified otherwise. Thus, the layering width is the maximum number of nodes (original and dummy) per layer. We have partitioned all DAGs into groups by node count. Each group covers an interval of size 5 on the $x$-axis. We display the average result for each group.

Figures 3(a) and (b) compare the width of the layerings taking into account the dummy nodes (i.e. each dummy node has width equal to one unit) and neglecting them (i.e. each dummy node has width equal to zero) respectively. In both cases $A_1$ gives the narrowest layerings which suggests that $A_1$ might be the best option if the width of the dummy nodes is considered less than or equal to one unit (which is a reasonable assumption). The height of the $A_1$ layerings (see Figure 4) is larger than the height of the other layerings. The height is the number of layers. It was expected that the narrower a layering, the larger is the

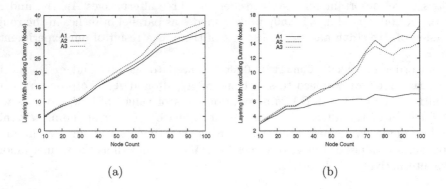

(a)                                                      (b)

**Fig. 3.** $A_1$, $A_2$, and $A_3$ compared: layering width (a) taking into account and (b) neglecting the contribution of the dummy nodes.

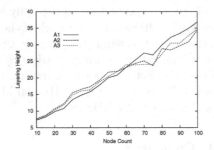

**Fig. 4.** $A_1$, $A_2$, and $A_3$ compared: layering height (number of layers).

number layers. Figure 5(a) shows the dummy node count divided by the total node count in a DAG. Figure 5(b) shows the edge density divided by the total edge count in a DAG. We can observe that the $A_1$ layerings have fewer dummy nodes and in general better edge density than the $A_2$ and the $A_3$ layerings.

Similarly, Figures 6(a) and (b) show the values of UBW and $c$ which lead to narrowest layerings. The simplest $A_1$ alternative finds narrowest layerings for considerably lower values of UBW and $c$ than $A_2$ and $A_3$. Moreover, those values of UBW and $c$ do not depend on the DAG size when $A_1$ is employed. The conclusion that we can make from these experiments is that the simplest alternative, $A_1$, is superior to the other two. It is enough to run MinWidth with $A_1$, UBW $= 1..4$ and $c = 1..2$ in order to achieve the narrowest possible layerings.

In any case MinWidth leads to layerings with a very high dummy node count. There is a simple heuristic that can be applied to a layering in order to reduce the dummy node count. It is the *Promotion* heuristic which works by iteratively moving (or promoting) nodes to upper layers if that movement decreases

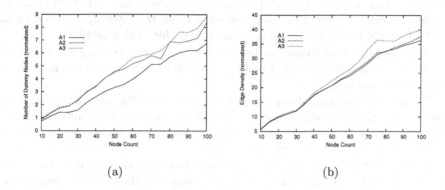

(a)                                                   (b)

**Fig. 5.** $A_1$, $A_2$, and $A_3$ compared: normalized values of (a) the dummy node count and (b) the edge density.

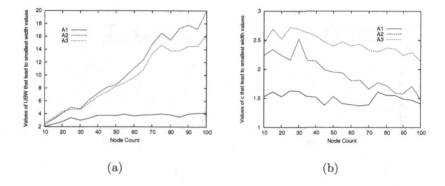

(a)                                                   (b)

**Fig. 6.** $A_1$, $A_2$, and $A_3$ compared: values of (a) UBW and (b) $c$ for which a narrowest layering was found.

the dummy node count [9]. The Promotion heuristic leads to close to the minimum dummy node count when applied to longest-path layerings. Since MinWidth is based on the longest-path algorithm we expected that the same Promotion heuristic might be successfully applied to MinWidth layerings as well. In the next section we compare MinWidth with $A_1$ followed by the Promotion heuristic to some well-known layering algorithms.

### 4.2   Effect of Promotion

Figures 6(a) and (b) suggest that when $A_1$ is employed it is enough to consider UBW $= 1..4$ and $c = 1..2$. Since MinWidth is very fast with fixed UBW and $c$, we can afford running it for relatively narrow ranges of UBW and $c$ values for better quality results. Thus, in a new series of experiments we run MinWidth with $A_1$

for UBW = 1..4 and $c = 1..2$, and choose the combination $(UBW, c)$ that leads to the narrowest layering. For convenience, we call the layering achieved by this method simply MinWidth layering in the remainder of this section.

We post-processed MinWidth layerings by applying to them the Promotion heuristic modified to perform a node promotion only if it does not increase the width of the layering.

We also run the longest-path algorithm and the Coffman-Graham algorithm followed by the same width-preserving node promotion. The Coffman-Graham algorithm takes an upper bound $m$ on the number of nodes in a layer as an input parameter [4]. Thus, we run it for $m = 1..n$, where $n$ is the number of nodes in the DAG, and chose the narrowest layering. We also run the network simplex algorithm of Gansner et al. [7] and compared the aesthetic properties of the four layering types: MinWidth, longest-path, Coffman-Graham and Gansner's network simplex. The results of the comparison are presented in Figures 7-10.

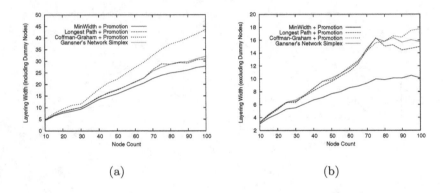

**Fig. 7.** Effect of promotion: layering width (a) taking into account and (b) neglecting the contribution of the dummy nodes.

It can be observed that the promotion heuristic is very efficient when applied after MinWidth. MinWidth leads to considerably narrower but taller layerings than the other three algorithms (see Figures 7(a) and (b)). It was expected that the narrower a layering, the larger is the number of layers. This can be confirmed in Figure 10(a).

The number of dummy nodes in the MinWidth layerings is close to the number of dummy nodes in the Coffman-Graham layerings and slightly higher than the number of dummy nodes in the longest-path and Gansner's layerings as it can be seen in Figure 8(a). However, Figure 8(b) shows that the MinWidth layerings have considerably lower edge density than the other layerings which means that they could possibly lead to clean drawings with small number of edge crossings. The number of edge crossings is widely accepted as one of the most important graph drawing aesthetic criteria [10].

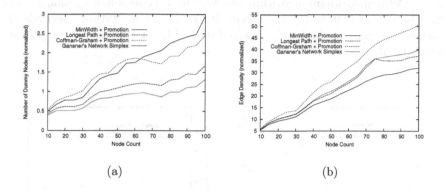

(a)                                    (b)

**Fig. 8.** Effect of promotion: normalized values of (a) the dummy node count and (b) the edge density.

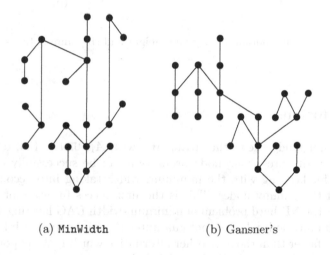

(a) MinWidth              (b) Gansner's

**Fig. 9.** Two layerings of the same DAG. The MinWidth layering is narrower than the Gansner's layering (assuming all DAG nodes and all dummy nodes have width one unit). All edges point downwards.

Figure 9 shows an example of the MinWidth layering of a DAG compared to the Gansner's layering of the same DAG. The DAG is taken from the Rome's graph dataset.

We run the second group of experiments on a single Pentium 4/2.4 GHz processor. The running times are presented in Figure 10(b). We observed that the average running time for MinWidth followed by promotion is up to 2 seconds for DAGs having no more than 75 nodes and it grows up to 6.2 seconds for DAGs with more than 75 and less than 100 nodes. The total running time for the Coffman-Graham algorithm was within 3 seconds and the longest-path algorithms was the fastest of the three with running time within 2 seconds. The Gansner's layerings (which we computed with ILOG CPLEX) are the fastest to be computed.

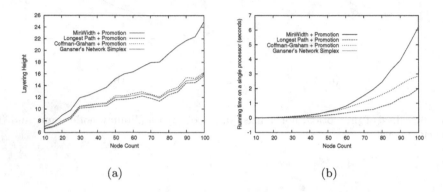

(a)                                                    (b)

**Fig. 10.** Effect of promotion: (a) layering height and (b) running times in seconds.

## 5    Conclusions

Our parameter study shows that MinWidth with $A_1$, UBW $= 1..4$, $c = 1..2$, and followed by width-preserving node promotion can be successfully employed as a heuristic for layering with the minimum width taking into account the contribution of the dummy nodes. This is the first successful attempt to design a heuristic for the NP-hard problem of minimum-width DAG layering with consideration of dummy nodes. It does not guarantee the minimum width but performs significantly faster than the only other alternative which is the exponential-time branch-and-cut algorithm of Healy and Nikolov.

The aesthetic properties of the MinWidth layerings compare well to the properties of the layerings constructed by the well-known layering algorithms. The MinWidth layerings have the lowest edge density which suggests that they could lead to clear and easy to comprehend drawings in the context of the STT method for hierarchical graph drawing. It has to be noted that the promotion heuristic slows down the computation significantly but the running time is still very acceptable for DAGs with up to 100 nodes.

The work we present can be continued by exploring other possibilities for the conditions we set up in MinWidth. However, we believe that MinWidth finds layerings which are narrow enough for practical applications. Further research could be related to the optimization of the running time of MinWidth and to experiments with larger DAGs and with DAGs with variable node widths.

# References

1. J. Branke, P. Eades, S. Leppert, and M. Middendorf. Width restricted layering of acyclic digraphs with consideration of dummy nodes. Technical Report No. 403, Intitute AIFB, University of Karlsruhe, 76128 Karlsruhe, Germany, 2001.
2. J. Branke, S. Leppert, M. Middendorf, and P. Eades. Width-restriced layering of acyclic digraphs with consideration of dummy nodes. *Information Processing Letters*, 81(2):59–63, January 2002.
3. M. J. Carpano. Automatic display of hierarchized graphs for computer aided decision analysis. *IEEE Transactions on Systems, Man and Cybernetics*, 10(11):705–715, 1980.
4. E. G. Coffman and R. L. Graham. Optimal scheduling for two processor systems. *Acta Informatica*, 1:200–213, 1972.
5. G. Di Battista, A. Garg, G. Liotta, R. Tamassia, E. Tassinari, and F. Vargiu. An experimental comparison of four graph drawing algorithms. *Computational Geometry: Theory and Applications*, 7:303–316, 1997.
6. P. Eades, X. Lin, and W. F. Smyth. A fast and effective heuristic for the feedback arc set problem. *Information Processing Letters*, 47:319–323, 1993.
7. E. R. Gansner, E. Koutsofios, S. C. North, and K.-P. Vo. A technique for drawing directed graphs. *IEEE Transactions on Software Engineering*, 19(3):214–230, March 1993.
8. P. Healy and N. S. Nikolov. A branch-and-cut approach to the directed acyclic graph layering problem. In M. Goodrich and S. Koburov, editors, *Graph Drawing: Proceedings of 10th International Symposium, GD 2002*, volume 2528 of *Lecture Notes in Computer Science*, pages 98–109. Springer-Verlag, 2002.
9. N.S. Nikolov and A. Tarassov. Graph layering by promotion of nodes. *Special issue of Discrete Applied Mathematics associated with the IV ALIO/EURO Workshop on Applied Combinatorial Optimization*, to appear.
10. H. C. Purchase, R. F. Cohen, and M. James. Validating graph drawing aesthetics. In F. J. Brandenburg, editor, *Graph Drawing: Symposium on Graph Drawing, GD '95*, volume 1027 of *Lecture Notes in Computer Science*, pages 435–446. Springer-Verlag, 1996.
11. K. Sugiyama and K. Misue. Graph drawing by the magneting spring model. *Journal of Visual Languages and Computing*, 6(3):217–231, 1995.
12. K. Sugiyama, S. Tagawa, and M. Toda. Methods for visual understanding of hierarchical system structures. *IEEE Transaction on Systems, Man, and Cybernetics*, 11(2):109–125, February 1981.
13. J. Utech, J. Branke, H. Schmeck, and P. Eades. An evolutionary algorithm for drawing directed graphs. In *Proceedings of the 1998 International Conference on Imaging Science, Systems, and Technology (CISST'98)*, pages 154–160, 1998.
14. J. N. Warfield. Crossing theory and hierarchy mapping. *IEEE Transactions on Systems, Man and Cybernetics*, 7(7):502–523, 1977.

# Author Index

# Lecture Notes in Computer Science

For information about Vols. 1–2951

please contact your bookseller or Springer-Verlag

Vol. 3004: J. Gottlieb, G.R. Raidl (Eds.), Evolutionary Computation in Combinatorial Optimization. X, 241 pages. 2004.

Vol. 3003: M. Keijzer, U.-M. O'Reilly, S.M. Lucas, E. Costa, T. Soule (Eds.), Genetic Programming. XI, 410 pages. 2004.

Vol. 3002: D.L. Hicks (Ed.), Metainformatics. X, 213 pages. 2004.

Vol. 3001: A. Ferscha, F. Mattern (Eds.), Pervasive Computing. XVII, 358 pages. 2004.

Vol. 2999: E.A. Boiten, J. Derrick, G. Smith (Eds.), Integrated Formal Methods. XI, 541 pages. 2004.

Vol. 2998: Y. Kameyama, P.J. Stuckey (Eds.), Functional and Logic Programming. X, 307 pages. 2004.

Vol. 2997: S. McDonald, J. Tait (Eds.), Advances in Information Retrieval. XIII, 427 pages. 2004.

Vol. 2996: V. Diekert, M. Habib (Eds.), STACS 2004. XVI, 658 pages. 2004.

Vol. 2995: C. Jensen, S. Poslad, T. Dimitrakos (Eds.), Trust Management. XIII, 377 pages. 2004.

Vol. 2994: E. Rahm (Ed.), Data Integration in the Life Sciences. X, 221 pages. 2004. (Subseries LNBI).

Vol. 2993: R. Alur, G.J. Pappas (Eds.), Hybrid Systems: Computation and Control. XII, 674 pages. 2004.

Vol. 2992: E. Bertino, S. Christodoulakis, D. Plexousakis, V. Christophides, M. Koubarakis, K. Böhm, E. Ferrari (Eds.), Advances in Database Technology - EDBT 2004. XVIII, 877 pages. 2004.

Vol. 2991: R. Alt, A. Frommer, R.B. Kearfott, W. Luther (Eds.), Numerical Software with Result Verification. X, 315 pages. 2004.

Vol. 2989: S. Graf, L. Mounier (Eds.), Model Checking Software. X, 309 pages. 2004.

Vol. 2988: K. Jensen, A. Podelski (Eds.), Tools and Algorithms for the Construction and Analysis of Systems. XIV, 608 pages. 2004.

Vol. 2987: I. Walukiewicz (Ed.), Foundations of Software Science and Computation Structures. XIII, 529 pages. 2004.

Vol. 2986: D. Schmidt (Ed.), Programming Languages and Systems. XII, 417 pages. 2004.

Vol. 2985: E. Duesterwald (Ed.), Compiler Construction. X, 313 pages. 2004.

Vol. 2984: M. Wermelinger, T. Margaria-Steffen (Eds.), Fundamental Approaches to Software Engineering. XII, 389 pages. 2004.

Vol. 2983: S. Istrail, M.S. Waterman, A. Clark (Eds.), Computational Methods for SNPs and Haplotype Inference. IX, 153 pages. 2004. (Subseries LNBI).

Vol. 2982: N. Wakamiya, M. Solarski, J. Sterbenz (Eds.), Active Networks. XI, 308 pages. 2004.

Vol. 2981: C. Müller-Schloer, T. Ungerer, B. Bauer (Eds.), Organic and Pervasive Computing – ARCS 2004. XI, 339 pages. 2004.

Vol. 2980: A. Blackwell, K. Marriott, A. Shimojima (Eds.), Diagrammatic Representation and Inference. XV, 448 pages. 2004. (Subseries LNAI).

Vol. 2979: I. Stoica, Stateless Core: A Scalable Approach for Quality of Service in the Internet. XVI, 219 pages. 2004.

Vol. 2978: R. Groz, R.M. Hierons (Eds.), Testing of Communicating Systems. XII, 225 pages. 2004.

Vol. 2977: G. Di Marzo Serugendo, A. Karageorgos, O.F. Rana, F. Zambonelli (Eds.), Engineering Self-Organising Systems. X, 299 pages. 2004. (Subseries LNAI).

Vol. 2976: M. Farach-Colton (Ed.), LATIN 2004: Theoretical Informatics. XV, 626 pages. 2004.

Vol. 2973: Y. Lee, J. Li, K.-Y. Whang, D. Lee (Eds.), Database Systems for Advanced Applications. XXIV, 925 pages. 2004.

Vol. 2972: R. Monroy, G. Arroyo-Figueroa, L.E. Sucar, H. Sossa (Eds.), MICAI 2004: Advances in Artificial Intelligence. XVII, 923 pages. 2004. (Subseries LNAI).

Vol. 2971: J.I. Lim, D.H. Lee (Eds.), Information Security and Cryptology -ICISC 2003. XI, 458 pages. 2004.

Vol. 2970: F. Fernández Rivera, M. Bubak, A. Gómez Tato, R. Doallo (Eds.), Grid Computing. XI, 328 pages. 2004.

Vol. 2968: J. Chen, S. Hong (Eds.), Real-Time and Embedded Computing Systems and Applications. XIV, 620 pages. 2004.

Vol. 2967: S. Melnik, Generic Model Management. XX, 238 pages. 2004.

Vol. 2966: F.B. Sachse, Computational Cardiology. XVIII, 322 pages. 2004.

Vol. 2965: M.C. Calzarossa, E. Gelenbe, Performance Tools and Applications to Networked Systems. VIII, 385 pages. 2004.

Vol. 2964: T. Okamoto (Ed.), Topics in Cryptology – CT-RSA 2004. XI, 387 pages. 2004.

Vol. 2963: R. Sharp, Higher Level Hardware Synthesis. XVI, 195 pages. 2004.

Vol. 2962: S. Bistarelli, Semirings for Soft Constraint Solving and Programming. XII, 279 pages. 2004.

Vol. 2961: P. Eklund (Ed.), Concept Lattices. IX, 411 pages. 2004. (Subseries LNAI).

Vol. 2960: P.D. Mosses (Ed.), CASL Reference Manual. XVII, 528 pages. 2004.

Vol. 2959: R. Kazman, D. Port (Eds.), COTS-Based Software Systems. XIV, 219 pages. 2004.

Vol. 2958: L. Rauchwerger (Ed.), Languages and Compilers for Parallel Computing. XI, 556 pages. 2004.

Vol. 2957: P. Langendoerfer, M. Liu, I. Matta, V. Tsaousidis (Eds.), Wired/Wireless Internet Communications. XI, 307 pages. 2004.

Vol. 2956: A. Dengel, M. Junker, A. Weisbecker (Eds.), Reading and Learning. XII, 355 pages. 2004.

Vol. 2954: F. Crestani, M. Dunlop, S. Mizzaro (Eds.), Mobile and Ubiquitous Information Access. X, 299 pages. 2004.

Vol. 2953: K. Konrad, Model Generation for Natural Language Interpretation and Analysis. XIII, 166 pages. 2004. (Subseries LNAI).

Vol. 2952: N. Guelfi, E. Astesiano, G. Reggio (Eds.), Scientific Engineering of Distributed Java Applications. X, 157 pages. 2004.